This book is in the

ADDISON-WESLEY SERIES IN MATHEMATICS

LYNN H. LOOMIS

Consulting Editor

Modern
Mathematical Analysis

Modern Mathematical Analysis

by

MURRAY H. PROTTER

University of California

Berkeley, California

and

CHARLES B. MORREY, JR.

University of California

Berkeley, California

ADDISON-WESLEY PUBLISHING COMPANY

Reading, Massachusetts

Menlo Park, California · London · Amsterdam · Sydney

ADDISON-WESLEY (CANADA) LIMITED

Don Mills, Ontario

Library of Congress Catalog Card No. 64-16907

ISBN 0-201-05995-9
  VWXYZ-MA-89876543

ADDISON-WESLEY PUBLISHING COMPANY, INC.
READING, MASSACHUSETTS · Palo Alto · London
NEW YORK · DALLAS · ATLANTA · BARRINGTON, ILLINOIS

ADDISON-WESLEY (CANADA) LIMITED
DON MILLS, ONTARIO

Preface

This book is designed for students who have already completed a one-year course in calculus and analytic geometry devoted to the study of functions of one variable. We assume that the student has learned the material in our text, *Calculus and Analytic Geometry, A First Course*, or in an equivalent text which covers approximately the same body of knowledge. Throughout this book we shall refer to theorems and examples in our previous text, using the term *First Course* in making such references. However, the topics in *First Course* are standard in character, and items such as Theorem of the Mean, Fundamental Theorem of Calculus, Chain Rule, and so forth may be found in almost all other elementary books on calculus and analytic geometry.

For most students the step from elementary calculus to advanced calculus is an abrupt and difficult one. By including in this book the more advanced portions of elementary calculus and the beginning portions of advanced calculus, we have attempted to make the transfer as smooth as possible. Thus the texts *First Course* and *Modern Mathematical Analysis* take the student through the first sixteen semester hours of college-level mathematics in a coherent, coordinated manner.

For some years it has been apparent that it is not only students of mathematics but also of engineering, physics, and chemistry who must master the elements of linear algebra as early as possible, preferably in the freshman or sophomore year. While it may be desirable to have a separate course in linear algebra, the curriculum at most colleges and universities will not allow a freshman or sophomore to take two mathematics courses simultaneously. It therefore seems natural to include linear algebra in the first two years of an analysis sequence; consequently we have devoted about one hundred and thirty pages to this topic (Chapters 6, 7, and 8).

From the logical point of view, it is possible to place linear algebra before the calculus. However, the theorems and proofs in linear algebra presented here require a degree of maturity which most beginning students do not possess. Furthermore, it is advantageous to place this subject matter as close as possible to its applications, which begin in the study of Fourier Series (Chapter 10).

Chapter 1 presents an introduction to solid analytic geometry, and Chapter 2 is devoted to a study of vectors in three-space. The elements of infinite series are taken up in Chapter 3, where the material covers all the usual topics which appear in an elementary calculus course.

Chapter 4 introduces partial differentiation and Chapter 5 discusses multiple integration, both topics being treated in a rather elementary manner. Later, after the study of vector spaces, transformations, and Jacobians, the rule for change of variables in a multiple integral is placed on a rigorous basis (Chapter 11).

Matrices, the solution of systems of linear equations, Cramer's Rule, and related topics are taken up in Chapter 6. The study of vector spaces and linear transformations between such spaces is undertaken in Chapter 7. Chapter 8 continues this study with a discussion of the eigenvalues of matrices. It is shown here that the eigenvalues of symmetric matrices are always real.

The more advanced topics of infinite series, those usually reserved for courses in advanced calculus, are given in Chapter 9. Chapter 10 applies the previous results on series to the study of Fourier Series.

Implicit function theorems are established in Chapter 11. The proofs of the elementary theorems are given in full, and the implicit function theorem for a general system is stated with care but not proved. Transformations, the Inversion Theorem, Jacobians, and the above-mentioned topic of change of variables in a multiple integral are included in this chapter.

The important subject of differentiation under the integral sign is frequently neglected in modern curricula in analysis. This topic, together with that of convergence of improper integrals, is treated in Chapter 12. In addition, the Gamma function is discussed, as are some of its properties.

Chapter 13 is devoted to vector field theory. At this point the student puts to good use his previous study of vectors in two and three dimensions and vector spaces. The elements of vector analysis and line integrals are included in this chapter.

It is possible to present proofs of Green's and Stokes' theorems with varying degrees of generality, depending on the assumptions made on the nature of the domain and on the smoothness of the functions involved. By employing the notion of a *partition of unity*, we obtain, by fairly elementary means, quite general proofs of these theorems. In fact, the proofs of Green's and Stokes' theorems given in Chapter 14 (and the Appendix) yield as a corollary a proof of Cauchy's theorem sufficiently general for use in the study of functions of a complex variable.

The final three chapters are devoted to an elementary study of ordinary differential equations. Chapter 15 treats the usual topics given in an unsophisticated first course on the subject. While the equations encountered here are rather special in character, we feel that the student needs this introduction if he intends to investigate the complicated, extensive body of analysis into which the field of differential equations has evolved. Chapter 16 discusses linear differential equations and systems with constant coefficients. In addition, a general existence theorem is stated without proof. The knowledge of eigenvalues of matrices which the student learned in Chapter 8 is exploited here. The final chapter, on the solution of differential equations by series, includes a brief study of Bessel's equation and Bessel functions.

It seems likely that this text contains more material than can be covered in eight semester hours of work. If the instructor wishes to delete material, he has a number of choices. For example, he may omit the chapters on differential equations, devoting an entire course to that subject later on. Or, if all the topics on differential equations are included, it is possible to save time by leaving out some of the text's more difficult proofs. These could be assigned as outside reading for the better students.

In the two volumes, *First Course* and *Modern Mathematical Analysis*, we have attempted to present those topics which we think a well-prepared college student should learn in his first two years of mathematics. It is clear that we have had to make some difficult and arbitrary choices in the selection of material. For example, there is no discussion of probability and statistics. It is our opinion that a beginning study of this subject requires an entire course, although one could argue that if linear algebra were omitted, the subject of probability could be discussed in a substantial manner. The subject of numerical analysis presents a second difficult problem. While this topic is touched briefly in many places, it is done so primarily to give the instructor a springboard for a lengthy digression, if he so wishes.

Berkeley, California   M.H.P.
February 1964   C.B.M., Jr.

Contents

CHAPTER 5.   MULTIPLE INTEGRATION

CHAPTER 6.   LINEAR ALGEBRA

CHAPTER 7.   VECTOR SPACES

CHAPTER 1

Solid Analytic Geometry

---

1. COORDINATES. THE DISTANCE FORMULA

In three-dimensional space, consider three mutually perpendicular lines which intersect in a point O. We designate these lines the *coordinate axes* and, starting from O, we set up number scales along each of them. If the positive directions of the x, y, and z axes are labeled x, y, and z, as shown in Fig. 1–1, we say the axes form a *right-handed system*. Figure 1–2 illustrates the axes in a *left-handed system*. We shall use a right-handed coordinate system throughout.

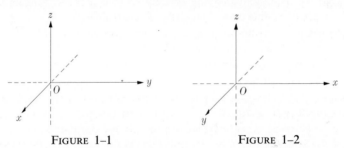

FIGURE 1–1   FIGURE 1–2

Any two intersecting lines in space determine a plane. A plane containing two of the coordinate axes is called a *coordinate plane*. Clearly, there are three such planes.

To each point P in three-dimensional space we can assign an ordered triple of numbers in the following way. Through P construct three planes, each parallel to one of the coordinate planes as shown in Fig. 1–3. We label the intersections of the planes through P with the coordinate axes Q, R, and S, as shown. Then,

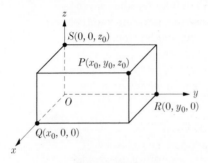

FIGURE 1–3

if Q is x_0 units from the origin O, R is y_0 units from O, and S is z_0 units from O, we assign to P the number triple (x_0, y_0, z_0) and say that the point P has *rectangular coordinates* (x_0, y_0, z_0). To each point in space there corresponds exactly one ordered number triple and, conversely, to each ordered number triple there is associated exactly one point in three-dimensional space. We have just described a *rectangular* or *cartesian* coordinate system. In Section 9 we shall discuss other coordinate systems.

In studying plane analytic geometry we saw that an equation such as

$$y = 3$$

represents all points lying on a line parallel to the x-axis and three units above it (Fig. 1–4). The equation

$$y = 3$$

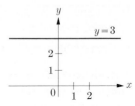

FIGURE 1–4

in the context of three-dimensional geometry represents something entirely different. The locus of points satisfying this equation is a plane parallel to the xz-plane (the xz-plane is the coordinate plane determined by the x-axis and the z-axis) and three units from it (Fig. 1–5). The plane represented by $y = 3$ is perpendicular to the y-axis and passes through the point $(0, 3, 0)$. Since there is exactly one plane which is perpendicular to a given line and which passes through a given point, we see that the locus of the equation $y = 3$ consists of one and only one such plane. Conversely, from the very definition of a rectangular coordinate system every point with y-coordinate 3 must lie in this plane. Equations such as $x = a$ or $y = b$ or $z = c$ always represent planes parallel to the coordinate planes.

We recall from Euclidean solid geometry that *any two nonparallel planes intersect in a straight line*. Therefore, the locus of all points which simultaneously satisfy the equations

$$x = a \qquad \text{and} \qquad y = b$$

is a line parallel to (or coincident with) the z-axis. Conversely, any such line is the locus of a pair of equations of the above form. Since the plane $x = a$ is parallel to the z-axis, and the plane $y = b$ is parallel to the z-axis, the line of intersection must be parallel to the z-axis also. (Corresponding statements hold with the axes interchanged.)

A plane separates three-dimensional space into two parts, each of which is called a **half-space.** The inequality

$$x > 5$$

represents all points with x-coordinate greater

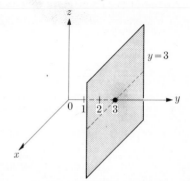

FIGURE 1–5

than 5.   The set of such points comprises a half-space. Two intersecting planes divide three-space into 4 regions which we call *infinite wedges*. Three intersecting planes divide space into 8 regions (or possibly fewer), four planes into 16 regions (or possibly fewer), and so on. The inequality

$$|y| \leq 4$$

represents all points between (and on) the planes $y = -4$ and $y = 4$. Regions in space defined by inequalities are more difficult to visualize than those in the plane. However, *polyhedral domains*—i.e., those determined by the intersection of a number of planes—are frequently simple enough to be sketched. A polyhedron with six faces in which opposite faces are congruent parallelograms is called a **parallelepiped.** Cubes and rectangular bins are particular cases of parallelepipeds.

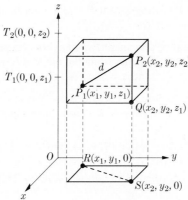

FIGURE 1–6

Theorem 1. *The distance d between the points $P_1(x_1, y_1, z_1)$ and $P_2(x_2, y_2, z_2)$ is*

$$d = \sqrt{(x_2 - x_1)^2 + (y_2 - y_1)^2 + (z_2 - z_1)^2}.$$

Proof. We make the construction shown in Fig. 1–6. By the Pythagorean theorem we have

$$d^2 = \overline{P_1Q}^2 + \overline{QP_2}^2.$$

Noting that $\overline{P_1Q} = \overline{RS}$, we use the formula for distance in the *xy*-plane to get

$$\overline{P_1Q}^2 = \overline{RS}^2 = (x_2 - x_1)^2 + (y_2 - y_1)^2.$$

Furthermore, since P_2 and Q are on a line parallel to the *z*-axis, we see that

$$\overline{QP_2}^2 = \overline{T_1T_2}^2 = (z_2 - z_1)^2.$$

Therefore

$$d^2 = (x_2 - x_1)^2 + (y_2 - y_1)^2 + (z_2 - z_1)^2.$$

The midpoint P of the line segment connecting the point $P_1(x_1, y_1, z_1)$ and $P_2(x_2, y_2, z_2)$ has coordinates $P(\overline{x}, \overline{y}, \overline{z})$ given by the formula

$$\overline{x} = \frac{x_1 + x_2}{2}, \qquad \overline{y} = \frac{y_1 + y_2}{2}, \qquad \overline{z} = \frac{z_1 + z_2}{2}.$$

If P_1 and P_2 lie in the xy-plane—i.e., if $z_1 = 0$ and $z_2 = 0$—then so does the midpoint, and we recognize the formula as the one we learned in plane analytic geometry. To establish the above formula we set up proportional triangles in complete analogy with the two-dimensional case. (See Exercise 24 at the end of this section.)

EXAMPLE 1. Find the coordinates of the point Q which divides the line segment from $P_1(1, 4, -2)$ to $P_2(-3, 6, 7)$ in the proportion 3 to 1.

Solution. The midpoint P of the segment P_1P_2 has coordinates $P(-1, 5, \frac{5}{2})$. When we find the midpoint of PP_2 we get $Q(-2, \frac{11}{2}, \frac{19}{4})$.

EXAMPLE 2. One endpoint of a segment P_1P_2 has coordinates $P_1(-1, 2, 5)$. The midpoint P is known to lie in the xz-plane, while the other endpoint is known to lie on the intersection of the planes $x = 5$ and $z = 8$. Find the coordinates of P and P_2.

Solution. For $P(\overline{x}, \overline{y}, \overline{z})$ we note that $\overline{y} = 0$, since P is in the xz-plane. Similarly, for $P_2(x_2, y_2, z_2)$ we have $x_2 = 5$ and $z_2 = 8$. From the midpoint formula we get

$$\overline{x} = \frac{-1 + 5}{2}, \qquad 0 = \overline{y} = \frac{2 + y_2}{2}, \qquad \overline{z} = \frac{5 + 8}{2}.$$

Therefore the points have coordinates $P(2, 0, \frac{13}{2})$, $P_2(5, -2, 8)$.

EXERCISES

In problems 1 through 5, find the lengths of the sides of triangle ABC and state whether the triangle is a right triangle, an isosceles triangle, or both.

1. $A(2, 1, 3)$, $B(3, -1, -2)$, $C(0, 2, -1)$ 2. $A(4, 3, 1)$, $B(2, 1, 2)$, $C(0, 2, 4)$
3. $A(3, -1, -1)$, $B(1, 2, 1)$, $C(6, -1, 2)$ 4. $A(1, 2, -3)$, $B(4, 3, -1)$, $C(3, 1, 2)$
5. $A(0, 0, 0)$, $B(4, 1, 2)$, $C(-5, -5, -1)$

In problems 6 through 9, find the midpoint of the segment joining the given points A, B.

6. $A(2, 1, 3)$, $B(-4, -1, 2)$ 7. $A(4, 6, 1)$, $B(2, -1, 3)$
8. $A(0, 2, -3)$, $B(1, 4, 6)$ 9. $A(-2, 0, 0)$, $B(0, 4, -1)$

In problems 10 through 13, find the lengths of the medians of the given triangles ABC.

10. $A(2, 1, 3)$, $B(3, -1, -2)$, $C(0, 2, -1)$ 11. $A(4, 3, 1)$, $B(2, 1, 2)$, $C(0, 2, 4)$
12. $A(3, -1, -1)$, $B(1, 2, 1)$, $C(6, -1, 2)$ 13. $A(1, 2, -3)$, $B(4, 3, -1)$, $C(3, 1, 2)$

14. One endpoint of a line segment is at $P(4, 6, -3)$ and the midpoint is at $Q(2, 1, 6)$. Find the other endpoint.

15. One endpoint of a line segment is at $P_1(-2, 1, 6)$ and the midpoint Q lies in the plane $y = 3$. The other endpoint, P_2, lies on the intersection of the planes $x = 4$ and $z = -6$. Find the coordinates of P_2 and Q.

In problems 16 through 19, determine whether or not the three given points lie on a line.

16. $A(1, -1, 2)$, $B(-1, -4, 3)$, $C(3, 2, 1)$ 17. $A(2, 3, 1)$, $B(4, 6, 5)$, $C(-2, -2, -7)$

18. $A(1, -1, 2)$, $B(3, 3, 4)$, $C(-2, -6, -1)$

19. $A(-4, 5, -6)$, $B(-1, 2, -1)$, $C(3, -3, 6)$

20. Describe the locus of points in space which satisfy the inequalities $-2 \le x < 3$.

21. Describe the locus of points in space which satisfy the relations $x = 2$, $z = -4$.

22. Describe the locus of points in space which satisfy the inequalities $x \ge 0$, $y \ge 0$, $z \ge 0$.

23. Describe the locus of points in space which satisfy the inequalities $x \ge 0$, $y > 0$, $z < 0$.

24. Derive the formula for determining the midpoint of a line segment.

25. The formula for the coordinates of a point $Q(x_0, y_0, z_0)$ which divides the line segment from $P_1(x_1, y_1, z_1)$ to $P_2(x_2, y_2, z_2)$ in the ratio p to q is

$$x_0 = \frac{px_2 + qx_1}{p + q}, \qquad y_0 = \frac{py_2 + qy_1}{p + q}, \qquad z_0 = \frac{pz_2 + qz_1}{p + q}.$$

Derive this formula.

26. Find the equation of the locus of all points equidistant from the points $(2, -1, 3)$ and $(3, 1, -1)$. Can you describe the locus?

27. Find the equation of the locus of all points equidistant from the points $(5, 1, 0)$ and $(2, -1, 4)$. Can you describe the locus?

28. The points $A(0, 0, 0)$, $B(1, 0, 0)$, $C(\frac{1}{2}, \frac{1}{2}, 1/\sqrt{2})$, $D(0, 1, 0)$ are the vertices of a four-sided figure. Show that $\overline{AB} = \overline{BC} = \overline{CD} = \overline{DA} = 1$. Prove that the figure is not a rhombus.

29. Prove that the diagonals joining opposite vertices of a rectangular parallelepiped (there are four of them which are interior to the parallelepiped) bisect each other.

2. DIRECTION COSINES AND NUMBERS

Consider a line L passing through the origin and place an arrow on it so that one of the two possible directions is distinguished (Fig. 1–7). We call such a line a *directed line*. If no arrow is placed, then L is called an *undirected line*. Denote by α, β, and γ the angles made by the directed line L and the positive directions of the x-, y-, and z-axes, respectively. We define these angles to be the **direction angles** of the directed line L. The undirected line L will have two possible sets of direction angles according as the arrow points in one direction or the other. The two sets are

$$\alpha, \beta, \gamma \qquad \text{and} \qquad 180 - \alpha, \quad 180 - \beta, \quad 180 - \gamma,$$

where α, β, γ are the direction angles of the directed line.

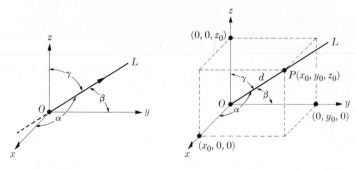

FIGURE 1–7   FIGURE 1–8

DEFINITION. If α, β, γ are direction angles of a line L, then $\cos \alpha$, $\cos \beta$, $\cos \gamma$ are called the **direction cosines** of the line L.

(unit vector)

Since $\cos (180 - \theta) = -\cos \theta$, we see that if λ, μ, ν are direction cosines of a directed line L, then λ, μ, ν and $-\lambda$, $-\mu$, $-\nu$ are the two sets of direction cosines of the undirected line L.

We shall show that the direction cosines of any line L satisfy the relation

$$\cos^2 \alpha + \cos^2 \beta + \cos^2 \gamma = 1.$$

Let $P(x_0, y_0, z_0)$ be a point on a line L which goes through the origin. Then the distance d of P from the origin is

$$d = \sqrt{x_0^2 + y_0^2 + z_0^2},$$

and (see Fig. 1–8) we have

$$\cos \alpha = \frac{x_0}{d}, \qquad \cos \beta = \frac{y_0}{d}, \qquad \cos \gamma = \frac{z_0}{d}.$$

Squaring and adding, we get the desired result.

To define the direction cosines of any line L in space, we simply consider the line L' parallel to L which passes through the origin, and assert that *by definition L has the same direction cosines as L'. Thus all parallel lines in space have the same direction cosines.*

DEFINITION. Two sets of number triples, a, b, c and a', b', c', neither all zero, are said to be **proportional** if there is a number k such that

$$a' = ka, \qquad b' = kb, \qquad c' = kc.$$

Remark. The number k may be positive or negative but not zero, since by hypothesis neither of the number triples is $0, 0, 0$. If none of the numbers a, b, and c is zero, we may write the proportionality relations as

$$\frac{a'}{a} = k, \qquad \frac{b'}{b} = k, \qquad \frac{c'}{c} = k$$

or, more simply,

$$\frac{a'}{a} = \frac{b'}{b} = \frac{c'}{c}.$$

DEFINITION. Suppose that a line L has direction cosines λ, μ, ν. Then a set of numbers a, b, c is called a *set of* **direction numbers** for L if a, b, c and λ, μ, ν are proportional. ~~not unit vector~~

A line L has unlimited sets of direction numbers.

Theorem 2. *If $P_1(x_1, y_1, z_1)$ and $P_2(x_2, y_2, z_2)$ are two points on a line L, then*

$$\lambda = \frac{x_2 - x_1}{d}, \qquad \mu = \frac{y_2 - y_1}{d}, \qquad \nu = \frac{z_2 - z_1}{d} \quad \text{unit vector}$$

is a set of direction cosines of L where d is the distance from P_1 to P_2.

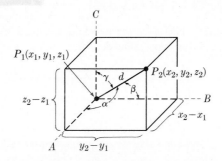

FIGURE 1–9

Proof. In Fig. 1–9 we note that the angles $\alpha, \beta,$ and γ are equal to the direction angles, since the lines P_1A, P_1B, P_1C are parallel to the coordinate axes. We read off from the figure that

$$\cos \alpha = \frac{x_2 - x_1}{d}, \qquad \cos \beta = \frac{y_2 - y_1}{d}, \qquad \cos \gamma = \frac{z_2 - z_1}{d},$$

which is the desired result.

Corollary 1. *If $P_1(x_1, y_1, z_1)$ and $P_2(x_2, y_2, z_2)$ are two points on a line L, then*

$$x_2 - x_1, \qquad y_2 - y_1, \qquad z_2 - z_1$$

is a set of direction numbers for L.

Multiplying λ, μ, ν of Theorem 2 by the constant d, we obtain the result of the Corollary.

EXAMPLE 1. Find direction numbers and direction cosines for the line L passing through the points $P_1(1, 5, 2)$ and $P_2(3, 7, -4)$.

Solution. From the Corollary, 2, 2, -6 is a set of direction numbers. We compute

$$d = \overline{P_1 P_2} = \sqrt{4 + 4 + 36} = \sqrt{44} = 2\sqrt{11},$$

and so

$$\frac{1}{\sqrt{11}}, \quad \frac{1}{\sqrt{11}}, \quad -\frac{3}{\sqrt{11}}$$

is a set of direction cosines. Since L is undirected, it has two such sets, the other being

$$-\frac{1}{\sqrt{11}}, \quad -\frac{1}{\sqrt{11}}, \quad \frac{3}{\sqrt{11}}.$$

EXAMPLE 2. Do the three points $P_1(3, -1, 4)$, $P_2(1, 6, 8)$, and $P_3(9, -22, -8)$ lie on the same straight line?

Solution. A set of direction numbers for the line L_1 through P_1 and P_2 is $-2, 7, 4$. A set of direction numbers for the line L_2 through P_2 and P_3 is $8, -28, -16$. Since the second set is proportional to the first (with $k = -4$), we conclude that L_1 and L_2 have the same direction cosines. Therefore the two lines are parallel. However, they have the point P_2 in common and so must coincide.

From Theorem 2 and the statements in Example 2, we easily obtain the next result.

Corollary 2. *A line L_1 is parallel to a line L_2 if and only if a set of direction numbers of L_1 is proportional to a set of direction numbers of L_2.*

The angle between two intersecting lines in space is defined in the same way as the angle between two lines in the plane. It may happen that two lines L_1 and L_2 in space are neither parallel nor intersecting. Such lines are said to be *skew* to each other. Nevertheless, the angle between L_1 and L_2 can still be defined. Denote by L_1' and L_2' the lines passing through the origin and parallel to L_1 and L_2, respectively. *The angle between L_1 and L_2 is defined to be the angle between the intersecting lines L_1' and L_2'.*

Theorem 3. *If L_1 and L_2 have direction cosines λ_1, μ_1, ν_1 and λ_2, μ_2, ν_2, respectively, and if θ is the angle between L_1 and L_2, then*

$$\cos \theta = \lambda_1 \lambda_2 + \mu_1 \mu_2 + \nu_1 \nu_2.$$

Proof. From the way we defined the angle between two lines we may consider L_1 and L_2 as lines passing through the origin. Let $P_1(x_1, y_1, z_1)$ be a point on L_1 and $P_2(x_2, y_2, z_2)$ a point on L_2 (see Fig. 1–10). Denote by d_1 the distance of P_1 from O, by d_2 the distance of P_2 from O; let $d = \overline{P_1 P_2}$. We apply the Law

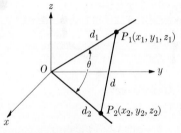

FIGURE 1–10

of Cosines to triangle OP_1P_2, getting

$$d^2 = d_1^2 + d_2^2 - 2d_1d_2 \cos \theta,$$

or

$$\cos \theta = \frac{d_1^2 + d_2^2 - d^2}{2d_1d_2}$$

and

$$\cos \theta = \frac{x_1^2 + y_1^2 + z_1^2 + x_2^2 + y_2^2 + z_2^2 - (x_2 - x_1)^2 - (y_2 - y_1)^2 - (z_2 - z_1)^2}{2d_1d_2}.$$

After simplification we get

$$\cos \theta = \frac{x_1x_2 + y_1y_2 + z_1z_2}{d_1d_2} = \frac{x_1}{d_1} \cdot \frac{x_2}{d_2} + \frac{y_1}{d_1} \cdot \frac{y_2}{d_2} + \frac{z_1}{d_1} \cdot \frac{z_2}{d_2}$$
$$= \lambda_1\lambda_2 + \mu_1\mu_2 + \nu_1\nu_2.$$

Corollary. *Two lines L_1 and L_2 with direction numbers a_1, b_1, c_1 and a_2, b_2, c_2, respectively, are perpendicular if and only if*

$$a_1a_2 + b_1b_2 + c_1c_2 = 0.$$

EXAMPLE 3. Find the cosine of the angle between the line L_1, passing through the points $P_1(1, 4, 2)$ and $P_2(3, -1, 3)$, and the line L_2, passing through the points $Q_1(3, 1, 2)$ and $Q_2(2, 1, 3)$.

Solution. A set of direction numbers for L_1 is 2, -5, 1. A set for L_2 is -1, 0, 1. Therefore direction cosines for the two lines are

$$L_1: \frac{2}{\sqrt{30}}, \frac{-5}{\sqrt{30}}, \frac{1}{\sqrt{30}}; \quad L_2: \frac{-1}{\sqrt{2}}, 0, \frac{1}{\sqrt{2}}.$$

We obtain

$$\cos \theta = -\frac{1}{\sqrt{15}} + 0 + \frac{1}{2\sqrt{15}} = -\frac{1}{2\sqrt{15}}.$$

We observe that two lines always have two possible supplementary angles of intersection. If $\cos \theta$ is negative, we have obtained the obtuse angle and, if it is positive, the acute angle of intersection.

EXERCISES

In problems 1 through 4, find a set of direction numbers and a set of direction cosines for the line passing through the given points.

1. $P_1(3, 1, 2)$, $P_2(2, 6, 0)$
2. $P_1(4, 1, 0)$, $P_2(0, 1, 2)$
3. $P_1(-5, -1, 4)$, $P_2(2, 0, 6)$
4. $P_1(-1, 5, 5)$, $P_2(3, 5, 5)$

In each of problems 5 through 8, a point P_1 and a set of direction numbers are given. Find another point which is on the line passing through P_1 and having the given set of direction numbers.

5. $P_1(2, 5, 1)$, direction numbers 3, 2, 5
6. $P_1(3, 0, -5)$, direction numbers 2, -1, 0
7. $P_1(0, 4, -3)$, direction numbers 0, 0, 5
8. $P_1(0, 0, 0)$, direction numbers 2, -4, 0

In each of problems 9 through 12, determine whether or not the three given points lie on a line.

9. $A(3, 1, 0)$, $B(2, 2, 2)$, $C(0, 4, 6)$
10. $A(2, -1, 1)$, $B(4, 1, -3)$, $C(7, 4, -9)$
11. $A(4, 2, -1)$, $B(2, 1, 1)$, $C(0, 0, 2)$
12. $A(5, 8, 6)$, $B(-2, -3, 1)$, $C(4, 2, 8)$

In each of problems 13 through 15, determine whether or not the line through the points P_1, P_2 is parallel to the line through the points Q_1, Q_2.

13. $P_1(4, 8, 0)$, $P_2(1, 2, 3)$; $Q_1(0, 5, 0)$, $Q_2(-3, -1, 3)$
14. $P_1(2, 1, 1)$, $P_2(3, 2, -1)$; $Q_1(0, 1, 4)$, $Q_2(2, 3, 0)$
15. $P_1(3, 1, 4)$, $P_2(-3, 2, 5)$; $Q_1(4, 6, 1)$, $Q_2(0, 5, 8)$

In each of problems 16 through 18, determine whether or not the line through the points P_1, P_2 is perpendicular to the line through the points Q_1, Q_2.

16. $P_1(2, 1, 3)$, $P_2(4, 0, 5)$; $Q_1(3, 1, 2)$, $Q_2(2, 1, 6)$
17. $P_1(2, -1, 0)$, $P_2(3, 1, 2)$; $Q_1(2, 1, 4)$, $Q_2(4, 0, 4)$
18. $P_1(0, -4, 2)$, $P_2(5, -1, 0)$; $Q_1(3, 0, 2)$, $Q_2(2, 1, 1)$

In each of problems 19 through 21, find $\cos \theta$ where θ is the angle between the line L_1 passing through P_1, P_2 and L_2 passing through Q_1, Q_2.

19. $P_1(2, 1, 4)$, $P_2(-1, 4, 1)$; $Q_1(0, 5, 1)$, $Q_2(3, -1, -2)$
20. $P_1(4, 0, 5)$, $P_2(-1, -3, -2)$; $Q_1(2, 1, 4)$, $Q_2(2, -5, 1)$
21. $P_1(0, 0, 5)$, $P_2(4, -2, 0)$; $Q_1(0, 0, 6)$, $Q_2(3, -2, 1)$

22. A *regular tetrahedron* is a 4-sided figure each side of which is an equilateral triangle. Find 4 points in space which are the vertices of a regular tetrahedron with each edge of length 2 units.

23. A *regular pyramid* is a 5-sided figure with a square base and sides consisting of 4 congruent isosceles triangles. If the base has a side of length 4 and if the height of the pyramid is 6 units, find the area of each of the triangular faces.

24. The points $P_1(1, 2, 3)$, $P_2(2, 1, 2)$, $P_3(3, 0, 1)$, $P_4(5, 2, 7)$ are the vertices of a plane quadrilateral. Find the coordinates of the midpoints of the sides. What kind of quadrilateral do these four midpoints form?

25. Prove that the four interior diagonals of a parallelepiped bisect each other.

3. EQUATIONS OF A LINE

In plane analytic geometry a single equation of the first degree,

$$Ax + By + C = 0,$$

is the equation of a line (so long as A and B are not both zero). *In the geometry of three dimensions such an equation represents a plane.* Although we shall postpone a systematic study of planes until the next section, we assert now that in three-dimensional geometry it is not possible to represent a line by a single first-degree equation.

A line in space is determined by two points. If $P_0(x_0, y_0, z_0)$ and $P_1(x_1, y_1, z_1)$ are given points, we seek an analytic method of representing the line L determined by these points. The result is obtained by solving a locus problem. A point $P(x, y, z)$ is on L if and only if the direction numbers determined by P and P_0 are proportional to those determined by P_1 and P_0 (Fig. 1–11). Calling the proportionality constant t, we see that the conditions are

$$x - x_0 = t(x_1 - x_0),$$
$$y - y_0 = t(y_1 - y_0),$$
$$z - z_0 = t(z_1 - z_0).$$

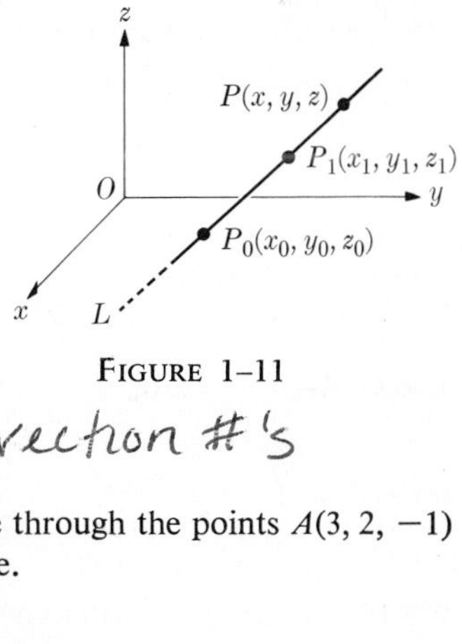

Thus we obtain the *two-point form of the parametric equations of a line*:

$$x = x_0 + (x_1 - x_0)t,$$
$$y = y_0 + (y_1 - y_0)t, \qquad (1)$$
$$z = z_0 + (z_1 - z_0)t.$$

FIGURE 1–11

direction #'s

EXAMPLE 1. Find the parametric equations of the line through the points $A(3, 2, -1)$ and $B(4, 4, 6)$. Locate three additional points on the line.

Solution. Substituting in (1) we obtain

$$x = 3 + t, \qquad y = 2 + 2t, \qquad z = -1 + 7t.$$

To get an additional point on the line we let $t = 2$ and obtain $P_1(5, 6, 13)$; $t = -1$ yields $P_2(2, 0, -8)$ and $t = 3$ gives $P_3(6, 8, 20)$.

Theorem 4. *The parametric equations of a line L through the point $P_0(x_0, y_0, z_0)$ with direction numbers a, b, c are given by*

$$x = x_0 + at, \qquad y = y_0 + bt, \qquad z = z_0 + ct. \qquad (2)$$

Proof. The point $P_1(x_0 + a, y_0 + b, z_0 + c)$ must be on L, since the direction numbers formed by P_0 and P_1 are just a, b, c. Using the two-point form (1) for the equations of a line through P_0 and P_1, we get (2) precisely.

EXAMPLE 2. Find the parametric equations of the line L through the point $A(3, -2, 5)$ with direction numbers 4, 0, -2. What is the relation of L to the coordinate planes?

Solution. Substituting in (2), we obtain

$$x = 3 + 4t, \qquad y = -2, \qquad z = 5 - 2t.$$

Since all points on the line must satisfy all three of the above equations, L must lie in the plane $y = -2$. This plane is parallel to the xz-plane. Therefore L is parallel to the xz-plane.

If none of the direction numbers is zero, the parameter t may be eliminated from the system of equations (2). We may write

$$\frac{x - x_0}{a} = \frac{y - y_0}{b} = \frac{z - z_0}{c} \tag{3}$$

for the equations of a line. For any value of t in (2) the ratios in (3) are equal. Conversely, if the ratios in (3) are all equal we may set the common value equal to t and (2) is satisfied.

If one of the direction numbers is zero, the form (3) may still be used if the zero in the denominator is interpreted properly. The equations

$$\frac{x - x_0}{a} = \frac{y - y_0}{b} = \frac{z - z_0}{0}$$

are understood to stand for the equations

$$\frac{x - x_0}{a} = \frac{y - y_0}{b} \qquad \text{and} \qquad z = z_0.$$

The system

$$\frac{x - x_0}{0} = \frac{y - y_0}{b} = \frac{z - z_0}{0}$$

stands for

$$x = x_0 \qquad \text{and} \qquad z = z_0.$$

We recognize these last two equations as those of planes parallel to coordinate planes. In other words, *a line is represented as the intersection of two planes.* This point will be discussed further in the next section.

The two-point form for the equations of a line also may be written *symmetrically.* We have

$$\frac{x - x_0}{x_1 - x_0} = \frac{y - y_0}{y_1 - y_0} = \frac{z - z_0}{z_1 - z_0}.$$

EXERCISES

In each of problems 1 through 4, find the equations of the line going through the given points. Find two additional points on each line.

1. $P_1(2, 3, 4)$, $P_2(-1, -3, 2)$
2. $P_1(1, -1, 3)$, $P_2(2, 1, 5)$
3. $P_1(1, 2, -1)$, $P_2(3, -1, 2)$
4. $P_1(-1, 2, 1)$, $P_2(3, 1, -1)$

In each of problems 5 through 9, find the equations of the line passing through the given point with the given direction numbers.

5. $P_1(1, 0, -1)$, direction numbers 2, 1, -3
6. $P_1(-2, 1, 3)$, direction numbers 3, -1, -2
7. $P_1(4, 0, 0)$, direction numbers 2, -1, -3
8. $P_1(1, 2, 0)$, direction numbers 0, 1, 3
9. $P_1(3, -1, -2)$, direction numbers 2, 0, 0

In each of problems 10 through 14, decide whether or not L_1 and L_2 are perpendicular.

10. L_1: $\dfrac{x-2}{2} = \dfrac{y+1}{-3} = \dfrac{z-1}{4}$; $\quad L_2$: $\dfrac{x-2}{-3} = \dfrac{y+1}{2} = \dfrac{z-1}{3}$

11. L_1: $\dfrac{x}{1} = \dfrac{y+1}{2} = \dfrac{z+1}{3}$; $\quad L_2$: $\dfrac{x-3}{0} = \dfrac{y+1}{-3} = \dfrac{z+4}{2}$

12. L_1: $\dfrac{x+2}{-1} = \dfrac{y-2}{2} = \dfrac{z+3}{3}$; $\quad L_2$: $\dfrac{x+2}{1} = \dfrac{y-2}{2} = \dfrac{z+3}{-1}$

13. L_1: $\dfrac{x+5}{4} = \dfrac{y-1}{3} = \dfrac{z+8}{5}$; $\quad L_2$: $\dfrac{x-4}{3} = \dfrac{y+7}{2} = \dfrac{z+4}{1}$

14. L_1: $\dfrac{x+1}{0} = \dfrac{y-2}{1} = \dfrac{z+8}{0}$; $\quad L_2$: $\dfrac{x-3}{1} = \dfrac{y+2}{0} = \dfrac{z-1}{0}$

15. Find the equations of the medians of the triangle with vertices at $A(4, 0, 2)$, $B(3, 1, 4)$, $C(2, 5, 0)$.

16. Find the points of intersection of the line

$$x = 3 + 2t, \qquad y = 7 + 8t, \qquad z = -2 + t$$

with each of the coordinate planes.

17. Find the points of intersection of the line

$$\frac{x+1}{-2} = \frac{y+1}{3} = \frac{z-2}{7}$$

with each of the coordinate planes.

18. Show that the following lines are coincident:

$$\frac{x-1}{2} = \frac{y+1}{-3} = \frac{z}{4}; \qquad \frac{x-5}{2} = \frac{y+7}{-3} = \frac{z-8}{4}.$$

19. Find the equations of the line through $(3, 1, 5)$ which is parallel to the line

$$x = 4 - t, \qquad y = 2 + 3t, \qquad z = -4 + t.$$

20. Find the equations of the line through $(3, 1, -2)$ which is perpendicular and intersects the line

$$\frac{x+1}{1} = \frac{y+2}{1} = \frac{z+1}{1}.$$

[*Hint:* Let (x_0, y_0, z_0) be the point of intersection and determine its coordinates.]

21. A triangle has vertices at $A(2, 1, 6)$, $B(-3, 2, 4)$ and $C(5, 8, 7)$. Perpendiculars are drawn from these vertices to the xz-plane. Locate the points A', B', and C' which are the intersections of the perpendiculars through A, B, C and the xz-plane. Find the equations of the sides of the triangle $A'B'C'$.

4. THE PLANE

Any three points not on a straight line determine a plane. While this characterization of a plane is quite simple, it is not convenient for beginning the study of planes. Instead we use the fact that *there is exactly one plane which passes through a given point and is perpendicular to a given line.*

Let $P_0(x_0, y_0, z_0)$ be a given point, and suppose that a given line L goes through the point $P_1(x_1, y_1, z_1)$ and has direction numbers A, B, C.

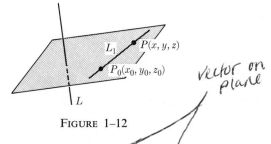

vector on plane

FIGURE 1–12

Theorem 5. *The equation of the plane passing through P_0 and perpendicular to L is*

normal

$$A(x - x_0) + B(y - y_0) + C(z - z_0) = 0.$$

Proof. We establish the result by solving a locus problem. Let $P(x, y, z)$ be a point on the locus (Fig. 1–12). From Euclidean geometry we recall that if a line L_1 through P_0 and P is perpendicular to L, then P must be in the desired plane. A set of direction numbers for the line L_1 is

$$x - x_0, \qquad y - y_0, \qquad z - z_0.$$

Since L has direction numbers A, B, C, we conclude that the two lines L and L_1 are perpendicular if and only if their direction numbers satisfy the relation

$$A(x - x_0) + B(y - y_0) + C(z - z_0) = 0,$$

which is the equation we seek.

Remark. Note that only the direction of L—and not the coordinates of P_1—enters the above equation. We obtain the same plane and the same equation if any line parallel to L is used in its stead.

EXAMPLE 1. Find the equation of the plane through the point $P_0(5, 2, -3)$ which is perpendicular to the line through the points $P_1(5, 4, 3)$ and $P_2(-6, 1, 7)$.

Solution. The line through the points P_1 and P_2 has direction numbers $-11, -3, 4$. The equation of the plane is

$$-11(x - 5) - 3(y - 2) + 4(z + 3) = 0$$

or

$$11x + 3y - 4z - 73 = 0.$$

All lines perpendicular to the same plane are parallel and therefore have proportional direction numbers.

DEFINITION. *A set of* **attitude numbers** *of a plane is any set of direction numbers of a line perpendicular to the plane.*

In Example 1 above, $11, 3, -4$ form a set of attitude numbers of the plane.

EXAMPLE 2. What are sets of attitude numbers for planes parallel to the coordinate planes?

Solution. A plane parallel to the *yz*-plane has an equation of the form

$$x - c = 0,$$

where c is a constant. A set of attitude numbers for this plane is $1, 0, 0$. A plane parallel to the *xz*-plane has attitude numbers $0, 1, 0$, and any plane parallel to the *xy*-plane has attitude numbers $0, 0, 1$.

Since lines perpendicular to the same or parallel planes are themselves parallel, we get at once the next theorem.

Theorem 6. *Two planes are parallel if and only if their attitude numbers are proportional.*

Theorem 7. *If A, B, and C are not all zero, the locus of an equation of the form*

$$Ax + By + Cz + D = 0 \qquad \text{normal} \qquad (1)$$

is a plane.

Proof. Suppose that $C \neq 0$, for example. Then the point $P_0(0, 0, -D/C)$ is on the locus, as its coordinates satisfy the above equation. Therefore we may write

$$A(x - 0) + B(y - 0) + C\left(z + \frac{D}{C}\right) = 0,$$

and the locus is the plane passing through P_0 perpendicular to any line with direction numbers A, B, C.

An equation of the plane through three points not on a line can be found by assuming that the plane has an equation of the form (1), substituting in turn the coordinates of the points, and solving simultaneously the three resulting equations. The fact that there are four constants, A, B, C, D, and only three equations is illusory, since we may divide through by one of them (say D) and obtain three equations in the unknowns A/D, B/D, C/D. This is equivalent to setting D (or one of the other constants) equal to some convenient value. An example illustrates the procedure.

EXAMPLE 3. Find an equation of the plane passing through the points $(2, 1, 3)$, $(1, 3, 2)$, $(-1, 2, 4)$.

Solution. Since the three points lie in the plane, each of them satisfies equation (1). We have

$$(2, 1, 3): \qquad 2A + B + 3C + D = 0,$$
$$(1, 3, 2): \qquad A + 3B + 2C + D = 0,$$
$$(-1, 2, 4): \qquad -A + 2B + 4C + D = 0.$$

Solving for A, B, C in terms of D, we obtain

$$A = -\tfrac{3}{25}D, \qquad B = -\tfrac{4}{25}D, \qquad C = -\tfrac{5}{25}D.$$

Setting $D = -25$, we get the equation

$$3x + 4y + 5z - 25 = 0.$$

EXERCISES

In each of problems 1 through 4, find the equation of the plane which passes through the given point P and has the given attitude numbers.

1. $P(0, 5, 6); 4, 2, -1$
2. $P(-3, 1, 4); 0, 2, 5$
3. $P(1, -2, -3); 4, 0, -1$
4. $P(2, 0, 6); 0, 0, -5$

In each of problems 5 through 8, find the equation of the plane which passes through the three points.

5. $(1, -2, 1), (2, 0, 3), (0, 1, -1)$
6. $(2, 2, 1), (-1, 2, 3), (3, -5, -2)$
7. $(3, -1, 2), (1, 2, -1), (2, 3, 1)$
8. $(-1, 3, 1), (2, 1, 2), (4, 2, -1)$

In each of problems 9 through 12, find the equation of the plane passing through P_1 and perpendicular to the line L_1.

9. $P_1(2, -1, 3);$ $L_1: x = -1 + 2t, \ y = 1 + 3t, \ z = -4t$
10. $P_1(1, 2, -3);$ $L_1: x = t, \ y = -2 - 2t, \ z = 1 + 3t$
11. $P_1(2, -1, -2);$ $L_1: x = 2 + 3t, \ y = 0, \ z = -1 - 2t$
12. $P_1(-1, 2, -3);$ $L_1: x = -1 + 5t, \ y = 1 + 2t, \ z = -1 + 3t$

In each of problems 13 through 16, find the equations of the line through P_1 and perpendicular to the given plane M_1.

13. $P_1(-2, 3, 1)$; M_1: $2x + 3y + z - 3 = 0$
14. $P_1(1, -2, -3)$; M_1: $3x - y - 2z + 4 = 0$
15. $P_1(-1, 0, -2)$; M_1: $x + 2z + 3 = 0$
16. $P_1(2, -1, -3)$; M_1: $x = 4$

In each of problems 17 through 20, find an equation of the plane through P_1 and parallel to the plane Φ.

17. $P_1(1, -2, -1)$; Φ: $3x + 2y - z + 4 = 0$
18. $P_1(-1, 3, 2)$; Φ: $2x + y - 3z + 5 = 0$
19. $P_1(2, -1, 3)$; Φ: $x - 2y - 3z + 6 = 0$
20. $P_1(3, 0, 2)$; Φ: $x + 2y + 1 = 0$

In each of problems 21 through 23, find equations of the line through P_1 parallel to the given line L.

21. $P_1(2, -1, 3)$; L: $\dfrac{x - 1}{3} = \dfrac{y + 2}{-2} = \dfrac{z - 2}{4}$

22. $P_1(0, 0, 1)$; L: $\dfrac{x + 2}{1} = \dfrac{y - 1}{3} = \dfrac{z + 1}{-2}$

23. $P_1(1, -2, 0)$; L: $\dfrac{x - 2}{2} = \dfrac{y + 2}{-1} = \dfrac{z - 3}{4}$

In each of problems 24 through 28, find the equation of the plane containing L_1 and L_2.

24. L_1: $\dfrac{x + 1}{2} = \dfrac{y - 2}{3} = \dfrac{z - 1}{1}$; L_2: $\dfrac{x + 1}{1} = \dfrac{y - 2}{-1} = \dfrac{z - 1}{2}$

25. L_1: $\dfrac{x - 1}{3} = \dfrac{y + 2}{2} = \dfrac{z - 2}{2}$; L_2: $\dfrac{x - 1}{1} = \dfrac{y + 2}{1} = \dfrac{z - 2}{0}$

26. L_1: $\dfrac{x + 2}{1} = \dfrac{y}{0} = \dfrac{z + 1}{2}$; L_2: $\dfrac{x + 2}{2} = \dfrac{y}{3} = \dfrac{z + 1}{1}$

27. L_1: $\dfrac{x}{2} = \dfrac{y - 1}{3} = \dfrac{z + 2}{-1}$; L_2: $\dfrac{x - 2}{2} = \dfrac{y + 1}{3} = \dfrac{z}{-1}$ $(L_1 \parallel L_2)$

28. L_1: $\dfrac{x - 2}{2} = \dfrac{y + 1}{-1} = \dfrac{z}{3}$; L_2: $\dfrac{x + 1}{2} = \dfrac{y}{-1} = \dfrac{z + 2}{3}$ $(L_1 \parallel L_2)$

In problems 29 and 30, find the equation of the plane through P_1 and the given line L.

29. $P_1(3, -1, 2)$; L: $\dfrac{x - 2}{2} = \dfrac{y + 1}{3} = \dfrac{z}{-2}$

30. $P_1(1, -2, 3)$; L: $x = -1 + t$, $y = 2 + 2t$, $z = 2 - 2t$

31. Show that the plane $2x - 3y + z - 2 = 0$ is parallel to the line

$$\frac{x - 2}{1} = \frac{y + 2}{1} = \frac{z + 1}{1}.$$

32. Show that the plane $5x - 3y - z - 6 = 0$ contains the line

$$x = 1 + 2t, \qquad y = -1 + 3t, \qquad z = 2 + t.$$

33. A plane has attitude numbers A, B, C, and a line has direction numbers a, b, c. What condition must be satisfied in order that the plane and line be parallel?

5. ANGLES. DISTANCE FROM A POINT TO A PLANE

The angle between two lines was defined in Section 2. We recall that if line L_1 has direction cosines λ_1, μ_1, ν_1 and line L_2 has direction cosines λ_2, μ_2, ν_2, then

$$\cos \theta = \lambda_1 \lambda_2 + \mu_1 \mu_2 + \nu_1 \nu_2,$$

where θ is the angle between L_1 and L_2.

DEFINITION. Let Φ_1 and Φ_2 be two planes, and let L_1 and L_2 be two lines which are perpendicular to Φ_1 and Φ_2, respectively. Then the **angle between Φ_1 and Φ_2** is, by definition, the angle between L_1 and L_2. Furthermore, we make the convention that we always select the acute angle between these lines as the angle between Φ_1 and Φ_2.

Theorem 8. *The angle θ between the planes $A_1x + B_1y + C_1z + D_1 = 0$ and $A_2x + B_2y + C_2z + D_2 = 0$ is given by*

$$\cos \theta = \frac{|A_1A_2 + B_1B_2 + C_1C_2|}{\sqrt{A_1^2 + B_1^2 + C_1^2} \; \sqrt{A_2^2 + B_2^2 + C_2^2}}.$$

angle between unit normal vector

Proof. From the definition of attitude numbers for a plane, we know that they are direction numbers of any line perpendicular to the plane. Converting to direction cosines, we get the above formula.

Corollary. *Two planes with attitude numbers A_1, B_1, C_1 and A_2, B_2, C_2 are perpendicular if and only if*

$$A_1A_2 + B_1B_2 + C_1C_2 = 0.$$

EXAMPLE 1. Find $\cos \theta$ where θ is the angle between the planes $3x - 2y + z = 4$ and $x + 4y - 3z - 2 = 0$.

Solution. Substituting in the formula of Theorem 8, we have

$$\cos \theta = \frac{|3 - 8 - 3|}{\sqrt{9 + 4 + 1}\sqrt{1 + 16 + 9}} = \frac{4}{\sqrt{91}}.$$

Two nonparallel planes intersect in a line. Every point on the line satisfies the equations of both planes and, conversely, every point which satisfies the equations of both planes must be on the line. *Therefore we may characterize any line in space by finding two planes which contain it.* Since every line has an unlimited number of planes which pass through it and since *any* two of them are sufficient to determine the line uniquely, we see that there is an unlimited number of ways of writing the equations of a line. The next example shows how to transform one representation into another.

EXAMPLE 2. The two planes

$$2x + 3y - 4z - 6 = 0 \quad \text{and} \quad 3x - y + 2z + 4 = 0$$

intersect in a line. (That is, the points which satisfy *both* equations determine the line.) Find a set of parametric equations of the line of intersection.

Solution. We solve the above equations for x and y in terms of z, getting

$$x = -\tfrac{2}{11}z - \tfrac{6}{11}, \qquad y = \tfrac{16}{11}z + \tfrac{26}{11}$$

and

$$\frac{x + \tfrac{6}{11}}{-\tfrac{2}{11}} = \frac{y - \tfrac{26}{11}}{\tfrac{16}{11}} = \frac{z}{1}.$$

We can therefore write

$$x = -\tfrac{6}{11} - \tfrac{2}{11}t, \qquad y = \tfrac{26}{11} + \tfrac{16}{11}t, \qquad z = t,$$

which are the desired parametric equations.

Three planes may be parallel, may pass through a common line, may have no common points, or may have a unique point of intersection. If they have a unique point of intersection, the intersection point may be found by solving simultaneously the three equations of the planes. If they have no common point, an attempt to solve simultaneously will fail. A further examination will show whether or not two or more of the planes are parallel.

EXAMPLE 3. Determine whether or not the planes Φ_1: $3x - y + z - 2 = 0$; Φ_2: $x + 2y - z + 1 = 0$; Φ_3: $2x + 2y + z - 4 = 0$ intersect. If so, find the point of intersection.

Solution. Eliminating z between Φ_1 and Φ_2, we have

$$4x + y - 1 = 0. \tag{1}$$

Eliminating z between Φ_2 and Φ_3, we find

$$3x + 4y - 3 = 0. \tag{2}$$

We solve equations (1) and (2) simultaneously to get

$$x = \tfrac{1}{13}, \qquad y = \tfrac{9}{13}.$$

Substituting in the equation for Φ_1, we obtain $z = \tfrac{32}{13}$. Therefore the single point of intersection of the three planes is $(\tfrac{1}{13}, \tfrac{9}{13}, \tfrac{32}{13})$.

EXAMPLE 4.   Find the point of intersection of the plane $3x - y + 2z - 3 = 0$ and the line

$$\frac{x + 1}{3} = \frac{y + 1}{2} = \frac{z - 1}{-2}.$$

Solution. We write the equations of the line in parametric form:

$$x = -1 + 3t, \qquad y = -1 + 2t, \qquad z = 1 - 2t.$$

The point of intersection is given by a value of t; call it t_0. This point must satisfy the equation of the plane. We have

$$3(-1 + 3t_0) - (-1 + 2t_0) + 2(1 - 2t_0) - 3 = 0,$$

or

$$t_0 = 1.$$

The desired point is $(2, 1, -1)$.

Theorem 9. *The distance d from the point $P_1(x_1, y_1, z_1)$ to the plane*

$$Ax + By + Cz + D = 0$$

is given by

$$d = \frac{|Ax_1 + By_1 + Cz_1 + D|}{\sqrt{A^2 + B^2 + C^2}}.$$

Proof. We write the equations of the line L through P_1 which is perpendicular to the plane. They are

$$L: x = x_1 + At, \qquad y = y_1 + Bt, \qquad z = z_1 + Ct.$$

Denote by (x_0, y_0, z_0) the intersection of L and the plane. Then

$$d^2 = (x_1 - x_0)^2 + (y_1 - y_0)^2 + (z_1 - z_0)^2. \tag{3}$$

Also (x_0, y_0, z_0) is on both the line and the plane. Therefore, we have for some value t_0

$$x_0 = x_1 + At_0,$$
$$y_0 = y_1 + Bt_0,$$
$$z_0 = z_1 + Ct_0$$

(4)

and

$$Ax_0 + By_0 + Cz_0 + D = 0$$
$$= A(x_1 + At_0) + B(y_1 + Bt_0) + C(z_1 + Ct_0) + D.$$

Thus, from (3) and (4), we write

$$d = \sqrt{A^2 + B^2 + C^2}\,|t_0|,$$

and now, inserting the relation

$$t_0 = \frac{-(Ax_1 + By_1 + Cz_1 + D)}{A^2 + B^2 + C^2}$$

in the preceding expression for d, we obtain the desired formula.

EXAMPLE 5. Find the distance from the point $(2, -1, 5)$ to the plane

$$3x + 2y - 2z - 7 = 0.$$

Solution.

$$d = \frac{|6 - 2 - 10 - 7|}{\sqrt{9 + 4 + 4}} = \frac{13}{\sqrt{17}}.$$

EXERCISES

In each of problems 1 through 4, find $\cos \theta$ where θ is the angle between the given planes.

1. $2x - y + 2z - 3 = 0$, $3x + 2y - 6z - 11 = 0$
2. $x + 2y - 3z + 6 = 0$, $x + y + z - 4 = 0$
3. $2x - y + 3z - 5 = 0$, $3x - 2y + 2z - 7 = 0$
4. $x + 4z - 2 = 0$, $y + 2z - 6 = 0$

In each of problems 5 through 8, find the equations in parametric form of the line of intersection of the given planes.

5. $3x + 2y - z + 5 = 0$, $2x + y + 2z - 3 = 0$
6. $x + 2y + 2z - 4 = 0$, $2x + y - 3z + 5 = 0$
7. $x + 2y - z + 4 = 0$, $2x + 4y + 3z - 7 = 0$
8. $2x + 3y - 4z + 7 = 0$, $3x - 2y + 3z - 6 = 0$

In each of problems 9 through 12, find the point of intersection of the given line and the given plane.

9. $3x - y + 2z - 5 = 0$, $\dfrac{x - 1}{2} = \dfrac{y + 1}{3} = \dfrac{z - 1}{-2}$

10. $2x + 3y - 4z + 15 = 0$, $\dfrac{x + 3}{2} = \dfrac{y - 1}{-2} = \dfrac{z + 4}{3}$

11. $x + 2z + 3 = 0$, $\dfrac{x + 1}{1} = \dfrac{y}{0} = \dfrac{z + 2}{2}$

12. $2x + 3y + z - 3 = 0$, $\dfrac{x + 2}{2} = \dfrac{y - 3}{3} = \dfrac{z - 1}{1}$

In each of problems 13 through 16, find the distance from the given point to the given plane.

13. $(2, 1, -1)$, $x - 2y + 2z + 5 = 0$

14. $(3, -1, 2)$, $3x + 2y - 6z - 9 = 0$

15. $(-1, 3, 2)$, $2x - 3y + 4z - 5 = 0$

16. $(0, 4, -3)$, $3y + 2z - 7 = 0$

17. Find the equation of the plane through the line

$$\frac{x + 1}{3} = \frac{y - 1}{2} = \frac{z - 2}{4}$$

which is perpendicular to the plane

$$2x + y - 3z + 4 = 0.$$

18. Find the equation of the plane through the line

$$\frac{x - 2}{2} = \frac{y - 2}{3} = \frac{z - 1}{-2}$$

which is parallel to the line

$$\frac{x + 1}{3} = \frac{y - 1}{2} = \frac{z + 1}{1}.$$

19. Find the equation of the plane through the line

$$\frac{x + 2}{3} = \frac{y}{-2} = \frac{z + 1}{2}$$

which is parallel to the line

$$\frac{x - 1}{2} = \frac{y + 1}{3} = \frac{z - 1}{4}.$$

20. Find the equations of any line through the point $(1, 4, 2)$ which is parallel to the plane

$$2x + y + z - 4 = 0.$$

21. Find the equation of the plane through $(3, 2, -1)$ and $(1, -1, 2)$ which is parallel to the line
$$\frac{x - 1}{3} = \frac{y + 1}{2} = \frac{z}{-2}.$$

In each of problems 22 through 26, find all the points of intersection of the three given planes. If the three planes pass through a line, find the equations of the line in parametric form.

22. $2x + y - 2z - 1 = 0,\quad 3x + 2y + z - 10 = 0,\quad x + 2y - 3z + 2 = 0$
23. $x + 2y + 3z - 4 = 0,\quad 2x - 3y + z - 2 = 0,\quad 3x + 2y - 2z - 5 = 0$
24. $3x - y + 2z - 4 = 0,\quad x + 2y - z - 3 = 0,\quad 3x - 8y + 7z + 1 = 0$
25. $2x + y - 2z - 3 = 0,\quad x - y + z + 1 = 0,\quad x + 5y - 7z - 3 = 0$
26. $x + 2y + 3z - 5 = 0,\quad 2x - y - 2z - 2 = 0,\quad x - 8y - 13z + 11 = 0.$

In each of problems 27 through 29, find the equations in parametric form of the line through the given point P_1 which intersects and is perpendicular to the given line L.

27. $P_1(3, -1, 2);\quad L: \dfrac{x - 1}{2} = \dfrac{y + 1}{-1} = \dfrac{z}{3}$

28. $P_1(-1, 2, 3);\quad L: \dfrac{x}{2} = \dfrac{y - 2}{0} = \dfrac{z + 3}{-3}$

29. $P_1(0, 2, 4);\quad L: \dfrac{x - 1}{3} = \dfrac{y - 2}{1} = \dfrac{z - 3}{4}$

30. (a) If $A_1x + B_1y + C_1z + D_1 = 0$ and $A_2x + B_2y + C_2z + D_2 = 0$ are two intersecting planes, what is the locus of all points which satisfy
$$A_1x + B_1y + C_1z + D_1 + k(A_2x + B_2y + C_2z + D_2) = 0,$$
where k is a constant? (b) Find the equation of the plane passing through the point $(2, 1, -3)$ and the intersection of the planes $3x + y - z - 2 = 0$, $2x + y + 4z - 1 = 0$.

6.   THE SPHERE. CYLINDERS

A **sphere** is the locus of all points at a given distance from a fixed point. The fixed point is called the **center** and the fixed distance is called the **radius.**

If the center is at the point (h, k, l), the radius is r, and (x, y, z) is any point on the sphere, then, from the formula for the distance between two points, we obtain the relation
$$(x - h)^2 + (y - k)^2 + (z - l)^2 = r^2. \tag{1}$$

Equation (1) is the *equation of a sphere.* If it is multiplied out and the terms collected we have the equivalent form
$$x^2 + y^2 + z^2 + Dx + Ey + Fz + G = 0. \tag{2}$$

EXAMPLE 1. Find the center and radius of the sphere with equation

$$x^2 + y^2 + z^2 + 4x - 6y + 9z - 6 = 0.$$

Solution. We complete the square by first writing

$$x^2 + 4x \qquad + y^2 - 6y \qquad + z^2 + 9z \qquad = 6;$$

then, adding the appropriate quantities to both sides, we have

$$x^2 + 4x + 4 + y^2 - 6y + 9 + z^2 + 9z + \tfrac{81}{4} = 6 + 4 + 9 + \tfrac{81}{4}$$

and

$$(x + 2)^2 + (y - 3)^2 + (z + \tfrac{9}{2})^2 = \tfrac{157}{4}.$$

The center is at $(-2, 3, -\tfrac{9}{2})$ and the radius is $\tfrac{1}{2}\sqrt{157}$.

EXAMPLE 2. Find the equation of the sphere which passes through $(2, 1, 3)$, $(3, 2, 1)$, $(1, -2, -3)$, $(-1, 1, 2)$.

Solution. Substituting these points in the form (2) above for the equation of a sphere, we obtain

$$
\begin{aligned}
(2, 1, 3): &\quad 2D + E + 3F + G = -14, \\
(3, 2, 1): &\quad 3D + 2E + F + G = -14, \\
(1, -2, -3): &\quad D - 2E - 3F + G = -14, \\
(-1, 1, 2): &\quad -D + E + 2F + G = -6.
\end{aligned}
$$

Solving these by elimination (first G, then D, then F) we obtain, successively,

$$D + E - 2F = 0, \qquad D + 3E + 6F = 0, \qquad 3D + F = -8,$$

and

$$2E + 8F = 0, \qquad -3E + 7F = -8; \qquad \text{and so} \qquad -38E = -64.$$

Therefore

$$E = \tfrac{32}{19}, \qquad F = -\tfrac{8}{19}, \qquad D = -\tfrac{48}{19}, \qquad G = -\tfrac{178}{19}.$$

The desired equation is

$$x^2 + y^2 + z^2 - \tfrac{48}{19}x + \tfrac{32}{19}y - \tfrac{8}{19}z - \tfrac{178}{19} = 0.$$

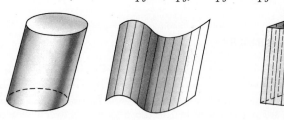

FIGURE 1–13

A **cylindrical surface** is a surface which consists of a collection of parallel lines. Each of the parallel lines is called a **generator** of the **cylinder** or cylindrical surface.

The customary right circular cylinder of elementary geometry is clearly a special case of the type of cylinder we are considering. Figure 1–13 shows some examples of cylindrical surfaces. Note that a plane is a cylinder.

Theorem 10. *An equation of the form*

$$f(x, y) = 0$$

is a cylindrical surface with generators all parallel to the z-axis. The surface intersects the xy-plane in the curve

$$f(x, y) = 0, \qquad z = 0.$$

A similar result holds with axes interchanged.

Proof. Suppose that x_0, y_0 satisfies $f(x_0, y_0) = 0$. Then any point (x_0, y_0, z) for $-\infty < z < \infty$ satisfies the same equation, since z is absent. Therefore the line parallel to the z-axis through $(x_0, y_0, 0)$ is a generator.

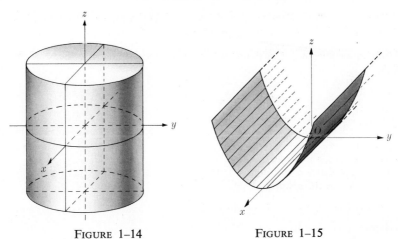

FIGURE 1–14   FIGURE 1–15

EXAMPLE 3. Describe and sketch the locus of the equation $x^2 + y^2 = 9$.

Solution. The locus is a right circular cylinder with generators parallel to the z-axis (Theorem 10). It is sketched in Fig. 1–14.

EXAMPLE 4. Describe and sketch the locus of the equation

$$y^2 = 4z.$$

Solution. According to Theorem 10 the locus is a cylinder with generators parallel to the x-axis. The intersection with the yz-plane is a parabola. The locus, called a parabolic cylinder, is sketched in Fig. 1–15.

EXERCISES

In each of problems 1 through 4, find the equation of the sphere with center C and radius r.

1. $C(2, 0, 1), r = 4$ 2. $C(3, -2, 1), r = 6$

3. $C(3, -2, 6), r = 7$ 4. $C(2, 1, -4), r = 3$

In each of problems 5 through 9, determine the locus of the equation. If it is a sphere find its center and radius.

5. $x^2 + y^2 + z^2 + 2x - 4z + 1 = 0$

6. $x^2 + y^2 + z^2 - 4x + 2y + 6z - 2 = 0$

7. $x^2 + y^2 + z^2 - 2x + 4y - 2z + 7 = 0$

8. $x^2 + y^2 + z^2 + 4x - 2y + 4z + 9 = 0$

9. $x^2 + y^2 + z^2 - 6x + 4y + 2z + 10 = 0$

10. Find the equation of the locus of all points which are twice as far from $A(3, -1, 2)$ as from $B(0, 2, -1)$.

11. Find the equation of the locus of all points which are three times as far from $A(2, 1, -3)$ as from $B(-2, -3, 5)$.

12. Find the equation of the locus of all points whose distances from the point $(0, 0, 4)$ are equal to their perpendicular distances from the xy-plane.

In each of problems 13 through 24, describe and sketch the locus of the given equation.

13. $x = 3$	14. $x^2 + y^2 = 16$	15. $2x + y = 3$
16. $x + 2z = 4$	17. $x^2 = 4z$	18. $z = 2 - y^2$
19. $4x^2 + y^2 = 16$	20. $4x^2 - y^2 = 16$	21. $y^2 + x^2 = 9$
22. $z^2 = 2 - 2x$	23. $x^2 + y^2 - 2x = 0$	24. $z^2 = y^2 + 4$

In each of problems 25 through 28, describe the curve of intersection, if any, of the given surface S and the given plane Φ.

25. $S: x^2 + y^2 + z^2 = 25;$ $\Phi: z = 3$

26. $S: 4x = y^2 + z^2;$ $\Phi: x = 4$

27. $S: x^2 + 2y^2 + 3z^2 = 12;$ $\Phi: y = 4$

28. $S: x^2 + y^2 = z^2;$ $\Phi: 2x + z = 4$

7. QUADRIC SURFACES

In the plane any equation of the form

$$Ax^2 + Bxy + Cy^2 + Dx + Ey + F = 0$$

is the representation of a curve. More specifically, we have found that circles, parabolas, ellipses, and hyperbolas, i.e., all conic sections, are represented by such second-degree equations.

In three-space the most general equation of the second degree in x, y, and z has the form

$$ax^2 + by^2 + cz^2 + dxy + exz + fyz + gx + hy + kz + l = 0, \quad (1)$$

where the quantities a, b, c, $\ldots$, l are positive or negative numbers or zero. The points in space satisfying such an equation all lie on a surface. Certain special cases, such as spheres and cylinders, were discussed in Section 6. Any second-degree equation which does not reduce to a cylinder, plane, line, or point corresponds to a surface which we call **quadric.** Quadric surfaces are classified into six types, and it can be shown that every second-degree equation which does not degenerate into a cylinder, a plane, etc., corresponds to one of these six types. The proof of this result involves the study of translation and rotation of coordinates in three-dimensional space, a topic beyond the scope of this book.

DEFINITIONS. The **x-, y-, and z-intercepts** of a surface are, respectively, the x-, y-, and z-coordinates of the points of intersection of the surface with the respective axes. When we are given an equation of a surface, we get the x-intercept by setting y and z equal to zero and solving for x. We proceed analogously for the y- and z-intercepts.

The **traces** of a surface on the coordinate planes are the curves of intersections of the surface with the coordinate planes. When we are given a surface, we obtain the trace on the xz-plane by first setting y equal to zero and then considering the resulting equation in x and z as the equation of a curve in the plane, as in plane analytic geometry. A **section of a surface by a plane** is the curve of intersection of the surface with the plane.

———————

EXAMPLE 1. Find the x-, y-, and z-intercepts of the surface

$$3x^2 + 2y^2 + 4z^2 = 12.$$

Describe the traces of this surface. Find the section of this surface by the plane $z = 1$ and by the plane $x = 3$.

Solution. We set $y = z = 0$, getting $3x^2 = 12$; the x-intercepts are at 2 and -2. Similarly, the y-intercepts are at $\pm\sqrt{6}$, the z-intercepts at $\pm\sqrt{3}$. To find the trace on the xy-plane, we set $z = 0$, getting

$$3x^2 + 2y^2 = 12 \quad \text{or} \quad \frac{x^2}{4} + \frac{y^2}{6} = 1.$$

We recognize this curve as an ellipse, with major semi-axis $\sqrt{6}$, minor semi-axis 2, foci at $(0, \sqrt{2}, 0)$, $(0, -\sqrt{2}, 0)$. Similarly, the trace on the xz-plane is the ellipse

$$\frac{x^2}{4} + \frac{z^2}{3} = 1,$$

and the trace on the yz-plane is the ellipse

$$\frac{y^2}{6} + \frac{z^2}{3} = 1.$$

The section of the surface by the plane $z = 1$ is the curve

$$3x^2 + 2y^2 + 4 = 12 \quad \text{or} \quad \frac{x^2}{8/3} + \frac{y^2}{4} = 1,$$

which we recognize as an ellipse. The section by the plane $x = 3$ is the curve

$$27 + 2y^2 + 4z^2 = 12 \quad \text{or} \quad 2y^2 + 4z^2 + 15 = 0.$$

Since the sum of three positive quantities can never be zero, we conclude that the plane $x = 3$ does not intersect the surface. The section is void.

DEFINITIONS. A surface is **symmetric with respect to the xy-plane** if and only if the point $(x, y, -z)$ lies on the surface whenever (x, y, z) does; *it is* **symmetric with respect to the x-axis** if and only if the point $(x, -y, -z)$ is on the locus whenever (x, y, z) is. Similar definitions are easily formulated for symmetry with respect to the remaining coordinate planes and axes.

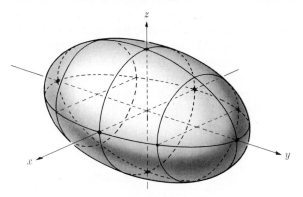

FIGURE 1–16

The notions of intercepts, traces, and symmetry are useful in the following description of the six types of quadric surfaces.

(i) An **ellipsoid** is the locus of an equation of the form

$$\frac{x^2}{A^2} + \frac{y^2}{B^2} + \frac{z^2}{C^2} = 1.$$

The surface is sketched in Fig. 1–16. The x-, y-, and z-intercepts are the numbers $\pm A$, $\pm B$, $\pm C$, respectively, and the traces on the xy-, xz-, and yz-planes are, respectively, the ellipses

$$\frac{x^2}{A^2} + \frac{y^2}{B^2} = 1, \quad \frac{x^2}{A^2} + \frac{z^2}{C^2} = 1, \quad \frac{y^2}{B^2} + \frac{z^2}{C^2} = 1.$$

Sections made by the planes $y = k$ (k a constant) are the similar ellipses

$$\frac{x^2}{A^2(1 - k^2/B^2)} + \frac{z^2}{C^2(1 - k^2/B^2)} = 1, \quad y = k, \quad -B < k < B.$$

Several such ellipses are drawn in Fig. 1–16.

If $A = B = C$, we obtain a sphere while, if two of the three numbers are equal, the surface is an *ellipsoid of revolution*, also called a **spheroid.** If, for example, $A = B$ and $C > A$, the surface is called a *prolate spheroid*, exemplified by a football. On the other hand, if $A = B$ and $C < A$, we have an *oblate spheroid*. The earth is approximately the shape of an oblate spheroid, with the section at the equator being circular and the distance between the North and South poles being smaller than the diameter of the equatorial circle.

(ii) An **elliptic hyperboloid of one sheet** is the locus of an equation of the form

$$\frac{x^2}{A^2} + \frac{y^2}{B^2} - \frac{z^2}{C^2} = 1.$$

A locus of such a surface is sketched in Fig. 1–17. The x-intercepts are at $\pm A$ and the y-intercepts at $\pm B$. As for the z-intercepts, we must solve the equation $-z^2/C^2 = 1$, which has no real solutions. Therefore the surface does not intersect the z-axis. The trace on the xy-plane is an ellipse, while the traces on the yz- and xz-planes are hyperbolas. The sections made by any plane $z = k$ are the ellipses

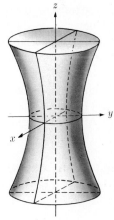

$$\frac{x^2}{A^2(1 + k^2/C^2)} + \frac{y^2}{B^2(1 + k^2/C^2)} = 1,$$

and the sections made by the planes $y = k$ are the hyperbolas

$$\frac{x^2}{A^2(1 - k^2/B^2)} - \frac{z^2}{C^2(1 - k^2/B^2)} = 1.$$

(iii) An **elliptic hyperboloid of two sheets** is the locus of an equation of the form

$$\frac{x^2}{A^2} - \frac{y^2}{B^2} - \frac{z^2}{C^2} = 1.$$

FIGURE 1–17

Such a locus is sketched in Fig. 1–18. We observe that we must have $|x| \geq A$, for otherwise the quantity $(x^2/A^2) < 1$ and the left side of the above equation will always be less than the right side. The x-intercepts are at $x = \pm A$. There are no y- and z-intercepts. The traces on the xz- and xy-planes are hyperbolas; there is no trace on the yz-plane. The sections made by the planes $x = k$ are the ellipses

$$\frac{y^2}{B^2(k^2/A^2 - 1)} + \frac{z^2}{C^2(k^2/A^2 - 1)} = 1, \qquad \text{if} \ \ |k| > A,$$

while the trace is void if $|k| < A$. The sections by the planes $y = k$ and $z = k$ are hyperbolas.

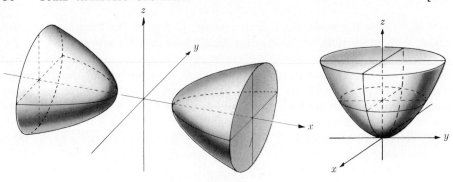

FIGURE 1–18   FIGURE 1–19

(iv) An **elliptic paraboloid** is the locus of an equation of the form

$$x^2/A^2 + y^2/B^2 = z.$$

A typical elliptic paraboloid is sketched in Fig. 1–19.

All three intercepts are 0; the traces on the yz- and xz-planes are parabolas, while the trace on the xy-plane consists of a single point, the origin. Sections made by planes $z = k$ are ellipses if $k > 0$, void if $k < 0$. Sections made by planes $x = k$ and $y = k$ are parabolas.

If $A = B$, we have a *paraboloid of revolution*, and the sections made by the planes $z = k, k > 0$ are circles. The reflecting surfaces of telescopes, automobile headlights, etc., are always paraboloids of revolution. (See Chapter 10, Section 2, of *First Course.**)

(v) A **hyperbolic paraboloid** is the locus of an equation of the form

$$x^2/A^2 - y^2/B^2 = z.$$

Such a locus is sketched in Fig. 1–20. As in the elliptic paraboloid, all intercepts are zero. The trace on the xz-plane is a parabola opening upward; the trace

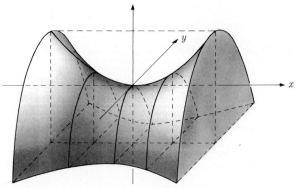

FIGURE 1–20

* Protter-Morrey, *Calculus with Analytic Goemetry, A First Course:* Reading, Mass., Addison-Wesley, 1963. Throughout the text we shall refer to this preceding volume as *First Course.*

on the yz-plane is a parabola opening downward; and the trace on the xy-plane is the pair of intersecting straight lines

$$y = \pm(B/A)\,x.$$

As Fig. 1–20 shows, the surface is "saddle-shaped"; sections made by planes $x = k$ are parabolas opening downward and those made by planes $y = k$ are parabolas opening upward. The sections made by planes $z = k$ are hyperbolas facing one way if $k < 0$ and the other way if $k > 0$. The trace on the xy-plane corresponds to $k = 0$ and, as we saw, consists of two intersecting lines.

  (vi) An **elliptic cone** is the locus of an equation of the form

$$\frac{x^2}{A^2} + \frac{y^2}{B^2} = \frac{z^2}{C^2}.$$

A typical cone of this type is shown in Fig. 1–21. Once again all intercepts are zero. The traces on the xz- and yz-planes are pairs of intersecting straight lines, while the trace on the xy-plane is a single point, the origin. Planes parallel to the coordinate planes yield sections which are the familiar conic sections of plane analytic geometry.

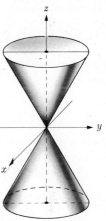

EXAMPLE 2. Name and sketch the locus of

$$4x^2 - 9y^2 + 8z^2 = 72.$$

Indicate a few sections parallel to the coordinate plane.

Solution. We divide by 72, getting

$$\frac{x^2}{18} - \frac{y^2}{8} + \frac{z^2}{9} = 1,$$

FIGURE 1–21

which is an elliptic hyperboloid of one sheet. The x-intercepts are $\pm3\sqrt{2}$, the z-intercepts are ±3, and there are no y-intercepts. The trace on the xy-plane is the hyperbola

$$\frac{x^2}{18} - \frac{y^2}{8} = 1,$$

the trace on the xz-plane is the ellipse

$$\frac{x^2}{18} + \frac{z^2}{9} = 1,$$

and the trace on the yz-plane is the hyperbola

$$\frac{z^2}{9} - \frac{y^2}{8} = 1.$$

The surface is sketched in Fig. 1–22.

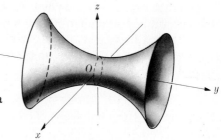

FIGURE 1–22

EXERCISES

Name and sketch the locus of each of the following equations.

1. $\dfrac{x^2}{16} + \dfrac{y^2}{9} + \dfrac{z^2}{4} = 1$ 2. $\dfrac{x^2}{9} + \dfrac{y^2}{12} + \dfrac{z^2}{9} = 1$ 3. $\dfrac{x^2}{16} + \dfrac{y^2}{20} + \dfrac{z^2}{20} = 1$

4. $\dfrac{x^2}{16} + \dfrac{y^2}{9} - \dfrac{z^2}{4} = 1$ 5. $\dfrac{x^2}{16} - \dfrac{y^2}{9} - \dfrac{z^2}{4} = 1$ 6. $\dfrac{x^2}{16} - \dfrac{y^2}{9} + \dfrac{z^2}{4} = 1$

7. $\dfrac{y^2}{9} + \dfrac{z^2}{4} = 1 + \dfrac{x^2}{16}$ 8. $-\dfrac{x^2}{16} - \dfrac{y^2}{9} + \dfrac{z^2}{4} = 1$ 9. $\dfrac{x}{4} = \dfrac{y^2}{4} + \dfrac{z^2}{9}$

10. $z = \dfrac{y^2}{4} - \dfrac{x^2}{9}$ 11. $y = \dfrac{x^2}{8} + \dfrac{z^2}{8}$ 12. $z^2 = 4x^2 + 4y^2$

13. $y^2 = x^2 + z^2$ 14. $z^2 = x^2 - y^2$ 15. $2x^2 + 6y^2 - 3z^2 = 8$

16. $4x^2 - 3y^2 + 2z^2 = 0$ 17. $3x = 2y^2 - 5z^2$ 18. $\dfrac{y}{5} = 8z^2 - 2x^2$

8. TRANSLATION OF AXES

Figure 1–23(a) shows a rectangular coordinate system in three dimensions. Suppose we introduce a second rectangular coordinate system, with axes x', y', and z' so located that the x'-axis is parallel to the x-axis and h units from it, the y'-axis is parallel to the y-axis and k units from it, and the z'-axis is parallel to the z-axis and l units from it (Fig. 1–23b). A point P in space will have coordinates in both systems. If its coordinates are (x, y, z) in the original system and (x', y', z') in the second system, the equations

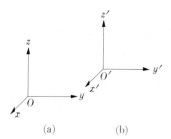

(a)   (b)

$$x' = x - h,$$
$$y' = y - k,$$
$$z' = z - l$$

FIGURE 1–23

hold. Two coordinate systems, xyz and $x'y'z'$, which satisfy these equations are said to be related by a *translation of axes*.

As in the case of plane analytic geometry (*First Course*, Chapter 10, Section 7), the method of translation of axes may be used to simplify second-degree equations, thereby making evident the nature of certain quadric surfaces. *The principal tool in this process is "completing the square."* We illustrate the method with several examples.

———————

EXAMPLE 1. Use a translation of coordinates to identify the quadric surface

$$x^2 + 4y^2 + 3z^2 + 2x - 8y + 9z = 10.$$

Sketch.

Solution. We write

$$x^2 + 2x \qquad + 4(y^2 - 2y \qquad) + 3(z^2 + 3z \qquad) = 10.$$

Completing the square, we obtain

$$(x + 1)^2 + 4(y - 1)^2 + 3(z + \tfrac{3}{2})^2 = 10 + 1 + 4 + \tfrac{27}{4}.$$

Introducing the translation of coordinates

$$x' = x + 1,$$
$$y' = y - 1,$$
$$z' = z + \tfrac{3}{2},$$

we find, for the equation of the surface,

$$x'^2 + 4y'^2 + 3z'^2 = \tfrac{87}{4},$$

which we recognize as an ellipsoid. The surface and both coordinate systems are sketched in Fig. 1–24.

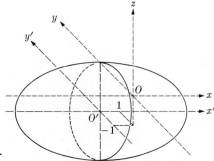

FIGURE 1–24

EXAMPLE 2. Use a translation of coordinates to identify the surface

$$2x^2 - 3y^2 + 6x - 12y - 4z = 0.$$

Sketch.

Solution. We write the equation in the form

$$2(x^2 + 3x \qquad) - 3(y^2 + 4y \qquad) = 4z$$

and complete the square, getting

$$2(x + \tfrac{3}{2})^2 - 3(y + 2)^2 = 4z + \tfrac{9}{2} - 12 = 4(z - \tfrac{15}{8}).$$

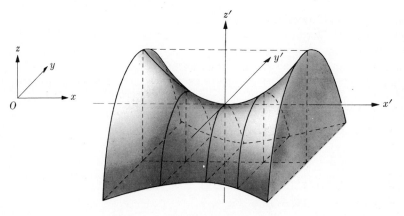

FIGURE 1–25

The translation of axes

$$x' = x + \tfrac{3}{2}, \qquad y' = y + 2, \qquad z' = z - \tfrac{15}{8},$$

yields the equation

$$2x'^2 - 3y'^2 = 4z'.$$

Dividing by 4, we get

$$\frac{x'^2}{2} - \frac{y'^2}{4/3} = z',$$

which we see is a hyperbolic paraboloid. The surface and both coordinate systems are sketched in Fig. 1–25.

EXAMPLE 3. Use a translation of coordinates to identify the surface

$$4x^2 - 3y^2 + 2z^2 + 8x + 18y - 8z = 15.$$

Sketch.

Solution. We write

$$4(x^2 + 2x \qquad) - 3(y^2 - 6y \qquad) + 2(z^2 - 4z \qquad) = 15.$$

Completing the square, we obtain

$$4(x + 1)^2 - 3(y - 3)^2 + 2(z - 2)^2 = 15 + 4 - 27 + 8 = 0.$$

The translation of axes,

$$x' = x + 1,$$
$$y' = y - 3,$$
$$z' = z - 2,$$

shows that the surface is an elliptic cone with equation

$$4x'^2 + 2z'^2 = 3y'^2.$$

The surface and coordinate systems are shown in Fig. 1–26.

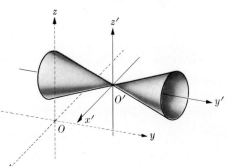

FIGURE 1–26

EXERCISES

In each of problems 1 through 12, employ a translation of coordinates to identify the locus. Sketch the surface, showing both coordinate systems.

1. $x^2 + y^2 + z^2 - 3x + 4y - 8z = 0$
2. $2x^2 + 4y^2 + 6z^2 - 2x + 4y - 15z = 10$
3. $x^2 + 2y^2 - z^2 + 2x + 4y - 6z = 18$
4. $x^2 + 4y^2 + z^2 - 3x + 2y - 4z + 9 = 0$
5. $x^2 + 8z^2 + 2x - 3y + 16z = 0$
6. $2x^2 + 3y^2 + 4x + 2z = 5$

7. $2y^2 - 3z^2 + 4x - 3y + 2z = 0$

8. $x^2 + y^2 - z^2 + 2x + 4y - 2z = 0$

9. $x^2 + y^2 - z^2 + 4x + 8y + 6z + 11 = 0$

10. $2x^2 - 5z^2 + 3x - 4y + 10z = 4$

11. $x^2 - y^2 + 2x - 3y + 4z = 0$

12. $x^2 - 2y^2 - 3z^2 + 4x - 6y + 8z = 2$

13. Use the rotation of coordinates in the plane

$$x = x' \cos \theta - y' \sin \theta, \qquad y = x' \sin \theta + y' \cos \theta,$$

to eliminate the xy-term in the equation of the surface

$$8x^2 - 4xy + 5y^2 + z^2 = 36.$$

Identify the surface and sketch.

14. Use the method of problem 13 to eliminate the xz-term in the equation

$$2x^2 + y^2 - 2z^2 + 3xz = 25.$$

Identify the surface and sketch.

15. Use a rotation of coordinates, as in problem 13, and a translation of coordinates to simplify and identify the surface

$$x^2 - 4xz + 2y^2 + 4z^2 - 3x + 2y - 4z = 5.$$

Sketch.

16. Use a rotation of coordinates, as in problem 13, to show that the equation

$$3x^2 = 2y + 2z$$

represents a parabolic cylinder.

9. OTHER COORDINATE SYSTEMS

In plane analytic geometry we employed a rectangular coordinate system for certain types of problems and a polar coordinate system for others. We saw that there are circumstances in which one system is more convenient than the other. A similar situation prevails in three-dimensional geometry, and we now take up systems of coordinates other than the rectangular one which we have studied exclusively so far. One such system, known as **cylindrical coordinates,** is described in the following way. A point P in space with rectangular coordinates (x, y, z) may also be located by replacing the x- and y-values with the corresponding polar coordinates r, θ and by allowing the z-value to remain unchanged. In other words, to each ordered number triple of the form (r, θ, z), there is associated a point in space. The transformation from cylindrical to rectangular coordinates is given by the equations

$$x = r \cos \theta, \qquad y = r \sin \theta, \qquad z = z.$$

The transformation from rectangular to cylindrical coordinates is given by

$$r^2 = x^2 + y^2, \qquad \tan \theta = \frac{y}{x}, \qquad z = z.$$

If the coordinates of a point are given in one system, the above equations show how to get the coordinates in the other. Figure 1–27 exhibits the relation between the two systems. It is always assumed that the origins of the systems coincide and that $\theta = 0$ corresponds to the xz-plane. We see that the locus $\theta = $ const consists of all points in a plane containing the z-axis. The locus $r = $ const consists of all points on a right circular cylinder with the z-axis as its central axis. (The term "cylindrical coordinates" comes from this fact.) The locus $z = $ const consists of all points in a plane parallel to the xy-plane.

EXAMPLE 1. Find the cylindrical coordinates of the points whose rectangular coordinates are $P(3, 3, 5)$, $Q(2, 0, -1)$, $R(0, 4, 4)$, $S(0, 0, 5)$, $T(2, 2\sqrt{3}, 1)$.

Solution. For the point P we have $r = \sqrt{9 + 9} = 3\sqrt{2}$, $\tan \theta = 1$, $\theta = \pi/4$, $z = 5$. Therefore the coordinates are $(3\sqrt{2}, \pi/4, 5)$. For Q we have $r = 2$, $\theta = 0$, $z = -1$. The coordinates are $(2, 0, -1)$. For R we get $r = 4$, $\theta = \pi/2$, $z = 4$. The result is $(4, \pi/2, 4)$. For S we see at once that the coordinates are $(0, \theta, 5)$ for any θ. For T we get $r = \sqrt{4 + 12} = 4$, $\tan \theta = \sqrt{3}$, $\theta = \pi/3$. The answer is $(4, \pi/3, 1)$.

Remark. Just as polar coordinates do not give a one-to-one correspondence between ordered number pairs and points in the plane, so cylindrical coordinates do not give a one-to-one correspondence between ordered number triples and points in space.

A **spherical coordinate system** is defined in the following way. A point P with rectangular coordinates (x, y, z) has spherical coordinates (ρ, θ, ϕ) where ρ is the distance of the point P from the origin, θ is the same quantity as in cylindrical coordinates, and ϕ is the angle that the line $\overline{OP}$ makes with the positive z-direction. Figure 1–28 exhibits the relation between rectangular and spherical coordinates.

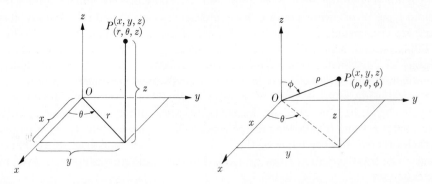

FIGURE 1–27   FIGURE 1–28

The transformation from spherical to rectangular coordinates is given by the equations

$$x = \rho \sin \phi \cos \theta, \qquad y = \rho \sin \phi \sin \theta, \qquad z = \rho \cos \phi.$$

The transformation from rectangular to spherical coordinates is given by

$$\rho^2 = x^2 + y^2 + z^2, \qquad \tan \theta = \frac{y}{x}, \qquad \cos \phi = \frac{z}{\sqrt{x^2 + y^2 + z^2}}.$$

We note that the locus ρ = const is a sphere with center at the origin (from which is derived the term "spherical coordinates"). The locus θ = const is a plane through the z-axis, as in cylindrical coordinates. The locus ϕ = const is a cone with vertex at the origin and angle opening 2ϕ (see Fig. 1–29).

EXAMPLE 2. Find an equation in spherical coordinates of the sphere

$$x^2 + y^2 + z^2 - 2z = 0.$$

Sketch the locus.

Solution. We have $\rho^2 = x^2 + y^2 + z^2$ and $z = \rho \cos \phi$. Therefore

$$\rho^2 - 2\rho \cos \phi = 0 \equiv \rho(\rho - 2 \cos \phi) = 0.$$

The locus of this equation is the locus of $\rho = 0$ and $\rho - 2 \cos \phi = 0$. The locus of $\rho = 0$ is on the locus of $\rho - 2 \cos \phi = 0$ (with $\phi = \pi/2$). Plotting the surface

$$\rho = 2 \cos \phi,$$

we get the surface shown in Fig. 1–30.

If ρ is constant, then the quantities (θ, ϕ) form a coordinate system on the surface of a sphere. Latitude and longitude on the surface of the earth also form a coordinate system. If we restrict θ so that $-\pi < \theta \le \pi$, then θ is called the

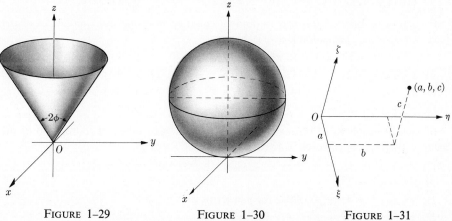

FIGURE 1–29   FIGURE 1–30   FIGURE 1–31

longitude of the point in spherical coordinates. If ϕ is restricted so that $0 \leq \phi \leq \pi$, then ϕ is called the *colatitude* of the point. That is, ϕ is $(\pi/2)$ — latitude, where latitude is taken in the ordinary sense—i.e., positive north of the equator and negative south of it.

Any three lines in space which do not lie in a plane and which intersect in a point may be used as the axes of a coordinate system. Taking the point of intersection as origin, we mark off scales on each axis and so obtain an oblique system of coordinates (see Fig. 1–31). Calling the axes ξ, η, and ζ, we locate a point (a, b, c) by starting at the origin, proceeding a distance a along the ξ-axis, then going parallel to the η-axis a distance b, and finally traversing c units in a direction parallel to the ζ-axis.

EXERCISES

1. Find a set of cylindrical coordinates for each of the points whose rectangular coordinates are

   (a) $(3, 3, 7)$,   (b) $(4, 8, 2)$,   (c) $(-2, 3, 1)$.

2. Find the rectangular coordinates of the points whose cylindrical coordinates are

   (a) $(2, \pi/3, 1)$,   (b) $(3, -\pi/4, 2)$,   (c) $(7, 2\pi/3, -4)$.

3. Find a set of spherical coordinates for each of the points whose rectangular coordinates are

   (a) $(2, 2, 2)$,   (b) $(2, -2, -2)$,   (c) $(-1, \sqrt{3}, 2)$.

4. Find the rectangular coordinates of the points whose spherical coordinates are

   (a) $(4, \pi/6, \pi/4)$,   (b) $(6, 2\pi/3, \pi/3)$,   (c) $(8, \pi/3, 2\pi/3)$.

5. Find a set of cylindrical coordinates for each of the points whose spherical coordinates are

   (a) $(4, \pi/3, \pi/2)$,   (b) $(2, 2\pi/3, 5\pi/6)$,   (c) $(7, \pi/2, \pi/6)$.

6. Find a set of spherical coordinates for each of the points whose cylindrical coordinates are

   (a) $(2, \pi/4, 1)$,   (b) $(3, \pi/2, 2)$,   (c) $(1, 5\pi/6, -2)$.

In each of problems 7 through 16, find an equation in cylindrical coordinates of the locus whose (x, y, z) equation is given. Sketch.

7. $x^2 + y^2 + z^2 = 9$ 8. $x^2 + y^2 + 2z^2 = 8$ 9. $x^2 + y^2 = 4z$

10. $x^2 + y^2 - 2x = 0$ 11. $x^2 + y^2 = z^2$ 12. $x^2 + y^2 + 2z^2 + 2z = 0$

13. $x^2 - y^2 = 4$ 14. $xy + z^2 = 5$ 15. $x^2 + y^2 - 4y = 0$

16. $x^2 + y^2 + z^2 - 2x + 3y - 4z = 0$

In each of problems 17 through 22, find an equation in spherical coordinates of the locus whose (x, y, z) equation is given. Sketch.

17. $x^2 + y^2 + z^2 - 4z = 0$ 18. $x^2 + y^2 + z^2 + 2z = 0$

19. $x^2 + y^2 = z^2$ 20. $x^2 + y^2 = 4$

21. $x^2 + y^2 = 4z + 4$ (Solve for ρ in terms of ϕ.)

22. $x^2 + y^2 - z^2 + z - y = 0$

10. LINEAR INEQUALITIES

Every plane divides three-dimensional space into two regions. If a plane is represented by the equation

$$Ax + By + Cz + D = 0,$$

then all points on one side of the plane satisfy the inequality

$$Ax + By + Cz + D > 0,$$

and all points on the other side satisfy the opposite inequality.

Two intersecting planes divide three-space into four regions, each of which may be described by a pair of linear inequalities. For example, the intersecting planes

$$\Phi_1: x - y + 2z - 4 = 0, \qquad \Phi_2: x + 2y - z + 2 = 0,$$

separate three-space into four regions which we label R_1, R_2, R_3, and R_4. Every point in R_1 satisfies the pair of inequalities

$$x - y + 2z - 4 > 0, \qquad x + 2y - z + 2 > 0;$$

every point in R_2 satisfies the pair of inequalities

$$x - y + 2z - 4 > 0, \qquad x + 2y - z + 2 < 0,$$

and so forth. Figure 1–32 shows, in a schematic way, how Φ_1 and Φ_2 divide space into four regions.

Two parallel planes divide three-dimensional space into three regions. For example, the parallel planes

$$2x + y - z + 1 = 0, \qquad 2x + y - z - 5 = 0,$$

separate space into three regions, R_1, R_2, and R_3, as shown in Fig. 1–33. Note that it is impossible for the two inequalities

$$2x + y - z < -1, \qquad 2x + y - z > 5,$$

to hold simultaneously, and so there is no region corresponding to this pair of inequalities.

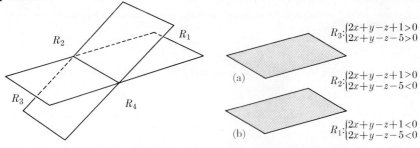

FIGURE 1–32   FIGURE 1–33

Three planes divide three-dimensional space into at most eight regions. If the planes are parallel or have a line in common, then the division is into fewer than eight regions. In fact, it is not difficult to see that three distinct planes always decompose the entire space into 4, 6, 7, or 8 regions, depending on the circumstances.

An enumeration of all possible types of regions into which four planes may separate space is lengthy and tedious. The student can verify that there can be at *most* $2^4 = 16$ regions. Similarly, it is not difficult to show that n planes can divide space into at most 2^n regions.

Rather than attempt to describe in a general way the manner in which a system of inequalities determines a region, we shall consider several special cases which are of particular interest.

The system of inequalities

$$x > 0, \quad y > 0, \quad z > 0$$

determines a region which we call the first octant. We recognize that $x = 0$ is the equation of the yz-plane. Similarly, $y = 0$ and $z = 0$ are the equations of the other two coordinate planes.

The four inequalities

$$x > 0, \quad y > 0, \quad z > 0,$$
$$2x + 2y + z - 2 < 0$$

enclose a region in the shape of a tetrahedron (Fig. 1–34).

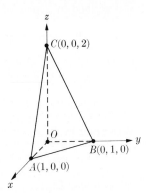

FIGURE 1–34

A region determined by a system of linear inequalities has a boundary consisting of three parts. First, there are the parts we call **faces,** which are composed of portions of planes, such as triangles, quadrilaterals, etc. Two adjoining faces have in common a straight line segment which we call an **edge.** Finally, the points which are the intersections of edges are called **vertices.**

In the example of the tetrahedron in Fig. 1–34, the four vertices are at $A(1, 0, 0)$, $B(0, 1, 0)$, $C(0, 0, 2)$, and the origin O.

To determine the vertices of the boundary of a region defined by a system of linear inequalities, we may proceed by a method which is an extension of the one described in *First Course*, Chapter 3, Section 5. We do so by finding the vertices in a particular example.

EXAMPLE. Determine the region in three-space (if any) which satisfies all the inequalities

$$L_1: x \geq 0, \quad L_2: y \geq 0, \quad L_3: z \geq 0,$$
$$L_4: -x - y - 2z + 4 \geq 0, \quad L_5: -x - y - 4z + 6 \geq 0.$$

Solution. We have a system of five linear inequalities, and we find all possible points of intersections of the five corresponding equations. Since three planes determine a point,

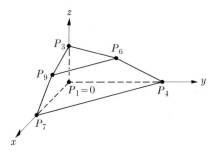

FIGURE 1–35

there are $5 \cdot 4 \cdot 3/1 \cdot 2 \cdot 3 = 10$ possible points of intersection. We find for these points of intersection:

$$P_1(0, 0, 0), \quad P_2(0, 0, 2), \quad P_3(0, 0, \tfrac{3}{2}), \quad P_4(0, 4, 0),$$
$$P_5(0, 6, 0), \quad P_6(0, 2, 1), \quad P_7(4, 0, 0), \quad P_8(6, 0, 0),$$
$$P_9(2, 0, 1), \quad P_{10}(\text{no solution}).$$

The point P_1 was obtained by solving simultaneously L_1, L_2, and L_3; the point P_2 by solving simultaneously L_1, L_2, and L_4; the point P_3 by solving L_1, L_2, and L_5; and so forth.

The coordinates of P_1 satisfy all five inequalities and, therefore, P_1 is a vertex of the region to be determined. The coordinates of P_2 do not satisfy L_5 and this point is rejected. Continuing, we see that the vertices of the region are the points P_1, P_3, P_4, P_6, P_7, and P_9 (Fig. 1–35). To determine an edge of the boundary, we check the vertices by pairs. The vertices P_1 and P_7 connect an edge, since P_1 was obtained from L_1, L_2, and L_3, while P_7 was obtained from L_2, L_3, and L_4. In general, if two vertices have in common two of the planes determining their intersection, then the line segment joining them is an edge. The edges of the region are shown in Fig. 1–35. We also see why P_{10} could not be obtained. The planes $z = 0$, $x + y + 2z - 4 = 0$, and $x + y + 4z - 6 = 0$ intersect in parallel lines. (Note that the line through P_6 and P_9 is parallel to the line through P_4 and P_7.)

EXERCISES

In each of the following problems, determine the region, if any, which satisfies the given system of inequalities.

1. $x \geq 0, \quad y \geq 0, \quad z \geq 0, \quad x + y + 2z - 4 \leq 0$

2. $x \geq 0, \quad y > 0, \quad z \geq 0, \quad 3x + 2y - 7 \leq 0$

3. $x > 0, \quad y > 0, \quad z < 0, \quad x + 2y - 2z - 4 \leq 0$

4. $x \geq 0, \quad y \geq 0, \quad z \geq 0, \quad 2x + y - 5 \leq 0, \quad z - 5 \leq 0$

5. $x \geq 0, \quad y \geq 0, \quad z \geq 0, \quad 3x + 2y - 9 \leq 0, \quad x \leq 2, \quad y \leq 2$

6. $x \geq 0, \quad y \geq 0, \quad z \geq 0, \quad x + y + 3z - 9 \leq 0, \quad x + y + 6z - 12 \leq 0,$
   $y \leq 15$

7. $x \leq 5, \quad z \geq -2, \quad y \leq 4, \quad x - y \leq 1, \quad z + x \geq 1, \quad y + z \geq 2$

8. $x \geq 0, \quad y \geq 0, \quad z \geq 0, \quad x + y + 2z - 1 \leq 0, \quad 3x + y \geq 12$

CHAPTER 2

Vectors in Three Dimensions

1. OPERATIONS WITH VECTORS

The development of vectors in three-dimensional space parallels the development in the plane, as given in Chapter 14 of *First Course*. (See footnote on page 30.) The student should review the material in that chapter, since the same notation and terminology will be employed here.

A *directed line segment* $\overrightarrow{AB}$ is defined as before, except that now the *base A* and the *head B* may be situated anywhere in three-space.

DEFINITION. A *vector* is the collection of all directed line segments having a given magnitude and a given direction.

Boldface letters will be used throughout for vectors. Any directed line segment in a collection forming a vector **v** is called a *representative* of the vector. The collection of directed line segments which defines a vector is called an *equivalence class* of directed line segments.

DEFINITIONS. The **length** of a vector is the common length of all its representative directed line segments. A **unit vector** is a vector of length one. Two vectors are said to be orthogonal (or perpendicular) if any representative of one vector is perpendicular to any representative of the other. The **zero vector,** denoted by **0**, is the class of directed line segments of zero length (i.e., simply points). We make the convention that the zero vector is orthogonal to all vectors.

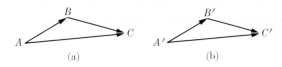

(a)   (b)

FIGURE 2–1

Vectors may be added to yield other vectors. Suppose **u** and **v** are vectors. To add **u** and **v**, we first select a representative of **u**, say $\overrightarrow{AB}$ as shown in Fig. 2–1. Next we take the particular representative of **v** which has its base at the point B and label it $\overrightarrow{BC}$. We then draw the directed line segment $\overrightarrow{AC}$. The sum **w** of **u** and **v** is the equivalence class of directed line segments of which $\overrightarrow{AC}$ is a representative. We write

$$\mathbf{u} + \mathbf{v} = \mathbf{w}.$$

It is important to note that we could have started with any representative of **u**, say $\overrightarrow{A'B'}$, in Fig. 2–1. Then we could have selected the representative of **v** with base at B'. The directed line segments $\overrightarrow{AC}$ and $\overrightarrow{A'C'}$ are representatives of the same vector, as is evident from elementary geometry.

The addition of vectors and the multiplication of vectors by numbers satisfy the same rules of arithmetic as do vectors in the plane. (See *First Course*, page 422.) These rules are:

$$\left.\begin{array}{l} \mathbf{u} + (\mathbf{v} + \mathbf{w}) = (\mathbf{u} + \mathbf{v}) + \mathbf{w} \\ c(d\mathbf{v}) = (cd)\mathbf{v} \end{array}\right\} \quad \text{Associative laws}$$

$$\mathbf{u} + \mathbf{v} = \mathbf{v} + \mathbf{u} \qquad \text{Commutative law}$$

$$\left.\begin{array}{l} (c + d)\mathbf{v} = c\mathbf{v} + d\mathbf{v} \\ c(\mathbf{u} + \mathbf{v}) = c\mathbf{u} + c\mathbf{v} \end{array}\right\} \quad \text{Distributive laws}$$

$$1 \cdot \mathbf{u} = \mathbf{u}, \qquad 0 \cdot \mathbf{u} = \mathbf{0}, \qquad (-1)\mathbf{u} = -\mathbf{u}.$$

Since $(-1)\mathbf{u} = -\mathbf{u}$, we can define *subtraction* of vectors in such a way that it is completely analogous to subtraction of vectors in the plane. In terms of representatives we construct the parallelogram, with $\overrightarrow{AB}$ and $\overrightarrow{DC}$ as two representatives of a vector **u** and $\overrightarrow{AD}$ and $\overrightarrow{BC}$ as representatives of a vector **v**. Then, as Fig. 2–2 shows, $\overrightarrow{AC}$ is a representative of the vector **w**, where

$$\mathbf{w} = \mathbf{u} + \mathbf{v},$$

and $\overrightarrow{BD}$ is a representative of the vector **t**, where

$$\mathbf{t} = \mathbf{v} + (-\mathbf{u}).$$

In other words,

$$\mathbf{t} = \mathbf{v} - \mathbf{u}.$$

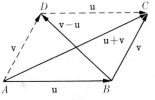

FIGURE 2–2

DEFINITIONS. Suppose that a rectangular coordinate system is given. We define the *vector* **i** to be a vector of unit length pointing in the direction of the positive x-axis. We define the *vector* **j** to be a vector of unit length pointing in the direction of the positive y-axis. We define the *vector* **k** to be a vector of unit length pointing in the direction of the positive z-axis.

Figure 2–3 shows a number of representatives of the vectors **i**, **j**, and **k**. The directed line segments $\overrightarrow{OA}$ and $\overrightarrow{DE}$ are representatives of **i**; the segments $\overrightarrow{OB}$ and $\overrightarrow{FG}$ are representatives of **j**; and the segments $\overrightarrow{OC}$ and $\overrightarrow{HJ}$ are representatives of **k**.

Theorem 1. *Every vector* **v** *in three-dimensional space may be written as a linear combination of* **i**, **j**, *and* **k**. *That is, there are numbers a, b, and c such that*

$$\mathbf{v} = a\mathbf{i} + b\mathbf{j} + c\mathbf{k}.$$

Proof. Select the representative $\overrightarrow{OA}$ of **v** which has its base at the origin (Fig. 2–4). Denote by $a\mathbf{i}$ the vector which has $\overrightarrow{OB}$ as representative, by $b\mathbf{j}$ the

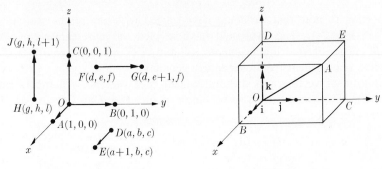

FIGURE 2–3   FIGURE 2–4

vector which has $\overrightarrow{OC}$ as representative, and by $c\mathbf{k}$ the vector which has $\overrightarrow{OD}$ as representative. Then, according to the law of addition of vectors, we see that $\mathbf{v} = a\mathbf{i} + b\mathbf{j} + c\mathbf{k}$.

Remark. The student should verify the result by taking a representative $\overrightarrow{FG}$ of $\mathbf{v}$ located in some general position in space.

EXAMPLE 1. A vector $\mathbf{v}$ has the directed line segment $\overrightarrow{AB}$ as one of its representatives. Express $\mathbf{v}$ as a linear combination of $\mathbf{i}$, $\mathbf{j}$, and $\mathbf{k}$, if A has coordinates $(3, -2, 4)$ and B has coordinates $(2, 1, 5)$.

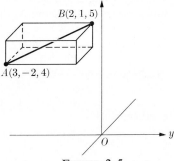

Solution. As shown in Fig. 2–5, the required quantities are $a = 2 - 3 = -1$, $b = 1 - (-2) = 3$, $c = 5 - 4 = 1$. Therefore,

$$\mathbf{v} = -\mathbf{i} + 3\mathbf{j} + \mathbf{k}.$$

FIGURE 2–5

Corollary. *If a representative $\overrightarrow{AB}$ of a vector $\mathbf{v}$ has for its coordinates (x_A, y_A, z_A) and (x_B, y_B, z_B), then we have*

$$\mathbf{v} = (x_B - x_A)\mathbf{i} + (y_B - y_A)\mathbf{j} + (z_B - z_A)\mathbf{k}.$$

The result is evident from the way in which Theorem 1 was proved, and Example 1 shows how it works in practice.

Theorem 2. *If $\mathbf{v}$ is any vector which, by Theorem 1, may be written in the form*

$$\mathbf{v} = a\mathbf{i} + b\mathbf{j} + c\mathbf{k},$$

then the length of $\mathbf{v}$, which we denote by $|\mathbf{v}|$, is given by

$$|\mathbf{v}| = \sqrt{a^2 + b^2 + c^2}.$$

Therefore, $\mathbf{v} = \mathbf{0}$ if and only if $a = b = c = 0$.

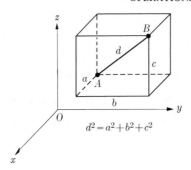

FIGURE 2–6

Proof. If $\overrightarrow{AB}$ is a representative of **v**, the result follows (Fig. 2–6) from the formula for the distance between two points in three-dimensional space.

EXAMPLE 2. Find the length of the vector $\mathbf{v} = 2\mathbf{i} - 3\mathbf{j} + 4\mathbf{k}$.

Solution. $|\mathbf{v}| = \sqrt{2^2 + 3^2 + 4^2} = \sqrt{29}$.

The next theorem is obtained at once from the laws of addition of vectors and multiplication of vectors by numbers.

Theorem 3. *If* $\mathbf{v} = a\mathbf{i} + b\mathbf{j} + c\mathbf{k}$ *and* $\mathbf{w} = d\mathbf{i} + e\mathbf{j} + f\mathbf{k}$, *then*

$$\mathbf{v} + \mathbf{w} = (a + d)\mathbf{i} + (b + e)\mathbf{j} + (c + f)\mathbf{k}.$$

Further, if h is any number, then

$$h\mathbf{v} = ha\mathbf{i} + hb\mathbf{j} + hc\mathbf{k}.$$

EXAMPLE 3. Given the vectors $\mathbf{u} = 3\mathbf{i} - 2\mathbf{j} + 4\mathbf{k}$ and $\mathbf{v} = 6\mathbf{i} - 4\mathbf{j} - 2\mathbf{k}$, express the vector $3\mathbf{u} - 2\mathbf{v}$ in terms of **i**, **j**, and **k**.

Solution. $3\mathbf{u} = 9\mathbf{i} - 6\mathbf{j} + 12\mathbf{k}$ and $-2\mathbf{v} = -12\mathbf{i} + 8\mathbf{j} + 4\mathbf{k}$. Adding these vectors, we get $3\mathbf{u} - 2\mathbf{v} = -3\mathbf{i} + 2\mathbf{j} + 16\mathbf{k}$.

DEFINITION. Let **v** be any vector except **0**. The *unit vector* **u** *in the direction of* **v** is defined by

$$\mathbf{u} = \left(\frac{1}{|\mathbf{v}|}\right)\mathbf{v}.$$

EXAMPLE 4. Given the vector $\mathbf{v} = 2\mathbf{i} - 3\mathbf{j} + \mathbf{k}$, find a unit vector in the direction of **v**.

Solution. We have $|v| = \sqrt{4 + 9 + 1} = \sqrt{14}$. The desired vector **u** is

$$\mathbf{u} = \frac{1}{\sqrt{14}}\mathbf{v} = \frac{2}{\sqrt{14}}\mathbf{i} - \frac{3}{\sqrt{14}}\mathbf{j} + \frac{1}{\sqrt{14}}\mathbf{k}.$$

EXAMPLE 5. Given the vector $\mathbf{v} = 2\mathbf{i} + 4\mathbf{j} - 3\mathbf{k}$, find the representative $\overrightarrow{AB}$ of **v** if the point A has coordinates $(2, 1, -5)$.

Solution. Denote the coordinates of B by x_B, y_B, z_B. Then we have

$$x_B - 2 = 2, \qquad y_B - 1 = 4, \qquad z_B + 5 = -3.$$

Therefore, $x_B = 4, y_B = 5, z_B = -8$.

EXERCISES

In problems 1 through 6, express **v** in terms of **i**, **j**, and **k**, given that the endpoints A and B of the representative $\overrightarrow{AB}$ of **v** have the given coordinates:

1. $A(3, 1, 0)$, $B(2, -1, 4)$
2. $A(0, 1, 5)$, $B(-2, 1, 4)$
3. $A(1, 6, 2)$, $B(1, -4, -5)$
4. $A(2, 2, 2)$, $B(0, 5, -1)$
5. $A(0, 0, 4)$, $B(4, 0, 0)$
6. $A(-1, -3, 5)$, $B(2, 1, -4)$

In problems 7 through 10, in each case find a unit vector **u** in the direction of **v**. Express **u** in terms of **i**, **j**, and **k**.

7. $\mathbf{v} = 3\mathbf{i} + 2\mathbf{j} - 4\mathbf{k}$
8. $\mathbf{v} = \mathbf{i} - \mathbf{j} + \mathbf{k}$
9. $\mathbf{v} = 2\mathbf{i} - 4\mathbf{j} - \mathbf{k}$
10. $\mathbf{v} = -2\mathbf{i} + 3\mathbf{j} + 5\mathbf{k}$

In problems 11 through 17, find the representative $\overrightarrow{AB}$ of the vector **v** from the information given.

11. $\mathbf{v} = 2\mathbf{i} + \mathbf{j} - 3\mathbf{k}$, $A(1, 2, -1)$
12. $\mathbf{v} = -\mathbf{i} + 3\mathbf{j} - 2\mathbf{k}$, $A(2, 0, 4)$
13. $\mathbf{v} = 3\mathbf{i} + 2\mathbf{j} - 4\mathbf{k}$, $B(2, 0, -4)$
14. $\mathbf{v} = -2\mathbf{i} + 4\mathbf{j} + \mathbf{k}$, $B(0, 0, -5)$
15. $\mathbf{v} = \mathbf{i} - 2\mathbf{j} + 2\mathbf{k}$; the midpoint of the segment AB has coordinates $(2, -1, 4)$.
16. $\mathbf{v} = 3\mathbf{i} + 4\mathbf{k}$; the midpoint of the segment AB has coordinates $(1, 2, -5)$.
17. $\mathbf{v} = -\mathbf{i} + \mathbf{j} - 2\mathbf{k}$; the point three-fourths of the distance from A to B has coordinates $(1, 0, 2)$.
18. Find a vector **u** in the direction of $\mathbf{v} = -\mathbf{i} + \mathbf{j} - \mathbf{k}$ and having half the length of **v**.
19. Given $\mathbf{u} = \mathbf{i} + 2\mathbf{j} - 4\mathbf{k}$, $\mathbf{v} = 3\mathbf{i} - 7\mathbf{j} + 5\mathbf{k}$, find $\mathbf{u} + \mathbf{v}$ in terms of **i**, **j**, and **k**. Sketch a figure.
20. Given that $\mathbf{u} = -3\mathbf{i} + 7\mathbf{j} - 4\mathbf{k}$, $\mathbf{v} = 2\mathbf{i} + \mathbf{j} - 6\mathbf{k}$, find $3\mathbf{u} - 7\mathbf{v}$ in terms of **i**, **j**, and **k**.
21. Let a and b be any real numbers. Show that the vector **k** is orthogonal to $a\mathbf{i} + b\mathbf{j}$.

2. THE INNER (SCALAR OR DOT) PRODUCT

Two vectors are said to be **parallel** or **proportional** when each is a scalar multiple of the other (and neither is zero). The representatives of parallel vectors are all parallel directed line segments.

By the **angle between two vectors v and w** (neither $= \mathbf{0}$), we mean the measure of the angle between any line containing a representative of **v** and an intersecting line containing a representative of **w** (Fig. 2–7). Two parallel vectors make an angle of 0 or π, depending on whether they are pointing in the same or opposite directions.

Theorem 4. *If θ is the angle between the vectors*

$$\mathbf{v} = a_1\mathbf{i} + a_2\mathbf{j} + a_3\mathbf{k}$$

and

$$\mathbf{w} = b_1\mathbf{i} + b_2\mathbf{j} + b_3\mathbf{k},$$

then

$$\cos\theta = \frac{a_1b_1 + a_2b_2 + a_3b_3}{|\mathbf{v}| \cdot |\mathbf{w}|}.$$

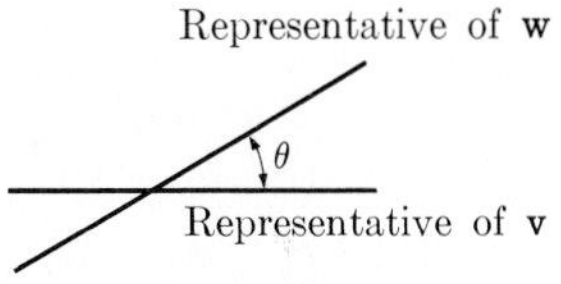

Representative of **w**

Representative of **v**

FIGURE 2–7

The proof is a straightforward extension of the proof of the analogous theorem in the plane (*First Course*, Theorem 4, page 428) and will therefore be omitted. The proof is also a direct consequence of Theorem 3 of Chapter 1.

EXAMPLE 1. Given the vectors $\mathbf{v} = 2\mathbf{i} + \mathbf{j} - 3\mathbf{k}$ and $\mathbf{w} = -\mathbf{i} + 4\mathbf{j} - 2\mathbf{k}$, find the cosine of the angle between **v** and **w**.

Solution. We have $|\mathbf{v}| = \sqrt{4 + 1 + 9} = \sqrt{14}$, $|\mathbf{w}| = \sqrt{1 + 16 + 4} = \sqrt{21}$.

Therefore

$$\cos\theta = \frac{-2 + 4 + 6}{\sqrt{14} \cdot \sqrt{21}} = \frac{8}{7\sqrt{6}}.$$

DEFINITIONS. Given the vectors **u** and **v**, we define the **inner (scalar or dot) product**

$$\mathbf{u} \cdot \mathbf{v}$$

by the formula

$$\mathbf{u} \cdot \mathbf{v} = |\mathbf{u}|\,|\mathbf{v}|\cos\theta,$$

where θ is the angle between the vectors. If either **u** or **v** is **0**, we define $\mathbf{u} \cdot \mathbf{v} = 0$. Two vectors **u** and **v** are **orthogonal** if and only if $\mathbf{u} \cdot \mathbf{v} = 0$.

Theorem 5. *The scalar product satisfies the laws*

(a) $\mathbf{u} \cdot \mathbf{v} = \mathbf{v} \cdot \mathbf{u};$ (b) $\mathbf{u} \cdot \mathbf{u} = |\mathbf{u}|^2.$

(c) *If* $\mathbf{u} = a_1\mathbf{i} + b_1\mathbf{j} + c_1\mathbf{k}$ *and* $\mathbf{v} = a_2\mathbf{i} + b_2\mathbf{j} + c_2\mathbf{k}$,   *then*

$$\mathbf{u} \cdot \mathbf{v} = a_1a_2 + b_1b_2 + c_1c_2.$$

Proof. Parts (a) and (b) are direct consequences of the definition; part (c) follows from Theorem 4, since

$$\mathbf{u} \cdot \mathbf{v} = |\mathbf{u}| \cdot |\mathbf{v}| \cos \theta = |\mathbf{u}| \cdot |\mathbf{v}| \frac{a_1a_2 + b_1b_2 + c_1c_2}{|\mathbf{u}| \cdot |\mathbf{v}|}.$$

Corollary. (a) *If c and d are any numbers and if* $\mathbf{u}$, $\mathbf{v}$, $\mathbf{w}$ *are any vectors, then*

$$\mathbf{u} \cdot (c\mathbf{v} + d\mathbf{w}) = c(\mathbf{u} \cdot \mathbf{v}) + d(\mathbf{u} \cdot \mathbf{w}).$$

(b) *We have*

$$\mathbf{i} \cdot \mathbf{i} = \mathbf{j} \cdot \mathbf{j} = \mathbf{k} \cdot \mathbf{k} = 1, \qquad \mathbf{i} \cdot \mathbf{j} = \mathbf{i} \cdot \mathbf{k} = \mathbf{j} \cdot \mathbf{k} = 0.$$

EXAMPLE 1. Find the scalar product of the vectors

$$\mathbf{u} = 3\mathbf{i} + 2\mathbf{j} - 4\mathbf{k} \qquad \text{and} \qquad \mathbf{v} = -2\mathbf{i} + \mathbf{j} + 5\mathbf{k}.$$

Solution. $\mathbf{u} \cdot \mathbf{v} = 3(-2) + 2 \cdot 1 + (-4)(5) = -24.$

EXAMPLE 2. Express $|3\mathbf{u} + 5\mathbf{v}|^2$ in terms of $|\mathbf{u}|^2$, $|\mathbf{v}|^2$, and $\mathbf{u} \cdot \mathbf{v}$.

Solution. $|3\mathbf{u} + 5\mathbf{v}|^2 = (3\mathbf{u} + 5\mathbf{v}) \cdot (3\mathbf{u} + 5\mathbf{v})$
$$= 9(\mathbf{u} \cdot \mathbf{u}) + 15(\mathbf{u} \cdot \mathbf{v}) + 15(\mathbf{v} \cdot \mathbf{u}) + 25(\mathbf{v} \cdot \mathbf{v})$$
$$= 9|\mathbf{u}|^2 + 30(\mathbf{u} \cdot \mathbf{v}) + 25|\mathbf{v}|^2.$$

DEFINITION. Let $\mathbf{v}$ and $\mathbf{w}$ be two vectors which make an angle θ. We denote by $|\mathbf{v}| \cos \theta$ the **projection of v along w**. We also call this quantity the **component of v along w**.

From the formula for $\cos \theta$, we may also write

$$|\mathbf{v}| \cos \theta = |\mathbf{v}| \frac{\mathbf{v} \cdot \mathbf{w}}{|\mathbf{v}| \cdot |\mathbf{w}|} = \frac{\mathbf{v} \cdot \mathbf{w}}{|\mathbf{w}|}.$$

EXAMPLE 3. Find the projection of $\mathbf{v} = -\mathbf{i} + 2\mathbf{j} + 3\mathbf{k}$ along $\mathbf{w} = 2\mathbf{i} - \mathbf{j} - 4\mathbf{k}$.

Solution. $\mathbf{v} \cdot \mathbf{w} = (-1)(2) + (2)(-1) + (3)(-4) = -16$; $|\mathbf{w}| = \sqrt{21}$. Therefore, the projection of $\mathbf{v}$ along $\mathbf{w} = -16/\sqrt{21}$.

An application of scalar product to mechanics occurs in the calculation of work done by a constant force $\mathbf{F}$ when its point of application moves along a segment from A to B. The **work done** in this case is defined as the product of the distance from A to B and the projection of $\mathbf{F}$ along $\mathbf{v}$ $(\overrightarrow{AB})$. We have

$$\text{Projection of } \mathbf{F} \text{ along } \mathbf{v} = \frac{\mathbf{F} \cdot \mathbf{v}}{|\mathbf{v}|};$$

since the distance from A to B is exactly $|\mathbf{v}|$, we conclude that

$$\text{Work done by } \mathbf{F} = \mathbf{F} \cdot \mathbf{v}.$$

EXAMPLE 4. Find the work done by the force

$$\mathbf{F} = 5\mathbf{i} - 3\mathbf{j} + 2\mathbf{k}$$

as its point of application moves from the point $A(2, 1, 3)$ to $B(4, -1, 5)$.

Solution. We have $\mathbf{v}(\overrightarrow{AB}) = (4 - 2)\mathbf{i} + (-1 - 1)\mathbf{j} + (5 - 3)\mathbf{k} = 2\mathbf{i} - 2\mathbf{j} + 2\mathbf{k}$. Therefore, work done $= 5 \cdot 2 + 3 \cdot 2 + 2 \cdot 2 = 20$.

Theorem 6. *If* $\mathbf{u}$ *and* $\mathbf{v}$ *are not* $\mathbf{0}$, *there is a unique number* k *such that* $\mathbf{v} - k\mathbf{u}$ *is orthogonal to* $\mathbf{u}$. *In fact,* k *can be found from the formula*

$$k = \frac{\mathbf{u} \cdot \mathbf{v}}{|\mathbf{u}|^2}.$$

Proof. $(\mathbf{v} - k\mathbf{u})$ is orthogonal to $\mathbf{u}$ if and only if $\mathbf{u} \cdot (\mathbf{v} - k\mathbf{u}) = 0$. But

$$\mathbf{u} \cdot (\mathbf{v} - k\mathbf{u}) = \mathbf{u} \cdot \mathbf{v} - k|\mathbf{u}|^2 = 0.$$

Therefore, selection of $k = \mathbf{u} \cdot \mathbf{v}/|\mathbf{u}|^2$ yields the result.

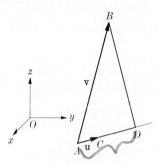

FIGURE 2–8

Figure 2–8 shows geometrically how k is to be selected. We drop a perpendicular from the head of $\mathbf{v}$ (point B) to the line containing $\mathbf{u}$ (point D). The directed segment $\overrightarrow{AD}$ gives the proper multiple of $\mathbf{u}(\overrightarrow{AC})$, and the directed segment $\overrightarrow{BD}$ represents the orthogonal vector.

EXAMPLE 5. Find a linear combination of $u = 2i + 3j - k$ and $v = i + 2j + k$ which is orthogonal to u.

Solution. We select $k = (2 + 6 - 1)/14 = \frac{1}{2}$, and the desired vector is $\frac{1}{2}j + \frac{3}{2}k$.

EXERCISES

In problems 1 through 5, find $\cos \theta$ where θ is the angle between the vectors u and v.

1. $u = 2i - 3j + k$, $v = -i + 2j + k$
2. $u = i + j - 4k$, $v = -2i + 3j - 4k$
3. $u = 2i + j + 5k$, $v = -6i + 2j - 4k$
4. $u = i + 2j - 3k$, $v = -2i - 4j + 6k$
5. $u = 2i + 3j + k$, $v = 2i + 4j - 12k$

In each of problems 6 through 10, find the projection of the vector v along u.

6. $u = 2i - 6j + 3k$, $v = i + 2j - 2k$
7. $u = 6i + 2j - 3k$, $v = -i + 8j + 4k$
8. $u = 12i + 3j + 4k$, $v = 4i + 8j + k$
9. $u = 3i + 5j - 4k$, $v = 4i - 3j + 5k$
10. $u = 2i - 5j + 3k$, $v = -i + 2j + 7k$

In each of problems 11 through 14, find the work done by the force F when its point of application moves from A to B.

11. $F = -32k$, $A: (-1, 1, 2)$, $B: (3, 2, -1)$
12. $F = 5i - 2j + 3k$, $A: (1, -2, 2)$, $B: (3, 1, -1)$
13. $F = -2i + 3j + 4k$, $A: (2, -1, -2)$, $B: (-1, 2, 3)$
14. $F = 3i - 2j - 3k$, $A: (-1, 2, 3)$, $B: (2, 1, -1)$

In each of problems 15 through 17, find a unit vector in the direction of u.

15. $u = 2i - 6j + 3k$ 16. $u = -i + 2k$ 17. $u = 3i - 2j + 7k$

In each of problems 18 through 21, find the value of k so that $v - ku$ is orthogonal to u. Also, find the value h so that $u - hv$ is orthogonal to v.

18. $u = 2i - j + 2k$, $v = 3i + j + 2k$
19. $u = 2i - 3j + 6k$, $v = 7i + 14k$
20. $u = 3i + 4j - 5k$, $v = 9i + 12j - 5k$
21. $u = i + 3j - 2k$, $v = 6i + 10j - 3k$
22. Write a detailed proof of Theorem 4.
23. Show that if u and v are any vectors ($\neq 0$), then u and v make equal angles with w if

$$w = \left(\frac{|v|}{|u| + |v|} \right) u + \left(\frac{|u|}{|u| + |v|} \right) v.$$

24. Show that if **u** and **v** are any vectors, the vectors $|\mathbf{v}|\mathbf{u} + |\mathbf{u}|\mathbf{v}$ and $|\mathbf{v}|\mathbf{u} - |\mathbf{u}|\mathbf{v}$ are orthogonal.

In each of problems 25 through 27, determine the relation between g and h so that $g\mathbf{u} + h\mathbf{v}$ is orthogonal to **w**.

25. $\mathbf{u} = 3\mathbf{i} - 2\mathbf{j} + \mathbf{k}, \qquad \mathbf{v} = \mathbf{i} + 2\mathbf{j} - 3\mathbf{k}, \qquad \mathbf{w} = -\mathbf{i} + \mathbf{j} + 2\mathbf{k}$

26. $\mathbf{u} = 2\mathbf{i} + \mathbf{j} - 2\mathbf{k}, \qquad \mathbf{v} = \mathbf{i} - \mathbf{j} + \mathbf{k}, \qquad \mathbf{w} = -\mathbf{i} + 2\mathbf{j} + 3\mathbf{k}$

27. $\mathbf{u} = \mathbf{i} + 2\mathbf{j} - 3\mathbf{k}, \qquad \mathbf{v} = 3\mathbf{i} + \mathbf{j} - \mathbf{k}, \qquad \mathbf{w} = 4\mathbf{i} - \mathbf{j} + 2\mathbf{k}$

In each of problems 28 through 30, determine g and h so that $\mathbf{w} - g\mathbf{u} - h\mathbf{v}$ is orthogonal to both **u** and **v**.

28. $\mathbf{u} = 2\mathbf{i} - \mathbf{j} + \mathbf{k}, \qquad \mathbf{v} = \mathbf{i} + \mathbf{j} + 2\mathbf{k}, \qquad \mathbf{w} = 2\mathbf{i} - \mathbf{j} + 4\mathbf{k}$

29. $\mathbf{u} = \mathbf{i} + \mathbf{j} - 2\mathbf{k}, \qquad \mathbf{v} = -\mathbf{i} + 2\mathbf{j} + 3\mathbf{k}, \quad \mathbf{w} = 5\mathbf{i} + 8\mathbf{k}$

30. $\mathbf{u} = 3\mathbf{i} - 2\mathbf{j}, \qquad\quad \mathbf{v} = 2\mathbf{i} - \mathbf{k}, \qquad\quad \mathbf{w} = 4\mathbf{i} - 2\mathbf{k}$

3. THE VECTOR OR CROSS PRODUCT

We saw in Section 2 that the scalar product of two vectors **u** and **v** is an operation which associates an ordinary number, or scalar, with each pair of vectors. The vector or cross product, on the other hand, is an operation which associates a vector with each ordered pair of vectors. However, before defining this operation, we shall discuss the notion of *right-handed* and *left-handed* triples of vectors.

Let **u** and **v** be two vectors (neither zero) which are not proportional. Then any two intersecting representatives determine a plane. Let **w** be a vector which is perpendicular to any such plane determined by **u** and **v**. We say that the triple, **u**, **v**, and **w**, forms a **right-handed triple** if the vectors are as shown in Fig. 2–9. They form a **left-handed triple** if they appear as in Fig. 2–10. A right-handed triple may be thought of as corresponding to a right-handed coordinate system with **u**, **v**, **w** corresponding to the *x*-, *y*-, and *z*-axes in orientation. A left-handed triple corresponds to a left-handed coordinate system.

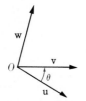

Right-handed triple

FIGURE 2–9

Left-handed triple

FIGURE 2–10

DEFINITION. Given the vectors **u** and **v**, we define the **vector** or **cross product** $\mathbf{u} \times \mathbf{v}$ as follows:

(i) If either **u** or **v** is **0**,

$$\mathbf{u} \times \mathbf{v} = \mathbf{0}.$$

(ii) If **u** is proportional to **v**,

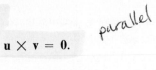

$$\mathbf{u} \times \mathbf{v} = \mathbf{0}.$$

(iii) If otherwise,

$$\mathbf{u} \times \mathbf{v} = \mathbf{w},$$

where **w** is a vector perpendicular to a plane determined by **u** and **v**; the magnitude of **w** is

$$|\mathbf{w}| = |\mathbf{u}|\,|\mathbf{v}|\sin\theta,$$

where θ is the angle between **u** and **v**; finally, the direction of **w** is selected so that **u**, **v**, and **w** form a right-handed triple.

Remark. We shall always select the unit vectors **i**, **j**, and **k** so as to form a right-handed triple. (We have done so up to now without pointing out this fact specifically.)

We obtain at once from the definition of vector product the simple relations:

$$\mathbf{i} \times \mathbf{i} = \mathbf{j} \times \mathbf{j} = \mathbf{k} \times \mathbf{k} = 0;$$
$$\mathbf{i} \times \mathbf{j} = -\mathbf{j} \times \mathbf{i} = \mathbf{k};$$
$$\mathbf{j} \times \mathbf{k} = -\mathbf{k} \times \mathbf{j} = \mathbf{i};$$
$$\mathbf{k} \times \mathbf{i} = -\mathbf{i} \times \mathbf{k} = \mathbf{j}.$$

Theorem 7. *If*

$$\mathbf{u} = a_1\mathbf{i} + a_2\mathbf{j} + a_3\mathbf{k} \quad and \quad \mathbf{v} = b_1\mathbf{i} + b_2\mathbf{j} + b_3\mathbf{k},$$

then

$$\mathbf{u} \times \mathbf{v} = (a_2b_3 - a_3b_2)\mathbf{i} + (a_3b_1 - a_1b_3)\mathbf{j} + (a_1b_2 - a_2b_1)\mathbf{k}. \qquad (1)$$

The proof will be omitted. (It may be found in Morrey, *University Calculus*, page 531.)

Corollary. *If* **u** *and* **v** *are any vectors, then*

(i) $\mathbf{u} \times \mathbf{v} = -\mathbf{v} \times \mathbf{u}$

(ii) $\mathbf{u} \times (\mathbf{v} + \mathbf{w}) = \mathbf{u} \times \mathbf{v} + \mathbf{u} \times \mathbf{w}.$

The results of this corollary are obtained by writing $\mathbf{u} = a_1\mathbf{i} + a_2\mathbf{j} + a_3\mathbf{k}$, $\mathbf{v} = b_1\mathbf{i} + b_2\mathbf{j} + b_3\mathbf{k}$, $\mathbf{w} = c_1\mathbf{i} + c_2\mathbf{j} + c_3\mathbf{k}$, and then applying the formula (1) for the cross product of two vectors.

EXAMPLE 1. Find the vector product $\mathbf{u} \times \mathbf{v}$ of $\mathbf{u} = 2\mathbf{i} - 3\mathbf{j} + \mathbf{k}$ and $\mathbf{v} = \mathbf{i} + \mathbf{j} - 2\mathbf{k}$.

Solution. Since $a_1 = 2, a_2 = -3, a_3 = 1, b_1 = 1, b_2 = 1, b_3 = -2$, we have

$$\mathbf{u} \times \mathbf{v} = (6 - 1)\mathbf{i} + (1 + 4)\mathbf{j} + (2 + 3)\mathbf{k} = 5\mathbf{i} + 5\mathbf{j} + 5\mathbf{k}.$$

Theorem 8. *The area of a parallelogram with adjacent sides $\overline{AB}$ and $\overline{AC}$ is given by*

$$|\mathbf{v}(\overrightarrow{AB}) \times \mathbf{v}(\overrightarrow{AC})|. \; = |\vec{w}|$$

The area of $\triangle ABC$ is then $\frac{1}{2}|\mathbf{v}(\overrightarrow{AB}) \times \mathbf{v}(\overrightarrow{AC})|$.

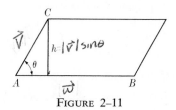

FIGURE 2–11

Proof. From Fig. 2–11, we see that the area of the parallelogram is $|AB| \cdot h = |AB| \cdot |AC| \sin \theta$. The result then follows from the definition of cross product.

———————

EXAMPLE 2. Find the area of $\triangle ABC$ with $A(-2, 1, 3)$, $B(1, -1, 1)$, $C(3, -2, 4)$.

Solution. We have $\mathbf{v}(\overrightarrow{AB}) = 3\mathbf{i} - 2\mathbf{j} - 2\mathbf{k}$, $\mathbf{v}(\overrightarrow{AC}) = 5\mathbf{i} - 3\mathbf{j} + \mathbf{k}$. From Theorem 8 we obtain

$$\mathbf{v}(\overrightarrow{AB}) \times \mathbf{v}(\overrightarrow{AC}) = -8\mathbf{i} - 13\mathbf{j} + \mathbf{k}$$

and

$$\tfrac{1}{2}|-8\mathbf{i} - 13\mathbf{j} + \mathbf{k}| = \tfrac{1}{2}\sqrt{64 + 169 + 1} = \tfrac{3}{2}\sqrt{26}.$$

———————

The vector product may be used to find the equation of a plane through three points. The next example illustrates the technique.

———————

EXAMPLE 3. Find the equation of the plane through the points $A(-1, 1, 2)$, $B(1, -2, 1)$, $C(2, 2, 4)$.

Solution. A vector normal to the plane will be perpendicular to both the vectors

$$\mathbf{v}(\overrightarrow{AB}) = 2\mathbf{i} - 3\mathbf{j} - \mathbf{k}$$

and

$$\mathbf{v}(\overrightarrow{AC}) = 3\mathbf{i} + \mathbf{j} + 2\mathbf{k}.$$

One such vector is the cross product

$$\mathbf{v}(\overrightarrow{AB}) \times \mathbf{v}(\overrightarrow{AC}) = -5\mathbf{i} - 7\mathbf{j} + 11\mathbf{k}.$$

Therefore the numbers $-5, -7, 11$ form a set of *attitude numbers* (see Chapter 1, Section 4) of the desired plane. Using $A(-1, 1, 2)$ as a point on the plane, we get for the equation

$$-5(x + 1) - 7(y - 1) + 11(z - 2) = 0$$

or

$$5x + 7y - 11z + 20 = 0.$$

EXAMPLE 4. Find the perpendicular distance between the skew lines

$$L_1: \frac{x + 2}{2} = \frac{y - 1}{3} = \frac{z + 1}{-1}, \qquad L_2: \frac{x - 1}{-1} = \frac{y + 1}{2} = \frac{z - 2}{4}.$$

Solution. The vector $\mathbf{v}_1 = 2\mathbf{i} + 3\mathbf{j} - \mathbf{k}$ is a vector along L_1. The vector $\mathbf{v}_2 = -\mathbf{i} + 2\mathbf{j} + 4\mathbf{k}$ is a vector along L_2. A vector perpendicular to both $\mathbf{v}_1$ and $\mathbf{v}_2$ (i.e., to both L_1 and L_2) is

$$\mathbf{v}_1 \times \mathbf{v}_2 = 14\mathbf{i} - 7\mathbf{j} + 7\mathbf{k}.$$

Call this common perpendicular $\mathbf{w}$. The desired length may be obtained as a *projection*. Select any point on L_1 (call it P_1) and any point on L_2 (call it P_2). Then the desired length is the projection of the vector $\mathbf{v}(\overrightarrow{P_1P_2})$ along $\mathbf{w}$. To get this, we select $P_1(-2, 1, -1)$ on L_1 and $P_2(1, -1, 2)$ on L_2; and so

$$\mathbf{v}(\overrightarrow{P_1P_2}) = 3\mathbf{i} - 2\mathbf{j} + 3\mathbf{k}.$$

Therefore,

$$\text{Projection of } \mathbf{v}(\overrightarrow{P_1P_2}) \text{ along } \mathbf{w} = \frac{\mathbf{v}(\overrightarrow{P_1P_2}) \cdot \mathbf{w}}{|\mathbf{w}|}$$

$$= \frac{3 \cdot 14 + (-2)(-7) + 3(7)}{7\sqrt{6}} = \frac{11}{\sqrt{6}}.$$

EXERCISES

In each of problems 1 through 6, find the cross product $\mathbf{u} \times \mathbf{v}$.

1. $\mathbf{u} = 2\mathbf{i} + \mathbf{j} + \mathbf{k}, \quad \mathbf{v} = -\mathbf{i} + 2\mathbf{j} + 3\mathbf{k}$
2. $\mathbf{u} = 2\mathbf{i} + \mathbf{j} - 3\mathbf{k}, \quad \mathbf{v} = 2\mathbf{i} + 2\mathbf{j} - 4\mathbf{k}$
3. $\mathbf{u} = \mathbf{i} + 3\mathbf{j}, \quad \mathbf{v} = 2\mathbf{i} - 5\mathbf{k}$
4. $\mathbf{u} = 3\mathbf{i} - 2\mathbf{j} + \mathbf{k}, \quad \mathbf{v} = 3\mathbf{j} + 4\mathbf{k}$
5. $\mathbf{u} = 2\mathbf{i} + 4\mathbf{j} - 3\mathbf{k}, \quad \mathbf{v} = -4\mathbf{i} - 8\mathbf{j} + 6\mathbf{k}$
6. $\mathbf{u} = 3\mathbf{i}, \quad \mathbf{v} = 2\mathbf{j} + \mathbf{k}$

In problems 7 through 11, find in each case the area of $\triangle ABC$ and the equation of the plane through A, B, and C.

7. $A(1, -2, 3), \quad B(3, 1, 2), \quad C(2, 3, -1)$
8. $A(3, 2, -2), \quad B(4, 1, 2), \quad C(1, 2, 3)$
9. $A(2, -1, 1), \quad B(3, 2, -1), \quad C(-1, 3, 2)$
10. $A(1, -2, 3), \quad B(2, -1, 1), \quad C(4, 2, -1)$
11. $A(-2, 3, 1), \quad B(4, 2, -2), \quad C(2, 0, 1)$

In problems 12 through 14, find in each case the perpendicular distance between the given lines.

12. $\dfrac{x + 1}{2} = \dfrac{y - 3}{-3} = \dfrac{z + 2}{4}; \quad \dfrac{x - 2}{3} = \dfrac{y + 1}{2} = \dfrac{z - 1}{5}$

13. $\dfrac{x - 1}{3} = \dfrac{y + 1}{2} = \dfrac{z - 1}{5}; \quad \dfrac{x + 2}{4} = \dfrac{y - 1}{3} = \dfrac{z + 1}{-2}$

14. $\dfrac{x + 1}{2} = \dfrac{y - 1}{-4} = \dfrac{z + 2}{3}; \quad \dfrac{x}{3} = \dfrac{y}{5} = \dfrac{z - 2}{-2}$

In problems 15 through 19, use vector methods to find, in each case, the equations in symmetric form of the line through the given point P and parallel to the two given planes.

15. $P(-1, 3, 2)$, $3x - 2y + 4z + 2 = 0$, $2x + y - z = 0$
16. $P(2, 3, -1)$, $x + 2y + 2z - 4 = 0$, $2x + y - 3z + 5 = 0$
17. $P(1, -2, 3)$, $3x + y - 2z + 3 = 0$, $2x + 3y + z - 6 = 0$
18. $P(-1, 0, -2)$, $2x + 3y - z + 4 = 0$, $3x - 2y + 2z - 5 = 0$
19. $P(3, 0, 1)$, $x + 2y = 0$, $3y - z = 0$

In problems 20 and 21, find in each case equations in symmetric form of the line of intersection of the given planes. Use the method of vector products.

20. $2(x - 1) + 3(y + 1) - 4(z - 2) = 0$
    $3(x - 1) - 4(y + 1) + 2(z - 2) = 0$
21. $3(x + 2) - 2(y - 1) + 2(z + 1) = 0$
    $(x + 2) + 2(y - 1) - 3(z + 1) = 0$

In each of problems 22 through 26, find an equation of the plane through the given point or points and parallel to the given line or lines.

22. $(1, 3, 2)$; $\dfrac{x + 1}{2} = \dfrac{y - 2}{-1} = \dfrac{z + 3}{3}$; $\dfrac{x - 2}{1} = \dfrac{y + 1}{-2} = \dfrac{z + 2}{2}$

23. $(2, -1, -3)$; $\dfrac{x - 1}{3} = \dfrac{y + 2}{2} = \dfrac{z}{-4}$; $\dfrac{x}{2} = \dfrac{y - 1}{-3} = \dfrac{z - 2}{2}$

24. $(2, 1, -2)$; $(1, -1, 3)$; $\dfrac{x + 1}{3} = \dfrac{y - 1}{2} = \dfrac{z - 2}{2}$

25. $(1, -2, 3)$; $(-1, 2, -1)$; $\dfrac{x - 2}{2} = \dfrac{y + 1}{3} = \dfrac{z - 1}{4}$

26. $(0, 1, 2)$; $(2, 0, 1)$; $\dfrac{x - 1}{3} = \dfrac{y + 1}{0} = \dfrac{z + 1}{1}$

In problems 27 through 29, find in each case the equation of the plane through the line L_1 which also satisfies the additional condition.

27. L_1: $\dfrac{x - 1}{2} = \dfrac{y + 1}{3} = \dfrac{z - 2}{1}$;   through $(2, 1, 1)$

28. L_1: $\dfrac{x - 2}{2} = \dfrac{y - 2}{3} = \dfrac{z - 1}{-2}$;   parallel to $\dfrac{x + 1}{3} = \dfrac{y - 1}{2} = \dfrac{z + 1}{1}$

29. L_1: $\dfrac{x + 1}{1} = \dfrac{y - 1}{2} = \dfrac{z - 2}{-2}$;   perpendicular to $2x + 3y - z + 4 = 0$

In problems 30 and 31, find the equation of the plane through the given points and perpendicular to the given planes.

30. $(1, 2, -1)$; $2x - 3y + 5z - 1 = 0$; $3x + 2y + 4z + 6 = 0$
31. $(-1, 3, 2)$; $(1, 6, 1)$; $3x - y + 4z - 7 = 0$

In problems 32 and 33, find equations in symmetric form of the line through the given point which is perpendicular to and intersects the given line. Use the cross product.

32. $(3, 3, -1)$; $\dfrac{x}{-1} = \dfrac{y-3}{1} = \dfrac{z+1}{1}$

33. $(3, -2, 0)$; $\dfrac{x-4}{3} = \dfrac{y-4}{-4} = \dfrac{z-5}{-1}$

34. Assuming Theorem 7, write out the details of the proof that

$$\mathbf{u} \times \mathbf{v} = -\mathbf{v} \times \mathbf{u}.$$

35. Assuming Theorem 7, write out the details of the proof that

$$\mathbf{u} \times (\mathbf{v} + \mathbf{w}) = \mathbf{u} \times \mathbf{v} + \mathbf{u} \times \mathbf{w}.$$

36. Show that if $\mathbf{u}$, $\mathbf{v}$, and $\mathbf{w}$ are any three vectors, then

$$(\mathbf{u} \times \mathbf{v}) \cdot \mathbf{w} = \mathbf{u} \cdot (\mathbf{v} \times \mathbf{w}).$$

4. DERIVATIVES OF VECTOR FUNCTIONS. SPACE CURVES. TANGENTS AND ARC LENGTH

In Chapter 14, Section 4, of *First Course*, we discussed two-dimensional vector functions and their derivatives. We now extend these definitions to vectors in space. The following theorem can be established in the same way as was the corresponding theorem in the two-dimensional case.

Theorem 9. *Suppose that*

$$\mathbf{f}(t) = f_1(t)\mathbf{i} + f_2(t)\mathbf{j} + f_3(t)\mathbf{k}$$

is a vector function and that the vector $\mathbf{c} = c_1\mathbf{i} + c_2\mathbf{j} + c_3\mathbf{k}$ *is a constant. Then*

(a) $\mathbf{f}(t) \to \mathbf{c}$ *as* $t \to a$ *if and only if*

$$f_1(t) \to c_1 \quad \text{and} \quad f_2(t) \to c_2 \quad \text{and} \quad f_3(t) \to c_3.$$

(b) $\mathbf{f}$ *is continuous at* a *if and only if* $f_1, f_2,$ *and* f_3 *are.*

(c) $\mathbf{f}'(t)$ *exists if and only if* $f_1'(t), f_2'(t),$ *and* $f_3'(t)$ *do.*

(d) *We have the formulas*

$$\mathbf{f}'(t) = f_1'(t)\mathbf{i} + f_2'(t)\mathbf{j} + f_3'(t)\mathbf{k}.$$

(e) *If* $\mathbf{v}(t) = a\mathbf{w}(t)$, *then* $\mathbf{v}'(t) = a\mathbf{w}'(t)$ *where* a *is a constant.*

(f) *If* $\mathbf{v}(t) = c(t)\mathbf{w}(t)$, *then*

$$\mathbf{v}'(t) = c(t)\mathbf{w}'(t) + c'(t)\mathbf{w}(t).$$

(g) *If* $\mathbf{v}(t) = \mathbf{w}(t)/c(t)$, *then*

$$\mathbf{v}'(t) = \frac{c(t)\mathbf{w}'(t) - c'(t)\mathbf{w}(t)}{[c(t)]^2}. \tag{1}$$

EXAMPLE 1. Given $\mathbf{f}(t) = 3t^2\mathbf{i} - 2t^3\mathbf{j} + (t^2 + 3)\mathbf{k}$, find $\mathbf{f}'(t), \mathbf{f}''(t), \mathbf{f}'''(t)$.

Solution. $\mathbf{f}'(t) = 6t\mathbf{i} - 6t^2\mathbf{j} + 2t\mathbf{k}$,
  $\mathbf{f}''(t) = 6\mathbf{i} - 12t\mathbf{j} + 2\mathbf{k}$,
  $\mathbf{f}'''(t) = -12\mathbf{j}$.

The next theorem shows how to differentiate functions involving scalar and vector products.

Theorem 10. *If* $\mathbf{u}(t)$ *and* $\mathbf{v}(t)$ *are differentiable, then the derivative of* $f(t) = \mathbf{u}(t) \cdot \mathbf{v}(t)$ *is given by the formula*

$$f'(t) = \mathbf{u}(t) \cdot \mathbf{v}'(t) + \mathbf{u}'(t) \cdot \mathbf{v}(t). \tag{2}$$

The derivative of $\mathbf{w}(t) = \mathbf{u}(t) \times \mathbf{v}(t)$ *is given by the formula*

$$\mathbf{w}'(t) = \mathbf{u}(t) \times \mathbf{v}'(t) + \mathbf{u}'(t) \times \mathbf{v}(t). \tag{3}$$

The proofs of formulas (2) and (3) follow from the corresponding differentiation formulas for ordinary functions. Note that in formula (3) it is essential to retain the order of the factors in each vector product.

EXAMPLE 2. Find the derivative

$$f'(t) \text{ of } f(t) = \mathbf{u}(t) \cdot \mathbf{v}(t) \qquad \text{and} \qquad \mathbf{w}'(t) \text{ of } \mathbf{w}(t) = \mathbf{u}(t) \times \mathbf{v}(t)$$

if

$\mathbf{u}(t) = (t + 3)\mathbf{i} + t^2\mathbf{j} + (t^3 - 1)\mathbf{k}$ and $\mathbf{v}(t) = 2t\mathbf{i} + (t^4 - 1)\mathbf{j} + (2t + 3)\mathbf{k}$.

Solution. By formula (2) we have

$$\begin{aligned}
f'(t) &= [(t + 3)\mathbf{i} + t^2\mathbf{j} + (t^3 - 1)\mathbf{k}] \cdot (2\mathbf{i} + 4t^3\mathbf{j} + 2\mathbf{k}) \\
&\quad + (\mathbf{i} + 2t\mathbf{j} + 3t^2\mathbf{k}) \cdot [2t\mathbf{i} + (t^4 - 1)\mathbf{j} + (2t + 3)\mathbf{k}] \\
&= (2t + 6) + 4t^5 + 2t^3 - 2 + 2t + 2t^5 - 2t + 6t^3 + 9t^2 \\
&= 6t^5 + 8t^3 + 9t^2 + 2t + 4.
\end{aligned}$$

According to (3), we have

$$\begin{aligned}
\mathbf{w}'(t) &= [(t + 3)\mathbf{i} + t^2\mathbf{j} + (t^3 - 1)\mathbf{k}] \times (2\mathbf{i} + 4t^3\mathbf{j} + 2\mathbf{k}) \\
&\quad + (\mathbf{i} + 2t\mathbf{j} + 3t^2\mathbf{k}) \times [2t\mathbf{i} + (t^4 - 1)\mathbf{j} + (2t + 3)\mathbf{k}].
\end{aligned}$$

Computing the two cross products on the right, we get

$$\mathbf{w}'(t) = (-7t^6 + 4t^3 + 9t^2 + 6t)\mathbf{i} + (8t^3 - 4t - 11)\mathbf{j} + (5t^4 + 12t^3 - 6t^2 - 1)\mathbf{k}.$$

———————————

Consider a rectangular coordinate system and a directed line segment from the origin O to a point P in space. As in two dimensions, we denote the vector $\mathbf{v}(\overrightarrow{OP})$ by $\mathbf{r}$. We define an **arc** C in space in a way completely analogous to that in which an arc in the plane was defined. (See *First Course*, page 388.) The vector equation

$$\mathbf{r}(t) = x(t)\mathbf{i} + y(t)\mathbf{j} + z(t)\mathbf{k}$$

is considered to be equivalent to the parametric equations

$$x = x(t), \qquad y = y(t), \qquad z = z(t),$$

which represent an arc C in space.

The definition of **length** of an arc C in three-space is identical with its definition in the plane. If we denote the length of an arc by $l(C)$ we get, in a way that is similar to the two-dimensional analysis, the formulas

$$l(C) = \int_a^b \sqrt{[x'(t)]^2 + [y'(t)]^2 + [z'(t)]^2}\, dt \qquad (4)$$

and

$$s'(t) = \sqrt{[x'(t)]^2 + [y'(t)]^2 + [z'(t)]^2} \qquad (5)$$

where $s(t)$ is an arc length function. We could also write the above formulas (4) and (5) in the vector form

$$l(C) = \int_a^b |\mathbf{r}'(t)|\, dt, \qquad s'(t) = |\mathbf{r}'(t)|.$$

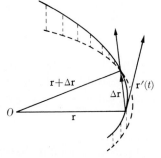

FIGURE 2–12

DEFINITIONS. If $\mathbf{r}'(t) \neq \mathbf{0}$, we define the vector $\mathbf{T}(t) = \mathbf{r}'(t)/|\mathbf{r}'(t)|$ as the **unit tangent vector** to the path corresponding to the value t. The line through the point P_0 corresponding to $\mathbf{r}(t_0)$ and parallel to $\mathbf{T}(t_0)$ is called the **tangent line to the arc** at t_0; the line directed in the same way as $\mathbf{T}(t_0)$ is called the **directed tangent line** at t_0 (Fig. 2–12).

———————————

EXAMPLE 3. The locus of the equations

$$x = a \cos t, \qquad y = a \sin t, \qquad z = bt \qquad (6)$$

is called a **helix**.

(a) Find $s'(t)$. (b) Find the length of that part of the helix for which $0 \le t \le 2\pi$. (c) Show that the unit tangent vector makes a constant angle with the z-axis.

Solution. (See Fig. 2–13 for the graph.)

(a) $x'(t) = -a \sin t$, $y'(t) = a \cos t$, $z'(t) = b$.
Therefore, (a) $s'(t) = \sqrt{a^2 + b^2}$.

(b) $l(C) = 2\pi\sqrt{a^2 + b^2}$.

(c) $\mathbf{T}(t) = \dfrac{-y\mathbf{i} + x\mathbf{j} + b\mathbf{k}}{\sqrt{a^2 + b^2}}$.

Letting ϕ be the angle between $\mathbf{T}$ and $\mathbf{k}$, we get

$$\cos \phi = b/\sqrt{a^2 + b^2}.$$

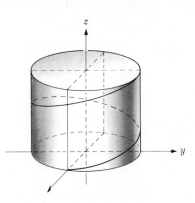

FIGURE 2–13

Remark. We note that the helix (6) winds
around the cylinder $x^2 + y^2 = a^2$.

DEFINITIONS. If t denotes time in the parametric equations of an arc $\mathbf{r}(t)$, then
$\mathbf{r}'(t)$ is the **velocity vector** $\mathbf{v}(t)$, and $\mathbf{r}''(t) = \mathbf{v}'(t)$ is called the **acceleration
vector.** The quantity $s'(t)$ (a scalar) is called the **speed** of the particle moving
according to the law

$$\mathbf{r}(t) = x(t)\mathbf{i} + y(t)\mathbf{j} + z(t)\mathbf{k}.$$

EXERCISES

In problems 1 through 6, find in each case the derivatives $\mathbf{f}'(t)$ and $\mathbf{f}''(t)$.

1. $\mathbf{f}(t) = (2t + 1)\mathbf{i} + t^2\mathbf{j} + (3t - 2)\mathbf{k}$

2. $\mathbf{f}(t) = (\cos t)\mathbf{i} + (2 \sin t)\mathbf{j} + (3t)\mathbf{k}$

3. $\mathbf{f}(t) = e^{2t}\mathbf{i} + e^{-2t}\mathbf{j} + 2\mathbf{k}$

4. $\mathbf{f}(t) = \dfrac{t}{t + 2}\mathbf{i} + \dfrac{t^2}{1 + t}\mathbf{j} + 3t^2\mathbf{k}$

5. $\mathbf{f}(t) = (\cos 2t)\mathbf{i} + \dfrac{1}{t}\mathbf{j} + (\sin 2t)\mathbf{k}$

6. $\mathbf{f}(t) = (\ln t)\mathbf{i} + e^{-t}\mathbf{j} + (t \ln t)\mathbf{k}$

In problems 7 through 9, find in each case either $f'(t)$ or $\mathbf{f}'(t)$, whichever is appropriate.

7. $f(t) = \mathbf{u}(t) \cdot \mathbf{v}(t)$, where

$$\mathbf{u}(t) = 3t\mathbf{i} + 2t^2\mathbf{j} + \frac{1}{t}\mathbf{k}, \qquad \mathbf{v} = t^2\mathbf{i} + \frac{1}{t}\mathbf{j} + t^3\mathbf{k}.$$

8. $\mathbf{f}(t) = \mathbf{u}(t) \times \mathbf{v}(t)$, where
$$\mathbf{u}(t) = (\cos t)\mathbf{i} + (\sin t)\mathbf{j} + t\mathbf{k}, \qquad \mathbf{v}(t) = (\sin t)\mathbf{i} + (\cos t)\mathbf{j} + t^2\mathbf{k}.$$

9. $f(t) = \mathbf{u}(t) \cdot [\mathbf{v}(t) \times \mathbf{w}(t)]$, where

$$\mathbf{u}(t) = t\mathbf{i} + (t + 1)\mathbf{k}, \qquad \mathbf{v}(t) = t^2\mathbf{j}, \qquad \mathbf{w}(t) = \frac{1}{t^2}\mathbf{k}.$$

In problems 10 through 13, find $l(C)$.

10. $C: x = t, \quad y = t^2/\sqrt{2}, \quad z = t^3/3; \quad\quad 0 \le t \le 2$

11. $C: x = t, \quad y = 3t^2/2, \quad z = 3t^3/2; \quad\quad 0 \le t \le 2$

12. $C: x = t, \quad y = \ln(\sec t + \tan t), \quad z = \ln \sec t; \quad\quad 0 \le t \le \pi/4$

13. $C: x = t \cos t, \quad y = t \sin t, \quad z = t; \quad\quad 0 \le t \le \pi/2$

In problems 14 through 16, the parameter t is the time in seconds. Taking s to be the length in feet, in each case find the velocity, speed, and acceleration.

14. $\mathbf{r}(t) = t^2\mathbf{i} + 2t\mathbf{j} + (t^3 - 1)\mathbf{k}$

15. $\mathbf{r}(t) = (t \sin t)\mathbf{i} + (t \cos t)\mathbf{j} + t\mathbf{k}$

16. $\mathbf{r}(t) = e^{3t}\mathbf{i} + e^{-3t}\mathbf{j} + te^{3t}\mathbf{k}$

5. TANGENTIAL AND NORMAL COMPONENTS. THE MOVING TRIHEDRAL

We consider the locus of the equation

$$\mathbf{r} = \mathbf{r}(t)$$

or

$$\mathbf{r}(t) = x(t)\mathbf{i} + y(t)\mathbf{j} + z(t)\mathbf{k}.$$

In the last section, we defined the *unit tangent vector* $\mathbf{T}(t)$ by the relation

$$\mathbf{T}(t) = \frac{\mathbf{r}'(t)}{|\mathbf{r}'(t)|},$$

which we could also write

$$\mathbf{T}(t) = \frac{\mathbf{r}'(t)}{s'(t)} = \frac{d\mathbf{r}}{ds}.$$

Now, since $\mathbf{T} \cdot \mathbf{T} = 1$ for all t, we differentiate this relation with respect to t to obtain

$$\mathbf{T}(t) \cdot \mathbf{T}'(t) + \mathbf{T}'(t) \cdot \mathbf{T}(t) = 2\mathbf{T}(t) \cdot \mathbf{T}'(t) = 0.$$

Therefore, the vector $\mathbf{T}'(t)$ is orthogonal to $\mathbf{T}(t)$. We define the vector $\kappa(t)$ by the relation

$$\kappa(t) = \frac{d\mathbf{T}}{ds} = \frac{\mathbf{T}'(t)}{s'(t)}.$$

Taking the scalar product of κ and $\mathbf{T}$, we see that

$$\kappa(t) \cdot \mathbf{T}(t) = \frac{1}{s'(t)}\mathbf{T}'(t) \cdot \mathbf{T}(t) = 0,$$

and so κ is orthogonal to $\mathbf{T}$. The vector κ is the vector rate of change of direction of the curve with respect to arc length. Bearing in mind the definition of curvature of a plane curve (see *First Course*, Chapter 12, Section 4, and Chapter 14, Section 6), we make the following definitions.

DEFINITIONS. The vector $\kappa(t)$, defined above, is called the **curvature vector** of the curve $\mathbf{r} = \mathbf{r}(t)$. The magnitude $|\kappa(t)| = \kappa(t)$ is called the curvature. If $\kappa(t) \neq 0$, we define the **principal normal vector** $\mathbf{N}(t)$ and the **binormal vector** $\mathbf{B}(t)$ by the relations

$$\mathbf{N}(t) = \frac{\kappa(t)}{|\kappa(t)|}, \qquad \mathbf{B}(t) = \mathbf{T}(t) \times \mathbf{N}(t).$$

The **center of curvature** $C(t)$ is defined by the equation

$$\mathbf{v}[\overrightarrow{OC}(t)] = \mathbf{r}(t) + \frac{1}{\kappa(t)} \mathbf{N}(t).$$

The **radius of curvature** is $R(t) = 1/\kappa(t)$. The **osculating plane** corresponding to t is the plane which contains the tangent line to the path and the center of curvature at t; it is defined only for those values of t for which $\kappa(t) \neq 0$.

———————

EXAMPLE 1. Given the curve

$$x = t, \qquad y = t^2, \qquad z = 1 + t^2,$$

find $\mathbf{T}(t)$ and $\kappa(t)$.

Solution. We may write

$$\mathbf{r}(t) = t\mathbf{i} + t^2\mathbf{j} + (1 + t^2)\mathbf{k}.$$

Therefore

$$\mathbf{r}'(t) = \mathbf{i} + 2t\mathbf{j} + 2t\mathbf{k}$$

and

$$|\mathbf{r}'(t)| = \sqrt{1 + 8t^2} = s'(t).$$

Since $\mathbf{T} = \mathbf{r}'(t)/|\mathbf{r}'(t)|$, we have

$$\mathbf{T}(t) = \frac{1}{\sqrt{1 + 8t^2}} (\mathbf{i} + 2t\mathbf{j} + 2t\mathbf{k}).$$

We may differentiate to get

$$\mathbf{T}'(t) = \frac{-8t}{(1 + 8t^2)^{3/2}} (\mathbf{i} + 2t\mathbf{j} + 2t\mathbf{k}) + \frac{1}{\sqrt{1 + 8t^2}} (2\mathbf{j} + 2\mathbf{k}).$$

Hence

$$\kappa(t) = \frac{\mathbf{T}'(t)}{s'(t)} = \frac{-8t}{(1 + 8t^2)^2} (\mathbf{i} + 2t\mathbf{j} + 2t\mathbf{k}) + \frac{1}{1 + 8t^2} (2\mathbf{j} + 2\mathbf{k})$$

$$= \frac{-8t}{(1 + 8t^2)^2} \mathbf{i} + \frac{2}{(1 + 8t^2)^2} \mathbf{j} + \frac{2}{(1 + 8t^2)^2} \mathbf{k}.$$

———————

It is clear from the definitions of $\mathbf{T}$ and $\mathbf{N}$ that they are unit vectors which are orthogonal. From the definition of cross product, we see at once that $\mathbf{B}$ is a unit vector orthogonal to both $\mathbf{T}$ and $\mathbf{N}$. The triple $\mathbf{T}(t)$, $\mathbf{N}(t)$, $\mathbf{B}(t)$ form a mutually orthogonal triple of unit vectors at each point, the **trihedral** at the point. (It is assumed that $\kappa(t) \neq 0$.) We may also write the formulas

$$\frac{d\mathbf{T}}{ds} = \kappa\mathbf{N} \quad \text{or} \quad \mathbf{T}'(t) = \kappa(t)s'(t)\,\mathbf{N}(t).$$

The equation of the osculating plane at $t = t_0$ is given by

$$b_1(x - x_0) + b_2(y - y_0) + b_3(z - z_0) = 0,$$

where (x_0, y_0, z_0) is the point on the curve corresponding to $t = t_0$ and $\mathbf{B}(t_0) = b_1\mathbf{i} + b_2\mathbf{j} + b_3\mathbf{k}$.

EXAMPLE 2. Given the helix

$$x = 4 \cos t, \qquad y = 4 \sin t, \qquad z = 2t$$

or, equivalently,

$$\mathbf{r}(t) = (4 \cos t)\mathbf{i} + (4 \sin t)\mathbf{j} + (2t)\mathbf{k},$$

find $\mathbf{T}$, $\mathbf{N}$, and $\mathbf{B}$ for $t = 2\pi/3$. Also find κ, the equation of the osculating plane, and the equations of the tangent line.

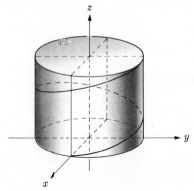

FIGURE 2–14

Solution. The curve is shown in Fig. 2–14. We have

$$x' = -4 \sin t, \qquad y' = 4 \cos t, \qquad z' = 2, \qquad s' = 2\sqrt{5},$$

$$\mathbf{T} = \frac{1}{2\sqrt{5}}(-y\mathbf{i} + x\mathbf{j} + 2\mathbf{k}), \qquad \mathbf{T}'(t) = \frac{1}{2\sqrt{5}}(-x\mathbf{i} - y\mathbf{j}).$$

We compute

$$\kappa(t) = \frac{|\mathbf{T}'(t)|}{s'(t)} = \frac{1}{5}$$

and, using the formula $\mathbf{T}'(t) = \kappa(t)s'(t)\mathbf{N}(t)$, we get

$$\mathbf{N}(t) = -(\cos t)\mathbf{i} - (\sin t)\mathbf{j}.$$

For $t = 2\pi/3$, $x_0 = -2$, $y_0 = 2\sqrt{3}$, $z_0 = 4\pi/3$, we obtain

$$\mathbf{T} = \frac{1}{\sqrt{5}}(-\sqrt{3}\,\mathbf{i} - \mathbf{j} + \mathbf{k}), \qquad \mathbf{N} = \frac{1}{2}\mathbf{i} - \frac{\sqrt{3}}{2}\mathbf{j},$$

$$\mathbf{B} = \frac{1}{2\sqrt{5}}(\sqrt{3}\,\mathbf{i} + \mathbf{j} + 4\mathbf{k}), \qquad \kappa\left(\frac{2\pi}{3}\right) = \frac{1}{5}.$$

The tangent line is

$$\frac{x+2}{\sqrt{3}} = \frac{y - 2\sqrt{3}}{1} = \frac{z - 4\pi/3}{-1},$$

and the equation of the osculating plane is

$$\sqrt{3}\,(x+2) + (y - 2\sqrt{3}) + 4(z - 4\pi/3) = 0.$$

DEFINITIONS. If a particle moves according to the law $\mathbf{r} = \mathbf{r}(t)$, the **tangential** and **normal components** of any vector are its components along $\mathbf{T}$ and $\mathbf{N}$, respectively.

Theorem 11. *If a particle moves according to the law* $\mathbf{r} = \mathbf{r}(t)$, *with t the time and* $\mathbf{T}(t)$, $\mathbf{N}(t)$, $R(t)$ *defined as above, then the acceleration vector* $\mathbf{a}(t) = \mathbf{v}'(t)$ *satisfies the equation*

$$\mathbf{a}(t) = s''(t)\mathbf{T}(t) + \frac{|\mathbf{v}(t)|^2}{R(t)}\mathbf{N}(t). \tag{1}$$

Proof. Since $\mathbf{v}(t) = \mathbf{r}'(t)$, we have, from the definition of $\mathbf{T}(t)$, that

$$\mathbf{v}(t) = s'(t)\mathbf{T}(t).$$

We differentiate this equation and obtain

$$\mathbf{a}(t) = s''(t)\mathbf{T}(t) + s'(t)\mathbf{T}'(t).$$

The relation $\mathbf{T}'(t) = \kappa(t)s'(t)\mathbf{N}(t)$ yields

$$\mathbf{a}(t) = s''(t)\mathbf{T}(t) + \kappa(t)[s'(t)]^2\mathbf{N}(t).$$

We observe that $|\mathbf{v}(t)| = s'(t)$ and $\kappa(t) = 1/R(t)$, and the proof is complete.

Remark. Since $\mathbf{T}$, $\mathbf{N}$, and $\mathbf{B}$ are mutually orthogonal vectors, formula (1) shows that $s''(t)$ and $|\mathbf{v}(t)|^2/R(t)$ are the tangential and normal components, respectively, of $\mathbf{a}(t)$.

EXAMPLE 3. A particle moves according to the law

$$x = t, \qquad y = t^2, \qquad z = t^3.$$

Find its vector velocity and acceleration, its speed, the unit vectors **T** and **N**, and the normal and tangential components of the acceleration vector, all at $t = 1$.

Solution. Since $\mathbf{r}(t) = t\mathbf{i} + t^2\mathbf{j} + t^3\mathbf{k}$, we have

$$\mathbf{v}(t) = \mathbf{r}'(t) = \mathbf{i} + 2t\mathbf{j} + 3t^2\mathbf{k},$$
$$\mathbf{a}(t) = \mathbf{v}'(t) = 2\mathbf{j} + 6t\mathbf{k}.$$

Also,

$$s'(t) = (1 + 4t^2 + 9t^4)^{1/2};$$
$$s''(t) = \frac{4t + 18t^3}{(1 + 4t^2 + 9t^4)^{1/2}}.$$

The path of the particle is sketched in Fig. 2–15. For $t = 1$,

$$\mathbf{v} = \mathbf{i} + 2\mathbf{j} + 3\mathbf{k},$$
$$\mathbf{a} = 2\mathbf{j} + 6\mathbf{k}.$$

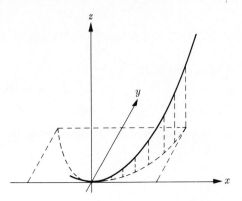

FIGURE 2–15

The speed is $s'(1) = \sqrt{14}$. We obtain

$$s''(1) = \frac{22}{\sqrt{14}}; \qquad \mathbf{T} = \frac{1}{\sqrt{14}}(\mathbf{i} + 2\mathbf{j} + 3\mathbf{k});$$

$$\mathbf{a} - s''\mathbf{T} = \frac{(s')^2}{R}\mathbf{N} = 2\mathbf{j} + 6\mathbf{k} - \frac{11}{7}(\mathbf{i} + 2\mathbf{j} + 3\mathbf{k})$$

$$= \frac{1}{7}(-11\mathbf{i} - 8\mathbf{j} + 9\mathbf{k}).$$

Therefore

$$|\mathbf{a} - s''\mathbf{T}| = \frac{1}{7}\sqrt{266} = \frac{14}{R};$$

$$\mathbf{N} = \frac{1}{\sqrt{266}}(-11\mathbf{i} - 8\mathbf{j} + 9\mathbf{k});$$

$$R = \frac{98}{\sqrt{266}}.$$

So far in this section we have shown how to obtain certain properties of curves in space by using tools of the calculus. The general study of the geometric properties of curves and surfaces in which methods of the calculus are employed is a branch of the subject known as *differential geometry*. The next theorem introduces a quantity τ known as the **torsion** of a curve. It is a remarkable fact that the curvature and torsion of a space curve describe it completely except for its position in space.

Theorem 12. *If* $\kappa(t) \neq 0$, *a quantity* $\tau(t)$ *exists such that*

$$\frac{d\mathbf{T}}{ds} = \kappa\mathbf{N}, \qquad \frac{d\mathbf{N}}{ds} = -\kappa\mathbf{T} - \tau\mathbf{B}, \qquad \frac{d\mathbf{B}}{ds} = \tau\mathbf{N}.$$

A proof is given in Morrey, *University Calculus*, page 544.

EXERCISES

In each of problems 1 through 6, find the unit tangent vector $\mathbf{T}(t)$.

1. $\mathbf{r}(t) = t^3\mathbf{i} + (1 - t)\mathbf{j} + (2t + 1)\mathbf{k}$

2. $\mathbf{r}(t) = (\sin t)\mathbf{i} + (\cos t)\mathbf{j} + 3t^4\mathbf{k}$

3. $x(t) = e^{-2t}, \qquad y(t) = e^{2t}, \qquad z(t) = 1 + t^2$

4. $\mathbf{r}(t) = \dfrac{t}{1 + t}\mathbf{i} + \dfrac{t^2}{1 + t}\mathbf{j} + \dfrac{1 - t}{1 + t}\mathbf{k}$

5. $x(t) = e^t \sin t, \quad y(t) = e^{2t} \cos t, \quad z(t) = e^{-t}$

6. $\mathbf{r}(t) = \ln (1 + t)\mathbf{i} + \dfrac{t}{1 + t^2}\mathbf{j} - 2t^3\mathbf{k}$

In each of problems 7 through 11, for the particular value of t given, find the vectors $\mathbf{T}, \mathbf{N}$, and $\mathbf{B}$; the curvature κ; the equations of the tangent line; and the equation of the osculating plane to the curves given.

7. $x = 1 + t, \qquad y = 3 - t, \qquad z = 2t + 4, \qquad t = 3$

8. $\mathbf{r}(t) = t\mathbf{i} + t^2\mathbf{j} + \frac{1}{3}t^3\mathbf{k}, \qquad t = 0$

9. $x = e^t \cos t, \qquad y = e^t \sin t, \qquad z = e^t, \qquad t = 0$

10. $x = \frac{1}{3}t^3, \qquad y = 2t, \qquad z = 2/t, \qquad t = 2$

11. $x = 2 \cosh (t/2), \qquad y = 2 \sinh (t/2), \qquad z = 2t, \qquad t = 0$

12. Given the path $x = e^t \cos t, y = e^t \sin t, z = e^t$, show that the path lies on the upper half of the cone $z^2 = x^2 + y^2$, and that the tangent vector $\mathbf{T}(t)$ cuts the generators of the cone at a constant angle and makes a constant angle with the z-axis for all t. Also, find $\mathbf{N}$ and κ in terms of t.

In problems 13 through 16, a particle moves according to the law given. In each case, find its vector velocity and acceleration, its speed, the radius of curvature of its path, the unit vectors $\mathbf{T}$ and $\mathbf{N}$, and the tangential and normal components of acceleration at the given time.

13. $x = t, \qquad y = \frac{3}{2}t^2, \qquad z = \frac{3}{2}t^3, \qquad t = 2$

14. $x = t, \qquad y = \ln (\sec t + \tan t), \qquad z = \ln \sec t, \qquad t = \pi/3$

15. $x = t^3/3, \qquad y = 2t, \qquad z = 2/t, \qquad t = 1$

16. $x = t \cos t, \qquad y = t \sin t, \qquad z = t, \qquad t = 0$

4

Elements of Infinite Series

1. INDETERMINATE FORMS

If $f(x)$ and $F(x)$ both approach 0 as x tends to a value a, the quotient

$$\frac{f(x)}{F(x)}$$

may approach a limit, may become infinite, or may fail to have any limit. We saw in the definition of derivative that it is the evaluation of just such expressions that leads to the usual differentiation formulas. We are aware that the expression

$$\frac{f(a)}{F(a)} = \frac{0}{0}$$

is in itself a meaningless one, and we use the term *indeterminate form* for the ratio 0/0.

If $f(x)$ and $F(x)$ both tend to infinity as x tends to a, the ratio $f(x)/F(x)$ may or may not tend to a limit. We use the same term, *indeterminate form*, for the expression ∞/∞, obtained by direct substitution of $x = a$ into the quotient $f(x)/F(x)$.

We recall the Theorem of the Mean, established in the study of differential calculus. (See *First Course*, Chapter 7, page 154.)

Theorem 1. (Theorem of the Mean.) *Suppose that f is continuous for $a \le x \le b$ and that $f'(x)$ exists for each x between a and b. Then there is an x_0 between a and b (that is, $a < x_0 < b$) such that*

$$\frac{f(b) - f(a)}{b - a} = f'(x_0).$$

Remark. Rolle's Theorem is the special case $f(a) = f(b) = 0$.

The evaluation of indeterminate forms requires an extension of the Theorem of the Mean which we now prove.

Theorem 2. (Generalized Theorem of the Mean.) *Suppose that f and F are continuous for $a \le x \le b$, and $f'(x)$ and $F'(x)$ exist for $a < x < b$ with $F'(x) \ne 0$ there. Then $F(b) - F(a) \ne 0$ and there is a number ξ with $a < \xi < b$ such that*

$$\frac{f(b) - f(a)}{F(b) - F(a)} = \frac{f'(\xi)}{F'(\xi)}. \tag{1}$$

Proof. The fact that $F(b) - F(a) \neq 0$ is obtained by applying the Theorem of the Mean (Theorem 1) to F. For then, $F(b) - F(a) = F'(x_0)(b - a)$ for some x_0 such that $a < x_0 < b$. By hypothesis, the right side is different from zero.

To prove the main part of the theorem, we define the function $\phi(x)$ by the formula

$$\phi(x) = f(x) - f(a) - \frac{f(b) - f(a)}{F(b) - F(a)}[F(x) - F(a)].$$

We compute $\phi(a)$, $\phi(b)$, and $\phi'(x)$, getting

$$\phi(a) = f(a) - f(a) - \frac{f(b) - f(a)}{F(b) - F(a)}[F(a) - F(a)] = 0,$$

$$\phi(b) = f(b) - f(a) - \frac{f(b) - f(a)}{F(b) - F(a)}[F(b) - F(a)] = 0,$$

$$\phi'(x) = f'(x) - \frac{f(b) - f(a)}{F(b) - F(a)} F'(x).$$

Applying the Theorem of the Mean (i.e., in the special form of Rolle's Theorem) to $\phi(x)$ in the interval (a, b), we find

$$\frac{\phi(b) - \phi(a)}{b - a} = 0 = \phi'(\xi) = f'(\xi) - \frac{f(b) - f(a)}{F(b) - F(a)} F'(\xi)$$

for some ξ between a and b. Dividing by $F'(\xi)$, we obtain formula (1).

The next theorem, known as **l'Hôpital's Rule,** is useful in the evaluation of indeterminate forms.

Theorem 3. (l'Hôpital's Rule.) *Suppose that*

$$\lim_{x \to a} f(x) = 0, \quad \lim_{x \to a} F(x) = 0, \quad and \quad \lim_{x \to a} \frac{f'(x)}{F'(x)} = L,$$

and that the hypotheses of Theorem 2 hold in some deleted interval about a. Then

$$\lim_{x \to a} \frac{f(x)}{F(x)} = \lim_{x \to a} \frac{f'(x)}{F'(x)} = L.$$

Proof. For some h we apply Theorem 2 in the interval $a < x < a + h$. Then

$$\frac{f(a + h) - f(a)}{F(a + h) - F(a)} = \frac{f(a + h)}{F(a + h)} = \frac{f'(\xi)}{F'(\xi)}, \quad a < \xi < a + h,$$

where we have taken $f(a) = F(a) = 0$. As h tends to 0, ξ tends to a, and so

$$\lim_{h \to 0} \frac{f(a + h)}{F(a + h)} = \lim_{\xi \to a} \frac{f'(\xi)}{F'(\xi)} = L.$$

A similar proof is valid for x on the interval $(a - h) < x < a$.

EXAMPLE 1. Evaluate

$$\lim_{x \to 3} \frac{x^3 - 2x^2 - 2x - 3}{x^2 - 9}.$$

Solution. We set $f(x) = x^3 - 2x^2 - 2x - 3$ and $F(x) = x^2 - 9$. We see at once that $f(3) = 0$ and $F(3) = 0$, and we have an indeterminate form. We calculate

$$f'(x) = 3x^2 - 6x - 2, \qquad F'(x) = 2x.$$

By Theorem 3 (l'Hôpital's Rule):

$$\lim_{x \to 3} \frac{f(x)}{F(x)} = \lim_{x \to 3} \frac{f'(x)}{F'(x)} = \frac{3(9) - 6(3) - 2}{2(3)} = \frac{7}{6}.$$

Remarks. It is essential that $f(x)$ and $F(x)$ *both* tend to zero as x tends to a before applying l'Hôpital's Rule. If either or both functions tend to finite limits $\neq 0$, or if one tends to zero and the other does not, then the limit of the quotient is found by the method of direct substitution as given in *First Course*, Chapter 5.

It may happen that $f'(x)/F'(x)$ is an indeterminate form as $x \to a$. Then l'Hôpital's Rule may be applied again, and the limit $f''(x)/F''(x)$ may exist as x tends to a. In fact, for some problems l'Hôpital's Rule may be required a number of times before the limit is actually determined. Example 3 below exhibits this point.

EXAMPLE 2. Evaluate

$$\lim_{x \to a} \frac{x^p - a^p}{x^q - a^q}, \qquad a \neq 0.$$

Solution. We set $f(x) = x^p - a^p$, $F(x) = x^q - a^q$. Then $f(a) = 0$, $F(a) = 0$. We compute $f'(x) = px^{p-1}$, $F'(x) = qx^{q-1}$. Therefore

$$\lim_{x \to a} \frac{f(x)}{F(x)} = \lim_{x \to a} \frac{f'(x)}{F'(x)} = \lim_{x \to a} \frac{px^{p-1}}{qx^{q-1}} = \frac{p}{q} a^{p-q}.$$

EXAMPLE 3. Evaluate

$$\lim_{x \to 0} \frac{x - \sin x}{x^3}.$$

Solution. We set $f(x) = x - \sin x$, $F(x) = x^3$. Since $f(0) = 0$, $F(0) = 0$, we apply l'Hôpital's Rule and get

$$\lim_{x \to 0} \frac{f(x)}{F(x)} = \lim_{x \to 0} \frac{1 - \cos x}{3x^2}.$$

But we note that $f'(0) = 0$, $F'(0) = 0$, and so we apply l'Hôpital's Rule again:

$$f''(x) = \sin x, \qquad F''(x) = 6x.$$

Hence

$$\lim_{x \to 0} \frac{f(x)}{F(x)} = \lim_{x \to 0} \frac{f''(x)}{F''(x)} \, .$$

Again we have an indeterminate form: $f''(0) = 0$, $F''(0) = 0$. We continue, to obtain $f'''(x) = \cos x$, $F'''(x) = 6$. Now we find that

$$\lim_{x \to 0} \frac{f(x)}{F(x)} = \lim_{x \to 0} \frac{f'''(x)}{F'''(x)} = \frac{\cos 0}{6} = \frac{1}{6} \, .$$

L'Hôpital's Rule can be extended to the case where *both* $f(x) \to \infty$ and $F(x) \to \infty$ as $x \to a$.

Theorem 4. (l'Hôpital's Rule.) *Suppose that*

$$\lim_{x \to a} f(x) = \infty, \qquad \lim_{x \to a} F(x) = \infty, \qquad and \qquad \lim_{x \to a} \frac{f'(x)}{F'(x)} = L.$$

Then

$$\lim_{x \to a} \frac{f(x)}{F(x)} = \lim_{x \to a} \frac{f'(x)}{F'(x)} = L.$$

This theorem is proved in Morrey, *University Calculus*, page 338.

EXAMPLE 4. Evaluate

$$\lim_{x \to 0} \frac{\ln x}{\operatorname{csch} x} \, .$$

Solution. We first note that x must tend to zero through positive values. In fact, all the theorems we have stated hold for one-sided limits as well as for ordinary limits. We set

$$f(x) = \ln x, \qquad F(x) = \operatorname{csch} x.$$

Then $f(x) \to -\infty$ and $F(x) \to +\infty$ as $x \to 0+$. Therefore

$$\lim_{x \to 0} \frac{f(x)}{F(x)} = \lim_{x \to 0} \frac{1/x}{-\operatorname{csch} x \coth x} = \lim_{x \to 0} \frac{-\sinh^2 x}{x \cosh x} \, .$$

We still have an indeterminate form, and we take derivatives again. Hence

$$\lim_{x \to 0} \frac{\ln x}{\operatorname{csch} x} = \lim_{x \to 0} \frac{-2 \sinh x \cosh x}{\cosh x + x \sinh x} = \frac{0}{1} = 0.$$

Remark. Theorems 3 and 4 are valid if $a = +\infty$ or $-\infty$. That is, if $f(\infty)/F(\infty)$ is indeterminate, then

$$\lim_{x \to \infty} \frac{f(x)}{F(x)} = \lim_{x \to \infty} \frac{f'(x)}{F'(x)} \, .$$

The next example exhibits this type of indeterminate form.

EXAMPLE 5. Evaluate

$$\lim_{x \to +\infty} \frac{8x}{e^x}.$$

Solution.

$$\lim_{x \to +\infty} \frac{8x}{e^x} = \lim_{x \to +\infty} \frac{8}{e^x} = 0.$$

Remarks. Indeterminate forms of the type $0 \cdot \infty$ or $\infty - \infty$ can often be evaluated by transforming the expression into a quotient of the form $0/0$ or ∞/∞. Limits involving exponential expressions may often be evaluated by taking logarithms. Of course, algebraic or trigonometric reductions may be made at any step. The next examples illustrate the procedure.

EXAMPLE 6. Evaluate

$$\lim_{x \to \pi/2} (\sec x - \tan x).$$

Solution. We employ trigonometric reduction to change $\infty - \infty$ into a standard form. We have

$$\lim_{x \to \pi/2} (\sec x - \tan x) = \lim_{x \to \pi/2} \frac{1 - \sin x}{\cos x} = \lim_{x \to \pi/2} \frac{-\cos x}{-\sin x} = 0.$$

EXAMPLE 7. Evaluate

$$\lim_{x \to 0} (1 + x)^{1/x}.$$

Solution. We have 1^∞, which is indeterminate. Set $y = (1 + x)^{1/x}$ and take logarithms. Then

$$\ln y = \ln (1 + x)^{1/x} = \frac{\ln (1 + x)}{x}.$$

By l'Hôpital's Rule,

$$\lim_{x \to 0} \frac{\ln (1 + x)}{x} = \lim_{x \to 0} \frac{1}{1 + x} = 1.$$

Therefore, $\lim_{x \to 0} \ln y = 1$, and we conclude that

$$\lim_{x \to 0} y = \lim_{x \to 0} (1 + x)^{1/x} = e.$$

EXERCISES

Evaluate the following limits:

1. $\displaystyle \lim_{x \to -2} \frac{2x^2 + 5x + 2}{x^2 - 4}$

2. $\displaystyle \lim_{x \to 2} \frac{x^3 - x^2 - x - 2}{x^3 - 8}$

3. $\displaystyle \lim_{x \to 1} \frac{x^3 - 3x + 2}{x^3 - x^2 - x + 1}$

4. $\displaystyle \lim_{x \to 2} \frac{x^4 - 3x^2 - 4}{x^3 + 2x^2 - 4x - 8}$

5. $\lim\limits_{x\to\infty} \dfrac{2x^3 - x^2 + 3x + 1}{3x^3 + 2x^2 - x - 1}$

6. $\lim\limits_{x\to 4} \dfrac{x^3 - 8x^2 + 2x + 1}{x^4 - x^2 + 2x - 3}$

7. $\lim\limits_{x\to\infty} \dfrac{x^3 - 3x + 1}{2x^4 - x^2 + 2}$

8. $\lim\limits_{x\to\infty} \dfrac{x^4 - 2x^2 - 1}{2x^3 - 3x^2 + 3}$

9. $\lim\limits_{x\to 0} \dfrac{\tan 3x}{\sin x}$

10. $\lim\limits_{x\to 0} \dfrac{\sin 7x}{x}$

11. $\lim\limits_{x\to 0} \dfrac{e^{2x} - 2x - 1}{1 - \cos x}$

12. $\lim\limits_{x\to 0} \dfrac{e^{3x} - 1}{1 - \cos x}$

13. $\lim\limits_{x\to 0} \dfrac{x - \sinh x}{(1 - \cosh x)^2}$

14. $\lim\limits_{x\to 0} \dfrac{\ln x}{e^x}$

15. $\lim\limits_{x\to\infty} \dfrac{\ln x}{x^h}, \quad h > 0$

16. $\lim\limits_{x\to 0} \dfrac{\ln (1 + 2x)}{3x}$

17. $\lim\limits_{x\to 0} \dfrac{3^x - 2^x}{x}$

18. $\lim\limits_{x\to 0} \dfrac{3^x - 2^x}{x^2}$

19. $\lim\limits_{x\to 0} \dfrac{3^x - 2^x}{\sqrt{x}}$

20. $\lim\limits_{\theta\to 0} \dfrac{\tanh 2\theta - 2\theta}{3\theta - \sinh 3\theta}$

21. $\lim\limits_{x\to \pi/2} \dfrac{1 - \sin x}{\cos x}$

22. $\lim\limits_{x\to 2} \dfrac{\sqrt{2x} - 2}{\ln (x - 1)}$

23. $\lim\limits_{x\to \pi/2} \dfrac{\ln \sin x}{1 - \sin x}$

24. $\lim\limits_{x\to \pi/2} \dfrac{\cos x}{\sin^2 x}$

25. $\lim\limits_{x\to\infty} \dfrac{x^3}{e^x}$

26. $\lim\limits_{x\to \pi/2} \dfrac{\tan x}{\ln \cos x}$

27. $\lim\limits_{x\to\infty} \dfrac{\sin x}{x}$

28. $\lim\limits_{x\to \pi/2} \dfrac{\sin x}{x}$

29. $\lim\limits_{x\to 0} \dfrac{x - \arctan x}{x - \sin x}$

30. $\lim\limits_{x\to 0} \sqrt{x} \ln x$

31. $\lim\limits_{x\to 0} x \cot x$

32. $\lim\limits_{x\to \pi/2} (x - \pi/2) \sec x$

33. $\lim\limits_{x\to\infty} \dfrac{\arctan x}{x}$

34. $\lim\limits_{\theta\to 0} \left(\csc \theta - \dfrac{1}{\theta} \right)$

35. $\lim\limits_{x\to 0} (\operatorname{csch} x - \coth x)$

36. $\lim\limits_{x\to 0} \left(\cot^2 x - \dfrac{1}{x^2} \right)$

37. $\lim\limits_{x\to 0} x^x$

38. $\lim\limits_{x\to 0} x^{4x}$

39. $\lim\limits_{x\to 0} (\sinh x)^{\tan x}$

40. $\lim\limits_{x\to\infty} \left(1 + \dfrac{k}{x} \right)^x$

41. $\lim\limits_{x\to 0} x^{(x^2)}$

42. $\lim\limits_{x\to 0} (\cot x)^x$

43. $\lim\limits_{x\to 0} x^{(1/\ln x)}$

44. $\lim\limits_{x\to\infty} \dfrac{x^p}{e^x}, \quad p > 0$

2. CONVERGENT AND DIVERGENT SERIES

In *First Course* (Chapter 5, Section 5, page 119), the idea of a sequence of numbers was introduced. We begin by repeating some of the material presented there. The numbers

$$b_1, b_2, b_3, \ldots, b_{12}, b_{13}, b_{14}$$

form a sequence of fourteen numbers. Since this set contains both a first and last element, the sequence is termed **finite**. In all other circumstances it is called **infinite**. The subscripts not only identify the location of each element but also serve to associate a positive integer with each member of the sequence. In other words, *a sequence is a function* with *domain* a portion (or all) of the positive integers and with *range* in the collection of real numbers.

If the domain is an infinite collection of positive integers, e.g., all positive integers, we write

$$a_1, a_2, \ldots, a_n, \ldots,$$

the final dots indicating the never-ending character of the sequence. Simple examples of infinite sequences are

$$1, \frac{1}{2}, \frac{1}{3}, \frac{1}{4}, \ldots, \frac{1}{n}, \ldots \tag{1}$$

$$\frac{1}{2}, \frac{2}{3}, \frac{3}{4}, \ldots, \frac{n}{n+1}, \ldots \tag{2}$$

$$2, 4, 6, \ldots, 2n, \ldots \tag{3}$$

DEFINITION. Given the infinite sequence

$$a_1, a_2, \ldots, a_n, \ldots,$$

we say that this **sequence has the limit** c if, for each $\epsilon > 0$, there is a positive integer N (the size of N depending on ϵ) such that

$$|a_n - c| < \epsilon \qquad \text{for all } n > N.$$

We also write $a_n \to c$ as $n \to \infty$ and, equivalently,

$$\lim_{n \to \infty} a_n = c.$$

In the sequence (1), we have

$$a_1 = 1, \quad a_2 = \frac{1}{2}, \quad \ldots, \quad a_n = \frac{1}{n}, \quad \ldots$$

and $\lim_{n \to \infty} a_n = 0$. The sequence (2) has the form

$$a_1 = \frac{1}{2}, \quad a_2 = \frac{2}{3}, \quad \ldots, \quad a_n = \frac{n}{n+1}, \quad \ldots$$

and $\lim_{n \to \infty} a_n = 1$. The sequence (3) does not tend to a limit.

An expression such as

$$u_1 + u_2 + u_3 + \cdots + u_{24}$$

is called a *finite series*. The *sum* of such a series is obtained by adding the 24 terms. We now extend the notion of a finite series by considering an expression of the form

$$u_1 + u_2 + u_3 + \cdots + u_n + \cdots$$

which is nonterminating and which we call an *infinite series*. Our first task is to give a meaning, if possible, to such an infinite succession of additions.

DEFINITION. Given the infinite series $u_1 + u_2 + u_3 + \cdots + u_n + \cdots$, the quantity $s_k = u_1 + u_2 + \cdots + u_k$ is called the **kth partial sum** of the series. That is,

$$s_1 = u_1, \qquad s_2 = u_1 + u_2, \qquad s_3 = u_1 + u_2 + u_3,$$

etc. Each partial sum is obtained simply by a *finite* number of additions.

DEFINITION. Given the series

$$u_1 + u_2 + u_3 + \cdots + u_n + \cdots \tag{4}$$

with the sequence of partial sums

$$s_1, s_2, s_3, \ldots, s_n, \ldots,$$

we define the **sum of the series** (4) to be

$$\lim_{n \to \infty} s_n \tag{5}$$

whenever the limit exists.

Using the $\sum$-notation for sum, we can also write

$$\sum_{n=1}^{\infty} u_n = \lim_{n \to \infty} s_n.$$

If the limit (5) does not exist, then the sum (4) **is not defined.**

DEFINITIONS. If the limit (5) exists, the series $\sum_{n=1}^{\infty} u_n$ is said to **converge** to that limit; otherwise the series is said to **diverge.**

The sequence of terms

$$a, ar, ar^2, ar^3, \ldots, ar^{n-1}, ar^n, \ldots$$

forms a *geometric progression.* Each term (except the first) is obtained by multiplication of the preceding term by r, the *common ratio.* The partial sums of the geometric series

$$a + ar + ar^2 + ar^3 + \cdots + ar^n + \cdots$$

are

$$s_1 = a,$$
$$s_2 = a + ar,$$
$$s_3 = a + ar + ar^2,$$
$$s_4 = a + ar + ar^2 + ar^3,$$

and, in general,

$$s_n = a(1 + r + r^2 + \cdots + r^{n-1}).$$

For example, with $a = 2$ and $r = \frac{1}{2}$,

$$s_n = 2\left(1 + \frac{1}{2} + \frac{1}{4} + \cdots + \frac{1}{2^{n-1}}\right).$$

The identity

$$(1 + r + r^2 + \cdots + r^{n-1})(1 - r) = 1 - r^n,$$

which may be verified by straightforward multiplication, leads to the formula

$$s_n = a\frac{1 - r^n}{1 - r}$$

for the nth partial sum. The example $a = 2$, $r = \frac{1}{2}$ gives

$$s_n = 2\frac{1 - 2^{-n}}{\frac{1}{2}} = 4 - \frac{1}{2^{n-2}}.$$

In general, we may write

$$s_n = a\frac{1 - r^n}{1 - r} = \frac{a}{1 - r} - \frac{a}{1 - r}r^n, \qquad r \neq 1. \tag{6}$$

The next theorem is a direct consequence of formula (6).

Theorem 5. *A geometric series*

$$a + ar + ar^2 + \cdots + ar^n + \cdots$$

converges if $-1 < r < 1$ *and diverges if* $|r| \geq 1$. *In the convergent case we have*

$$\sum_{n=1}^{\infty} ar^{n-1} = \frac{a}{1 - r}. \tag{7}$$

Proof. From (6) we see that $r^n \to 0$ if $|r| < 1$, yielding (7); also, $r^n \to \infty$ if $|r| > 1$. For $r = 1$, the partial sum s_n is na, and s_n does not tend to a limit as $n \to \infty$. If $r = -1$, the partial sum s_n is a if n is odd and 0 if n is even.

The next theorem is useful in that it exhibits a limitation on the behavior of the terms of a convergent series.

Theorem 6. *If the series*

$$\sum_{k=1}^{\infty} u_k = u_1 + u_2 + u_3 + \cdots + u_n + \cdots \tag{8}$$

converges, then

$$\lim_{n \to \infty} u_n = 0.$$

Proof. Writing

$$s_n = u_1 + u_2 + \cdots + u_n,$$
$$s_{n-1} = u_1 + u_2 + \cdots + u_{n-1},$$

we have, by subtraction, $u_n = s_n - s_{n-1}$. Letting c denote the sum of the series, we see that $s_n \to c$ as $n \to \infty$; also, $s_{n-1} \to c$ as $n \to \infty$. Therefore

$$\lim_{n \to \infty} u_n = \lim_{n \to \infty} (s_n - s_{n-1}) = \lim_{n \to \infty} s_n - \lim_{n \to \infty} s_{n-1} = c - c = 0.$$

Remark. The converse of Theorem 6 is not necessarily true. Later we shall show (by example) that it is possible both for u_n to tend to 0 and for the series to diverge.

Corollary. *If u_n does not tend to zero as $n \to \infty$, then the series $\sum_{n=1}^{\infty} u_n$ is divergent.*

Convergent series may be added, subtracted, and multiplied by constants, as the next theorem shows.

Theorem 7. *If $\sum_{n=1}^{\infty} u_n$ and $\sum_{n=1}^{\infty} v_n$ both converge and c is any number, then the series*

$$\sum_{n=1}^{\infty} (cu_n), \qquad \sum_{n=1}^{\infty} (u_n + v_n), \qquad \sum_{n=1}^{\infty} (u_n - v_n)$$

all converge, and

$$\sum_{n=1}^{\infty} (cu_n) = c \sum_{n=1}^{\infty} u_n,$$

$$\sum_{n=1}^{\infty} (u_n \pm v_n) = \sum_{n=1}^{\infty} u_n \pm \sum_{n=1}^{\infty} v_n.$$

Proof. For each n, we have the following equalities for the partial sums:

$$\sum_{j=1}^{n} (cu_j) = c \sum_{j=1}^{n} u_j; \qquad \sum_{j=1}^{n} (u_j \pm v_j) = \sum_{j=1}^{n} u_j \pm \sum_{j=1}^{n} v_j.$$

The results follow from the theorems on limits of sequences, as given in *First Course,* Chapter 5.

EXAMPLE. Express the repeating decimal $A = 0.151515\ldots$ as the ratio of two integers.

Solution. We write A in the form of a geometric series:

$$A = 0.15(1 + 0.01 + (0.01)^2 + (0.01)^3 + \cdots),$$

in which $a = 0.15$ and $r = 0.01$. This series is convergent and has sum

$$s = \frac{0.15}{1 - 0.01} = \frac{0.15}{0.99} = \frac{5}{33} = A.$$

EXERCISES

In problems 1 through 6, express each repeating decimal as the ratio of two integers.

1. $0.535353\ldots$

2. $0.012012012\ldots$

3. $463.546354635463\ldots$

4. $22.818181\ldots$

5. $3.7217217217\ldots$

6. $27.5431313131\ldots$

7. Find the sum of the geometric series if $a = 3$, $r = -\frac{1}{3}$.

8. The first term of a geometric series is 3 and the fifth term is $\frac{16}{27}$. Find the sum of the infinite series.

In problems 9 through 12, write the first five terms of each of the series given. Use the Corollary to Theorem 6 to show that the series is divergent.

9. $\displaystyle\sum_{n=1}^{\infty} \frac{2n}{3n + 5}$

10. $\displaystyle\sum_{n=1}^{\infty} \frac{n^2 - 2n + 3}{2n^2 + n + 1}$

11. $\displaystyle\sum_{n=1}^{\infty} (-1)^{n+1} \frac{e^n}{n^3}$

12. $\displaystyle\sum_{n=1}^{\infty} \frac{n^2 + n + 2}{\log (n + 1)}$

In problems 13 through 16, *assume* that the series

$$\sum_{n=1}^{\infty} \frac{1}{n^2}, \qquad \sum_{n=1}^{\infty} \frac{1}{n^3}, \qquad \sum_{n=1}^{\infty} \frac{1}{n^4}$$

all converge. In each case use Theorem 7 to show that the given series is convergent.

13. $\displaystyle\sum_{n=1}^{\infty} \frac{3n + 2}{n^3}$

14. $\displaystyle\sum_{n=1}^{\infty} \frac{n - 2}{n^3}$

15. $\displaystyle\sum_{n=1}^{\infty} \frac{3n^2 + 4}{n^4}$

16. $\displaystyle\sum_{n=1}^{\infty} \frac{3n^2 - 2n + 4}{n^4}$

17. Suppose that the series $\sum_{k=1}^{\infty} u_k$ converges. Show that any series obtained from this one by deleting a finite number of terms also converges. [*Hint:* Since $s_n = \sum_{k=1}^{n} u_k$ converges to a limit, find the value to which S_n, the partial sums of the deleted series, must tend.]

3. SERIES OF POSITIVE TERMS

Except in very special cases, it is not possible to tell if a series converges by finding whether or not s_n, the nth partial sum, tends to a limit. (The geometric series, however, is one of the special cases where it *is* possible.) In this section, we present some *indirect* tests for convergence and divergence which apply only to series with positive (or at least nonnegative) terms. That is, we assume throughout this section that $u_n \geq 0$ for $n = 1, 2, \ldots$. Tests for series with terms which may be positive or negative will be discussed in the following sections.

Theorem 8. *Suppose that $u_n \geq 0$, $n = 1, 2, \ldots$ and $s_n = \sum_{k=1}^{n} u_k$ is the nth partial sum. Then, either* (a), *there is a number M such that all the $s_n \leq M$, in which case the series $\sum_{k=1}^{\infty} u_k$ converges to a value $s \leq M$, or else* (b), $s_n \to +\infty$ *and the series diverges.*

Proof. By subtraction, we have

$$u_n = s_n - s_{n-1} \geq 0,$$

and so the s_n form an increasing (or at least nondecreasing) sequence. If all $s_n \leq M$, then by Axiom C (*First Course*, page 122), which says that a nondecreasing bounded sequence has a limit, we conclude that $s_n \to s \leq M$. Thus part (a) of the theorem is established. If there is no such M, then for each number E, no matter how large, there must be an $s_n > E$; and all s_m with $m > n$ are greater than or equal to s_n. This is another way of saying $s_n \to +\infty$.

The next theorem is one of the most useful tests for deciding convergence and divergence of series.

Theorem 9. (Comparison Test.) *Suppose that all $u_n \geq 0$.* (a) *If $\sum_{n=1}^{\infty} a_n$ is a convergent series and $u_n \leq a_n$ for all n, then $\sum_{n=1}^{\infty} u_n$ is convergent and*

$$\sum_{n=1}^{\infty} u_n \leq \sum_{n=1}^{\infty} a_n.$$

(b) *If $\sum_{n=1}^{\infty} a_n$ is a divergent series of nonnegative terms and $u_n \geq a_n$ for all n, then $\sum_{n=1}^{\infty} u_n$ diverges.*

Proof. We let

$$s_n = u_1 + u_2 + \cdots + u_n, \qquad S_n = a_1 + a_2 + \cdots + a_n$$

be the nth partial sums. The s_n, S_n are both nondecreasing sequences. In case (a), we let S be the limit of S_n and, since

$$s_n \leq S_n \leq S$$

for every n, we apply Theorem 8 to conclude that s_n converges. In case (b), we have $S_n \to +\infty$ and $s_n \geq S_n$ for every n. Hence, $s_n \to +\infty$.

Remarks. In order to apply the Comparison Test, the student must show either (a), that the terms u_n of the given series are $\leq a_n$ where $\sum_{n=1}^{\infty} a_n$ is a *known* convergent series or (b), that each $u_n \geq a_n$ where $\sum_{n=1}^{\infty} a_n$ is a *known* divergent series. In all other cases, no conclusion can be drawn.

For the Comparison Test to be useful, we must have at hand as large a number as possible of series (of positive terms) about whose convergence and divergence we are fully informed. Then, when confronted with a new series of positive terms, we shall have available a body of series for comparison purposes. So far, the only series which we have shown to be convergent are the geometric series with $r < 1$, and the only series which we have shown to be divergent are those in which u_n does not tend to zero. We now study the convergence and divergence of a few special types of series in order to obtain material which can be used for the Comparison Test.

DEFINITION. If n is a positive integer, we define $n!$ (read *n factorial*) $= 1 \cdot 2 \cdots n$; it is convenient to define $0! = 1$.

For example, $5! = 1 \cdot 2 \cdot 3 \cdot 4 \cdot 5 = 120$, etc. We see that

$$(n + 1)! = (n + 1) \cdot n!, \qquad n \geq 0.$$

EXAMPLE 1. Test the series

$$\sum_{n=1}^{\infty} \frac{1}{n!}$$

for convergence or divergence.

Solution. Writing the first few terms, we obtain

$$\frac{1}{1!} = \frac{1}{1}, \qquad \frac{1}{2!} = \frac{1}{1 \cdot 2}, \qquad \frac{1}{3!} = \frac{1}{1 \cdot 2 \cdot 3}, \qquad \frac{1}{4!} = \frac{1}{1 \cdot 2 \cdot 3 \cdot 4}.$$

Since each factor except 1 and 2 in $n!$ is larger than 2, we have the inequalities

$$n! \geq 2^{n-1} \qquad \text{and} \qquad \frac{1}{n!} \leq \frac{1}{2^{n-1}}.$$

The series $\sum_{n=1}^{\infty} a_n$ with $a_n = 1/2^{n-1}$ is a geometric series with $r = \frac{1}{2}$, and is therefore convergent. Hence, by the Comparison Test, $\sum_{n=1}^{\infty} 1/n!$ converges.

Remark. Since any *finite* number of terms at the beginning of a series does not affect convergence or divergence, the comparison between u_n and a_n in Theorem 9 is not required for all n. It is required for all n *except a finite number.*

The next theorem gives us an entire collection of series useful for comparison purposes.

Theorem 10. (The *p*-series.) *The series*

$$\sum_{n=1}^{\infty} \frac{1}{n^p},$$

known as the p-series, is convergent if p > 1 and divergent if p ≤ 1.

The proof of this theorem is deferred until later in the section. We note that it is not necessary that *p* be an integer.

EXAMPLE 2. Test the series

$$\sum_{n=1}^{\infty} \frac{1}{n(n+1)}$$

for convergence or divergence.

Solution. For each *n* we have

$$\frac{1}{n(n+1)} \leq \frac{1}{n^2}.$$

Since $\sum_{n=1}^{\infty} 1/n^2$ is a *p*-series with $p = 2$ and so converges, we are in a position to use the Comparison Test.

$$\sum_{n=1}^{\infty} \frac{1}{n(n+1)}$$

converges.

EXAMPLE 3. Test the series

$$\sum_{n=1}^{\infty} \frac{1}{n+10}$$

for convergence or divergence.

Solution 1. Writing out a few terms, we have

$$\tfrac{1}{11} + \tfrac{1}{12} + \tfrac{1}{13} + \tfrac{1}{14} + \cdots,$$

and we see that the series is just like $\sum_{n=1}^{\infty} 1/n$, except that the first ten terms are missing. According to the Remark before Theorem 10, we may compare the given series with the *p*-series for $p = 1$. The comparison establishes divergence.

Solution 2. We have, for every $n \geq 1$,

$$n + 10 \leq 11n,$$

and so

$$\frac{1}{n+10} \geq \frac{1}{11n}.$$

The series

$$\sum_{n=1}^{\infty} \frac{1}{11n} = \frac{1}{11} \sum_{n=1}^{\infty} \frac{1}{n}$$

is divergent (*p*-series with $p = 1$) and, therefore, the given series diverges.

The next theorem yields another test which is used frequently in conjunction with the Comparison Test.

Theorem 11. (Integral Test.) *Assume that f is a continuous, nonnegative, and nonincreasing function defined for all $x \geq 1$. That is, we suppose that*

$$f(x) \geq 0, \ (nonnegative)$$

and

$$f(x) \geq f(y) \quad for \ x \leq y \ (nonincreasing).$$

Suppose that $\sum_{n=1}^{\infty} u_n$ is a series with

$$u_n = f(n) \quad for \ each \ n \geq 1.$$

Then (a) $\sum_{n=1}^{\infty} u_n$ is convergent if the improper integral $\int_1^{\infty} f(x)\,dx$ is convergent and, conversely (b), the improper integral converges if the series does.

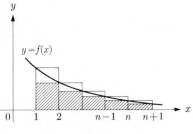

FIGURE 3–1

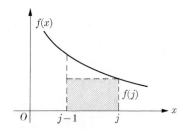

FIGURE 3–2

Proof. (See Fig. 3–1.) (a) Suppose first that the improper integral is convergent. Then, since $f(x) \geq f(j)$ for $x \leq j$, we see that

$$\int_{j-1}^{j} f(x)\,dx \geq f(j), \tag{1}$$

a fact verified by noting in Fig. 3–2 that $f(j)$ is the shaded area and the integral is the area under the curve. We define

$$a_j = \int_{j-1}^{j} f(x)\,dx,$$

and then

$$a_1 + \int_1^{n} f(x)\,dx = a_1 + a_2 + \cdots + a_n,$$

where we have set $a_1 = f(1) = u_1$. By hypothesis, $\int_1^{\infty} f(x)\,dx$ is finite, and so the series $\sum_{n=1}^{\infty} a_n$ is convergent. Since $f(j) = u_j$, the inequality $a_j \geq u_j$ is a restatement of (1), and now the comparison theorem applies to yield the result.

(b) Suppose now that $\sum_{n=1}^{\infty} u_n$ converges. Let

$$v_n = \int_n^{n+1} f(x)\, dx \le f(n) = u_n, \qquad n = 1, 2, \ldots,$$

the inequality holding since $f(x) \le f(n)$ for $n \le x$ (Fig. 3–3). Because $f(x) \ge 0$, each $v_n \ge 0$, and so $\sum_{n=1}^{\infty} v_n$ converges to some number S. That is,

$$\sum_{k=1}^{n} v_k = \int_1^{n+1} f(x)\, dx \le S$$

for every n. Let ϵ be any positive number. There is an N such that

$$S - \epsilon < \int_1^{n+1} f(x)\, dx \le S \qquad \text{for all} \quad n \ge N.$$

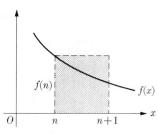

FIGURE 3–3

Since $f(x) \ge 0$, we see that

$$S - \epsilon < \int_1^{n+1} f(x)\, dx \le \int_1^{X} f(x)\, dx \le S \qquad \text{if} \quad X \ge N + 1.$$

Hence the improper integral converges.

We shall now employ the Integral Test to establish the convergence and divergence of the p-series.

Proof of Theorem 10. We define the function $f(x) = 1/x^p$, which satisfies all the conditions of the integral test if $p > 0$. We have

$$\int_1^{\infty} \frac{dx}{x^p} = \lim_{t \to \infty} \int_1^{t} \frac{dx}{x^p} = \lim_{t \to \infty} \left(\frac{t^{1-p} - 1}{1 - p} \right) \qquad \text{for} \quad p \ne 1.$$

The limit exists for $p > 1$ and fails to exist for $p < 1$. As for the case $p = 1$, we have

$$\int_1^{t} \frac{dx}{x} = \ln t,$$

which tends to ∞ as $t \to \infty$. Thus Theorem 10 is established.

EXAMPLE 4. Test the series

$$\sum_{n=1}^{\infty} \frac{1}{(n+1) \ln (n+1)}$$

for convergence or divergence.

Solution. Let

$$f(x) = \frac{1}{(x+1) \ln (x+1)}$$

and note that all conditions for the Integral Test are fulfilled. We obtain

$$\int_{1}^{t} \frac{dx}{(x+1) \ln (x+1)} = \int_{2}^{t+1} \frac{du}{u \ln u} = \int_{2}^{t+1} \frac{d(\ln u)}{\ln u}$$

$$= \ln [\ln (t+1)] - \ln (\ln 2).$$

The expression on the right diverges as $t \to \infty$ and, therefore, the given series is divergent.

————————

The particular *p*-series with $p = 1$, known as the *harmonic series*, is interesting, as it appears to be on the borderline between convergence and divergence of the various *p*-series. It is an example of a series in which the general term u_n tends to zero while the series diverges. (See page 75.) We can prove divergence of the harmonic series without recourse to the Integral Test. To do so, we write

$$\tfrac{1}{1} + \tfrac{1}{2} + (\tfrac{1}{3} + \tfrac{1}{4}) + (\tfrac{1}{5} + \tfrac{1}{6} + \tfrac{1}{7} + \tfrac{1}{8})$$

$$\text{(2)}$$

$$+ (\tfrac{1}{9} + \tfrac{1}{10} + \tfrac{1}{11} + \tfrac{1}{12} + \tfrac{1}{13} + \tfrac{1}{14} + \tfrac{1}{15} + \tfrac{1}{16}) + \text{(next 16 terms)} + \cdots.$$

We have the obvious inequalities

$$\tfrac{1}{3} + \tfrac{1}{4} > \tfrac{1}{4} + \tfrac{1}{4} = \tfrac{1}{2},$$

$$\tfrac{1}{5} + \tfrac{1}{6} + \tfrac{1}{7} + \tfrac{1}{8} > \tfrac{1}{8} + \tfrac{1}{8} + \tfrac{1}{8} + \tfrac{1}{8} = \tfrac{1}{2},$$

$$\tfrac{1}{9} + \tfrac{1}{10} + \tfrac{1}{11} + \tfrac{1}{12} + \tfrac{1}{13} + \tfrac{1}{14} + \tfrac{1}{15} + \tfrac{1}{16}$$

$$> \tfrac{1}{16} + \tfrac{1}{16} + \tfrac{1}{16} + \tfrac{1}{16} + \tfrac{1}{16} + \tfrac{1}{16} + \tfrac{1}{16} + \tfrac{1}{16} = \tfrac{1}{2},$$

and so forth.

In other words, each set of terms in a set of parentheses in series (2) is larger than $\tfrac{1}{2}$. By taking a sufficient number of parentheses, we can make the partial sum s_n of the harmonic series as large as we please. Therefore the series diverges.

————————

EXAMPLE 5. Show that the series

$$\sum_{n=1}^{\infty} \frac{1}{3n-2}$$

diverges.

Solution. We may use the Integral Test, or observe that

$$\sum_{n=1}^{\infty} \frac{1}{3n-2} = \frac{1}{3}\sum_{n=1}^{\infty} \frac{1}{(n-\frac{2}{3})} \quad \text{and} \quad \frac{1}{n-\frac{2}{3}} \geq \frac{1}{n} \quad \text{for all} \quad n \geq 1.$$

The comparison test shows divergence. The divergence may also be shown directly by recombination of terms, as in the harmonic series.

EXERCISES

Test the following series for convergence or divergence.

1. $\displaystyle\sum_{n=1}^{\infty} \frac{1}{n\sqrt{n}}$

2. $\displaystyle\sum_{n=1}^{\infty} \frac{1}{\sqrt{n}}$

3. $\displaystyle\sum_{n=1}^{\infty} \frac{1}{(n+1)(n+2)}$

4. $\displaystyle\sum_{n=1}^{\infty} \frac{n+1}{n\sqrt{n}}$

5. $\displaystyle\sum_{n=1}^{\infty} \frac{2n+3}{n^2+3n+2}$

6. $\displaystyle\sum_{n=1}^{\infty} \frac{1}{n\cdot 2^n}$

7. $\displaystyle\sum_{n=1}^{\infty} \frac{n-1}{n^3}$

8. $\displaystyle\sum_{n=1}^{\infty} \frac{n^2+3n-6}{n^4}$

9. $\displaystyle\sum_{n=1}^{\infty} \frac{1}{\sqrt{n(n+1)}}$

10. $\displaystyle\sum_{n=1}^{\infty} \frac{1}{2n+3}$

11. $\displaystyle\sum_{n=1}^{\infty} \frac{1}{n+100}$

12. $\displaystyle\sum_{n=1}^{\infty} \frac{2}{2^n+3}$

13. $\displaystyle\sum_{n=1}^{\infty} \frac{1}{n^2+2}$

14. $\displaystyle\sum_{n=1}^{\infty} \frac{2^n}{1000}$

15. $\displaystyle\sum_{n=1}^{\infty} \frac{2n+5}{n^3}$

16. $\displaystyle\sum_{n=1}^{\infty} \frac{n-1}{n^2}$

17. $\displaystyle\sum_{n=1}^{\infty} \frac{\ln n}{n}$

18. $\displaystyle\sum_{n=1}^{\infty} \frac{1}{(n+1)[\ln(n+1)]^2}$

19. $\displaystyle\sum_{n=1}^{\infty} \frac{\ln(n+1)}{(n+1)^3}$

20. $\displaystyle\sum_{n=1}^{\infty} \frac{n!}{1\cdot 3\cdot 5\cdot 7\ldots(2n-1)}$

21. $\displaystyle\sum_{n=1}^{\infty} \frac{n}{e^n}$

22. $\displaystyle\sum_{n=1}^{\infty} \frac{1}{\sqrt{n^2+1}}$

23. $\displaystyle\sum_{n=1}^{\infty} \frac{n+1}{n\cdot 2^n}$

24. $\displaystyle\sum_{n=1}^{\infty} \frac{n^4}{e^n}$

25. $\displaystyle\sum_{n=1}^{\infty} \frac{n}{2^n}$

26. $\displaystyle\sum_{n=1}^{\infty} \frac{n^2}{2^n}$

27. $\displaystyle\sum_{n=1}^{\infty} \frac{\ln n}{n^2}$ 28. $\displaystyle\sum_{n=1}^{\infty} \frac{\ln n}{n\sqrt{n}}$

29. $\displaystyle\sum_{n=1}^{\infty} \frac{n^4}{n!}$ 30. $\displaystyle\sum_{n=1}^{\infty} \frac{(2n)!}{(3n)!}$

4. SERIES OF POSITIVE AND NEGATIVE TERMS

In this section we establish three theorems which serve as important tests for the convergence and divergence of series whose terms are not necessarily positive.

Theorem 12. *If $\sum_{n=1}^{\infty} |u_n|$ converges, then $\sum_{n=1}^{\infty} u_n$ converges, and*

$$\left| \sum_{n=1}^{\infty} u_n \right| \leq \sum_{n=1}^{\infty} |u_n|.$$

Proof. We define the numbers $v_1, v_2, \ldots, v_n, \ldots$ by the relations

$$v_n = \begin{cases} u_n & \text{if } u_n \text{ is nonnegative,} \\ 0 & \text{if } u_n \text{ is negative.} \end{cases}$$

In other words, the series $\sum_{n=1}^{\infty} v_n$ consists of all the nonnegative entries in $\sum_{n=1}^{\infty} u_n$. Similarly, we define the sequence w_n by

$$w_n = \begin{cases} 0 & \text{if } u_n \text{ is nonnegative,} \\ -u_n & \text{if } u_n \text{ is negative.} \end{cases}$$

The w_n are all positive, and we have

$$v_n + w_n = |u_n|, \qquad v_n - w_n = u_n \tag{1}$$

for each n. Since $v_n \leq |u_n|$ and $w_n \leq |u_n|$, and since, by hypothesis, $\sum_{n=1}^{\infty} |u_n|$ converges, we may apply the comparison test to conclude that the series

$$\sum_{n=1}^{\infty} v_n, \qquad \sum_{n=1}^{\infty} w_n$$

converge. Also,

$$\left| \sum_{n=1}^{\infty} u_n \right| = \left| \left(\sum_{n=1}^{\infty} v_n \right) - \left(\sum_{n=1}^{\infty} w_n \right) \right|$$

$$\leq \sum_{n=1}^{\infty} v_n + \sum_{n=1}^{\infty} w_n = \sum_{n=1}^{\infty} |u_n|.$$

Remark. Theorem 12 shows that if the series of absolute values $\sum_{n=1}^{\infty} |u_n|$ is convergent, then the series itself also converges. The converse is not necessarily true. We give an example below (page 86) in which $\sum_{n=1}^{\infty} u_n$ converges while $\sum_{n=1}^{\infty} |u_n|$ diverges.

DEFINITIONS. A series $\sum_{n=1}^{\infty} u_n$ which is such that $\sum_{n=1}^{\infty} |u_n|$ converges is said to be **absolutely convergent.** However, if $\sum_{n=1}^{\infty} u_n$ converges and $\sum_{n=1}^{\infty} |u_n|$ diverges, then the series $\sum_{n=1}^{\infty} u_n$ is said to be **conditionally convergent.**

The next theorem yields a test for series whose terms are alternately positive and negative. Since the hypotheses are rather stringent, the test can be used only under special circumstances.

Theorem 13. (Alternating Series Theorem.) *Suppose that the numbers $u_1, u_2, \ldots,$ $u_n, \ldots$ satisfy the hypotheses:*

(i) *the u_n are alternately positive and negative,*

(ii) $|u_{n+1}| < |u_n|$ *for every n, and*

(iii) $\lim_{n \to \infty} u_n = 0.$

Then $\sum_{n=1}^{\infty} u_n$ is convergent. Furthermore, if the sum is denoted by s, then s lies between the partial sums s_n and s_{n+1} for each n.

Proof. Assume that u_1 is positive. (If it is not, we can consider the series beginning with u_2, since discarding a finite number of terms does not affect convergence.) Therefore, all u_k with odd subscripts are positive and all u_k with even subscripts are negative. We state this fact in the form

$$u_{2n-1} > 0, \qquad u_{2n} < 0$$

for each *n*. We now write

$$s_{2n} = (u_1 + u_2) + (u_3 + u_4) + (u_5 + u_6) + \cdots + (u_{2n-1} + u_{2n}).$$

Since $|u_{2k}| < u_{2k-1}$ for each k, we know that each quantity in the parentheses is positive. Hence s_{2n} increases for all *n*. On the other hand,

$$s_{2n} = u_1 + (u_2 + u_3) + (u_4 + u_5) + \cdots + (u_{2n-2} + u_{2n-1}) + u_{2n}.$$

Each quantity in parentheses in the above expression is negative, and so is u_{2n}. Therefore $s_{2n} < u_1$ for all *n*. We conclude that s_{2n} is an increasing sequence bounded by the number u_1. It must tend to a limit (Axiom C).

We apply similar reasoning to s_{2n-1}. We have

$$s_{2n-1} + u_{2n} = s_{2n}$$

and, since $u_{2n} < 0$, we have $s_{2n-1} > s_{2n}$ for every *n*. Therefore for $n > 1$, s_{2n-1} is bounded from below by $s_2 = u_1 + u_2$. Also,

$$s_{2n+1} = s_{2n-1} + (u_{2n} + u_{2n+1}).$$

Since the quantity in parentheses on the right is negative,

$$s_{2n-1} > s_{2n+1};$$

FIGURE 3–4

in other words, the partial sums with odd subscripts form a decreasing bounded sequence (Fig. 3–4). The limits approached by s_{2n} and s_{2n-1} must be the same, since by hypothesis (iii),

$$u_{2n} = s_{2n} - s_{2n-1} \to 0.$$

If s is the limit, we see that any even sum is less than or equal to s, while any odd sum is greater than or equal to s.

EXAMPLE 1. Test the series

$$\sum_{n=1}^{\infty} \frac{(-1)^{n+1}}{n}$$

for convergence or divergence. If it is convergent, determine whether it is conditionally convergent or absolutely convergent.

Solution. We set $u_n = (-1)^{n+1}/n$ and observe that the three hypotheses of Theorem 13 hold; i.e., the terms alternate in sign, $1/(n + 1) < 1/n$ for each n, and $\lim_{n\to\infty} (-1)^n/n = 0$. Therefore the given series converges. However, the series $\sum_{n=1}^{\infty} |u_n|$ is the harmonic series

$$\sum_{n=1}^{\infty} \frac{1}{n},$$

which is divergent. Therefore the original series is conditionally convergent.

The next test is one of the most useful for determining absolute convergence of series.

Theorem 14. (Ratio Test.) *Suppose that in the series $\sum_{n=1}^{\infty} u_n$ every $u_n \neq 0$ and that*

$$\lim_{n\to\infty} \left| \frac{u_{n+1}}{u_n} \right| = \rho \quad or \quad \left| \frac{u_{n+1}}{u_n} \right| \to +\infty.$$

Then

(i) *if $\rho < 1$, the series $\sum_{n=1}^{\infty} u_n$ converges absolutely;*

(ii) *if $\rho > 1$, or if $|u_{n+1}/u_n| \to +\infty$, the series diverges;*

(iii) *if $\rho = 1$, the test gives no information.*

Proof. (i) Suppose that $\rho < 1$. Choose any ρ' such that $\rho < \rho' < 1$. Then, since

$$\lim_{n\to\infty} \left| \frac{u_{n+1}}{u_n} \right| = \rho,$$

there must be a sufficiently large N for which

$$\left|\frac{u_{n+1}}{u_n}\right| < \rho', \qquad \text{for all} \quad n \geq N.$$

Then we obtain

$$|u_{N+1}| < \rho'|u_N|, \qquad |u_{N+2}| < \rho'|u_{N+1}|, \qquad |u_{N+3}| < \rho'|u_{N+2}|, \qquad \text{etc.}$$

By substitution we find

$$|u_{N+2}| < \rho'^2|u_N|, \qquad |u_{N+3}| < \rho'^3|u_N|, \qquad |u_{N+4}| < \rho'^4|u_N|, \qquad \text{etc.}$$

and, in general,

$$|u_{N+k}| < (\rho')^k|u_N| \qquad \text{for } k = 1, 2, \ldots. \tag{2}$$

The series

$$\sum_{k=1}^{\infty} |u_N|(\rho')^k = |u_N| \sum_{k=1}^{\infty} (\rho')^k$$

is a geometric series with ratio less than 1 and hence convergent. From (2) and the Comparison Test, we conclude that

$$\sum_{k=1}^{\infty} |u_{N+k}| \tag{3}$$

converges. Since (3) differs from the series

$$\sum_{n=1}^{\infty} |u_n|$$

in only a finite number of terms (N, to be exact), statement (i) of the theorem is established.

(ii) Suppose that $\rho > 1$ or $|u_{n+1}/u_n| \to +\infty$. There is an N such that

$$\frac{|u_{n+1}|}{|u_n|} > 1 \qquad \text{for all} \quad n \geq N.$$

By induction $|u_n| > |u_N|$ for all $n > N$. Therefore u_n does not tend to zero, and the series diverges.

To establish (iii), we exhibit two cases in which $\rho = 1$, one of them corresponding to a divergent series, the other to a convergent series. The p-series

$$\sum_{n=1}^{\infty} \frac{1}{n^p}$$

has $u_n = 1/n^p$. Therefore

$$\left|\frac{u_{n+1}}{u_n}\right| = \frac{n^p}{(n+1)^p} = 1 \bigg/ \left(1 + \frac{1}{n}\right)^p.$$

Since for any p

$$\lim_{n \to \infty} 1 \bigg/ \left(1 + \frac{1}{n}\right)^p = 1 = \rho,$$

we see that if $p > 1$ the series converges (and $\rho = 1$), while if $p \leq 1$, the series diverges (and $\rho = 1$).

Remarks. A good working procedure for a student is to try first the ratio test for convergence or divergence. If the limit ρ turns out to be 1, some other test must then be tried. The integral test is one possibility. (We observe that the integral test establishes convergence and divergence for the p-series, while the ratio test fails.) When the terms have alternating signs, the Alternating Series Theorem is suggested. In addition, we may also try comparison theorems.

In the statement of Theorem 14, it may appear at first glance that all possible situations for ρ have been considered. That is not the case, since it may happen that

$$\left|\frac{u_{n+1}}{u_n}\right|$$

does not tend to any limit and does not tend to $+\infty$. In such circumstances, more sophisticated ratio tests are available—ones which, however, are beyond the scope of this course.

———————

EXAMPLE 2. Test for absolute convergence:

$$\sum_{n=1}^{\infty} \frac{2^n}{n!}.$$

Solution. Applying the ratio test, we have

$$u_{n+1} = \frac{2^{n+1}}{(n+1)!}, \qquad u_n = \frac{2^n}{n!},$$

and

$$\left|\frac{u_{n+1}}{u_n}\right| = \frac{2^{n+1}}{(n+1)!} \cdot \frac{n!}{2^n} = \frac{2}{n+1}.$$

Therefore

$$\lim_{n \to \infty} \left|\frac{u_{n+1}}{u_n}\right| = 0 = \rho.$$

The series converges absolutely.

EXAMPLE 3. Test for absolute and conditional convergence:

$$\sum_{n=1}^{\infty} \frac{(-1)^n n}{2^n}.$$

Solution. We have

$$u_{n+1} = (-1)^{n+1} \frac{n+1}{2^{n+1}}, \qquad u_n = \frac{(-1)^n n}{2^n},$$

and therefore

$$\left|\frac{u_{n+1}}{u_n}\right| = \frac{n+1}{2^{n+1}} \cdot \frac{2^n}{n} = \frac{1}{2}\left(\frac{n+1}{n}\right) = \frac{1}{2}\left(\frac{1+1/n}{1}\right).$$

Hence

$$\lim_{n\to\infty}\left|\frac{u_{n+1}}{u_n}\right| = \frac{1}{2} = \rho.$$

The series converges absolutely.

EXAMPLE 4. Test for absolute convergence:

$$\sum_{n=1}^{\infty}\frac{(2n)!}{n^{100}}.$$

Solution. We have

$$u_n = \frac{(2n)!}{n^{100}} \quad \text{and} \quad u_{n+1} = \frac{[2(n+1)]!}{(n+1)^{100}}.$$

Therefore

$$\left|\frac{u_{n+1}}{u_n}\right| = \frac{(2n+2)!}{(n+1)^{100}} \cdot \frac{n^{100}}{(2n)!} = (2n+1)(2n+2)\left(\frac{n}{1+n}\right)^{100}$$

$$= (2n+1)(2n+2)\left(\frac{1}{1+1/n}\right)^{100}.$$

Hence

$$\lim_{n\to\infty}\left|\frac{u_{n+1}}{u_n}\right| = +\infty,$$

and the series is divergent.

EXERCISES

Test each of the following series for convergence or divergence. If the series is convergent, determine whether it is absolutely or conditionally so.

1. $\displaystyle\sum_{n=1}^{\infty}\frac{n!}{10^n}$

2. $\displaystyle\sum_{n=1}^{\infty}\frac{10^n}{n!}$

3. $\displaystyle\sum_{n=1}^{\infty}\frac{(-1)^{n-1}n!}{10^n}$

4. $\displaystyle\sum_{n=1}^{\infty}\frac{(-1)^{n-1}10^n}{n!}$

5. $\displaystyle\sum_{n=1}^{\infty}n\left(\frac{3}{4}\right)^n$

6. $\displaystyle\sum_{n=1}^{\infty}n^2\left(\frac{3}{4}\right)^n$

7. $\displaystyle\sum_{n=1}^{\infty}\frac{(-1)^n}{\sqrt{n}}$

8. $\displaystyle\sum_{n=1}^{\infty}\frac{(-1)^n}{n^p}, \quad 0 < p < 1$

9. $\displaystyle\sum_{n=1}^{\infty}\frac{(-1)^n}{n\sqrt{n}}$

10. $\displaystyle\sum_{n=1}^{\infty}\frac{(-1)^{n+1}(n-1)}{n^2+1}$

11. $\displaystyle\sum_{n=1}^{\infty}\frac{(-1)^n n^2}{2^n}$

12. $\displaystyle\sum_{n=1}^{\infty}\frac{(-1)^{n+1}(n-1)^2}{n^3}$

13. $\displaystyle\sum_{n=1}^{\infty} \frac{(-1)^{n-1}(4/3)^n}{n^2}$

14. $\displaystyle\sum_{n=1}^{\infty} \frac{(-1)^n(3/2)^2}{n^4}$

15. $\displaystyle\sum_{n=1}^{\infty} \frac{(-5)^{n-1}}{n \cdot n!}$

16. $\displaystyle\sum_{n=1}^{\infty} \frac{(-2)^{n-1} \cdot (n+1)}{(2n)!}$

17. $\displaystyle\sum_{n=1}^{\infty} \frac{(-1)^{n-1}n!}{1 \cdot 3 \cdot 5 \cdots (2n-1)}$

18. $\displaystyle\sum_{n=1}^{\infty} \frac{(-1)^{n-1}(n!)^2 \cdot 2^n}{(2n)!}$

19. $\displaystyle\sum_{n=1}^{\infty} \frac{(-1)^{n-1}(n+1)}{n\sqrt{n}}$

20. $\displaystyle\sum_{n=1}^{\infty} \frac{(n!)^2 5^n}{(2n)!}$

21. $\displaystyle\sum_{n=1}^{\infty} \frac{(-1)^n 2 \cdot 4 \cdot 6 \cdots (2n)}{1 \cdot 4 \cdot 7 \cdots (3n-2)}$

22. $\displaystyle\sum_{n=1}^{\infty} \frac{(-1)^{n+1} 3^{n+1}}{2^{4n}}$

23. $\displaystyle\sum_{n=1}^{\infty} \frac{(-1)^{n-1}n}{n+1}$

24. $\displaystyle\sum_{n=1}^{\infty} \frac{(-1)^n(n-2)}{n^{7/4}}$

25. $\displaystyle\sum_{n=1}^{\infty} \frac{(-1)^n(6n^2 - 9n + 4)}{n^3}$

26. $\displaystyle\sum_{n=1}^{\infty} \frac{(-1)^{n+1}}{(n+1)\ln(n+1)}$

27. $\displaystyle\sum_{n=1}^{\infty} \frac{(-1)^{n+1}\ln(n+1)}{n+1}$

28. $\displaystyle\sum_{n=1}^{\infty} \frac{(-1)^{n-1}\ln n}{n^2}$

5. POWER SERIES

A **power series** is a series of the form

$$c_0 + c_1(x-a) + c_2(x-a)^2 + \cdots + c_n(x-a)^n + \cdots,$$

in which a and the c_i, $i = 0, 1, 2, \ldots$, are constants. If a particular value is given to x, we then obtain an infinite series of numbers of the type we have been considering. The special case $a = 0$ occurs frequently, in which case the series becomes

$$c_0 + c_1 x + c_2 x^2 + c_3 x^3 + \cdots + c_n x^n + \cdots.$$

Most often, we use the $\sum$-notation, writing

$$\sum_{n=0}^{\infty} c_n(x-a)^n \qquad \text{and} \qquad \sum_{n=0}^{\infty} c_n x^n.$$

If a power series converges for certain values of x, we may define a function of x by setting

$$f(x) = \sum_{n=0}^{\infty} c_n(x-a)^n \qquad \text{or} \qquad g(x) = \sum_{n=0}^{\infty} c_n x^n$$

for those values of x. We shall see that all the functions we have studied can be represented by convergent power series (with certain exceptions for the value a).

The Ratio Test may be used to determine when a power series converges. We begin with several examples.

EXAMPLE 1. Find the values of x for which the series

$$\sum_{n=1}^{\infty} \frac{1}{n} x^n$$

converges.

Solution. We apply the Ratio Test, noting that

$$u_n = \frac{1}{n} x^n, \qquad u_{n+1} = \frac{1}{n+1} x^{n+1}.$$

Then

$$\left| \frac{u_{n+1}}{u_n} \right| = \frac{|x|^{n+1}}{n+1} \cdot \frac{n}{|x|^n} = |x| \frac{n}{n+1}.$$

It is important to observe that x remains *unaffected* as $n \to \infty$. Hence

$$\lim_{n \to \infty} \left| \frac{u_{n+1}}{u_n} \right| = \lim_{n \to \infty} |x| \frac{n}{n+1} = |x| \lim_{n \to \infty} \frac{1}{1 + 1/n} = |x|.$$

That is, $\rho = |x|$ in the Ratio Test.

We conclude: (a) the series converges if $|x| < 1$; (b) the series diverges if $|x| > 1$; (c) if $|x| = 1$, the Ratio Test gives no information. The last case corresponds to $x = \pm 1$, and we may try other methods for these two series, which are

$$\sum_{n=1}^{\infty} \frac{1}{n} \quad \text{(if } x = 1\text{)} \quad \text{and} \quad \sum_{n=1}^{\infty} \frac{(-1)^n}{n} \quad \text{(if } x = -1\text{)}.$$

The first series above is the divergent harmonic series. The second series converges by the Alternating Series Theorem. Therefore the given series converges for $-1 \leq x < 1$.

EXAMPLE 2. Find the values of x for which the series

$$\sum_{n=1}^{\infty} \frac{(-1)^n (x + 1)^n}{2^n n^2}$$

converges.

Solution. We have

$$u_n = \frac{(-1)^n (x + 1)^n}{2^n n^2}, \qquad u_{n+1} = \frac{(-1)^{n+1}(x + 1)^{n+1}}{2^{n+1}(n + 1)^2}.$$

Therefore

$$\left| \frac{u_{n+1}}{u_n} \right| = \frac{|x + 1|^{n+1}}{2^{n+1}(n + 1)^2} \cdot \frac{2^n n^2}{|x + 1|^n} = \frac{1}{2} |x + 1| \left(\frac{n}{n + 1} \right)^2$$

and

$$\lim_{n \to \infty} \frac{|u_{n+1}|}{|u_n|} = \frac{1}{2} |x + 1| \lim_{n \to \infty} \left(\frac{1}{1 + 1/n} \right)^2 = \frac{|x + 1|}{2}.$$

According to the Ratio Test: (a) the series converges if $\frac{1}{2}|x + 1| < 1$; (b) the series diverges if $\frac{1}{2}|x + 1| > 1$; (c) if $|x + 1| = 2$, the test fails.

The inequality $|x + 1| < 2$ may be written

$$-2 < x + 1 < 2 \quad \text{or} \quad -3 < x < 1,$$

and the series converges in this interval, while it diverges for x outside this interval. The values $x = -3$ and $x = 1$ remain for consideration. The corresponding series are

$$\sum_{n=1}^{\infty} \frac{(-1)^n(-2)^n}{2^n n^2} = \sum_{n=1}^{\infty} \frac{1}{n^2} \quad \text{and} \quad \sum_{n=1}^{\infty} \frac{(-1)^n 2^n}{2^n n^2} = \sum_{n=1}^{\infty} \frac{(-1)^n}{n^2}.$$

Both series converge absolutely by the p-series test. The original series converges for x in the interval $-3 \le x \le 1$.

EXAMPLE 3. Find the values of x for which the series

$$\sum_{n=0}^{\infty} \frac{(-1)^n x^n}{n!}$$

converges.

Solution. We have

$$u_n = \frac{(-1)^n x^n}{n!}, \qquad u_{n+1} = \frac{(-1)^{n+1} x^{n+1}}{(n + 1)!},$$

and

$$\left| \frac{u_{n+1}}{u_n} \right| = \frac{|x|^{n+1}}{(n + 1)!} \cdot \frac{n!}{|x|^n} = |x| \frac{1}{n + 1}.$$

Hence $\rho = 0$, regardless of the value of $|x|$. The series converges for all values of x; that is, $-\infty < x < \infty$.

EXAMPLE 4. Find the values of x for which the series

$$\sum_{n=0}^{\infty} \frac{(-1)^n n! x^n}{10^n}$$

converges.

Solution. We have

$$u_n = \frac{(-1)^n n! x^n}{10^n}, \qquad u_{n+1} = \frac{(-1)^{n+1}(n + 1)! x^{n+1}}{10^{n+1}}$$

and, if $x \ne 0$,

$$\left| \frac{u_{n+1}}{u_n} \right| = \frac{|x|^{n+1}(n + 1)!}{10^{n+1}} \cdot \frac{10^n}{|x|^n n!} = |x| \cdot \frac{n + 1}{10}.$$

Therefore, if $x \ne 0$, $|u_{n+1}/u_n| \to \infty$ and the series diverges. The series converges only for $x = 0$.

The convergence properties of the most general power series are illustrated in the examples above. However, the proof of the theorem which states this fact (given below in Theorem 16) is beyond the scope of this text. In all the examples above we see that it always happened that $|u_{n+1}/u_n|$ tended to a limit or to $+\infty$. The examples are deceptive, since there are cases in which $|u_{n+1}/u_n|$ may neither tend to a limit nor tend to $+\infty$.

Lemma. *If the series $\sum_{n=0}^{\infty} u_n$ converges, there is a number M such that $|u_n| \leq M$ for every n.*

Proof. By Theorem 6 we know that $\lim_{n \to \infty} u_n = 0$. From the definition of a limit, there must be a number N such that

$$|u_n| < 1 \quad \text{for all} \quad n > N$$

(by taking $\epsilon = 1$ in the definition of limit). We define M to be the largest of the numbers

$$|u_0|, \quad |u_1|, \quad |u_2|, \quad \ldots, \quad |u_N|, \quad 1,$$

and the result is established.

Theorem 15. *If the series $\sum_{n=0}^{\infty} a_n x^n$ converges for some $x_1 \neq 0$, then the series converges absolutely for all x for which $|x| < |x_1|$, and there is a number M such that*

$$|a_n x^n| \leq M \left| \frac{x}{x_1} \right|^n$$

for all n.

Proof. Since the series $\sum_{n=0}^{\infty} a_n x_1^n$ converges, we know from the Lemma above that there is a number M such that

$$|a_n x_1^n| \leq M \quad \text{for all } n.$$

Then

$$|a_n x^n| = \left| a_n x_1^n \frac{x^n}{x_1^n} \right| = |a_n x_1^n| \cdot \left| \frac{x}{x_1} \right|^n \leq M \left| \frac{x}{x_1} \right|^n.$$

The series

$$\sum_{n=0}^{\infty} M \left| \frac{x}{x_1} \right|^n$$

is a geometric series with ratio less than 1, and so convergent. Hence, by the Comparison Test, the series

$$\sum_{n=0}^{\infty} a_n x^n$$

converges absolutely.

Remark. Theorem 15 may be established for series of the form $\sum_{n=0}^{\infty} a_n (x - a)^n$ in a completely analogous manner.

Theorem 16. *Let $\sum_{n=0}^{\infty} a_n(x - a)^n$ be any given power series. Then either*

(i) *the series converges only for $x = a$;*
(ii) *the series converges for all values of x; or*
(iii) *there is a number $R > 0$ such that the series converges for all x for which $|x - a| < R$ and diverges for all x for which $|x - a| > R$.*

We omit the proof. (See, however, Morrey, *University Calculus*, page 464.) The consequence (iii) in Theorem 16 states that there is an *interval of convergence* $-R < x - a < R$ or $a - R < x < a + R$. Nothing is stated about what happens when $x = a - R$ or $a + R$. These *endpoint* problems must be settled on a case-by-case basis. The alternatives (i) and (ii) correspond to $R = 0$ and $R = +\infty$, respectively.

EXERCISES

In problems 1 through 27, find the values of x for which the following power series converge. Include a discussion of the endpoints.

1. $\displaystyle\sum_{n=0}^{\infty} x^n$

2. $\displaystyle\sum_{n=0}^{\infty} (-1)^n x^n$

3. $\displaystyle\sum_{n=0}^{\infty} (2x)^n$

4. $\displaystyle\sum_{n=0}^{\infty} \left(\frac{1}{4}x\right)^n$

5. $\displaystyle\sum_{n=0}^{\infty} (-1)^n(n + 1)x^n$

6. $\displaystyle\sum_{n=1}^{\infty} \frac{(x - 1)^n}{3^n n^2}$

7. $\displaystyle\sum_{n=1}^{\infty} \frac{(x - 1)^n}{2^n n^3}$

8. $\displaystyle\sum_{n=1}^{\infty} \frac{(-1)^{n+1}(x - 2)^n}{n\sqrt{n}}$

9. $\displaystyle\sum_{n=1}^{\infty} \frac{(x + 2)^n}{\sqrt{n}}$

10. $\displaystyle\sum_{n=0}^{\infty} \frac{(10x)^n}{n!}$

11. $\displaystyle\sum_{n=0}^{\infty} \frac{x^n}{(2n)!}$

12. $\displaystyle\sum_{n=0}^{\infty} \frac{n!(x + 1)^n}{5^n}$

13. $\displaystyle\sum_{n=0}^{\infty} \frac{(-1)^n(3/2)^n x^n}{n + 1}$

14. $\displaystyle\sum_{n=0}^{\infty} \frac{(2n)!x^n}{n!}$

15. $\displaystyle\sum_{n=1}^{\infty} n^2(x - 1)^n$

16. $\displaystyle\sum_{n=1}^{\infty} \frac{n(x + 2)^n}{2^n}$

17. $\displaystyle\sum_{n=1}^{\infty} \frac{(-1)^{n-1}(x + 4)^n}{3^n \cdot n^2}$

18. $\displaystyle\sum_{n=1}^{\infty} \frac{n!(x - 3)^n}{1 \cdot 3 \cdot 5 \cdots (2n - 1)}$

19. $\displaystyle\sum_{n=1}^{\infty} \frac{(-1)^{n+1}(n!)^2(x - 2)^n}{2^n(2n)!}$

20. $\displaystyle\sum_{n=1}^{\infty} \frac{n!(x - 1)^n}{4^n \cdot 1 \cdot 3 \cdot 5 \cdots (2n - 1)}$

21. $\displaystyle\sum_{n=1}^{\infty} \frac{(-1)^{n-1} n!(3/2)^n x^n}{1 \cdot 3 \cdot 5 \cdots (2n-1)}$

22. $\displaystyle\sum_{n=0}^{\infty} \frac{(-1)^n 3^{n+1} x^n}{2^{3n}}$

23. $\displaystyle\sum_{n=1}^{\infty} \frac{(n-2)x^n}{n^2}$

24. $\displaystyle\sum_{n=0}^{\infty} \frac{(6n^2 + 3n + 1)x^n}{2^n(n+1)^3}$

25. $\displaystyle\sum_{n=1}^{\infty} \frac{(-1)^{n-1} x^n}{(n+1)\ln(n+1)}$

26. $\displaystyle\sum_{n=1}^{\infty} \frac{\ln(n+1)3^n(x-1)^n}{n+1}$

27. $\displaystyle\sum_{n=1}^{\infty} \frac{(-1)^{n-1}(\ln n)2^n x^n}{3^n n^2}$

28. Prove Theorem 15 for series of the form

$$\sum_{n=0}^{\infty} a_n(x-a)^n.$$

29. (a) Find the interval of convergence of the series

$$\sum_{n=1}^{\infty} \frac{1 \cdot 3 \cdot 5 \cdots (2n-1)}{2 \cdot 4 \cdot 6 \cdots (2n)} x^n.$$

(b) Show that the series in (a) is identical with the series

$$\sum_{n=1}^{\infty} \frac{(2n)!}{2^{2n}(n!)^2} x^n.$$

30. Find the interval of convergence of the series

$$\sum_{n=1}^{\infty} \frac{1 \cdot 3 \cdot 5 \cdots (2n-1)(x-2)^n}{2^n \cdot 1 \cdot 4 \cdot 7 \cdots (3n-2)}.$$

31. Find the interval of convergence of the binomial series

$$1 + \sum_{n=1}^{\infty} \frac{m(m-1) \cdots (m-n+1)}{n!} x^n; \qquad m \text{ fixed.}$$

6. TAYLOR'S SERIES

Suppose that a power series

$$\sum_{n=0}^{\infty} a_n(x-a)^n$$

converges in some interval $-R < x - a < R$ $(R > 0)$. Then the sum of the series has a value for each x in this interval and so defines a function of x. We can therefore write

$$f(x) = a_0 + a_1(x-a) + a_2(x-a)^2 + a_3(x-a)^3 + \cdots,$$

$$a - R < x < a + R.$$
(1)

We ask the question: What is the relationship between the coefficients a_1, a_2, a_3, ..., a_n, ... and the function f?

We shall proceed naïvely, as if the right side of (1) were a polynomial. Setting $x = a$, we find at once that

$$f(a) = a_0.$$

We differentiate (1) (as if the right side were a polynomial) and get

$$f'(x) = a_1 + 2a_2(x - a) + 3a_3(x - a)^2 + 4a_4(x - a)^3 + \cdots.$$

For $x = a$, we find that

$$f'(a) = a_1.$$

We continue both differentiating and setting $x = a$, to obtain

$$f''(x) = 2a_2 + 3 \cdot 2a_3(x - a) + 4 \cdot 3a_4(x - a)^2 + 5 \cdot 4a_5(x - a)^3 + \cdots,$$

$$f''(a) = 2a_2 \quad \text{or} \quad a_2 = \frac{f''(a)}{2!},$$

$$f'''(x) = 3 \cdot 2a_3 + 4 \cdot 3 \cdot 2a_4(x - a) + 5 \cdot 4 \cdot 3a_5(x - a)^2$$
$$+ 6 \cdot 5 \cdot 4a_6(x - a)^3 + \cdots,$$

$$f'''(a) = 3 \cdot 2a_3 \quad \text{or} \quad a_3 = \frac{f'''(a)}{3!},$$

and so forth. The pattern is now clear. The general formula for the coefficients a_0, a_1, a_2, ..., a_n, ... is

$$a_n = \frac{f^{(n)}(a)}{n!}.$$

In Section 8, it will be shown that all of the above steps are legitimate so long as the series is convergent in some positive interval. Substituting the formulas for the coefficients a_n into the power series, we obtain

$$f(x) = \sum_{n=0}^{\infty} \frac{f^{(n)}(a)}{n!} (x - a)^n. \tag{2}$$

DEFINITION. The right side of Eq. (2) is called the **Taylor series for f about the point a** or the **expansion of f into a power series about** a.

For the special case $a = 0$, the Taylor series is

$$f(x) = \sum_{n=0}^{\infty} \frac{f^{(n)}(0)}{n!} x^n. \tag{3}$$

The right side of (3) is called the **Maclaurin series for f**.

EXAMPLE 1. Assuming that $f(x) = \sin x$ is given by its Maclaurin series, expand $\sin x$ into such a series.

Solution. We have

$$f(x) = \sin x, \qquad f(0) = 0,$$
$$f'(x) = \cos x, \qquad f'(0) = 1,$$
$$f''(x) = -\sin x, \qquad f''(0) = 0,$$
$$f^{(3)}(x) = -\cos x, \qquad f^{(3)}(0) = -1,$$
$$f^{(4)}(x) = \sin x, \qquad f^{(4)}(0) = 0.$$

It is clear that $f^{(5)} = f', f^{(6)} = f''$, etc., so that the sequence $0, 1, 0, -1, 0, 1, 0, -1, \ldots$ repeats itself indefinitely. Therefore, from (3) we obtain

$$\sin x = x - \frac{x^3}{3!} + \frac{x^5}{5!} - \frac{x^7}{7!} + \frac{x^9}{9!} \cdots$$
$$= \sum_{k=0}^{\infty} \frac{(-1)^k x^{2k+1}}{(2k+1)!}. \qquad (4)$$

Remark. It may be verified (by the Ratio Test, for example) that the series (4) converges for all values of x.

EXAMPLE 2. Expand the function

$$f(x) = \frac{1}{x}$$

into a Taylor series about $x = 1$, assuming that such an expansion is valid.

Solution. We have

$$f(x) = x^{-1} \qquad\qquad f(1) = 1$$
$$f'(x) = (-1)x^{-2} \qquad\qquad f'(1) = -1$$
$$f''(x) = (-1)(-2)x^{-3} \qquad\qquad f''(1) = (-1)^2 \cdot 2!$$
$$f^{(3)}(x) = (-1)(-2)(-3)x^{-4} \qquad\qquad f^{(3)}(1) = (-1)^3 \cdot 3!$$
$$f^{(n)}(x) = (-1)(-2)\cdots(-n)x^{-n-1} \qquad f^{(n)}(1) = (-1)^n \cdot n!$$

Therefore from (2) with $a = 1$, we obtain

$$f(x) = \frac{1}{x} = \sum_{n=0}^{\infty} (-1)^n (x-1)^n. \qquad (5)$$

Remark. The series (5) converges for $|x - 1| < 1$ or $0 < x < 2$, as may be confirmed by the Ratio Test.

Examples 1 and 2 have meaning only if it is known that the functions are representable by means of power series. There are examples of functions for which it is possible to compute all the quantities $f^{(n)}(x)$ at a given value a, and yet the Taylor series about a will not represent the function. (See Morrey, *University Calculus*, p. 467. See also Exercise 33 below.)

EXAMPLE 3. Compute the first six terms of the Maclaurin expansion of the function

$$f(x) = \tan x,$$

assuming that such an expansion is valid.

Solution. We have

$$f(x) = \tan x, \qquad f(0) = 0$$
$$f'(x) = \sec^2 x, \qquad f'(0) = 1$$
$$f''(x) = 2 \sec^2 x \tan x, \qquad f''(0) = 0$$
$$f^{(3)}(x) = 2 \sec^4 x + 4 \sec^2 x \tan^2 x, \qquad f^{(3)}(0) = 2$$
$$f^{(4)}(x) = 8 \tan x \sec^2 x (2 + 3 \tan^2 x), \qquad f^{(4)}(0) = 0$$
$$f^{(5)}(x) = 48 \tan^2 x \sec^4 x + 8 \sec^2 x (2 + 3 \tan^2 x)(\sec^2 x + 2 \tan^2 x), \qquad f^{(5)}(0) = 16$$

Therefore

$$f(x) = \tan x = x + \frac{x^3}{3} + \frac{2x^5}{15} + \cdots.$$

Remark. Example 3 shows that the general pattern for the successive derivatives may not always be readily discernible. Examples 1 and 2, on the other hand, show how the general formula for the nth derivative may be arrived at simply.

EXERCISES

In problems 1 through 16, find the Taylor (Maclaurin if $a = 0$) series for each function f about the given value of a.

1. $f(x) = e^x, \quad a = 0$
2. $f(x) = \cos x, \quad a = 0$
3. $f(x) = \ln (1 + x), \quad a = 0$
4. $f(x) = \ln (1 + x), \quad a = 1$
5. $f(x) = (1 - x)^{-2}, \quad a = 0$
6. $f(x) = (1 - x)^{-1/2}, \quad a = 0$
7. $f(x) = (1 + x)^{1/2}, \quad a = 0$
8. $f(x) = e^x, \quad a = 1$
9. $f(x) = \ln x, \quad a = 3$
10. $f(x) = \sin x, \quad a = \pi/4$
11. $f(x) = \cos x, \quad a = \pi/3$
12. $f(x) = \sin x, \quad a = 2\pi/3$
13. $f(x) = \sqrt{x}, \quad a = 4$
14. $f(x) = \sin (x + \frac{1}{2}), \quad a = 0$
15. $f(x) = \cos (x + \frac{1}{2}), \quad a = 0$
16. $f(x) = x^m, \quad a = 1$

In each of problems 17 through 31, find the first few terms of the Taylor expansion about the given value of a. Carry out the process to include the term $(x - a)^n$ for the given integer n.

17. $f(x) = e^{-x^2}, \quad a = 0, \quad n = 4$
18. $f(x) = xe^x, \quad a = 0, \quad n = 4$
19. $f(x) = \dfrac{1}{1 + x^2}, \quad a = 0, \quad n = 4$
20. $f(x) = \arctan x, \quad a = 0, \quad n = 5$
21. $f(x) = e^x \cos x, \quad a = 0, \quad n = 4$
22. $f(x) = \dfrac{1}{\sqrt{1 - x^2}}, \quad a = 0, \quad n = 4$

23. $f(x) = \arcsin x$, $a = 0$, $n = 5$
24. $f(x) = \tanh x$, $a = 0$, $n = 5$
25. $f(x) = \ln \sec x$, $a = 0$, $n = 6$
26. $f(x) = \sec x$, $a = 0$, $n = 4$
27. $f(x) = \operatorname{sech} x$, $a = 0$, $n = 4$
28. $f(x) = \csc x$, $a = \pi/2$, $n = 4$
29. $f(x) = \sec x$, $a = \pi/3$, $n = 3$
30. $f(x) = \ln \sin x$, $a = \pi/4$, $n = 4$
31. $f(x) = \tan x$, $a = 0$, $n = 8$ (See Example 3.)
32. (a) Given the polynomial

$$f(x) = 3 + 2x - x^2 + 4x^3 - 2x^4, \tag{6}$$

show that f may be written in the form

$$f(x) = a_0 + a_1(x - 1) + a_2(x - 1)^2 + a_3(x - 1)^3 + a_4(x - 1)^4.$$

[*Hint:* Use the Taylor expansion (2) and (6) to get each a_i.]

*(b) Given the same polynomial in two forms,

$$f(x) = \sum_{k=0}^{N} a_k(x - a)^k, \qquad f(x) = \sum_{k=0}^{n} b_k(x - b)^k,$$

express each b_i in terms of a, b, and the a_i.

*33. (a) Given the function (see Fig. 3–5)

$$F(x) = \begin{cases} e^{-1/x^2}, & x \neq 0, \\ 0, & x = 0, \end{cases}$$

use l'Hôpital's Rule to show that

$$F'(0) = 0.$$

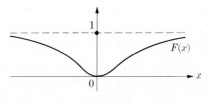

FIGURE 3–5

(b) Show that $F^{(n)}(0) = 0$ for every positive integer n.

(c) What can be said about the Taylor series for F?

7. TAYLOR'S THEOREM WITH REMAINDER

If a function f possesses only a finite number—say n—of derivatives, then it is clear that it is not possible to represent it by a Taylor series, since the coefficients $a_k = f^{(k)}(a)/k!$ cannot be computed beyond a_n. In such cases it is still possible to obtain a *finite version* of a Taylor expansion.

Suppose that $f(x)$ possesses n continuous derivatives in some interval about the value a. Then it is always possible to write

$$f(x) = f(a) + \frac{f'(a)}{1!}(x - a) + \frac{f^{(2)}(a)}{2!}(x - a)^2 + \cdots$$

$$+ \frac{f^{(n)}(a)}{n!}(x - a)^n + R_n \tag{1}$$

for x in this interval. The right side consists of a polynomial in x of degree n and a *remainder R_n* about which, as yet, we have no knowledge. The content of *Taylor's Theorem* concerns the character of R_n. This theorem is not only of great theoretical value but may also be used in approximations and numerical computations.

Theorem 17. (Taylor's Theorem with Derivative Form of Remainder.) *Suppose that $f, f', f^{(2)}, \ldots, f^{(n)}, f^{(n+1)}$ are all continuous in some interval containing a and b. Then there is a number ξ between a and b such that*

$$f(b) = f(a) + \frac{f'(a)}{1!}(b - a) + \frac{f^{(2)}(a)}{2!}(b - a)^2 + \cdots + \frac{f^{(n)}(a)}{n!}(b - a)^n$$

$$+ \frac{f^{(n+1)}(\xi)(b - a)^{n+1}}{(n + 1)!}.$$

That is, the remainder R_n is given by the formula

$$R_n = \frac{f^{(n+1)}(\xi)(b - a)^{n+1}}{(n + 1)!}. \tag{2}$$

Remarks. (i) We see that R_n depends on both b and a, and we write, in general, $R_n = R_n(a, b)$, a function of two variables.

(ii) If we take the special case $n = 0$, we obtain $f(b) = f(a) + f'(\xi)(b - a)$, which we recognize as the Theorem of the Mean.

Proof. The proof makes use of Rolle's Theorem. We create a function $\phi(x)$ which is zero at a and b and so, by Rolle's Theorem, there must be a number ξ between a and b where $\phi'(\xi) = 0$. The algebra is lengthy, and the student should write out the details for the cases $n = 1, 2, 3$ in order to grasp the essence of the proof. We use the form (1) for $x = b$ and write

$$f(b) = f(a) + \frac{f'(a)}{1!}(b - a) + \frac{f^{(2)}(a)(b - a)^2}{2!} + \cdots$$

$$+ \frac{f^{(n)}(a)(b - a)^n}{n!} + R_n(a, b);$$

we wish to find $R_n(a, b)$. We define the function

$$\phi(x) = f(b) - f(x) - \frac{f'(x)(b - x)}{1!} - \frac{f^{(2)}(x)(b - x)^2}{2!} - \frac{f^{(3)}(x)(b - x)^3}{3!}$$

$$- \cdots - \frac{f^{(n-1)}(x)(b - x)^{n-1}}{(n - 1)!} - \frac{f^{(n)}(x)(b - x)^n}{n!} - R_n(a, b)\frac{(b - x)^{n+1}}{(b - a)^{n+1}}.$$

The function ϕ was concocted in such a way that $\phi(a) = 0$ and $\phi(b) = 0$, facts which are easily checked by straight substitution. We compute the derivative

$\phi'(x)$ (using the formula for the derivative of a product wherever necessary):

$$\phi'(x) = -f'(x) + f'(x) - \frac{f^{(2)}(x)(b-x)}{1!} + \frac{2f^{(2)}(x)(b-x)}{2!}$$

$$- \frac{f^{(3)}(x)(b-x)^2}{2!} + \frac{3f^{(3)}(x)(b-x)^2}{3!} - \frac{f^{(4)}(x)(b-x)^3}{3!} + \cdots$$

$$- \frac{f^{(n+1)}(x)(b-x)^n}{n!} + \frac{R_n(a,b)(n+1)(b-x)^n}{(b-a)^{n+1}}.$$

Amazingly, all the terms cancel except the last two, and we find

$$\phi'(x) = - \frac{f^{(n+1)}(x)(b-x)^n}{n!} + R_n(a,b)(n+1)\frac{(b-x)^n}{(b-a)^{n+1}}.$$

Using Rolle's Theorem, we know there must be a value ξ between a and b such that $\phi'(\xi) = 0$. Therefore we get

$$0 = - \frac{f^{(n+1)}(\xi)(b-\xi)^n}{n!} + R_n(a,b)(n+1)\frac{(b-\xi)^n}{(b-a)^{n+1}}$$

or, upon solving for $R_n(a,b)$, the formula (2) exactly.

Remarks. (i) If we know that $f(x)$ has continuous derivatives of all orders and if $R_n(a,b) \to 0$ as $n \to \infty$, then we can establish the validity of the Taylor series.

(ii) In any case, R_n is a measure of how much f differs from a certain polynomial of degree n. If R_n is small, then the polynomial may be used for approximations.

When we use Taylor's Theorem in the computation of functions from the approximating polynomial, errors may arise from two sources: the error R_n, made above by neglecting the powers of $(b-a)$ beyond the nth; and the "round-off error" made by expressing each term in decimal form. If we wish to compute the value of some function $f(b)$ to an accuracy of four decimal places, it is essential to be able to say for certain that $f(b)$ is between some decimal fraction with four decimals -0.00005 and the same decimal fraction $+0.00005$. Time is saved by computing each term to two decimals more than are required. Frequently R_n is close to the value of the first term in the series omitted, and this fact can be used as a guide in choosing the number of terms. Although we do not know R_n exactly, we can often show that there are two numbers m and M with

$$m \le f^{(n+1)}(x) \le M \qquad \text{for } all \ x \text{ between } a \text{ and } b.$$

Then we get for $R_n(a,b)$ the inequality

$$\frac{m(b-a)^{n+1}}{(n+1)!} \le R_n(a,b) \le \frac{M(b-a)^{n+1}}{(n+1)!}.$$

EXAMPLE 1. Compute $(1.1)^{1/5}$ to an accuracy of four decimal places.

Solution. The key to the solution, using Taylor's Theorem, is the fact that we can set

$$f(x) = (1 + x)^{1/5}, \quad a = 0, \quad b = 0.1.$$

Then

$f(x) = (1 + x)^{1/5},$	$f(a) = 1$	$=$	1.0000 00
$f'(x) = \dfrac{1}{5}(1 + x)^{-4/5},$	$f'(a)(b - a) = \dfrac{1}{5}(0.1)^1$	$=$	0.0200 00
$f''(x) = -\dfrac{4}{25}(1 + x)^{-9/5},$	$\dfrac{f''(a)(b - a)^2}{2!} = -\dfrac{2}{25}(0.1)^2$	$= -0.0008$ 00	
$f'''(x) = \dfrac{36}{125}(1 + x)^{-14/5},$	$\dfrac{f'''(a)(b - a)^3}{3!} = \dfrac{6}{125}(0.1)^3$	$=$	0.0000 48

For all x between 0 and 1 we have

$$0 < (1 + x)^{-14/5} < 1.$$

Therefore we can estimate R_n for $n = 2$:

$$0 < R_2 = \frac{36}{125}(1 + \xi)^{-14/5}\frac{(b - a)^3}{3!} < \frac{6}{125}(0.1)^3 = 0.0000 \ \ 48.$$

Adding the terms in the Taylor expansion through $n = 2$, we get

$$(1.1)^{1/5} = 1.0192, \text{ approximately};$$

in fact, a more precise statement is

$$1.0192 < (1.1)^{1/5} < 1.0192 \ \ 48.$$

Remark. In Example 1, we could have selected $f(x) = x^{1/5}$ with $a = 1, b = 1.1$. The result is the same.

EXAMPLE 2. Compute $\sqrt[3]{7}$ to an accuracy of four decimal places.

Solution. Set

$$f(x) = x^{1/3}, \quad a = 8, \quad b = 7.$$

Then $b - a = -1$ and

$f(x) = x^{1/3},$	$f(a) = 2$	$=$	2.0000 00
$f'(x) = \dfrac{1}{3}x^{-2/3},$	$f'(a)(b - a) = -\dfrac{1}{12}$	$= -0.0833$ 33	
$f''(x) = -\dfrac{2}{9}x^{-5/3},$	$\dfrac{f''(a)(b - a)^2}{2!} = -\dfrac{1}{9 \cdot 2^5}$	$= -0.0034$ 72	
$f^{(3)}(x) = \dfrac{10}{27}x^{-8/3},$	$\dfrac{f'''(a)(b - a)^3}{3!} = -\dfrac{5}{81 \cdot 2^8}$	$= -0.0002$ 41	
$f^{(4)}(x) = \dfrac{-80}{81}x^{-11/3},$	$\dfrac{f^{(4)}(a)(b - a)^4}{4!} = -\dfrac{5}{243 \cdot 2^{10}}$	$= -0.0000$ 20	

It would appear to be sufficient to use only the terms through $(b - a)^3$. However, by computing the sum of the decimal fractions given, we obtain 1.9129 54. But the next term is -0.0000 20 which, if included, would reduce the value to 1.9129 34. If we stopped with the $(b - a)^3$ term and rounded off, we would obtain 1.9130 whereas, if we keep the next term and round off, we obtain 1.9129. So the term in $(b - a)^4$ should be retained, and the remainder R_4 must be estimated. We have

$$f^{(5)}(x) = \frac{880}{243} x^{-14/3} = \frac{880 x^{1/3}}{243 x^5}.$$

Since we are concerned with the interval $7 < x < 8$, we see that $x^{1/3} < 2$ and $x^5 > (49)(343)$. Hence

$$0 < \frac{f^{(5)}(x)}{5!} < \frac{1760}{(243)(49)(343)(120)} < \frac{1}{270,000} < 0.000004.$$

Since $(b - a)^5 = -1$, we conclude that

$$-0.000004 < R_4 < 0$$

and that

$$\sqrt[3]{7} = 1.9129$$

to the required accuracy. Actually, if we merely keep an extra decimal in each term retained, we see that

$$\sqrt[3]{7} = 1.91293$$

to five decimals.

Remarks. In Example 1 there was no round-off error, since each decimal fraction gave the exact value of the corresponding term. This was not true in Example 2, however. In general, the round-off error in each term may be as much as $\frac{1}{2}$ in the last decimal place retained. Round-off errors may tend to cancel each other if there is a large number of computations in a given problem.

One may ask why two additional decimal places were kept in the above examples. Why not one or four or seven? It is clear that if several thousand additions are made the round-off error may be much larger. In computations with high-speed electronic computers the round-off error may be serious, since computations frequently run into the millions. The study of how such questions are handled is a part of the subject known as numerical analysis, a topic which has come under intensive examination because of the capabilities of high-speed computers. We have here an example of how the presence of computing machines has given rise to an entire complex of purely mathematical questions—some of which have been answered while many are still unsolved.

The remainder $R_n(a, b)$ in Taylor's Theorem may be given in many forms. The next theorem, stated without proof, gives the remainder in the form of an integral.

Theorem 18. (Taylor's Theorem with Integral Form of Remainder.) *Suppose that $f, f', f^{(2)}, \ldots, f^{(n)}, f^{(n+1)}$ are all continuous in some interval containing a*

and b. Then $f(x)$ *may be written in the form*

$$f(b) = f(a) + \frac{f'(a)}{1!} (b - a) + \frac{f^{(2)}(a)(b - a)^2}{2!} + \cdots$$

$$+ \frac{f^{(n)}(a)(b - a)^n}{n!} + R_n$$

where

$$R_n = \frac{1}{n!} \int_a^b f^{(n+1)}(t)(b - t)^n \, dt.$$

EXAMPLE 3. Write $\ln (1 + x)$ as a polynomial of the third degree, and estimate the remainder R_n for $0 < x < \frac{1}{2}$.

Solution. We select

$$f(x) = \ln (1 + x), \qquad a = 0, \qquad b = x.$$

Then

$$f'(x) = \frac{1}{1 + x}, \qquad\qquad f'(0) = 1,$$

$$f''(x) = - \frac{1}{(1 + x)^2}, \qquad f''(0) = -1,$$

$$f^{(3)}(x) = \frac{2}{(1 + x)^3}, \qquad f^{(3)}(0) = 2,$$

$$f^{(4)}(x) = - \frac{6}{(1 + x)^4}.$$

Therefore

$$\ln (1 + x) = x - \frac{x^2}{2} + \frac{x^3}{3} + R_n,$$

with

$$R_n = - \frac{6}{6} \int_0^x \frac{1}{(1 + t)^4} (x - t)^3 \, dt.$$

A simple estimate replaces $(1 + t)^{-4}$ by its smallest value 1, and we find

$$|R_n| < \int_0^{1/2} \left(\frac{1}{2} - t \right)^3 dt = \frac{1}{64}.$$

EXERCISES

In each of problems 1 through 20, compute the given quantities to the specified number of decimal places. Make sure of your accuracy by using Taylor's Theorem with Remainder. Use the fact that $2 < e < 4$ wherever necessary.

1. $e^{-0.2}$,   5 decimals

2. $e^{-0.4}$,   4 decimals

3. $e^{0.2}$,   5 decimals

4. $\sin (0.5)$,   5 decimals

5. $\cos (0.5)$,   5 decimals

6. $\tan (0.1)$,   3 decimals

7. ln (1.2),   4 decimals   8. ln (0.9),   5 decimals

9. e^{-1},   5 decimals   10. e,   5 decimals

11. $(1.08)^{1/4}$,   5 decimals   12. $(0.92)^{1/4}$,   5 decimals

13. $(0.91)^{1/3}$,   5 decimals   14. $(0.90)^{1/5}$,   5 decimals

15. $(30)^{1/5}$,   5 decimals   16. $(15)^{1/4}$,   5 decimals

17. $(0.8)^{1/5}$,   5 decimals   18. $(65)^{1/6}$,   5 decimals

19. ln (0.8),   5 decimals   20. ln (0.6),   3 decimals

Given that $1° = \pi/180$ radians $= 0.0174533$ radians and $5° = \pi/36$ radians $= 0.0872655$ radians, compute each of the following to the number of decimal places required.

21. sin 1°,   6 decimals   22. sin 5°,   5 decimals   23. cos 5°,   5 decimals

8. DIFFERENTIATION AND INTEGRATION OF SERIES

In Section 6 we developed the formula for the Taylor series of a function and, in so doing, we ignored the validity of the manipulations which were performed. Now we shall establish the theorems which verify the correctness of the results already obtained.

Theorem 19. *If $R > 0$ and the series*

$$\sum_{n=0}^{\infty} a_n x^n \tag{1}$$

converges for $|x| < R$, then the series obtained from (1) by term-by-term differentiation converges absolutely for $|x| < R$.

Proof. Term-by-term differentiation of (1) yields

$$\sum_{n=1}^{\infty} n a_n x^{n-1}. \tag{2}$$

Choose any value x such that $|x| < R$ and choose x_1 so that $|x| < |x_1| < R$. According to the Lemma of Section 5 (page 93), there is a positive number M with the property that

$$|a_n x_1^n| \leq M \qquad \text{for all } n.$$

We have the relation

$$|n a_n x^{n-1}| = \left| n a_n \frac{x^{n-1}}{x_1^n} \cdot x_1^n \right| \leq n \frac{M}{x_1} \left| \frac{x}{x_1} \right|^{n-1},$$

and now we can apply the comparison test to the series (2). The series

$$\frac{M}{x_1} \sum_{n=1}^{\infty} n \left| \frac{x}{x_1} \right|^{n-1}$$

converges by the Ratio Test, since $\rho = |x/x_1| < 1$; hence, so does the series (2). Since x was any number in the interval $(-R, R)$, the interval of convergence of (2) is the same as that of (1).

Corollary.   *Under the hypotheses of Theorem* 19, *the series* (1) *may be differentiated any number of times and each of the differentiated series converges for* $|x| < R$.

Remarks.   (i) The Corollary is obtained by induction, since each differentiated series has the same radius of convergence as the one before.   (ii) The results of Theorem 19 and the Corollary are valid for a series of the form

$$\sum_{n=0}^{\infty} a_n(x - a)^n,$$

which converges for $|x - a| < R$ so long as $R > 0$.   The proof is the same. (iii) The quantity R may be $+\infty$, in which case the series and its derived ones converge for all values of x.

Theorem 20.   *If* $R > 0$ *and* f *is defined by*

$$f(x) = \sum_{n=0}^{\infty} a_n x^n \qquad for \ |x| < R,$$

then f *is continuous for* $|x| < R$.

Proof.   Let x_0 be any number such that $-R < x_0 < R$; we wish to show that f is continuous at x_0.   In other words we must show that

$$f(x) \to f(x_0) \quad \text{as} \quad x \to x_0.$$

We have

$$|f(x) - f(x_0)| = \left| \sum_{n=0}^{\infty} a_n(x^n - x_0^n) \right|$$

$$\leq \sum_{n=0}^{\infty} |a_n| \, |x^n - x_0^n|.$$

We apply the Theorem of the Mean to the function $g(x) = x^n$; that is, the relation

$$g(x) - g(x_0) = g'(\xi)(x - x_0), \qquad (\xi \text{ is between } x \text{ and } x_0),$$

applied to the function $g(x) = x^n$, is

$$x^n - x_0^n = n\xi_n^{n-1}(x - x_0), \qquad (\xi_n \text{ between } x \text{ and } x_0).$$

The subscript n has been put on ξ to identify the particular exponent of the function x^n.   We may also write

$$|x^n - x_0^n| = n|\xi_n|^{n-1}|x - x_0|.$$

Thus we find

$$|f(x) - f(x_0)| \leq \sum_{n=0}^{\infty} n|a_n| \, |\xi_n|^{n-1} |x - x_0|.$$

So long as x is in the interval of convergence there is an x_1 such that $|\xi_n| < x_1$ for all n. We deduce that

$$|f(x) - f(x_0)| \leq |x - x_0| \sum_{n=0}^{\infty} n|a_n| x_1^{n-1}.$$

Now we apply Theorem 19 to conclude that the series on the right converges; call its sum K. Then

$$|f(x) - f(x_0)| \leq |x - x_0| \cdot K.$$

As x tends to x_0 the quantity on the right tends to zero, and so $f(x)$ tends to $f(x_0)$. That is, $f(x)$ is continuous at x_0.

Theorem 21. *Suppose that $R > 0$ and*

$$f(x) = \sum_{n=0}^{\infty} a_n x^n \tag{3}$$

converges for $|x| < R$. We define

$$F(x) = \int_0^x f(t) \, dt.$$

Then

$$F(x) = \sum_{n=0}^{\infty} a_n \frac{x^{n+1}}{n+1} \tag{4}$$

holds for $|x| < R$.

Proof. Let x be any number such that $-R < x < R$. Choose x_1 so that $|x| < |x_1| < R$. We note that for any n

$$\int_0^x a_n t^n \, dt = \frac{a_n x^{n+1}}{n+1}.$$

Therefore

$$F(x) - \sum_{n=0}^{N} a_n \frac{x^{n+1}}{n+1} = \int_0^x \left[f(t) - \sum_{n=0}^{N} a_n t^n \right] dt. \tag{5}$$

But now

$$f(t) - \sum_{n=0}^{N} a_n t^n = \sum_{n=0}^{\infty} a_n t^n - \sum_{n=0}^{N} a_n t^n = \sum_{n=N+1}^{\infty} a_n t^n,$$

and for all t such that $|\xi_n| < x_1 < R$,

$$\left| f(t) - \sum_{n=0}^{N} a_n t^n \right| \leq \sum_{n=N+1}^{\infty} |a_n| \, |x|^n. \tag{6}$$

Since the series (3) converges absolutely at x, the right side of (6)—being the remainder—tends to zero as $N \to \infty$. We conclude from (5) that

$$\left| F(x) - \sum_{n=0}^{N} a_n \frac{x^{n+1}}{n+1} \right| \leq \int_0^x \left(\sum_{n=N+1}^{\infty} |a_n|\,|x|^n \right) dt = \left(\sum_{n=N+1}^{\infty} |a_n|\,|x|^n \right) \cdot x.$$

If N tends to ∞, the right side above tends to zero and the left side above yields (4).

EXAMPLE 1. Assuming that the function $f(x) = \sin x$ is given by the series

$$\sin x = x - \frac{x^3}{3!} + \frac{x^5}{5!} - \cdots + \frac{(-1)^n x^{2n+1}}{(2n+1)!} + \cdots,$$

find the Taylor series for $\cos x$.

Solution. Applying the Ratio Test to the series

$$\sum_{n=0}^{\infty} \frac{(-1)^n x^{2n+1}}{(2n+1)!},$$

we see that it converges for all values of x. We define

$$F(x) = \int_0^x \sin t\, dt = -\cos x + 1.$$

Integrating the above series for $\sin x$ term by term, we obtain

$$F(x) = 1 - \cos x = \sum_{n=0}^{\infty} \frac{(-1)^n x^{2n+2}}{(2n+2)!}$$

or

$$\cos x = 1 - \frac{x^2}{2!} + \frac{x^4}{4!} - \frac{x^6}{6!} + \cdots + \frac{(-1)^n x^{2n}}{(2n)!} + \cdots = \sum_{n=0}^{\infty} \frac{(-1)^n x^{2n}}{(2n)!}.$$

The next theorem relates the derivative of a function given by a series with the term-by-term differentiation of the series.

Theorem 22. *If $R > 0$ and*

$$f(x) = \sum_{n=0}^{\infty} a_n x^n \qquad \text{for } |x| < R, \tag{7}$$

then $f(x)$ has continuous derivatives of all orders for $|x| < R$ which are given there by series obtained by successive term-by-term differentiations of (7).

Proof. The fact that

$$\sum_{n=1}^{\infty} n a_n\, x^{n-1} \tag{8}$$

converges for $|x| < R$ was shown in Theorem 19. We must show that $g(x) = \sum_{n=1}^{\infty} na_n x^{n-1}$ is the derivative of f. Theorem 20 establishes the fact that g is continuous. Then, integrating the series (8) term by term we get, on the one hand,

$$a_0 + \int_0^x g(t)\, dt$$

and, on the other, the series for $f(x)$. That is,

$$f(x) = a_0 + \int_0^x g(t)\, dt.$$

The Fundamental Theorem of the Calculus then asserts that

$$f'(x) = g(x),$$

which is the result we wished to establish.

Remark. The result of Theorem 22 holds equally well for functions f given by series of the form

$$f(x) = \sum_{n=0}^{\infty} a_n(x - a)^n,$$

which converge for $|x - a| < R$ with $R > 0$.

EXAMPLE 2. Assuming that the expansion of $f(x) = \sinh x$ is given by

$$\sinh x = x + \frac{x^3}{3!} + \frac{x^5}{5!} + \cdots + \frac{x^{2n-1}}{(2n-1)!} + \cdots,$$

valid for all x, obtain an expansion for $\cosh x$.

Solution. Differentiating term by term, we find

$$\cosh x = 1 + \frac{x^2}{2!} + \frac{x^4}{4!} + \cdots + \frac{x^{2n}}{(2n)!} + \cdots.$$

We now make use of the above theorems and the fact that the function $1/(1 + x)$ may be expanded in the simple geometric series

$$\frac{1}{1 + x} = \sum_{n=0}^{\infty} (-1)^n x^n, \qquad |x| < 1$$

to obtain additional series expansions.

Theorem 23.

$$\ln (1 + x) = \sum_{n=1}^{\infty} \frac{(-1)^{n-1} x^n}{n} \qquad for\ |x| < 1.$$

Proof. Letting $F(x) = \ln(1 + x)$ and differentiating, we find

$$F'(x) = f(x) = \frac{1}{1 + x} = \sum_{n=0}^{\infty} (-1)^n x^n, \qquad |x| < 1.$$

Now, by Theorem 21, we may integrate term by term and get

$$F(x) = \int_0^x f(t)\, dt = \sum_{n=0}^{\infty} \frac{(-1)^n x^{n+1}}{n + 1},$$

which is identical with the statement of the theorem.

EXAMPLE 3. Find the Maclaurin series for

$$f(x) = \frac{1}{(1 - x)^2}.$$

Solution. If we define $F(x) = (1 - x)^{-1}$, we see that $F'(x) = f(x)$.

Now

$$F(x) = \sum_{n=0}^{\infty} x^n, \qquad |x| < 1$$

[which we can obtain from the expansion for $(1 + x)^{-1}$ if we replace x by $-x$]. Therefore

$$f(x) = \frac{1}{(1 - x)^2} = \sum_{n=1}^{\infty} nx^{n-1} = \sum_{k=0}^{\infty} (k + 1)x^k, \qquad |x| < 1.$$

EXAMPLE 4. Find the Maclaurin expansion for $f(x) = (1 + x^2)^{-1}$.

Solution. The geometric series

$$\frac{1}{1 + u} = \sum_{n=0}^{\infty} (-1)^n u^n, \qquad |u| < 1$$

after substitution of x^2 for u, becomes

$$\frac{1}{1 + x^2} = \sum_{n=0}^{\infty} (-1)^n x^{2n},$$

valid for $x^2 < 1$. The inequality $x^2 < 1$ is equivalent to the inequality $|x| < 1$.

EXERCISES

In each of problems 1 through 13, find the Taylor or Maclaurin series for the functions f, assuming that the Taylor or Maclaurin series for the functions given in Sections 6 and 7 are known.

1. $f(x) = \begin{cases} (\sin x)/x, & x \neq 0 \\ 1, & x = 0 \end{cases}$ 2. $f(x) = \begin{cases} (e^x - 1)/x, & x \neq 0 \\ 1, & x = 0 \end{cases}$

3. $f(x) = \begin{cases} (1 - \cos x)/x, & x \neq 0 \\ 0, & x = 0 \end{cases}$

4. $f(x) = \ln(x + \sqrt{1 + x^2}) = \text{argsinh } x$

5. $f(x) = \arcsin x$ 6. $f(x) = \arctan x$

7. $f(x) = \text{argtanh } x$ 8. $f(x) = (1 + x)^{-2}$

9. $f(x) = \ln \dfrac{1 + x}{1 - x}$ 10. $f(x) = (1 - x)^{-3}$

11. $f(x) = x(1 + x^2)^{-2}$ 12. $f(x) = x \ln(1 + x^2)$

13. $f(x) = \sin^2 x$

9. VALIDITY OF TAYLOR EXPANSIONS AND COMPUTATIONS WITH SERIES

The theorems of this section state that certain functions are truly represented by their Taylor series.

Theorem 24. *For any values of a and x, we have*

$$e^x = e^a \sum_{n=0}^{\infty} \frac{(x - a)^n}{n!} \, ; \tag{1}$$

i.e., *the Taylor series for e^x about $x = a$ converges to e^x for any a and x.*

Proof. For simplicity, set $a = 0$, the proof being analogous when $a \neq 0$. If we let $f(x) = e^x$, then $f^{(n)}(x) = e^x$ for all n. Now, using Taylor's Theorem with Remainder, we have

$$e^x = \sum_{k=0}^{n} \frac{x^k}{k!} + R_n,$$

where

$$R_n = \frac{e^\xi x^{n+1}}{(n + 1)!},$$

with ξ between 0 and x. If x is positive, then $e^\xi < e^x$ while, if x is negative, then $e^\xi < e^0 = 1$. In either case,

$$R_n \leq C \frac{|x|^{n+1}}{(n + 1)!}, \tag{2}$$

where C is the larger of 1 and e^x but is *independent of n*. If the right side of (2) is the general term of a series then, by the Ratio Test, that series is convergent for all x. The general term of any convergent series must tend to zero, and so

$$C \frac{|x|^{n+1}}{(n + 1)!} \to 0 \quad \text{as} \quad n \to \infty \text{ for each } x.$$

We conclude that $R_n \to 0$ as $n \to \infty$, and so (1) is established.

Theorem 25. *The following functions are given by their Maclaurin series*:

$$\sin x = \sum_{n=0}^{\infty} \frac{(-1)^n x^{2n+1}}{(2n+1)!},$$

$$\cos x = \sum_{n=0}^{\infty} \frac{(-1)^n x^{2n}}{(2n)!},$$

$$\sinh x = \sum_{n=0}^{\infty} \frac{x^{2n+1}}{(2n+1)!},$$

$$\cosh x = \sum_{n=0}^{\infty} \frac{x^{2n}}{(2n)!}.$$

The proofs of these results follow the same outline as the proof of Theorem 24 and are left as exercises for the student at the end of this section.

Theorem 26. (Binomial Theorem.) *For each real number m, we have*

$$(1 + x)^m = 1 + \sum_{n=1}^{\infty} \frac{m(m-1)(m-2)\cdots(m-n+1)}{n!} x^n \quad \text{for } |x| < 1.$$

Proof. To show that the series on the right converges absolutely for $|x| < 1$, we apply the Ratio Test:

$$\left| \frac{u_{n+1}}{u_n} \right|$$

$$= \left| \frac{m(m-1)\cdots(m-n+1)(m-n)}{(n+1)!} \cdot \frac{n!}{m(m-1)\cdots(m-n+1)} \cdot \frac{x^{n+1}}{x^n} \right|$$

$$= \frac{|m-n|}{n+1} |x| = \frac{|1 - m/n|}{1 + 1/n} |x|.$$

The quantity on the right tends to $|x|$ as $n \to \infty$, and so the series converges for $|x| < 1$. We define

$$f(x) = 1 + \sum_{n=1}^{\infty} \frac{m(m-1)\cdots(m-n+1)}{n!} x^n,$$

and we wish to show that $f(x) = (1 + x)^m$ if $|x| < 1$. Employing Theorem 22, we get $f'(x)$ by term-by-term differentiation of the series for f. We have

$$f'(x) = m + \sum_{n=2}^{\infty} \frac{m(m-1)\cdots(m-n+1)}{(n-1)!} x^{n-1}. \tag{3}$$

Multiplying both sides of (3) by x, we get

$$xf'(x) = \sum_{n=1}^{\infty} n \frac{m(m-1)\cdots(m-n+1)}{n!} x^n. \tag{4}$$

We add (3) and (4) to obtain

$$(1 + x)f'(x) = m\left\{1 + \sum_{n=1}^{\infty} \frac{m(m - 1) \cdots (m - n + 1)}{n!} x^n\right\} = mf(x).$$

The derivative of $\ln f(x)$ is given by

$$\frac{d}{dx} \ln f(x) = \frac{f'(x)}{f(x)} = \frac{m}{1 + x}.$$

On the other hand,

$$\frac{d}{dx} \ln (1 + x)^m = m \frac{d}{dx} \ln (1 + x) = \frac{m}{1 + x}.$$

Since $f(x)$ and $(1 + x)^m$ have the same derivative, and since $f(0) = 1$, we conclude finally that

$$f(x) = (1 + x)^m, \qquad |x| < 1.$$

EXAMPLE 1. Write the first 5 terms of the series expansion for $(1 + x)^{3/2}$.

Solution. We have $m = \frac{3}{2}$, and therefore

$$(1 + x)^{3/2} = 1 + \frac{\frac{3}{2}}{1!}x + \frac{(\frac{3}{2})(\frac{1}{2})}{2!}x^2 + \frac{(\frac{3}{2})(\frac{1}{2})(-\frac{1}{2})}{3!}x^3 + \frac{(\frac{3}{2})(\frac{1}{2})(-\frac{1}{2})(-\frac{3}{2})}{4!}x^4$$

$$= 1 + \frac{3}{2}x + \frac{3}{8}x^2 - \frac{3}{2^3 \cdot 3!}x^3 + \frac{3^2}{2^4 \cdot 4!}x^4 - \cdots.$$

EXAMPLE 2. Write the binomial series for $(1 + x)^7$.

Solution. We have

$$(1 + x)^7 = 1 + \sum_{n=1}^{\infty} \frac{7(6) \cdots (7 - n + 1)}{n!} x^n.$$

We now observe that beginning with $n = 8$ all the terms have a zero in the numerator. Therefore,

$$(1 + x)^7 = 1 + \sum_{n=1}^{7} \frac{7(6) \cdots (7 - n + 1)}{n!} x^n$$

$$= 1 + 7x + \frac{7 \cdot 6}{2!}x^2 + \frac{7 \cdot 6 \cdot 5}{3!}x^3 + \cdots + \frac{7!}{7!}x^7.$$

For m a positive integer, the binomial series always terminates after a finite number of terms.

EXAMPLE 3. Compute

$$\int_0^{0.5} e^{x^2} \, dx$$

to an accuracy of five decimal places.

Solution. We have

$$e^u = \sum_{n=0}^{\infty} \frac{u^n}{n!} \qquad \text{for all } u,$$

and so

$$e^{x^2} = \sum_{n=0}^{\infty} \frac{x^{2n}}{n!} \qquad \text{for all } x.$$

By Theorem 21 we can integrate term by term to get

$$\int_0^x e^{x^2}\, dx = \sum_{n=0}^{\infty} \frac{x^{2n+1}}{n!(2n+1)}.$$

For $x = 0.5$, we now compute

$$x \;=\; 0.5 \;=\; \frac{1}{2} \;=\; 0.50000 \;\; 00,$$

$$\frac{x^3}{1!\cdot 3} = \frac{(0.5)^3}{3} = \frac{1}{24} = 0.04166 \;\; 67^{-},$$

$$\frac{x^5}{2!\cdot 5} = \frac{(0.5)^5}{10} = \frac{1}{320} = 0.00312 \;\; 50,$$

$$\frac{x^7}{3!\cdot 7} = \frac{(0.5)^7}{42} = \frac{1}{5376} = 0.00018 \;\; 60^{+},$$

$$\frac{x^9}{4!\cdot 9} = \frac{(0.5)^9}{216} = \frac{1}{110{,}592} = 0.00000 \;\; 90^{+},$$

Sum of the right-hand column $\qquad\qquad = 0.54498 \;\; 67.$

If we wish to stop at this point we must estimate the error made by neglecting all the remaining terms. The remainder is

$$\sum_{n=5}^{\infty} \frac{x^{2n+1}}{n!(2n+1)} \le \frac{x^{11}}{5!\cdot 11}\left(1 + \frac{x^2}{6} + \frac{x^4}{6^2} + \frac{x^6}{6^3} + \cdots\right)$$

$$\le \frac{x^{11}}{1320}\,\frac{1}{(1 - x^2/6)} = \frac{24}{23}\cdot\frac{1}{1320}\cdot\frac{1}{2048}$$

$$\le 0.00000 \;\; 04.$$

Therefore

$$\int_0^{0.5} e^{x^2}\, dx = \begin{cases} 0.54499 \text{ to an accuracy of 5 decimals} \\ 0.544987 \text{ to an accuracy of 6 decimals} \end{cases}.$$

EXERCISES

In each of problems 1 through 7, write the beginning of the binomial series for the given expression to the required number of terms.

1. $(1 + x)^{-3/2}$,   5 terms

2. $(1 - x)^{1/2}$,   4 terms

3. $(1 + x^2)^{-2/3}$,   5 terms

4. $(1 - x^2)^{-1/2}$,   6 terms

5. $(1 + x^3)^7$, all terms   6. $(5 + x)^{1/2}$, 4 terms   7. $(3 + \sqrt{x})^{-3}$, 5 terms

In each of problems 8 through 17, compute the value of the definite integral to the number of decimal places specified. Estimate the remainder.

8. $\int_0^1 \sin(x^2)\, dx$,   5 decimals

9. $\int_0^1 \cos(x^2)\, dx$,   5 decimals

10. $\int_0^1 e^{-x^2}\, dx$,   5 decimals

11. $\int_0^{0.5} \dfrac{dx}{1 + x^3}$,   5 decimals

12. $\int_0^1 \dfrac{\sin x}{x}\, dx$,   5 decimals

13. $\int_0^1 \dfrac{e^x - 1}{x}\, dx$,   5 decimals

14. $\int_0^{0.5} \dfrac{dx}{\sqrt{1 + x^3}}$,   5 decimals

15. $\int_0^{1/3} \dfrac{dx}{\sqrt[3]{1 + x^2}}$,   5 decimals

16. $\int_0^{1/3} \dfrac{dx}{\sqrt[3]{1 - x^2}}$,   5 decimals

17. $\int_0^{0.5} \dfrac{dx}{\sqrt{1 - x^3}}$,   5 decimals

18. Use the series for $\ln \dfrac{1 + x}{1 - x}$ to find $\ln 1.5$ to 5 decimals of accuracy.

19. Same as Problem 18, to find $\ln 2$.

20. Prove that
$$\sin x = \sum_{n=0}^{\infty} \frac{(-1)^n x^{2n+1}}{(2n + 1)!}$$

21. Prove that
$$\cos x = \sum_{n=0}^{\infty} \frac{(-1)^n x^{2n}}{(2n)!}$$

22. Prove that
$$\sinh x = \sum_{n=0}^{\infty} \frac{x^{2n+1}}{(2n + 1)!}$$

23. Prove that
$$\cosh x = \sum_{n=0}^{\infty} \frac{x^{2n}}{(2n)!}$$

24. If $f(x) = \sin x$ show, by induction, that $f^{(k)}(x) = \sin(x + \tfrac{1}{2}k\pi)$.

25. Use the result of Problem 24 to show that for all a and x the Taylor series for $\sin x$ is given by
$$\sin x = \sum_{n=0}^{\infty} \frac{\sin(a + \tfrac{1}{2}n\pi)}{n!} (x - a)^n.$$

26. If $f(x) = \cos x$ show, by induction, that $f^{(k)}(x) = \cos(x + \tfrac{1}{2}k\pi)$.

27. Use the result of Problem 26 to show that for all a and x the Taylor series for $\cos x$ is given by
$$\cos x = \sum_{n=0}^{\infty} \frac{\cos(a + \tfrac{1}{2}n\pi)}{n!} (x - a)^n.$$

10. ALGEBRAIC OPERATIONS WITH SERIES

In previous sections we discussed the question of term-by-term differentiation and integration of series. Now we turn to the question of multiplication and division of power series.

Suppose we are given two power series

$$\sum_{n=0}^{\infty} a_n x^n = a_0 + a_1 x + a_2 x^2 + \cdots + a_n x^n + \cdots,$$

$$\sum_{n=0}^{\infty} b_n x^n = b_0 + b_1 x + b_2 x^2 + \cdots + b_n x^n + \cdots.$$

Without considering questions of convergence, we multiply the two series by following the rules for multiplying two polynomials. We obtain the successive lines, each obtained by multiplying an element of the second series with all the terms of the first series:

$$b_0: \quad a_0 b_0 + a_1 b_0 x + a_2 b_0 x^2 + \cdots + a_n b_0 x^n + \cdots,$$

$$b_1 x: \quad a_0 b_1 x + a_1 b_1 x^2 + \cdots + a_{n-1} b_1 x^n + a_n b_1 x^{n+1} + \cdots,$$

$$b_2 x^2: \quad a_0 b_2 x^2 + \cdots + a_{n-2} b_2 x^n + a_{n-1} b_2 x^{n+1} + a_n b_2 x^{n+2} + \cdots,$$

$$\vdots$$

$$b_n x^n: \quad a_0 b_n x^n + a_1 b_n x^{n+1} + a_2 b_n x^{n+2} + \cdots.$$

Adding the columns, we obtain the power series

$$a_0 b_0 + (a_1 b_0 + a_0 b_1)x + (a_2 b_0 + a_1 b_1 + a_0 b_2)x^2$$
$$+ (a_3 b_0 + a_2 b_1 + a_1 b_2 + a_0 b_3)x^3 + \cdots$$
$$+ (a_n b_0 + a_{n-1} b_1 + \cdots + a_0 b_n)x^n + \cdots.$$

The technique for computing the coefficients of any term is easy to determine. The subscripts of the a's decrease by one as the subscripts of the b's increase, the total always remaining the same.

DEFINITION. Given the series

$$\sum_{n=0}^{\infty} a_n x^n, \quad \sum_{n=0}^{\infty} b_n x^n,$$

we define the **Cauchy Product** to be the series

$$c_0 + c_1 x + c_2 x^2 + \cdots + c_n x^n + \cdots$$

where

$$c_n = a_n b_0 + a_{n-1} b_1 + \cdots + a_0 b_n = \sum_{k=0}^{\infty} a_{n-k} b_k.$$

EXAMPLE 1. Given the Maclaurin series for e^x and cos x, find the first seven terms of the Cauchy Product of these two series—i.e., the terms through x^6.

Solution. We have

$$e^x = 1 + x + \frac{x^2}{2} + \frac{x^3}{6} + \frac{x^4}{24} + \frac{x^5}{120} + \frac{x^6}{720} + \cdots,$$

$$\cos x = 1 \quad - \frac{x^2}{2} \quad + \frac{x^4}{24} \quad - \frac{x^6}{720} + \cdots.$$

We multiply term by term to obtain

$$1 + x + \frac{x^2}{2} + \frac{x^3}{6} + \frac{x^4}{24} + \frac{x^5}{120} + \frac{x^6}{720}$$

$$- \frac{x^2}{2} - \frac{x^3}{2} - \frac{x^4}{4} - \frac{x^5}{12} - \frac{x^6}{48}$$

$$+ \frac{x^4}{24} + \frac{x^5}{24} + \frac{x^6}{48}$$

$$- \frac{x^6}{720}$$

$$\text{Cauchy Product} = 1 + x - \frac{x^3}{3} - \frac{x^4}{6} - \frac{x^5}{30} + 0 \cdot x^6 + \cdots.$$

Theorem 27. *If*

$$f(x) = \sum_{n=0}^{\infty} a_n x^n, \qquad g(x) = \sum_{n=0}^{\infty} b_n x^n,$$

both converge for $|x| < R$; *then the Cauchy Product of the two series converges to* $f(x) \cdot g(x)$ *for* $|x| < R$.

The proof of this theorem is given in Morrey, *University Calculus*, page 711.

On the basis of Theorem 27, we see that, in Example 1 above, the Cauchy Product is actually the Maclaurin expansion of the function $e^x \cos x$, valid for all x.

Now we apply the process of long division to two series as if they were polynomials. We write

$$
\begin{array}{r}
c_0 + \quad c_1 x \\
\hline
b_0 + b_1 x + b_2 x^2 + \cdots \,\overline{)\,a_0 + \quad a_1 x + \quad a_2 x^2 + \cdots} \\
c_0 b_0 + c_0 b_1 x + c_0 b_2 x^2 + \cdots \\
\hline
+ \, (a_1 - c_0 b_1)x + (a_2 - c_0 b_2)x^2 + \cdots \\
c_1 b_0 \, x + \qquad\quad c_1 b_1 \, x^2 + \cdots \\
\hline
+ \, (a_2 - c_0 b_2 - c_1 b_1)x^2 + \cdots \\
c_2 b_0 \, x^2 + \cdots
\end{array}
$$

In order for the division process to proceed, we must have (assuming $b_0 \neq 0$)

$$a_0 = c_0 b_0 \quad \text{or} \quad c_0 = a_0/b_0,$$
$$a_1 - c_0 b_1 = c_1 b_0 \quad \text{or} \quad c_1 = (a_1 - c_0 b_1)/b_0,$$
$$a_2 - c_0 b_2 - c_1 b_1 = c_2 b_0 \quad \text{or} \quad c_2 = (a_2 - c_0 b_2 - c_1 b_1)/b_0,$$

etc. By induction it can be established that the Cauchy Product of the quotient series with the divisor series yields the dividend series.

Theorem 28. *Under the hypotheses of Theorem 27, the quotient series converges to $f(x)/g(x)$ for $|x| < T$ for some $T > 0$ so long as $b_0 \neq 0$.*

The proof is omitted.

EXAMPLE 2. By division of the Maclaurin series for $\sin x$ by the one for $\cos x$, find the terms up to x^5 in the Maclaurin series for $\tan x$.

Solution. We have

$$
\begin{array}{r}
x + \dfrac{x^3}{3} + \dfrac{2x^5}{15} \\[2mm]
\left(1 - \dfrac{x^2}{2} + \dfrac{x^4}{24} - \cdots\right) \overline{\Big) \; x - \dfrac{x^3}{6} + \dfrac{x^5}{120} - \cdots} \\[2mm]
x - \dfrac{x^3}{2} + \dfrac{x^5}{24} - \cdots \\[2mm]
\hline
+ \dfrac{x^3}{3} - \dfrac{x^5}{30} + \cdots \\[2mm]
+ \dfrac{x^3}{3} - \dfrac{x^5}{6} + \cdots \\[2mm]
\hline
\dfrac{2x^5}{15} + \cdots
\end{array}
$$

EXERCISES

In each of problems 1 through 20, find the Maclaurin series to the number of terms given by the index n.

1. $\dfrac{\sin x}{1 + x}$, $\quad n = 5$

2. $\dfrac{\sin x}{1 - x}$, $\quad n = 5$

3. $\dfrac{\cos x}{1 + x}$, $\quad n = 5$

4. $\dfrac{\cos x}{1 - x}$, $\quad n = 5$

5. $\sqrt{1 + x} \ln (1 + x)$, $\quad n = 5$

6. $\sqrt{1 - x} \ln (1 - x)$, $\quad n = 5$

7. $\dfrac{\ln (1 + x)}{\sqrt{1 + x}}$, $\quad n = 5$

8. $\dfrac{\ln (1 - x)}{\sqrt{1 + x}}$, $\quad n = 5$

9. $e^x \sec x, \quad n = 5$

10. $e^{-x} \tan x, \quad n = 3$

11. $\dfrac{e^x}{\sqrt{1 + x^2}}, \quad n = 5$

12. $\dfrac{e^{-x}}{\sqrt{1 + x^2}}, \quad n = 5$

13. $\dfrac{\arctan x}{1 + x}, \quad n = 5$

14. $\dfrac{1 - x}{1 - x^3} = \dfrac{1}{1 + x + x^2}, \quad \text{all } n$

15. $\dfrac{\arcsin x}{\cosh x}, \quad n = 5$

16. $\operatorname{sech} x, \quad n = 6$

17. $(1 + x)^{1/3}(1 + x^2)^{4/3}, \quad n = 4$

18. $\sin^2 x \cos x, \quad n = 4$

19. $(1 + x^2)^{3/2}(1 - x^3)^{-1/2}, \quad n = 4$

20. $\tanh^2 x, \quad n = 5$

CHAPTER 4

Partial Differentiation

1. LIMITS AND CONTINUITY. PARTIAL DERIVATIVES

A symbolic expression of the form

$$z = F(x, y),$$

where x and y are independent variables, indicates that the dependent variable z is a function of *both* the independent variables. A function of three variables is written

$$w = G(x, y, z),$$

where x, y, and z are independent variables and w is the dependent variable. We can go on and consider functions of four, of five, or of any number of independent variables. If the exact number of independent variables is n, we usually write

$$y = f(x_1, x_2, \ldots, x_n),$$

in which $x_1, x_2, \ldots, x_n$ are n independent variables and y is the dependent variable.

We now give a precise definition of a function of two variables. (See, for example, *First Course*, page 184.)

DEFINITION. Consider a collection of ordered pairs (A, w) where the elements A are themselves ordered pairs of real numbers and the elements w are real numbers. If no two members of the collection have the same item A as a first element —i.e., if it can never happen that there are two members (A_1, w_1) and (A_1, w_2) with $w_1 \neq w_2$—then we call this collection **a function of two variables.** The totality of possible ordered pairs A is called the **domain** of the function. The totality of possible values for w is called the **range** of the function.

Remark. The definition of a *function of three variables* is precisely the same as that for a function of two variables except that the elements A are ordered triples rather than ordered pairs. A *function of n variables* is defined by considering the elements A to be ordered n-tuples of real numbers.

We are now ready to define a limit of a function of two variables. Let f be a function of the two independent variables x, y. We wish to examine the behavior of f as $(x, y) \to (a, b)$—i.e., as $x \to a$ and $y \to b$.

120

DEFINITION. We say that $f(x, y)$ **tends to the number** L **as** (x, y) **tends to** (a, b), and we write

$$f(x, y) \to L \quad \text{as} \quad (x, y) \to (a, b)$$

if and only if for each $\epsilon > 0$ there is a $\delta > 0$ such that

$$|f(x, y) - L| < \epsilon$$

whenever

$$|x - a| < \delta \quad \text{and} \quad |y - b| < \delta \quad \text{and} \quad (x, y) \neq (a, b).$$

A geometric interpretation of this definition is exhibited in Fig. 4–1. The definition asserts that whenever (x, y) are in the shaded square, as shown, then the function values, which we represent by z, must lie in the rectangular box of height 2ϵ between the values $L - \epsilon$ and $L + \epsilon$. This interpretation is an extension of the one given in *First Course*, Chapter 5, page 97, for functions of one variable.

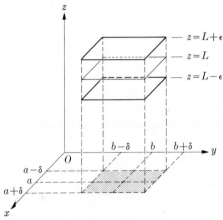

FIGURE 4–1

Remark. It is not necessary that (a, b) be in the domain of f. That is, a limit may exist with $f(a, b)$ being undefined.

DEFINITION. We say that f **is continuous at** (a, b) if and only if

(i) $f(a, b)$ is defined, and

(ii) $f(x, y) \to f(a, b)$ as $(x, y) \to (a, b)$.

Definitions of limits and continuity for functions of three, four, and more variables are completely similar.

Functions of two or more variables do not have ordinary derivatives of the type we studied for functions of one variable. If f is a function of two variables, say x and y, then for each *fixed* value of y, f is a function of a single variable x. The derivative with respect to x (keeping y fixed) is then called the partial derivative with respect to x. For x fixed and y varying, we also obtain a partial derivative with respect to y.

DEFINITIONS. We define **the partial derivatives of a function** f of the variables x and y by

$$f_{,1}(x, y) = \lim_{h \to 0} \frac{f(x + h, y) - f(x, y)}{h},$$

$$f_{,2}(x, y) = \lim_{h \to 0} \frac{f(x, y + h) - f(x, y)}{h},$$

where y is kept fixed in the first limit and x is kept fixed in the second. If F is a function of the three variables x, y, and z, we define the *partial derivatives* $F_{,1}$, $F_{,2}$, and $F_{,3}$ by

$$F_{,1}(x, y, z) = \lim_{h \to 0} \frac{F(x + h, y, z) - F(x, y, z)}{h}, \quad (y, z \text{ fixed})$$

$$F_{,2}(x, y, z) = \lim_{h \to 0} \frac{F(x, y + h, z) - F(x, y, z)}{h}, \quad (x, z \text{ fixed})$$

$$F_{,3}(x, y, z) = \lim_{h \to 0} \frac{F(x, y, z + h) - F(x, y, z)}{h}, \quad (x, y \text{ fixed}).$$

In other words, **to find the partial derivative** $F_{,1}$ **of a function of three variables, regard** y **and** z **as constants and find the usual derivative with respect to** x**; the derivatives** $F_{,2}$ **and** $F_{,3}$ **are found correspondingly. The same procedure applies for functions of any number of variables.**

EXAMPLE 1. Given $f(x, y) = x^3 + 7x^2y + 8y^3 + 3x - 2y + 7$, find $f_{,1}$ and $f_{,2}$.

Solution. Keeping y fixed and differentiating with respect to x, we find that

$$f_{,1} = 3x^2 + 14xy + 3.$$

Similarly, keeping x fixed, we get

$$f_{,2} = 7x^2 + 24y^2 - 2.$$

EXAMPLE 2. Given $f(x, y, z) = x^3 + y^3 + z^3 + 3xyz$, find $f_{,1}(x, y, z)$, $f_{,2}(x, y, z)$, and $f_{,3}(x, y, z)$.

Solution.

$$f_{,1}(x, y, z) = 3x^2 + 3yz, \qquad f_{,2}(x, y, z) = 3y^2 + 3xz, \qquad f_{,3}(x, y, z) = 3z^2 + 3xy.$$

Remarks on notation. The notation presented here is not classical, although it is coming into use more and more. The chief advantage of our notation is that it is independent of the letters used. A more common notation for $f_{,1}$ is

$$f_x.$$

It has the disadvantage of implying that the independent variable must be the letter x. Another common symbol is

$$\frac{\partial f}{\partial x}$$

(read partial of f with respect to x). This notation has the double disadvantage of using the letter x and of giving the impression (incorrectly) that the derivative is a fraction with ∂f and ∂x having independent meanings (which they do not). If we write $z = f(x, y)$, still another symbol for partial derivative is the expression

$$\frac{\partial z}{\partial x}.$$

Because of the multiplicity of symbols for partial derivatives used in texts on mathematics and various related branches of technology, it is important that the student familiarize himself with all of them.

EXAMPLE 3. Given $f(x, y) = e^{xy} \cos x \sin y$, find $f_x(x, y)$ and $f_y(x, y)$.

Solution. We have

$$f_x(x, y) = e^{xy} \sin y(-\sin x) + \cos x \sin y e^{xy} \cdot y$$
$$= e^{xy} \sin y(y \cos x - \sin x),$$

and

$$f_y(x, y) = e^{xy} \cos x(x \sin y + \cos y).$$

EXAMPLE 4. Given that $z = x \arctan (y/x)$, find $\partial z/\partial x$ and $\partial z/\partial y$.

Solution. We have

$$\frac{\partial z}{\partial x} = x \cdot \frac{-(y/x^2)}{1 + (y^2/x^2)} + \arctan (y/x)$$
$$= \arctan (y/x) - \frac{xy}{x^2 + y^2},$$

and

$$\frac{\partial z}{\partial y} = x \frac{1/x}{1 + (y^2/x^2)}$$
$$= \frac{x^2}{x^2 + y^2}.$$

EXAMPLE 5. Given $f(x, y) = x^2 - 3xy + 2x - 3y + 5$, find $f_{,1}(2, 3)$.

Solution. We have

$$f_{,1}(x, y) = 2x - 3y + 2, \qquad f_{,1}(2, 3) = -3.$$

EXERCISES

In each of problems 1 through 12, find $f_{,1}(x, y)$ and $f_{,2}(x, y)$.

1. $f(x, y) = x^2 + 2xy^2 - 2x$ 2. $f(x, y) = x^3 y^2 + x^2 y^3 + 3$

3. $f(x, y) = x^3 y - 3x^2 y^2 + 2xy$ 4. $f(x, y) = \sqrt{x^2 + 1} + y^3$

5. $f(x, y) = \sqrt{x^2 + y^2}$ 6. $f(x, y) = \dfrac{xy}{x^2 + y^2}$

7. $f(x, y) = \ln(x^2 + y^2)$ 8. $f(x, y) = \ln \sqrt{x^2 + y^2}$

9. $f(x, y) = \arctan \dfrac{y}{x}$ 10. $f(x, y) = \arcsin \dfrac{x}{1 + y}$

11. $f(x, y) = xye^{x^2 + y^2}$ 12. $f(x, y) = \cos(xe^y)$

In each of problems 13 through 18, find $f_{,1}$ and $f_{,2}$ at the values indicated.

13. $f(x, y) = x \arcsin(x - y)$, $x = 1$, $y = 2$

14. $f(u, v) = e^{uv} \sec \left(\dfrac{u}{v} \right)$, $u = v = 3$

15. $f(x, z) = e^{\sin x} \tan xz$, $x = \dfrac{\pi}{4}$, $z = 1$

16. $f(t, u) = \dfrac{\cos 2tu}{t^2 + u^2}$, $t = 0$, $u = 1$

17. $f(y, x) = x^{xy}$, $x = y = 2$ 18. $f(s, t) = s^t + t^s$, $s = t = 3$

In each of problems 19 through 22, find $f_{,1}(x, y, z)$, $f_{,2}(x, y, z)$ and $f_{,3}(x, y, z)$.

19. $f(x, y, z) = x^2 y - 2x^2 z + 3xyz - y^2 z + 2xz^2$

20. $f(x, y, z) = \dfrac{xyz}{x^2 + y^2 + z^2}$ 21. $f(x, y, z) = e^{xyz} \sin xy \cos 2xz$

22. $f(x, y, z) = x^2 + z^2 + y^3 + 2x - 3y + 4z$

In problems 23 through 26 find in each case the indicated partial derivative.

23. $w = \ln \left(\dfrac{xy}{x^2 + y^2} \right)$; $\dfrac{\partial w}{\partial x}, \dfrac{\partial w}{\partial y}$

24. $w = (r^2 + s^2 + t^2) \cosh rst$; $\dfrac{\partial w}{\partial r}, \dfrac{\partial w}{\partial t}$

25. $w = e^{\sin(y/x)}$; $\dfrac{\partial w}{\partial y}, \dfrac{\partial w}{\partial x}$

26. $w = (\sec tu) \arcsin tv$; $\dfrac{\partial w}{\partial t}, \dfrac{\partial w}{\partial u}, \dfrac{\partial w}{\partial v}$

2. IMPLICIT DIFFERENTIATION

An equation involving $x, y,$ and z establishes a relation among the variables. If we can solve for z in terms of x and y, then we may have one or more functions determined by the relation. For example, the equation

$$2x^2 + y^2 + z^2 - 16 = 0 \tag{1}$$

may be solved for z to give

$$z = \pm\sqrt{16 - 2x^2 - y^2}. \tag{2}$$

If one or more functions are determined by a relation, it is possible to compute partial derivatives implicitly in a way that is completely similar to the methods used for ordinary derivatives. (See *First Course*, Chapter 6, page 142.)

For example, in Equation (1) above, considering x and y as independent variables with z as the dependent variable, we can compute $\partial z/\partial x$ directly from (1) without resorting to (2). We keep y fixed and, in (1), differentiate implicitly with respect to x, getting

$$4x + 2z\frac{\partial z}{\partial x} = 0 \quad \text{and} \quad \frac{\partial z}{\partial x} = -\frac{2x}{z}.$$

Further examples exhibit the method.

EXAMPLE 1. Suppose that x, y, and z are variables and that z is a function of x and y which satisfies

$$x^3 + y^3 + z^3 + 3xyz = 5.$$

Find $\partial z/\partial x$ and $\partial z/\partial y$.

Solution. Holding y constant and differentiating z with respect to x implicitly, we obtain

$$3x^2 + 3z^2\frac{\partial z}{\partial x} + 3xy\frac{\partial z}{\partial x} + 3yz = 0.$$

Therefore

$$\frac{\partial z}{\partial x} = -\frac{x^2 + yz}{xy + z^2}.$$

Holding x constant and differentiating with respect to y, we get

$$3y^2 + 3z^2\frac{\partial z}{\partial y} + 3xy\frac{\partial z}{\partial y} + 3xz = 0$$

and

$$\frac{\partial z}{\partial y} = -\frac{y^2 + xz}{xy + z^2}.$$

The same technique works with equations relating four or more variables, as the next example shows.

EXAMPLE 2. If r, s, t, and w are variables and if w is a function of r, s, and t which satisfies

$$e^{rt} - 2se^w + wt - 3w^2r = 5,$$

find $\partial w/\partial r$, $\partial w/\partial s$, and $\partial w/\partial t$.

Solution. To find $\partial w/\partial r$, we keep s and t fixed and differentiate implicitly with respect to r. The result is

$$te^{rt} - 2se^w \frac{\partial w}{\partial r} + t \frac{\partial w}{\partial r} - 3w^2 - 6rw \frac{\partial w}{\partial r} = 0$$

or

$$\frac{\partial w}{\partial r} = \frac{3w^2 - te^{rt}}{t - 2se^w - 6rw}.$$

Keeping r and t fixed, we obtain

$$-2e^w - 2se^w \frac{\partial w}{\partial s} + t \frac{\partial w}{\partial s} - 6rw \frac{\partial w}{\partial s} = 0$$

and

$$\frac{\partial w}{\partial s} = \frac{2e^w}{t - 2se^w - 6rw}.$$

Similarly, with r and s fixed the result is

$$re^{rt} - 2se^w \frac{\partial w}{\partial t} + w + t \frac{\partial w}{\partial t} - 6rw \frac{\partial w}{\partial t} = 0$$

or

$$\frac{\partial w}{\partial t} = \frac{w + re^{rt}}{2se^w - t + 6rw}.$$

EXERCISES

In each of problems 1 through 12, assume that w is a function of all other variables. Find the partial derivatives as indicated in each case.

1. $3x^2 + 2y^2 + 6w^2 - x + y - 12 = 0;$ $\quad \dfrac{\partial w}{\partial x}, \dfrac{\partial w}{\partial y}$

2. $x^2 + y^2 + w^2 + 3xy - 2xw + 3yw = 36;$ $\quad \dfrac{\partial w}{\partial x}, \dfrac{\partial w}{\partial y}$

3. $x^2 - 2xy + 2xw + 3y^2 + w^2 = 21;$ $\quad \dfrac{\partial w}{\partial x}, \dfrac{\partial w}{\partial y}$

4. $x^2y - x^2w - 2xy^2 - yw^2 + w^3 = 7;$ $\quad \dfrac{\partial w}{\partial x}, \dfrac{\partial w}{\partial y}$

5. $w - (r^2 + s^2) \cosh rw = 0;$ $\quad \dfrac{\partial w}{\partial r}, \dfrac{\partial w}{\partial s}$

6. $w - e^{w \sin(y/x)} = 1;$ $\quad \dfrac{\partial w}{\partial x}, \dfrac{\partial w}{\partial y}$

7. $e^{xyw} \sin xy \cos 2xw - 4 = 0;$ $\quad \dfrac{\partial w}{\partial x}, \dfrac{\partial w}{\partial y}$

8. $w^2 - 3xw - \ln \left(\dfrac{xy}{x^2 + y^2} \right) = 0;$ $\quad \dfrac{\partial w}{\partial x}, \dfrac{\partial w}{\partial y}$

9. $xyz + x^2z + xzw - yzw + yz^2 - w^3 = 3;$ $\quad \dfrac{\partial w}{\partial x}, \dfrac{\partial w}{\partial y}, \dfrac{\partial w}{\partial z}$

10. $r^2 + 3s^2 - 2t^2 + 6tw - 8w^2 + 12sw^3 = 4;$ $\dfrac{\partial w}{\partial r}, \dfrac{\partial w}{\partial s}, \dfrac{\partial w}{\partial t}$

11. $we^{zw} - ye^{yw} + e^{xy} = 1;$ $\dfrac{\partial w}{\partial x}, \dfrac{\partial w}{\partial y}$

12. $x^3 + 3x^2 y + 2z^2 t - 4zt^3 + 7xw - 8yw^2 + w^4 = 5;$ $\dfrac{\partial w}{\partial x}, \dfrac{\partial w}{\partial y}, \dfrac{\partial w}{\partial z}, \dfrac{\partial w}{\partial t}$

3. THE CHAIN RULE

The Chain Rule is one of the most effective devices for calculating ordinary derivatives. (See *First Course*, Chapter 6, page 133.) In this section we show how to extend the Chain Rule for the computation of partial derivatives. The basis of the Rule in the case of functions of one variable is the Fundamental Lemma on Differentiation, which we now recall.

Theorem 1. *If F has a derivative at a value a so that $F'(a)$ exists, then*

$$F(a + h) - F(a) = [F'(a) + G(h)]h$$

where $G(h)$ tends to zero as h tends to zero and $G(0) = 0$.

The proof of this theorem is on page 133 of *First Course*.
The above theorem has a natural generalization for functions of two variables.

Theorem 2. (Fundamental Lemma on Differentiation.) *Suppose that f is a continuous function of two variables (say x and y) and that $f_{,1}$ and $f_{,2}$ are continuous at (x_0, y_0). Then there are two functions, $G_1(h, k)$ and $G_2(h, k)$ continuous at $(0, 0)$ with $G_1(0, 0) = G_2(0, 0) = 0$, such that*

$$f(x_0 + h, y_0 + k) - f(x_0, y_0) = f_{,1}(x_0, y_0)h + f_{,2}(x_0, y_0)k$$
$$+ G_1(h, k)h + G_2(h, k)k. \quad (1)$$

Proof. The proof depends on writing the left side of (1) in a more complicated way:

$$f(x_0 + h, y_0 + k) - f(x_0, y_0) = [f(x_0 + h, y_0 + k) - f(x_0 + h, y_0)]$$
$$+ [f(x_0 + h, y_0) - f(x_0, y_0)]. \quad (2)$$

Figure 4–2 shows the points at which f is evaluated in (2) (h and k are taken to be positive in the figure). We apply the Theorem of the Mean to each of the quantities in brackets in (2) above. The results are

$$\frac{f(x_0 + h, y_0 + k) - f(x_0 + h, y_0)}{(y_0 + k) - y_0} = f_{,2}(x_0 + h, \eta) \quad (3)$$

and

$$\frac{f(x_0 + h, y_0) - f(x_0, y_0)}{(x_0 + h) - x_0} = f_{,1}(\xi, y_0), \quad (4)$$

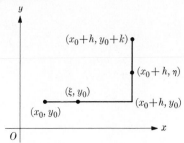

FIGURE 4–2

where η is between y_0 and $y_0 + k$ and ξ is between x_0 and $x_0 + h$. Typical locations for ξ and η are shown in Fig. 4–2. Substituting (3) and (4) into the right side of (2), we find

$$f(x_0 + h, y_0 + k) - f(x_0, y_0) = f_{,1}(\xi, y_0)h + f_{,2}(x_0 + h, \eta)k. \qquad (5)$$

The quantities G_1 and G_2 are defined by

$$G_1 = f_{,1}(\xi, y_0) - f_{,1}(x_0, y_0),$$
$$G_2 = f_{,2}(x_0 + h, \eta) - f_{,2}(x_0, y_0).$$

Multiplying the expression for G_1 by h and that for G_2 by k and inserting the result in the right side of (5), we obtain the statement of the theorem. Since $\xi \to x_0$ as $h \to 0$ and $\eta \to y_0$ as $k \to 0$, it follows (since $f_{,1}$ and $f_{,2}$ are continuous) that G_1 and G_2 tend to zero as h and k tend to zero. Thus G_1 and G_2 are continuous at $(0, 0)$ if we define them as being equal to zero there.

Theorem 3. (Chain Rule.) *Suppose that $z = f(x, y)$ is continuous and that $f_{,1}$, $f_{,2}$ are continuous. Assume that $x = x(r, s)$ and $y = y(r, s)$ are functions of r and s such that $x_{,1}, x_{,2}, y_{,1}, y_{,2}$ all exist. Then z is a function of r and s and the following formulas hold:*

$$\left. \begin{aligned} \frac{\partial f}{\partial r} &= \left(\frac{\partial f}{\partial x}\right)\left(\frac{\partial x}{\partial r}\right) + \left(\frac{\partial f}{\partial y}\right)\left(\frac{\partial y}{\partial r}\right) \\ \frac{\partial f}{\partial s} &= \left(\frac{\partial f}{\partial x}\right)\left(\frac{\partial x}{\partial s}\right) + \left(\frac{\partial f}{\partial y}\right)\left(\frac{\partial y}{\partial s}\right) \end{aligned} \right\} \qquad (6)$$

Proof. The first formula will be established; the second is proved similarly. We use the Δ notation. A change Δr in r induces a change Δx in x and a change Δy in y. That is,

$$\Delta x = x(r + \Delta r, s) - x(r, s), \qquad \Delta y = y(r + \Delta r, s) - y(r, s).$$

The function f, thought of as a function of r and s, has the partial derivatives

$$\frac{\partial f}{\partial r} = \lim_{\Delta r \to 0} \frac{\Delta f}{\Delta r}$$

where Δf, the change in f due to the change Δr in r, is given by

$$\Delta f = f(x + \Delta x, y + \Delta y) - f(x, y).$$

The Fundamental Lemma on differentiation with $h = \Delta x$ and $k = \Delta y$ reads

$$\Delta f = \frac{\partial f}{\partial x}\Delta x + \frac{\partial f}{\partial y}\Delta y + G_1 \Delta x + G_2 \Delta y,$$

where we have changed notation by using $\partial f/\partial x$ in place of $f_{,1}(x, y)$ and $\partial f/\partial y$ for $f_{,2}(x, y)$. We divide the equation above by Δr:

$$\frac{\Delta f}{\Delta r} = \frac{\partial f}{\partial x}\frac{\Delta x}{\Delta r} + \frac{\partial f}{\partial y}\frac{\Delta y}{\Delta r} + G_1 \frac{\Delta x}{\Delta r} + G_2 \frac{\Delta y}{\Delta r}.$$

Letting Δr tend to zero and remembering that $G_1 \to 0$, $G_2 \to 0$, we obtain the desired formula.

Remarks. (i) The formulas (6) may be written in various notations. Two common expressions are

$$\left.\begin{aligned} \frac{\partial z}{\partial r} &= \left(\frac{\partial z}{\partial x}\right)\left(\frac{\partial x}{\partial r}\right) + \left(\frac{\partial z}{\partial y}\right)\left(\frac{\partial y}{\partial r}\right) \\ \frac{\partial z}{\partial s} &= \left(\frac{\partial z}{\partial x}\right)\left(\frac{\partial x}{\partial s}\right) + \left(\frac{\partial z}{\partial y}\right)\left(\frac{\partial y}{\partial s}\right) \end{aligned}\right\} \tag{7}$$

and

$$\left.\begin{aligned} f_r &= f_x x_r + f_y y_r \\ f_s &= f_x x_s + f_y y_s \end{aligned}\right\} \tag{8}$$

(ii) To use our preferred notation we introduce the symbol $g(r, s)$ to represent the function f considered as a function of r and s. The formulas expressing the Chain Rule are then

$$\left.\begin{aligned} g_{,1}(r, s) &= f_{,1}(x, y)x_{,1}(r, s) + f_{,2}(x, y)y_{,1}(r, s) \\ g_{,2}(r, s) &= f_{,1}(x, y)x_{,2}(r, s) + f_{,2}(x, y)y_{,2}(r, s) \end{aligned}\right\} \tag{9}$$

(iii) For functions of one variable, the Chain Rule is easily remembered as the rule which allows us to think of derivatives as fractions. The formula

$$\frac{dy}{dx} = \frac{dy}{du}\cdot\frac{du}{dx}$$

is an example. The symbol du has a meaning of its own. To attempt to draw such an analogy with the Chain Rule for Partial Derivatives leads to disaster. Formulas (6) are the ones we usually employ in the applications. The parentheses around the individual terms are used to indicate the inseparable nature of each item. Actually, the forms (8) and (9) for the same formulas avoid the danger of treating partial derivatives as fractions.

EXAMPLE 1.   Suppose that $f(x, y) = x^3 + y^3$, $x = 2r + s$, $y = 3r - 2s$.   Find $\partial f / \partial r$ and $\partial f / \partial s$.

Solution. We can employ the Chain Rule and obtain

$$\frac{\partial f}{\partial x} = 3x^2, \qquad \frac{\partial f}{\partial y} = 3y^2$$

$$\frac{\partial x}{\partial r} = 2, \qquad \frac{\partial x}{\partial s} = 1, \qquad \frac{\partial y}{\partial r} = 3, \qquad \frac{\partial y}{\partial s} = -2.$$

Therefore

$$\frac{\partial f}{\partial r} = (3x^2)(2) + (3y^2)(3) = 6x^2 + 9y^2 = 6(2r + s)^2 + 9(3r - 2s)^2,$$

$$\frac{\partial f}{\partial s} = (3x^2)(1) + (3y^2)(-2) = 3x^2 - 6y^2 = 3(2r + s)^2 - 6(3r - 2s)^2.$$

In Theorem 3 (the Chain Rule), the variables r and s are independent variables; we denote the variables x and y **intermediate variables** and, of course, z is the dependent variable.   The formulas we derived extend easily to any number of independent variables and any number of intermediate variables.   For example, if

$$w = f(x, y, z)$$

and if

$$x = x(r, s), \quad y = y(r, s), \quad z = z(r, s),$$

then

$$\frac{\partial f}{\partial r} = \left(\frac{\partial f}{\partial x}\right)\left(\frac{\partial x}{\partial r}\right) + \left(\frac{\partial f}{\partial y}\right)\left(\frac{\partial y}{\partial r}\right) + \left(\frac{\partial f}{\partial z}\right)\left(\frac{\partial z}{\partial r}\right),$$

and there is a similar formula for $\partial f / \partial s$. The case of four intermediate variables and one independent variable—that is,

$$w = f(x, y, u, v), \quad x = x(t), \quad y = y(t), \quad u = u(t), \quad v = v(t)$$

—leads to the formula

$$\frac{df}{dt} = \frac{\partial f}{\partial x}\frac{dx}{dt} + \frac{\partial f}{\partial y}\frac{dy}{dt} + \frac{\partial f}{\partial u}\frac{du}{dt} + \frac{\partial f}{\partial v}\frac{dv}{dt}.$$

The ordinary d is used for derivatives with respect to t, since w, x, y, u, and v are all functions of the single variable t.

Remark. As an aid in remembering the Chain Rule, we note that *there are as many terms in the formula as there are intermediate variables.*

EXAMPLE 2.   If $z = 2x^2 + xy - y^2 + 2x - 3y + 5$, $x = 2s - t$, $y = s + t$, find $\partial z / \partial t$.

Solution. We use the Chain Rule:

$$\frac{\partial z}{\partial x} = 4x + y + 2, \quad \frac{\partial z}{\partial y} = x - 2y - 3, \quad \frac{\partial x}{\partial t} = -1, \quad \frac{\partial y}{\partial t} = 1.$$

Therefore

$$\frac{\partial z}{\partial t} = (4x + y + 2)(-1) + (x - 2y - 3)(1) = -3x - 3y - 5$$
$$= -3(2s - t) - 3(s + t) - 5$$
$$= -9s - 5.$$

EXAMPLE 3. Given $w = x^2 + 3y^2 - 2z^2 + 4x - y + 3z - 1$, $x = t^2 - 2t + 1$, $y = 3t - 2$, $z = t^2 + 4t - 3$, find dw/dt when $t = 2$.

Solution. Employing the Chain Rule, we find

$$\frac{\partial w}{\partial x} = 2x + 4, \quad \frac{\partial w}{\partial y} = 6y - 1, \quad \frac{\partial w}{\partial z} = -4z + 3,$$
$$\frac{dx}{dt} = 2t - 2, \quad \frac{dy}{dt} = 3, \quad \frac{dz}{dt} = 2t + 4.$$

Therefore

$$\frac{dw}{dt} = (2x + 4)(2t - 2) + (6y - 1)(3) + (-4z + 3)(2t + 4).$$

When $t = 2$, we have $x = 1$, $y = 4$, $z = 9$, and so

$$\frac{dw}{dt} = (6)(2) + (23)(3) + (-33)(8) = -183.$$

EXERCISES

In each of problems 1 through 12, use the Chain Rule to obtain the indicated partial derivative.

1. $f(x, y) = x^2 + y^2$; $x = s - 2t$, $y = 2s + t$; $\quad \dfrac{\partial f}{\partial s}, \dfrac{\partial f}{\partial t}$

2. $f(x, y) = x^2 - xy - y^2$; $x = s + t$, $y = -s + t$; $\quad \dfrac{\partial f}{\partial s}, \dfrac{\partial f}{\partial t}$

3. $f(x, y) = x^2 + y^2$; $x = s^2 - t^2$, $y = 2st$; $\quad \dfrac{\partial f}{\partial s}, \dfrac{\partial f}{\partial t}$

4. $f(x, y) = \dfrac{x}{x^2 + y^2}$; $x = s \cos t$, $y = s \sin t$; $\quad \dfrac{\partial f}{\partial s}, \dfrac{\partial f}{\partial t}$

5. $f(x, y) = \dfrac{x}{\sqrt{x^2 + y^2}}$; $x = 2s - t$, $y = s + 2t$; $\quad \dfrac{\partial f}{\partial s}, \dfrac{\partial f}{\partial t}$

6. $f(x, y) = e^x \cos y$; $x = s^2 - t^2$, $y = 2st$; $\quad \dfrac{\partial f}{\partial s}, \dfrac{\partial f}{\partial t}$

7. $f(x, y, z) = x^2 + y^2 + z^2 + 3xy - 2xz + 4$; $x = 3s + t$, $y = 2s - t$, $z = s + 2t$; $\quad \dfrac{\partial f}{\partial s}, \dfrac{\partial f}{\partial t}$

8. $f(x, y, z) = x^3 + 2y^3 + z^3;\ x = s^2 - t^2,\ y = s^2 + t^2,\ z = 2st;\quad \dfrac{\partial f}{\partial s}, \dfrac{\partial f}{\partial t}$

9. $f(x, y, z) = x^2 - y^2 + 2z^2;\ x = r^2 + 1,\ y = r^2 - 2r + 1,\ z = r^2 - 2;\ \dfrac{df}{dr}$

10. $f(x, y) = \dfrac{x - y}{1 + x^2 + y^2};\ x = r + 3s - t,\ y = r - 2s + 3t;\quad \dfrac{\partial f}{\partial r}, \dfrac{\partial f}{\partial s}, \dfrac{\partial f}{\partial t}$

11. $f(u) = u^3 + 2u^2 - 3u + 1;\ u = r^2 - s^2 + t^2;\quad \dfrac{\partial f}{\partial r}, \dfrac{\partial f}{\partial s}, \dfrac{\partial f}{\partial t}$

12. $f(x, y, u, v) = x^2 + y^2 - u^2 - v^2 + 3x - 2y + u - v;$

$x = 2r + s - t,\ y = r - 2s + t,\ u = 3r - 2s + t,\ v = r - s - t;$

$\dfrac{\partial f}{\partial r}, \dfrac{\partial f}{\partial s}, \dfrac{\partial f}{\partial t}$

In problems 13 through 20, use the Chain Rule to find the indicated derivatives at the values given.

13. $z = x^2 - y^2;\ x = r \cos \theta,\ y = r \sin \theta;\quad \dfrac{\partial z}{\partial r}, \dfrac{\partial z}{\partial \theta}$ where $r = \sqrt{2},\ \theta = \dfrac{\pi}{4}$

14. $w = x^2 + y^2 - z^2;\ x = 1 - t^2,\ y = 2t + 3,\ z = t^2 + t,\ \dfrac{dw}{dt}$ where $t = -1$

15. $w = xy + yz + zx;\ x = t \cos t,\ y = t \sin t,\ z = t;\quad \dfrac{dw}{dt}$ where $t = \dfrac{\pi}{4}$

16. $z = u^3 + 2u - 3;\ u = s^2 + t^2 - 4;\quad \dfrac{\partial z}{\partial s}, \dfrac{\partial z}{\partial t}$ where $s = 1$ and $t = 2$

17. $z = \dfrac{xy}{x^2 + y^2};\ x = r \cos \theta,\ y = r \sin \theta;\quad \dfrac{\partial z}{\partial r}, \dfrac{\partial z}{\partial \theta}$ where $r = 3,\ \theta = \dfrac{\pi}{6}$

18. $f = e^{xyz} \cos xyz;\ x = r^2 + t^2,\ y = t^2 + r,\ z = r + t;$

$\dfrac{\partial f}{\partial x}, \dfrac{\partial f}{\partial y}, \dfrac{\partial f}{\partial z},\ r = 2,\ t = -2$

19. $w = x^3 + y^3 + z^3 - u^2 - v^2;\ x = r^2 + s^2 + t^2,\ y = r^2 + s^2 - t^2,$

$z = r^2 - s^2 - t^2,\ u = r^2 + t^2,\ v = r^2 - s^2;$

$\dfrac{\partial w}{\partial s}, \dfrac{\partial w}{\partial t}$, where $r = 1,\ s = 0,\ t = -1$

20. $w = x^4 - y^4 - z^4;\ x = 5r + 3s - 2t + u - v,$

$y = 2r - 4s + t - u^2 + v^2;\ z = s^3 - 2t^2 + 3v^2;$

$\dfrac{\partial w}{\partial s}, \dfrac{\partial w}{\partial v}$ where $r = 1,\ s = -1,\ t = 0,\ u = 3,\ v = -2$

4. APPLICATIONS OF THE CHAIN RULE

The Chain Rule may be employed profitably in many types of applications. These are best illustrated with examples, and we shall begin with two problems in related rates, a topic discussed in *First Course* (Chapter 7, page 193).

EXAMPLE 1. At a certain instant the altitude of a right circular cone is 30 in. and is increasing at the rate of 2 in./sec. At the same instant, the radius of the base is 20 in. and is increasing at the rate of 1 in./sec. At what rate is the volume increasing at that instant? (See Fig. 4–3.)

Solution. The volume V is given by

$$V = \tfrac{1}{3}\pi r^2 h,$$

with r and h functions of the time t. We can apply the Chain Rule to obtain

$$\frac{dV}{dt} = \frac{\partial V}{\partial r}\frac{dr}{dt} + \frac{\partial V}{\partial h}\frac{dh}{dt}$$

$$= \frac{2}{3}\pi rh\frac{dr}{dt} + \frac{1}{3}\pi r^2 \frac{dh}{dt}.$$

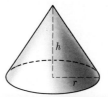

FIGURE 4–3

At the given instant,

$$\frac{dV}{dt} = \frac{2}{3}\pi(20)(30)(1) + \frac{1}{3}\pi(20)^2(2) = \frac{2000\pi}{3}\text{ in}^3/\text{sec}.$$

EXAMPLE 2. The base B of a trapezoid increases in length at the rate of 2 in./sec and the base b decreases in length at the rate of 1 in./sec. If the altitude h is increasing at the rate of 3 in./sec, how rapidly is the area A changing when $B = 30$ in., $b = 50$ in., and $h = 10$ in.? (See Fig. 4–4.)

Solution. The area A is given by

$$A = \tfrac{1}{2}(B + b)h,$$

with B, b, and h functions of time. We apply the Chain Rule to get

$$\frac{dA}{dt} = \frac{\partial A}{\partial B}\frac{dB}{dt} + \frac{\partial A}{\partial b}\frac{db}{dt} + \frac{\partial A}{\partial h}\frac{dh}{dt}$$

$$= \frac{1}{2}h\frac{dB}{dt} + \frac{1}{2}h\frac{db}{dt} + \frac{1}{2}(B + b)\frac{dh}{dt}$$

$$= (5)(2) + (5)(-1) + (40)(3) = 125\text{ in}^2/\text{sec}.$$

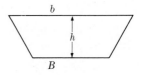

FIGURE 4–4

Note that since b is decreasing, db/dt is negative.

———

The next example shows that a clear understanding of the symbolism in partial differentiation is required in many applications.

———

EXAMPLE 3. Suppose that $z = f(x + at)$ and a is constant. Show that

$$\frac{\partial z}{\partial t} = a\frac{\partial z}{\partial x}.$$

Solution. We observe that f is a function of *one argument* (in which, however, two variables occur in a particular combination). We let

$$u = x + at$$

and, if we now write

$$z = f(u), \qquad u = x + at,$$

we recognize the applicability of the Chain Rule. Therefore

$$\frac{\partial z}{\partial x} = \frac{dz}{du}\frac{\partial u}{\partial x} = f'(u)\cdot 1,$$

$$\frac{\partial z}{\partial t} = \frac{dz}{du}\frac{\partial u}{\partial t} = f'(u)\cdot a.$$

We conclude that

$$\frac{\partial z}{\partial t} = a\,\frac{\partial z}{\partial x}.$$

EXAMPLE 4. An airplane is traveling directly east at 300 mi/hr and is climbing at the rate of 600 ft/min. At a certain instant, the airplane is 12,000 ft above ground and 5 mi directly west of an observer on the ground. How fast is the distance changing between the airplane and the observer at this instant?

Solution. Referring to Fig. 4–5, with the observer at O and the airplane at A, we see that x, y, and s are functions of the time t. The distance s between the airplane and the observer is given by

$$s = (x^2 + y^2)^{1/2},$$

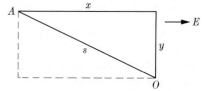

and we wish to find ds/dt. Using the Chain Rule, we get

$$\frac{ds}{dt} = \frac{\partial s}{\partial x}\frac{dx}{dt} + \frac{\partial s}{\partial y}\frac{dy}{dt}$$

$$= \frac{x}{\sqrt{x^2 + y^2}}\frac{dx}{dt} + \frac{y}{\sqrt{x^2 + y^2}}\frac{dy}{dt}.$$

FIGURE 4–5

From the given data we see that, at the instant in question,

$$y = 12{,}000, \; x = 26{,}400, \; \frac{dx}{dt} = -440 \text{ ft/sec and } \frac{dy}{dt} = +10 \text{ ft/sec.}$$

Therefore

$$\frac{ds}{dt} = \frac{26{,}400}{1000\sqrt{840.96}}(-440) + \frac{12{,}000}{1000\sqrt{840.96}}(10)$$

$$= -\frac{(264)(44)}{\sqrt{840.96}} + \frac{120}{\sqrt{840.96}} = -396^{+} \text{ ft/sec.}$$

The negative sign indicates that the airplane is approaching the observer.

EXERCISES

1. Find the rate at which the lateral area of the cone in Example 1 is increasing at the given instant.

2. At a certain instant a right circular cylinder has radius of base 10 in. and altitude 15 in. At this instant the radius is decreasing at the rate of 5 in./sec and the altitude is increasing at the rate of 4 in./sec. How rapidly is the volume changing at this moment?

3. A gas obeys the law $pv = RT(R = \text{const})$. At a certain instant while the gas is being compressed, $v = 15$ ft^3, $p = 25$ lb/in^2, v is decreasing at the rate of 3 ft^3/min, and p is increasing at the rate of $6\frac{2}{3}$ lb/in^2/min. Find dT/dt. (Answer in terms of R.)

4. (a) In problem 2, find how rapidly the lateral surface area of the cylinder is changing at the same instant. (b) What would the result be for the area A consisting of the top and bottom of the cylinder as well as the lateral surface?

5. At a certain instant of time, the angle A of a triangle ABC is 60° and increasing at the rate of 5°/sec, the side $\overline{AB}$ is 10 in. and increasing at the rate of 1 in./sec, and side $\overline{AC}$ is 16 in. and decreasing at the rate of $\frac{1}{2}$ in./sec. Find the rate of change of side $\overline{BC}$.

6. A point moves along the surface

$$z = x^2 + 2y^2 - 3x + y$$

in such a way that $dx/dt = 3$ and $dy/dt = 2$. Find how z changes with time when $x = 1$, $y = 4$.

7. Water is leaking out of a conical tank at the rate of 0.5 ft^3/min. The tank is also stretching in such a way that, while it remains conical, the distance across the top at the water surface is increasing at the rate of 0.2 ft/min. How fast is the height h of water changing at the instant when $h = 10$ and the volume of water is 75 cu. ft?

8. A rectangular bin is changing in size in such a way that its length is increasing at the rate of 3 in./sec, its width is decreasing at the rate of 2 in./sec, and its height is increasing at the rate of 1 in./sec. (a) How fast is the volume changing at the instant when the length is 15, the width is 10, and the height is 8? (b) How fast is the total surface area changing at the same instant?

9. (a) Given $z = f(y/x)$. Find $\partial z/\partial x$ and $\partial z/\partial y$ in terms of $f'(y/x)$ and x and y. [Hint. Let $u = y/x$.] (b) Show that $x\,(\partial z/\partial x) + y\,(\partial z/\partial y) = 0$.

10. (a) Given that $w = f(y - x - t, z - y + t)$. By letting $u = y - x - t$, $v = z - y + t$, find $\partial w/\partial x, \partial w/\partial y, \partial w/\partial z, \partial w/\partial t$ in terms of $f_{,1}$ and $f_{,2}$. (b) Show that

$$\frac{\partial w}{\partial x} + 2\frac{\partial w}{\partial y} + \frac{\partial w}{\partial z} + \frac{\partial w}{\partial t} = 0.$$

11. Suppose that $z = f(x, y)$ and $x = r\cos\theta$, $y = r\sin\theta$. (a) Express $\partial z/\partial r$ and $\partial z/\partial\theta$ in terms of $\partial z/\partial x$ and $\partial z/\partial y$. (b) Show that

$$\left(\frac{\partial z}{\partial r}\right)^2 + \frac{1}{r^2}\left(\frac{\partial z}{\partial\theta}\right)^2 = \left(\frac{\partial z}{\partial x}\right)^2 + \left(\frac{\partial z}{\partial y}\right)^2.$$

12. Suppose that $z = f(x, y)$, $x = e^s \cos t$, $y = e^s \sin t$. Show that

$$\left(\frac{\partial z}{\partial s}\right)^2 + \left(\frac{\partial z}{\partial t}\right)^2 = e^{2s}\left[\left(\frac{\partial z}{\partial x}\right)^2 + \left(\frac{\partial z}{\partial y}\right)^2\right].$$

13. Suppose that $u = f(x + at, y + bt)$, with a and b constants. Show that

$$\frac{\partial u}{\partial t} = a\frac{\partial u}{\partial x} + b\frac{\partial u}{\partial y}.$$

14. Given that

$$f(x, y) = \frac{x + y}{x^2 - xy + y^2},$$

show that

$$xf_{,1} + yf_{,2} = -f.$$

15. Given that $f(x, y) = x^2 - y^2 + xy \ln(y/x)$, show that $xf_{,1} + yf_{,2} = 2f$.

16. If in Example 4 a second observer is situated at a point O', 12 miles west of the one at O, find the rate of change of the distance between A and O' at the same instant.

17. If in Example 2 one of the acute angles is held constant at $60°$, find the rate of change of the perimeter of the trapezoid at the instant in question.

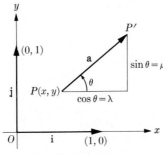

FIGURE 4–6

5. DIRECTIONAL DERIVATIVES. GRADIENT

The partial derivative of a function with respect to x may be considered as the derivative in the x-direction; the partial derivative with respect to y is the derivative in the y-direction. We now show how we may define the derivative in *any direction*. To see this, we consider a function $f(x, y)$ and a point $P(x, y)$ in the xy-plane. A particular direction is singled out by specifying the angle θ which a line through P makes with the positive x-axis (Fig. 4–6). We may also prescribe the direction by drawing the directed line segment $\overrightarrow{PP'}$ of *unit* length, as shown, and defining the vector **a** by the relation

$$\mathbf{a} = \lambda\mathbf{i} + \mu\mathbf{j},$$

with $\lambda = \cos\theta$, $\mu = \sin\theta$, and **i** and **j** the customary unit vectors. We note that **a** is a vector of unit length. The vector **a** determines the same direction as the angle θ.

DEFINITION. Let f be a function of two variables. We define the **directional derivative** $D_a f$ **of** f **in the direction of a** by

$$D_a f(x, y) = \lim_{h \to 0} \frac{f(x + \lambda h, y + \mu h) - f(x, y)}{h}$$

whenever the limit exists.

Remark. We note that, when $\theta = 0$, then $\lambda = 1$, $\mu = 0$, and the direction is the positive x-direction. The directional derivative is exactly $\partial f / \partial x$. Similarly, if $\theta = \pi/2$, we have $\lambda = 0$, $\mu = 1$, and the directional derivative is $\partial f / \partial y$.

The working formula for directional derivatives is established in the next theorem.

Theorem 4. *If $f(x, y)$ and its partial derivatives are continuous and*

$$\mathbf{a} = (\cos \theta)\mathbf{i} + (\sin \theta)\mathbf{j},$$

then

$$D_a f(x, y) = f_{,1}(x, y) \cos \theta + f_{,2}(x, y) \sin \theta.$$

Proof. We define the function $g(s)$ by

$$g(s) = f(x + s \cos \theta, \ y + s \sin \theta),$$

in which we keep x, y, and θ fixed and allow s to vary. The Chain Rule now yields
$g'(s) = f_{,1}(x + s \cos \theta, \ y + s \sin \theta) \cos \theta + f_{,2}(x + s \cos \theta, \ y + s \sin \theta) \sin \theta$
and

$$g'(0) = f_{,1}(x, y) \cos \theta + f_{,2}(x, y) \sin \theta.$$

Since, by definition, $g'(0)$ is precisely $D_a f(x, y)$, the result is established.

EXAMPLE 1.   Given $f(x, y) = x^2 + 2y^2 - 3x + 2y$, find the directional derivative of f in the direction $\theta = \pi/6$. What is the value of this derivative at the point $(2, -1)$?

Solution. We compute
$$f_{,1} = 2x - 3, \qquad f_{,2} = 4y + 2.$$
Therefore
$$D_a f = (2x - 3) \cdot \tfrac{1}{2}\sqrt{3} + (4y + 2) \cdot \tfrac{1}{2}.$$

In particular, when $x = 2$ and $y = -1$, we obtain

$$D_a f = \tfrac{1}{2}\sqrt{3} - 1.$$

An alternate notation for directional derivative for functions of two variables is the symbol

$$d_\theta f(x, y),$$

in which θ is the angle the direction makes with the x-axis. For a fixed value of

x and y, the directional derivative is a function of θ. It is an ordinary problem in maxima and minima to find the value of θ which makes the directional derivative at a given point the largest or the smallest. The next example shows the method.

EXAMPLE 2. Given $f(x, y) = x^2 - xy - y^2$, find $d_\theta f(x, y)$ at the point $(2, -3)$. For what value of θ does $d_\theta(2, -3)$ take on its maximum value?

Solution. We have

$$f_{,1}(x, y) = 2x - y, \qquad f_{,2}(x, y) = -x - 2y.$$

Therefore for $x = 2, y = -3$,

$$d_\theta f(2, -3) = 7 \cos \theta + 4 \sin \theta.$$

To find the maximum of this function of θ we differentiate

$$k(\theta) = 7 \cos \theta + 4 \sin \theta$$

and set the derivative equal to zero. We get

$$k'(\theta) = -7 \sin \theta + 4 \cos \theta = 0 \qquad \text{or} \qquad \tan \theta = \tfrac{4}{7}.$$

The result is $\cos \theta = \pm 7/\sqrt{65}$, $\sin \theta = \pm 4/\sqrt{65}$, and

$$\tan \theta = \tfrac{4}{7}, \quad \theta = \begin{cases} 29°45' \text{ approximately,} \\ 209°45' \text{ approximately.} \end{cases}$$

It is clear by substitution that the first choice for θ makes $k(\theta)$ a maximum, while the second makes it a minimum.

The definition of directional derivative for functions of two variables has a natural extension to functions of three variables. In three dimensions, a direction is determined by a set of direction cosines λ, μ, ν or, equivalently, a vector

$$\mathbf{a} = \lambda \mathbf{i} + \mu \mathbf{j} + \nu \mathbf{k}.$$

We recall that $\lambda^2 + \mu^2 + \nu^2 = 1$, and so $\mathbf{a}$ is a **unit** vector.

DEFINITION. We define the **directional derivative $D_\mathbf{a}f$ of $f(x, y, z)$ in the direction of a** by

$$D_\mathbf{a}f(x, y, z) = \lim_{h \to 0} \frac{f(x + \lambda h, y + \mu h, z + \nu h) - f(x, y, z)}{h}$$

whenever the limit exists.

The proof of the next theorem is entirely analogous to the proof of Theorem 4.

Theorem 5. *If* $f(x, y, z)$ *and its partial derivatives are continuous and* $\mathbf{a} = \lambda\mathbf{i} + \mu\mathbf{j} + \nu\mathbf{k}$ *is a unit vector, then*

$$D_\mathbf{a}f(x, y, z) = \lambda f_{,1}(x, y, z) + \mu f_{,2}(x, y, z) + \nu f_{,3}(x, y, z).$$

EXAMPLE 3. Find the directional derivative of $f(x, y, z) = x^2 + y^2 + z^2 - 3xy + 2xz - yz$ at the point $(1, 2, -1)$.

Solution. We have

$$f_{,1}(x, y, z) = 2x - 3y + 2z; \quad f_{,2}(x, y, z) = 2y - 3x - z; \quad f_{,3}(x, y, z) = 2z + 2x - y.$$

Denoting the direction by $\mathbf{a} = \lambda\mathbf{i} + \mu\mathbf{j} + \nu\mathbf{k}$, we get

$$D_\mathbf{a}f(1, 2, -1) = -6\lambda + 2\mu - 2\nu.$$

EXAMPLE 4. Given the function

$$f(x, y, z) = xe^{yz} + ye^{xz} + ze^{xy},$$

find the directional derivative at $P(1, 0, 2)$ in the direction going from P to $P'(5, 3, 3)$.

Solution. A set of direction numbers for the line through P and P' is 4, 3, 1. The corresponding direction cosines are $4/\sqrt{26}$, $3/\sqrt{26}$, $1/\sqrt{26}$, which we denote by λ, μ, ν. The direction $\mathbf{a}$ is given by

$$\mathbf{a} = \frac{4}{\sqrt{26}}\mathbf{i} + \frac{3}{\sqrt{26}}\mathbf{j} + \frac{1}{\sqrt{26}}\mathbf{k}.$$

We find that

$$\begin{array}{lll} f_{,1}(x, y, z) = e^{yz} + yze^{xz} + zye^{xy} & \text{and} & f_{,1}(1, 0, 2) = 1, \\ f_{,2}(x, y, z) = xze^{yz} + e^{xz} + xze^{xy} & \text{and} & f_{,2}(1, 0, 2) = 4 + e^2, \\ f_{,3}(x, y, z) = xye^{yz} + xye^{xz} + e^{xy} & \text{and} & f_{,3}(1, 0, 2) = 1. \end{array}$$

Therefore

$$D_\mathbf{a}f(1, 0, 2) = \frac{4}{\sqrt{26}} + (4 + e^2)\frac{3}{\sqrt{26}} + \frac{1}{\sqrt{26}} = \frac{17 + 3e^2}{\sqrt{26}}.$$

As the next definition shows, the *gradient* of a function is a vector containing the partial derivatives of the function.

DEFINITIONS. (i) If $f(x, y)$ has partial derivatives, we define the **gradient vector**

$$\mathbf{grad}\, f(x, y) = f_{,1}(x, y)\mathbf{i} + f_{,2}(x, y)\mathbf{j}.$$

(ii) If $g(x, y, z)$ has partial derivatives, we define

$$\mathbf{grad}\, g(x, y, z) = g_{,1}(x, y, z)\mathbf{i} + g_{,2}(x, y, z)\mathbf{j} + g_{,3}(x, y, z)\mathbf{k}.$$

The symbol ∇, an inverted delta, is called "del" and is a common one used to denote the gradient. We will frequently write ∇f for **grad** f.

If **b** and **c** are two vectors, we recall that the scalar product **b** · **c** of **b** $= b_1\mathbf{i} + b_2\mathbf{j} + b_3\mathbf{k}$ and **c** $= c_1\mathbf{i} + c_2\mathbf{j} + c_3\mathbf{k}$ is given by

$$\mathbf{b} \cdot \mathbf{c} = b_1 c_1 + b_2 c_2 + b_3 c_3.$$

For two-dimensional vectors the result is the same, with $b_3 = c_3 = 0$.

We recognize that if **a** is a *unit* vector so that $\mathbf{a} = \lambda\mathbf{i} + \mu\mathbf{j} + \nu\mathbf{k}$, then we have the formula

$$D_{\mathbf{a}}f = \lambda f_{,1} + \mu f_{,2} + \nu f_{,3} = \mathbf{a} \cdot \nabla f.$$

Looked at another way, the scalar product of **a** and ∇f is given by

$$\mathbf{a} \cdot \nabla f = |\mathbf{a}|\,|\nabla f|\cos\phi = D_{\mathbf{a}}f,$$

where ϕ is the angle between the **a** and ∇f.

From the above formula we can conclude that $D_{\mathbf{a}}f$ *is a maximum when ϕ is zero*—i.e., when **a** *is in the direction of* the **grad** f.

EXAMPLE 5. Given the function

$$f(x, y, z) = x^3 + 2y^3 + z^3 - 4xyz,$$

find the maximum value of $D_{\mathbf{a}}f$ at the point $P = (-1, 1, 2)$.

Solution. We have

$$f_{,1} = 3x^2 - 4yz; \quad f_{,2} = 6y^2 - 4xz; \quad f_{,3} = 3z^2 - 4xy.$$

Therefore

$$\nabla f(-1, 1, 2) = -5\mathbf{i} + 14\mathbf{j} + 16\mathbf{k},$$

and a unit vector **a** in the direction of ∇f is

$$\mathbf{a} = -\frac{5}{3\sqrt{53}}\mathbf{i} + \frac{14}{3\sqrt{53}}\mathbf{j} + \frac{16}{3\sqrt{53}}\mathbf{k}.$$

The maximum value of $D_{\mathbf{a}}f$ is given by

$$D_{\mathbf{a}}f = -5\left(\frac{-5}{3\sqrt{53}}\right) + 14\left(\frac{14}{3\sqrt{53}}\right) + 16\left(\frac{16}{3\sqrt{53}}\right) = 3\sqrt{53}.$$

EXERCISES

In problems 1 through 6, find in each case $d_0 f(x, y)$ at the given point.

1. $f(x, y) = x^2 + y^2$; $(3, 4)$
2. $f(x, y) = x^3 + y^3 - 3x^2y - 3xy^2$; $(1, -2)$

3. $f(x, y) = \arctan(y/x);$ (4, 3)   4. $f(x, y) = \sin xy;$ $(2, \pi/4)$

5. $f(x, y) = e^x \cos y;$ $(0, \pi/3)$ 6. $f(x, y) = (\sin x)^{xy};$ $(\pi/2, 0)$

In each of problems 7 through 10 find the value of $d_\theta f(x, y)$ at the given point. Also, find the value of θ which makes $d_\theta f$ a maximum at this point. Express your answer in terms of $\sin \theta$ and $\cos \theta$.

7. $f(x, y) = x^2 + y^2 - 2x + 3y;$ $(2, -1)$ 8. $f(x, y) = \arctan(x/y);$ (3, 4)

9. $f(x, y) = e^x \sin y;$ $(0, \pi/6)$ 10. $f(x, y) = (\sin y)^{xy};$ $(0, \pi/2)$

In each of problems 11 through 14, find $D_\mathbf{a} f$ at the given point.

11. $f(x, y, z) = x^2 + xy - xz + y^2 - z^2;$ $(2, 1, -2)$

12. $f(x, y, z) = x^2 y + xze^y - xye^z;$ $(-2, 3, 0)$

13. $f(x, y, z) = \cos xy + \sin xz;$ $(0, 2, -1)$

14. $f(x, y, z) = \ln(x + y + z) - xyz;$ $(-1, 2, 1)$

In problems 15 through 18, in each case find $D_\mathbf{a} f$ at the given point P when $\mathbf{a}$ is the given unit vector.

15. $f(x, y, z) = x^2 + 2xy - y^2 + xz + z^2;$ $P(2, 1, 1);$ $\mathbf{a} = \frac{1}{3}\mathbf{i} - \frac{2}{3}\mathbf{j} + \frac{2}{3}\mathbf{k}$

16. $f(x, y, z) = x^2 y + xye^z - 2xze^y;$ $P(1, 2, 0);$ $\mathbf{a} = \frac{2}{7}\mathbf{i} - \frac{3}{7}\mathbf{j} + \frac{6}{7}\mathbf{k}$

17. $f(x, y, z) = \sin xz + \cos xy;$ $P(0, -1, 2);$ $\mathbf{a} = \dfrac{1}{\sqrt{6}}\mathbf{i} - \dfrac{1}{\sqrt{6}}\mathbf{j} + \dfrac{2}{\sqrt{6}}\mathbf{k}$

18. $f(x, y, z) = \tan xyz + \sin xy - \cos xz;$ $P(0, 1, 1);$

$$\mathbf{a} = \dfrac{1}{\sqrt{26}}\mathbf{i} + \dfrac{3}{\sqrt{26}}\mathbf{j} + \dfrac{4}{\sqrt{26}}\mathbf{k}$$

19. The temperature at any point of a rectangular plate in the xy-plane is given by the formula $T = 50(x^2 - y^2)$ (in degrees centigrade). Find $d_\theta T(4, 3)$, and find $\tan \theta$ when $d_\theta T(4, 3) = 0$. Find also the slope of the curve $T = $ const which passes through that point.

In each of problems 20 through 23, find ∇f at the given point.

20. $f(x, y) = x^3 - 2x^2 y + xy^2 - y^3;$ $P(3, -2)$

21. $f(x, y) = \ln(x^2 + y^2 + 1) + e^{2xy};$ $P(0, -2)$

22. $f(x, y, z) = \sin xy + \sin xz + \sin yz;$ $P(1, 2, -1)$

23. $f(x, y, z) = xze^{xy} + yze^{xz} + xye^{yz};$ $P(-1, 2, 1)$

In each of problems 24 through 27, find $D_\mathbf{a} f$ at the given point P where $\mathbf{a}$ is a unit vector in the direction $\overrightarrow{PP'}$. Also, find at P the value of $D_\mathbf{\bar{a}} f$ where $\mathbf{\bar{a}}$ is a unit vector such that $D_\mathbf{\bar{a}} f$ is a maximum.

24. $f(x, y, z) = x^2 + 3xy + y^2 + z^2;$ $P(1, 0, 2);$ $P'(-1, 3, 4)$

25. $f(x, y, z) = e^x \cos y + e^y \sin z;$ $P(2, 1, 0);$ $P'(-1, 2, 2)$

26. $f(x, y, z) = \ln(x^2 + y^2) + e^z;$ $P(0, 1, 0);$ $P'(-4, 2, 3)$

27. $f(x, y, z) = x \cos y + y \cos z + z \cos x;$ $P(2, 1, 0);$ $P'(1, 4, 2)$

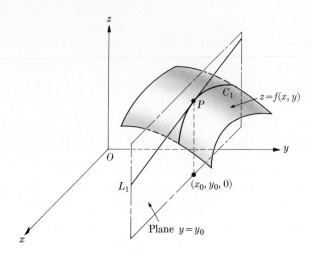

FIGURE 4–7

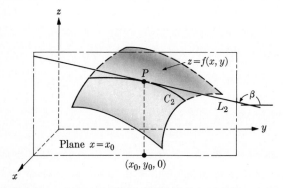

FIGURE 4–8

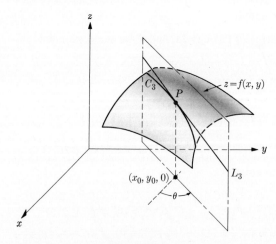

FIGURE 4–9

6. GEOMETRIC INTERPRETATION OF PARTIAL DERIVATIVES. TANGENT PLANES

From the geometric point of view, a function of one variable represents a curve in the plane. The derivative at a point on the curve is the slope of the line tangent to the curve at this point. A function of two variables $z = f(x, y)$ represents a surface in three-dimensional space. If x_0, y_0 are the coordinates of a point in the xy-plane, then $P(x_0, y_0, z_0)$, with $z_0 = f(x_0, y_0)$, is a point on the surface. Consider the vertical plane $y = y_0$, as shown in Fig. 4–7. This plane cuts the surface $z = f(x, y)$ in a curve C_1 which contains the point P. From the definition of partial derivative we see that

$$f_x(x_0, y_0)$$

is the slope of the line tangent to the curve C_1 at the point P. This line is labeled L_1 in Fig. 4–7 and, of course, is in the plane $y = y_0$. In a completely analogous way we construct a plane $x = x_0$ intersecting the surface with equation $z = f(x, y)$ in a curve C_2. The partial derivative

$$f_y(x_0, y_0)$$

is the slope of the line tangent to C_2 at the point P. The line is denoted L_2 in Fig. 4–8, and its slope is the tangent of the angle β, as shown.

The directional derivative $d_\theta f(x_0, y_0)$ has a similar interpretation. We construct the vertical plane through $(x_0, y_0, 0)$ which makes an angle θ with the positive x-direction. Such a plane is shown in Fig. 4–9. The curve C_3 is the intersection of this plane with the surface, and the line L_3 is the line tangent to C_3 at P. Then $d_\theta f(x_0, y_0)$ is the slope of L_3.

According to Theorem 2, we may write the formula

$$f(x, y) - f(x_0, y_0) = f_{,1}(x_0, y_0)(x - x_0) + f_{,2}(x_0, y_0)(y - y_0)$$
$$+ G_1 \cdot (x - x_0) + G_2 \cdot (y - y_0),$$

in which we take

$$x - x_0 = h \quad \text{and} \quad y - y_0 = k.$$

Since G_1 and G_2 tend to zero as $(x, y) \to (x_0, y_0)$, the next definition has an intuitive geometric meaning.

DEFINITION. The plane whose equation is

$$z - z_0 = m_1(x - x_0) + m_2(y - y_0)$$

where

$$z_0 = f(x_0, y_0), \qquad m_1 = f_{,1}(x_0, y_0), \qquad m_2 = f_{,2}(x_0, y_0)$$

is called the **tangent plane** to the surface $z = f(x, y)$ at (x_0, y_0).

Remarks. (i) According to the geometric interpretation of partial derivative which we gave, it is easy to verify that the tangent plane contains the lines L_1

and L_2, tangents to C_1 and C_2, respectively.   (ii) From the definition of tangent plane we observe at once that

$$f_{,1}(x_0, y_0), \qquad f_{,2}(x_0, y_0), \qquad -1$$

is a set of attitude numbers for the plane.   (Attitude numbers were defined on page 15.)   The relation between attitude numbers and direction numbers leads to the next definition.

DEFINITION.   The line with equations

$$\frac{x - x_0}{m_1} = \frac{y - y_0}{m_2} = \frac{z - z_0}{-1},$$

with $z_0 = f(x_0, y_0)$, $m_1 = f_{,1}(x_0, y_0)$, $m_2 = f_{,2}(x_0, y_0)$, is called the **normal line** to the surface at the point $P(x_0, y_0, z_0)$.   Clearly, the normal line is perpendicular to the tangent plane.

Remarks.   In addition to $f_{,1}(x_0, y_0)$ we shall use the symbols

$$f_x(x_0, y_0), \qquad \frac{\partial f(x_0, y_0)}{\partial x}, \qquad \text{and} \qquad \frac{\partial z}{\partial x}\bigg|_{\substack{x=x_0 \\ y=y_0}}$$

for m_1, and analogous notations for m_2.

EXAMPLE 1.   Find the equation of the tangent plane and the equations of the normal line to the surface

$$z = x^2 + xy - y^2$$

at the point where $x = 2$, $y = -1$.

Solution.   We have $z_0 = f(2, -1) = 1$; $f_{,1}(x, y) = 2x + y$, $f_{,2}(x, y) = x - 2y$. Therefore $m_1 = 3$, $m_2 = 4$, and the desired equation for the tangent plane is

$$z - 1 = 3(x - 2) + 4(y + 1).$$

The equations of the normal line are

$$\frac{x - 2}{3} = \frac{y + 1}{4} = \frac{z - 1}{-1}.$$

If the equation of the surface is given in implicit form it is possible to use the methods of Section 2 to find $\partial z/\partial x$ and $\partial z/\partial y$ at the desired point.   The next example shows how we obtain the equations of the tangent plane and normal line under such circumstances.

EXAMPLE 2.   Find the equations of the tangent plane and normal line at $(3, -1, 2)$ to the locus of

$$xy + yz + xz - 1 = 0.$$

Solution. Holding y constant and differentiating with respect to x, we get

$$y + y\frac{\partial z}{\partial x} + z + x\frac{\partial z}{\partial x} = 0 \quad \text{and} \quad \frac{\partial z}{\partial x} = -\frac{y + z}{y + x} = -\frac{1}{2} = m_1.$$

Similarly, holding x constant, we obtain

$$x + y\frac{\partial z}{\partial y} + z + x\frac{\partial z}{\partial y} = 0, \quad \frac{\partial z}{\partial y} = -\frac{x + z}{x + y} = -\frac{5}{2} = m_2.$$

The equation of the tangent plane is

$$z - 2 = -\tfrac{1}{2}(x - 3) - \tfrac{5}{2}(y + 1) \quad \text{or} \quad x + 5y + 2z - 2 = 0.$$

The normal line has equations

$$\frac{x - 3}{-1/2} = \frac{y + 1}{-5/2} = \frac{z - 2}{-1} \quad \text{or} \quad \frac{x - 3}{1} = \frac{y + 1}{5} = \frac{z - 2}{2}.$$

The methods of the calculus of functions of several variables enable us to establish purely geometric facts, as the next example shows.

EXAMPLE 3. Show that any line normal to the sphere

$$x^2 + y^2 + z^2 = a^2$$

always passes through the center of the sphere.

Solution. Since the sphere has center at $(0, 0, 0)$, we must show that the equations of the line are satisfied for $x = y = z = 0$. Let (x_0, y_0, z_0) be a point on the sphere. Differentiating implicitly, we find

$$\frac{\partial z}{\partial x} = -\frac{x}{z}, \quad \frac{\partial z}{\partial y} = -\frac{y}{z}.$$

The normal line to the sphere at (x_0, y_0, z_0) has equations

$$\frac{x - x_0}{-x_0/z_0} = \frac{y - y_0}{-y_0/z_0} = \frac{z - z_0}{-1}. \tag{1}$$

Letting $x = y = z = 0$, we get an identity for (1); hence the line passes through the origin.

Let the equations

$$z = f(x, y) \quad \text{and} \quad z = g(x, y)$$

represent surfaces which intersect in a curve C. The *tangent line* to the intersection at a point P on C is by definition the line of intersection of the tangent planes to f and g at P (Fig. 4–10).

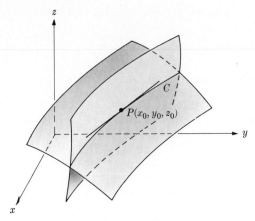

FIGURE 4–10

We can use vector algebra in the following way to find the equations of this tangent line. If P has coordinates (x_0, y_0, z_0), then

$$\mathbf{u} = f_x(x_0, y_0)\mathbf{i} + f_y(x_0, y_0)\mathbf{j} + (-1)\mathbf{k}$$

is a vector perpendicular to the plane tangent to f at P. Similarly, we have

$$\mathbf{v} = g_x(x_0, y_0)\mathbf{i} + g_y(x_0, y_0)\mathbf{j} + (-1)\mathbf{k}$$

is a vector perpendicular to the plane tangent to g at P. The line of intersection of these tangent planes is perpendicular to both $\mathbf{u}$ and $\mathbf{v}$. We recall from the study of vectors that, if $\mathbf{u}$ and $\mathbf{v}$ are nonparallel vectors, the vector $\mathbf{u} \times \mathbf{v}$ is perpendicular to both $\mathbf{u}$ and $\mathbf{v}$. Defining the vector $\mathbf{w} = a\mathbf{i} + b\mathbf{j} + c\mathbf{k}$ by the relation

$$\mathbf{w} = \mathbf{u} \times \mathbf{v},$$

we see that the equations of the line of intersection of the tangent plane are

$$\frac{x - x_0}{a} = \frac{y - y_0}{b} = \frac{z - z_0}{c}.$$

The next example shows how the method works.

———————

EXAMPLE 4. Find the equations of the line tangent to the intersection of the surfaces

$$z = f(x, y) = x^2 + 2y^2, \qquad z = g(x, y) = 2x^2 - 3y^2 + 1$$

at the point $(2, 1, 6)$.

Solution. We have

$$f_x = 2x, \quad f_y = 4y; \qquad g_x = 4x, \quad g_y = -6y.$$

Therefore

$$\mathbf{u} = 4\mathbf{i} + 4\mathbf{j} - \mathbf{k}; \qquad \mathbf{v} = 8\mathbf{i} - 6\mathbf{j} - \mathbf{k}; \qquad \mathbf{u} \times \mathbf{v} = -10\mathbf{i} - 4\mathbf{j} - 56\mathbf{k}.$$

The desired equations are

$$\frac{x-2}{10} = \frac{y-1}{4} = \frac{z-6}{56}.$$

EXERCISES

In problems 1 through 14, find in each case the equation of the tangent plane and the equations of the normal line to the given surface at the given point.

1. $z = x^2 + 2y^2$; $(2, -1, 6)$
2. $z = 3x^2 - y^2 - 2$; $(-1, 2, -3)$
3. $z = xy$; $(2, -1, -2)$
4. $z = x^2y^2$; $(-2, 2, 16)$
5. $z = e^x \sin y$; $(1, \pi/2, e)$
6. $z = e^{2x} \cos 3y$; $(1, \pi/3, -e^2)$
7. $z = \ln \sqrt{x^2 + y^2}$; $(-3, 4, \ln 5)$
8. $x^2 + 2y^2 + 3z^2 = 6$; $(1, 1, -1)$
9. $x^2 + 2y^2 - 3z^2 = 3$; $(2, 1, -1)$
10. $x^2 + 3y^2 - z^2 = 0$; $(2, -2, 4)$
11. $x^2 + z^2 = 25$; $(4, -2, -3)$
12. $xy + yz + xz = 1$; $(2, 3, -1)$
13. $x^{1/2} + y^{1/2} + z^{1/2} = 6$; $(4, 1, 9)$
14. $y^{1/2} + z^{1/2} = 7$; $(3, 16, 9)$
15. Show that the equation of the plane tangent at (x_1, y_1, z_1) to the ellipsoid

$$\frac{x^2}{A^2} + \frac{y^2}{B^2} + \frac{z^2}{C^2} = 1 \qquad \text{is} \qquad \frac{x_1 x}{A^2} + \frac{y_1 y}{B^2} + \frac{z_1 z}{C^2} = 1.$$

16. Show that every plane tangent to the cone

$$x^2 + y^2 = z^2$$

passes through the origin.

17. Show that every line normal to the cone

$$z^2 = 3x^2 + 3y^2$$

intersects the z-axis.

18. Show that the sum of the squares of the intercepts of any plane tangent to the surface

$$x^{2/3} + y^{2/3} + z^{2/3} = a^{2/3}$$

is constant.

In problems 19 through 22, find the equations of the line tangent to the intersection of the two surfaces at the given point.

19. $z = x^2 + y^2$, $z = 2x + 4y + 20$; $(4, -2, 20)$
20. $z = \sqrt{x^2 + y^2}$, $z = 2x - 3y - 13$; $(3, -4, 5)$
21. $z = x^2$, $z = 25 - y^2$; $(4, -3, 16)$
22. $z = \sqrt{25 - 9x^2}$, $z = e^{xy} + 3$; $(1, 0, 4)$

7. THE TOTAL DIFFERENTIAL. APPROXIMATION

The differential of a function of one variable is a function of two variables selected in a special way. We recall that if $y = f(x)$, then the quantity df, called the differential of f, is defined by the relation

$$df = f'(x)h,$$

where h and x are independent variables. (See *First Course*, Chapter 7, page 184.)

Let f be a function of several variables; the next definition is the appropriate one for generalizing the notion of differential to such functions.

DEFINITIONS. The **total differential** of $f(x, y)$ is the function df of four variables x, y, h, k given by the formula

$$df(x, y, h, k) = f_{,1}(x, y)h + f_{,2}(x, y)k.$$

If F is a function of three variables—say x, y, and z—we define the **total differential** as the function of six variables x, y, z, h, k, l given by

$$dF(x, y, z, h, k, l) = F_{,1}(x, y, z)h + F_{,2}(x, y, z)k + F_{,3}(x, y, z)l.$$

A quantity related to the total differential is the difference of a function at two nearby values. As is customary, we use Δ notation and, for functions of two variables, we define the quantity Δf by the relation

$$\Delta f \equiv \Delta f(x, y, h, k) = f(x + h, y + k) - f(x, y).$$

Here Δf is a function of four variables, as is df. If f depends on x, y, and z, then Δf is a function of six variables defined by

$$\Delta f \equiv \Delta f(x, y, z, h, k, l) = f(x + h, y + k, z + l) - f(x, y, z).$$

———————

EXAMPLE 1. Given the function

$$f(x, y) = x^2 + xy - 2y^2 - 3x + 2y + 4,$$

find $df(a, b, h, k)$ and $\Delta f(a, b, h, k)$ with $a = 3$, $b = 1$.

Solution. We have

$$f(3, 1) = 7; \quad f_{,1}(x, y) = 2x + y - 3; \quad f_{,2}(x, y) = x - 4y + 2;$$
$$f_{,1}(3, 1) = 4; \quad f_{,2}(3, 1) = 1.$$

Also

$$f(3 + h, 1 + k) = (3 + h)^2 + (3 + h)(1 + k) - 2(1 + k)^2 - 3(3 + h)$$
$$+ 2(1 + k) + 4 = h^2 + hk - 2k^2 + 4h + k + 7.$$

Therefore

$$df(3, 1, h, k) = 4h + k; \quad \Delta f(3, 1, h, k) = 4h + k + h^2 + hk - 2k^2.$$

The close relationship between df and Δf is exhibited in Theorem 2. Equation (1) of that theorem may be written in the form

$$\Delta f(x_0, y_0, h, k) = df(x_0, y_0, h, k) + G_1(h, k)h + G_2(h, k)k.$$

The conclusion of the theorem implies that

$$\frac{\Delta f - df}{|h| + |k|} \to 0 \quad \text{as} \quad h, k \to 0,$$

since both G_1 and G_2 tend to zero with h and k. In many problems Δf is difficult to calculate, while df is easy. If h and k are both "small," we can use df as an approximation to Δf. The next example shows the technique.

———

EXAMPLE 2. Find, approximately, the value of $\sqrt{(5.98)^2 + (8.01)^2}$.

Solution. We consider the function

$$z = f(x, y) = \sqrt{x^2 + y^2},$$

and we wish to find $f(5.98, 8.01)$. We see at once that $f(6, 8) = 10$; hence we may write

$$f(5.98, 8.01) = f(6, 8) + \Delta f,$$

where Δf is defined as above with $x_0 = 6$, $y_0 = 8$, $h = -0.02$, $k = 0.01$. The approximation consists of replacing Δf by df. We have

$$\frac{\partial z}{\partial x} = \frac{x}{\sqrt{x^2 + y^2}}, \qquad \frac{\partial z}{\partial y} = \frac{y}{\sqrt{x^2 + y^2}},$$

and so

$$df(6, 8, -0.02, 0.01) = \tfrac{6}{10}(-0.02) + \tfrac{8}{10}(0.01) = -0.004.$$

We conclude that $\sqrt{(5.98)^2 + (8.01)^2} = 10 - 0.004 = 9.996$, approximately. The true value is $9.9959992+$.

———

As in the case of functions of one variable, the symbolism for the total differential may be used as an aid in differentiation. We let $z = f(x, y)$ and employ the symbols

$$dz \text{ for } df, \qquad dx \text{ for } h, \qquad \text{and} \qquad dy \text{ for } k.$$

As in the case of one variable, there is a certain ambiguity, since dz has a precise definition as the total differential, while dx and dy are used as independent variables. The next theorem shows how the Chain Rule comes to our rescue and removes all difficulties when dx and dy are in turn functions of other variables (i.e., dx and dy are what we call intermediate variables).

Theorem 6. *Suppose that* $z = f(x, y)$, *and* x *and* y *are functions of some other variables. Then*

$$dz = \frac{\partial z}{\partial x} dx + \frac{\partial z}{\partial y} dy.$$

The result for $w = F(x, y, z)$ *is similar. That is, the formula*

$$dw = \frac{\partial w}{\partial x} dx + \frac{\partial w}{\partial y} dy + \frac{\partial w}{\partial z} dz$$

holds when x, y, *and* z *are either independent or intermediate variables.*

Proof. We establish the result for $z = f(x, y)$ with x and y functions of two variables, say r and s. The proof in all other cases is analogous. We may write

$$x = x(r, s), \qquad y = y(r, s),$$

and then

$$z = f(x, y) = f[x(r, s), y(r, s)] = g(r, s).$$

The definition of total differential yields

$$dz = g_{,1}(r, s)h + g_{,2}(r, s)k,$$
$$dx = x_{,1}(r, s)h + x_{,2}(r, s)k,$$
$$dy = y_{,1}(r, s)h + y_{,2}(r, s)k.$$

According to the Chain Rule, we have

$$g_{,1}(r, s) = \frac{\partial f}{\partial r} = \frac{\partial f}{\partial x}\frac{\partial x}{\partial r} + \frac{\partial f}{\partial y}\frac{\partial y}{\partial r}$$
$$= f_{,1}(x, y)x_{,1}(r, s) + f_{,2}(x, y)y_{,1}(r, s);$$
$$g_{,2}(r, s) = \frac{\partial f}{\partial s} = \frac{\partial f}{\partial x}\frac{\partial x}{\partial s} + \frac{\partial f}{\partial y}\frac{\partial y}{\partial s}$$
$$= f_{,1}(x, y)x_{,2}(r, s) + f_{,2}(x, y)y_{,2}(r, s).$$

Substituting the above expressions for $g_{,1}$ and $g_{,2}$ into that for dz, we obtain

$$dz = \frac{\partial f}{\partial x}[x_{,1}(r, s)h + x_{,2}(r, s)k] + \frac{\partial f}{\partial y}[y_{,1}(r, s)h + y_{,2}(r, s)k]$$

or

$$dz = \frac{\partial f}{\partial x} dx + \frac{\partial f}{\partial y} dy.$$

We recognize this last formula as the statement of the theorem.

───────────

EXAMPLE 3. Given

$$z = e^x \cos y + e^y \sin x, \qquad x = r^2 - t^2, \qquad y = 2rt,$$

find $dz(r, t, h, k)$ in two ways and verify that the results coincide.

Solution. We have, by one method,

$$dz = \frac{\partial z}{\partial x} dx + \frac{\partial z}{\partial y} dy$$

$$= (e^x \cos y + e^y \cos x) dx + (-e^x \sin y + e^y \sin x) dy;$$

$$dx = 2rh - 2tk;$$

$$dy = 2th + 2rk.$$

Therefore

$$dz = (e^x \cos y + e^y \cos x)(2rh - 2tk) + (-e^x \sin y + e^y \sin x)(2th + 2rk)$$

$$= 2[e^x \cos y + e^y \cos x)r + (e^y \sin x - e^x \sin y)t]h$$

$$+ 2[(-e^x \cos y - e^y \cos x)t + (e^y \sin x - e^x \sin y)r]k. \qquad (1)$$

On the other hand,

$$dz = \frac{\partial z}{\partial r} h + \frac{\partial z}{\partial t} k \qquad (2)$$

and, using the Chain Rule,

$$\frac{\partial z}{\partial r} = \frac{\partial z}{\partial x} \frac{\partial x}{\partial r} + \frac{\partial z}{\partial y} \frac{\partial y}{\partial r}, \qquad \frac{\partial z}{\partial t} = \frac{\partial z}{\partial x} \frac{\partial x}{\partial t} + \frac{\partial z}{\partial y} \frac{\partial y}{\partial t}.$$

We compute the various quantities in the two formulas above and find that

$$\frac{\partial z}{\partial r} = (e^x \cos y + e^y \cos x)(2r) + (e^y \sin x - e^x \sin y)(2t),$$

$$\frac{\partial z}{\partial t} = (e^x \cos y + e^y \cos x)(-2t) + (e^y \sin x - e^x \sin y)(2r).$$

Substituting these expressions in (2), we get (1) precisely.

EXERCISES

In each of problems 1 through 6, find $df(x, y, h, k)$ and $\Delta f(x, y, h, k)$ for the given values of x, y, h, and k.

1. $f(x, y) = x^2 - xy + 2y^2$, $x = 2$, $y = -1$, $h = -0.01$, $k = 0.02$
2. $f(x, y) = 2x^2 + 3xy - y^2$, $x = 1$, $y = 2$, $h = 0.02$, $k = -0.01$
3. $f(x, y) = \sin xy + \cos (x + y)$, $x = \pi/6$, $y = 0$, $h = 2\pi$, $k = 3\pi$
4. $f(x, y) = e^{xy} \sin (x + y)$, $x = \pi/4$, $y = 0$, $h = -\pi/2$, $k = 4\pi$
5. $f(x, y) = x^3 - 3xy + y^3$, $x = -2$, $y = 1$, $h = -0.03$, $k = -0.02$
6. $f(x, y) = x^2y - 2xy^2 + 3x$, $x = 1$, $y = 1$, $h = 0.02$, $k = 0.01$

In each of problems 7 through 10, find $df(x, y, z, h, k, l)$ and $\Delta f(x, y, z, h, k, l)$ for the given values of x, y, z, h, k, and l.

7. $f(x, y, z) = x^2 - 2y^2 + z^2 - xz$; $(x, y, z) = (2, -1, 3)$;
   $(h, k, l) = (0.01, -0.02, 0.03)$

8. $f(x, y, z) = xy - xz + yz + 2x - 3y + 1$; $(x, y, z) = (2, 0, -3)$;
   $(h, k, l) = (0.1, -0.2, 0.1)$

9. $f(x, y, z) = x^2y - xyz + z^3$; $(x, y, z) = (1, 2, -1)$;
   $(h, k, l) = (-0.02, 0.01, 0.02)$

10. $f(x, y, z) = \sin(x + y) - \cos(x - z) + \sin(y + 2z)$;
    $(x, y, z) = (\pi/3, \pi/6, 0)$; $(h, k, l) = (\pi/4, \pi/2, 2\pi)$

11. We define the **approximate percentage error** of a function f (see, for example, *First Course*, page 188) by the relation

$$\text{Approximate percentage error} = 100\,\frac{df}{f}.$$

    Find the approximate percentage error if $f(x, y, z) = 3x^3y^7z^4$.

12. Find the approximate percentage error (see problem 11) if $f = kx^my^nz^p$
    (k = const).

13. A crate has square ends, 11.98 in. on each side, and has a length of 30.03 in. Find its approximate volume, using differentials.

14. Use differentials to find the approximate value of $\sqrt{(5.02)^2 + (11.97)^2}$.

15. The legs of a right triangle are measured and found to be 6.0 and 8.0 in., with a possible error of 0.1 in. Find approximately the maximum possible value of the error in computing the hypotenuse. What is the maximum approximate percentage error? (See problem 11.)

16. Find in degrees the maximum possible approximate error in the computed value of the smaller acute angle in the triangle of problem 15.

17. The diameter and height of a right circular cylinder are found by measurement to be 8.0 and 12.5 in., respectively, with possible errors of 0.05 in. in each measurement. Find the maximum possible approximate error in the computed volume.

18. A right circular cone is measured, and the radius of the base is 12.0 in. with the height 16.0 in. If the possible error in each measurement is 0.06, find the maximum possible error in the computed volume. What is the maximum possible approximate error in the lateral surface area?

19. By measurement, a triangle is found to have two sides of length 50 in. and 70 in.; the angle between them is 30°. If there are possible errors of $\frac{1}{2}\%$ in the measurements of the sides and $\frac{1}{2}$ degree in that of the angle, find the maximum approximate percentage error in the measurement of the area. (See problem 11.)

20. Use differentials to find the approximate value of

$$\sqrt{(3.02)^2 + (1.99)^2 + (5.97)^2}.$$

21. Use differentials to find the approximate value of

$$[(3.01)^2 + (3.98)^2 + (6.02)^2 + 5(1.97)^2]^{-1/2}.$$

In each of problems 22 through 26, find dz in two ways in terms of the independent variables.

22. $z = x^2 + xy - y^2$; $x = r^2 + 2s^2$, $y = rs + 2$

23. $z = 2x^2 + 3xy + y^2$; $x = t^3 + 2t - 1$, $y = t^2 + t - 3$

24. $z = x^3 + y^3 - x^2y$; $x = r + 2s - t$, $y = r - 3s + 2t$

25. $z = u^2 + 2v^2 - x^2 + 3y^2$; $u = r^2 - s^2$, $v = r^2 + s^2$, $x = 2rs$, $y = 2r/s$

26. $z = u^3 + v^3 + w^3$; $u = r^2 + s^2 + t^2$, $v = r^2 - s^2 + t^2$, $w = r^2 + s^2 - t^2$

8. APPLICATIONS OF THE TOTAL DIFFERENTIAL

Once we clearly understand the concept of function we are able to use the notation of the total differential to obtain a number of useful differentiation formulas.

One of the simplest formulas, which we now develop, uses the fact that *the total differential of a constant is zero.* Suppose that x and y are related by some equation such as

$$f(x, y) = 0.$$

If we think of y as a function of x, we can compute the derivative dy/dx by implicit methods in the usual way. However, we may also use an alternate procedure. Since $f = 0$, the differential df also vanishes. Therefore we can write

$$df = \frac{\partial f}{\partial x} dx + \frac{\partial f}{\partial y} dy = 0$$

or

$$\frac{dy}{dx} = - \frac{\partial f/\partial x}{\partial f/\partial y} \quad \left(\text{if } \frac{\partial f}{\partial y} \neq 0 \right). \tag{1}$$

EXAMPLE 1. Use the methods of partial differentiation to compute dy/dx if

$$x^4 + 3x^2y^2 - y^4 + 2x - 3y = 5.$$

Solution. Setting

$$f(x, y) = x^4 + 3x^2y^2 - y^4 + 2x - 3y - 5 = 0,$$

we find

$$f_x = 4x^3 + 6xy^2 + 2, \qquad f_y = 6x^2y - 4y^3 - 3.$$

Therefore, using (1) above,

$$\frac{dy}{dx} = - \frac{4x^3 + 6xy^2 + 2}{6x^2y - 4y^3 - 3}.$$

Of course, the same result is obtained by the customary process of implicit differentiation.

The above method for ordinary differentiation may be extended to yield partial derivatives. Suppose x, y, and z are connected by a relation of the form

$$F(x, y, z) = 0,$$

and we imagine that z is a function of x and y. That is, we make the assumption that it is possible to solve for z in terms of x and y even though we have no intention of doing so; in fact, we may find it exceptionally difficult (if not down-right impossible) to perform the necessary steps. If z is a function of x and y, then we have the formula for the total differential:

$$dz = \frac{\partial z}{\partial x} dx + \frac{\partial z}{\partial y} dy. \tag{2}$$

On the other hand, since $F = 0$, the differential dF is also. Therefore

$$dF = F_x \, dx + F_y \, dy + F_z \, dz = 0.$$

Solving for dz, we get

$$dz = \left(- \frac{F_x}{F_z}\right) dx + \left(- \frac{F_y}{F_z}\right) dy, \tag{3}$$

assuming that $F_z \neq 0$. Comparing Eq. (2) and (3), it is possible to conclude that

$$\frac{\partial z}{\partial x} = - \frac{F_x}{F_z} \quad \text{and} \quad \frac{\partial z}{\partial y} = - \frac{F_y}{F_z}. \tag{4}$$

(The validity of this conclusion is established in Morrey, *University Calculus*, page 568.) We exhibit the technique in the next example.

EXAMPLE 2. Use formulas (4) to find $\partial z/\partial x$ and $\partial z/\partial y$ if

$$e^{xy} \cos z + e^{-xz} \sin y + e^{yz} \cos x = 0.$$

Solution. We set

$$F(x, y, z) = e^{xy} \cos z + e^{-xz} \sin y + e^{yz} \cos x,$$

and compute

$$F_x = ye^{xy} \cos z - ze^{-xz} \sin y - e^{yz} \sin x,$$
$$F_y = xe^{xy} \cos z + e^{-xz} \cos y + ze^{yz} \cos x,$$
$$F_z = -e^{xy} \sin z - xe^{-xz} \sin y + ye^{yz} \cos x.$$

Therefore

$$\frac{\partial z}{\partial x} = - \frac{ye^{xy} \cos z - ze^{-xz} \sin y - e^{yz} \sin x}{-e^{xy} \sin z - xe^{-xz} \sin y + ye^{yz} \cos x},$$

$$\frac{\partial z}{\partial y} = - \frac{xe^{xy} \cos z + e^{-xz} \cos y + ze^{yz} \cos x}{-e^{xy} \sin z - xe^{-xz} \sin y + ye^{yz} \cos x}.$$

Remarks. (i) Note that we also could have found the derivatives by the implicit methods described in Section 2. (ii) Formulas similar to (4) may be established for a single relation with any number of variables. For example, if we are given $G(x, y, u, v, w) = 0$ and we assume w is a function of the remaining variables with $G_w \neq 0$, then

$$\frac{\partial w}{\partial x} = - \frac{G_x}{G_w}, \quad \frac{\partial w}{\partial y} = - \frac{G_y}{G_w}, \quad \frac{\partial w}{\partial u} = - \frac{G_u}{G_w}, \quad \frac{\partial w}{\partial v} = - \frac{G_v}{G_w}.$$

A more complicated application of differentials is exhibited in the derivation of the next set of formulas. Suppose, for example, that x, y, u, v are related by *two* equations, so that

$$F(x, y, u, v) = 0 \quad \text{and} \quad G(x, y, u, v) = 0.$$

If we could solve one of them for, say u, and substitute in the other, we would get a single equation for x, y, v. Then, solving for v, we would find that v is a function of x and y. Similarly, we might find u as a function of x and y. Of course, all this work is purely fictitious, since we have no intention of carrying out such a process. In fact, it may be impossible. The main point is that we know that under appropriate circumstances the process is *theoretically feasible*. (This fact is discussed in Chapter II.) Therefore it makes sense to write the symbols

$$\frac{\partial u}{\partial x}, \quad \frac{\partial u}{\partial y}, \quad \frac{\partial v}{\partial x}, \quad \frac{\partial v}{\partial y} \tag{5}$$

whenever $u = u(x, y)$ and $v = v(x, y)$. Furthermore, the selection of u and v in terms of x and y is arbitrary. We could equally well attempt to solve for v and x in terms of u and y or for any two of the variables in terms of the remaining two variables.

The problem we pose is one of determining the quantities in (5) without actually finding the functions $u(x, y)$ and $v(x, y)$. We use the total differential. Since $F = 0$ and $G = 0$, so are dF and dG. We have

$$dF = F_x \, dx + F_y \, dy + F_u \, du + F_v \, dv = 0,$$
$$dG = G_x \, dx + G_y \, dy + G_u \, du + G_v \, dv = 0.$$

We write these equations,

$$F_u \, du + F_v \, dv = -F_x \, dx - F_y \, dy, \quad G_u \, du + G_v \, dv = -G_x \, dx - G_y \, dy,$$

and consider du and dv as unknowns with everything else known. Solving two equations in two unknowns is easy. We obtain

$$du = \frac{G_x F_v - G_v F_x}{F_u G_v - F_v G_u} \, dx + \frac{G_y F_v - G_v F_y}{F_u G_v - F_v G_u} \, dy, \tag{6}$$

$$dv = \frac{G_u F_x - G_x F_u}{F_u G_v - F_v G_u} \, dx + \frac{G_u F_y - G_y F_u}{F_u G_v - F_v G_u} \, dy. \tag{7}$$

(It is assumed that $F_u G_v - F_v G_u \neq 0$.) On the other hand, we know that if $u = u(x, y)$, $v = v(x, y)$, then

$$du = \frac{\partial u}{\partial x} \, dx + \frac{\partial u}{\partial y} \, dy, \tag{8}$$

$$dv = \frac{\partial v}{\partial x} \, dx + \frac{\partial v}{\partial y} \, dy. \tag{9}$$

Therefore, comparing (6) with (8) and (7) with (9), we find

$$\frac{\partial u}{\partial x} = \frac{G_x F_v - G_v F_x}{F_u G_v - F_v G_u},$$ (10)

and similar formulas for $\partial u/\partial y$, $\partial v/\partial x$, and $\partial v/\partial y$. If F and G are given, the right side of (10) is computable.

EXAMPLE 3. Given the relations for x, y, u, v:

$$u^2 - uv - v^2 + x^2 + y^2 - xy = 0,$$
$$uv - x^2 + y^2 = 0,$$

and assuming that $u = u(x, y)$, $v = v(x, y)$, find $\partial u/\partial x$, $\partial u/\partial y$, $\partial v/\partial x$, and $\partial v/\partial y$.

Solution. We could find F_x, F_y, ..., G_u, G_v and then substitute for the coefficients in (6) and (7) to obtain the result. Instead we make use of the fact that we can treat differentials both as independent variables and as total differentials, with no fear of difficulty (because of the Chain Rule). Taking such differentials in each of the equations given, we get

$$2u \, du - u \, dv - v \, du - 2v \, dv + 2x \, dx + 2y \, dy - x \, dy - y \, dx = 0,$$
$$v \, du + u \, dv - 2x \, dx + 2y \, dy = 0.$$

Solving the two equations simultaneously for du and dv in terms of dx and dy, we obtain

$$du = \frac{uy + 4xv}{2(u^2 + v^2)} \, dx + \frac{ux - 4y(u + v)}{2(u^2 + v^2)} \, dy,$$

$$dv = \frac{4xu - yv}{2(u^2 + v^2)} \, dx + \frac{4y(v - u) - xv}{2(u^2 + v^2)} \, dy.$$

From these equations we read off the results. For example,

$$\frac{\partial u}{\partial y} = \frac{ux - 4y(u + v)}{2(u^2 + v^2)},$$

and there are corresponding expressions for $\partial u/\partial x$, $\partial v/\partial x$, $\partial v/\partial y$.

EXERCISES

In each of problems 1 through 7, find the derivative dy/dx by the methods of partial differentiation.

1. $x^2 + 3xy - 4y^2 + 2x - 6y + 7 = 0$
2. $x^3 + 3x^2y - 4xy^2 + y^3 - x^2 + 2y - 1 = 0$
3. $\ln(1 + x^2 + y^2) + e^{xy} = 5$
4. $x^4 - 3x^2y^2 + y^4 - x^2y + 2xy^2 = 3$ 5. $e^{xy} + \sin xy + 1 = 0$
6. $xe^y + ye^x + \sin(x + y) - 2 = 0$ 7. $\arctan(y/x) + (x^2 + y^2)^{3/2} = 2$

In each of problems 8 through 12, assume that w is a function of the remaining variables. Find the partial derivatives as indicated by the method of Example 2.

8. $x^2 + y^2 + w^2 - 3xyw - 4 = 0$; $\dfrac{\partial w}{\partial y}$

9. $x^3 + 3x^2 w - y^2 w + 2yw^2 - 3w + 2x = 8$; $\dfrac{\partial w}{\partial x}$

10. $e^{xy} + e^{yw} - e^{xw} + xyw = 4$; $\dfrac{\partial w}{\partial y}$

11. $\sin(xyw) + x^2 + y^2 + w^2 = 3$; $\dfrac{\partial w}{\partial x}$

12. $(w^2 - y^2)(w^2 + x^2)(x^2 - y^2) = 1$; $\dfrac{\partial w}{\partial y}$

In problems 13 and 14, use the methods of this section to find the partial derivatives as indicated.

13. $x^2 + y^2 - z^2 - w^2 + 3xy - 2xz + 4xw - 3zw + 2x - 3y = 0$; $\dfrac{\partial w}{\partial y}$

14. $x^2 y^2 z^2 w^2 + x^2 z^2 w^4 - y^4 w^4 + x^6 w^2 - 2y^3 w^3 = 8$; $\dfrac{\partial w}{\partial z}$

If $F(x, y, z) = 0$ and $G(x, y, z) = 0$, then we may consider z and y as functions of the single variable x; that is, $z = z(x)$, $y = y(x)$. Using differentials, we obtain

$$F_x\, dx + F_y\, dy + F_z\, dz = 0, \qquad G_x\, dx + G_y\, dy + G_z\, dz = 0,$$

and so we can get the ordinary derivatives dz/dx and dy/dx. Use this method in problems 15 through 18 to obtain these derivatives.

15. $z = x^2 + y^2$, $y^2 = 4x + 2z$

16. $x^2 - y^2 + z^2 = 7$, $2x + 3y + 4z = 15$

17. $2x^2 + 3y^2 + 4z^2 = 12$, $x = yz$

18. $xyz = 5$, $x^2 + y^2 - z^2 = 16$

In each of problems 19 through 23, find $\partial u/\partial x$, $\partial u/\partial y$, $\partial v/\partial x$, and $\partial v/\partial y$ by the method of differentials.

19. $x = u^2 - v^2$, $y = 2uv$

20. $x = u + v$, $y = uv$

21. $u + v - x^2 = 0$, $u^2 - v^2 - y = 0$

22. $u^3 + xv^2 - xy = 0$, $u^2 y + v^3 + x^2 - y^2 = 0$

23. $u^2 + v^2 + x^2 - y^2 = 4$, $u^2 - v^2 - x^2 - y^2 = 1$

24. Given that $F(x, y, z) = 0$, show that by considering each of the variables in turn as the dependent variable, the relation

$$\frac{\partial x}{\partial y} \cdot \frac{\partial y}{\partial z} \cdot \frac{\partial z}{\partial x} = -1$$

holds.

25. Given that $x = f(u, v)$, $y = g(u, v)$, find $\partial u/\partial x$, $\partial u/\partial y$, $\partial v/\partial x$, $\partial v/\partial y$ in terms of u and v and the derivatives of f and g.

26. Given that $F(u, v, x, y, z) = 0$ and $G(u, v, x, y, z) = 0$, assume that $u = u(x, y, z)$ and $v = v(x, y, z)$ and find formulas for $\partial u/\partial x$, $\partial v/\partial x$, $\ldots$, $\partial v/\partial z$ in terms of the derivatives of F and G.

9. SECOND AND HIGHER DERIVATIVES

If f is a function of two variables—say x and y—then $f_{,1}$ and $f_{,2}$ are also functions of the same two variables. When we differentiate $f_{,1}$ and $f_{,2}$, we obtain second partial derivatives. The *second partial derivatives of* $f_{,1}$ are defined by the formulas

$$f_{,1,1}(x, y) = \lim_{h \to 0} \frac{f_{,1}(x + h, y) - f_{,1}(x, y)}{h},$$

$$f_{,1,2}(x, y) = \lim_{k \to 0} \frac{f_{,1}(x, y + k) - f_{,1}(x, y)}{k}.$$

The derivatives of $f_{,2}$ are defined by similar expressions. We observe that, if f is a function of two variables, there are four second partial derivatives.

There is a multiplicity of notations for partial derivatives which at times may lead to confusion. For example, if we write $z = f(x, y)$, then the following five symbols all have the same meaning:

$$f_{,1,1}; \qquad \frac{\partial^2 z}{\partial x^2}; \qquad \frac{\partial^2 f}{\partial x^2}; \qquad f_{xx}; \qquad z_{xx}.$$

For other partial derivatives we have the variety of expressions:

$$f_{,1,2} = f_{xy} = \frac{\partial}{\partial y}\left(\frac{\partial z}{\partial x}\right) = \frac{\partial^2 z}{\partial y\,\partial x} = \frac{\partial^2 f}{\partial y\,\partial x} = z_{xy},$$

$$f_{,2,1} = f_{yx} = \frac{\partial}{\partial x}\left(\frac{\partial z}{\partial y}\right) = \frac{\partial^2 z}{\partial x\,\partial y} = \frac{\partial^2 f}{\partial x\,\partial y} = z_{yx},$$

$$f_{,1,2,1} = \frac{\partial}{\partial x}\left(\frac{\partial^2 z}{\partial y\,\partial x}\right) = \frac{\partial^3 z}{\partial x\,\partial y\,\partial x} = \frac{\partial^3 f}{\partial x\,\partial y\,\partial x} = f_{xyx} = z_{xyx},$$

and so forth. Note that, in the subscript notation, symbols such as f_{xyy} or z_{xyy} mean that the order of partial differentiation is taken from left to right—that is, first with respect to x and then twice with respect to y. On the other hand, the symbol

$$\frac{\partial^3 z}{\partial x\,\partial y\,\partial y}$$

asserts that we first take two derivatives with respect to y and then one with respect to x. The denominator symbol and the subscript symbol are the reverse of each other.

EXAMPLE 1. Given $z = x^3 + 3x^2y - 2x^2y^2 - y^4 + 3xy$, find

$$\frac{\partial z}{\partial x}, \quad \frac{\partial z}{\partial y}, \quad \frac{\partial^2 z}{\partial x^2}, \quad \frac{\partial^2 z}{\partial x \, \partial y}, \quad \frac{\partial^2 z}{\partial y \, \partial x}, \quad \frac{\partial^2 z}{\partial y^2}.$$

Solution. We have

$$\frac{\partial z}{\partial x} = 3x^2 + 6xy - 4xy^2 + 3y; \qquad \frac{\partial z}{\partial y} = 3x^2 - 4x^2y - 4y^3 + 3x;$$

$$\frac{\partial^2 z}{\partial x^2} = 6x + 6y - 4y^2; \qquad \frac{\partial^2 z}{\partial y^2} = -4x^2 - 12y^2;$$

$$\frac{\partial^2 z}{\partial y \, \partial x} = 6x - 8xy + 3; \qquad \frac{\partial^2 z}{\partial x \, \partial y} = 6x - 8xy + 3.$$

In the example above, it is not accidental that $\partial^2 z/\partial y \, \partial x = \partial^2 z/\partial x \, \partial y$, as the next theorem shows.

Theorem 7. *Assume that $f(x, y), f_{,1}, f_{,2}, f_{,1,2}$, and $f_{,2,1}$ are all continuous at (x_0, y_0). Then*

$$f_{,1,2}(x_0, y_0) = f_{,2,1}(x_0, y_0).$$

In traditional notation, the formula reads

$$f_{xy}(x_0, y_0) = f_{yx}(x_0, y_0).$$

(*The order of partial differentiation may be reversed without affecting the result.*)

Proof. The result is obtained by use of a quantity we call the *double difference*, denoted by $\Delta_2 f$, and defined by the formula

$$\Delta_2 f = [f(x_0 + h, y_0 + h) - f(x_0 + h, y_0)] - [f(x_0, y_0 + h) - f(x_0, y_0)]. \quad (1)$$

We shall show that, as h tends to zero, the quantity $\Delta_2 f/h^2$ tends to $f_{xy}(x_0, y_0)$. On the other hand, we shall also show that the same quantity tends to $f_{yx}(x_0, y_0)$. The principal tool is the repeated application of the Theorem of the Mean. (See page 66.) We may write $\Delta_2 f$ in a more transparent way by defining

$$\phi(s) = f(x_0 + s, y_0 + h) - f(x_0 + s, y_0), \qquad (2)$$
$$\chi(t) = f(x_0 + h, y_0 + t) - f(x_0, y_0 + t). \qquad (3)$$

(The quantities x_0, y_0, h are considered fixed in the definition of ϕ and χ.) Then straight substitution in (1) shows that

$$\Delta_2 f = \phi(h) - \phi(0) \qquad (4)$$

and

$$\Delta_2 f = \chi(h) - \chi(0). \qquad (5)$$

We apply the Theorem of the Mean in (4) and (5), getting two expressions for $\Delta_2 f$. They are

$$\Delta_2 f = \phi'(s_1) \cdot h \qquad \text{with } 0 < s_1 < h,$$
$$\Delta_2 f = \chi'(t_1) \cdot h \qquad \text{with } 0 < t_1 < h.$$

The derivatives $\phi'(s_1)$ and $\chi'(t_1)$ are easily computed from (2) and (3). We obtain

$$\phi'(s_1) = f_{,1}(x_0 + s_1, y_0 + h) - f_{,1}(x_0 + s_1, y_0),$$
$$\chi'(t_1) = f_{,2}(x_0 + h, y_0 + t_1) - f_{,2}(x_0, y_0 + t_1),$$

and the two expressions for $\Delta_2 f$ yield

$$\frac{1}{h}\Delta_2 f = [f_{,1}(x_0 + s_1, y_0 + h) - f_{,1}(x_0 + s_1, y_0)], \tag{6}$$

$$\frac{1}{h}\Delta_2 f = [f_{,2}(x_0 + h, y_0 + t_1) - f_{,2}(x_0, y_0 + t_1)]. \tag{7}$$

In (6) the Theorem of the Mean may be applied to the expression on the right with respect to $y_0 + h$ and y_0. We get

$$\frac{1}{h}\Delta_2 f = f_{,1,2}(x_0 + s_1, y_0 + t_2)h \qquad \text{with} \qquad 0 < t_2 < h. \tag{8}$$

Similarly, the Theorem of the Mean may be applied in (7) to the expression on the right with respect to $x_0 + h$ and x_0. The result is

$$\frac{1}{h}\Delta_2 f = f_{,2,1}(x_0 + s_2, y_0 + t_1)h \qquad \text{with} \qquad 0 < s_2 < h. \tag{9}$$

Dividing by h in (8) and (9), we find that

$$\frac{1}{h^2}\Delta_2 f = f_{,1,2}(x_0 + s_1, y_0 + t_2) = f_{,2,1}(x_0 + s_2, y_0 + t_1).$$

Letting h tend to zero and noticing that s_1, s_2, t_1, and t_2 all tend to zero with h, we obtain the result.

Corollary 1. *If f is a function of any number of variables and s and t are any two of them, then*

$$f_{st} = f_{ts}.$$

For example, if the function is $f(x, y, s, t, u, v)$, then

$$f_{xt} = f_{tx}, \qquad f_{yu} = f_{uy}, \qquad f_{yv} = f_{vy}, \qquad \text{etc.}$$

The proof of the corollary is identical with the proof of the theorem.

Corollary 2. *For derivatives of the third, fourth, or any order, it does not matter in what order the differentiations with respect to the various variables are performed. For instance,*

$$\frac{\partial^4 z}{\partial x \, \partial x \, \partial y \, \partial y} = \frac{\partial^4 z}{\partial x \, \partial y \, \partial x \, \partial y} = \frac{\partial^4 z}{\partial x \, \partial y \, \partial y \, \partial x}$$

$$= \frac{\partial^4 z}{\partial y \, \partial x \, \partial x \, \partial y} = \frac{\partial^4 z}{\partial y \, \partial x \, \partial y \, \partial x} = \frac{\partial^4 z}{\partial y \, \partial y \, \partial x \, \partial x}.$$

Remarks. (i) It is true that there are functions for which f_{xy} is not equal to f_{yx}. Of course, the hypotheses of Theorem 7 are violated for such functions. (ii) All the functions we have considered thus far and all the functions we shall consider from now on will always satisfy the hypotheses of Theorem 7. Therefore the order of differentiation will be reversible throughout.

EXAMPLE 2. Given $u = e^x \cos y + e^y \sin z$, find all first partial derivatives and verify that

$$\frac{\partial^2 u}{\partial x \, \partial y} = \frac{\partial^2 u}{\partial y \, \partial x}, \qquad \frac{\partial^2 u}{\partial x \, \partial z} = \frac{\partial^2 u}{\partial z \, \partial x}, \qquad \frac{\partial^2 u}{\partial y \, \partial z} = \frac{\partial^2 u}{\partial z \, \partial y}.$$

Solution. We have

$$\frac{\partial u}{\partial x} = e^x \cos y; \qquad \frac{\partial u}{\partial y} = -e^x \sin y + e^y \sin z; \qquad \frac{\partial u}{\partial z} = e^y \cos z.$$

Therefore

$$\frac{\partial^2 u}{\partial y \, \partial x} = -e^x \sin y = \frac{\partial^2 u}{\partial x \, \partial y},$$

$$\frac{\partial^2 u}{\partial z \, \partial x} = 0 \qquad\quad = \frac{\partial^2 u}{\partial x \, \partial z},$$

$$\frac{\partial^2 u}{\partial z \, \partial y} = e^y \cos z \quad\; = \frac{\partial^2 u}{\partial y \, \partial z}.$$

EXAMPLE 3. Suppose that $u = F(x, y, z)$ and $z = f(x, y)$. Obtain a formula for $\partial^2 u / \partial x^2$ in terms of the derivatives of F (that is, F_x, F_y, F_z, F_{xx}, etc.) and the derivatives of f (or, equivalently, z). That is, in the expression for F we consider x, y, z *intermediate variables*, while in the expression for f we consider x and y *independent variables*.

Solution. We apply the Chain Rule to F to obtain $\partial u / \partial x$ with x as an independent variable. We get

$$\frac{\partial u}{\partial x} = F_x \frac{\partial x}{\partial x} + F_y \frac{\partial y}{\partial x} + F_z \frac{\partial z}{\partial x}.$$

Since x and y are independent, $\partial y / \partial x = 0$; also, $\partial x / \partial x = 1$. Therefore

$$\frac{\partial u}{\partial x} = F_x + F_z \frac{\partial z}{\partial x}.$$

In order to differentiate a second time, we must recognize that F_x and F_z are again functions of the three intermediate variables. We find that

$$\frac{\partial^2 u}{\partial x^2} = F_{xx}\frac{\partial x}{\partial x} + F_{xy}\frac{\partial y}{\partial x} + F_{xz}\frac{\partial z}{\partial x} + \frac{\partial z}{\partial x}\left(F_{zx}\frac{\partial x}{\partial x} + F_{zy}\frac{\partial y}{\partial x} + F_{zz}\frac{\partial z}{\partial x}\right) + F_z\frac{\partial^2 z}{\partial x^2}.$$

The result is

$$\frac{\partial^2 u}{\partial x^2} = F_{xx} + 2F_{xz}\frac{\partial z}{\partial x} + F_{zz}\left(\frac{\partial z}{\partial x}\right)^2 + F_z\frac{\partial^2 z}{\partial x^2}.$$

EXERCISES

In each of problems 1 through 8, verify that $f_{,1,2} = f_{,2,1}$.

1. $f(x, y) = x^2 - 2xy - 3y^2$
2. $f(x, y) = 3x^2 + 4xy + 2y^2 - 3x + 7y - 6$
3. $f(x, y) = x^3 - x^2y + 2xy^2$
4. $f(x, y) = x^4 + 4x^3y - 3x^2y^2 + 6xy^3 + 9y^4$
5. $f(r, s) = e^{rs}\sin r \cos s$
6. $f(u, v) = e^{2u}\cos v + e^{3v}\sin u$
7. $f(s, t) = \arctan(t/s)$
8. $f(x, z) = \ln\dfrac{1 + x}{1 + z} - e^{xz}$

In each of problems 9 through 13, verify that $u_{xy} = u_{yx}$ and $u_{xz} = u_{zx}$.

9. $u = \ln\sqrt{x^2 + y^2 + z^2}$ 10. $u = \ln(x + \sqrt{y^2 + z^2})$
11. $u = x^3 + y^3 + z^3 - 3xyz$
12. $u = e^{xy} + e^{2xz} - e^{3yz}$
13. $u = e^{xy}/\sqrt{x^2 + z^2}$

14. Given that $u = 1/\sqrt{x^2 + y^2 + z^2}$, verify that

$$\frac{\partial^2 u}{\partial x^2} + \frac{\partial^2 u}{\partial y^2} + \frac{\partial^2 u}{\partial z^2} = 0.$$

15. Given that $u = xe^x\cos y$, verify that

$$\frac{\partial^4 u}{\partial x^4} + 2\frac{\partial^4 u}{\partial x^2\,\partial y^2} + \frac{\partial^4 u}{\partial y^4} = 0.$$

In each of problems 16 through 19, r and s are independent variables. Find $\partial^2 z/\partial r^2$ by (a) the Chain Rule and by (b) finding z in terms of r and s first.

16. $z = x^2 - xy - y^2$, $x = r + s$, $y = s - r$
17. $z = x^2 - y^2$, $x = r\cos s$, $y = r\sin s$
18. $z = x^3 - y^3$, $x = 2r - s$, $y = s + 2r$
19. $z = x^2 - 2xy - y^2$, $x = r^2 - s^2$, $y = 2rs$

20. Given that $u = F(x, y, z)$ and $z = f(x, y)$, find $\partial^2 u/\partial y\, \partial x$ with all variables as in Example 3.

21. Given that $u = F(x, y, z)$ and $z = f(x, y)$, find $\partial^2 u/\partial y^2$ with all variables as in Example 3.

22. If $u = F(x, y)$, $y = f(x)$, find $d^2 u/dx^2$.

23. Given that $u = f(x + 2y) + g(x - 2y)$, show that

$$u_{xx} - \tfrac{1}{4} u_{yy} = 0.$$

24. Given that $u = F(x, y)$, $x = r\cos\theta$, $y = r\sin\theta$, find $\partial^2 u/\partial r^2$, r and θ being independent variables.

25. Given $u = F(x, y)$, $x = f(r, s)$, $y = g(r, s)$, find $\partial^2 u/\partial r\, \partial s$, r and s being independent variables.

26. If $F(x, y) = 0$, find $d^2 y/dx^2$ in terms of partial derivatives of F.

27. Given that $u = F(x, y)$, $x = e^s \cos t$, $y = e^s \sin t$, use the Chain Rule to show that

$$\frac{\partial^2 u}{\partial s^2} + \frac{\partial^2 u}{\partial t^2} = e^{2s}\left(\frac{\partial^2 u}{\partial x^2} + \frac{\partial^2 u}{\partial y^2}\right),$$

where s and t are independent variables and x and y are intermediate variables.

28. Given $V = F(x, y)$, $x = \tfrac{1}{2}r(e^s + e^{-s})$, $y = \tfrac{1}{2}r(e^s - e^{-s})$, show that

$$V_{xx} - V_{yy} = V_{rr} + \frac{1}{r}\,V_r + \frac{1}{r^2}\,V_{ss}.$$

*29. If $u = f(x - ut)$, show that $u_t + uu_x = 0$.

10. TAYLOR'S THEOREM WITH REMAINDER

Taylor's theorem for functions of one variable was established in Chapter 3 on page 100. There we found that if $F(x)$ has $n + 1$ derivatives in an interval containing a value x_0, then we can obtain the expansion

$$F(x) = F(x_0) + F'(x_0)(x - x_0) + \cdots + \frac{F^{(n)}(x_0)(x - x_0)^n}{n!} + R_n, \quad (1)$$

where the remainder R_n is given by the formula

$$R_n = \frac{F^{(n+1)}(\xi)(x - x_0)^{n+1}}{(n + 1)!},$$

with ξ some number between x_0 and x.

Taylor's theorem in several variables is a generalization of the expansion (1). We carry out the procedure for a function of two variables $f(x, y)$, the process for functions of more variables being completely analogous. Consider the function

$$\phi(t) = f(x + \lambda t, y + \mu t),$$

in which the quantities x, y, λ, and μ are temporarily kept constant. Then ϕ is a function of the single variable t, and we may compute its derivative. Using the Chain Rule, we obtain

$$\phi'(t) = f_{,1}(x + \lambda t, y + \mu t)\lambda + f_{,2}(x + \lambda t, y + \mu t)\mu.$$

It is convenient to use the more suggestive notation

$$\phi'(t) = f_x(x + \lambda t, y + \mu t)\lambda + f_y(x + \lambda t, y + \mu t)\mu.$$

In fact, we will simplify matters further by omitting the arguments in f. We write

$$\phi'(t) = f_x\lambda + f_y\mu.$$

It is important to compute second, third, fourth, etc., derivatives of ϕ. We do so by applying the Chain Rule repeatedly. The result is

$$\phi''(t) = \lambda^2 f_{xx} + 2\lambda\mu f_{xy} + \mu^2 f_{yy},$$
$$\phi^{(3)}(t) = \lambda^3 f_{xxx} + 3\lambda^2\mu f_{xxy} + 3\lambda\mu^2 f_{xyy} + \mu^3 f_{yyy},$$
$$\phi^{(4)}(t) = \lambda^4 f_{xxxx} + 4\lambda^3\mu f_{xxxy} + 6\lambda^2\mu^2 f_{xxyy} + 4\lambda\mu^3 f_{xyyy} + \mu^4 f_{yyyy}.$$

Examining the pattern in each of the above derivatives, we see that the coefficients in ϕ'' are formed by the **symbolic** expression

$$\left(\lambda\frac{\partial}{\partial x} + \mu\frac{\partial}{\partial y}\right)^2 f,$$

provided that the exponent applied to a partial derivative is interpreted as repeated differentiation instead of multiplication. Using this new symbolism we can easily write any derivative of ϕ. The kth derivative is

$$\phi^{(k)}(t) = \left(\lambda\frac{\partial}{\partial x} + \mu\frac{\partial}{\partial y}\right)^k f. \tag{2}$$

For instance, with $k = 7$ we obtain

$$\phi^{(7)}(t) = \lambda^7 f_{xxxxxxx} + 7\lambda^6\mu f_{xxxxxxy} + \frac{7 \cdot 6}{1 \cdot 2}\lambda^5\mu^2 f_{xxxxxyy} + \cdots + \mu^7 f_{yyyyyyy}.$$

Of course all derivatives are evaluated at $(x + \lambda t, y + \mu t)$. The validity of (2) may be established by induction. (See Morrey, *University Calculus*, pp. 574 and 695.)

Before stating Taylor's theorem for functions of two variables, we must introduce still more symbols. The quantity

$$\sum_{1 \le r+s \le p} (\quad)$$

means that the sum of the terms in parentheses is taken over all possible combinations of r and s which add up to a number between 1 and p. Neither r nor s is allowed to be negative. For example, if $p = 3$ the combinations are

$$(r = 0, s = 1), \quad (r = 0, s = 2), \quad (r = 0, s = 3)$$
$$(r = 1, s = 0), \quad (r = 1, s = 1), \quad (r = 1, s = 2)$$
$$(r = 2, s = 0), \quad (r = 2, s = 1)$$
$$(r = 3, s = 0).$$

The symbol

$$\sum_{r+s=p} (\quad)$$

means that the sum is taken over all possible nonnegative combinations of r and s which add up to p exactly. For instance, if $p = 3$, then the combinations are

$$(r = 0, s = 3), \quad (r = 1, s = 2), \quad (r = 2, s = 1), \quad (r = 3, s = 0).$$

Theorem 8. (Taylor's Theorem.) *Suppose that f is a function of two variables and that f and all of its partial derivatives of order up to $p + 1$ are continuous in a neighborhood of the point (a, b). Then we have the expansion*

$$f(x, y) = f(a, b) + \sum_{1 \le r+s \le p} \frac{\partial^{r+s} f(a, b)}{\partial x^r \, \partial y^s} \frac{(x - a)^r}{r!} \cdot \frac{(y - b)^s}{s!} + R_p, \quad (3)$$

where the remainder R_p is given by the formula

$$R_p = \sum_{r+s=p+1} \frac{\partial^{r+s} f(\xi, \eta)}{\partial x^r \, \partial y^s} \frac{(x - a)^r}{r!} \frac{(y - b)^s}{s!},$$

with the value (ξ, η) situated on the line segment joining the points (a, b) to (x, y). (See Fig. 4–11.)

Proof. We let $d = \sqrt{(x - a)^2 + (y - b)^2}$ and define

$$\lambda = \frac{x - a}{d}, \quad \mu = \frac{y - b}{d}.$$

The function

$$\phi(t) = f(a + \lambda t, b + \mu t), \quad 0 \le t \le d$$

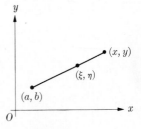

FIGURE 4–11

may be differentiated according to the rules described at the beginning of the section. Taylor's theorem for $\phi(t)$ (a function of *one* variable) taken about $t = 0$ and evaluated at d yields

$$\phi(d) = \phi(0) + \phi'(0)d + \phi''(0)\frac{d^2}{2!} + \cdots$$
$$+ \frac{\phi^{(p)}(0) \, d^p}{p!} + \frac{\phi^{(p+1)}(\tau) \, d^{p+1}}{(p + 1)!}. \quad (4)$$

The kth derivative of ϕ evaluated at 0 is given by

$$\phi^k(0) = \left(\lambda \frac{\partial}{\partial x} + \mu \frac{\partial}{\partial y}\right)^k f(a, b).$$

We now recall the binomial formula:

$$(A + B)^k = A^k + \frac{k}{1} A^{k-1}B + \frac{k(k-1)}{1 \cdot 2} A^{k-2}B^2 + \cdots + B^k$$

$$= \sum_{q=0}^{k} \frac{k!}{q!(k-q)!} A^{k-q}B^q.$$

Applying the binomial formula to the symbolic expression for $\phi^{(k)}(0)$, we obtain

$$\phi^{(k)}(0) = \sum_{q=0}^{k} \frac{k!}{q!(k-q)!} \frac{\partial^k f(a, b)}{\partial x^{k-q} \, \partial y^q} \lambda^{k-q}\mu^q. \tag{5}$$

Noting that $\phi(d) = f(a + \lambda d, b + \mu d) = f(x, y)$ and that $\phi(0) = f(a, b)$, we find, upon substitution of (5) into (4):

$$f(x, y) = f(a, b) + \frac{\partial f(a, b)}{\partial x}(x - a) + \frac{\partial f(a, b)}{\partial y}(y - b)$$

$$+ \frac{\partial^2 f(a, b)}{\partial x^2} \frac{(x - a)^2}{2!} + \frac{\partial^2 f(a, b)}{\partial x \, \partial y}(x - a)(y - b)$$

$$+ \frac{\partial^2 f(a, b)}{\partial y^2} \frac{(y - b)^2}{2!} + \cdots,$$

which is precisely the formula in the statement of the theorem. The remainder term shows that if $0 < \tau < d$, then

$$\xi = a + \frac{(x - a)}{d}\tau,$$

$$\eta = b + \frac{(y - b)}{d}\tau,$$

which places (ξ, η) on the line segment joining (a, b) and (x, y).

Remarks. (i) Taylor's formula [as (3) is usually called] is also often written in the form

$$f(x, y) = f(a, b)$$

$$+ \sum_{q=1}^{p} \frac{1}{q!} \left[\sum_{r=0}^{q} \frac{q!}{(q - r)!r!} \frac{\partial^q f(a, b)}{\partial x^{q-r} \, \partial y^r}(x - a)^{q-r}(y - b)^r \right] + R_p.$$

(ii) For some functions f, if we let p tend to infinity we may find that $R_p \to 0$. We thereby obtain a representation of f as an infinite series in x and y. Such a series is called a **double series,** and we say that f is **expanded about the point** (a, b).

For functions of three variables we obtain a triple sum, the Taylor formula in this case being

$$f(x, y, z) = f(a, b, c)$$
$$+ \sum_{1 \le r+s+t \le p} \frac{\partial^{r+s+t} f(a, b, c)}{\partial x^r \, \partial y^s \, \partial z^t} \frac{(x-a)^r}{r!} \frac{(y-b)^s}{s!} \frac{(z-c)^t}{t!} + R_p,$$

with

$$R_p = \sum_{r+s+t=p+1} \frac{\partial^{p+1} f(\xi, \eta, \zeta)}{\partial x^r \, \partial y^s \, \partial z^t} \frac{(x-a)^r}{r!} \frac{(y-b)^s}{s!} \frac{(z-x)^t}{t!},$$

where (ξ, η, ζ) is on the line segment joining (a, b, c) and (x, y, z).

EXAMPLE 1. Expand $x^2 y$ about the point $(1, -2)$ out to and including the terms of the second degree. Find R_2.

Solution. Setting $f(x, y) = x^2 y$, we obtain

$$f_x = 2xy, \quad f_y = x^2, \quad f_{xx} = 2y, \quad f_{xy} = 2x, \quad f_{yy} = 0,$$
$$f_{xxx} = 0, \quad f_{xxy} = 2, \quad f_{xyy} = f_{yyy} = 0.$$

Noting that $f(1, -2) = -2$, we find

$$x^2 y = -2 - 4(x-1) + (y+2) + \frac{1}{2!}[-4(x-1)^2 + 4(x-1)(y+2)] + R_2,$$

with

$$R_2 = \frac{1}{3!} \, 3 \cdot 2(x-1)^2 (y+2) = (x-1)^2 (y+2).$$

EXAMPLE 2. Given $f(x, y, z) = e^x \cos y + e^y \cos z + e^z \cos x$. Define

$$\phi(t) = f(x + \lambda t, y + \mu t, z + \nu t).$$

Find $\phi'(0)$ and $\phi''(0)$ in terms of $x, y, z, \lambda, \mu, \nu$.

Solution. We have

$$\phi'(0) = f_{,1}(x, y, z)\lambda + f_{,2}(x, y, z)\mu + f_{,3}(x, y, z)\nu.$$

Computing the derivatives, we obtain

$$\phi'(0) = (e^x \cos y - e^z \sin x)\lambda + (e^y \cos z - e^x \sin y)\mu + (e^z \cos x - e^y \sin z)\nu.$$

The formula for $\phi''(0)$ is

$$\phi''(0) = \left(\lambda \frac{\partial}{\partial x} + \mu \frac{\partial}{\partial y} + \nu \frac{\partial}{\partial z} \right)^2 f$$
$$= \lambda^2 (e^x \cos y - e^z \cos x) + \mu^2 (e^y \cos z - e^x \cos y)$$
$$+ \nu^2 (e^z \cos x - e^y \cos z) + 2\lambda\mu(-e^x \sin y)$$
$$+ 2\lambda\nu(-e^z \sin x) + 2\mu\nu(-e^y \sin z).$$

EXERCISES

1. Expand $x^3 + xy^2$ about the point $(2, 1)$.

2. Expand $x^4 + x^2y^2 - y^4$ about the point $(1, 1)$ out to terms of the second degree. Find the form of R_2.

3. Find the expansion of $\sin (x + y)$ about $(0, 0)$ out to and including the terms of the third degree in (x, y). Compare the result with that which you get by writing $\sin u \approx u - \frac{1}{6}u^3$ and setting $u = x + y$.

4. Find the expansion of $\cos (x + y)$ about $(0, 0)$ out to and including terms of the fourth degree in (x, y). Compare the result with that which you get by writing $\cos u = 1 - \frac{1}{2}u^2 + \frac{1}{24}u^4$ and setting $u = x + y$.

5. Find the expansion of e^{x+y} about $(0, 0)$ out to and including the terms of the third degree in (x, y). Compare the result with that which you get by setting $e^u \approx 1 + u + \frac{1}{2}u^2 + \frac{1}{6}u^3$, and then setting $u = x + y$. Next compare the result with that obtained by multiplying the series for e^x by that for e^y and keeping terms up to and including the third degree.

6. Find the expansion of $\sin x \sin y$ about $(0, 0)$ out to and including the terms of the fourth degree in (x, y). Compare the result with that which you get by multiplying the series for $\sin x$ and $\sin y$.

7. Do the same as problem 6 for $\cos x \cos y$.

8. Expand $e^x \arctan y$ about $(1, 1)$ out to and including the terms of the second degree in $(x - 1)$ and $(y - 1)$.

9. Expand $x^2 + 2xy + yz + z^2$ about $(1, 1, 0)$.

10. Expand $x^3 + x^2y - yz^2 + z^3$ about $(1, 0, 1)$ out to and including the terms of the second degree in $(x - 1)$, y, and $(z - 1)$.

11. If $f(x, y) = x^2 + 4xy + y^2 - 6x$ and $\phi(t) = f(x + \lambda t, y + \mu t)$, find $\phi''(0)$ when $x = -1$ and $y = 2$. Is $\phi''(0) > 0$ when $(x, y) = (-1, 2)$ if λ and μ are related so that $\lambda^2 + \mu^2 = 1$?

12. If $f(x, y) = x^3 + 3xy^2 - 3x^2 - 3y^2 + 4$ and $\phi(t) = f(x + \lambda t, y + \mu t)$, find $\phi''(0)$ for $(x, y) = (2, 0)$. Show that $\phi''(0) > 0$ for all λ and μ such that $\lambda^2 + \mu^2 = 1$.

*13. (a) Write the appropriate expansion formula using binomial coefficients for

$$(A + B + C)^k,$$

with k a positive integer. (b) If $f(x, y, z)$ and $\phi(t) = f(x + \lambda t, y + \mu t, z + \nu t)$ are sufficiently differentiable, show the relationship between the symbolic expression

$$\left(\lambda \frac{\partial}{\partial x} + \mu \frac{\partial}{\partial y} + \nu \frac{\partial}{\partial z} \right)^k f(x + \lambda t, y + \mu t, z + \nu t) \qquad \text{and} \qquad \phi^{(k)}(t).$$

14. Write Taylor's formula for a function $f(x, y, u, v)$ of four variables expanded about the point a, b, c, d. How many second derivative terms are there? Third derivative terms?

11. MAXIMA AND MINIMA

One of the principal applications of differentiation of functions of one variable occurs in the study of maxima and minima. In Chapter 7 of *First Course* we derived various tests using first and second derivatives which enable us to determine relative maxima and minima of functions of a single variable. These tests are useful for graphing functions, for solving problems involving related rates, and for attacking a variety of geometrical and physical problems. (See *First Course*, Chapter 7, Sections 3, 4, 6, and 7.)

The study of maxima and minima for functions of two, three, or more variables has its basis in the following theorem, which is stated without proof.

Theorem 9. *Let R be a region in the xy-plane with the boundary curve of R considered as part of R also* (Fig. 4–12). *If f is a function of two variables defined and continuous in R, then there is* (at least) *one point in R where f takes on a maximum value and there is* (at least) *one point in R where f takes on a minimum value.*

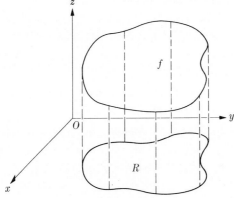

FIGURE 4–12

Remarks. (i) Theorem 9 is a straightforward generalization of Theorem 1 given in *First Course* on page 147 (Extreme Value Theorem). (ii) Analogous theorems may be stated for functions of three, four, or more variables. (iii) The maximum and minimum may occur on the boundary of R. Thus, as in the case of one variable where the interval must be *closed*, the region R *must contain its boundary* in order to ensure the validity of the result.

DEFINITION. A function $f(x, y)$ is said to have a **relative maximum** at (x_0, y_0) if there is some region containing (x_0, y_0) *in its interior* such that

$$f(x, y) \leq f(x_0, y_0)$$

for all (x, y) in this region. More precisely, there must be some positive number δ (which may be "small") such that the above inequality holds for all (x, y) in the square

$$|x - x_0| < \delta, \qquad |y - y_0| < \delta.$$

A similar definition holds for **relative minimum** when the inequality $f(x, y) \geq f(x_0, y_0)$ is satisfied in a square about (x_0, y_0). The above definitions are easily extended to functions of three, four, or more variables.

Theorem 10. *Suppose that $f(x, y)$ is defined in a region R containing (x_0, y_0) in its interior. Suppose that $f_{,1}(x_0, y_0)$ and $f_{,2}(x_0, y_0)$ are defined and that*

$$f(x, y) \leq f(x_0, y_0)$$

for all (x, y) in R; that is, $f(x_0, y_0)$ is a relative maximum. Then

$$f_{,1}(x_0, y_0) = f_{,2}(x_0, y_0) = 0.$$

Proof. We show that $f_{,1}(x_0, y_0) = 0$, the proof for $f_{,2}$ being analogous. By definition,

$$f_{,1}(x_0, y_0) = \lim_{h \to 0} \frac{f(x_0 + h, y_0) - f(x_0, y_0)}{h}.$$

By hypothesis,

$$f(x_0 + h, y_0) - f(x_0, y_0) \leq 0$$

for all h sufficiently small so that $(x_0 + h, y_0)$ is in R. If h is positive, then

$$\frac{f(x_0 + h, y_0) - f(x_0, y_0)}{h} \leq 0,$$

and as $h \to 0$ we conclude that $f_{,1}$ must be nonpositive. On the other hand, if $h < 0$, then

$$\frac{f(x_0 + h, y_0) - f(x_0, y_0)}{h} \geq 0,$$

since division of both sides of an inequality by a negative number reverses its direction. Letting $h \to 0$, we conclude that $f_{,1}$ is nonnegative. A quantity which is both nonnegative and nonpositive vanishes.

Corollary. *The same result holds at a relative minimum.*

DEFINITION. A value (x_0, y_0) at which both $f_{,1}$ and $f_{,2}$ vanish is called a **critical point of f**.

Discussion. The conditions that $f_{,1}$ and $f_{,2}$ vanish at a point are *necessary* conditions for a relative maximum or a relative minimum. It is easy to find a function for which $f_{,1}$ and $f_{,2}$ vanish at a point, with the function having neither a relative maximum nor a relative minimum at that point. A critical point at which f is neither a maximum nor a minimum may be a "**saddle point**." A simple example of a function which has such a point is given by

$$f(x, y) = x^2 - y^2.$$

Note that for any x, y, a, b the relation $\lambda^2 + \mu^2 = 1$ prevails. The Taylor expansion (1) now becomes

$$f(x, y) - f(a, b) = \tfrac{1}{2}r^2(A\lambda^2 + 2B\lambda\mu + C\mu^2 + r\rho), \qquad (2)$$

where

$$\rho = \frac{1}{3}\left(\frac{\partial^3 f}{\partial x^3}\lambda^3 + 3\frac{\partial^3 f}{\partial x^2\,\partial y}\lambda^2\mu + 3\frac{\partial^2 f}{\partial x\,\partial y^2}\lambda\mu^2 + \frac{\partial^3 f}{\partial y^3}\mu^3\right)_{\substack{x=\xi \\ y=\eta}}.$$

The quantity ρ is bounded since, by hypothesis, f has continuous third derivatives. The behavior of $f(x, y) - f(a, b)$ is determined completely by the size of $r\rho$ and the size of the quadratic expression

$$A\lambda^2 + 2B\lambda\mu + C\mu^2, \qquad (3)$$

with $\lambda^2 + \mu^2 = 1$. If

$$B^2 - AC < 0 \qquad \text{and} \qquad A > 0,$$

then there are no real roots to (3) and it has a positive minimum value. (Call it m.) Now, selecting r so small that $r\rho$ is negligible compared with m, we deduce that the right side of (2) is always positive if (x, y) is sufficiently close to (a, b). Hence

$$f(x, y) - f(a, b) > 0$$

and f is a minimum at (a, b). We have just established part (i) of the theorem. By the same argument, if

$$B^2 - AC < 0 \qquad \text{and} \qquad A < 0,$$

then (3) is always negative and

$$f(x, y) - f(a, b) < 0$$

for (x, y) near (a, b). Thus the statement of (ii) follows. Part (iii) results when $B^2 - AC > 0$, in which case (3) (and therefore (2)) is sometimes positive and sometimes negative. Then f can have neither a maximum nor a minimum at (a, b) and the surface $z = f(x, y)$ can be shown to be saddle-shaped near (a, b). Part (iv) is provided for completeness.

EXAMPLE 2. Test for relative maxima and minima the function f defined by

$$f(x, y) = x^3 + 3xy^2 - 3x^2 - 3y^2 + 4.$$

Solution. We have

$$f_{,1} = 3x^2 + 3y^2 - 6x \qquad \text{and} \qquad f_{,2} = 6xy - 6y.$$

We set these equations equal to zero and solve simultaneously. Writing

$$x^2 + y^2 - 2x = 0, \qquad y(x - 1) = 0,$$

we see that the second equation vanishes only when $y = 0$ or $x = 1$. If $y = 0$, the first equation gives $x = 0$ and 2; if $x = 1$, the first equation gives $y = \pm 1$. The critical points are

$$(0, 0), \quad (2, 0), \quad (1, 1), \quad (1, -1).$$

To apply the Second Derivative Test, we compute

$$A = f_{,1,1} = 6x - 6, \qquad B = f_{,1,2} = 6y, \qquad C = f_{,2,2} = 6x - 6.$$

At $(0, 0)$: $AC - B^2 > 0$ and $A < 0$,   a maximum.

At $(2, 0)$: $AC - B^2 > 0$ and $A > 0$,   a minimum.

At $(1, 1)$: $AC - B^2 < 0$,   saddle point.

At $(1, -1)$: $AC - B^2 < 0$,   saddle point.

EXAMPLE 3. Find the dimensions of the rectangular box, open at the top, which has maximum volume if the surface area is 12.

Solution. Let V be the volume of the box; let (x, y) be the horizontal directions and z the height. Then

$$V = xyz,$$

and the surface area is given by

$$xy + 2xz + 2yz = 12.$$

Solving this equation for z and substituting its value in the expression for V, we get

$$V = \frac{xy(12 - xy)}{2(x + y)} = \frac{12xy - x^2y^2}{2(x + y)}.$$

The domains for x, y, z are restricted by the inequalities

$$x > 0, \qquad y > 0, \qquad xy < 12.$$

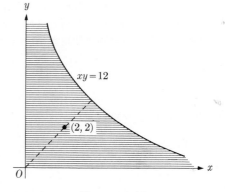

FIGURE 4–14

In other words, x and y must lie in the shaded region shown in Fig. 4–14. To find the critical points, we compute

$$V_x = \frac{y^2(12 - x^2 - 2xy)}{2(x + y)^2}, \qquad V_y = \frac{x^2(12 - y^2 - 2xy)}{2(x + y)^2}$$

and set these expressions equal to zero. We obtain (excluding $x = y = 0$)

$$x^2 + 2xy = 12, \qquad y^2 + 2xy = 12.$$

Subtracting, we find that $x = \pm y$. If $x = y$, then the positive solution is $x = y = 2$. We reject $x = -y$, since both quantities must be positive. From the formula for surface area we conclude that $z = 1$ when $x = y = 2$. From geometrical considerations, we conclude that these are the dimensions which give a maximum volume.

Remarks. (i) The determination of maxima and minima hinges on our ability to solve the two simultaneous equations in two unknowns resulting when we set

$f_x = 0$ and $f_y = 0$. In Example 1 these equations are linear and so quite easy to solve. In Examples 2 and 3, however, the equations are nonlinear, and there are no routine methods for solving nonlinear simultaneous equations. Elementary courses in algebra usually avoid such topics, and the student is left to his own devices. The only general rule we can state is: try to solve one of the equations for one of the unknowns in terms of the other. Substitute this value in the second equation and try to find all solutions of the second equation. Otherwise use trickery and guesswork. In actual practice, systems of nonlinear equations may be solved by a variety of numerical techniques. The remarkable achievements of the electronic computer come to the fore in such problems. (ii) Definitions of critical point, relative maximum and minimum, etc., for functions of three, four, and more variables are simple extensions of the two-variable case. If $f(x, y, z)$ has first derivatives, then a point where

$$f_x = 0, \qquad f_y = 0, \qquad f_z = 0$$

is a **critical point**. We obtain such points by solving simultaneously three equations in three unknowns. To obtain the critical points for functions of n variables, we set all n first derivatives equal to zero and solve simultaneously the n equations in n unknowns. Extensions of the Second Derivative Test for functions of three or more variables are given in advanced courses.

EXERCISES

In each of problems 1 through 13, test the functions f for relative maxima and minima.

1. $f(x, y) = x^2 + 2y^2 - 4x + 4y - 3$
2. $f(x, y) = x^2 - y^2 + 2x - 4y - 2$
3. $f(x, y) = x^2 + 2xy + 3y^2 + 2x + 10y + 9$
4. $f(x, y) = x^2 - 3xy + y^2 + 13x - 12y + 13$
5. $f(x, y) = y^3 + x^2 - 6xy + 3x + 6y - 7$
6. $f(x, y) = x^3 + y^2 + 2xy + 4x - 3y - 5$
7. $f(x, y) = 3x^2y + x^2 - 6x - 3y - 2$
8. $f(x, y) = xy + 4/x + 2/y$ 9. $f(x, y) = \sin x + \sin y + \sin(x + y)$
10. $f(x, y) = x^3 - 6xy + y^3$ 11. $f(x, y) = 8^{2/3} - x^{2/3} - y^{2/3}$
12. $f(x, y) = e^x \cos y$ 13. $f(x, y) = e^{-x} \sin^2 y$

In each of problems 14 through 17, find the critical points.

14. $f(x, y, z) = x^2 + 2y^2 + z^2 - 6x + 3y - 2z - 5$
15. $f(x, y, z) = x^2 + y^2 - 2z^2 + 3x + y - z - 2$
16. $f(x, y, z) = x^2 + y^2 + z^2 + 2xy - 3xz + 2yz - x + 3y - 2z - 5$
17. $f(x, y, z, t) = x^2 + y^2 + z^2 - t^2 - 2xy + 4xz + 3xt - 2yt + 4x - 5y - 3$
18. In three-dimensional space find the minimum distance from the origin to the plane

$$3x + 4y + 2z = 6.$$

19. In the plane find the minimum distance from the point $(-1, -3)$ to the line $x + 3y = 7$.

20. In three-dimensional space find the minimum distance from the point $(-1, 3, 2)$ to the plane
$$x + 3y - 2z = 8.$$

21. In three-dimensional space find the minimum distance from the origin to the cone
$$z^2 = (x - 1)^2 + (y - 2)^2.$$

22. For a package to go by parcel post, the sum of the length and girth (perimeter of cross-section) must not exceed 100 in. Find the dimensions of the package of largest volume which can be sent; assume the package has the shape of a rectangular box.

23. Find the dimensions of the rectangular parallelepiped of maximum volume with edges parallel to the axes which can be inscribed in the ellipsoid
$$\frac{x^2}{9} + \frac{y^2}{4} + \frac{z^2}{16} = 1.$$

24. Find the shape of the closed rectangular box of largest volume with a surface area of 16 sq in.

25. The base of an open rectangular box costs half as much per square foot as the sides. Find the dimensions of the box of largest volume which can be made for D dollars.

*26. The cross-section of a trough is an isosceles trapezoid (see Fig. 4–15). If the trough is made by bending up the sides of a strip of metal 18 in. wide, what should the dimensions be in order for the area of the cross-section to be a maximum? Choose h and l as independent variables.

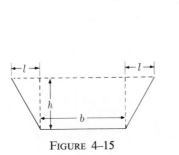 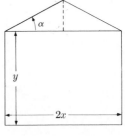

FIGURE 4–15   FIGURE 4–16

*27. A pentagon is composed of a rectangle surmounted by an isosceles triangle (see Fig. 4–16). If the pentagon has a given perimeter P, find the dimensions for maximum area. Choose variables as indicated in Fig. 4–16.

12. MAXIMA AND MINIMA; LAGRANGE MULTIPLIERS

In Example 2 of Section 11 (page 173), we solved the problem of finding the relative maxima and minima of the function
$$f(x, y) = x^3 + 3xy^2 - 3x^2 - 3y^2 + 4. \tag{1}$$

In Example 1 of the same section (page 171), we solved the problem of finding the minimum of the function

$$f(x, y, z) = x^2 + y^2 + z^2, \tag{2}$$

subject to the condition that (x, y, z) is on the plane

$$2x + 3y - z - 1 = 0. \tag{3}$$

The problem of finding the critical points of (1) is quite different from that of finding those of (2) because, in the latter case, the additional condition (3) is attached. This distinction leads to the following definitions.

DEFINITIONS. The problem of finding maxima and minima of a function of several variables [such as (1) above] without added conditions is called a problem in **free maxima and minima.** When a condition such as (3) is imposed on a function such as (2) above, the problem of determining the maximum and minimum of that function is called a problem in **constrained maxima and minima.** The added condition is called a **side condition.** Problems in maxima and minima may have one or more side conditions. When side conditions occur, they are usually crucial. For example, the minimum of the function f given by (2) without a side condition is obviously zero.

While the problem of minimizing (2) with the side condition (3) has already been solved, we shall do it again by a new and interesting method. This method, due to Lagrange, changes a problem in constrained maxima and minima to a problem in free maxima and minima.

We first introduce a new variable, traditionally denoted by λ, and form the function

$$F(x, y, z, \lambda) = (x^2 + y^2 + z^2) + \lambda(2x + 3y - z - 1).$$

The problem of finding the critical points of (2) with side condition (3) can be shown to be equivalent (under rather general circumstances) to that of finding the critical points of F considered as a function of the *four* variables x, y, z, λ. (See the end of this section.) We proceed by computing F_x, F_y, F_z, and F_λ and setting each of these expressions equal to zero. We obtain

$$F_x = 2x + 2\lambda = 0, \qquad F_y = 2y + 3\lambda = 0,$$
$$F_z = 2z - \lambda = 0, \qquad F_\lambda = 2x + 3y - z - 1 = 0.$$

Note that the equation $F_\lambda = 0$ is precisely the side condition (3). That is, any solution to the problem will automatically satisfy the side condition. We solve these equations simultaneously by writing

$$x = -\lambda, \qquad y = -\tfrac{3}{2}\lambda, \qquad z = \tfrac{1}{2}\lambda,$$
$$2(-\lambda) + 3(-\tfrac{3}{2}\lambda) - (\tfrac{1}{2}\lambda) - 1 = 0,$$

and we get $\lambda = -\frac{1}{7}$, $x = \frac{1}{7}$, $y = \frac{3}{14}$, $z = -\frac{1}{14}$. The solution satisfies $F_\lambda = 0$ and so is on the plane (2).

The general method, known as the *method of Lagrange multipliers*, may be stated as follows: In order to find the critical points of a function

$$f(x, y, z)$$

subject to the side condition

$$\phi(x, y, z) = 0,$$

form the function

$$F(x, y, z, \lambda) = f(x, y, z) + \lambda\phi(x, y, z)$$

and find the critical points of F considered as a function of the four variables x, y, z, λ.

The method is quite general in that several "multipliers" may be introduced if there are several side conditions. To find the critical points of

$$f(x, y, z),$$

subject to the conditions

$$\phi_1(x, y, z) = 0 \quad \text{and} \quad \phi_2(x, y, z) = 0. \tag{4}$$

form the function

$$F(x, y, z, \lambda_1, \lambda_2) = f(x, y, z) + \lambda_1\phi_1(x, y, z) + \lambda_2\phi_2(x, y, z)$$

and find the critical points of F as a function of the five variables x, y, z, λ_1, and λ_2.

We shall exhibit the method by working several examples.

———————

EXAMPLE 1. Find the minimum of the function

$$f(x, y) = x^2 + 2y^2 + 2xy + 2x + 3y,$$

subject to the condition that x and y satisfy the equation

$$x^2 - y = 1.$$

Solution. We form the function

$$F(x, y, \lambda) = (x^2 + 2y^2 + 2xy + 2x + 3y) + \lambda(x^2 - y - 1).$$

Then

$$F_x = 2x + 2y + 2 + 2x\lambda = 0,$$
$$F_y = 4y + 2x + 3 - \lambda = 0,$$
$$F_\lambda = x^2 - y - 1 = 0.$$

Substituting $y = x^2 - 1$ in the first two equations, we get

$$x + x^2 - 1 + 1 + \lambda x = 0, \qquad 4x^2 - 4 + 2x + 3 = \lambda.$$

Solving, we obtain

$$x = 0, \qquad y = -1, \qquad \lambda = -1,$$

and

$$x = -\tfrac{3}{4}, \qquad y = -\tfrac{7}{16}, \qquad \lambda = -\tfrac{1}{4}.$$

Evaluating f at these points, we find that a lower value occurs when $x = -\tfrac{3}{4}, y = -\tfrac{7}{16}$. From geometrical considerations we conclude that f is a minimum at this value.

Remarks. We could have solved this problem as a simple maximum and minimum problem by substituting $y = x^2 - 1$ in the equation for f and finding the critical points of the resulting function of the single variable x. However, in some problems the side condition may be so complicated that we cannot easily solve for one of the variables in terms of the others, although it may be possible to do so theoretically. It is in such cases that the power of the method of Lagrange multipliers becomes apparent. The system of equations obtained by setting the first derivatives equal to zero may be solvable even though the side condition alone may not be. The next example illustrates this point.

EXAMPLE 2. Find the critical values of

$$f(x, y) = x^2 + y^2, \tag{5}$$

subject to the condition that

$$x^3 + y^3 - 6xy = 0. \tag{6}$$

Solution. We form the function

$$F(x, y, \lambda) = x^2 + y^2 + \lambda(x^3 + y^3 - 6xy)$$

and obtain the derivatives

$$F_x = 2x + 3x^2\lambda - 6y\lambda = 0,$$
$$F_y = 2y + 3y^2\lambda - 6x\lambda = 0,$$
$$F_\lambda = x^3 + y^3 - 6xy = 0.$$

Solving simultaneously, we find from the first two equations that

$$\lambda = \frac{-2x}{3x^2 - 6y}, \qquad \lambda = \frac{-2y}{3y^2 - 6x}, \qquad \text{and} \qquad x(3y^2 - 6x) = y(3x^2 - 6y).$$

The equations

$$x^2y - xy^2 + 2x^2 - 2y^2 = 0,$$
$$x^3 + y^3 - 6xy = 0$$

may be solved simultaneously by a trick. Factoring the first equation, we see that

$$(x - y)(2x + 2y + xy) = 0$$

and $x = y$ is a solution. When $x = y$, the second equation yields

$$2x^3 - 6x^2 = 0, \qquad x = 0, 3.$$

We discard the complex solutions obtained by setting $2x + 2y + xy = 0$. The values $x = 0$, $y = 0$ clearly yield a minimum, while from geometric considerations (Fig. 4–17), the point $x = 3$, $y = 3$ corresponds to a relative maximum. There is no true maximum of f, since $x^2 + y^2$ (the distance from the origin to the curve) grows without bound if either x or y does.

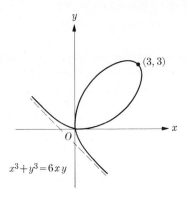

FIGURE 4–17

Note that it is not easy to solve Eq. (6) for either x or y and substitute in (5) to get a function of one variable. Therefore the methods of one-dimensional calculus are not readily usable in this problem.

The next example illustrates the technique when there are two side conditions.

EXAMPLE 3. Find the minimum of the function

$$f(x, y, z, t) = x^2 + 2y^2 + z^2 + t^2,$$

subject to the conditions

$$x + 3y - z + \ t = 2, \tag{7}$$
$$2x - \ y + z + 2t = 4. \tag{8}$$

Solution. We form the function

$$F(x, y, z, t, \lambda_1, \lambda_2) = (x^2 + 2y^2 + z^2 + t^2) + \lambda_1(x + 3y - z + t - 2)$$
$$+ \lambda_2(2x - y + z + 2t - 4).$$

We have

$$F_x = 2x + \lambda_1 + 2\lambda_2 = 0, \qquad F_t = 2t + \lambda_1 + 2\lambda_2 = 0,$$
$$F_y = 4y + 3\lambda_1 - \lambda_2 = 0, \qquad F_{\lambda_1} = x + 3y - z + t - 2 = 0,$$
$$F_z = 2z - \lambda_1 + \lambda_2 = 0, \qquad F_{\lambda_2} = 2x - y + z + 2t - 4 = 0.$$

Solving these six linear equations in six unknowns is tedious but routine. We obtain

$$x = \tfrac{67}{69}, \qquad y = \tfrac{6}{69}, \qquad z = \tfrac{14}{69}, \qquad t = \tfrac{67}{69}.$$

The corresponding values of λ_1 and λ_2 are: $\lambda_1 = -26/69$, $\lambda_2 = -54/69$.

The validity of the method of Lagrange multipliers hinges on the ability to solve an equation for a side condition such as

$$\phi(x, y, z) = 0 \tag{9}$$

for one of the unknowns in terms of the other two. Theorems which state when such a process can be performed (theoretically, that is, not actually) are called *implicit function theorems* and are studied in Chapter 11. Suppose we wish to find the critical values of

$$f(x, y, z)$$

with the side condition

$$\phi(x, y, z) = 0.$$

If we assume that (x_0, y_0, z_0) is the point where f has its critical value, and if we assume that $\phi_z(x_0, y_0, z_0) \neq 0$, then it is possible to establish the validity of the method of Lagrange multipliers. It can be shown that if

$$\phi_z(x_0, y_0, z_0) \neq 0,$$

then we may solve the equation $\phi(x, y, z) = 0$ for z in terms of x and y, so that $z = g(x, y)$. We now set

$$H(x, y) = f[x, y, g(x, y)],$$

and we note that H has a critical point at (x_0, y_0). Therefore

$$H_x = f_x + f_z g_x = 0, \qquad H_y = f_y + f_z g_y = 0. \tag{10}$$

But, by differentiating (9) implicitly, we obtain

$$\frac{\partial z}{\partial x} = g_x = -\frac{\phi_x}{\phi_z}, \qquad \frac{\partial z}{\partial y} = g_y = -\frac{\phi_y}{\phi_z}, \tag{11}$$

$$[\text{since } \phi_z \neq 0 \text{ near } (x_0, y_0, z_0)].$$

Substituting (11) into (10), we find

$$f_x - \frac{f_z}{\phi_z}\phi_x = 0 \qquad \text{and} \qquad f_y - \frac{f_z}{\phi_z}\phi_y = 0.$$

We add to these equations the obvious identity

$$f_z - \frac{f_z}{\phi_z}\phi_z = 0,$$

and then we set $\lambda_0 = -f_z(x_0, y_0, z_0)/\phi_z(x_0, y_0, z_0)$. In this way we obtain the equations

$$f_x + \lambda_0\phi_x = 0, \qquad f_y + \lambda_0\phi_y = 0, \qquad f_z + \lambda_0\phi_z = 0, \qquad \phi = 0,$$

which are just the equations satisfied at a critical point of $F = f + \lambda\phi$. The proof when there are more side conditions is similar but somewhat more complicated.

EXERCISES

Solve the following problems by the method of Lagrange multipliers.

1. Find the minimum of $f(x, y, z) = x^2 + y^2 + z^2$ subject to the condition that $x + 3y - 2z = 4$.

2. Find the minimum of $f(x, y, z) = 3x^2 + 2y^2 + 4z^2$ subject to the condition that $2x + 4y - 6z + 5 = 0$.

3. Find the minimum of $f(x, y, z) = x^2 + y^2 + z^2$ subject to the condition that $ax + by + cz = d$.

4. Find the minimum of $f(x, y, z) = ax^2 + by^2 + cz^2$ subject to the condition that $dx + ey + gz + h = 0$ (a, b, c positive).

5. Find the minimum of $f(x, y, z) = x^2 + y^2 + z^2$ if (x, y, z) is on the line of intersection of the planes

$$x + 2y + z - 1 = 0, \qquad 2x - y - 3z - 4 = 0.$$

6. Find the minimum of $f(x, y, z) = 2x^2 + y^2 + 3z^2$ if (x, y, z) is on the line of intersection of the planes

$$2x + y - 3z = 4, \qquad x - y + 2z = 6.$$

7. Find the point on the curve $x^2 + 2xy + 2y^2 = 100$ which is closest to the origin.

8. Find the relative maxima and minima of the function $f(x, y, z) = x^3 + y^3 + z^3$ where (x, y, z) is on the plane $x + y + z = 4$.

9. Find the dimensions of the rectangular box, open at the top, which has maximum volume if the surface area is 12. (Compare with Example 3, page 174.)

10. A tent is made in the form of a cylinder surmounted by a cone (Fig. 4–18). If the cylinder has radius 5 and the total surface area is 100, find the height H of the cylinder and the height h of the cone which make the volume a maximum.

11. A container is made of a right circular cylinder with radius 5 and with a conical cap at each end. If the volume is given, find the height H of the cylinder and the height h of each of the conical caps which together make the total surface area as small as possible.

FIGURE 4–18

12. Find the minimum of the function

$$f(x, y, z, t) = x^2 + y^2 + z^2 + t^2$$

subject to the condition $3x + 2y - 4z + t = 2$.

13. Find the minimum of the function

$$f(x, y, z, t) = x^2 + y^2 + z^2 + t^2$$

subject to the conditions

$$x + y - z + 2t = 2, \qquad 2x - y + z + 3t = 3.$$

14. Find the minimum of the function

$$f(x, y, z, t) = 2x^2 + y^2 + z^2 + 2t^2$$

subject to the conditions

$$x + y + z + 2t = 1, \qquad 2x + y - z + 4t = 2, \qquad x - y + z - t = 4.$$

15. Find the points on the curve $x^4 + y^4 + 3xy = 2$ which are closest to the origin; find those which are farthest from the origin.

16. Find three critical points of the function $x^4 + y^4 + z^4 + 3xyz$ subject to the condition that (x, y, z) is on the plane $x + y + z = 3$. Can you identify these points?

17. Work Exercise 22 of Section 11 by the method of Lagrange multipliers.

18. Find the dimensions of the rectangular parallelepiped of maximum volume with edges parallel to the axes which can be inscribed in the ellipsoid

$$\frac{x^2}{a^2} + \frac{y^2}{b^2} + \frac{z^2}{c^2} = 1.$$

19. If the base of an open rectangular box costs three times as much per square foot as the sides, find the dimensions of the box of largest volume which can be made for D dollars.

20. Find and identify the critical points of the function

$$f(x, y, z) = 2x^2 + y^2 + z^2$$

subject to the condition that (x, y, z) is on the surface $x^2yz = 1$.

21. Find the critical points of the function $f(x, y, z) = x^a y^b z^c$ if $x + y + z = A$, where a, b, c, A are given positive numbers.

13.   EXACT DIFFERENTIALS

In Section 7 we saw that the differential of a function $f(x, y)$ is given by

$$df = \frac{\partial f}{\partial x} dx + \frac{\partial f}{\partial y} dy. \tag{1}$$

The quantity df is a function of four variables, since $\partial f/\partial x$ and $\partial f/\partial y$ are functions of x and y and dx and dy are additional independent variables. It turns out that expressions of the form

$$P(x, y)\, dx + Q(x, y)\, dy$$

occur frequently in problems in engineering and physics. It is natural to ask when such an expression is the total differential of a function f. For example, if we are given

$$(3x^2 + 2y)\, dx + (2x - 3y^2)\, dy,$$

we may guess (correctly) that the function $f(x, y) = x^3 + 2xy - y^3$ has the above expression as its total differential, df. On the other hand, if we are given

$$(2x^2 - 3y)\, dx + (2x - y^3)\, dy, \tag{2}$$

then it can be shown that *there is no function f whose total differential is the expression (2).*

DEFINITION. If there is a function $f(x, y)$ such that

$$df = P(x, y)\, dx + Q(x, y)\, dy$$

for all (x, y) in some region and for all values of dx and dy, we say that

$$P(x, y)\, dx + Q(x, y)\, dy$$

is an **exact differential.** If there is a function $F(x, y, z)$ such that

$$dF = P(x, y, z)\, dx + Q(x, y, z)\, dy + R(x, y, z)\, dz$$

for all (x, y, z) in some region and for all values of dx, dy, and dz, we say that $P\, dx + Q\, dy + R\, dz$ is an **exact differential.** For functions with any number of variables the extension is immediate.

The next theorem gives a precise criterion for determining when a differential expression is an exact differential.

Theorem 12. *Suppose that $P(x, y)$, $Q(x, y)$, $\partial P/\partial y$, $\partial Q/\partial x$ are continuous in a rectangle S. Then the expression*

$$P(x, y)\, dx + Q(x, y)\, dy \tag{3}$$

is an exact differential for (x, y) in the region S if and only if

$$\frac{\partial P}{\partial y} = \frac{\partial Q}{\partial x} \qquad \text{for all} \qquad (x, y) \text{ in } S. \tag{4}$$

Proof. The theorem has two parts: we must show (a), that if (3) is an exact differential, then (4) holds; and (b), that if (4) holds, then the expression (3) is an exact differential.

To establish (a) we start with the assumption that there is a function f such that

$$df = P\,dx + Q\,dy,$$

and so $\partial f/\partial x = P(x, y)$ and $\partial f/\partial y = Q(x, y)$. We differentiate and obtain

$$\frac{\partial^2 f}{\partial y\,\partial x} = \frac{\partial P}{\partial y}$$

and

$$\frac{\partial^2 f}{\partial x\,\partial y} = \frac{\partial Q}{\partial x}.$$

Now Theorem 7 of Section 9, which states that the order of differentiation is immaterial, may be invoked to conclude that (4) holds.

To prove (b) we assume that (4) holds, and we must construct a function f such that df is equal to the differential expression (3). That is, we must find a function f such that

$$\frac{\partial f}{\partial x} = P(x, y) \qquad \text{and} \qquad \frac{\partial f}{\partial y} = Q(x, y). \tag{5}$$

Let (a, b) be a point of S; suppose we try to solve these two partial differential equations for the function f. We integrate the first with respect to x, getting

$$f(x, y) = C(y) + \int_a^x P(\xi, y)\,d\xi$$

where, instead of a "constant" of integration, we get a function of the remaining variable. Letting $x = a$, we find that $f(a, y) = C(y)$, and we can write

$$f(x, y) = f(a, y) + \int_a^x P(\xi, y)\,d\xi. \tag{6}$$

Setting $x = a$ in the second equation of (5) and integrating with respect to y, we obtain

$$f(a, y) = C_1 + \int_b^y Q(a, \eta)\,d\eta;$$

letting $y = b$, we conclude that

$$f(a, y) = f(a, b) + \int_b^y Q(a, \eta)\,d\eta.$$

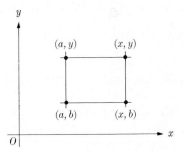

FIGURE 4–19

Substitution of this expression for $f(a, y)$ into the equation (6) yields (Fig. 4–19)

$$f(x, y) = f(a, b) + \int_b^y Q(a, \eta)\,d\eta + \int_a^x P(\xi, y)\,d\xi. \tag{7}$$

We may repeat the entire process by integrating with respect to y first and with respect to x second. The three equations are

$$f(x, y) = f(x, b) + \int_b^y Q(x, \eta)\, d\eta, \qquad f(x, b) = f(a, b) + \int_a^x P(\xi, b)\, d\xi,$$

and

$$f(x, y) = f(a, b) + \int_a^x P(\xi, b)\, d\xi + \int_b^y Q(x, \eta)\, d\eta. \tag{8}$$

The two expressions for f given by (7) and (8) will be identical if and only if (after subtraction) the equation

$$\int_a^x [P(\xi, y) - P(\xi, b)]\, d\xi = \int_b^y [Q(x, \eta) - Q(a, \eta)]\, d\eta \tag{9}$$

holds. To establish (9), we start with the observation that

$$P(\xi, y) - P(\xi, b) = \int_b^y \frac{\partial P(\xi, \eta)}{\partial y}\, d\eta = \int_b^y \frac{\partial Q(\xi, \eta)}{\partial x}\, d\eta,$$

where, for the first time, we have used the hypothesis that $\partial P/\partial y = \partial Q/\partial x$. Therefore, upon integration,

$$\int_a^x [P(\xi, y) - P(\xi, b)]\, d\xi = \int_a^x \left[\int_b^y \frac{\partial Q(\xi, \eta)}{\partial x}\, d\eta \right] d\xi.$$

It will be shown in Chapter 5, Section 2 that the order of integrations in the term on the right may be interchanged, so that

$$\int_a^x [P(\xi, y) - P(\xi, b)]\, d\xi = \int_b^y \left[\int_a^x \frac{\partial Q(\xi, \eta)}{\partial x}\, d\xi \right] d\eta$$

$$= \int_b^y [Q(x, \eta) - Q(a, \eta)]\, d\eta.$$

But this equality is (9) precisely; the theorem is established when we observe that as a result of (7) or (8), the relations

$$\frac{\partial f}{\partial x} = P \qquad \text{and} \qquad \frac{\partial f}{\partial y} = Q$$

hold.

The proof of Theorem 12 contains in it the method for finding the function f when it exists. Examples illustrate the technique.

EXAMPLE 1. Show that

$$(3x^2 + 6y)\, dx + (3y^2 + 6x)\, dy$$

is an exact differential, and find the function f of which it is the total differential.

Solution. Setting $P = 3x^2 + 6y$, $Q = 3y^2 + 6x$, we obtain

$$Q_x = 6, \qquad P_y = 6,$$

so that $P\,dx + Q\,dy$ is an exact differential. We write (as in the proof of the theorem)

$$f_x = 3x^2 + 6y$$

and integrate to get

$$f = x^3 + 6xy + C(y).$$

We differentiate with respect to y. We find

$$f_y = 6x + C'(y),$$

and this expression must be equal to Q. Therefore

$$6x + C'(y) = 3y^2 + 6x$$

or

$$C'(y) = 3y^2, \quad C(y) = y^3 + C_1.$$

Thus

$$f(x, y) = x^3 + 6xy + y^3 + C_1.$$

A constant of integration will always appear in the integration of exact differentials.

EXAMPLE 2. Show that

$$(e^x \cos y - e^y \sin x)\,dx + (e^y \cos x - e^x \sin y)\,dy$$

is an exact differential, and find the function f of which it is the differential.

Solution. Setting $P = e^x \cos y - e^y \sin x$, $Q = e^y \cos x - e^x \sin y$, we have

$$\frac{\partial P}{\partial y} = -e^x \sin y - e^y \sin x = \frac{\partial Q}{\partial x},$$

and the differential is exact. Integrating $f_x = P$, we get

$$f(x, y) = e^x \cos y + e^y \cos x + C(y).$$

Differentiating with respect to y, we find

$$f_y = -e^x \sin y + e^y \cos x + C'(y) = Q = e^y \cos x - e^x \sin y.$$

Therefore, $C'(y) = 0$ and C is a constant. The function f is given by

$$f(x, y) = e^x \cos y + e^y \cos x + C.$$

The next theorem is an extension of Theorem 12 to functions of three variables.

Theorem 13. *Suppose that P(x, y, z), Q(x, y, z), R(x, y, z) are continuous in some rectangular parallelepiped S. Then*

$$P(x, y, z)\, dx\ +\ Q(x, y, z)\, dy\ +\ R(x, y, z)\, dz$$

is an exact differential on S if and only if

$$\frac{\partial P}{\partial y} = \frac{\partial Q}{\partial x}, \qquad \frac{\partial P}{\partial z} = \frac{\partial R}{\partial x}, \qquad \frac{\partial Q}{\partial z} = \frac{\partial R}{\partial y}.$$

It is assumed that all the above partial derivatives are continuous functions of (x, y, z) on S.

The proof of this theorem follows the lines (and uses the proof) of Theorem 12. It may be found in Morrey, *University Calculus*, page 586.

The next example shows how to integrate an exact differential in three variables.

EXAMPLE 3. Determine whether or not

$$(3x^2 - 4xy + z^2 + yz - 2)\, dx + (xz - 6y^2 - 2x^2)\, dy + (9z^2 + 2xz + xy + 6z)\, dz$$

is an exact differential and, if so, find the function f of which it is the total differential.

Solution. Setting P, Q, R equal to the coefficients of dx, dy, and dz, respectively, we obtain

$$P_y = -4x + z = Q_x, \qquad P_z = 2z + y = R_x, \qquad Q_z = x = R_y.$$

Therefore $P\, dx + Q\, dy + R\, dz$ is an exact differential, and we proceed to find f. Writing $f_x = P$, we integrate to get

$$f(x, y, z) = x^3 - 2x^2y + xz^2 + xyz - 2x + C(y, z).$$

We differentiate with respect to y:

$$f_y(x, y, z) = -2x^2 + xz + C_y(y, z) = Q = xz - 6y^2 - 2x^2.$$

Hence

$$C_y(y, z) = -6y^2$$

and, upon integration with respect to y,

$$C(y, z) = -2y^3 + C_1(z).$$

We may write

$$f(x, y, z) = x^3 - 2x^2y + xz^2 + xyz - 2x - 2y^3 + C_1(z),$$

and we wish to find $C_1(z)$. We differentiate f with respect to z:

$$f_z = 2xz + xy + C_1'(z) = R = 9z^2 + 2xz + xy + 6z.$$

We obtain

$$C_1'(z) = 9z^2 + 6z \qquad \text{and} \qquad C_1(z) = 3z^3 + 3z^2 + C_2.$$

Therefore

$$f(x, y, z) = x^3 - 2y^3 + 3z^3 - 2x^2y + xz^2 + xyz + 3z^2 - 2x + C_2.$$

EXERCISES

In each of problems 1 through 18, determine which of the differentials are exact. In case a differential is exact, find the functions of which it is the total differential.

1. $(x^3 + 3x^2y) dx + (x^3 + y^3) dy$

2. $(2x + 3y) dx + (3x + 2y) dy$

3. $\left(2y - \dfrac{1}{x}\right) dx + \left(2x + \dfrac{1}{y}\right) dy$

4. $(x^2 + 2xy) dx + (y^3 - x^2) dy$

5. $x^2 \sin y\, dx + x^2 \cos y\, dy$

6. $\dfrac{x^2 + y^2}{2y^2} dx - \dfrac{x^3}{3y^3} dy$

7. $2xe^{x^2} \sin y\, dx + e^{x^2} \cos y\, dy$

8. $(ye^{xy} + 3x^2) dx + (xe^{xy} - \cos y) dy$

9. $\dfrac{x\, dy - y\, dx}{x^2 + y^2},\ x > 0$

10. $(2x \ln y) dx + \dfrac{x^2}{y} dy$

11. $(x + \cos x \tan y) dx + (y + \tan x \cos y) dy$

12. $\dfrac{1}{y} e^{2x/y} dx - \dfrac{1}{y^3} e^{2x/y}(y + 2x) dy$

13. $(3x^2 \ln y - x^3) dx + \dfrac{3x^2}{y} dy$

14. $\dfrac{x\, dx}{\sqrt{x^2 + y^2}} + \left(\dfrac{y}{\sqrt{x^2 + y^2}} - 2\right) dy$

15. $(2x - y + 3z) dx + (3y + 2z - x) dy + (2x + 3y - z) dz$

16. $(2xy + z^2) dx + (2yz + x^2) dy + (2xz + y^2) dz$

17. $(e^x \sin y \cos z) dx + (e^x \cos y \cos z) dy - (e^x \sin y \sin z) dz$

18. $\left(\dfrac{1}{y^2} - \dfrac{y}{x^2z} - \dfrac{z}{x^2y}\right) dx + \left(\dfrac{1}{xz} - \dfrac{x}{y^2z} - \dfrac{z}{xy^2}\right) dy + \left(\dfrac{1}{xy} - \dfrac{x}{yz^2} - \dfrac{y}{xz^2}\right) dz$

*19. (a) Given the differential expression

$$P\, dx + Q\, dy + R\, dz + S\, dt,$$

where P, Q, R, S are functions of x, y, z, t, state a theorem which is a plausible generalization of Theorem 13 in order to decide when the above expression is exact. (b) Use the result of part (a) to show that the following expression is exact. Find the function f of which it is the total differential.

$$(3x^2 + 2z + 3) dx + (2y - t - 2) dy + (3z^2 + 2x) dz + (4 - 3t^2 - y) dt.$$

14. DEFINITION OF A LINE INTEGRAL

Let C be an arc in the plane extending from the point $A(a, b)$ to the point $B(c, d)$, as shown in Fig. 4–20. Suppose that $f(x, y)$ is a continuous function defined in a region which contains the arc C in its interior. We make a decomposition of the arc C by introducing $n - 1$ points between A and B along C. We label these points $P_1, P_2, \ldots, P_{n-2}, P_{n-1}$, and set $A = P_0$, $B = P_n$. Denote the coordinates of the point P_i by (x_i, y_i), $i = 0, 1, 2, \ldots, n$. (See Fig. 4–21). Between each two successive points of the subdivision we select a point on the curve. Call these points $Q_1, Q_2, \ldots, Q_n$, and denote the coordinates of Q_i by (ξ_i, η_i), $i = 1, 2,$ $\ldots, n$. This selection may be made in any manner whatsoever so long as Q_i is on the part of C between P_{i-1} and P_i (Fig. 4–22).

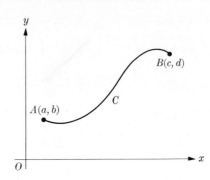

FIGURE 4–20

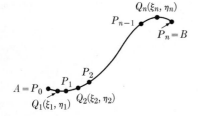 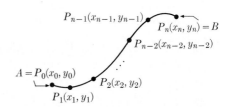

FIGURE 4–21   FIGURE 4–22

We form the sum

$$f(\xi_1, \eta_1)(x_1 - x_0) + f(\xi_2, \eta_2)(x_2 - x_1) + \cdots + f(\xi_n, \eta_n)(x_n - x_{n-1}),$$

or, written more compactly,

$$\sum_{i=1}^{n} f(\xi_i, \eta_i)(x_i - x_{i-1}) = \sum_{i=1}^{n} f(\xi_i, \eta_i)\,\Delta_i x. \tag{1}$$

As in the case of subdivisions of an interval along the x-axis (see *First Course*, page 213), we define the *norm of the subdivision* $P_0, P_1, P_2, \ldots, P_n$ of the curve C to be the maximum distance between any two successive points of the subdivision. We denote the norm by $\|\Delta\|$.

DEFINITION. Suppose there is a number L with the following property: for each $\epsilon > 0$ there is a $\delta > 0$ such that

$$\left| \sum_{i=1}^{n} f(\xi_i, \eta_i)(x_i - x_{i-1}) - L \right| < \epsilon$$

for every subdivision with $\|\Delta\| < \delta$ and for any choices of the (ξ_i, η_i) as described above. Then we say that **the line integral of f with respect to x along the curve C exists and its value is L.** There are a number of symbols for this line integral such as

$$\int_C f(x, y)\, dx \quad \text{and} \quad (C)\int_A^B f(x, y)\, dx. \tag{2}$$

Note that the value of the integral will depend, in general, not only on f and the points A and B but also on the particular arc C selected.

The expression (1) is one of two types of sums which are commonly formed in line integrations. We also introduce the sum

$$\sum_{i=1}^n f(\xi_i, \eta_i)(y_i - y_{i-1})$$

in which the points (ξ_i, η_i) are selected as before. The limit, if it exists, (as $\|\Delta\| \to 0$) is the line integral

$$(C)\int_A^B f(x, y)\, dy, \tag{3}$$

and will generally have a value different from (2).

In Section 16, we shall give an application of line integration to a physical problem.

If the arc C happens to be a segment of the x-axis, then the line integral $\int_C f(x, y)\, dx$ reduces to an ordinary integral. To see this we note that in the approximating sums all the $\eta_i = 0$. Therefore we have

not true in all cases

$$(C)\int_A^B f(x, y)\, dx = \int_a^c f(x, 0)\, dx.$$

On the other hand, when C is a segment of the x-axis, the integral $\int_C f(x, y)\, dy$ always vanishes, since in each approximating sum $y_i - y_{i-1} = 0$ for every i.

Simple properties of line integrals, analogous to those for ordinary integrals, may be derived directly from the definition. For example, if the arc C is traversed in the opposite direction, the line integral changes sign. That is,

$$(C)\int_A^B f(x, y)\, dx = -(C)\int_B^A f(x, y)\, dx.$$

If C_1 is an arc extending from A_1 to A_2 and C_2 is an arc extending from A_2 to A_3, then

$$(C_1)\int_{A_1}^{A_2} f(x, y)\, dx + (C_2)\int_{A_2}^{A_3} f(x, y)\, dx = (C_1 + C_2)\int_{A_1}^{A_3} f(x, y)\, dx, \tag{4}$$

where the symbol $C_1 + C_2$ has the obvious meaning. As in the case of ordinary integrals, line integrals satisfy the additive property:

$$\int_C [f(x, y) + g(x, y)]\, dx = \int_C f(x, y)\, dx + \int_C g(x, y)\, dx.$$

Statements similar to those above hold for integrals of the type $\int_C f(x, y)\, dy$.

There is one more type of line integral which we can define. If the arc C and the function f are as before and if s denotes arc length along C measured from the point A to the point B, we can define the *line integral with respect to the arc length* s. We use the symbol

$$(C)\int_A^B f(x, y)\, ds$$

for this line integral. If C is given in the form $y = g(x)$, we use the relation $ds = [1 + (g'(x))^2]^{1/2}\, dx$ to define:

$$(C)\int_A^B f(x, y)\, ds = (C)\int_A^B f(x, y)\sqrt{1 + (g'(x))^2}\, dx,$$

in which the right-hand side has already been defined. If the curve C is in the form $x = h(y)$, we may write

$$(C)\int_A^B f(x, y)\, ds = (C)\int_A^B f(x, y)\sqrt{1 + (h'(y))^2}\, dy.$$

If C is in neither the form $y = g(x)$ nor the form $x = h(y)$, it may be broken up into a sum of arcs, each one of which does have the appropriate functional behavior. Then the integrals over each piece may be calculated and the results added.

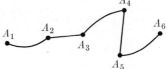

FIGURE 4–23

For ordinary integrals we stated a simple theorem to the effect that if a function f is continuous on an interval $[a, b]$, then it is integrable there. (See *First Course*, page 218.) It can be shown that if $f(x, y)$ is continuous and if the arc C is rectifiable (has finite length), then the line integrals exist. We shall consider throughout only functions and arcs which are sufficiently smooth so that the line integrals always exist. It is worth remarking that if C consists of a collection of smooth arcs joined together (Fig. 4–23), then because of (4) the line integral along C exists as the sum of the line integrals taken along each of the pieces.

Line integrals in three dimensions may be defined similarly to the way they were defined in the plane. An arc C joining the points A and B in three-space may be given either parametrically by three equations,

$$x = x(t), \qquad y = y(t), \qquad z = z(t), \qquad t_0 \le t \le t_1,$$

or nonparametrically by two equations,

$$y = g_1(x), \qquad z = g_2(x).$$

If $f(x, y, z)$ is a function defined along C, then a subdivision of the arc C leads to a sum of the form

$$\sum_{i=1}^{n} f(\xi_i, \eta_i, \zeta_i) (x_i - x_{i-1})$$

which, in turn, is an approximation to the line integral

$$\int_C f(x, y, z) \, dx.$$

Line integrals such as $\int_C f(x, y, z) \, dy$, $\int_C f(x, y, z) \, dz$ are defined similarly.

15. CALCULATION OF LINE INTEGRALS

In the study of integration of functions of one variable, we saw that the definition of integral (*First Course*, page 213) turned out to be fairly worthless as a tool for computing the value of any specific integral. While we did have a certain amount of practice in calculating areas by sums (*First Course*, page 203), the methods we employed for performing integration most often used certain properties of integrals, special formulas for antiderivatives, and so forth (*First Course*, page 223).

The situation with line integrals is similar. In the last section we defined various types of line integrals, and now we shall exhibit methods for calculating the value of these integrals when the curve C and the function f are specifically given. It is an interesting fact that *all such integrals may be reduced to ordinary integrations of the type we have already studied.* Once the reduction is made, the problem becomes routine and all the formulas we learned for evaluation of integrals may be used.

The next theorem establishes the rule for reducing a line integration to an ordinary integration of a function of a single variable.

Theorem 14. *Let C be a rectifiable arc given in the form*

$$x = x(t), \qquad y = y(t), \qquad t_0 \leq t \leq t_1, \tag{1}$$

so that the point $A(a, b)$ corresponds to t_0, and $B(c, d)$ corresponds to t_1. Suppose $f(x, y)$ is a continuous function along C, and $x'(t)$, $y'(t)$ are continuous. Then

$$(C)\int_A^B f(x, y) \, dx = \int_{t_0}^{t_1} f[x(t), y(t)]x'(t) \, dt,$$

$$(C)\int_A^B f(x, y) \, dy = \int_{t_0}^{t_1} f[x(t), y(t)]y'(t) \, dt,$$

$$(C)\int_A^B f(x, y) \, ds = \int_{t_0}^{t_1} f[x(t), y(t)]\sqrt{(x'(t))^2 + (y'(t))^2} \, dt.$$

For a proof of this theorem, see Morrey, *University Calculus*, page 590. A similar theorem is valid for line integrals in three-space.

Corollary. *If the arc C is in the form $y = g(x)$, then*

$$(C)\int_A^B f(x, y)\, dx = \int_a^c f[x, g(x)]\, dx.$$

For, if $y = g(x)$, then x may be used as a parameter in place of t in (1) and the corollary is a restatement of the theorem. Similar statements may be made if C is given by an equation of the type $x = h(y)$.

EXAMPLE 1. Evaluate the integrals

$$\int_C (x^2 - y^2)\, dx - \int_C 2xy\, dy$$

where C is the arc (Fig. 4–24):

$$x = t^2 - 1, \qquad y = t^2 + t + 2, \qquad 0 \le t \le 1.$$

According to Theorem 14, we compute

$$x'(t) = 2t, \qquad y'(t) = 2t + 1$$

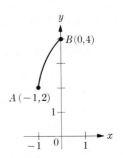

FIGURE 4–24

and make the appropriate substitutions. We get

$$\int_C (x^2 - y^2)\, dx = \int_0^1 \left[(t^2 - 1)^2 - (t^2 + t + 2)^2 \right] \cdot 2t\, dt,$$

$$-\int_C 2xy\, dy = -2\int_0^1 (t^2 - 1)(t^2 + t + 2)(2t + 1)\, dt.$$

Multiplying out the integrands, we find

$$\int_C (x^2 - y^2)\, dx = 2\int_0^1 (-2t^3 - 7t^2 - 4t - 3)t\, dt,$$

$$-2\int_C xy\, dy = -2\int_0^1 (t^4 + t^3 + t^2 - t - 2)(2t + 1)\, dt.$$

The integration is now routine, and the final result is

$$\int_C [(x^2 - y^2)\, dx - 2xy\, dy] = -2\int_0^1 (2t^5 + 5t^4 + 10t^3 + 3t^2 - 2t - 2)\, dt$$

$$= -\frac{11}{3}.$$

EXAMPLE 2. Evaluate the integral

$$\int_C (x^2 - 3xy + y^3)\, dx$$

where C is the arc

$$y = 2x^2, \qquad 0 \le x \le 2.$$

Solution. We have

$$\int_C (x^2 - 3xy + y^3)\, dx = \int_0^2 [x^2 - 3x(2x^2) + (2x^2)^3]\, dx$$

$$= \left[\frac{x^3}{3} - \frac{3}{2}x^4 + \frac{8}{7}x^7\right]_0^2 = \frac{2624}{21}.$$

EXAMPLE 3. Evaluate

$$\int_C y\, ds$$

where C is the arc

$$y = \sqrt{x}, \qquad 0 \le x \le 6.$$

Solution. We have

$$ds = \sqrt{1 + \left(\frac{dy}{dx}\right)^2}\, dx = \frac{1}{2}\sqrt{\frac{1 + 4x}{x}}\, dx,$$

and therefore

$$\int_C y\, ds = \frac{1}{2}\int_0^6 \sqrt{x}\sqrt{\frac{1 + 4x}{x}}\, dx = \frac{1}{8}\int_0^6 \sqrt{1 + 4x}\, d(1 + 4x)$$

$$= \left[\frac{1}{8}\cdot\frac{2}{3}(1 + 4x)^{3/2}\right]_0^6 = \frac{31}{3}.$$

The next example shows how we evaluate integrals when the arc C consists of several pieces.

EXAMPLE 4. Evaluate

$$\int_C [(x + 2y)\, dx + (x^2 - y^2)\, dy],$$

where C is the line segment C_1 from $(0,0)$ to $(1,0)$ followed by the line segment C_2 from $(1,0)$ to $(1,1)$ (Fig. 4–25).

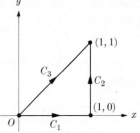

FIGURE 4–25

Solution. Along C_1 we have $x = x$, $y = 0$, $0 \le x \le 1$, so $dx = dx$, $dy = 0$, and

$$\int_{C_1} [(x + 2y)\, dx + (x^2 - y^2)\, dy] = \int_0^1 x\, dx = \frac{1}{2}.$$

Along C_2 we have $x = 1$, $y = y$, and so $dx = 0$, $dy = dy$. We obtain

$$\int_{C_2} [(x + 2y)\, dx + (x^2 - y^2)\, dy] = \int_0^1 (1 - y^2)\, dy = \frac{2}{3}.$$

Therefore

$$\int_C [(x + 2y)\, dx + (x^2 - y^2)\, dy] = \frac{1}{2} + \frac{2}{3} = \frac{7}{6}.$$

EXAMPLE 5. Evaluate the integral of Example 4 where the arc C is now the line segment C_3 from $(0, 0)$ to $(1, 1)$. (See Fig. 4–25.)

Solution. Along C_3 we have $y = x$, and so $dy = dx$. Therefore

$$\int_{C_3} [(x + 2y)\, dx + (x^2 - y^2)\, dy] = \int_0^1 3x\, dx = \frac{3}{2}.$$

The next example illustrates the method for evaluation of line integrals in three space.

EXAMPLE 6. Evaluate the integral

$$\int_C [(x^2 + y^2 - z^2)\, dx + yz\, dy + (x - y)\, dz]$$

where C is the arc

$$x = t^2 + 2, \qquad y = 2t - 1, \qquad z = 2t^2 - t, \qquad 0 \le t \le 1. \tag{2}$$

Solution. We substitute for x, y, z from (2) and insert the values $dx = 2t\, dt$, $dy = 2\, dt$, $dz = (4t - 1)\, dt$, to obtain

$$\int_0^1 \{[(t^2 + 2)^2 + (2t - 1)^2 - (2t^2 - t)^2]2t\, dt + (2t - 1)(2t^2 - t)2\, dt$$
$$+ (t^2 - 2t + 3)(4t - 1)\, dt\}.$$

Upon multiplying out all the terms and performing the resulting routine integration we get the value $263/30$.

EXERCISES

In each of problems 1 through 10, evaluate $\int_C P\, dx + Q\, dy$ and draw a sketch of the arc C.

1. $\int_C [(x + y)\, dx + (x - y)\, dy]$ where C is the segment from $(0, 0)$ to $(2, 1)$.

2. $\int_C [(x + y)\, dx + (x - y)\, dy]$ where C consists of the segment from $(0, 0)$ to $(2, 0)$ followed by that from $(2, 0)$ to $(2, 1)$.

3. $\int_C [(x^2 - 2y)\, dx + (2x + y^2)\, dy]$ where C is the arc of $y^2 = 4x - 1$ going from $(\frac{1}{2}, -1)$ to $(\frac{5}{4}, 2)$.

4. $\int_C [(x^2 - 2y)\, dx + (2x + y^2)\, dy]$ where C is the segment going from $(\frac{1}{2}, -1)$ to $(\frac{5}{4}, 2)$.

5. $\int_C [y\, dx + (x^2 + y^2)\, dy]$ where C is the arc of the circle $y = +\sqrt{4 - x^2}$ from $(-2, 0)$ to $(0, 2)$.

6. $\int_C [y\, dx + (x^2 + y^2)\, dy]$ where C consists of the line segment from $(-2, 0)$ to $(0, 0)$ followed by that from $(0, 0)$ to $(0, 2)$.

7. $\int_C \left(\frac{x^2}{\sqrt{x^2 - y^2}}\, dx + \frac{2y}{4x^2 + y^2}\, dy \right)$

where C is the arc $y = \frac{1}{2}x^2$ from $(0, 0)$ to $(2, 2)$.

8. $\int_C \left(\dfrac{x^2}{\sqrt{x^2 - y^2}} \, dx + \dfrac{2y}{4x^2 + y^2} \, dy \right)$

where C consists of the line segment from $(0, 0)$ to $(2, 0)$, followed by the line segment from $(2, 0)$ to $(2, 2)$.

9. $\int_C \left(\dfrac{-y}{x\sqrt{x^2 - y^2}} \, dx + \dfrac{1}{\sqrt{x^2 - y^2}} \, dy \right)$

where C is the arc of $x^2 - y^2 = 9$ from $(3, 0)$ to $(5, 4)$.

10. Same integral as in problem 9, where C consists of the line segment from $(3, 0)$ to $(5, 0)$, followed by the line segment from $(5, 0)$ to $(5, 4)$.

11. Calculate $\int_C \sqrt{x + (3y)^{5/3}} \, ds$ where C is the arc $y = \frac{1}{3}x^3$ going from $(0, 0)$ to $(3, 9)$.

12. Calculate $\int_C \sqrt{x + 3y} \, ds$ where C is the straight line segment going from $(0, 0)$ to $(3, 9)$.

13. Calculate $\int_C y^2 \sin^3 x \sqrt{1 + \cos^2 x} \, ds$ where C is the arc $y = \sin x$ going from $(0, 0)$ to $(\pi/2, 1)$.

14. Calculate $\int_C (2x^2 + 3y^2 - xy) \, ds$ where C is the arc

$$\left. \begin{array}{l} x = 3 \cos t \\ y = 3 \sin t \end{array} \right\} \quad 0 \le t \le \frac{\pi}{4}.$$

15. Calculate $\int_C x^2 \, ds$ where C is the arc $x = 2y^{3/2}$ going from $(2, 1)$ to $(16, 4)$.

16. Calculate $\int_C [(x^2 + y^2) \, dx + (x^2 - y^2) \, dy]$ where C is the arc

$$\left. \begin{array}{l} x = t^2 + 3 \\ y = t - 1 \end{array} \right\} \quad 1 \le t \le 2.$$

17. Calculate $\int_C [\sin x \, dy + \cos y \, dx]$ where C is the arc

$$\left. \begin{array}{l} x = t^2 + 3 \\ y = 2t^2 - 1 \end{array} \right\} \quad 0 \le t \le 2.$$

18. Calculate $\int_C [(x - y) \, dx + (y - z) \, dy + (z - x) \, dz]$ where C is the line segment extending from $(1, -1, 2)$ to $(2, 3, 1)$.

19. Calculate $\int_C [(x^2 - y^2) \, dx + 2xz \, dy + (xy - yz) \, dz]$ where C is the line segment

$$\left. \begin{array}{l} x = 2t - 1 \\ y = t + 1 \\ z = t - 2 \end{array} \right\} \quad 0 \le t \le 3.$$

20. Calculate $\int_C [(x - y + z) \, dx + (y + z - x) \, dy + (z + x - y) \, dz]$ where C consists of straight line segments connecting the points $(1, -1, 2)$, $(2, -1, 2)$, $(2, 3, 2)$, and $(2, 3, 1)$, in that order.

21. Calculate

$$\int_C \frac{x\,dx + y\,dy + z\,dz}{x^2 + y^2 + z^2}$$

where C is the arc

$$x = 2t, \qquad y = 2t + 1, \qquad z = t^2 + t,$$

joining the points $(0, 1, 0)$ and $(2, 3, 2)$.

22. Same as problem 21, where C is the straight line segment joining $(0, 1, 0)$ and $(2, 3, 2)$.

23. Evaluate

$$\int_C \frac{y\,dx + x\,dy}{\sqrt{x^2 + y^2}}$$

where C is the *closed curve*

$$\left.\begin{matrix} x = \cos t \\ y = \sin t \end{matrix}\right\} - \pi \le t \le \pi.$$

24. Evaluate

$$\int_C \frac{-y\,dx + x\,dy}{\sqrt{x^2 + y^2}}$$

where C is the same curve as in problem 23.

16.   PATH–INDEPENDENT LINE INTEGRALS. WORK

In general, the value of a line integral depends on the integrand, on the two endpoints, and on the arc connecting these endpoints. However, there are special circumstances when the value of a line integral depends solely on the integrand and endpoints but *not* on the arc on which the integration is performed. When such conditions prevail, we say that the integral is **independent of the path.** The next theorem establishes the connection between path-independent integrals and exact differentials. (See Section 13.)

Theorem 15.   *Suppose that $P(x, y)\,dx + Q(x, y)\,dy$ is an exact differential. That is, there is a function $f(x, y)$ with*

$$df = P\,dx + Q\,dy.$$

Let C be an arc given parametrically by

$$x = x(t), \qquad y = y(t), \qquad t_0 \le t \le t_1$$

where $x'(t), y'(t)$ are continuous. Then

$$\int_C (P\,dx + Q\,dy) = f[x(t_1), y(t_1)] - f[x(t_0), y(t_0)].$$

Thus the integral depends only on the endpoints and not on the arc C joining them.

Proof. We define the function $F(t)$ by

$$F(t) = f[x(t), y(t)], \qquad t_0 \leq t \leq t_1.$$

We use the Chain Rule to calculate the derivative:

$$F'(t) = f_{,1}x'(t) + f_{,2}y'(t)$$

and

$$F'(t) = P[x(t), y(t)]x'(t) + Q[x(t), y(t)]y'(t). \tag{1}$$

Integrating both sides of (1) with respect to t and employing Theorem 14, we conclude that

$$F(t_1) - F(t_0) = \int_C P \, dx + Q \, dy.$$

The result follows when we note that

$$F(t_1) = f[x(t_1), y(t_1)] \qquad \text{and} \qquad F(t_0) = f[x(t_0), y(t_0)].$$

Corollary. *If $P \, dx + Q \, dy + R \, dz$ is an exact differential, then*

$$\int_C (P \, dx + Q \, dy + R \, dz) = f[x(t_1), y(t_1), z(t_1)] - f[x(t_0), y(t_0), z(t_0)]$$

where $df = P \, dx + Q \, dy + R \, dz$ and the parametric equations of C are:

$$\{x(t), y(t), z(t)\}, \qquad t_0 \leq t \leq t_1.$$

EXAMPLE 1. Show that the integrand of

$$\int_C [(2x + 3y) \, dx + (3x - 2y) \, dy]$$

is an exact differential and find the value of the integral over any arc C going from the point $(1, 3)$ to the point $(-2, 5)$.

Solution. Setting $P = 2x + 3y$, $Q = 3x - 2y$, we have

$$\frac{\partial P}{\partial y} = 3 = \frac{\partial Q}{\partial x}.$$

By Theorem 12, the integrand is an exact differential. Using the methods of Section 13 for integrating exact differentials, we find that

$$f(x, y) = x^2 + 3xy - y^2 + C_1.$$

Therefore

$$\int_C [(2x + 3y) \, dx + (3x - 2y) \, dy] = f(-2, 5) - f(1, 3) = -52.$$

Notice that in the evaluation process the constant C_1 disappears.

EXAMPLE 2. Show that the integrand of

$$\int_C (3x^2 + 6xy)\, dx + (3x^2 - 3y^2)\, dy$$

is an exact differential, and find the value of the integral over any arc C going from the point $(1, 1)$ to the point $(2, 3)$.

Solution. Setting $P = 3x^2 + 6xy$, $Q = 3x^2 - 3y^2$, we have

$$\frac{\partial P}{\partial y} = 6x = \frac{\partial Q}{\partial x}.$$

Instead of finding the function f with the property that $df = P\, dx + Q\, dy$, we may pick *any* simple path joining $(1, 1)$ and $(2, 3)$ and evaluate the integral along that path. We select the horizontal path C_1 from $(1, 1)$ to $(2, 1)$, followed by the vertical path from $(2, 1)$ to $(2, 3)$, as shown in Fig. 4–26. The result is

$$\int_{C_1} (3x^2 + 6x)\, dx + \int_{C_2} (12 - 3y^2)\, dy = \left[x^3 + 3x^2\right]_1^2 + \left[12y - y^3\right]_1^3 = 14.$$

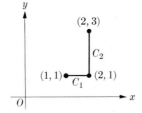

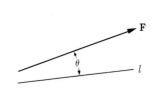

FIGURE 4–26   FIGURE 4–27

An application of line integrals occurs in the determination of the work done by a force acting on a particle in motion along a path. We recall that we considered the work done in the case of the motion of objects along a straight line. (See *First Course*, page 244.) The basic formula depends on the elementary idea that if the force F is constant and if the distance the particle moves (along a straight line) is d, then the work W is given by the relation

$$W = Fd.$$

For motion in the plane, *force* is a vector quantity, since it has both *magnitude* and, at the point of application, a *direction* along which it acts. We employ the representation

$$\mathbf{F} = P\mathbf{i} + Q\mathbf{j}$$

for a force vector in the plane. Suppose a particle is constrained to move in the direction of a straight line l. If the motion is caused by a force $\mathbf{F}$, then the only portion of $\mathbf{F}$ which has any effect on the motion is the component of $\mathbf{F}$ in the direction of l. We recall that this quantity is called the **projection** of $\mathbf{F}$ on l (Fig. 4–27), and is given by

$$|\mathbf{F}|\cos\theta,$$

where θ is the angle between the direction of $\mathbf{F}$ and the direction of l.

If the particle moves along l a distance d and if the force $\mathbf{F}$ is constant then, according to the elementary principle, the work done is

$$W = |\mathbf{F}| \cos \theta \cdot d.$$

We may represent the quantity above in another way. Let $\mathbf{r}$ be a vector of length d in the direction of l. Then we may write $|\mathbf{r}| = d$ and

$$W = |\mathbf{F}| \cdot |\mathbf{r}| \cos \theta.$$

We recognize this quantity as the inner, or scalar, product of the vectors $\mathbf{F}$ and $\mathbf{r}$. That is,

$$W = \mathbf{F} \cdot \mathbf{r}, \tag{2}$$

where $\mathbf{r}$ is a vector of the appropriate length in the direction of motion. Note that work is a scalar quantity.

Using (2) as the basic formula for motion in the plane, we proceed to define work when the force is variable and the motion is along a curved path. We write

$$\mathbf{F} = P(x, y)\mathbf{i} + Q(x, y)\mathbf{j}$$

for a force which may vary from point to point. We let C be an arc and suppose that a particle is constrained so that it must move along C (as a bead on a wire); we suppose the motion is caused by a force $\mathbf{F}$. We let the arc C be given by the equations

$$x = x(t), \qquad y = y(t), \qquad a \leq t \leq b,$$

and make a subdivision $a = t_0 < t_1 < \cdots < t_{n-1} < t_n = b$ of the interval $a \leq t \leq b$. We thus get a subdivision $P_0, P_1, \ldots, P_{n-1}, P_n$ of the arc C. We now replace each subarc by a straight line segment and *assume* that the force is approximately constant along each such subarc. Denoting the ith subarc by $P_{i-1}P_i$ and the force along this subarc by $\mathbf{F}_i$, we obtain, for the work done along this arc, the approximate quantity (Fig. 4–28),

$$W_i = |\mathbf{F}_i| \cos \theta_i \sqrt{[x(t_i) - x(t_{i-1})]^2 + [y(t_i) - y(t_{i-1})]^2}.$$

If we introduce the vector

$$\Delta_i x \mathbf{i} + \Delta_i y \mathbf{j}$$

and write $\mathbf{F}_i = P_i \mathbf{i} + Q_i \mathbf{j}$ then, taking (2) into account, the work is

$$W_i = P_i \Delta_i x + Q_i \Delta_i y.$$

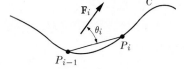

FIGURE 4–28

The total work done is the sum of the amounts of work done on the individual subarcs. (See *First Course*, page 244, PRINCIPLE 1.) We have, approximately,

$$W = \sum_{i=1}^{n} P[x(\tau_i), y(\tau_i)] \Delta_i x + Q[x(\tau_i), y(\tau_i)] \Delta_i y.$$

Proceeding to the limit, we define the **total work** by the formula

$$W = \int_C (P\,dx + Q\,dy).$$

For motion in three dimensions with the force given by a vector

$$\mathbf{F} = P(x, y, z)\mathbf{i} + Q(x, y, z)\mathbf{j} + R(x, y, z)\mathbf{k},$$

work is defined by the formula

$$W = \int_C (P\,dx + Q\,dy + R\,dz).$$

EXAMPLE 3. A particle moves along the curve $y = x^2$ from the point $(1, 1)$ to the point $(3, 9)$. If the motion is caused by the force $\mathbf{F} = (x^2 - y^2)\mathbf{i} + x^2 y\mathbf{j}$ applied to the particle, find the total work done.

Solution. Using the formula for work, we have

$$W = \int_C (x^2 - y^2)\,dx + x^2 y\,dy.$$

Employing the normal methods for calculating such integrals, we find

$$W = \int_1^3 (x^2 - x^4)\,dx + x^2(x^2)(2x)\,dx$$

$$= \left[\frac{1}{3}x^6 - \frac{1}{5}x^5 + \frac{1}{3}x^3\right]_1^3 = \frac{3044}{15}.$$

If the arc is measured in inches and the force in pounds, the total work is $\frac{3044}{15}$ in.-lb.

EXERCISES

In each of problems 1 through 11, show that the integrand is an exact differential and evaluate the integral.

1. $\int_C [(x^2 + 2y)\,dx + (2y + 2x)\,dy]$ where C is any arc from $(2, 1)$ to $(4, 2)$.

2. $\int_C [(3x^2 + 4xy - 2y^2)\,dx + (2x^2 - 4xy - 3y^2)\,dy]$ where C is any arc from $(1, 1)$ to $(3, 2)$.

3. $\int_C (e^x \cos y\,dx - e^x \sin y\,dy)$ where C is any arc from $(1, 0)$ to $(0, 1)$.

4. $\int_C \left[\left(\frac{2xy^2}{1 + x^2} + 3\right)dx + (2y \ln(1 + x^2) - 2)\,dy\right]$

   where C is any arc from $(0, 2)$ to $(5, 1)$.

5. $\int_C \left[\frac{y^2}{(x^2 + y^2)^{3/2}}\,dx - \frac{xy}{(x^2 + y^2)^{3/2}}\,dy\right]$

   where C is any arc from $(4, 3)$ to $(-3, 4)$ which does not pass through the origin.

6. $\displaystyle\int_C \left[\frac{x}{\sqrt{1+x^2+y^2}}\, dx + \frac{y}{\sqrt{1+x^2+y^2}}\, dy \right]$

where C is any arc from $(-2,-2)$ to $(4,1)$.

7. $\int_C \{[ye^{xy}(\cos xy - \sin xy) + \cos x]\, dx + [xe^{xy}(\cos xy - \sin xy) + \sin y]\, dy\}$
where C is any arc from $(0,0)$ to $(3,-2)$.

8. $\int_C [(2x - 2y + z + 2)\, dx + (2y - 2x - 1)\, dy + (-2z + x)\, dz]$ where C is any arc from $(1,0,2)$ to $(3,-1,4)$.

9. $\int_C [(2x + y - z)\, dx + (-2y + x + 2z + 3)\, dy + (4z - x + 2y - 2)\, dz]$
where C is any arc from $(0,2,-1)$ to $(1,-2,4)$.

10. $\int_C [(3x^2 - 3yz + 2xz)\, dx + (3y^2 - 3xz + z^2)\, dy + (3z^2 - 3xy + x^2 + 2yz)\, dz]$
where C is any arc from $(-1,2,3)$ to $(3,2,-1)$.

11. $\int_C [(yze^{xyz}\cos x - e^{xyz}\sin x + y\cos xy + z\sin xz)\, dx$
$\qquad\qquad + (xze^{xyz}\cos x + x\cos xy)\, dy + (xye^{xyz}\cos x + x\sin xz)\, dz]$

where C is any arc from $(0,0,0)$ to $(-1,-2,-3)$.

12. A particle is moving along the path

$$x = t + 1, \qquad y = 2t^2 + t + 2$$

from the point $(1,2)$ to the point $(2,5)$, subject to the force $\mathbf{F} = (x^2 + y)\mathbf{i} + 2xy\mathbf{j}$. Find the total work done.

13. A particle is moving in the xy-plane along a straight line from the point $A(a,b)$ to the point $B(c,d)$, subject to the force

$$\mathbf{F} = \frac{-x}{x^2 + y^2}\mathbf{i} - \frac{y}{x^2 + y^2}\mathbf{j}.$$

Find the work done. Show that the work done is the same if a different path between A and B is selected. (The path does not go through the origin.)

14. A particle moves in the xy-plane along the straight line connecting $A(a,b)$ and $B(c,d)$, subject to the force

$$\mathbf{F} = \frac{-x}{(x^2 + y^2)^{3/2}}\mathbf{i} + \frac{-y}{(x^2 + y^2)^{3/2}}\mathbf{j}.$$

Find the work done. Show that the work done is the same if a different path is selected joining the points A and B but not going through the origin.

15. A particle moves along the straight line (in three space) joining the points $A(a,b,c)$ and $B(d,e,f)$, subject to the force

$$\mathbf{F} = \frac{-x}{(x^2 + y^2 + z^2)^{3/2}}\mathbf{i} - \frac{y}{(x^2 + y^2 + z^2)^{3/2}}\mathbf{j} - \frac{z}{(x^2 + y^2 + z^2)^{3/2}}\mathbf{k}.$$

Find the work done. Show that the work done is the same if a different path is selected joining the points A and B but not going through the origin.

16. Same as problem 15, with

$$\mathbf{F} = \frac{x}{(x^2 + y^2 + z^2)^2}\mathbf{i} + \frac{y}{(x^2 + y^2 + z^2)^2}\mathbf{j} + \frac{z}{(x^2 + y^2 + z^2)^2}\mathbf{k}.$$

CHAPTER **5**

Multiple Integration

1. DEFINITION OF THE DOUBLE INTEGRAL

Let F be a region of area A situated in the xy-plane. We shall always assume that a region includes its boundary curve. Such regions are sometimes called *closed regions* in analogy with closed intervals on the real line—that is, ones which include their endpoints. We subdivide the xy-plane into rectangles by drawing lines parallel to the coordinate axes. These lines may or may not be equally spaced (Fig. 5–1). Starting in some convenient place (such as the upper left-hand corner of F), we systematically number all the rectangles *lying entirely within F*. Suppose there are n such and we label them $r_1, r_2, \ldots, r_n$. We use the symbols $A(r_1), A(r_2), \ldots, A(r_n)$ for the areas of these rectangles. The collection of n rectangles $\{r_1, r_2, \ldots, r_n\}$ is called a **subdivision** Δ of F. The **norm of the subdivision,** denoted as usual by $\|\Delta\|$, is the length of the diagonal of the largest rectangle in the subdivision Δ.

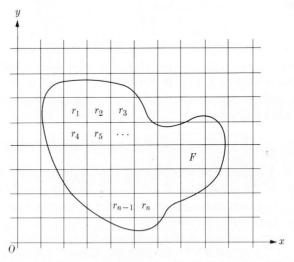

FIGURE 5–1

Suppose that $f(x, y)$ is a function defined for all (x, y) in the region F. The definition of the *double integral of f over the region F* is similar to the definition of the integral for functions of one variable. (See *First Course*, page 213.) Select arbitrarily a point in each of the rectangles of the subdivision Δ, denoting the

204

coordinates of the point in the rectangle r_i by (ξ_i, η_i). (See Fig. 5–2.) Now form the sum

$$f(\xi_1, \eta_1)A(r_1) + f(\xi_2, \eta_2)A(r_2) + \cdots + f(\xi_n, \eta_n)A(r_n)$$

or, more compactly,

$$\sum_{i=1}^{n} f(\xi_i, \eta_i)A(r_i). \qquad (1)$$

This sum is an approximation to the double integral we shall define. Sums such as (1) may be formed for subdivisions with any positive norm and with the ith point (ξ_i, η_i) chosen in any way whatsoever in the rectangle r_i.

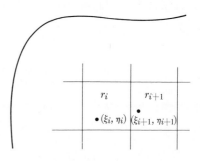

FIGURE 5–2

DEFINITION. We say that **a number L is the limit of sums of type** (1) and write

$$\lim_{||\Delta|| \to 0} \sum_{i=1}^{n} f(\xi_i, \eta_i)A(r_i) = L$$

if the number L has the property: for each $\epsilon > 0$ there is a $\delta > 0$ such that

$$\left| \sum_{i=1}^{n} f(\xi_i, \eta_i)A(r_i) - L \right| < \epsilon$$

for every subdivision Δ with $||\Delta|| < \delta$ and for all possible choices of the points (ξ_i, η_i) in the rectangles r_i.

It can be shown that if the number L exists, then it must be unique.

DEFINITION. If f is defined in a region F and the number L defined above exists, we say that f **is integrable over** F and write

$$\iint_F f(x, y)\, dA.$$

We also call the expression above the **double integral of f over F.**

The double integral has a geometric interpretation in terms of the volume of a solid. We recall the methods of finding volumes of solids of revolution developed in *First Course*, on page 482. Now we shall discuss the notion of volume in somewhat more detail. The definition of volume depends on (i), the definition of the volume of a cube—namely, length times width times height, and (ii), a limiting process.

Let S be a solid in three-space. We divide all of space into cubes by constructing planes parallel to the coordinate planes at a distance apart of $1/2^n$ units, with n some positive integer. In such a network, the cubes are of three kinds: type (1), those cubes entirely within S; type (2), those cubes partly in S and partly outside S; and type (3), those cubes entirely outside S (Fig. 5–3). We define

$$V_n^-(S) = \frac{1}{8^n} \text{ times the number of cubes of type (1)},$$

$$V_n^+(S) = V_n^-(S) + \frac{1}{8^n} \text{ times the number of cubes of type (2)}.$$

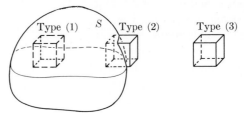

FIGURE 5–3

(Compare this discussion with that for area, in *First Course*, page 200 ff.) Intuitively we expect that, however the volume of S is defined, the number $V_n^-(S)$ would be smaller than the volume, while the number $V_n^+(S)$ would be larger. It can be shown that, as n increases, $V_n^-(S)$ gets larger or at least does not decrease, while $V_n^+(S)$ gets smaller or is at least nonincreasing. Obviously,

$$V_n^-(S) \leq V_n^+(S),$$

always. Since bounded increasing sequences and bounded decreasing sequences tend to limits, the following definitions are appropriate.

DEFINITIONS. The **inner volume** of a solid S, denoted $V^-(S)$, is $\lim_{n \to \infty} V_n^-(S)$. The **outer volume,** denoted $V^+(S)$, is $\lim_{n \to \infty} V_n^+(S)$. A set of points S in three-space has a **volume** whenever $V^-(S) = V^+(S)$. This common value is denoted by $V(S)$ and is called the volume of S.

Remark. It is not difficult to construct point sets for which $V^-(S) \neq V^+(S)$. For example, take S to be all points (x, y, z) such that x, y, and z are rational and $0 \leq x \leq 1, 0 \leq y \leq 1, 0 \leq z \leq 1$. The student can verify that $V_n^-(S) = 0$ for every n, while $V_n^+(S) = 1$ for every n.

If S_1 and S_2 are two solids with no points in common, it can be shown, as expected, that $V(S_1 + S_2) = V(S_1) + V(S_2)$. Also, the subdivision of all of space into cubes is not vital. Rectangular parallelepipeds would do equally well, with the formula for the volume of a rectangular parallelepiped taken as length times width times height.

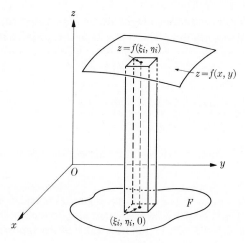

FIGURE 5–4

The volume of a solid is intimately connected with the double integral in the same way that the area of a region is connected with the single integral. (See *First Course*, page 213.) We now exhibit this connection.

Suppose that $f(x, y)$ is a positive function defined for (x, y) in some region F (Fig. 5–4). An item in the sum (1) approximating the double integral is

$$f(\xi_i, \eta_i)A(r_i),$$

which we recognize as the volume of the rectangular column of height $f(\xi_i, \eta_i)$ and area of base $A(r_i)$ (Fig. 5–4). The sum of such columns is an approximation to the volume of the cylindrical solid bounded by the surface $z = f(x, y)$, the plane figure F, and lines parallel to the z-axis through the boundary of F (Fig. 5–5). It can be shown that, with appropriate hypotheses on the function f, the double integral

$$\iint_F f(x, y)\, dA$$

measures the "volume under the surface" in the same way that a single integral

$$\int_a^b f(x)\, dx$$

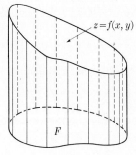

FIGURE 5–5

measures the area under the curve. An outline of a proof of the following theorem is given in Morrey, *University Calculus*, page 596.*

---

* Actually it is shown there that the rectangles r_i may be replaced by any figures F_i (i.e., any regions which have area) in the sums $\sum f(\xi_i, \eta_i)A(F_i)$ approximating the integral.

Theorem 1. *If $f(x, y)$ is continuous for (x, y) in a closed region F, then f is integrable over F. Furthermore, if $f(x, y) > 0$ for (x, y) in F, then*

$$V(S) = \iint\limits_{F} f(x, y)\, dA,$$

where $V(S)$ is the volume of the solid defined by

$$(x, y) \text{ in } F \text{ and } 0 \leq z \leq f(x, y).$$

Methods for the evaluation of double integrals are discussed in Section 3.

EXAMPLE. Given $f(x, y) = 1 + xy$ and the region F bounded by the lines $y = 0$, $y = x$, and $x = 1$ (Fig. 5–6), let Δ be the subdivision formed by the lines $x = 0, 0.2$, $0.5, 0.8, 1$ and $y = 0, 0.2, 0.5, 0.7, 1$. Find the value of the approximating sum

$$\sum_{i=1}^{n} f(\xi_i, \eta_i) A(r_i)$$

to the double integral

$$\iint\limits_{F} f(x, y)\, dA$$

if the points (ξ_i, η_i) are selected at the centers of the rectangles.

Solution. Referring to Fig. 5–6, we see that there are 6 rectangles in the subdivision which we label $r_1, r_2, \ldots, r_6$, as shown. We compute:

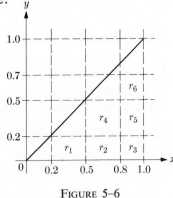

$$A(r_1) = 0.06, \quad f(0.35, \ 0.1) = 1.035$$
$$A(r_2) = 0.06, \quad f(0.65, \ 0.1) = 1.065$$
$$A(r_3) = 0.04, \quad f(0.9, \ \ 0.1) = 1.090$$
$$A(r_4) = 0.09, \quad f(0.65, 0.35) = 1.2375$$
$$A(r_5) = 0.06, \quad f(0.9, \ 0.35) = 1.315$$
$$A(r_6) = 0.04, \quad f(0.9, \ \ 0.6) = 1.540$$

Multiplying and adding, we find that

$$\sum_{i=1}^{6} f(\xi_i, \eta_i) A(r_i) = 0.421475 \quad \text{(Answer).}$$

FIGURE 5–6

EXERCISES

In each of problems 1 though 10, calculate the sum $\sum_{i=1}^{n} f(\xi_i, \eta_i) A(r_i)$ for the subdivision Δ of the region F formed by the given lines and with the points (ξ_i, η_i) selected as directed in each case.

1. $f(x, y) = x^2 + 2y^2$; F is the rectangle $0 \leq x \leq 1, 0 \leq y \leq 1$. The subdivision Δ is: $x = 0, 0.4, 0.8, 1$; $y = 0, 0.3, 0.7, 1$. For each i the point (ξ_i, η_i) is taken at the center of the rectangle r_i.

2. Same as problem 1, with (ξ_i, η_i) taken at the point of r_i which is closest to the origin.

3. $f(x, y) = 1 + x^2 - y^2$; F is the triangular region formed by the lines $y = 0$, $y = x$, $x = 2$. The subdivision Δ is: $x = 0, 0.5, 1, 1.6, 2$; $y = 0, 0.6, 1, 1.5, 2$. For each i, the point (ξ_i, η_i) is taken at the center of the rectangle r_i.

4. Same as problem 3, with (ξ_i, η_i) taken at the point of r_i which is closest to the origin.

5. Same as problem 3, with (ξ_i, η_i) selected on the lower edge of r_i, midway between the vertical subdivision lines.

6. $f(x, y) = x^2 - 2xy + 3x - 2y$; F is the trapezoid bounded by the lines $x = 0$, $x = 2$, $y = 0$, $y = x + 1$. The subdivision Δ is: $x = 0, 0.4, 1, 1.5, 2$; $y = 0$, $0.6, 1, 1.4, 1.8, 2, 3$. For each i the point (ξ_i, η_i) is taken at the center of the rectangle r_i.

7. Same as problem 6, with (ξ_i, η_i) taken at the point of r_i farthest from the origin.

8. Same as problem 6, with (ξ_i, η_i) taken at the point of r_i closest to the origin.

9. Same as problem 6, with the subdivision Δ: $x = 0, 0.4, 0.8, 1, 1.5, 1.8, 2$; $y = 0$, $0.3, 0.6, 1, 1.4, 1.6, 1.8, 2, 3$. Can any statement be made comparing the results of problems 6 and 9 with $\iint_F f(x, y)\, dA$?

10. We have

$$f(x, y) = \frac{x - y}{1 + x + y};$$

F is the region bounded by the line $y = 0$ and the curve $y = 2x - x^2$. The subdivision Δ is: $x = 0, 0.5, 1.0, 1.5, 2$; $y = 0, 0.2, 0.4, 0.6, 0.8, 1$. For each i, the point (ξ_i, η_i) is taken at the center of the rectangle r_i.

2. PROPERTIES OF THE DOUBLE INTEGRAL

In analogy with the properties of the definite integral of functions of one variable (*First Course*, page 219), we state several basic properties of the double integral. The simplest properties are given in the two following theorems.

Theorem 2. *If c is any number and f is integrable over a closed region F, then cf is integrable and*

$$\iint_F cf(x, y)\, dA = c \iint_F f(x, y)\, dA.$$

Theorem 3. *If f and g are integrable over a closed region F, then*

$$\iint_F [f(x, y) + g(x, y)]\, dA = \iint_F f(x, y)\, dA + \iint_F g(x, y)\, dA.$$

The result holds for the sum of any finite number of integrable functions. The proofs of Theorems 2 and 3 are obtained directly from the definition.

Theorem 4. *Suppose that f is integrable over a closed region F and*

$$m \leq f(x, y) \leq M \text{ for all } (x, y) \text{ in } F.$$

Then, if A(F) denotes the area of F, we have

$$mA(F) \leq \iint_F f(x, y) \, dA \leq MA(F).$$

The proof of Theorem 4 follows exactly the same pattern as does the proof in the one-variable case. (See *First Course*, page 220, Theorem 6.)

Theorem 5. *If f and g are integrable over F and $f(x, y) \leq g(x, y)$ for all (x, y) in F, then*

$$\iint_F f(x, y) \, dA \leq \iint_F g(x, y) \, dA.$$

The proof is established by the same argument used in the one-variable case (*First Course*, page 221, Theorem 7).

Theorem 6. *If the closed region F is decomposed into regions F_1 and F_2 and if f is continuous over F, then*

$$\iint_F f(x, y) \, dA = \iint_{F_1} f(x, y) \, dA + \iint_{F_2} f(x, y) \, dA.$$

The proof depends on the definition of double integral and on the basic theorems on limits.

EXERCISES

In problems 1 through 7, use Theorem 4 to find in each case the largest and smallest values the given double integrals can possibly have.

1. $\iint_F xy \, dA$ where F is the region bounded by the lines $x = 0$, $y = 0$, $x = 2$, $y = x + 3$.

2. $\iint_F (x^2 + y^2) \, dA$ where F is the region bounded by the lines $x = -2$, $x = 3$, $y = x + 2$, $y = -2$.

3. $\iint_F (1 + 2x^2 + y^2) \, dA$ where F is the region bounded by the lines $x = -3$, $x = 3$, $y = 4$, $y = -4$.

4. $\iint_F y^4 \, dA$ where F is the region bounded by the line $y = 0$ and the curve $y = 2x - x^2$.

5. $\iint_F (x - y) \, dA$ where F is the region enclosed in the circle $x^2 + y^2 = 9$.

6. $\iint_F 1/(1 + x^2 + y^2) \, dA$ where F is the region enclosed in the ellipse $4x^2 + 9y^2 = 36$.

7. $\iint_F \sqrt{1 + x^2 + y^2} \, dA$ where F is the region bounded by the curves $y = 3x - x^2$ and $y = x^2 - 3x$.

8. Write out a proof of Theorem 3.　　　9. Write out a proof of Theorem 4.

10. Write out a proof of Theorem 5.　　11. Write out a proof of Theorem 6.

3. EVALUATION OF DOUBLE INTEGRALS. ITERATED INTEGRALS

The definition of the double integral is useless as a tool for evaluation in any particular case. Of course, it may happen that the function $f(x, y)$ and the region F are particularly simple, so that the limit of the sum $\sum_{i=1}^{n} f(\xi_i, \eta_i) A(r_i)$ can be found directly. However, such limits cannot generally be found. As in the case of ordinary integrals and line integrals, it is important to develop simple and routine methods for determining the value of a given double integral. In this section we show how the evaluation of a double integral may be performed by successive evaluations of single integrals. In other words, we reduce the problem to one we have already studied extensively. The student will recall that the evaluation of line integrals was reduced to known techniques for single integrals in a similar way.

Let F be the rectangle with sides $x = a$, $x = b$, $y = c$, $y = d$, as shown in Fig. 5–7. Suppose that $f(x, y)$ is continuous for (x, y) in F. We form the ordinary integral with respect to x,

$$\int_a^b f(x, y)\, dx,$$

in which we keep y fixed when performing the integration. Of course, the value of the above integral will depend on the value of y used, and so we may write

$$A(y) = \int_a^b f(x, y)\, dx.$$

The function $A(y)$ is defined for $c \leq y \leq d$ and, in fact, it can be shown that if $f(x, y)$ is continuous on F, then $A(y)$ is continuous on $[c, d]$. The integral of $A(y)$ may be computed, and we write

$$\int_c^d A(y)\, dy = \int_c^d \left[\int_a^b f(x, y)\, dx \right] dy. \tag{1}$$

We could start the other way around by fixing x and forming the integral

$$B(x) = \int_c^d f(x, y)\, dy.$$

Then

$$\int_a^b B(x)\, dx = \int_a^b \left[\int_c^d f(x, y)\, dy \right] dx. \tag{2}$$

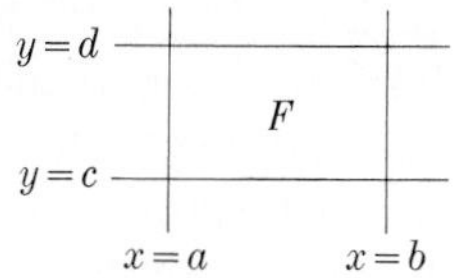

FIGURE 5–7

Note that the integrals are computed *successively*; in (1) we first integrate with respect to x (keeping y constant) and then with respect to y; in (2) we first integrate with respect to y (keeping x constant) and then with respect to x.

DEFINITION. The integrals

$$\int_c^d \left[\int_a^b f(x, y)\, dx \right] dy, \qquad \int_a^b \left[\int_c^d f(x, y)\, dy \right] dx$$

are called the **iterated integrals** of f. The terms *repeated* integrals and *successive* integrals are also used.

NOTATION. The brackets in iterated integrals are unwieldy, and we will write

$$\int_c^d \int_a^b f(x, y)\, dx\, dy \qquad \text{to mean} \qquad \int_c^d \left[\int_a^b f(x, y)\, dx \right] dy,$$

$$\int_a^b \int_c^d f(x, y)\, dy\, dx \qquad \text{to mean} \qquad \int_a^b \left[\int_c^d f(x, y)\, dy \right] dx.$$

Iterated integrals are computed in the usual way, as the next example shows.

EXAMPLE 1. Evaluate

$$\int_1^4 \int_{-2}^3 (x^2 - 2xy^2 + y^3)\, dx\, dy.$$

Solution. Keeping y fixed, we have

$$\int_{-2}^3 (x^2 - 2xy^2 + y^3)\, dx = \left[\frac{1}{3}x^3 - x^2 y^2 + y^3 x \right]_{-2}^3$$

$$= 9 - 9y^2 + 3y^3 - \left(-\frac{8}{3} - 4y^2 - 2y^3 \right)$$

$$= \frac{35}{3} - 5y^2 + 5y^3.$$

Therefore

$$\int_1^4 \int_{-2}^3 (x^2 - 2xy^2 + y^3)\, dx\, dy = \int_1^4 \left(\frac{35}{3} - 5y^2 + 5y^3 \right) dy$$

$$= \left[\frac{35}{3} y - \frac{5}{3} y^3 + \frac{5}{4} y^4 \right]_1^4 = \frac{995}{4}.$$

Iterated integrals may be defined over regions F which have curved boundaries. This situation is more complicated than the one just discussed. Consider a region F such as that shown in Fig. 5–8, in which the boundary consists of the lines $x = a$, $x = b$, and the graphs of the functions $p(x)$ and $q(x)$ with $p(x) \le q(x)$ for $a \le x \le b$. We may define

$$\int_a^b \int_{p(x)}^{q(x)} f(x, y)\, dy\, dx,$$

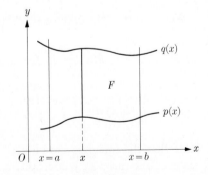

FIGURE 5–8

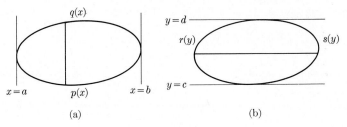

FIGURE 5–9

in which we first integrate (for fixed x) from the lower curve to the upper curve, i.e., along a typical line as shown in Fig. 5–8; then we integrate with respect to x over all such typical segments from a to b.

More generally, iterated integrals may be defined over a region F such as the one shown in Fig. 5–9(a). Integrating first with respect to y, we have

$$\int_a^b \int_{p(x)}^{q(x)} f(x, y) \, dy \, dx.$$

On the other hand, the integral taken first with respect to x requires that we represent F as shown in Fig. 5–9(b). Then we have

$$\int_c^d \int_{r(y)}^{s(y)} f(x, y) \, dx \, dy.$$

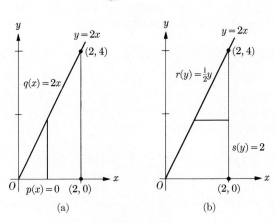

FIGURE 5–10

EXAMPLE 2. Given the function $f(x, y) = xy$ and the triangular region F bounded by the lines $y = 0$, $y = 2x$, $x = 2$ [Fig. 5–10(a)], find the value of both iterated integrals.

Solution. Referring to Fig. 5–10(a), we see that for

$$\int_a^b \int_{p(x)}^{q(x)} xy \, dy \, dx,$$

we have $p(x) = 0$, $q(x) = 2x$, $a = 0$, $b = 2$. Therefore

$$\int_0^2 \int_0^{2x} xy \, dy \, dx = \int_0^2 \left[\frac{1}{2} xy^2 \right]_0^{2x} dx$$

$$= \int_0^2 2x^3 \, dx = \left[\frac{1}{2} x^4 \right]_0^2 = 8.$$

Integrating with respect to x first [Fig. 5–10(b)], we have

$$\int_c^d \int_{r(y)}^{s(y)} xy \, dx \, dy$$

with

$$r(y) = \frac{1}{2} y, \quad s(y) = 2, \quad c = 0, \quad d = 4.$$

Therefore

$$\int_0^4 \int_{y/2}^2 xy \, dx \, dy = \int_0^4 \left[\frac{1}{2} x^2 y \right]_{y/2}^2 dy$$

$$= \int_0^4 \left(2y - \frac{1}{8} y^3 \right) dy = \left[y^2 - \frac{1}{32} y^4 \right]_0^4 = 8.$$

It is not accidental that the two integrals in Example 2 have the same value. The next theorem describes the general situation.

Theorem 7. *Suppose F is a closed region consisting of all (x, y) such that*

$$a \leq x \leq b, \qquad p(x) \leq y \leq q(x),$$

where p and q are continuous and $p(x) \leq q(x)$ for $a \leq x \leq b$. Suppose that $f(x, y)$ is continuous for (x, y) in F. Then

$$\iint_F f(x, y) \, dA = \int_a^b \int_{p(x)}^{q(x)} f(x, y) \, dy \, dx.$$

The corresponding result holds if the closed region F has the representation

$$c \leq y \leq d, \qquad r(y) \leq x \leq s(y)$$

where $r(y) \leq s(y)$ for $c \leq y \leq d$. In such a case,

$$\iint_F f(x, y) \, dA = \int_c^d \int_{r(y)}^{s(y)} f(x, y) \, dx \, dy.$$

In other words, both iterated integrals, when computable, are equal to the double integral and therefore equal to each other.

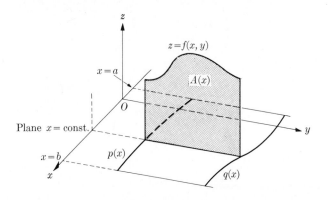

FIGURE 5–11

Partial proof. We shall discuss the first result, the second being similar. Suppose first that $f(x, y)$ is positive. A plane $x = $ const intersects the surface $f(x, y)$ in a curve (Fig. 5–11). The area under this curve in the $x = $ const plane is shown as a shaded region. Denoting the area of this region by $A(x)$, we have the formula

$$A(x) = \int_{p(x)}^{q(x)} f(x, y) \, dy.$$

It can be shown that $A(x)$ is continuous. Furthermore, it can also be shown (Morrey, *University Calculus*, pp. 599–601) that if $A(x)$ is integrated between $x = a$ and $x = b$, the volume V under the surface $f(x, y)$ is swept out. We recall that a similar argument was used in obtaining volumes of revolution by the disc method. (See *First Course*, page 482.) The double integral yields the volume under the surface, and so we write

$$V = \iint_F f(x, y) \, dA.$$

On the other hand, we obtain the volume by integrating $A(x)$; that is,

$$V = \int_a^b A(x) \, dx = \int_a^b \int_{p(x)}^{q(x)} f(x, y) \, dy \, dx.$$

If $f(x, y)$ is not positive but is bounded from below by the plane $z = c$, then subtraction of the volume of the cylinder of height c and cross-section F leads to the same result.

Remarks. We have considered two ways of expressing a region F in the xy-plane. They are

$$a \le x \le b, \qquad p(x) \le y \le q(x) \tag{3}$$

and

$$c \le y \le d, \qquad r(y) \le x \le s(y). \tag{4}$$

It frequently happens that a region F is expressible more simply in one of the

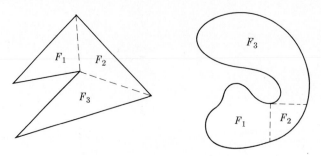

FIGURE 5–12

above forms than in the other. In doubtful cases, a sketch of F may show which is simpler and, therefore, which of the iterated integrals is evaluated more easily.

A region F may not be expressible in either the form (3) or the form (4). In such cases, F may sometimes be subdivided into a number of regions, each having one of the two forms. The integrations are then performed for each subregion and the results added. Figure 5–12 gives examples of how the subdivision process might take place.

EXAMPLE 3. Evaluate $\iint_F x^2 y^2\, dA$ where F is the figure bounded by the lines $y = 1$, $y = 2$, $x = 0$, and $x = y$ (Fig. 5–13).

Solution. The closed region F is the set of (x, y) such that

$$1 \le y \le 2, \qquad 0 \le x \le y.$$

We use Theorem 7 and evaluate the iterated integral, to find

$$\iint_F x^2 y^2\, dA = \int_1^2 \int_0^y x^2 y^2\, dx\, dy = \int_1^2 \left[\frac{1}{3} x^3 y^2\right]_0^y dy$$

$$= \frac{1}{3} \int_1^2 y^5\, dy = \frac{7}{2}.$$

Note that in the above example the iterated integral in the other order is a little more difficult, since the curves $p(x)$, $q(x)$ are

$$p(x) = \begin{cases} 1 \text{ for } 0 \le x \le 1 \\ x \text{ for } 1 \le x \le 2 \end{cases}, \qquad q(x) = 2, \quad 0 \le x \le 2.$$

The evaluation would have to take place in two parts, so that

$$\iint_F x^2 y^2\, dA = \int_0^1 \int_1^2 x^2 y^2\, dy\, dx + \int_1^2 \int_x^2 x^2 y^2\, dy\, dx.$$

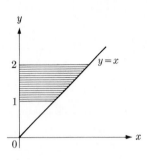

FIGURE 5–13

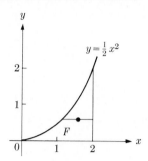

FIGURE 5–14

EXAMPLE 4. Evaluate

$$\int_0^2 \int_0^{x^2/2} \frac{x}{\sqrt{1 + x^2 + y^2}}\, dy\, dx.$$

Solution. Carrying out the integration first with respect to y is possible but difficult and leads to a complicated integral for x. Therefore we shall try to express the integral as an iterated integral in the opposite order and use Theorem 7. We construct the region F as shown in Fig. 5–14. The region is expressed by the inequalities

$$0 \le x \le 2 \quad \text{and} \quad 0 \le y \le \tfrac{1}{2}x^2.$$

However, it is also expressed by the inequalities

$$0 \le y \le 2 \quad \text{and} \quad \sqrt{2y} \le x \le 2.$$

Therefore, integrating with respect to x first, we have

$$\int_0^2 \int_0^{x^2/2} \frac{x}{\sqrt{1 + x^2 + y^2}}\, dy\, dx$$

$$= \iint_F \frac{x}{\sqrt{1 + x^2 + y^2}}\, dA = \int_0^2 \int_{\sqrt{2y}}^2 \frac{x}{\sqrt{1 + x^2 + y^2}}\, dx\, dy$$

$$= \int_0^2 [\sqrt{1 + x^2 + y^2}]_{\sqrt{2y}}^2\, dy = \int_0^2 [\sqrt{5 + y^2} - (1 + y)]\, dy$$

$$= [\tfrac{5}{2} \ln (y + \sqrt{y^2 + 5}) + \tfrac{1}{2}y\sqrt{y^2 + 5} - y - \tfrac{1}{2}y^2]_0^2$$

$$= \tfrac{5}{2} \ln 5 + 3 - 4 - \tfrac{5}{2} \ln \sqrt{5} = -1 + \tfrac{5}{4} \ln 5.$$

The next example shows how the volume of a solid may be found by iterated integration.

EXAMPLE 5. Let S be the solid bounded by the surface $z = xy$, the cylinders $y = x^2$ and $y^2 = x$, and the plane $z = 0$. Find the volume $V(S)$.

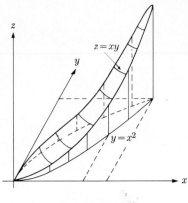

FIGURE 5–15

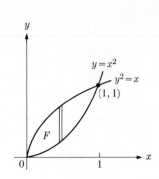

FIGURE 5–16

Solution. The solid S is shown in Fig. 5–15. It consists of all points "under" the surface $z = xy$, bounded by the cylinders, and "above" the xy-plane. The region F in the xy-plane is bounded by the curves $y = x^2$, $y^2 = x$ and is shown in Fig. 5–16. Therefore

$$V(S) = \iint_F xy \, dA = \int_0^1 \int_{x^2}^{\sqrt{x}} xy \, dy \, dx = \int_0^1 \left[\frac{xy^2}{2} \right]_{x^2}^{\sqrt{x}} dx$$

$$= \frac{1}{2} \int_0^1 (x^2 - x^5) \, dx = \frac{1}{12}.$$

If a solid S is bounded by two surfaces of the form $z = f(x, y)$ and $z = g(x, y)$ with $f(x, y) \leq g(x, y)$, then the volume between the surfaces may be found as a double integral, and that integral in turn may be evaluated by iterated integrals. The closed region F over which the integration is performed is found by the projection onto the xy-plane of the curve of intersection of the two surfaces. To find this projection we merely set

$$f(x, y) = g(x, y)$$

and trace this curve in the xy-plane. The next example shows the method.

EXAMPLE 6. Find the volume bounded by the surfaces

$$z = x^2 \quad \text{and} \quad z = 4 - x^2 - y^2.$$

Solution. A portion of the solid S (the part corresponding to $y \leq 0$) is shown in Fig. 5–17. We set $x^2 = 4 - x^2 - y^2$ and find that the closed region F in the xy-plane is the ellipse (Fig. 5–18)

$$\frac{x^2}{2} + \frac{y^2}{4} = 1.$$

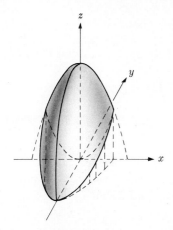

FIGURE 5–17

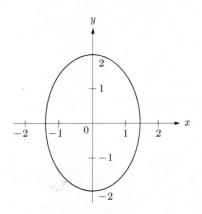

FIGURE 5–18

Note that the surface $z = 4 - x^2 - y^2 \equiv g(x, y)$ is above the surface $z = x^2 \equiv f(x, y)$ for (x, y) inside the above ellipse. Therefore

$$V(S) = \iint\limits_{F} (4 - y^2 - x^2 - x^2)\, dA$$

$$= \int_{-2}^{2} \int_{-\sqrt{4-y^2}/\sqrt{2}}^{+\sqrt{4-y^2}/\sqrt{2}} (4 - y^2 - 2x^2)\, dx\, dy$$

$$= \int_{-2}^{2} \frac{2\sqrt{2}}{3} (4 - y^2)^{3/2}\, dy = \frac{4\sqrt{2}}{3} \int_{0}^{2} (4 - y^2)^{3/2}\, dy$$

$$= \frac{64\sqrt{2}}{3} \int_{0}^{\pi/2} \cos^4 \theta\, d\theta = \frac{16\sqrt{2}}{3} \int_{0}^{\pi/2} (1 + 2\cos 2\theta + \cos^2 2\theta)\, d\theta$$

$$= \frac{8\pi\sqrt{2}}{3} + \left[\frac{16\sqrt{2}}{3} \sin 2\theta\right]_{0}^{\pi/2} + \frac{8\sqrt{2}}{3} \int_{0}^{\pi/2} (1 + \cos 4\theta)\, d\theta = 4\pi\sqrt{2}.$$

EXERCISES

In problems 1 through 10, evaluate the iterated integrals as indicated. Sketch the region F in the xy-plane over which the integration is taken.

1. $\int_{1}^{4} \int_{2}^{5} (x^2 - y^2 + xy - 3)\, dx\, dy$ 2. $\int_{0}^{2} \int_{-3}^{2} (x^3 + 2x^2 y - y^3 + xy)\, dy\, dx$

3. $\int_{1}^{4} \int_{\sqrt{x}}^{x^2} (x^2 + 2xy - 3y^2)\, dy\, dx$ 4. $\int_{0}^{1} \int_{x^3}^{x^2} (x^2 - xy)\, dy\, dx$

5. $\int_{2}^{3} \int_{1+y}^{\sqrt{y}} (x^2 y + xy^2)\, dx\, dy$ 6. $\int_{-2}^{2} \int_{-\sqrt{4-x^2}}^{+\sqrt{4-x^2}} y\, dy\, dx$

7. $\displaystyle\int_{-3}^{3}\int_{-\sqrt{18-2y^2}}^{+\sqrt{18-2y^2}} x \, dx \, dy$

8. $\displaystyle\int_{-3}^{3}\int_{z^2}^{18-z^2} xy^3 \, dy \, dx$

9. $\displaystyle\int_{0}^{2}\int_{x^2}^{2x^2} x \cos y \, dy \, dx$

10. $\displaystyle\int_{1}^{2}\int_{x^3}^{4x^3} \frac{1}{y} \, dy \, dx$

In problems 11 through 17, evaluate the double integrals as indicated. Sketch the region F.

11. $\displaystyle\iint_F (x^2 + y^2) \, dA$, $F: 0 \le y \le 2, \, y^2 \le x \le 4$

12. $\displaystyle\iint_F x \cos y \, dA$, F bounded by the curve $y = x^2$ and the lines $y = 0, \, x = \sqrt{\pi/2}$

13. $\displaystyle\iint_F \frac{x}{x^2 + y^2} \, dA$, F bounded by $y = 0, \, y = x, \, x = 1,$ and $x = \sqrt{3}$

14. $\displaystyle\iint_F \ln y \, dA$, F bounded by $y = 1, \, y = x - 1,$ and $x = 3$

15. $\displaystyle\iint_F \frac{x}{\sqrt{1 - y^2}} \, dA$, F bounded by $x = 0, \, y = 0, \, y = \frac{1}{2},$ and $y = x$

16. $\displaystyle\iint_F \frac{x}{\sqrt{x^2 + y^2}} \, dA$, F bounded by $y = x, \, y = 1,$ and $x = 2$

17. $\displaystyle\iint_F \frac{1}{y^2} e^{x/\sqrt{y}} \, dA$, F bounded by $x = 1, \, y = 2,$ and $y = x^2 (x \ge 1)$

In each of problems 18 through 22, (a) sketch the domain over which the integration is performed; (b) write the equivalent iterated integral in the reverse order; (c) evaluate the integral obtained in (b).

18. $\displaystyle\int_{1}^{2}\int_{1}^{x} \frac{x^2}{y^2} \, dy \, dx$

19. $\displaystyle\int_{-2}^{2}\int_{-\sqrt{4-x^2}}^{+\sqrt{4-x^2}} xy \, dy \, dx$

20. $\displaystyle\int_{0}^{a}\int_{0}^{\sqrt{a^2-x^2}} (a^2 - y^2)^{3/2} \, dy \, dx$

21. $\displaystyle\int_{0}^{1}\int_{y}^{1} \sqrt{1 + x^2} \, dx \, dy$

22. $\displaystyle\int_{0}^{1}\int_{\sqrt{x}}^{1} \sqrt{1 + y^3} \, dy \, dx$

In each of problems 23 through 32, find the volume $V(S)$ of the solid described.

23. S is bounded by the surfaces $z = 0, z = x,$ and $y^2 = 2 - x$.

24. S is bounded by the planes $z = 0, y = 0, y = x, x + y = 2,$ and $x + y + z = 3$.

25. S is bounded by the surfaces $x = 0, z = 0, y^2 = 4 - x,$ and $z = y + 2$.

26. S is bounded by the surfaces $x^2 + z^2 = 4, y = 0,$ and $x + y + z = 3$.

27. S is bounded by the surfaces $y^2 = z, y = z^3, z = x,$ and $y^2 = 2 - x$.

28. S is bounded by the coordinate planes and the surface $x^{1/2} + y^{1/2} + z^{1/2} = a^{1/2}$.

29. S is bounded by the surfaces $y = x^2$ and $z^2 = 4 - y$.

30. S is bounded by the surfaces $y^2 = x$, $x + y = 2$, $x + z = 0$, and $z = x + 1$.

31. S is bounded by the surfaces $x^2 = y + z$, $y = 0$, $z = 0$, and $x = 2$.

32. S is bounded by the surfaces $y^2 + z^2 = 2x$ and $y = x - \frac{3}{2}$.

4. AREA, DENSITY, AND MASS

The double integral of a nonnegative function $z = f(x, y)$ taken over a region F may be interpreted as a volume. The value of such an integral is the volume of the cylinder having generators parallel to the z-axis and situated between the surface $z = f(x, y)$ and the region F in the xy-plane.

If we select for the surface f the particularly simple function $z = 1$, then the volume V is given by the formula

$$V = \iint_F 1 \, dA.$$

On the other hand, the volume of a right cylinder of cross-section F and height 1 is

$$V = A(F) \cdot 1.$$

(See Fig. 5–19.) Therefore

$$A(F) = \iint_F dA.$$

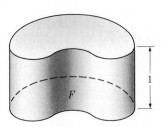

FIGURE 5–19

We see that *the double integral of the function 1 taken over F is precisely the area of F.* By Theorem 7, we conclude that the iterated integral of the function 1 also yields the area of F.

EXAMPLE 1. Use iterated integration to find the area of the region F bounded by the curves $y = x^4$ and $y = 4 - 3x^2$.

Solution. The region F is shown in Fig. 5–20. One of the iterated integrals for the area is

$$A(F) = \int_{-1}^{1} \int_{x^4}^{4-3x^2} dy \, dx,$$

and its evaluation gives

$$A(F) = \int_{-1}^{1} [y]_{x^4}^{4-3x^2} \, dx = \int_{-1}^{1} (4 - 3x^2 - x^4) \, dx$$

$$= \left[4x - x^3 - \frac{1}{5} x^5 \right]_{-1}^{1} = \frac{28}{5}.$$

Note that the iterated integral in the other direction is more difficult to evaluate.

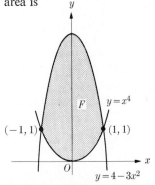

FIGURE 5–20

If a flat object is made of an extremely thin uniform material, then the mass of the object is just a multiple of the area of the plane region on which the object rests. (The multiple depends on the units used.) If a thin object (resting on the xy-plane) is made of a nonuniform material, then the mass* of the object may be expressed in terms of the density $\rho(x, y)$ of the material at any point. It is assumed that the material is uniform in the z-direction. Letting F denote the region occupied by the object, we decompose F into rectangles $r_1, r_2, \ldots, r_n$ in the usual way. Then an approximation to the mass of the ith rectangle is given by

$$\rho(\xi_i, \eta_i)A(r_i),$$

where $A(r_i)$ is the area of r_i and (ξ_i, η_i) is a point in r_i. The total mass of F is approximated by

$$\sum_{i=1}^{n} \rho(\xi_i, \eta_i)A(r_i),$$

and when we proceed to the limit in the customary manner, the mass $M(F)$ is

$$M(F) = \iint_F \rho(x, y)\, dA.$$

In other words, the double integral is a useful device for finding the mass of a thin object with variable density.

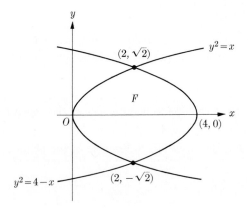

FIGURE 5–21

EXAMPLE 2. A thin object occupies the region F bounded by the curves $y^2 = x$ and $y^2 = 4 - x$. The density is given by $\rho(x, y) = 1 + 2x + y$. Find the total mass.

Solution. We have

$$M(F) = \iint_F (1 + 2x + y)\, dA.$$

_____

* For a precise definition of mass in terms of set functions, see Morrey, *University Calculus*, page 610.

Sketching the region F (Fig. 5–21), we obtain for $M(F)$ the iterated integral

$$M(F) = \int_{-\sqrt{2}}^{\sqrt{2}} \int_{y^2}^{4-y^2} (1 + 2x + y)\, dx\, dy$$

$$= \int_{-\sqrt{2}}^{\sqrt{2}} [x + x^2 + xy]_{y^2}^{4-y^2}\, dy$$

$$= \int_{-\sqrt{2}}^{\sqrt{2}} (20 + 4y - 10y^2 - 2y^3)\, dy$$

$$= \left[20y + 2y^2 - \frac{10}{3} y^3 - \frac{1}{2} y^4 \right]_{-\sqrt{2}}^{\sqrt{2}} = \frac{80}{3} \sqrt{2}.$$

EXERCISES

In each of problems 1 through 5 use iterated integration to find the area of the given region F. Subdivide F and do each part separately whenever necessary.

1. F is bounded by $y = x^3$ and $y = \sqrt{x}$.

2. F is determined by the inequalities

$$y \geq 1, \qquad y \geq x, \qquad y^2 \leq 4x.$$

3. F is determined by the inequalities

$$xy \leq 4, \qquad y \leq x, \qquad 27y \geq 4x^2.$$

4. F is determined by the inequalities

$$y^2 \leq x, \qquad y^2 \leq 6 - x, \qquad y \leq x - 2.$$

5. F is determined by the inequalities

$$x^2 + y^2 \leq 9, \qquad y \leq x + 3, \qquad y \leq -x.$$

In each of problems 6 through 14, find the mass of the given region F.

6. F is the interior of the circle $x^2 + y^2 = 64$; $\rho = x^2 + y^2$.

7. F is bounded by the curves $y = x^2$ and $y^2 = x$; $\rho = 3y$.

8. F is bounded by $y = x^2$ and $y = x + 2$; $\rho = x^2 y$.

9. F is bounded by $y = x^3$ and $y = \sqrt{x}$; $\rho = 2x$.

10. F is bounded by $x + y = 5$ and $xy = 4$; $\rho = 4y$.

11. F is bounded by $y^2 = x$ and $x = y + 2$; $\rho = x^2 y^2$.

12. F is the triangle with vertices at $(0, 0)$, $(a, 0)$, (b, c), $a > b > 0$, $c > 0$; $\rho = 2x$.

13. F is bounded by $y = 0$ and $y = \sqrt{a^2 - x^2}$; $\rho = 3y$.

14. F is a rectangle with vertices at $(0, 0)$, $(a, 0)$, (a, b), $(0, b)$; $\rho = 3x/(1 + x^2 y^2)$.

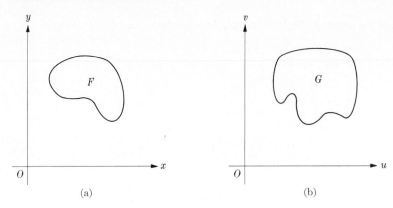

FIGURE 5–22

5. TRANSFORMATIONS AND MAPPINGS IN THE PLANE

Consider a pair of equations

$$u = f(x, y), \qquad v = g(x, y) \tag{1}$$

defined for (x, y) in some region F in the xy-plane. Letting (u, v) denote rectangular coordinates in another plane (Fig. 5–22), we see that the above pair of equations assigns to each point of F a point in the uv-plane. The totality of points so determined in the uv-plane (denoted by G) is called the **image** of F under the **transformation** or **mapping** defined by (1).

It sometimes happens that the system (1) can be solved for x and y in terms of u and v. In such a case we write

$$x = p(u, v), \qquad y = q(u, v),$$

although the domain and image of this *inverse* transformation may not correspond precisely to the regions G and F of the original transformation.

One of the simplest types of transformations is determined by the equations

$$u = x + h, \qquad v = y + k$$

and is called a **translation** mapping. The inverse transformation is

$$x = u - h, \qquad y = v - k.$$

Another simple mapping is a **rotation** mapping (see *First Course*, page 322), given by the equations

$$u = x \cos \theta + y \sin \theta, \qquad v = -x \sin \theta + y \cos \theta, \tag{2}$$

with the inverse transformation given by

$$x = u \cos \theta - v \sin \theta, \qquad y = u \sin \theta + v \cos \theta. \tag{3}$$

The transformation

$$u = ax + by, \qquad v = cx + dy, \tag{4}$$

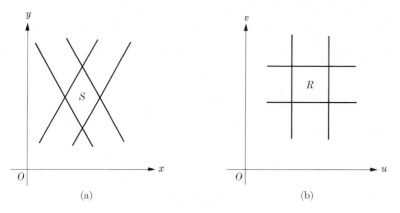

FIGURE 5–23

with a, b, c, d constants is called a **linear transformation.** If $ad - bc \neq 0$, then we may solve for x, y in terms of u, v uniquely. We note that a rectangle R formed by lines parallel to the axes in the uv-plane corresponds to a parallelogram S in the xy-plane (Fig. 5–23). In general, the areas of R and S will be different.

The transformation

$$u = \sqrt{x^2 + y^2}, \quad v = \arctan \frac{y}{x}, \quad x > 0 \tag{5}$$

has the inverse mapping

$$x = u \cos v, \quad y = u \sin v; \quad u > 0, \quad -\frac{\pi}{2} < v < \frac{\pi}{2}. \tag{6}$$

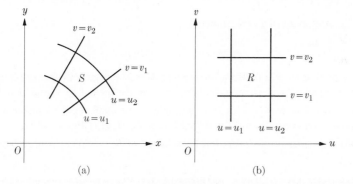

FIGURE 5–24

Figure 5–24 shows how a rectangle R in the uv-plane corresponds to a region S in the xy-plane bounded by two circular arcs and two rays through the origin.

EXAMPLE. Given the transformation (5). Find the area S in the xy-plane corresponding to the rectangle R in the uv-plane determined by the lines $u = u_1$, $u = u_2$, $v = v_1$, $v = v_2$ with $u_2 > u_1$, $v_2 > v_1$.

Solution. The area of a sector of a circle with angle opening θ (measured in radians) and radius r is given by the elementary formula $\frac{1}{2}\theta r^2$. The lines $v = v_1$ and $v = v_2$ determine an angle of $v_2 - v_1$ radians in the xy-plane. Therefore the area $A(S)$ is given by

$$A(S) = \tfrac{1}{2}(v_2 - v_1)u_2^2 - \tfrac{1}{2}(v_2 - v_1)u_1^2 = \tfrac{1}{2}(v_2 - v_1)(u_2^2 - u_1^2)$$
$$= \tfrac{1}{2}(v_2 - v_1)(u_2 - u_1)(u_2 + u_1) = \tfrac{1}{2}(u_2 + u_1)A(R).$$

In order to identify the plane in which the areas are computed, we write

$$A_{x,y}(S)$$

for the area of a region S in the xy-plane. The above relation then becomes

$$A_{x,y}(S) = \tfrac{1}{2}(u_1 + u_2)A_{u,v}(R).$$

Some mappings, such as translations and rotations, transform a region with a given area into a region with the same area. Such mappings are called **area-preserving.** A mapping which leaves the distance between any two points unchanged is called **isometric.**

EXERCISES

1. Show that translation and rotation mappings are both area-preserving and isometric.

2. Show that the transformation

   $$u = x \cos \theta + y \sin \theta + h, \qquad v = -x \sin \theta + y \cos \theta + k$$

   is both area-preserving and isometric.

3. The mapping
   $$u = \alpha x, \qquad v = \beta y,$$

   with $\alpha > 0, \beta > 0$ is called a *stretching*. If S is a region in the xy-plane and R its image under the stretching, find the relation between $A_{x,y}(S)$ and $A_{u,v}(R)$.

4. If $ad - bc \neq 0$ in (4) and R is a rectangle in the uv-plane, S its image under the mapping inverse to (4), find the formula relating $A_{x,y}(S)$ and $A_{u,v}(R)$.

*5. Show that a linear transformation (4) with $ad - bc = 1$ is area-preserving.

6. EVALUATION OF DOUBLE INTEGRALS BY POLAR COORDINATES

The polar coordinates (r, θ) of a point in the plane are related to the rectangular coordinates (x, y) of the same point by the equations

$$x = r \cos \theta, \qquad y = r \sin \theta, \qquad r > 0. \tag{1}$$

We recall that certain problems concerned with finding areas by integration are solved more easily in polar coordinates than in rectangular coordinates. (See *First Course*, page 416.) The same situation prevails in problems involving double integration.

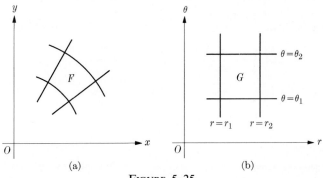

FIGURE 5–25

Instead of considering (1) as a means of representing a point in two different coordinate systems, we interpret the equations as a mapping between the xy-plane and the $r\theta$-plane. In the context of mappings as discussed in Section 5, we draw the $r\theta$-plane as shown in Fig. 5–25, treating $r = 0$ and $\theta = 0$ as perpendicular straight lines. A rectangle G in the $r\theta$-plane bounded by the lines $r = r_1$, $r = r_2$, and $\theta = \theta_1$, $\theta = \theta_2$ (θ in radians) with $2\pi > \theta_2 > \theta_1 \geq 0$, $r_2 > r_1 > 0$ has an image F in the xy-plane bounded by two circular arcs and two rays. (See the example in Section 5.) For the area of F we have

$$A_{x,y}(F) = \tfrac{1}{2}(r_2^2 - r_1^2)(\theta_2 - \theta_1).$$

This area may be written as an iterated integral. A simple calculation shows that

$$A_{x,y}(F) = \int_{\theta_1}^{\theta_2}\left[\int_{r_1}^{r_2} r\,dr\right]d\theta.$$

Because double integrals and iterated integrals are equivalent for evaluation purposes, we can also write

$$A_{x,y}(F) = \iint_G r\,dA_{r,\theta}, \tag{2}$$

where $dA_{r,\theta}$ is an element of area in the $r\theta$-plane, r and θ being treated as rectangular coordinates.

More generally, it can be shown (Morrey, *University Calculus*, page 672) that if G is *any region* in the $r\theta$-plane and F is its image under the transformation (1), then the area of F may be found by formula (2). Thus, areas of regions may be determined by expressing the double integral in polar coordinates as in (2) and then evaluating the double integral by iterated integrals in the usual way.

EXAMPLE 1. A region F above the x-axis is bounded on the left by the line $y = -x$, and on the right by the curve $C: x^2 + y^2 = 3\sqrt{x^2 + y^2} - 3x$, as shown in Fig. 5–26. Find its area.

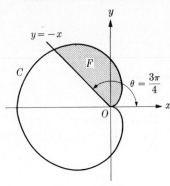

FIGURE 5–26

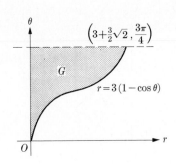

FIGURE 5–27

Solution. We employ polar coordinates to describe the region. The curve C is the cardioid $r = 3(1 - \cos \theta)$, and the line $y = -x$ is the ray $\theta = 3\pi/4$. The region F is described by the inequalities

$$0 \leq \theta \leq \frac{3\pi}{4}$$

and

$$0 \leq r \leq 3(1 - \cos \theta).$$

Employing the $r\theta$-plane, we see that the region G shown in Fig. 5–27 is the image of F under the mapping (1). Therefore, for the area $A(F)$ we obtain

$$A(F) = \iint_F dA_{x,y} = \iint_G r\, dA_{r,\theta} = \int_0^{3\pi/4} \int_0^{3(1-\cos \theta)} r\, dr\, d\theta$$

$$= \int_0^{3\pi/4} \frac{1}{2} [r^2]_0^{3(1-\cos \theta)}\, d\theta = \frac{9}{2} \int_0^{3\pi/4} (1 - \cos \theta)^2\, d\theta.$$

To perform the integration we multiply out and find that

$$A(F) = \frac{9}{2} \int_0^{3\pi/4} (1 - 2 \cos \theta + \cos^2 \theta)\, d\theta$$

$$= \frac{9}{2} \left[\theta - 2 \sin \theta + \frac{1}{2} \theta + \frac{1}{4} \sin 2\theta \right]_0^{3\pi/4}$$

$$= \frac{9}{8} \left(\frac{9}{2} \pi - 4\sqrt{2} - 1 \right).$$

The transformation of regions from the xy-plane to the $r\theta$-plane is useful because general double integrals as well as areas may be evaluated by means of polar coordinates. The theoretical basis for the method is the Fundamental Lemma on Integration, a proof of which may be found in Morrey, *University Calculus*, page 611.

Theorem 8. (Fundamental Lemma on Integration.) *Assume that f and g are continuous on some region F. Then for each ε > 0 there is a δ > 0 such that*

$$\left| \sum_{i=1}^{n} f_i g_i A(F_i) - \iint_{F} f(x, y) g(x, y) \, dA \right| < \epsilon$$

for every subdivision $F_1, F_2, \ldots, F_n$ of F with norm less than δ and any numbers $f_1, f_2, \ldots, f_n, g_1, g_2, \ldots, g_n$ where each f_i and each g_i is between the minimum and maximum values of f and g, respectively, on F_i.

The fundamental lemma is the basis for the next theorem, the proof of which we sketch. (See Morrey, *University Calculus*, page 613.)

Theorem 9. *Suppose F and G are regions related according to the mapping $x = r \cos \theta$, $y = r \sin \theta$ and $f(x, y)$ is continuous on F. Then the function $g(r, \theta) = f(r \cos \theta, r \sin \theta)$ is defined and continuous on G and*

$$\iint_{F} f(x, y) \, dA_{x,y} = \iint_{G} g(r, \theta) r \, dA_{r,\theta}.$$

Sketch of proof. Consider a subdivision of G into "figures" $G_1, \ldots, G_n$. (See the discussion of volume in Section 1.) Let (r_i, θ_i) be in G_i for each i, and let (ξ_i, η_i) and F_i be the respective images of (r_i, θ_i) and G_i. Then $(F_1, \ldots, F_n)$ is a subdivision of F. From the expression for area in the xy-plane as an integral, we obtain

$$A_{x,y}(F_i) = \iint_{G_i} r \, dA_{r,\theta}.$$

Using Theorem 4 concerning bounds for integrals, we obtain

$$\iint_{G_i} r \, dA_{r,\theta} = \bar{r}_i \, A_{r,\theta}(G_i),$$

where $\bar{r}_i$ is between the minimum and maximum of r on G_i. Thus

$$\sum_{i=1}^{n} f(\xi_i, \eta_i) A_{x,y}(F_i) = \sum_{i=1}^{n} g(r_i, \theta_i) \cdot \bar{r}_i \cdot A_{r,\theta}(G_i).$$

The theorem follows by letting the norms of the subdivisions → 0, using the fundamental lemma to evaluate the limit of the sum on the right.

EXAMPLE 2. Use polar coordinates to evaluate

$$\iint_{F} \sqrt{x^2 + y^2} \, dA_{x,y},$$

where F is the circular area bounded by $x^2 + y^2 = 2x$.

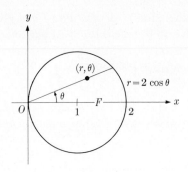

FIGURE 5–28

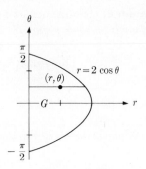

FIGURE 5–29

Solution. *F* may be described in polar coordinates by the inequalities

$$-\frac{\pi}{2} \le \theta \le \frac{\pi}{2}, \qquad 0 \le r \le 2\cos\theta.$$

(See Fig. 5–28.) Figure 5–29 shows *G*, the image of *F* in the $r\theta$-plane. Therefore

$$\iint_F \sqrt{x^2 + y^2}\, dA_{x,y} = \iint_G r \cdot r\, dA_{r,\theta} = \int_{-\pi/2}^{\pi/2} \int_0^{2\cos\theta} r^2\, dr\, d\theta$$

$$= \int_{-\pi/2}^{\pi/2} \frac{8}{3}\cos^3\theta\, d\theta = \frac{16}{3}\int_0^{\pi/2}(1 - \sin^2\theta)\cos\theta\, d\theta = \frac{32}{9}.$$

Although the construction of the region *G* in the $r\theta$-plane is helpful in understanding the transformation (1), it is not necessary for determining the limits in the iterated integrals. The limits of integration in polar coordinates may be found by using rectangular and polar coordinates in the same plane and using a sketch of the region *F* to read off the limits for *r* and θ.

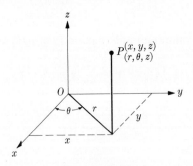

FIGURE 5–30

Double integrals are useful for finding volumes bounded by surfaces. Cylindrical coordinates (r, θ, z) are a natural extension to three-space of polar coordinates in the plane. The *z*-direction is selected as in rectangular coordinates, as shown in Fig. 5–30. If a closed surface in space is expressed in cylindrical coordinates, we may find the volume enclosed by this surface by evaluating a double integral in polar coordinates. An example illustrates the method.

EXAMPLE 3. A region S is bounded by the surfaces $x^2 + y^2 - 2x = 0, 4z = x^2 + y^2$, $z^2 = x^2 + y^2$. Use cylindrical coordinates to find the volume $V(S)$.

Solution. In cylindrical coordinates, the paraboloid $4z = x^2 + y^2$ has equation $4z = r^2$; the cylinder $x^2 + y^2 - 2x = 0$ has equation $r = 2\cos\theta$; and the cone $z^2 = x^2 + y^2$ has equation $z^2 = r^2$. The region is shown in Fig. 5–31, and we note that the projection of S on the xy-plane is precisely the plane region F of Example 2. We obtain

$$V(S) = \iint\limits_{F} \left[\sqrt{x^2 + y^2} - \frac{x^2 + y^2}{4}\right] dA_{x,y}$$

$$= \iint\limits_{G} \left(r - \frac{1}{4}r^2\right) r\, dA_{r,\theta}$$

$$= \int_{-\pi/2}^{\pi/2} \int_{0}^{2\cos\theta} \left(r^2 - \frac{1}{4}r^3\right) dr\, d\theta$$

$$= \int_{-\pi/2}^{\pi/2} \left(\frac{8}{3}\cos^3\theta - \cos^4\theta\right) d\theta$$

$$= \frac{32}{9} - \frac{1}{2}\int_{0}^{\pi/2} \left(1 + 2\cos 2\theta + \frac{1 + \cos 4\theta}{2}\right) d\theta = \frac{32}{9} - \frac{3\pi}{8}.$$

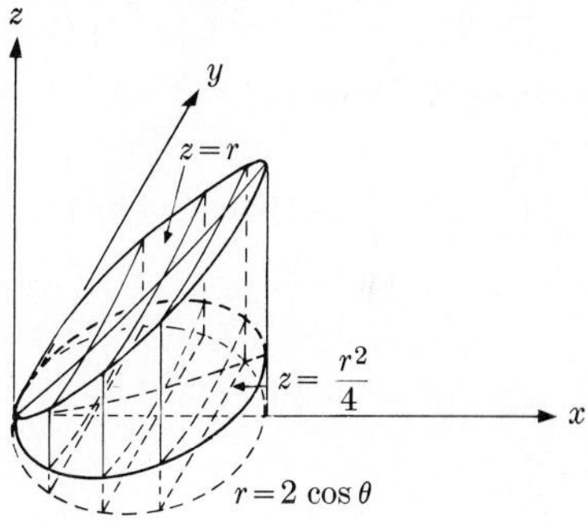

FIGURE 5–31

EXERCISES

In each of problems 1 through 7, evaluate the given integral by first expressing it as a double integral and then changing to polar coordinates.

1. $\displaystyle\int_{0}^{2}\int_{0}^{\sqrt{4-y^2}} \sqrt{x^2 + y^2}\, dx\, dy$

2. $\displaystyle\int_{-2}^{2}\int_{-\sqrt{4-x^2}}^{\sqrt{4-x^2}} e^{-(x^2+y^2)}\, dy\, dx$

3. $\displaystyle\int_{-\sqrt{\pi}}^{\sqrt{\pi}}\int_{-\sqrt{\pi-y^2}}^{\sqrt{\pi-y^2}} \sin(x^2 + y^2)\, dx\, dy$

4. $\displaystyle\int_{0}^{4}\int_{-\sqrt{4x-x^2}}^{\sqrt{4x-x^2}} \sqrt{x^2 + y^2}\, dy\, dx$

5. $\displaystyle\int_{-2}^{2}\int_{2-\sqrt{4-x^2}}^{2+\sqrt{4-x^2}} \sqrt{16 - x^2 - y^2}\, dy\, dx$

6. $\displaystyle\int_{0}^{2}\int_{0}^{x} (x^2 + y^2)\, dy\, dx$

7. $\displaystyle\int_{0}^{1}\int_{y}^{\sqrt{y}} (x^2 + y^2)^{-1/2}\, dx\, dy$

In each of problems 8 through 10, use polar coordinates to find the area of the region given.

8. The region inside the circle $x^2 + y^2 - 8y = 0$ and outside the circle $x^2 + y^2 = 9$.

9. The region bounded by $y^2 = 4x$ and $y = \frac{1}{2}x$.

10. The region interior to the curve $(x^2 + y^2)^3 = 16x^2$.

In each of problems 11 through 26, find the volume of S.

11. S is the set bounded by the surfaces $z = 0$, $2z = x^2 + y^2$, and $x^2 + y^2 = 4$.

12. S is the set bounded by the cone $z^2 = x^2 + y^2$ and the cylinder $x^2 + y^2 = 4$.

13. S is the set cut from a sphere of radius 4 by a cylinder of radius 2 whose axis is a diameter of the sphere.

14. S is the set above the cone $z^2 = x^2 + y^2$ and inside the sphere

$$x^2 + y^2 + z^2 = a^2.$$

15. S is the set bounded by the cone $z^2 = x^2 + y^2$ and the cylinder

$$x^2 + y^2 - 2y = 0.$$

16. S is the set bounded by the sphere $x^2 + y^2 + z^2 = 4$ and the cylinder

$$x^2 + y^2 = 2x.$$

17. S is the set bounded by the cone $z^2 = x^2 + y^2$ and the paraboloid

$$3z = x^2 + y^2.$$

18. S is bounded by the surfaces $z = 0$, $2z = x^2 + y^2$, and $2y = x^2 + y^2$.

19. S is bounded by the cylinder $x^2 + y^2 = 4$ and the hyperboloid

$$x^2 + y^2 - z^2 = 1.$$

20. S is bounded by the cone $z^2 = x^2 + y^2$ and the cylinder $r = 1 + \cos \theta$.

21. S is bounded by the surfaces $z = x$ and $2z = x^2 + y^2$.

22. S is bounded by the surfaces $z = 0$, $z = x^2 + y^2$, and $r = 2(1 + \cos \theta)$.

23. S is inside the sphere $x^2 + y^2 + z^2 = a^2$ and inside the cylinder erected on one loop of the curve $z = 0$, $r = a \cos 2\theta$.

24. S is inside the sphere $x^2 + y^2 + z^2 = 4$ and inside the cylinder erected on one loop of the curve $z = 0$, $r^2 = 4 \cos 2\theta$.

25. S is bounded by the surfaces $z^2 = x^2 + y^2$, $y = 0$, $y = x$, and $x = a$.

*26. S is bounded by the surfaces $z^2 = x^2 + y^2$ and $x - 2z + 2 = 0$.

7.   MOMENT OF INERTIA AND CENTER OF MASS

Consider the idealized situation in which an object of mass m occupies a single point. Let L be a line which we designate as an axis.

DEFINITION.   The **moment of inertia of a particle of mass m about the axis L** is mr^2, where r is the perpendicular distance of the object from the axis (Fig. 5–32). If we have a system of particles $m_1, m_2, \ldots, m_n$ at perpendicular distances,

FIGURE 5-32

respectively, of $r_1, r_2, \ldots, r_n$ from the axis L, then the **moment of inertia of the system,** I, is given by

$$I = m_1 r_1^2 + m_2 r_2^2 + \cdots + m_n r_n^2$$

$$= \sum_{i=1}^{n} m_i r_i^2.$$

Let F be an object made of thin material occupying a region in the xy-plane (Fig. 5-33). We wish to define the moment of inertia of F about an axis L. The axis L may be any line in three-dimensional space. We proceed as in the definition of integration. First we make a subdivision of the plane into rectangles or squares. We designate the rectangles either wholly or partly in F by $F_1, F_2, \ldots, F_n$. Since the object F may be of irregular shape and of variable density, the mass of the subregions may not be calculable exactly. We select a point in each subregion F_i and denote its coordinates (ξ_i, η_i). We assume that the entire mass of F_i, denoted $m(F_i)$, is concentrated at the point (ξ_i, η_i). Letting r_i be the perpendicular distance of (ξ_i, η_i) from the line L, we form the sum

$$\sum_{i=1}^{n} m(F_i) r_i^2.$$

DEFINITION. If the above sums tend to a limit (called I) as the norms of the subdivisions tend to zero, and if this limit is independent of the manner in which the (ξ_i, η_i) are selected within the F_i, then we say that I is the **moment of inertia of the mass distribution about the axis** L.

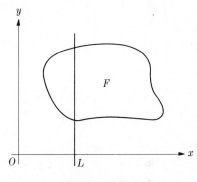

FIGURE 5-33

The above definition of moment of inertia leads in a natural way to the next theorem.

Theorem 10. *Given a mass distribution occupying a region F in the xy-plane and having a continuous density $\rho(x, y)$. Then the moment of inertia about the y-axis, I_1, is given by*

$$I_1 = \iint_F x^2 \rho(x, y) \, dA.$$

Similarly, the moments of inertia about the x-axis and the z-axis are, respectively,

$$I_2 = \iint_F y^2 \rho(x, y) \, dA, \qquad I_3 = \iint_F (x^2 + y^2) \rho(x, y) \, dA.$$

The proof depends on the Fundamental Lemma of Integration. (See Exercise 29 at the end of this section.)

Corollary. *The moments of inertia of F about the lines* $L_1: x = a, z = 0$; $L_2: y = b, z = 0$; $L_3: x = a, y = b$ *are, respectively,*

$$I_1^a = \iint_F (x - a)^2 \rho(x, y) \, dA, \qquad I_2^b = \iint_F (y - b)^2 \rho(x, y) \, dA,$$

$$I_3^{a,b} = \iint_F [(x - a)^2 + (y - b)^2] \rho(x, y) \, dA.$$

EXAMPLE 1. Find the moment of inertia about the x-axis of the homogeneous plate bounded by the line $y = 0$ and $y = 4 - x^2$ (Fig. 5-34).

Solution. According to Theorem 10, we have

$$I_2 = \iint_F y^2 \rho \, dA = \rho \iint_F y^2 \, dy \, dx = \rho \int_{-2}^{2} \int_{0}^{4-x^2} y^2 \, dy \, dx = \frac{\rho}{3} \int_{-2}^{2} (4 - x^2)^3 \, dx$$

$$= \frac{\rho}{3} \int_{-2}^{2} (64 - 48x^2 + 12x^4 - x^6) \, dx = \frac{4096\rho}{105}.$$

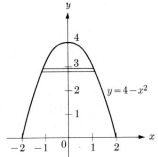

FIGURE 5-34

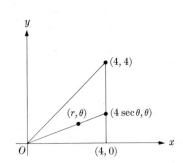

FIGURE 5-35

EXAMPLE 2. Find the moment of inertia about the z-axis of the homogeneous triangular plate bounded by the lines $y = 0$, $y = x$, and $x = 4$.

Solution 1. We have

$$I_3 = \iint_F (x^2 + y^2) \rho \, dA = \rho \int_0^4 \int_0^x (x^2 + y^2) \, dy \, dx$$

$$= \rho \int_0^4 \left[x^2 y + \frac{1}{3} y^3 \right]_0^x dx = \frac{4\rho}{3} \int_0^4 x^3 \, dx = \frac{256\rho}{3}.$$

Solution 2. We may introduce polar coordinates as shown in Fig. 5-35. Then

$$I_3 = \iint_F (x^2 + y^2) \rho \, dA_{x,y} = \rho \iint_G r^2 \cdot r \, dr \, d\theta = \rho \int_0^{\pi/4} \int_0^{4 \sec \theta} r^3 \, dr \, d\theta$$

$$= 64\rho \int_0^{\pi/4} \sec^4 \theta \, d\theta = 64\rho \left[\tan \theta + \frac{1}{3} \tan^3 \theta \right]_0^{\pi/4} = \frac{256\rho}{3}.$$

The moment of inertia about the z-axis of two-dimensional objects in the xy-plane is called the **polar moment of inertia.** Since the combination $x^2 + y^2 = r^2$ is always present in calculating polar moments, a change to polar coordinates is frequently indicated.

A quantity known as the radius of gyration is intimately connected with the notion of moment of inertia. It is defined in terms of the total mass m of an object and the moment of inertia I about a specific axis.

DEFINITION. The **radius of gyration of an object about an axis** L is that number R such that

$$R^2 = \frac{I}{m},$$

where I is the moment of inertia about L, and m is the total mass of the object.

If we imagine the total mass of a body as concentrated at one point which is at distance R from the axis L, then the moment of inertia of this idealized "point mass" will be the same as the moment of inertia of the original body.

EXAMPLE 3. Find the radius of gyration for the problem in Example 1. Do the same for Example 2.

Solution. The mass m of the homogeneous plate in Example 1 is its area multiplied by ρ. We have

$$m = \rho \int_{-2}^{2} \int_{0}^{4-x^2} dy\, dx = \rho \int_{-2}^{2} (4 - x^2)\, dx = \rho \left[4x - \frac{x^3}{3} \right]_{-2}^{2} = \frac{32\rho}{3}.$$

Therefore

$$R = \sqrt{\frac{I}{m}} = \left(\frac{4096\rho}{105} \cdot \frac{3}{32\rho} \right)^{1/2} = 8\sqrt{\frac{2}{35}}.$$

The mass of the triangular plate in Example 2 is 8ρ, and so

$$R = \sqrt{\frac{I}{m}} = \left(\frac{256\rho}{3} \cdot \frac{1}{8\rho} \right)^{1/2} = 4\sqrt{\frac{2}{3}}.$$

The center of mass of an object was defined in *First Course*, on page 508 ff. To calculate the center of mass we make use of the *moment of a mass* m with respect to one of the coordinate axes. We recall that if particles of masses $m_1, m_2, \ldots, m_n$ are situated at the points $(x_1, y_1), (x_2, y_2), \ldots, (x_n, y_n)$, respectively, then the **algebraic moment** (sometimes called *first moment* or simply *moment*) of this system about the y-axis is

$$m_1 x_1 + m_2 x_2 + \cdots + m_n x_n = \sum_{i=1}^{n} m_i x_i.$$

Its algebraic moment about the x-axis is

$$\sum_{i=1}^{n} m_i y_i.$$

More generally, the algebraic moments about the line $x = a$ and about the line $y = b$ are, respectively,

$$\sum_{i=1}^{n} m_i(x_i - a) \quad \text{and} \quad \sum_{i=1}^{n} m_i(y_i - b)$$

We now define the moment of a thin object occupying a region F in the xy-plane.

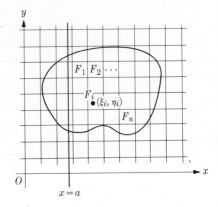

FIGURE 5–36

DEFINITION. Assume that a thin mass occupies a region F in the xy-plane. Let $F_1, F_2, \ldots, F_n$ be a subdivision of F as shown in Fig. 5–36. Choose a point (ξ_i, η_i) in each F_i and replace the mass in F_i by a particle of mass $m(F_i)$ located at (ξ_i, η_i). The n idealized masses have moment

$$\sum_{i=1}^{n} (\xi_i - a)m(F_i)$$

about the line $x = a$. If the above sum tends to a limit M_1 as the norms of the subdivision tend to zero and for any choices of the points (ξ_i, η_i) in F_i, then we define the limit M_1 as the **moment of the mass distribution about the line** $x = a$. An analogous definition for the limit M_2 of sums of the form

$$\sum_{i=1}^{n} (\eta_i - b)m(F_i)$$

yields the first moment about the line $y = b$.

The definition of first moment and the Fundamental Lemma on Integration yield the next theorem.

Theorem 11. *If a distribution of mass over a region F in the xy-plane has a continuous density $\rho(x, y)$, then the moments M_1 and M_2 of F about the lines $x = a$ and $y = b$ are given by the formulas*

$$M_1 = \iint_F (x - a)\rho(x, y)\, dA, \qquad M_2 = \iint_F (y - b)\rho(x, y)\, dA.$$

Corollary. *Given a distribution of mass over a region F in the xy-plane as in Theorem 11, then there are unique values of a and b (denoted $\bar{x}$ and $\bar{y}$, respectively)* *such that* $M_1 = M_2 = 0$. *In fact, the values of* $\bar{x}$ *and* $\bar{y}$ *are given by*

$$\bar{x} = \frac{\iint\limits_{F} x\rho(x, y)\, dA}{m(F)}, \qquad \bar{y} = \frac{\iint\limits_{F} y\rho(x, y)\, dA}{m(F)}, \qquad where \quad m(F) = \iint\limits_{F} \rho(x, y)\, dA.$$

Proof. If we set $M_1 = 0$, we get

$$0 = \iint\limits_{F} (x - a)\rho(x, y)\, dA = \iint\limits_{F} x\rho(x, y)\, dA - a\iint\limits_{F} \rho(x, y)\, dA.$$

Since $m(F) = \iint\limits_{F} \rho(x, y)\, dA$, we find for the value of a:

$$a = \frac{\iint\limits_{F} x\rho(x, y)\, dA}{m(F)} = \bar{x}.$$

The value $\bar{y}$ is found similarly.

DEFINITION. The point $(\bar{x}, \bar{y})$ is called the **center of mass** of the distribution over F.

EXAMPLE 4. Find the center of mass of the region F bounded by $y = x^3$ and $y = \sqrt{x}$ if the density of F is given by $\rho = 3x$.

Solution. (See Fig. 5–37.) For the first moments, we have

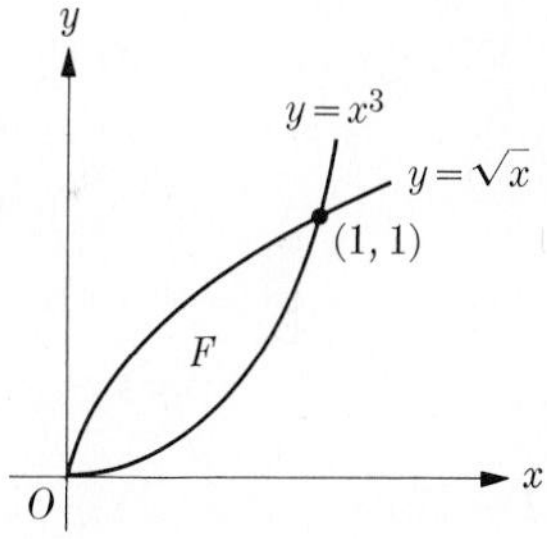

$$M_1 = \iint\limits_{F} x\rho\, dA = 3\int_0^1 \int_{x^3}^{\sqrt{x}} x^2\, dy\, dx$$

$$= 3\int_0^1 x^2 [y]_{x^3}^{\sqrt{x}}\, dx$$

$$= 3\int_0^1 (x^{5/2} - x^5)\, dx = \frac{5}{14},$$

$$M_2 = \iint\limits_{F} y\rho\, dA = 3\int_0^1 \int_{x^3}^{\sqrt{x}} yx\, dy\, dx = \frac{3}{2}\int_0^1 x[y^2]_{x^3}^{\sqrt{x}}\, dx$$

$$= \frac{3}{2}\int_0^1 (x^2 - x^7)\, dx = \frac{5}{16},$$

$$m(F) = \iint\limits_{F} \rho\, dA = 3\int_0^1 \int_{x^3}^{\sqrt{x}} x\, dy\, dx = 3\int_0^1 x[y]_{x^3}^{\sqrt{x}}\, dx$$

$$= 3\int_0^1 (x^{3/2} - x^4)\, dx = \frac{3}{5}.$$

FIGURE 5–37

Therefore

$$\bar{x} = \frac{M_1}{m(F)} = \frac{5}{14} \cdot \frac{5}{3} = \frac{25}{42}; \qquad \bar{y} = \frac{M_2}{m(F)} = \frac{5}{16} \cdot \frac{5}{3} = \frac{25}{48}.$$

EXAMPLE 5. Find the center of mass of a plate in the form of a circular sector of radius a and central angle 2α if its thickness is proportional to its distance from the center of the circle from which the sector is taken.

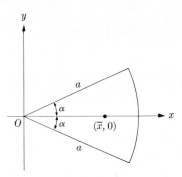

FIGURE 5–38

Solution. Select the sector so that the vertex is at the origin and the x-axis bisects the region (Fig. 5–38). Then the density is given by $\rho = kr$, where k is a proportionality constant. By symmetry we have $\bar{y} = 0$. Using polar coordinates, we obtain

$$M_1 = \iint_F xkr \, dA_{x,y} = k \int_{-\alpha}^{\alpha} \int_0^a r^2 \cos\theta \, r \, dr \, d\theta$$

$$= \frac{ka^4}{4} \int_{-\alpha}^{\alpha} \cos\theta \, d\theta = \frac{1}{2} ka^4 \sin\alpha,$$

$$m(F) = k \iint_F r \, dA_{x,y} = k \int_{-\alpha}^{\alpha} \int_0^a r^2 \, dr \, d\theta = \frac{2}{3} ka^3 \alpha.$$

Therefore

$$\bar{x} = \frac{M_1}{m(F)} = \frac{3a \sin\alpha}{4\alpha}.$$

EXERCISES

In each of problems 1 through 7, find the moment of inertia and the radius of gyration about the given axis of the plate whose density and bounding curves are given.

1. F is the square with vertices $(0, 0)$, $(a, 0)$, (a, a), $(0, a)$, $\rho =$ constant; y-axis.

2. F is the triangle with vertices $(0, 0)$, $(a, 0)$, (b, c), with $a > 0$, $c > 0$, $\rho =$ constant; x-axis.

3. F is bounded by $y = \sqrt{x}$ and $y = x^2$, $\rho =$ constant; y-axis.

4. F is bounded by $x + y = 5$, $xy = 4$, $\rho = ky$; x-axis.

5. F is bounded by $y = 0$ and $y = \sqrt{a^2 - x^2}$; $\rho = ky$; x-axis.

6. F is bounded by $x + y = 5$, $xy = 4$, $\rho = $ constant; x-axis.

7. F is bounded by $y = 0$ and the arch of $y = \sin x$ for $0 \leq x \leq \pi$, $\rho = $ constant; y-axis.

In each of problems 8 through 17, find the moment of inertia about the given axis of the plate whose density and bounding curves are given.

8. F is bounded by $x^2 + y^2 = a^2$, $\rho = k\sqrt{x^2 + y^2}$; z-axis.

9. F is bounded by $y = x^2$ and $y = x + 2$, $\rho = $ constant; x-axis.

10. F is the square with vertices $(0, 0)$, $(a, 0)$, (a, a), $(0, a)$, $\rho = $ constant; z-axis.

11. F is bounded by $y = x^2$ and $y^2 = x$, $\rho = ky$; y-axis.

12. F is bounded by $y = x^2$ and $y = x + 2$, $\rho = $ constant; axis is line $y = 4$.

13. F is bounded by $x^2 + y^2 = a^2$, $\rho = k\sqrt{x^2 + y^2}$; x-axis.

14. F is bounded by $r = 2a \cos \theta$, $\rho = kr$; z-axis.

15. F is bounded by one loop of $r^2 = a^2 \cos 2\theta$, $\rho = $ constant; z-axis.

16. F is bounded by one loop of $r^2 = a^2 \cos 2\theta$, $\rho = $ constant; x-axis.

17. F is in the first quadrant, inside $r = 1$, and bounded by $r = 1$, $\theta = r$ and $\theta = \pi/2$, $\rho = $ constant; z-axis.

In each of problems 18 through 28, find the center of mass of the plate described.

18. F is bounded by $x + y = 5$ and $xy = 4$, $\rho = ky$.

19. F is bounded by $y^2 = x$ and $x = y + 2$, $\rho = kx$.

20. F is the triangle with vertices at $(0, 0)$, $(a, 0)$, (b, c) with $0 < b < a$, $0 < c$, $\rho = kx$.

21. F is bounded by $y = x^2$ and $y^2 = x$, $\rho = ky$.

22. F is bounded by $y = x^2$ and $y = x + 2$, $\rho = $ constant.

23. F is the square with vertices at $(0, 0)$, $(a, 0)$, (a, a), $(0, a)$, $\rho = k(x^2 + y^2)$.

24. F is the triangle with vertices at $(0, 0)$, $(1, 0)$, $(1, 1)$, $\rho = kr^2$.

25. F is bounded by the cardioid $r = 2(1 + \cos \theta)$, $\rho = $ constant.

26. F is bounded by one loop of the curve $r = 2 \cos 2\theta$, $\rho = $ constant.

27. F is bounded by $3x^2 + 4y^2 = 48$ and $(x - 2)^2 + y^2 = 1$, $\rho = $ constant.

28. F is bounded by one loop of the curve $r^2 = a^2 \cos 2\theta$, $\rho = $ constant.

29. The *Theorem of the Mean* for double integrals states that *if f is integrable over a region F of area $A(F)$ and if $m \leq f(x, y) \leq M$ for all (x, y) on F, then there is a number $\bar{f}$ between m and M such that*

$$\iint_F f(x, y)\, dA = \bar{f} A(F).$$

Use the Fundamental Lemma of Integration and the Theorem of the Mean to establish Theorem 10. Use the idea of the proof of Theorem 9.

30. Show that if a mass distribution F lies between the lines $x = a$ and $x = b$, then $a \leq \bar{x} \leq b$. Similarly, if F lies between the lines $y = c$ and $y = d$, then $c \leq \bar{y} \leq d$.

31. Let $F_1, F_2, \ldots, F_n$ be regions no two of which have any points in common, and let $(\bar{x}_1, \bar{y}_1), (\bar{x}_2, \bar{y}_2), \ldots, (\bar{x}_n, \bar{y}_n)$ be their respective centers of mass. Denote the mass of F_i by m_i. If F is the region containing all the points in every F_i, show that the center of mass $(\bar{x}, \bar{y})$ of F is given by

$$\bar{x} = \frac{m_1\bar{x}_1 + m_2\bar{x}_2 + \cdots + m_n\bar{x}_n}{m_1 + m_2 + \cdots + m_n}, \qquad \bar{y} = \frac{m_1\bar{y}_1 + m_2\bar{y}_2 + \cdots + m_n\bar{y}_n}{m_1 + m_2 + \cdots + m_n}.$$

32. Show that if F is symmetric with respect to the x-axis and $\rho(x, -y) = \rho(x, y)$ for all (x, y) on F, then $\bar{y} = 0$. A similar result holds for symmetry with respect to the y-axis.

8.   SURFACE AREA

To define surface area we employ a procedure similar to that used for defining area in the plane. First, we define surface area in the simplest case and, second, we employ a limiting process for the definition of surface area of a general curved surface.

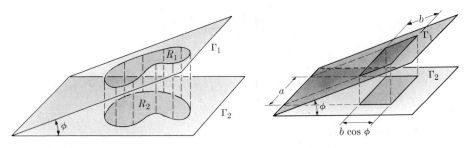

FIGURE 5–39   FIGURE 5–40

Suppose two planes Γ_1 and Γ_2 intersect at an angle ϕ (Fig. 5–39). From each point of R_1, a region in the plane Γ_1, we drop a perpendicular to the plane Γ_2. The set of points of intersection of these perpendiculars with Γ_2 forms a region which we denote R_2. The set R_2 is called the *projection* of R_1 *on* Γ_2. We shall now determine the relationship between the area $A(R_1)$ of R_1 and the area $A(R_2)$ of R_2. If R_1 is a rectangle—the simplest possible case—the problem may be solved by elementary geometry. For convenience, we select the rectangle in Γ_1 so that one side is parallel to the line of intersection of the two planes (Fig. 5–40). Let the lengths of the sides of the rectangle be a and b, as shown. The projection of the rectangle in Γ_1 onto Γ_2 is a rectangle, as the student may easily verify. The lengths of the sides of the rectangle in Γ_2 are a and $b \cos \phi$. The area $A_1 = ab$ of the rectangle in Γ_1 and the area $A_2 = ab \cos \phi$ of the rectangle in Γ_2 satisfy the relation

$$A_2 = A_1 \cos \phi. \tag{1}$$

Equation (1) is the basis of the next useful result.

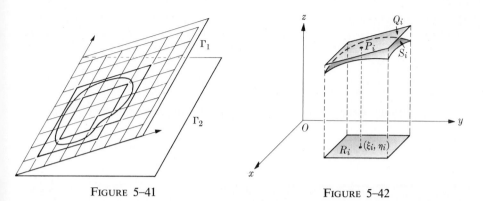

FIGURE 5–41   FIGURE 5–42

Lemma. Let R_1 be a region in a plane Γ_1 and let R_2 be the projection of R_1 onto a plane Γ_2. Then

$$A(R_2) = A(R_1) \cos \phi, \tag{2}$$

where ϕ is the angle between the planes Γ_1 and Γ_2.

This lemma is proved by subdividing the plane Γ_1 into a network of rectangles and observing that these rectangles project onto rectangles in Γ_2 with areas related by equation (1). Since the areas of R_1 and R_2 are obtained as limits of sums of the areas of rectangles, formula (2) holds in the limit (Fig. 5–41). (The details of this argument are given in Morrey, *University Calculus*, page 623.) We observe that if the planes are parallel, then ϕ is zero, the region and its projection are congruent, and the areas are equal. If the planes are perpendicular, then $\phi = \pi/2$, the projection of R_1 degenerates into a line segment, and $A(R_2)$ vanishes. Thus formula (2) is valid for all angles ϕ such that $0 \le \phi \le \pi/2$.

Suppose we have a surface S represented by an equation

$$z = f(x, y)$$

for (x, y) on some region F in the xy-plane. We shall consider only functions f which have continuous first partial derivatives for all (x, y) on F.

To define the area of the surface S we begin by subdividing the xy-plane into a rectangular mesh. Suppose R_i, a rectangle of the subdivision, is completely contained in F. Select a point (ξ_i, η_i) in R_i. This selection may be made in any manner whatsoever. The point $P_i(\xi_i, \eta_i, \zeta_i)$, with $\zeta_i = f(\xi_i, \eta_i)$, is on the surface S. Construct the plane tangent to the surface S at P_i (Fig. 5–42). Planes parallel to the z-axis and through the edges of R_i cut out a portion (denoted S_i) of the surface, and they cut out a quadrilateral, denoted Q_i, from the tangent plane. The projection of Q_i on the xy-plane is R_i. If the definition of surface area is to satisfy our intuition, then the area of S_i must be close to the area of Q_i whenever the subdivision in the xy-plane is sufficiently fine. However, Q_i is a plane region, and its area can be found exactly. We recall from Chapter 4 (page 143) that we can determine the equation of a plane tangent to a surface $z = f(x, y)$

at a given point on the surface. Such a determination is possible because the quantities

$$f_{,1}(\xi_i, \eta_i), \qquad f_{,2}(\xi_i, \eta_i), \qquad -1$$

form a set of attitude numbers for the tangent plane at the point (ξ_i, η_i, ζ_i) where $\zeta_i = f(\xi_i, \eta_i)$.

On page 18 we showed that the formula for the angle between two planes is

$$\cos \phi = \frac{|a_1 b_1 + a_2 b_2 + a_3 b_3|}{\sqrt{a_1^2 + a_2^2 + a_3^2} \sqrt{b_1^2 + b_2^2 + b_3^2}},$$

where a_1, a_2, a_3 and b_1, b_2, b_3 are sets of attitude numbers of the two planes. We now find the cosine of the angle between the plane tangent to the surface and the xy-plane. Letting ϕ denote the angle between the tangent plane and the xy-plane and recalling that the xy-plane has attitude numbers $0, 0, -1$, we get

$$\cos \phi = \frac{0 \cdot f_{,1} + 0 \cdot f_{,2} + 1 \cdot 1}{\sqrt{1 + f_{,1}^2 + f_{,2}^2}} = (1 + f_{,1}^2 + f_{,2}^2)^{-1/2}.$$

According to the above lemma, we have

$$A(R_i) = A(Q_i) \cos \phi$$

or

$$A(Q_i) = A(R_i)\sqrt{1 + f_{,1}^2(\xi_i, \eta_i) + f_{,2}^2(\xi_i, \eta_i)}.$$

We add all expressions of the above type for rectangles R_i which are in F. We obtain the sum

$$\sum_{i=1}^{n} A(Q_i) = \sum_{i=1}^{n} A(R_i)\sqrt{1 + f_{,1}^2(\xi_i, \eta_i) + f_{,2}^2(\xi_i, \eta_i)}, \qquad (3)$$

and we expect that this sum is a good approximation to the (as yet undefined) surface area if the norm of the rectangular subdivision in the xy-plane is sufficiently small.

DEFINITION. If the limit of the sums (3) exists as the norms of the subdivisions tend to zero and for arbitrary selections of the values (ξ_i, η_i) in R_i, then we say that the surface $z = f(x, y)$ has **surface area.** The **value of the surface area** $A(S)$ of S is the limit of the sum (3).

Theorem 12. *The sums in (3) tend to*

$$\iint_F \sqrt{1 + [f_{,1}(x, y)]^2 + [f_{,2}(x, y)]^2} \, dA$$

whenever the first derivatives $f_{,1}$ and $f_{,2}$ are continuous on F.

This theorem is an immediate consequence of Theorem 1 and the fact that $\sqrt{1 + f_{,1}^2 + f_{,2}^2}$ is continuous if $f_{,1}$ and $f_{,2}$ are. The integration formula of the theorem may be used to calculate surface area, as the next examples show.

EXAMPLE 1. Find the area of the surface $z = \frac{2}{3}(x^{3/2} + y^{3/2})$ situated above the square $F: 0 \le x \le 1, 0 \le y \le 1$ (Fig. 5–43).

Solution. Setting $z = f(x, y)$, we have $f_{,1} = x^{1/2}, f_{,2} = y^{1/2}$, and

$$A(S) = \iint_F (1 + x + y)^{1/2}\, dA = \int_0^1 \int_0^1 (1 + x + y)^{1/2}\, dy\, dx.$$

Therefore

$$A(S) = \int_0^1 \frac{2}{3}[(1 + x + y)^{3/2}]_0^1\, dx = \frac{2}{3}\int_0^1 [(2 + x)^{3/2} - (1 + x)^{3/2}]\, dx$$

$$= \frac{4}{15}[(2 + x)^{5/2} - (1 + x)^{5/2}]_0^1 = \frac{4}{15}(1 + 9\sqrt{3} - 8\sqrt{2}).$$

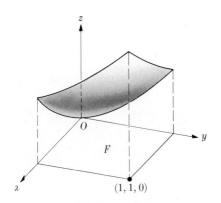

FIGURE 5–43

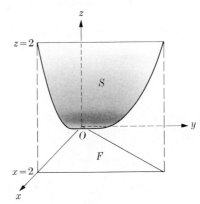

FIGURE 5–44

EXAMPLE 2. Find the area of the part of the cylinder $z = \frac{1}{2}x^2$ cut out by the planes $y = 0, y = x$, and $x = 2$.

Solution. See Figs. 5–44 and 5–45, which show the surface S and the projection F. We have $\partial z/\partial x = x, \partial z/\partial y = 0$. Therefore

$$A(S) = \iint_F \sqrt{1 + x^2}\, dA = \int_0^2 \int_0^x \sqrt{1 + x^2}\, dy\, dx$$

$$= \int_0^2 x\sqrt{1 + x^2}\, dx = \frac{1}{3}[(1 + x^2)^{3/2}]_0^2$$

$$= \frac{1}{3}(5\sqrt{5} - 1).$$

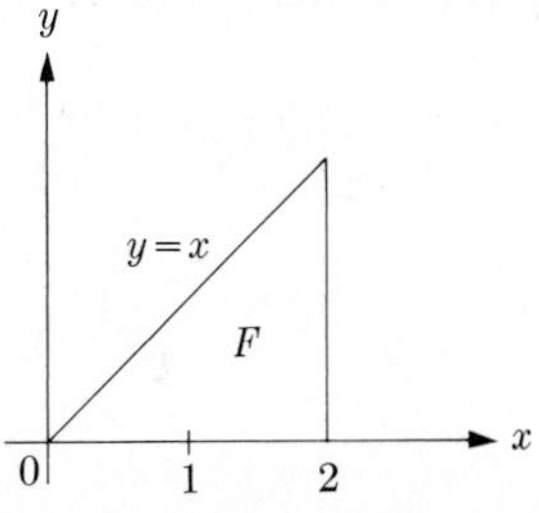

FIGURE 5–45

The next example shows that it is sometimes useful to use polar coordinates for the evaluation of the double integral.

EXAMPLE 3.   Find the surface area of the part of the sphere $x^2 + y^2 + z^2 = a^2$ cut out by the vertical cylinder erected on one loop of the curve whose equation in polar coordinates is $r = a \cos 2\theta$.

Solution. (See Fig. 5–46.) The surface consists of two parts, one above and one below the xy-plane, symmetrically placed. The area of the upper half will be found. We have

$$z = \sqrt{a^2 - x^2 - y^2},$$

$$\frac{\partial z}{\partial x} = \frac{-x}{\sqrt{a^2 - x^2 - y^2}},$$

$$\frac{\partial z}{\partial y} = \frac{-y}{\sqrt{a^2 - x^2 - y^2}}.$$

Therefore (Fig. 5–46), we obtain

$$A(S) = \iint\limits_{F} \frac{a}{\sqrt{a^2 - x^2 - y^2}} \, dA_{x,y}$$

$$= \int_{-\pi/4}^{\pi/4} \int_{0}^{a \cos 2\theta} \frac{ar}{\sqrt{a^2 - r^2}} \, dr \, d\theta.$$

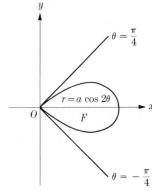

FIGURE 5–46

This integral is an improper integral, but it can be shown to be convergent. Taking this fact for granted, we get

$$A(S) = 2a \int_{0}^{\pi/4} \left[-\sqrt{a^2 - r^2} \right]_{0}^{a \cos 2\theta} d\theta$$

$$= 2a^2 \int_{0}^{\pi/4} (1 - \sin 2\theta) \, d\theta = \frac{1}{2} a^2 (\pi - 2).$$

The total surface area is $a^2(\pi - 2)$.

If the given surface is of the form $y = f(x, z)$ or $x = f(y, z)$, we get similar formulas for the surface area. These are

$$A(S) = \iint\limits_{F} \sqrt{1 + \left(\frac{\partial z}{\partial x}\right)^2 + \left(\frac{\partial z}{\partial y}\right)^2} \, dA_{x,y} \qquad \text{if} \quad z = f(x, y),$$

$$A(S) = \iint\limits_{F} \sqrt{1 + \left(\frac{\partial y}{\partial x}\right)^2 + \left(\frac{\partial y}{\partial z}\right)^2} \, dA_{x,z} \qquad \text{if} \quad y = f(x, z),$$

$$A(S) = \iint\limits_{F} \sqrt{1 + \left(\frac{\partial x}{\partial y}\right)^2 + \left(\frac{\partial x}{\partial z}\right)^2} \, dA_{y,z} \qquad \text{if} \quad x = f(y, z).$$

EXERCISES

In each of the problems 1 through 18, find the area of the surface described.

1. The portion of the surface $z = \frac{2}{3}(x^{3/2} + y^{3/2})$ situated above the triangle

$$F: \{0 \le x \le y, 0 \le y \le 1\}.$$

2. The portion of the plane $x/a + y/b + z/c = 1$ in the first octant ($a > 0, b > 0, c > 0$).

3. The part of the cylinder $x^2 + z^2 = a^2$ inside the cylinder $x^2 + y^2 = a^2$.

4. The part of the cylinder $x^2 + z^2 = a^2$ above the square $|x| \le \frac{1}{2}a, |y| \le \frac{1}{2}a$.

5. The part of the cone $z^2 = x^2 + y^2$ inside the cylinder $x^2 + y^2 = 2x$.

6. The part of the cone $z^2 = x^2 + y^2$ above the figure bounded by one loop of the curve $r^2 = 4 \cos 2\theta$.

7. The part of the cone $x^2 = y^2 + z^2$ between the cylinder $y^2 = z$ and the plane $y = z - 2$.

8. The part of the cone $y^2 = x^2 + z^2$ cut off by the plane $2y = (x + 2)\sqrt{2}$.

9. The part of the cone $x^2 = y^2 + z^2$ inside the sphere $x^2 + y^2 + z^2 = 2z$.

10. The part of the surface $z = xy$ inside the cylinder $x^2 + y^2 = a^2$.

11. The part of the surface $4z = x^2 - y^2$ above the region bounded by the curve $r^2 = 4 \cos \theta$.

12. The part of the surface of a sphere of radius $2a$ inside a cylinder of radius a if the center of the sphere is on the surface of the cylinder.

13. The part of the surface of a sphere of radius a, center at the origin, inside the cylinder erected on one loop of the curve $r = a \cos 3\theta$.

14. The part of the sphere $x^2 + y^2 + z^2 = 4z$ inside the paraboloid $x^2 + y^2 = z$.

15. The part of the cylinder $y^2 + z^2 = 2z$ cut off by the cone $x^2 = y^2 + z^2$.

16. The part of the cylinder $x^2 + y^2 = 2ax$ inside the sphere $x^2 + y^2 + z^2 = 4a^2$.

17. The part of the cylinder $y^2 + z^2 = 4a^2$ above the xy-plane and bounded by the planes $y = 0, x = a$, and $y = x$.

18. The part of the paraboloid $y^2 + z^2 = 4ax$ cut off by the cylinder $y^2 = ax$ and the plane $x = 3a$.

19. (a) Use elementary geometry (and trigonometry) to establish equation (2) for an arbitrary triangle. (b) Use the result of (a) to establish equation (2) for an arbitrary polygon.

9. VOLUMES OF SOLIDS OF REVOLUTION

We previously developed methods for finding the volume of certain solids of revolution. These techniques were applicable whenever the resulting process reduced to a single integration. (See *First Course*, page 482 ff.) Now that areas may be determined in a more general way by double integrations, we can calculate the volume of a greater variety of solids of revolution. The basic tool is the Theorem of Pappus which we now state. (See also *First Course*, page 523.)

Theorem 13. (Theorem of Pappus.) *If a plane figure F lies on one side of a line L in its plane, the volume of the set S generated by revolving F around L is equal to the product of $A(F)$, the area of F, and the length of the path described by the centroid of F; in other words, if F is in the xy-plane and L is the x-axis (see Fig. 5–47), then*

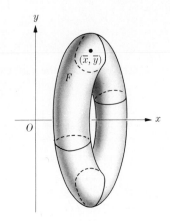

$$V(S) = 2\pi\bar{y}A(F) = \iint_F 2\pi y\, dA. \qquad (1)$$

The second equality in (1) above follows from the definition of $\bar{y}$. In case F is a rectangle of the form $a \le x \le b$, $c \le y \le d$, where $c \ge 0$, then S is just a circular ring of altitude $b - a$, inner radius of base c (if $c > 0$) and outer radius d. Thus

FIGURE 5–47

$$V(S) = \pi(d^2 - c^2)(b - a) = 2\pi\left(\frac{c+d}{2}\right)A(F) = 2\pi\bar{y}A(F)$$
$$= \iint_F 2\pi y\, dA.$$

The theorem is proved in general by subdividing F and noting that

$$\sum_{i=1}^{n} V(S_i) = \sum_{i=1}^{n} \iint_{F_i} 2\pi y\, dA = \iint_{F_n^*} 2\pi y\, dA,$$

where F_n^* is the union of the F_i; we then pass to the limit. Additional details may be found in Morrey, *University Calculus*, page 627.

EXAMPLE 1. The region F bounded by the curves $y = x^3$ and $y = \sqrt{x}$ is revolved about the x-axis. Find the volume generated.

Solution. See Fig. 5–48. We have

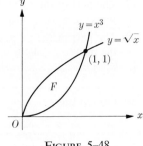

$$V(S) = 2\pi \iint_F y\, dA = 2\pi \int_0^1 \int_{x^3}^{\sqrt{x}} y\, dy\, dx$$

$$= \pi \int_0^1 (x - x^6)\, dx = \frac{5}{14}\pi.$$

FIGURE 5–48

The Theorem of Pappus in this more general form is especially useful whenever the transformation to polar coordinates is appropriate. The next example illustrates this point.

EXAMPLE 2. Find the volume of the set S generated by revolving around the x-axis the upper half F of the area bounded by the cardioid $r = a(1 + \cos \theta)$. (See Fig. 5–49.)

Solution. We have

$$V(S) = \iint_F 2\pi y \, dA = \iint_F 2\pi r \sin \theta \, dA_{x,y}$$

$$= 2\pi \int_0^\pi \int_0^{a(1+\cos \theta)} r^2 \sin \theta \, dr \, d\theta$$

$$= \frac{2}{3}\pi a^3 \int_0^\pi (1 + \cos \theta)^3 \sin \theta \, d\theta$$

$$= \frac{\pi a^3}{6}\left[-(1 + \cos \theta)^4\right]_0^\pi = \frac{8\pi a^3}{3}.$$

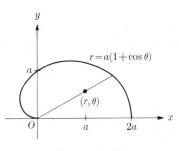

$r = a(1 + \cos \theta)$

(r, θ)

FIGURE 5–49

The formula analogous to (1) for revolving a region F in the xy-plane about the y-axis is

$$V(S) = 2\pi \bar{x} A(F) = 2\pi \iint_F x \, dA.$$

EXERCISES

In each of problems 1 through 12, find the volume of the set obtained by revolving the region described about the axis indicated. Sketch the region.

1. The upper half of the ellipse $x^2/a^2 + y^2/b^2 = 1$; the x-axis.

2. The region in the first quadrant inside $3x^2 + 4y^2 = 48$ and outside $(x - 2)^2 + y^2 = 1$; the x-axis.

3. The region satisfying the inequalities $xy \le 4$, $y \le x$, $27y \ge 4x^2$; the x-axis.

4. The upper half of the figure bounded by the right-hand loop of the curve $r^2 = a^2 \cos \theta$; the x-axis.

5. The upper half of the right-hand loop of the curve $r = a \cos^2 \theta$; the x-axis.

6. The region inside $x^2 + y^2 = 64$ and outside $x^2 + y^2 = 8x$; the line $x = 8$.

7. The loop of $r^3 = \sin 2\theta$ in the first quadrant; the y-axis.

8. The upper half of the area outside the circle $r = 4$ and inside the limaçon $r = 3 + 2 \cos \theta$; the x-axis.

9. The upper half of the area to the right of the line $x = \frac{3}{2}$ and inside the cardioid $r = 2(1 + \cos \theta)$; the x-axis.

10. The upper half of the area to the right of the parabola $r = 9/(1 + \cos \theta)$ and inside the cardioid $r = 4(1 + \cos \theta)$; the x-axis.

11. The right-hand horizontal loop of the curve $r = 2 \cos 2\theta$; the y-axis. Find the centroid of that loop.

12. The upper half of the right-hand loop of the curve $r^2 = a^2 \cos 2\theta$; the x-axis.

10. THE TRIPLE INTEGRAL

The definition of the triple integral parallels that of the double integral. In the simplest case, we consider a rectangular box R bounded by the six planes $x = a_0$, $x = a_1$, $y = b_0$, $y = b_1$, $z = c_0$, $z = c_1$ (Fig. 5–50). Let $f(x, y, z)$ be a function of three variables defined for (x, y, z) in R. We subdivide the entire three-dimensional space into rectangular boxes by constructing planes parallel to the coordinate planes. Let $B_1, B_2, \ldots, B_n$ be those boxes of the subdivision which contain points of R. Denote by $V(B_i)$ the volume of the ith box, B_i. We select a point $P_i(\xi_i, \eta_i, \zeta_i)$ in B_i; this selection may be made in any manner whatsoever. The sum

$$\sum_{i=1}^{n} f(\xi_i, \eta_i, \zeta_i) V(B_i)$$

is an approximation to the triple integral. The *norm of the subdivision* is the length of the longest diagonal of the boxes $B_1, B_2, \ldots, B_n$. If the above sums tend to a limit as the norms of the subdivisions tend to zero and for any choices of the points P_i, we call this limit the **triple integral of f over** R. The expression

$$\iiint\limits_{R} f(x, y, z)\, dV$$

is used to represent this limit.

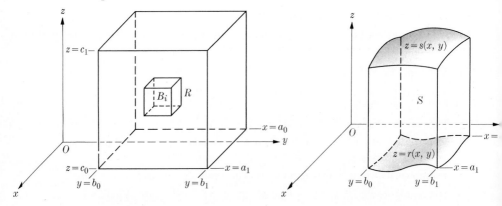

FIGURE 5–50   FIGURE 5–51

Just as the double integral is equal to a twice-iterated integral, so the triple integral has the same value as a threefold iterated integral. In the case of the rectangular box R, we obtain

$$\iiint\limits_{R} f(x, y, z)\, dV = \int_{a_0}^{a_1} \left\{ \int_{b_0}^{b_1} \left[\int_{c_0}^{c_1} f(x, y, z)\, dz \right] dy \right\} dx.$$

Suppose a region S is bounded by the planes $x = a_0$, $x = a_1$, $y = b_0$, $y = b_1$, and by the surfaces $z = r(x, y)$, $z = s(x, y)$, as shown in Fig. 5–51. The triple integral may be defined in the same way as for a rectangular box R, and once again it is equal to the iterated integral. We have

$$\iiint\limits_{S} f(x, y, z)\, dV = \int_{a_0}^{a_1} \left\{ \int_{b_0}^{b_1} \left[\int_{r(x,y)}^{s(x,y)} f(x, y, z)\, dz \right] dy \right\} dx.$$

We state without proof the following theorem, which applies in the general case.

Theorem 14. *Suppose that S is a region defined by the inequalities*

$$S: a \le x \le b, \qquad p(x) \le y \le q(x), \qquad r(x, y) \le z \le s(x, y),$$

where the functions p, q, r, and s are continuous. If f is a continuous function on S, then

$$\iiint\limits_{S} f(x, y, z)\, dV = \int_{a}^{b} \left\{ \int_{p(x)}^{q(x)} \left[\int_{r(x,y)}^{s(x,y)} f(x, y, z)\, dz \right] dy \right\} dx.$$

The iterated integrations are performed in turn by holding all variables constant except the one being integrated. Brackets and braces in multiple integrals will be omitted unless there is danger of confusion.

EXAMPLE 1. Evaluate the iterated integral

$$\int_{0}^{3} \int_{0}^{6-2z} \int_{0}^{4-(2/3)y-(4/3)z} yz\, dx\, dy\, dz.$$

Solution. We have

$$\int_{0}^{3} \int_{0}^{6-2z} \int_{0}^{4-(2/3)y-(4/3)z} yz\, dx\, dy\, dz$$

$$= \int_{0}^{3} \int_{0}^{6-2z} [xyz]_{0}^{4-(2/3)y-(4/3)z}\, dy\, dz$$

$$= \int_{0}^{3} \int_{0}^{6-2z} yz\left(4 - \frac{2}{3}y - \frac{4}{3}z\right) dy\, dz$$

$$= \int_{0}^{3} \left[2zy^2 - \frac{2}{9}zy^3 - \frac{2}{3}y^2 z^2 \right]_{0}^{6-2z} dz$$

$$= \int_{0}^{3} \left[\left(2z - \frac{2}{3}z^2\right)(6 - 2z)^2 - \frac{2}{9}z(6 - 2z)^3 \right] dz$$

$$= \frac{1}{9} \int_{0}^{3} z(6 - 2z)^3\, dz.$$

The integration may be performed by the substitution $u = 6 - 2z$. The result is $54/5$.

The determination of the limits of integration is the principal difficulty in re-ducing a triple integral to an iterated integral. The student who works a large number of problems will develop good powers of visualization of three-dimen-sional figures. There is no simple mechanical technique for determining the limits of integration in the wide variety of problems we encounter. The next examples illustrate the process.

EXAMPLE 2. Evaluate

$$\iiint\limits_{S} x \, dV,$$

where S is the region bounded by the surfaces $y = x^2$, $y = x + 2$, $4z = x^2 + y^2$, and $z = x + 3$.

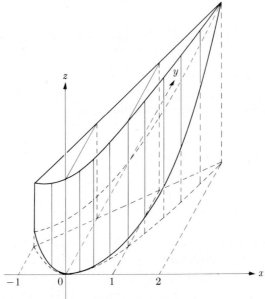

FIGURE 5–52

Solution. To transform the triple integral into an iterated integral, we must determine the limits of integration. The region S is sketched in Fig. 5–52. The projection of S on the xy-plane is the region F bounded by the curves $y = x^2$ and $y = x + 2$, as shown in Fig. 5–53. From this projection, the region rises with vertical walls, bounded from below by the paraboloid $z = \frac{1}{4}(x^2 + y^2)$ and above by the plane $z = x + 3$. Since F is described by the inequalities

$$-1 \le x \le 2, \qquad x^2 \le y \le x + 2,$$

we have

$$S: \begin{cases} -1 \le x \le 2, \\ x^2 \le y \le x + 2, \\ \frac{1}{4}(x^2 + y^2) \le z \le x + 3. \end{cases}$$

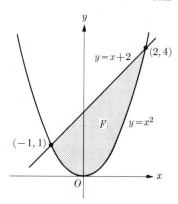

FIGURE 5–53

Therefore

$$\iiint_S x\, dV = \int_{-1}^{2} \int_{x^2}^{x+2} \int_{(x^2+y^2)/4}^{x+3} x\, dz\, dy\, dx$$

$$= \int_{-1}^{2} \int_{x^2}^{x+2} \left[x^2 + 3x - \frac{1}{4}(x^3 + xy^2) \right] dy\, dx$$

$$= \int_{-1}^{2} \left\{ \left(3x + x^2 - \frac{1}{4}x^3 \right)(2 + x - x^2) - \frac{x}{12} \left[(2 + x)^3 - x^6 \right] \right\} dx$$

$$= \frac{837}{160}.$$

In the case of double integrals there are two possible orders of integration, one of them often being easier to calculate than the other. In the case of triple integrals there are six possible orders of integration. It becomes a matter of practice and trial and error to find which order is the most convenient.

The limits of integration may sometimes be found by projecting the region on one of the coordinate planes and then finding the equations of the "bottom" and "top" surfaces. This method was used in Example 2. If part of the boundary is a cylinder perpendicular to one of the coordinate planes, that fact can be used to determine the limits of integration.

EXAMPLE 3. Express the integral

$$I = \iiint_S f(x, y, z)\, dV$$

as an iterated integral in six different ways if S is the region bounded by the surfaces

$$z = 0, \qquad z = x, \qquad \text{and} \qquad y^2 = 4 - 2x.$$

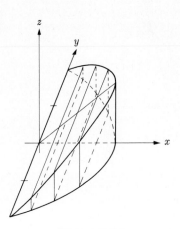

FIGURE 5–54

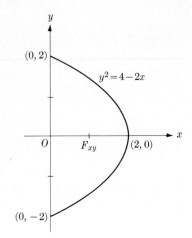

FIGURE 5–55

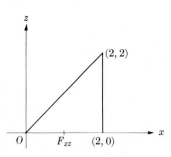

FIGURE 5–56

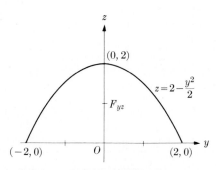

FIGURE 5–57

Solution. The region S is shown in Fig. 5–54. The projection of S on the xy-plane is the two-dimensional region F_{xy} bounded by $x = 0$ and $y^2 = 4 - 2x$, as shown in Fig. 5–55. Therefore the integral may be written

$$I = \int_0^2 \int_{-\sqrt{4-2x}}^{+\sqrt{4-2x}} \int_0^x f(x, y, z) \, dz \, dy \, dx$$

$$= \int_{-2}^2 \int_0^{2-(1/2)y^2} \int_0^x f(x, y, z) \, dz \, dx \, dy.$$

The projection of S on the xz-plane is the triangular region bounded by the curves $z = 0$, $z = x$, and $x = 2$, as shown in Fig. 5–56. The iterated integral in this case becomes

$$I = \int_0^2 \int_0^x \int_{-\sqrt{4-2x}}^{+\sqrt{4-2x}} f(x, y, z) \, dy \, dz \, dx$$

$$= \int_0^2 \int_z^2 \int_{-\sqrt{4-2x}}^{+\sqrt{4-2x}} f(x, y, z) \, dy \, dx \, dz.$$

The projection of S on the yz-plane is the plane region bounded by $z = 0$ and $z = 2 - \frac{1}{2}y^2$ (Fig. 5–57). Then I takes the form

$$I = \int_{-2}^{2} \int_{0}^{2-(1/2)y^2} \int_{z}^{2-(1/2)y^2} f(x, y, z)\, dx\, dz\, dy$$

$$= \int_{0}^{2} \int_{-\sqrt{4-2z}}^{+\sqrt{4-2z}} \int_{z}^{2-(1/2)y^2} f(x, y, z)\, dx\, dy\, dz.$$

EXERCISES

In each of problems 1 through 6, find the value of the iterated integral.

1. $\displaystyle\int_{0}^{1} \int_{0}^{x} \int_{0}^{x-y} x\, dz\, dy\, dx$

2. $\displaystyle\int_{-1}^{1} \int_{0}^{1-y^2} \int_{-\sqrt{x}}^{\sqrt{x}} 2y^2\sqrt{x}\, dz\, dx\, dy$

3. $\displaystyle\int_{0}^{1} \int_{y^2}^{\sqrt{y}} \int_{0}^{y+z} xy\, dx\, dz\, dy$

4. $\displaystyle\int_{0}^{4} \int_{0}^{\sqrt{16-x^2}} \int_{0}^{\sqrt{16-x^2-y^2}} (x + y + z)\, dz\, dy\, dx$

5. $\displaystyle\int_{0}^{2} \int_{0}^{\sqrt{4-z^2}} \int_{0}^{2-z} z\, dx\, dy\, dz$

6. $\displaystyle\int_{0}^{1} \int_{0}^{x} \int_{0}^{y} \frac{1 + \sqrt[3]{z}}{\sqrt{z}}\, dz\, dy\, dx$

In problems 7 through 17, evaluate

$$\iiint_{S} f(x, y, z)\, dV$$

where S is bounded by the given surfaces and f is the given function.

7. $z = 0,\ y = 0,\ y = x,\ x + y = 2,\ x + y + z = 3;\ f(x, y, z) = x$

8. $x = 0,\ x = \sqrt{a^2 - y^2 - z^2};\ f(x, y, z) = x$

9. $z = 0,\ x^2 + z = 1,\ y^2 + z = 1,\ f(x, y, z) = z^2$

10. $x^2 + z^2 = a^2,\ y^2 + z^2 = a^2,\ f(x, y, z) = x^2 + y^2$

11. $x = 0,\ y = 0,\ z = 0,\ (x/a) + (y/b) + (z/c) = 1,\ (a, b, c > 0);\ f(x, y, z) = z$

12. $y = z^2,\ y^2 = z,\ x = 0,\ x = y - z^2;\ f(x, y, z) = y + z^2$

13. $x = 0,\ y = 0,\ z = 0,\ x^{1/2} + y^{1/2} + z^{1/2} = a^{1/2};\ f(x, y, z) = z$

14. $x = 0,\ y = 0,\ z = 0,\ y^2 = 4 - z,\ x = y + 2;\ f(x, y, z) = x^2$

15. $z = x^2 + y^2,\ z = 27 - 2x^2 - 2y^2;\ f(x, y, z) = 1$

16. $z^2 = 4ax,\ x^2 + y^2 = 2ax;\ f(x, y, z) = 1$

*17. $y^2 + z^2 = 4ax,\ y^2 = ax,\ x = 3a;\ f(x, y, z) = x^2$

In problems 18 through 22, express each iterated integral as a triple integral by describing the set S over which the integration is performed. Sketch the set S and then express the iterated integral in two orders differing from the original. Do not evaluate the integrals.

18. $\displaystyle\int_{0}^{1} \int_{0}^{x} \int_{0}^{x-y} x\, dz\, dy\, dx$

19. $\displaystyle\int_{-1}^{1} \int_{0}^{1-y^2} \int_{-\sqrt{x}}^{\sqrt{x}} 2y^2\sqrt{x}\, dz\, dx\, dy$

20. $\int_0^1 \int_{y^2}^{\sqrt{y}} \int_0^{y+z} xy \, dx \, dz \, dy$ 21. $\int_{-2}^2 \int_0^{4-y^2} \int_0^{y+2} (y^2 + z^2) \, dz \, dx \, dy$

22. $\int_0^1 \int_{x^2}^{\sqrt{x}} \int_0^{y-x^2} f(x, y, z) \, dz \, dy \, dx$

23. Express the integral of $f(x, y, z)$ over the region S bounded by the surface $z = \sqrt{16 - x^2 - y^2}$ and the plane $z = 2$ in 6 ways.

24. Express the integral

$$\int_0^2 \int_0^z \int_0^x (x^2 + y^2 + z^2) \, dy \, dx \, dz$$

in 5 additional ways.

11. MASS OF A SOLID. TRIPLE INTEGRALS IN CYLINDRICAL AND SPHERICAL COORDINATES

From the definition of triple integral we see that if $f(x, y, z) \equiv 1$, then the triple integral taken over a region S is precisely the volume $V(S)$. More generally, if a solid object occupies a region S, and if the density at any point is given by $\delta(x, y, z)$, then the total mass, $m(S)$, is given by the triple integral

$$m(S) = \iiint_S \delta(x, y, z) \, dV.$$

Notation. For the remainder of this chapter the symbol δ will be used for density. The quantity ρ, which we previously used for density, will denote one of the variables in spherical coordinates.

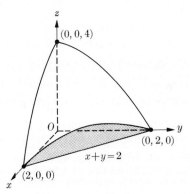

FIGURE 5–58

EXAMPLE 1. The solid S in the first octant is bounded by the surfaces $z = 4 - x^2 - y^2$, $z = 0$, $x + y = 2$, $x = 0$, $y = 0$. The density is given by $\delta(x, y, z) = 2z$. Find the total mass.

Solution. We have (Fig. 5–58)

$$m(S) = \iiint\limits_{S} 2z \, dV = 2\int_0^2 \int_0^{2-x} \int_0^{4-x^2-y^2} z \, dz \, dy \, dx$$

$$= \int_0^2 \int_0^{2-x} (4 - x^2 - y^2)^2 \, dy \, dz$$

$$= \int_0^2 \int_0^{2-x} (16 + x^4 + y^4 - 8x^2 - 8y^2 + 2x^2y^2) \, dy \, dz$$

$$= \int_0^2 \left[(4 - x^2)^2 y - \frac{2}{3}(4 - x^2)y^3 + \frac{1}{5}y^5 \right]_0^{2-x} dx$$

$$= \int_0^2 \left[2(4 - x^2)^2 - x(4 - x^2)^2 - \frac{2}{3}(4 - x^2)(2 - x)^3 + \frac{1}{5}(2 - x)^5 \right] dx.$$

The above integral, a polynomial in x, can be evaluated. The answer is $704/45$.

We found that certain double integrals are easy to evaluate if a polar coordinate system is used. Similarly, there are triple integrals which, although difficult to evaluate in rectangular coordinates, are simple integrations when transformed into other systems. The most useful transformations are those to cylindrical and spherical coordinates. (See Chapter 1, page 36.)

Cylindrical coordinates consist of polar coordinates in the plane and a z-coordinate as in a rectangular system. The transformation from rectangular to cylindrical coordinates is

$$x = r \cos \theta, \qquad y = r \sin \theta, \qquad z = z. \tag{1}$$

A region S in (x, y, z) space corresponds to a region U in (r, θ, z) space. The volume of S, $V(S)$, may be found in terms of a triple integral in (r, θ, z) space by the formula (Fig. 5–59)

$$V(S) = \iiint\limits_{U} r \, dV_{r\theta z}.$$

This formula is a natural extension of the formula relating area in rectangular and in polar coordinates. (See page 227.)

More generally, if $f(x, y, z)$ is a continuous function, and if we define

$$g(r, \theta, z) = f(r \cos \theta, r \sin \theta, z),$$

then we have the following relationship between triple integrals:

$$\iiint\limits_{S} f(x, y, z) \, dV_{xyz} = \iiint\limits_{U} g(r, \theta, z) r \, dV_{r\theta z}.$$

A triple integral in cylindrical coordinates may be evaluated by iterated integrals. We write

$$\iiint g(r,\, \theta,\, z) r \, dV_{r\theta z} = \iiint g(r,\, \theta,\, z) r \, dr \, d\theta \, dz$$

and, as before, there are five other orders of integration possible. Once again the major problem is the determination of the limits of integration. For this purpose it is helpful to superimpose cylindrical coordinates on a rectangular system, sketch the surface, and read off the limits of integration. The next example shows the method.

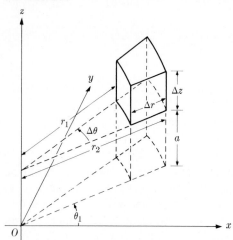

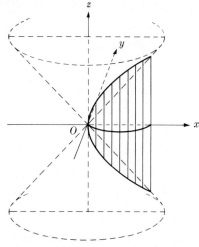

FIGURE 5–59   FIGURE 5–60

EXAMPLE 2. Find the mass of the solid bounded by the cylinder $x^2 + y^2 = ax$ and the cone $z^2 = x^2 + y^2$ if the density $\delta = k\sqrt{x^2 + y^2}$ (Fig. 5–60).

Solution. We change to cylindrical coordinates. The cylinder is $r = a \cos \theta$ and the cone is $z^2 = r^2$. The density is kr. The region S corresponds to the region U given by

$$U: \quad -\frac{\pi}{2} \le \theta \le \frac{\pi}{2}, \quad 0 \le r \le a \cos \theta, \quad -r \le z \le r.$$

Therefore

$$m(S) = \iiint_S k\sqrt{x^2 + y^2} \, dV_{xyz} = k \iiint_U r \cdot r \cdot dV_{r\theta z}$$

$$= k \int_{-\pi/2}^{\pi/2} \int_0^{a \cos \theta} \int_{-r}^{r} r^2 \, dz \, dr \, d\theta$$

$$= 2k \int_{-\pi/2}^{\pi/2} \int_0^{a \cos \theta} r^3 \, dr \, d\theta = \frac{1}{2} ka^4 \int_{-\pi/2}^{\pi/2} \cos^4 \theta \, d\theta$$

$$= \frac{1}{8} ka^4 \int_{-\pi/2}^{\pi/2} \left(1 + 2 \cos 2\theta + \frac{1 + \cos 4\theta}{2} \right) d\theta = \frac{3k\pi a^4}{16}.$$

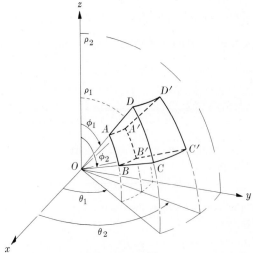

FIGURE 5–61

The transformation from rectangular to spherical coordinates is given by the equations (see page 37)

$$x = \rho \cos \theta \sin \phi, \qquad y = \rho \sin \theta \sin \phi, \qquad z = \rho \cos \phi. \tag{2}$$

The region U determined by the inequalities

$$\rho_1 \le \rho \le \rho_2, \qquad \theta_1 \le \theta \le \theta_2, \qquad \phi_1 \le \phi \le \phi_2$$

corresponds to a rectangular box in (ρ, θ, ϕ) space. We wish to find the volume of the solid S in (x, y, z) space which corresponds to U under the transformation (2). Referring to Fig. 5–61, we see that S is a region such as $ABCDA'B'C'D'$ between the spheres $\rho = \rho_1$ and $\rho = \rho_2$, between the planes $\theta = \theta_1$ and $\theta = \theta_2$, and between the cones $\phi = \phi_1$ $(OADD'A')$ and $\phi = \phi_2$ $(OBCC'B')$. The region S is obtained by sweeping the plane region F ($ABCD$ shown in Fig. 5–62) through an angle $\Delta\theta = \theta_2 - \theta_1$. The Theorem of Pappus (page 246) for determining volumes of solids of revolution applies also for areas swept through any angle about an axis. We obtain

$$V(S) = \Delta\theta \iint\limits_{F} x \, dA_{zx}.$$

Since ρ, ϕ are polar coordinates in the zx-plane (Fig. 5–62), we have $x = \rho \sin \phi$ and $dA_{zx} = \rho \, dA_{\rho\phi} = \rho \, d\rho \, d\phi$. Therefore

$$V(S) = \Delta\theta \int_{\phi_1}^{\phi_2} \int_{\rho_1}^{\rho_2} \rho^2 \sin \phi \, d\rho \, d\phi$$

$$= \int_{\theta_1}^{\theta_2} \int_{\phi_1}^{\phi_2} \int_{\rho_1}^{\rho_2} \rho^2 \sin \phi \, d\rho \, d\phi \, d\theta$$

$$= \iiint\limits_{U} \rho^2 \sin \phi \, dV_{\rho\theta\phi}.$$

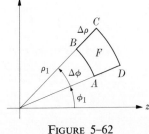

FIGURE 5–62

More generally, if $f(x, y, z)$ is continuous on a region S and if

$$g(\rho, \theta, \phi) = f(\rho \cos \theta \sin \phi, \rho \sin \theta \sin \phi, \rho \cos \phi),$$

then the triple integral of f may be transformed according to the formula

$$\iiint_S f(x, y, z)\, dV_{xyz} = \iiint_U g(\rho, \theta, \phi)\rho^2 \sin \phi \, dV_{\rho\theta\phi}$$

Once again, the triple integral is evaluated by iterated integrations. The next example illustrates the process.

EXAMPLE 3.   Find the volume above the cone $z^2 = x^2 + y^2$ and inside the sphere $x^2 + y^2 + z^2 = 2az$ (Fig. 5–63).

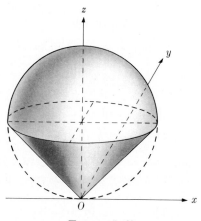

FIGURE 5–63

Solution. In spherical coordinates the cone and sphere have the equations

$$\phi = \frac{\pi}{4} \qquad \text{and} \qquad \rho = 2a \cos \phi,$$

respectively.   Therefore

$$V(S) = \iiint_S dV_{xyz} = \iiint_U \rho^2 \sin \phi \, dV_{\rho\theta\phi}$$

$$= \int_0^{\pi/4} \int_0^{2a \cos \phi} \int_0^{2\pi} \rho^2 \sin \phi \, d\theta \, d\rho \, d\phi$$

$$= 2\pi \int_0^{\pi/4} \int_0^{2a \cos \phi} \rho^2 \sin \phi \, d\rho \, d\phi = \frac{16a^3\pi}{3} \int_0^{\pi/4} \cos^3 \phi \sin \phi \, d\phi$$

$$= \frac{4a^3\pi}{3} \left[-\cos^4 \phi\right]_0^{\pi/4} = \pi a^3.$$

EXERCISES

In each of problems 1 through 16, find the mass of the solid having the given density δ and bounded by the surfaces whose equations are given.

1. $z^2 = x^2 + y^2$, $x^2 + y^2 + z^2 = a^2$, above the cone, $\delta =$ const.

2. The rectangular parallelepiped bounded by $x = -a$, $x = a$, $y = -b$, $y = b$, $z = -c$, $z = c$, $\delta = k(x^2 + y^2 + z^2)$.

3. $x^2 + y^2 + z^2 = a^2$, $x^2 + y^2 + z^2 = b^2$, $a < b$, $\delta = k\sqrt{x^2 + y^2 + z^2}$.

4. The rectangular parallelepiped bounded by $x = 0$, $x = 2a$, $y = 0$, $y = 2b$, $z = 0$, $z = 2c$, $\delta = k(x^2 + y^2 + z^2)$.

5. $x^2 + y^2 = a^2$, $x^2 + y^2 + z^2 = 4a^2$; $\delta = kz^2$; outside the cylinder.

6. The tetrahedron bounded by the coordinate planes and $x + y + z = 1$; $\delta = kxyz$.

7. $x^2 + y^2 = 2ax$, $x^2 + y^2 + z^2 = 4a^2$; $\delta = k(x^2 + y^2)$.

8. $z^2 = 25(x^2 + y^2)$, $z = x^2 + y^2 + 4$; $\delta =$ const; above the paraboloid.

9. $z^2 = x^2 + y^2$, $x^2 + y^2 + z^2 = 2az$; above the cone; $\delta = kz$.

10. Interior of $x^2 + y^2 + z^2 = a^2$; $\delta = k(x^2 + y^2 + z^2)^n$, n a positive number.

11. $x^2 + y^2 = az$, $x^2 + y^2 + z^2 = 2az$; above the paraboloid; $\delta =$ const.

12. $2z = x^2 + y^2$, $z = 2x$; $\delta = k\sqrt{x^2 + y^2}$.

13. $x^2 + y^2 + z^2 = a^2$, $r^2 = a^2 \cos 2\theta$ (cylindrical coordinates); $\delta =$ const.

14. $x^2 + y^2 + z^2 = 4az$, $z = 3a$, above the plane; $\delta = k\sqrt{x^2 + y^2 + z^2}$.

15. $z^2 = x^2 + y^2$, $(x^2 + y^2)^2 = a^2(x^2 - y^2)$; $\delta = k\sqrt{x^2 + y^2}$.

*16. $z^2 = x^2 + y^2$, $x^2 + y^2 + z^2 = 2ax$; $\delta =$ const; above the cone.

12. MOMENT OF INERTIA. CENTER OF MASS

The definition of moment of inertia of a solid body is similar to the definition of moment of inertia of a plane region (page 232). The following definition is basic for the material of this section.

DEFINITION. Suppose that a solid body occupies a region S and let L be any line in three-space. We make a subdivision of space into rectangular boxes and let $S_1, S_2, \ldots, S_n$ be those boxes which contain points of S. For each i, select any point $P_i(\xi_i, \eta_i, \zeta_i)$ in S_i. If the sums

$$\sum_{i=1}^{n} r_i^2 m(S_i) \qquad (r_i = \text{distance of } P_i \text{ from } L)$$

tend to a limit I as the norms of the subdivisions tend to zero, and for any choices of the P_i, then I is called the **moment of inertia of the solid S about L**.

It can be shown that if a solid S has continuous density $\delta(x, y, z)$, then the moments of inertia I_x, I_y, and I_z about the x, y, and z axes, respectively, are given by the triple integrals

$$I_x = \iiint\limits_{S} (y^2 + z^2)\,\delta(x, y, z)\,dV,$$

$$I_y = \iiint\limits_{S} (x^2 + z^2)\,\delta(x, y, z)\,dV,$$

$$I_z = \iiint\limits_{S} (x^2 + y^2)\,\delta(x, y, z)\,dV.$$

If a solid S has a density $\delta(x, y, z)$ and a mass $m(S)$, the point $(\bar{x}, \bar{y}, \bar{z})$, defined by the formulas

$$\bar{x} = \frac{\displaystyle\iiint\limits_{S} x\,\delta(x, y, z)\,dV}{m(S)}, \qquad \bar{y} = \frac{\displaystyle\iiint\limits_{S} y\,\delta(x, y, z)\,dV}{m(S)}, \qquad \bar{z} = \frac{\displaystyle\iiint\limits_{S} z\,\delta(x, y, z)\,dV}{m(S)},$$

is called the **center of mass of** S.

In determining the center of mass, it is helpful to take into account all available symmetries. The following rules are noted:

(a) *If S is symmetric in the xy-plane and $\delta(x, y, -z) = \delta(x, y, z)$, then $\bar{z} = 0$. A similar result holds for other coordinate planes.*

(b) *If S is symmetric in the x-axis and $\delta(x, -y, -z) = \delta(x, y, z)$, then $\bar{y} = \bar{z} = 0$. A similar result holds for the other axes.*

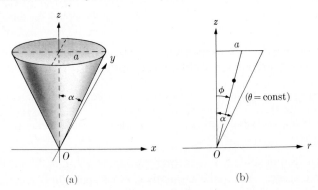

FIGURE 5–64

EXAMPLE 1. Find the moment of inertia of a homogeneous solid cone of base radius a and altitude h about a line through the vertex and perpendicular to the axis.

Solution. We take the vertex at the origin, the z-axis as the axis of the cone, and the x-axis as the line L about which the moment of inertia is to be computed (Fig. 5–64). Let $\alpha = \arctan(a/h)$ be half the angle opening of the cone. In spherical coordinates we get

$$I = \iiint\limits_{S} (y^2 + z^2)\,dV_{xyz} = \int_0^\alpha \int_0^{h\,\sec\,\phi} \int_0^{2\pi} \rho^2\,(\sin^2\phi\,\sin^2\theta + \cos^2\phi)\rho^2\,\sin\phi\,d\theta\,d\rho\,d\phi.$$

Since

$$\int_0^{2\pi} \sin^2 \theta \, d\theta = \pi,$$

we obtain

$$I = \pi \int_0^{\alpha} \int_0^{h \sec \phi} \rho^4 (1 + \cos^2 \phi) \sin \phi \, d\rho \, d\phi$$

$$= \frac{\pi h^5}{5} \int_0^{\alpha} \left[(\cos \phi)^{-5} + (\cos \phi)^{-3} \right] \sin \phi \, d\phi$$

$$= \frac{\pi h^5}{5} \left[\frac{\sec^4 \alpha - 1}{4} + \frac{\sec^2 \alpha - 1}{2} \right] = \frac{\pi h a^2}{20} (4h^2 + a^2),$$

since $\tan \alpha = a/h$.

EXAMPLE 2. Find the center of gravity of a solid hemisphere of radius a which has density proportional to the distance from the center.

Solution. We select the hemisphere so that the plane section is in the xy-plane and the z-axis is an axis of symmetry (Fig. 5–65). Then $\bar{x} = \bar{y} = 0$. Changing to spherical coordinates, we have

$$m(S) = \int_0^{2\pi} \int_0^{\pi/2} \int_0^{a} k\rho \cdot \rho^2 \sin \phi \, d\rho \, d\phi \, d\theta = \frac{1}{2} \pi k a^4$$

and

$$\bar{z} = \frac{\displaystyle\int_0^{2\pi} \int_0^{\pi/2} \int_0^{a} \rho \cos \phi \cdot k\rho \cdot \rho^2 \sin \phi \, d\rho \, d\phi \, d\theta}{\frac{1}{2}\pi k a^4}$$

$$= \frac{2}{\pi a^4} \cdot \frac{1}{5} a^5 \cdot 2\pi \int_0^{\pi/2} \cos \phi \sin \phi \, d\phi$$

$$= \frac{4}{5} a \left[\frac{\sin^2 \phi}{2} \right]_0^{\pi/2} = \frac{2a}{5}.$$

The center of gravity is at $(0, 0, 2a/5)$.

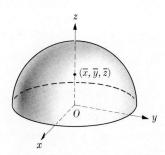

FIGURE 5–65

EXERCISES

In each of problems 1 through 15, find the moment of inertia about the given axis of the solid having the specified density δ and bounded by the surfaces as described.

1. A cube of side a; $\delta = $ const; about an edge.

2. A cube of side a; $\delta = $ const; about a line parallel to an edge, at distance 2 from it, and in a plane of one of the faces.

3. Bounded by $x = 0$, $y = 0$, $z = 0$, $x + z = a$, $y = z$; $\delta = kx$; about the x-axis.

4. $x^2 + y^2 = a^2$; $x^2 + z^2 = a^2$; $\delta = $ const; about the z-axis.

5. $z = x$, $y^2 = 4 - 2z$, $x = 0$; $\delta = $ const; about the z-axis.

6. $x = 0$, $y = 0$, $z^2 = 1 - x - y$; $\delta = $ const; about the z-axis.

7. $z^2 = y^2(1 - x^2)$, $y = 1$; δ = const; about the x-axis.

8. $x^2 + y^2 = a^2$, $x^2 + y^2 = b^2$, $z = 0$, $z = h$; $\delta = k\sqrt{x^2 + y^2}$; about the x-axis ($a < b$).

9. $x^2 + y^2 + z^2 = a^2$, $x^2 + y^2 + z^2 = b^2$; $\delta = k\sqrt{x^2 + y^2 + z^2}$; about the z-axis ($a < b$).

10. $\rho = 4$, $\rho = 5$, $z = 1$, $z = 3$ $(1 \leq z \leq 3)$; δ = const; about the z-axis.

11. $z^2 = x^2 + y^2$, $(x^2 + y^2)^2 = a^2(x^2 - y^2)$; $\delta = k\sqrt{x^2 + y^2}$; about the z-axis.

12. $z = 0$, $z = \sqrt{a^2 - x^2 - y^2}$; $\delta = kz$; about the x-axis.

13. $x^2/a^2 + y^2/b^2 + z^2/c^2 = 1$; δ = const; about the z-axis. (Divide the integral into two parts.)

14. $(r - b)^2 + z^2 = a^2$ $(0 < a < b)$; $\delta = kz$; about the z-axis. Assume $z \geq 0$.

15. $\rho = b$, $\rho = c$, $r = a$ $(a < b < c)$; outside $r = a$, between $\rho = b$ and $\rho = c$; δ = const; about the z-axis.

In each of problems 16 through 32, find the center of mass of the solid having the given density and bounded by the surfaces as described.

16. $x = 0$, $y = 0$, $z = 0$, $x/a + y/b + z/c = 1$; δ = const.

17. $x = 0$, $y = 0$, $z = 0$, $x + z = a$, $y = z$; $\delta = kx$.

18. $x^2 + y^2 = a^2$, $x^2 + z^2 = a^2$, δ = const; the portion where $x \geq 0$.

19. $z = x$, $z = -x$, $y^2 = 4 - 2x$; δ = const.

20. $z^2 = y^2(1 - x^2)$, $y = 1$; δ = const.

21. $z = 0$, $x^2 + z = 1$, $y^2 + z = 1$; δ = const.

22. $x = 0$, $y = 0$, $z = 0$, $x^{1/2} + y^{1/2} + z^{1/2} = a^{1/2}$; δ = const.

23. $y^2 + z^2 = 4ax$, $y^2 = ax$, $x = 3a$; δ = const. (Inside $y^2 = ax$.)

24. $z^2 = 4ax$, $x^2 + y^2 = 2ax$; δ = const.

25. $z^2 = x^2 + y^2$, $x^2 + y^2 + z^2 = a^2$, above the cone; δ = const.

26. $z^2 = x^2 + y^2$, $x^2 + y^2 = 2ax$; $\delta = k(x^2 + y^2)$.

27. $z^2 = x^2 + y^2$, $x^2 + y^2 + z^2 = 2az$, above the cone; $\delta = kz$.

28. $x^2 + y^2 = az$, $x^2 + y^2 + z^2 = 2az$, above the paraboloid; δ = const.

29. $x^2 + y^2 + z^2 = 4az$, $z = 3a$, above the plane; $\delta = k\sqrt{x^2 + y^2 + z^2}$.

30. $\rho = 4$, $\rho = 5$, $z = 1$, $z = 3$ $(1 \leq z \leq 3)$; δ = const.

*31. $z^2 = x^2 + y^2$, $(x^2 + y^2)^2 = a^2(x^2 - y^2)$; $\delta = k\sqrt{x^2 + y^2}$; the part for which $x \geq 0$.

*32. $z^2 = x^2 + y^2$, $x^2 + y^2 + z^2 = 2ax$, above the cone; δ = const.

Linear Algebra

1. SOLUTION OF SYSTEMS OF EQUATIONS IN n VARIABLES BY ELIMINATION

The algebraic problem of solving two simultaneous linear equations in two unknowns is equivalent to the geometric problem of finding the point of intersection of two lines in a plane. We know that a solution can always be found except when the lines are parallel. The algebraic problem of solving three simultaneous linear equations in three unknowns is equivalent to the problem of finding the point of intersection of three planes in space. A solution can be obtained under most circumstances. There are, however, a number of exceptional cases which occur when two or three of the planes are parallel, when the planes intersect in three parallel lines, and so forth. One of the simplest techniques for solving systems of two or three simultaneous linear equations is the method of elimination which the student learned in high school and which was discussed in *First Course*, pages 57 and 58. The process, which works whenever there is a solution, is easily extended to equations in n unknowns with $n > 3$. We illustrate with an example.

EXAMPLE 1. Solve the following system of simultaneous equations:

$$
\begin{aligned}
2x - y + 3z - 2w &= 4, \\
x + 7y + z - w &= 2, \\
3x + 5y - 5z + 3w &= 0, \\
4x - 3y + 2z - w &= 5.
\end{aligned}
$$

Solution. We multiply the second equation by 2 and subtract it from the first, getting

$$-15y + z = 0.$$

Next we multiply the second equation by 3 and subtract it from the third equation; then we multiply the second equation by 4 and subtract it from the fourth equation. In this way we obtain the system

$$
\begin{aligned}
-15y + z &= 0, \\
-16y - 8z + 6w &= -6, \\
-31y - 2z + 3w &= -3,
\end{aligned}
$$

in which only the variables y, z, and w are present. This reduced system may be solved simultaneously by elimination. The quantity w may be eliminated from the second and third equations to give the system

$$-15y + z = 0,$$
$$46y - 4z = 0,$$

which has the solution $y = z = 0$. The value of w may be found from the second or third equation of the preceding set by setting $y = z = 0$. We get $w = -1$. Finally, inserting $y = z = 0$ and $w = -1$ in any of the four original equations, we obtain $x = 1$. We may verify the result in two ways: (1) by substituting $x = 1$, $y = 0$, $z = 0$, $w = -1$ in all four of the original equations and checking the equality, or (2) by noticing that in the elimination process each step is reversible. Therefore, we can begin with the equations $x = 1$, $y = 0$, $z = 0$, $w = -1$ and derive the system we set out to solve.

EXAMPLE 2. Solve the simultaneous system:

$$x - 2y + z - 4w + t = 14,$$
$$2x + 2y - z + w + 3t = 0,$$
$$-x + y + 2z - w + 2t = 8,$$
$$3x - y + 2z + 2w - 2t = 5,$$
$$x + 3y + 2z + 2w - 3t = -7.$$

Solution. We solve the first equation for x in terms of the other unknowns, getting

$$x = 14 + 2y - z + 4w - t.$$

Next we substitute this value of x in each of the remaining four equations to obtain the system of four equations in four unknowns:

$$6y - 3z + 9w + t = -28,$$
$$-y + 3z - 5w + 3t = 22,$$
$$5y - z + 14w - 5t = -37,$$
$$5y + z + 6w - 4t = -21.$$

Solving the second of these equations for y in terms of the remaining quantities, we get

$$y = -22 + 3z - 5w + 3t.$$

Substitution of this value of y in the remaining three equations yields

$$15z - 21w + 19t = 104,$$
$$14z - 11w + 10t = 73,$$
$$16z - 19w + 11t = 89.$$

These three equations in the unknowns z, w, t may be reduced by solving the second equation for t in terms of z and w and then substituting in the first and third equations. The result is

$$116z + w = 347,$$
$$2z - 23w = 29.$$

Finally, we solve the first equation above for w in terms of z and substitute in the second equation to get $z = 3$. Now we can select the following equations from the various systems above:

$$
\begin{aligned}
x - 2y + t - 4w + z &= 14, \\
-y + 3t - 5w + 3z &= 22, \\
10t - 11w + 14z &= 73, \\
w + 116z &= 347, \\
z &= 3.
\end{aligned}
$$

A system of this type is called a **triangular form** and may be solved almost by inspection by starting with the last equation and substituting successively in each of the preceding equations. The solution is $x = 1$, $y = -2$, $z = 3$, $w = -1$, $t = 2$. The triangular system is completely equivalent to the original system in that each may be derived from the other.

———————

In both examples it turned out that there is exactly one solution to the system of equations. The question of determining *when* a system of n equations has any solution at all or when it may have many solutions is discussed in Section 6. The more complicated problem of solutions of systems in which the number of equations differs from the number of unknowns is treated in Section 8.

EXERCISES

In the following problems, solve the systems of simultaneous equations, and check the answers.

1. $\begin{aligned}
x - y - 2z &= -2 \\
3x - 2y - 4z &= -5 \\
2y + 3z &= 2
\end{aligned}$

2. $\begin{aligned}
2x + y + 3z &= 6 \\
2x + z &= 5 \\
-7x + 2y + z &= -15
\end{aligned}$

3. $\begin{aligned}
2x + y - z &= -1 \\
3x - 2y + z &= 7 \\
x + 2y + 2z &= 3
\end{aligned}$

4. $\begin{aligned}
2x - 3y - 2z &= -10 \\
3x + 2y - 4z &= -3 \\
4x - 2y + 3z &= -5
\end{aligned}$

5. $\begin{aligned}
x - y + 2z + w &= -3 \\
2x + y - z - w &= 8 \\
2x + 2y + 3z + w &= 1 \\
-x + y - 2z + 3w &= -5
\end{aligned}$

6. $\begin{aligned}
2x - 3y + z - w &= 7 \\
x + 2y - z + 2w &= 1 \\
-x + y + 2z + w &= 4 \\
3x - 2y + z - 2w &= 3
\end{aligned}$

7. $\begin{aligned}
3x + y - 2z + w &= 5 \\
2x - y + z - 2w &= 2 \\
-2x + 2y - z + 2w &= -1 \\
-3x - 2y - 3z + w &= -3
\end{aligned}$

8. $\begin{aligned}
2x - y + 3z + w &= 6 \\
x + 2y - z + 2w &= -9 \\
-x + 3y - 2z - 3w &= -1 \\
2x + y + z - 2w &= 9
\end{aligned}$

9. $\begin{aligned}
2x + 3y - 2z + 2w &= 14 \\
3x - 2y + 3z + 4w &= 4 \\
-2x + 2y + 2z - 3w &= -6 \\
3x + 2y - 2z - 2w &= 5
\end{aligned}$

10. $\begin{aligned}
3x - 2y + 2z + 3w &= 15 \\
-2x + 2y + 3z - 2w &= -2 \\
2x + 3y - 2z + 2w &= -1 \\
4x - 2y - 3z + 3w &= 7
\end{aligned}$

11. $2x - y + 3z - w + t = 0$
     $x + y + 2z + 2w - 3t = 13$
     $2x + 3y - 2z + 4w - t = 15$
     $-x - 3y + 4z - 2w + t = -7$
     $4x + y - 5z + 2w - 3t = 19$

12. $x + 2y - 3z + 4w - 2t = -1$
     $-x - y - z + 4w - 4t = -19$
     $2x + 4y - 3z + w = 14$
     $x + 3z + 5w - t = -9$
     $2x + 3y + 4z - 2w + 3t = 17$

2. MATRICES

A **matrix** is a rectangular array of numbers enclosed in parentheses. For example, the array

$$\begin{pmatrix} 3 & 2 & 0 & -1 \\ 1 & 4 & 6 & 2 \\ -3 & 1 & 3 & 0 \end{pmatrix}$$

is a matrix with three rows and four columns. A matrix with m rows and n columns is written

$$\begin{pmatrix} a_{11} & a_{12} & a_{13} & \cdots & a_{1n} \\ a_{21} & a_{22} & a_{23} & \cdots & a_{2n} \\ \vdots & & & & \vdots \\ a_{m1} & a_{m2} & a_{m3} & \cdots & a_{mn} \end{pmatrix}$$

The individual entries in the matrix are called its **elements;** the quantity a_{ij} in the above matrix is the element in the ith row and jth column. The subscripts used to indicate the elements will always denote the row first and the column second. If the number of rows of a matrix is the same as the number of columns, it is said to be **square.** We shall use capital letters such as A, B, C, ... to denote matrices. The corresponding lower-case letters with subscripts, such as a_{ij}, b_{ij}, c_{ij}, etc., will be used to denote the elements. We use the expression "m by n matrix" and write "$m \times n$ matrix" for a matrix with m rows and n columns.

A matrix with one row and n columns is called a **row vector.** For example, the matrix

$$(3, \quad -1, \quad 2, \quad 0, \quad 5, \quad 4)$$

is a row vector with 6 columns. A matrix with m rows and one column is called a **column vector.** The array

$$\begin{pmatrix} -3 \\ 0 \\ 2 \\ 4 \\ 5 \end{pmatrix}$$

is a column vector with 5 rows. We say that two matrices are of the **same size** if and only if they have the same number of rows and the same number of columns.

The most important properties of matrices are contained in the rules of operation which we now define.

RULE I. (Multiplication by a constant.) *If A is a matrix and c is a number, then cA is the matrix obtained by multiplying each element of A by the number c. For example, if c* = 3 *and*

$$A = \begin{pmatrix} 3 & -2 & 0 \\ 1 & 4 & 2 \\ -1 & -3 & 1 \end{pmatrix}, \quad then \quad 3A = \begin{pmatrix} 9 & -6 & 0 \\ 3 & 12 & 6 \\ -3 & -9 & 3 \end{pmatrix}$$

We use the symbol $-A$ for the matrix $(-1)A$.

RULE II. (Addition of matrices.) *If A and B are of the same size and have elements a_{ij} and b_{ij}, respectively, we define their* **sum** $A + B$ *as the matrix C with elements c_{ij} such that*

$$c_{ij} = a_{ij} + b_{ij}$$

for each i and j.

For example, if

$$A = \begin{pmatrix} 3 & 2 & -1 \\ 4 & 6 & 0 \end{pmatrix}, \quad B = \begin{pmatrix} -2 & 1 & 5 \\ 0 & 4 & -2 \end{pmatrix},$$

then

$$C = A + B = \begin{pmatrix} 1 & 3 & 4 \\ 4 & 10 & -2 \end{pmatrix}$$

It is important to remember that addition of matrices can be defined only for matrices of the *same size.* Subtraction of two matrices is defined in terms of addition. We have $A - B = A + (-1)B$.

RULE III. (Equality of matrices.) *Two matrices are equal if and only if all their corresponding elements are equal. That is, in order that $A = B$ we must have A and B the same size and $a_{ij} = b_{ij}$ for each i and j; in other words A and B denote the same matrix.*

EXAMPLE 1. Solve the following matrix equation for A:

$$2A - 3\begin{pmatrix} 2 & -1 & 3 \\ -3 & 2 & 1 \end{pmatrix} = \begin{pmatrix} -2 & 3 & -3 \\ 1 & 4 & 3 \end{pmatrix}$$

Solution. In order to make sense, A must be a 2×3 matrix. We may use the rules for adding matrices to write

$$2A = \begin{pmatrix} -2 & 3 & -3 \\ 1 & 4 & 3 \end{pmatrix} + 3\begin{pmatrix} 2 & -1 & 3 \\ -3 & 2 & 1 \end{pmatrix}$$

$$= \begin{pmatrix} -2 & 3 & -3 \\ 1 & 4 & 3 \end{pmatrix} + \begin{pmatrix} 6 & -3 & 9 \\ -9 & 6 & 3 \end{pmatrix} = \begin{pmatrix} 4 & 0 & 6 \\ -8 & 10 & 6 \end{pmatrix}$$

Therefore

$$A = \begin{pmatrix} 2 & 0 & 3 \\ -4 & 5 & 3 \end{pmatrix}$$

We define the **zero matrix** as the matrix with all elements zeros.

RULE IV. (Multiplication of matrices.) *Let A be an m × n matrix and B an n × p matrix. The product AB is that m × p matrix C with elements c_{ij} given by*

$$c_{ij} = \sum_{k=1}^{n} a_{ik}b_{kj}, \qquad i = 1, 2, \ldots, m; \qquad j = 1, 2, \ldots, p.$$

It is extremely important to note that the product of two matrices is defined *only* when the number of columns in the first matrix is equal to the number of rows in the second matrix.

EXAMPLE 2. Compute AB, given that

$$A = \begin{pmatrix} 2 & -1 & 3 \\ 1 & -2 & -1 \end{pmatrix}, \qquad B = \begin{pmatrix} 3 & -1 \\ 1 & 2 \\ -1 & 1 \end{pmatrix}$$

Solution.

$$AB = \begin{pmatrix} 2 \cdot 3 + (-1) \cdot 1 + 3 \cdot (-1) & 2 \cdot (-1) + (-1) \cdot 2 + 3 \cdot 1 \\ 1 \cdot 3 + (-2) \cdot 1 + (-1)(-1) & 1 \cdot (-1) + (-2) \cdot 2 + (-1) \cdot 1 \end{pmatrix}$$

$$= \begin{pmatrix} 2 & -1 \\ 2 & -6 \end{pmatrix}$$

Remarks. Note that A is a 2 × 3 matrix and B is a 3 × 2 matrix. Therefore the product can be formed, and the result is a 2 × 2 matrix. The product BA can also be formed, the result being a 3 × 3 matrix. It is clear that $AB \neq BA$, since AB is a 2 × 2 matrix and BA is a 3 × 3 matrix. In general, *multiplication of matrices is not commutative.* This is true even for square matrices. We observe that if

$$A = \begin{pmatrix} 1 & 2 \\ -1 & 1 \end{pmatrix} \qquad \text{and} \qquad B = \begin{pmatrix} 2 & 0 \\ 1 & -1 \end{pmatrix},$$

then

$$AB = \begin{pmatrix} 4 & -2 \\ -1 & -1 \end{pmatrix}, \qquad BA = \begin{pmatrix} 2 & 4 \\ 2 & 1 \end{pmatrix}$$

The student can easily establish the following result, which is stated in the form of a theorem.

Theorem 1. (a) *If A is an m × n matrix and B and C are n × p matrices, then*

$$A(B + C) = AB + AC.$$

(b) *If A and B are m × n matrices and C is an n × p matrix, then*

$$(A + B)C = AC + BC.$$

(c) *If A is an m $\times$ n matrix, B is an n $\times$ p matrix, and c is a number, then*

$$(cA)B = A(cB) = c(AB).$$

If the matrix C is the product of the matrices A and B, the elements of C may be expressed in terms of the inner or scalar product of two vectors. The element in the ith row and jth column of C is the inner product of the ith row vector of A with the jth column vector of B. The formula

$$c_{ij} = a_{i1}b_{1j} + a_{i2}b_{2j} + \cdots + a_{in}b_{nj}$$

verifies this fact.

A system of linear equations is easily written in matrix form. For example, the system

$$\begin{aligned}
a_{11}x_1 + a_{12}x_2 + \cdots + a_{1n}x_n &= b_1 \\
a_{21}x_1 + a_{22}x_2 + \cdots + a_{2n}x_n &= b_2 \\
&\vdots \\
a_{m1}x_1 + a_{m2}x_2 + \cdots + a_{mn}x_n &= b_m
\end{aligned}$$

has m equations and the n unknowns $x_1, x_2, \ldots, x_n$; it may be written in the form

$$AX = B,$$

where A is the $m \times n$ matrix with elements a_{ij} and where X and B are the column vectors with n rows and m rows, respectively:

$$X = \begin{pmatrix} x_1 \\ x_2 \\ \vdots \\ x_n \end{pmatrix}, \qquad B = \begin{pmatrix} b_1 \\ b_2 \\ \vdots \\ b_m \end{pmatrix}$$

A precise definition of an $m \times n$ matrix may be given in terms of functions. The domain of a matrix function consists of all ordered pairs (i, j) of integers with $1 \le i \le m$ and $1 \le j \le n$. The range of the function is the real number system. The particular nature of the function is determined by the rules of operation which we described in terms of rectangular arrays.

EXERCISES

In problems 1 through 7, solve for A.

1. $A = 2\begin{pmatrix} 1 & -2 \\ 2 & 3 \end{pmatrix} - 3\begin{pmatrix} 2 & 1 \\ -1 & 4 \end{pmatrix}$

2. $A = 3\begin{pmatrix} 2 & -3 & 1 \\ 0 & 1 & -1 \end{pmatrix} - 2\begin{pmatrix} 1 & -1 & 2 \\ 2 & 3 & -1 \end{pmatrix}$

3. $2A - \begin{pmatrix} 1 & 2 & 3 \\ 2 & -1 & 2 \end{pmatrix} = 3\begin{pmatrix} 3 & 0 & 5 \\ 2 & 1 & 4 \end{pmatrix}$

4. $3A + \begin{pmatrix} 1 & -1 \\ 2 & 3 \\ -1 & 2 \end{pmatrix} = 2\begin{pmatrix} 2 & 1 \\ 1 & -3 \\ 1 & 4 \end{pmatrix}$

5. $\begin{pmatrix} 2 & -1 & 3 \\ -1 & 3 & 2 \\ 1 & 2 & -1 \end{pmatrix} - 2A = \begin{pmatrix} 2 & 3 & 1 \\ 3 & -1 & 2 \\ 1 & -2 & 1 \end{pmatrix}$

6. $\begin{pmatrix} 2 & 1 \\ 0 & 3 \end{pmatrix} A = \begin{pmatrix} 3 & 0 \\ -3 & -6 \end{pmatrix}$

7. $A\begin{pmatrix} 3 & 1 \\ -2 & 2 \end{pmatrix} = \begin{pmatrix} 5 & 7 \\ -5 & 9 \end{pmatrix}$

8. If A is a matrix and c and d are numbers, show that $c(dA) = (cd)A$.

9. If A and B are matrices of the same size and c and d are numbers, show that

$$(c + d)A = cA + dA \quad \text{and} \quad c(A + B) = cA + cB.$$

In problems 10 through 13, solve simultaneously for A and B.

10. $A - 2B = \begin{pmatrix} 1 & 2 \\ -1 & 1 \end{pmatrix}, \quad A - B = \begin{pmatrix} 2 & 1 \\ 1 & -1 \end{pmatrix}$

11. $A + 2B = \begin{pmatrix} 2 & 1 & 0 \\ 1 & -1 & 2 \end{pmatrix}, \quad 2A + 3B = \begin{pmatrix} 1 & 2 & -1 \\ 2 & 0 & 1 \end{pmatrix}$

12. $A - B = \begin{pmatrix} 1 & -2 \\ -1 & 3 \end{pmatrix}, \quad A + B = \begin{pmatrix} 3 & 0 \\ 3 & 1 \end{pmatrix}$

13. $2A - B = \begin{pmatrix} 3 & -3 & 0 \\ 3 & 3 & 2 \end{pmatrix}, \quad -A + 2B = \begin{pmatrix} 0 & 3 & 3 \\ -3 & 0 & -4 \end{pmatrix}$

In problems 14 through 21, compute AB when possible and compute BA when possible.

14. $A = \begin{pmatrix} 2 & -1 \\ 1 & 2 \end{pmatrix}, \quad B = \begin{pmatrix} 1 & 0 \\ 2 & 1 \end{pmatrix}$ 15. $A = \begin{pmatrix} 1 & -1 \\ 2 & 3 \end{pmatrix}, \quad B = \begin{pmatrix} 2 & -1 \\ 1 & 3 \end{pmatrix}$

16. $A = \begin{pmatrix} 1 & 0 \\ 0 & 0 \end{pmatrix}, \quad B = \begin{pmatrix} 0 & 0 \\ 1 & 1 \end{pmatrix}$ 17. $A = \begin{pmatrix} 1 & 0 \\ 0 & 1 \end{pmatrix}, \quad B = \begin{pmatrix} 0 & 1 \\ 0 & 1 \end{pmatrix}$

18. $A = \begin{pmatrix} 1 & 0 & 2 \\ -2 & 1 & 3 \end{pmatrix}, \quad B = \begin{pmatrix} 2 & 1 \\ -1 & 1 \\ 1 & 2 \end{pmatrix}$

19. $A = \begin{pmatrix} 1 & 2 & -1 \\ -2 & 1 & 3 \end{pmatrix}, \quad B = \begin{pmatrix} 2 & -1 & 1 \\ 1 & 3 & -2 \\ -1 & 2 & 1 \end{pmatrix}$

20. $A = \begin{pmatrix} 2 & 1 \\ -1 & 2 \\ -2 & 3 \end{pmatrix}, \quad B = \begin{pmatrix} 1 & 2 & -1 & -2 \\ -1 & 3 & 1 & -1 \end{pmatrix}$

21. $A = \begin{pmatrix} 1 & 1 & 0 \\ 0 & 0 & 0 \\ 0 & 1 & 0 \end{pmatrix}$, $\quad B = \begin{pmatrix} 0 & 0 & 0 \\ 0 & 0 & 0 \\ 1 & 0 & 0 \end{pmatrix}$

22. Prove Theorem 1.

23. Let

$$A = \begin{pmatrix} 0 & 0 \\ 0 & 0 \end{pmatrix} \quad \text{and} \quad B = \begin{pmatrix} 1 & 0 \\ 0 & 1 \end{pmatrix}$$

Show that A and B commute with all 2×2 matrices. Are there any other matrices which commute with all 2×2 matrices?

24. Find two 3×3 matrices A and B with the property that $AB = 0$, and $BA = 0$ and $A \neq 0$, $B \neq 0$.

3. MATRICES, CONTINUED. DOUBLE SUMS AND DOUBLE SEQUENCES

If A is an $m \times n$ matrix, the **transpose of A** is that $n \times m$ matrix obtained from A by interchanging its rows and columns. We use the symbol A^t for the transpose of A. The element in the ith row and jth column of A^t, denoted a^t_{ij}, is given by

$$a^t_{ij} = a_{ji}, \quad i = 1, 2, \ldots, n; \quad j = 1, 2, \ldots, m.$$

For example, if

$$A = \begin{pmatrix} 2 & -1 & 3 \\ 1 & -2 & -1 \end{pmatrix}, \quad \text{then} \quad A^t = \begin{pmatrix} 2 & 1 \\ -1 & -2 \\ 3 & -1 \end{pmatrix}$$

The following theorem follows from the definition of transpose.

Theorem 2. (a) *If A and B are $m \times n$ matrices and c is a constant, then*

$$(A + B)^t = A^t + B^t, \quad (cA)^t = cA^t.$$

(b) *If A is an $m \times n$ matrix and B is an $n \times p$ matrix, then*

$$(AB)^t = B^t A^t.$$

Proof. Part (a) can be performed by the student. (See problem 8 at the end of this section.) To prove (b), we let $C = AB$. Then C is an $m \times p$ matrix, and C^t is a $p \times m$ matrix. Using a_{ij}, b_{ij}, and c_{ij} for elements of A, B, and C, respectively, we have

$$c_{ij} = \sum_{k=1}^{n} a_{ik}b_{kj}, \quad i = 1, 2, \ldots, m; \quad j = 1, 2, \ldots, p.$$

Also

$$c^t_{ij} = c_{ji} = \sum_{k=1}^{n} a_{jk}b_{ki}, \quad i = 1, 2, \ldots, p; \quad j = 1, 2, \ldots, m.$$

We must show that the elements of B^tA^t are precisely c_{ij}^t. We let $D = B^tA^t$. Since B^t is a $p \times n$ matrix and since A^t is an $n \times m$ matrix, the matrix D is $p \times m$, and we have

$$b_{ij}^t = b_{ji}, \qquad i = 1, 2, \ldots, p, \qquad j = 1, 2, \ldots, n,$$
$$a_{ij}^t = a_{ji}, \qquad i = 1, 2, \ldots, n, \qquad j = 1, 2, \ldots, m,$$
$$d_{ij} = \sum_{k=1}^{n} b_{ik}^t a_{kj}^t = \sum_{k=1}^{n} b_{ki} a_{jk} = \sum_{k=1}^{n} a_{jk} b_{ki} = c_{ji} = c_{ij}^t.$$

DEFINITIONS. If A is a *square matrix*, we call those elements of the form a_{ii} **diagonal elements,** and we call the totality of diagonal elements the **diagonal** of the matrix. A **diagonal matrix** is one in which all elements are zero, except possibly those on the diagonal. That is,

$$a_{ij} = 0 \quad \text{if} \quad i \neq j.$$

If, in a diagonal matrix A, we have $a_{ii} = 1$ for all i, the matrix is called the **identity matrix** and is denoted by I. A square matrix is said to be **triangular** if and only if all the elements on one side of the diagonal are zero, i.e.,

$$a_{ij} = 0 \quad \text{for} \quad i > j \qquad \text{or} \qquad a_{ij} = 0 \quad \text{for} \quad i < j.$$

Examples of 4×4 diagonal, identity, and triangular matrices are shown below.

$$D = \begin{pmatrix} 3 & 0 & 0 & 0 \\ 0 & 2 & 0 & 0 \\ 0 & 0 & 1 & 0 \\ 0 & 0 & 0 & 4 \end{pmatrix} \qquad\qquad I = \begin{pmatrix} 1 & 0 & 0 & 0 \\ 0 & 1 & 0 & 0 \\ 0 & 0 & 1 & 0 \\ 0 & 0 & 0 & 1 \end{pmatrix}$$

Diagonal   Identity

$$T_1 = \begin{pmatrix} 3 & -1 & 0 & 4 \\ 0 & 2 & 1 & 5 \\ 0 & 0 & 3 & 4 \\ 0 & 0 & 0 & 5 \end{pmatrix} \qquad\qquad T_2 = \begin{pmatrix} 2 & 0 & 0 & 0 \\ 1 & 3 & 0 & 0 \\ 2 & 5 & 1 & 0 \\ 0 & 4 & 6 & 2 \end{pmatrix}$$

$a_{ij} = 0$ if $i > j$ $a_{ij} = 0$ if $i < j$

It is useful to introduce the **Kronecker delta** symbol δ_{ij}, which is defined by the relation

$$\delta_{ij} = \begin{cases} 1 & \text{if} \quad i = j, \\ 0 & \text{if} \quad i \neq j. \end{cases}$$

If B is a square matrix with elements b_{ij}, then the relation $b_{ij} = \delta_{ij}$ implies that the matrix B is the identity matrix I. We also notice that the Kronecker delta allows simplifications of the type

$$\sum_{j=1}^{n} \delta_{ij} x_j = x_i, \qquad \sum_{i=1}^{n} \delta_{ij} y_i = y_j.$$

Theorem 3. (a) *If A and B are each n × n diagonal matrices, then AB = BA and this is a diagonal matrix.* (b) *If A is any n × n matrix and I is the n × n identity matrix, then IA = AI = A.*

The proofs are left to the reader. (See problems 9 and 10 at the end of this section.)

We recall that a sequence is a succession of numbers such as

$$a_1, \quad a_2, \quad a_3, \quad \ldots, \quad a_{14}.$$

If there is a first and last element, the sequence is called *finite*; otherwise it is *infinite*. We may consider a sequence as a function with domain consisting of a portion of the integers (i.e., the subscripts) and with the range (the numbers a_i) consisting of a portion of the real number system.

A **double sequence** is a function the domain of which is some set S of ordered pairs (i, j) of integers and with range consisting of a portion of the real number system. A matrix is a special type of double sequence in which the domain S consists of pairs (i, j) in which $1 \le i \le m$ and $1 \le j \le n$. In general, the domain S of a double sequence is not restricted in this way; in fact, S may be infinite, in which case we say that the double sequence is **infinite.**

A finite **double sum** is an expression of the form

$$\sum_{(i,j)\in S} a_{ij},$$

in which the a_{ij} form a finite double sequence with domain S and with the sum extending over all elements of S. The symbol $(i, j) \in S$, which we read "(i, j) belongs to S," indicates this fact. The general commutative property for addition of ordinary numbers shows that the order in which we add the terms of a finite double sum is irrelevant.

Suppose the domain of a double sequence consists of all pairs (i, j) for which $1 \le i \le m$ and $1 \le j \le n$. Then we can easily conclude that

$$\sum_{(i,j)\in S} a_{ij} = \sum_{i=1}^{m}\left[\sum_{j=1}^{n} a_{ij}\right] = \sum_{j=1}^{n}\left[\sum_{i=1}^{m} a_{ij}\right]. \tag{1}$$

The latter two sums are called **iterated sums,** since we sum first with respect to one index and then with respect to the other. The fact that double sums and iterated sums are identical leads to many conveniences, and we shall use this fact on a number of occasions. For example, it follows that

$$\left(\sum_{i=1}^{m} a_i\right)\left(\sum_{j=1}^{n} b_j\right) = \sum_{(i,j)\in S} a_i b_j, \qquad S:\{1 \le i \le m, \; 1 \le j \le n\}.$$

A convenient notation consists of the symbol $1 \le i < j \le n$ used to represent the set S, which consists of all (i, j) for which $1 \le i \le n, 1 \le j \le n$ and $i < j$.

Note that i cannot have the value n, since i must be strictly less than j and j can be at most n. Similarly, j cannot have the value 1. We write

$$\sum_{1 \leq i < j \leq n} a_{ij} \qquad \text{to mean} \qquad \sum_{(i,j) \in S} a_{ij},$$

in which S is the set described above. We may also conclude on the basis of the equivalence of double sums and iterated sums that

$$\sum_{1 \leq i < j \leq n} a_{ij} = \sum_{i=1}^{n-1} \sum_{j=i+1}^{n} a_{ij}. \tag{2}$$

It can also be shown that

$$\sum_{1 \leq i < j \leq n} a_{ij} = \sum_{j=2}^{n} \sum_{i=1}^{j-1} a_{ij}. \tag{3}$$

The set S for $n = 6$ is shown schematically below.

j

6	x	x	x	x	x	
5	x	x	x	x		
4	x	x	x			
3	x	x				
2	x					
1			$1 \leq i < j \leq 6$			

$$\begin{array}{cccccc} 1 & 2 & 3 & 4 & 5 & 6 \end{array} \qquad i$$

EXAMPLE 1. Write out the iterated sum

$$\sum_{i=1}^{4} \sum_{j=1}^{i} a_{ij}.$$

Solution. We have

$$\sum_{i=1}^{4} \sum_{j=1}^{i} a_{ij} = \left[\sum_{j=1}^{1} a_{1j} \right] + \left[\sum_{j=1}^{2} a_{2j} \right] + \left[\sum_{j=1}^{3} a_{3j} \right] + \left[\sum_{j=1}^{4} a_{4j} \right]$$

$$= a_{11} + (a_{21} + a_{22}) + (a_{31} + a_{32} + a_{33})$$

$$+ (a_{41} + a_{42} + a_{43} + a_{44}).$$

EXAMPLE 2. Verify formulas (2) and (3) for $n = 3$.

Solution. The symbol

$$\sum_{1 \leq i < j \leq 3} a_{ij} \qquad \text{means} \qquad \sum_{(i,j) \in S} a_{ij},$$

where S consists of all (i, j) for which $1 \leq i \leq 3, 1 \leq j \leq 3$ and $i < j$. Therefore

$$\sum_{1 \leq i < j \leq 3} a_{ij} = a_{12} + a_{13} + a_{23}.$$

From formulas (2) and (3) we have

$$\sum_{i=1}^{2} \sum_{j=i+1}^{3} a_{ij} = \sum_{j=2}^{3} a_{1j} + \sum_{j=3}^{3} a_{2j} = (a_{12} + a_{13}) + a_{23},$$

$$\sum_{j=2}^{3} \sum_{i=1}^{j-1} a_{ij} = \sum_{i=1}^{1} a_{i2} + \sum_{i=1}^{2} a_{i3} = a_{12} + (a_{13} + a_{23}).$$

Theorem 4. (Associative Law for Multiplication of Matrices.) *Suppose that A, B, and C are $m \times n$, $n \times p$, and $p \times q$ matrices, respectively. Then*

$$(AB)C = A(BC).$$

Proof. We define $D = AB$ and $F = DC = (AB)C$. Also, we define $E = BC$ and $G = AE = A(BC)$. We must show that $F = G$. Using lower-case letters for the elements of the matrix with the corresponding capital letters, we have

$$d_{ij} = \sum_{k=1}^{n} a_{ik} b_{kj}, \qquad f_{ij} = \sum_{l=1}^{p} d_{il} c_{lj}.$$

We substitute the expression for d_{ij} on the left into that for f_{ij} (changing subscripts in the process), and we find

$$f_{ij} = \sum_{l=1}^{p} \sum_{k=1}^{n} a_{ik} b_{kl} c_{lj}.$$

Similarly, we may write

$$e_{ij} = \sum_{l=1}^{p} b_{il} c_{lj}, \qquad g_{ij} = \sum_{k=1}^{n} a_{ik} e_{kj}.$$

A straight substitution shows that the expressions for f_{ij} and g_{ij} are identical.

EXERCISES

In each of problems 1 through 6, compute the quantities $(AB)^t$ and $B^t A^t$ for the given matrices A and B. Verify that the expressions are equal.

1. $A = \begin{pmatrix} 2 & 3 & 1 \\ -1 & 0 & 2 \end{pmatrix}, \quad B = \begin{pmatrix} 1 & 1 \\ 0 & 1 \\ 1 & 1 \end{pmatrix}$

2. $A = \begin{pmatrix} 3 & 1 \\ 2 & -1 \end{pmatrix}, \quad B = \begin{pmatrix} 3 & 2 \\ -1 & -2 \end{pmatrix}$

3. $A = \begin{pmatrix} 1 & 2 & -1 \\ 3 & -2 & 1 \end{pmatrix}$, $B = \begin{pmatrix} 2 & 1 & 0 \\ -1 & 1 & 2 \\ 0 & 1 & 0 \end{pmatrix}$

4. $A = \begin{pmatrix} 4 \\ 1 \\ 2 \\ 3 \end{pmatrix}$, $B = \begin{pmatrix} 2 & 5 & -1 & 4 \end{pmatrix}$

5. $A = \begin{pmatrix} 3 & -1 & 2 & 5 \\ 4 & 1 & 2 & 4 \\ -1 & 0 & 2 & 0 \\ 6 & 1 & 5 & 4 \end{pmatrix}$, $B = \begin{pmatrix} 3 & -2 & 1 & 4 \\ -2 & 0 & 5 & -2 \\ 1 & 5 & 1 & 3 \\ 4 & -2 & 3 & 2 \end{pmatrix}$

6. $A = \begin{pmatrix} 3 & 2 & -1 & 4 \\ 1 & 6 & 8 & 5 \end{pmatrix}$, $B = \begin{pmatrix} 4 & 2 \\ 1 & 6 \\ 8 & 5 \\ -1 & 0 \end{pmatrix}$

7. Show that $IA = A$ if I is the $m \times m$ identity and A is any $m \times n$ matrix.

8. Prove part (a) of Theorem 2.

9. Prove part (a) of Theorem 3.

10. Prove part (b) of Theorem 3.

11. Verify formula (1) of this section for $m = 2$ and $n = 3$.

12. Verify formula (1) of this section for $m = 4$ and $n = 3$.

13. Verify the formula

$$\left(\sum_{i=1}^{m} a_i \right) \left(\sum_{j=1}^{n} b_j \right) = \sum_{(i,j)\in S} a_i b_j, \qquad S: 1 \le i \le m; \quad 1 \le j \le n$$

for $m = 3$ and $n = 4$.

14. Verify the formula

$$\sum_{1 \le i < j \le n} a_{ij} = \sum_{i=1}^{n-1} \sum_{j=i+1}^{n} a_{ij}$$

for $n = 4$.

15. Verify the formula

$$\sum_{1 \le i < j \le n} a_{ij} = \sum_{j=2}^{n} \sum_{i=1}^{j-1} a_{ij}$$

for $n = 5$.

In each of problems 16 through 18, write out and evaluate the following double sums.

16. $\displaystyle\sum_{(i,j)\in S} (i + j)$, $\quad S: 1 \le i \le 3, \quad 1 \le j \le 4$.

17. $\displaystyle\sum_{(i,j)\in S} ij$, $\quad S = $ all (i, j) with $1 \le i \le 4, 1 \le j \le 4, j \le i$.

18. $\displaystyle\sum_{(i,j)\in S} (3i + 2j)$, $\quad S = $ all (i, j) with $0 \le i \le 4, 0 \le j \le 4, 0 \le i + j \le 4$.

In each of problems 19 through 23, write out and evaluate the given iterated sums.

19. $\displaystyle\sum_{i=1}^{4}\sum_{j=1}^{3}(i-j)$

20. $\displaystyle\sum_{i=1}^{4}\sum_{j=1}^{5-i}(i^2 j)$

21. $\displaystyle\sum_{i=0}^{3}\sum_{j=i+1}^{i+3}(i+j+1)$

22. $\displaystyle\sum_{i=-1}^{4}\sum_{j=0}^{5}\frac{i}{j+2}$

23. $\displaystyle\sum_{i=0}^{3}\sum_{j=0}^{i+1} ij$

24. Write the iterated sum in Example 1 as an iterated sum in the other order (i.e., with j in the "outside" summation).

25. Write the double sum in problem 18 as an iterated sum with j in the outside summation.

26. In the sum in problem 18, let $p = i + j$ be the index in the outer sum and verify that

$$\sum_{(i,j)\in S}(3i+2j) = \sum_{p=0}^{4}\sum_{j=0}^{p}[3(p-j)+2j] = \sum_{p=0}^{4}\sum_{j=0}^{p}(3p-j).$$

4. DETERMINANTS

With each $n \times n$ square matrix we associate a number called its **determinant.** If the matrix is A, we denote its determinant by **det** A. If the matrix A is written out as a square array, its determinant is denoted by the same array between vertical bars. For example, if

$$A = \begin{pmatrix} 2 & -1 & 3 \\ 3 & 1 & 2 \\ -1 & 2 & -3 \end{pmatrix}, \quad\text{then}\quad \det A = \begin{vmatrix} 2 & -1 & 3 \\ 3 & 1 & 2 \\ -1 & 2 & -3 \end{vmatrix}$$

DEFINITIONS. The **order** of an $n \times n$ square matrix is the integer n. A **submatrix** of a given matrix is any matrix obtained by deleting certain rows and columns from the original matrix and consolidating the remaining elements.

For example, the matrix

$$\begin{pmatrix} 2 & 0 & 1 & 4 \\ -3 & \frac{1}{2} & 2 & 5 \\ \sqrt{6} & 7 & 8 & 1 \\ 0 & 5 & 4 & -6 \end{pmatrix}$$

has the submatrix

$$\begin{pmatrix} 2 & 0 & 4 \\ \sqrt{6} & 7 & 1 \\ 0 & 5 & -6 \end{pmatrix},$$

obtained by deleting the second row and third column. Another submatrix, obtained by striking out the third and fourth rows and the fourth column, is

$$\begin{pmatrix} 2 & 0 & 1 \\ -3 & \frac{1}{2} & 2 \end{pmatrix}$$

A *determinant* is a function the domain of which is the collection of square matrices; its range is the real number system. To define the determinant of a matrix, we proceed inductively with respect to its order n. For $n = 1$, we define

$$\det (a_{11}) = a_{11}.$$

Assuming, for $n > 1$, that we have defined determinants of order $\leq n - 1$, we define those of order n by the formula

$$\det A \equiv \begin{vmatrix} a_{11} & a_{12} & \cdots & a_{1n} \\ a_{21} & a_{22} & \cdots & a_{2n} \\ \vdots & & & \vdots \\ a_{n1} & a_{n2} & \cdots & a_{nn} \end{vmatrix} = \sum_{i=1}^{n} (-1)^{i+n} a_{in} M_{in}. \tag{1}$$

In this formula M_{in} is the determinant of the $(n - 1) \times (n - 1)$ submatrix of A, obtained by deleting its ith row and nth column. The determinant M_{in} is called the **minor** of the element a_{in}.

To illustrate formula (1), we obtain for $n = 2$

$$\begin{vmatrix} a_{11} & a_{12} \\ a_{21} & a_{22} \end{vmatrix} = (-1)^{1+2} a_{12} a_{21} + (-1)^{2+2} a_{22} a_{11} = a_{11} a_{22} - a_{12} a_{21};$$

for $n = 3$,

$$\begin{vmatrix} a_{11} & a_{12} & a_{13} \\ a_{21} & a_{22} & a_{23} \\ a_{31} & a_{32} & a_{33} \end{vmatrix} = (-1)^{1+3} a_{13} \begin{vmatrix} a_{21} & a_{22} \\ a_{31} & a_{32} \end{vmatrix} + (-1)^{2+3} a_{23} \begin{vmatrix} a_{11} & a_{12} \\ a_{31} & a_{32} \end{vmatrix}$$

$$+ (-1)^{3+3} a_{33} \begin{vmatrix} a_{11} & a_{12} \\ a_{21} & a_{22} \end{vmatrix}$$

$$= a_{13}(a_{21} a_{32} - a_{31} a_{22}) - a_{23}(a_{11} a_{32} - a_{31} a_{12})$$

$$+ a_{33}(a_{11} a_{22} - a_{21} a_{12}).$$

In formula (1) it is convenient to consolidate the quantity $(-1)^{i+n}$ and the minor M_{in}. We define the **cofactor** A_{ij} of the element a_{ij} in $\det A$ of (1) by the relation

$$A_{ij} = (-1)^{i+j} M_{ij}. \tag{2}$$

In terms of cofactors we obtain the basic expansion theorem for determinants:

Theorem 5. *If A is an $n \times n$ matrix with $n \geq 2$, then*

(a) $\det A = \displaystyle\sum_{i=1}^{n} a_{ij} A_{ij}$ *for each fixed j, $1 \leq j \leq n$;*

$$\text{(b) } \det A = \sum_{j=1}^{n} a_{ij}A_{ij} \text{ for each fixed } i, \quad 1 \leq i \leq n.$$

We postpone the proof until the next section.

The formula in (a) is called the **expansion of det A according to its jth column;** that in (b) is called the **expansion of det A according to its ith row.** *From* (2) *it follows that the signs preceding M_{ij} alternate as one proceeds along a row or column, so that* (2) *is needed only to get the first sign correct.*

EXAMPLE 1. Evaluate the following determinant by expanding it according to its second column and then evaluating the 2×2 determinants

$$\det A = \begin{vmatrix} 1 & -2 & 3 \\ 2 & 1 & -1 \\ -2 & -1 & 2 \end{vmatrix}$$

Solution. Expanding according to the second column, we have

$$\det A = -(-2)\begin{vmatrix} 2 & -1 \\ -2 & 2 \end{vmatrix} + 1\begin{vmatrix} 1 & 3 \\ -2 & 2 \end{vmatrix} - (-1)\begin{vmatrix} 1 & 3 \\ 2 & -1 \end{vmatrix}$$

$$= 2(4 - 2) + 1(2 + 6) + 1(-1 - 6) = 5.$$

EXAMPLE 2. Write out the expansion of the determinant

$$\det A = \begin{vmatrix} -1 & 2 & 3 & -4 \\ 4 & 2 & 0 & 1 \\ -1 & 1 & 2 & 3 \\ -5 & 1 & 6 & 2 \end{vmatrix}$$

according to the fourth column. Do not evaluate.

Solution.

$$\det A = -(-4)\begin{vmatrix} 4 & 2 & 0 \\ -1 & 1 & 2 \\ -5 & 1 & 6 \end{vmatrix} + 1\begin{vmatrix} -1 & 2 & 3 \\ -1 & 1 & 2 \\ -5 & 1 & 6 \end{vmatrix} - 3\begin{vmatrix} -1 & 2 & 3 \\ 4 & 2 & 0 \\ -5 & 1 & 6 \end{vmatrix}$$

$$+ 2\begin{vmatrix} -1 & 2 & 3 \\ 4 & 2 & 0 \\ -1 & 1 & 2 \end{vmatrix}$$

EXERCISES

In each of problems 1 through 4, evaluate the given determinant by expanding it according to (a) the second row, and (b) the third column.

1. $\begin{vmatrix} 3 & 2 & -1 \\ -1 & 0 & 1 \\ 2 & 1 & -2 \end{vmatrix}$

2. $\begin{vmatrix} 1 & 0 & 3 \\ 2 & -1 & -2 \\ 1 & 3 & 2 \end{vmatrix}$

3. $\begin{vmatrix} 2 & 3 & 1 \\ 1 & 2 & -2 \\ -2 & 1 & 3 \end{vmatrix}$

4. $\begin{vmatrix} 1 & \frac{1}{2} & 5 \\ 2 & 1 & 0 \\ 3 & -6 & \frac{1}{3} \end{vmatrix}$

In problems 5 through 7, expand in each case according to the second row. Do not evaluate.

5. $\begin{vmatrix} 2 & 0 & 2 & 3 \\ -1 & 3 & 6 & -1 \\ 1 & -1 & -2 & 4 \\ 0 & 4 & 8 & 2 \end{vmatrix}$

6. $\begin{vmatrix} 0 & 1 & -1 & 2 \\ 4 & 4 & -4 & 2 \\ 3 & -1 & 2 & 3 \\ 1 & -2 & 3 & 4 \end{vmatrix}$

7. $\begin{vmatrix} 8 & 0 & 2 & 5 \\ 6 & -1 & -1 & 4 \\ 0 & 2 & 5 & 1 \\ 4 & 4 & 0 & 0 \end{vmatrix}$

In each of problems 8 through 10, expand according to the fourth column. Do not evaluate.

8. $\begin{vmatrix} 6 & 1 & 2 & 0 \\ 0 & 1 & -1 & -1 \\ -1 & -1 & 3 & 2 \\ 0 & 2 & 5 & 1 \end{vmatrix}$

9. $\begin{vmatrix} -1 & 1 & 6 & -3 & -5 \\ -2 & 2 & 0 & -1 & 6 \\ 1 & 0 & 6 & 2 & 1 \\ -3 & 3 & 2 & 4 & 1 \\ 3 & 1 & 4 & 1 & 2 \end{vmatrix}$

10. $\begin{vmatrix} 3 & -1 & 2 & 4 & 6 \\ 7 & 8 & 2 & 0 & 5 \\ 4 & -1 & -2 & 0 & \frac{1}{2} \\ 6 & 2 & 8 & 5 & 1 \\ 7 & 4 & 3 & 0 & 5 \end{vmatrix}$

11. Prove the identity

$$\begin{vmatrix} 1 & a & a^2 \\ 1 & b & b^2 \\ 1 & c & c^2 \end{vmatrix} = (b - a)(c - a)(c - b).$$

12. Show that in the plane the equation of a line through the points (x_0, y_0) and (x_1, y_1) is given by

$$\begin{vmatrix} x & y & 1 \\ x_0 & y_0 & 1 \\ x_1 & y_1 & 1 \end{vmatrix} = 0.$$

13. Expand a general 4×4 determinant (a) according to the third column, and (b) according to the second row. Do not evaluate.

14. Show that if $\mathbf{u} = a_1\mathbf{i} + a_2\mathbf{j} + a_3\mathbf{k}$ and $\mathbf{v} = b_1\mathbf{i} + b_2\mathbf{j} + b_3\mathbf{k}$ then (formally)

$$\mathbf{u} \times \mathbf{v} = \begin{vmatrix} \mathbf{i} & \mathbf{j} & \mathbf{k} \\ a_1 & a_2 & a_3 \\ b_1 & b_2 & b_3 \end{vmatrix}$$

5. PROPERTIES OF DETERMINANTS

The evaluation of determinants of high order by expansion in rows or columns is a tedious and lengthy process. We now establish a number of important theorems which lead to rapid methods of evaluating determinants.

Theorem 6. *If A is an $n \times n$ matrix, then*

$$\det A^t = \det A.$$

Proof. We proceed by induction on the order n. For $n = 1$ the result is obvious. Now let $n > 1$ and denote $B = A^t$. Assume that the result holds for all matrices of order $\leq n - 1$. We wish to show that it holds for a matrix of order n. Using lower-case letters in the usual way, we write

$$b_{ij} = a^t_{ij} = a_{ji}.$$

The cofactor of b_{ij} is denoted B_{ij} and the cofactor of a_{ji} is the determinant A_{ji}. The determinants A_{ji} and B_{ij} come from two $(n - 1) \times (n - 1)$ matrices, each of which is the transpose of the other. According to the induction hypothesis,

$$B_{ij} = A_{ji}.$$

Expanding $\det B$ according to its ith row, we obtain

$$\det B = \sum_{j=1}^{n} b_{ij}B_{ij} = \sum_{j=1}^{n} a_{ji}A_{ji} = \det A.$$

Theorem 7. (a) *If all the elements in the kth row or kth column $(1 \leq k \leq n)$ of a matrix A are zero, then $\det A = 0$.*
(b) *If a matrix A' is obtained from A by multiplying the elements of the kth row or column by a constant c, then*

$$\det A' = c \det A.$$

(c) *If each element a_{kj} of the kth row of a matrix A equals $a'_{kj} + a''_{kj}$, then*

$$\det A = \det A' + \det A'',$$

where A' and A'' are obtained from A by replacing a_{kj} by a'_{kj} and by a''_{kj}, respectively. The analogous result holds for the kth column.

Proof. (a) Expanding according to the kth row (or kth column), we introduce the factor zero in each term.
(b) Expanding according to the kth row (or kth column), we introduce the factor c in each term of the expansion.
(c) Expanding according to the kth row and setting $a_{kj} = a'_{kj} + a''_{kj}$, we get expansions of A' and A'' in terms of the kth row.

As an example of part (c) of the theorem, we have

$$
\begin{vmatrix} a_{11} & a_{12} & a_{13} \\ a_{21} & a_{22} & a_{23} \\ a'_{31} + a''_{31} & a'_{32} + a''_{32} & a'_{33} + a''_{33} \end{vmatrix} = \begin{vmatrix} a_{11} & a_{12} & a_{13} \\ a_{21} & a_{22} & a_{23} \\ a'_{31} & a'_{32} & a'_{33} \end{vmatrix} + \begin{vmatrix} a_{11} & a_{12} & a_{13} \\ a_{21} & a_{22} & a_{23} \\ a''_{31} & a''_{32} & a''_{33} \end{vmatrix}
$$

Theorem 8. (a) *If A' is obtained from A by interchanging two rows or two columns, then*

$$\det A' = -\det A.$$

(b) *If two rows or two columns of A are proportional, then* $\det A = 0$.

Proof. (a) We proceed by induction on n. If $n = 1$, there is nothing to prove. For $n = 2$, the result follows at once by inspection of the formula

$$
\begin{vmatrix} a_{11} & a_{12} \\ a_{21} & a_{22} \end{vmatrix} = a_{11}a_{22} - a_{21}a_{12}.
$$

We let $n > 2$, suppose that two rows are interchanged, and assume that the result holds for $n - 1$. Now we expand A according to the ith row, where the ith row is *not* one of those being interchanged. Then $a'_{ij} = a_{ij}$, and each cofactor A'_{ij} is obtained from A_{ij} by interchanging two rows. Invoking the induction hypothesis, we have

$$A'_{ij} = -A_{ij}.$$

Therefore

$$\det A' = \sum_{j=1}^{n} a'_{ij}A'_{ij} = -\sum_{j=1}^{n} a_{ij}A_{ij} = -\det A.$$

The proof is identical for the case when two columns are interchanged.

(b) If two rows (or columns) are proportional, then either one row consists of zeros so that the determinant is zero, or else one row (or column) is a constant c times the other row (or column). Using part (b) of Theorem 7, we see that the determinant $D = cD'$ where D' has two identical rows (or columns). Interchanging the two identical rows and employing part (a) (which was just established), we find $D' = -D'$, so that D' is zero; therefore $D = 0$.

By combining part (c) of Theorem 7 and part (b) of Theorem 8, we obtain the next extremely useful result.

Theorem 9. *If A' is obtained from A by multiplying the kth row by the constant c and adding the result to the ith row where $i \neq k$, then*

$$\det A' = \det A.$$

The same result holds for two columns.

Proof. The element a'_{ij} of A' is of the form $a_{ij} + ca_{kj}$, so that [by Theorem 7(c)]

$$\det A' = \det A + c \det A'',$$

where A'' is obtained from A by replacing the ith row by the kth row. But then the ith and kth rows of A'' are identical, so that $\det A'' = 0$. The same proof holds for columns.

Corollary. *If A' is obtained from A by multiplying the kth row by c_i and adding the result to the ith row for $i = 1, 2, \ldots, k - 1, k + 1, \ldots, n$ in turn, then $\det A' = \det A$. The same is true for columns.*

Proof. Each step of the process is one for which Theorem 9 applies and which leaves $\det A$ unchanged. Therefore a succession of such steps will not alter the determinant.

The next example shows how to use the results of this section to simplify and evaluate a determinant.

———

EXAMPLE 1. Simplify by using the corollary, and use the expansion theorem to evaluate the determinant

$$D = \begin{vmatrix} 2 & -1 & 1 & 0 \\ -3 & 0 & 1 & -2 \\ 1 & 1 & -1 & 1 \\ 2 & -1 & 5 & -1 \end{vmatrix}$$

Solution. By adding the third row to the first and fourth rows in turn, we find that

$$D = \begin{vmatrix} 3 & 0 & 0 & 1 \\ -3 & 0 & 1 & -2 \\ 1 & 1 & -1 & 1 \\ 3 & 0 & 4 & 0 \end{vmatrix}$$

Expanding according to the second column, we obtain

$$D = (-1) \begin{vmatrix} 3 & 0 & 1 \\ -3 & 1 & -2 \\ 3 & 4 & 0 \end{vmatrix}$$

Multiplying the first row by 2 and adding the result to the second row, we get

$$D = - \begin{vmatrix} 3 & 0 & 1 \\ 3 & 1 & 0 \\ 3 & 4 & 0 \end{vmatrix} = - \begin{vmatrix} 3 & 1 \\ 3 & 4 \end{vmatrix} = -9.$$

The 2×2 determinant was obtained by expansion according to the third column.

EXAMPLE 2. Show that

$$D = \begin{vmatrix} x & x^2 & x^3 \\ y & y^2 & y^3 \\ z & z^2 & z^3 \end{vmatrix} = xyz(y - x)(z - y)(z - x).$$

Solution. We may factor out an x from the first row, a y from the second row, and a z from the third row, to obtain

$$D = xyz \begin{vmatrix} 1 & x & x^2 \\ 1 & y & y^2 \\ 1 & z & z^2 \end{vmatrix}$$

Subtracting the third row from the first and second in turn, and then expanding in terms of the first column, we get

$$D = xyz \begin{vmatrix} 0 & x-z & x^2-z^2 \\ 0 & y-z & y^2-z^2 \\ 1 & z & z^2 \end{vmatrix} = xyz \begin{vmatrix} x-z & x^2-z^2 \\ y-z & y^2-z^2 \end{vmatrix}$$

$$= xyz(x-z)(y-z) \begin{vmatrix} 1 & x+z \\ 1 & y+z \end{vmatrix} = xyz(y-x)(z-x)(z-y).$$

The next theorem is employed in Section 6.

Theorem 10. *For any square matrix A, we have*

$$\text{(a)} \quad \sum_{i=1}^{n} a_{ij}A_{ik} = \delta_{jk} (\det A),$$

$$\text{(b)} \quad \sum_{k=1}^{n} a_{ik}A_{jk} = \delta_{ij} (\det A).$$

Proof. (a) If $j = k$, then $\delta_{jk} = 1$ and the formula (a) is the statement of the expansion theorem (Theorem 5) according to the kth column. If $j \neq k$, then $\delta_{jk} = 0$ and the right side of (a) is zero. As for the left side of (a), we introduce the matrix A' obtained from A by replacing the kth column with the jth column. Then the left side of (a) is the expansion of A' according to the jth column. But since A' has two columns alike (jth and kth), we have $\det A' = 0$, and so the left side of (a) vanishes. The proof of (b) is the same.

We conclude this section with a proof of Theorem 5, stated in Section 4.

Proof of Theorem 5. Our first aim is to establish the formula

$$\det A = \sum_{i=1}^{n} a_{ij}A_{ij} \quad \text{for each fixed } j, \quad 1 \leq j \leq n. \tag{1}$$

If $j = n$, the above formula is just the definition of $\det A$. It must be shown that all other expansions, $1 \leq j \leq n - 1$, yield the same result.

We proceed by induction on n. If $n = 1$, there is nothing to prove. If $n = 2$, the result is easily verified by inspection of the expansion of a 2×2 matrix. So we let $n > 2$ and assume the result is true for all determinants of order $\leq n - 1$.

Choose j, $1 \leq j \leq n - 1$, and let $M_{ik,jn}$ denote the determinant formed from A by deleting its ith and kth rows and its jth and nth columns. $M_{ik,jn}$ is the determinant of an $(n - 2) \times (n - 2)$ matrix. Recalling that M_{in} is the minor of the element a_{in}, we obtain from the definition of determinant

$$D \equiv \det A = \sum_{i=1}^{n} (-1)^{i+n} a_{in} M_{in}. \tag{2}$$

Now, using our induction hypothesis, we may expand each M_{in} according to its jth column. The determinant of A is shown below with lines through the ith row and nth column so that the remaining terms form M_{in}.

$$\begin{vmatrix} a_{11} & a_{12} & \cdots & a_{1j} & \cdots & a_{1n} \\ \vdots & & & & & \\ a_{k1} & a_{k2} & \cdots & a_{kj} & \cdots & a_{kn} \\ \vdots & & & & & \\ a_{i1} & a_{i2} & & a_{ij} & & a_{in} \\ \vdots & & & & & \\ a_{n1} & a_{n2} & \cdots & a_{nj} & \cdots & a_{nn} \end{vmatrix}$$

If $k < i$ (as shown above), then a_{kp} is in the kth row of M_{in} but, if $k > i$, then each a_{kp} is in the $(k - 1)$st row of M_{in}. Therefore, if $1 < i < n$, we have the expansion

$$M_{in} = \sum_{k=1}^{i-1} (-1)^{k+j} a_{kj} M_{ik,jn} + \sum_{k=i+1}^{n} (-1)^{k-1+j} a_{kj} M_{ik,jn}. \tag{3}$$

If $i = 1$, the first sum in (3) is missing while, if $i = n$, the second sum is absent. Substituting M_{in} from (3) into (2), we see that

$$D = \sum_{i=2}^{n} \sum_{k=1}^{i-1} (-1)^{i+k+j+n} a_{in} a_{kj} M_{ik,jn} - \sum_{i=1}^{n-1} \sum_{k=i+1}^{n} (-1)^{i+k+j+n} a_{in} a_{kj} M_{ik,jn}. \tag{4}$$

If we interchange the indices i and k in the first sum and make use of the double-sum notation of Section 3, we can combine the two iterated sums in (4) into the single double sum

$$D = \sum_{1 \le i < k \le n} (-1)^{i+k+j+n} (a_{kn} a_{ij} - a_{in} a_{kj}) M_{ik,jn}. \tag{5}$$

The relation $M_{ik,jn} = M_{ki,jn}$, which follows directly from the definition of these four-subscript determinants, was used to obtain (5).

To establish (1), it is sufficient to show that the expansion

$$\sum_{k=1}^{n} (-1)^{k+j} a_{kj} M_{kj} \tag{6}$$

is equal to (5). To do so, we expand M_{kj} according to its $(n - 1)$st column, noting that the $(n - 1)$st column of M_{kj} is the nth column of D with a_{kn} removed. If $i < k$, then the elements a_{ip} are in the ith row of M_{kj} but, if $i > k$, then the a_{ip} are in the $(i - 1)$st row of M_{kj}. If we then write out the expansion of M_{kj}, collect terms, and substitute in (6), we find an expression which is identical with (5).

The proof of the formula

$$\det A = \sum_{k=1}^{n} a_{ik} A_{ik} \quad \text{for each fixed } i, \quad 1 \leq i \leq n,$$

is obtained by first showing that an expansion according to the nth row is equal to the expansion according to the nth column. This proof is omitted. Then the proof that the expansion according to two different rows yields the same result is analogous to the proof for columns.

EXERCISES

In each of problems 1 through 12, simplify and evaluate the determinant.

1. $\begin{vmatrix} 2 & -1 & 3 \\ 2 & -1 & 1 \\ 1 & 3 & -2 \end{vmatrix}$

2. $\begin{vmatrix} 3 & 1 & -1 \\ 1 & 3 & -2 \\ -2 & 1 & 3 \end{vmatrix}$

3. $\begin{vmatrix} 1 & 2 & 3 \\ 1 & -1 & 1 \\ 2 & 4 & -1 \end{vmatrix}$

4. $\begin{vmatrix} 5 & 0 & -1 & 1 \\ 1 & -1 & 1 & -2 \\ 4 & -4 & 2 & 1 \\ -2 & 1 & 1 & 0 \end{vmatrix}$

5. $\begin{vmatrix} 2 & 3 & 6 & 3 \\ 0 & 1 & 3 & 1 \\ -1 & -2 & 0 & 4 \\ 1 & 2 & 4 & -1 \end{vmatrix}$

6. $\begin{vmatrix} 2 & -1 & 3 & 2 \\ 1 & -2 & -2 & 3 \\ -1 & 3 & 2 & 0 \\ 3 & 2 & -2 & 1 \end{vmatrix}$

7. $\begin{vmatrix} -2 & 3 & 2 & 4 \\ -3 & 1 & -2 & 3 \\ 2 & 2 & 3 & -2 \\ 4 & -3 & 2 & 1 \end{vmatrix}$

8. $\begin{vmatrix} 1 & 2 & -3 & 4 \\ 3 & -4 & 2 & -1 \\ 2 & -2 & 3 & 4 \\ 1 & 2 & -2 & 3 \end{vmatrix}$

9. $\begin{vmatrix} 3 & 2 & -1 & 4 \\ 2 & -3 & 4 & 1 \\ -4 & 2 & 0 & 3 \\ 2 & 4 & -1 & 2 \end{vmatrix}$

10. $\begin{vmatrix} 2 & -1 & 3 & 4 & 0 \\ -1 & 2 & 1 & 0 & -1 \\ -3 & 0 & 0 & 1 & 0 \\ 0 & 1 & 0 & -1 & 2 \\ 0 & -2 & 0 & 2 & -1 \end{vmatrix}$

11. $\begin{vmatrix} 1 & 2 & 0 & -1 & 2 \\ 2 & 3 & -1 & 0 & 1 \\ 0 & -1 & 2 & 4 & -2 \\ -1 & 0 & 4 & -1 & 0 \\ 1 & 2 & -1 & 0 & 1 \end{vmatrix}$

12. $\begin{vmatrix} 2 & -1 & 0 & 4 & 1 & -3 \\ 2 & 1 & 2 & 1 & 3 & 2 \\ 4 & -1 & 0 & 2 & -2 & 3 \\ 1 & 5 & 4 & 0 & 2 & 0 \\ 6 & 2 & 1 & 4 & -3 & 0 \\ -1 & 2 & 5 & -3 & 4 & 2 \end{vmatrix}$

13. Show that

$$\begin{vmatrix} 1 & x & x^2 & x^3 \\ 1 & y & y^2 & y^3 \\ 1 & z & z^2 & z^3 \\ 1 & w & w^2 & w^3 \end{vmatrix} = (x - y)(y - z)(z - w)(x - z)(x - w)(y - w).$$

[*Hint:* Use the relation $a^3 - b^3 = (a - b)(a^2 + ab + b^2)$.]

14. Show that the equation of a plane through the three points $P_0(x_0, y_0, z_0)$, $P_1(x_1, y_1, z_1)$, and $P_2(x_2, y_2, z_2)$ is given by

$$\begin{vmatrix} 1 & x & y & z \\ 1 & x_0 & y_0 & z_0 \\ 1 & x_1 & y_1 & z_1 \\ 1 & x_2 & y_2 & z_2 \end{vmatrix} = 0.$$

15. Given the matrix

$$A = \begin{pmatrix} 3 & -1 & 2 \\ 1 & 4 & -1 \\ 2 & 0 & 5 \end{pmatrix},$$

verify Theorem 10 by computing

$$a_{11}A_{13} + a_{21}A_{23} + a_{31}A_{33},$$

and showing that the result vanishes.

16. Given the matrices

$$A = \begin{pmatrix} 1 & 2 & -1 \\ 3 & 1 & 1 \\ -2 & 0 & 5 \end{pmatrix}, \quad B = \begin{pmatrix} 3 & -1 & 2 \\ 4 & 0 & 1 \\ -2 & 1 & 5 \end{pmatrix}$$

Show that det $(AB) = $ (det A) (det B).

6. CRAMER'S RULE

With the aid of determinants, we are able to give a formula for the solution of n simultaneous linear equations in n unknowns. The resulting theorem is known as **Cramer's Rule.**

Theorem 11. *If* det $A \neq 0$, *the system of equations*

$$\begin{aligned} a_{11}x_1 + a_{12}x_2 + \cdots + a_{1n}x_n &= b_1 \\ a_{21}x_1 + a_{22}x_2 + \cdots + a_{2n}x_n &= b_2 \\ &\vdots \\ a_{n1}x_1 + a_{n2}x_2 + \cdots + a_{nn}x_n &= b_n \end{aligned} \qquad (1)$$

has a unique solution given by

$$x_k = \frac{1}{D} \begin{vmatrix} a_{11} & \cdots & a_{1,k-1} & b_1 & a_{1,k+1} & \cdots & a_{1n} \\ \vdots & & & & & & \vdots \\ a_{n1} & \cdots & a_{n,k-1} & b_n & a_{n,k+1} & \cdots & a_{nn} \end{vmatrix}, \quad \begin{aligned} D &= \det A, \\ k &= 1, 2, \ldots, n, \end{aligned} \qquad (2)$$

where, if $k = 1$ *or* n, *the column of* b's *is in the first or nth column, respectively.*

Proof. We must show that (1) implies (2) and that (2) implies (1). Suppose first that $x_1, x_2, \ldots, x_n$ are numbers which satisfy (1). Multiply the ith equation of (1) by A_{ik}, the cofactor of a_{ik}, where k is a fixed number between 1 and n. We get

$$\sum_{j=1}^{n} a_{ij}A_{ik}x_j = b_iA_{ik}.$$

Now we add all such equations; that is, we sum on the index i. After interchanging the order of the iterated sum, we obtain

$$\sum_{j=1}^{n} \left[\sum_{i=1}^{n} a_{ij}A_{ik} \right] x_j = \sum_{i=1}^{n} b_iA_{ik}. \tag{3}$$

The sum on the right in (3) is the expansion according to the kth column of the determinant displayed in (2). As for the left side of (3), the quantity in brackets is precisely the expression which appears in Theorem 10 of the last section and is equal to δ_{jk} (det A). Therefore, using the property of the Kronecker δ, we find that

$$\sum_{j=1}^{n} \delta_{jk} (\text{det } A)x_j = Dx_k = \sum_{i=1}^{n} b_iA_{ik},$$

and the formula (2) follows.

If we assume formula (2) holds, we see that

$$Dx_k = \sum_{i=1}^{n} b_iA_{ik}.$$

We multiply this expression by a_{jk} and sum with respect to k to get (after interchanging the order of summation)

$$D \sum_{k=1}^{n} a_{jk}x_k = \sum_{i=1}^{n} \left[\sum_{k=1}^{n} a_{jk}A_{ik} \right] b_i.$$

Using Theorem 10 of Section 5 again, we obtain

$$D \sum_{k=1}^{n} a_{jk}x_k = \sum_{i=1}^{n} \delta_{ij} (\text{det } A)b_i = Db_j,$$

from which (1) follows.

EXAMPLE. Solve, using Cramer's Rule,

$$3x_1 - 2x_2 + 4x_3 = 5,$$
$$x_1 + x_2 + 3x_3 = 2,$$
$$-x_1 + 2x_2 - x_3 = 1.$$

Solution. We have

$$D = \begin{vmatrix} 3 & -2 & 4 \\ 1 & 1 & 3 \\ -1 & 2 & -1 \end{vmatrix} = \begin{vmatrix} 5 & -2 & 10 \\ 0 & 1 & 0 \\ -3 & 2 & -7 \end{vmatrix} = \begin{vmatrix} 5 & 10 \\ -3 & -7 \end{vmatrix} = -5,$$

$$x_1 = -\tfrac{1}{5}\begin{vmatrix} 5 & -2 & 4 \\ 2 & 1 & 3 \\ 1 & 2 & -1 \end{vmatrix} = -\tfrac{1}{5}\begin{vmatrix} 9 & 0 & 10 \\ 2 & 1 & 3 \\ -3 & 0 & -7 \end{vmatrix} = -\tfrac{1}{5}\begin{vmatrix} 9 & 10 \\ -3 & -7 \end{vmatrix} = \tfrac{33}{5}.$$

In a similar way, we find that

$$x_2 = -\tfrac{1}{5}\begin{vmatrix} 3 & 5 & 4 \\ 1 & 2 & 3 \\ -1 & 1 & -1 \end{vmatrix} = \tfrac{13}{5}, \qquad x_3 = -\tfrac{1}{5}\begin{vmatrix} 3 & -2 & 5 \\ 1 & 1 & 2 \\ -1 & 2 & 1 \end{vmatrix} = -\tfrac{12}{5}.$$

Cramer's Rule is useful in many theoretical investigations, since it gives an explicit formula for the solution. However, because of the enormous work of evaluating $n + 1$ determinants when n is large, Cramer's Rule is rather poor for numerical computations. Furthermore, there are many chances for error in the large number of multiplications and additions which must be performed. Techniques for solving linear systems which proceed by iteration and elimination are computationally superior not only because there are fewer arithmetical operations, but also because numerical errors are frequently self-correcting.

EXERCISES

Solve the following systems, using Cramer's Rule.

1. $2x_1 - x_2 + 3x_3 = 1$
$3x_1 + x_2 - x_3 = 2$
$x_1 + 2x_2 + 3x_3 = -6$

2. $2x_1 - x_2 + x_3 = -3$
$x_1 + 3x_2 - 2x_3 = 0$
$x_1 - x_2 + x_3 = -2$

3. $x_1 + 3x_2 - 2x_3 = 4$
$-2x_1 + x_2 + 3x_3 = 2$
$2x_1 + 4x_2 - x_3 = -1$

4. $2x_1 + x_2 - x_3 = 1$
$x_1 - 2x_2 + 3x_3 = 0$
$2x_1 - 3x_2 + 4x_3 = 0$

5. $2x_1 - x_2 + 2x_3 = 11$
$x_1 + 2x_2 - x_3 = -3$
$3x_1 - 2x_2 - 3x_3 = -1$

6. $2x_1 - x_2 - 2x_3 = 0$
$-x_1 + 2x_2 - 3x_3 = 11$
$3x_1 - 2x_2 + 4x_3 = -15$

7. $x_1 - 2x_2 + 2x_3 = -1$
$2x_1 - 3x_2 - 3x_3 = 1$
$3x_1 + x_2 + 2x_3 = 3$

8. $x_1 \qquad + x_3 - 2x_4 = 3$
$x_2 + 2x_3 - x_4 = 2$
$2x_1 + 3x_2 - 2x_3 \qquad = -1$
$x_1 - x_2 \qquad - 4x_4 = 0$

9. $3x_1 + 2x_2 \qquad - 4x_4 = 0$
$x_2 - 2x_3 + x_4 = -1$
$2x_1 + 3x_2 \qquad = 1$
$x_1 \qquad + 4x_3 - 2x_4 = 2$

10. $2x_1 + x_2 - 2x_3 + 3x_4 - 4x_5 = 0$
$4x_1 - x_2 + x_3 - 3x_4 + 2x_5 = 1$
$-2x_1 + x_2 + 2x_3 + 6x_4 - 2x_5 = -2$
$-4x_1 + 3x_2 - 5x_3 - 6x_4 + 4x_5 = 13$
$6x_1 - 3x_2 + 4x_3 + 9x_4 - 6x_5 = -13$

7. THE RANK OF A MATRIX. ELEMENTARY TRANSFORMATIONS

Cramer's Rule applies when the number of equations is the same as the number of unknowns and when the determinant of the coefficients is not zero. In order to treat systems in which the number of equations is different from the number of unknowns, it is necessary to introduce a quantity called the *rank* of a matrix.

DEFINITIONS. A square matrix is said to be **nonsingular** if its determinant is not zero. The **rank** of an $m \times n$ matrix is the largest integer r for which a nonsingular $r \times r$ submatrix exists. The rank of any **matrix of zeros** is zero.

For example, the matrix

$$\begin{pmatrix} 4 & 0 & 0 & 0 \\ 1 & 0 & 3 & 0 \\ 2 & 0 & 0 & 0 \end{pmatrix}$$

is of rank 2, since the 2×2 submatrix

$$\begin{pmatrix} 4 & 0 \\ 1 & 3 \end{pmatrix},$$

obtained by deleting the third row and the second and fourth columns, is nonsingular, and since every 3×3 submatrix has zero determinant.

We note that the rank of an $m \times n$ matrix can never exceed the smaller of the numbers m and n.

DEFINITION. An **elementary transformation** of a matrix is a process of obtaining a second matrix from the given matrix in one of the following ways:
    (a) interchanging two rows or two columns,
    (b) multiplying a row or column by a nonzero constant,
    (c) multiplying one row (or column) by a constant and adding it to another row (or column).

We observe that transformations of type (c) are just those which we used for simplifying and evaluating determinants.

Theorem 12. *If A' is obtained from A by an elementary transformation, the rank of A' equals the rank of A.*

Proof. For transformations of type (a) and (b), the result is immediate from the definition. To prove the result for type (c), let r be the rank of A. If A is an $m \times n$ matrix and r is the smaller of m and n (i.e., the rank of A is as large as possible), then

$$\text{rank } A' \leq \text{rank } A. \tag{1}$$

We show first that (1) holds regardless of the rank of A. Suppose that r is smaller than m and n; let D' be a $k \times k$ submatrix of A' with $k > r$. To be specific, suppose that A' is obtained from A by multiplying the first row of A by c and adding the result to the second row. If D' contains both the first and second rows of A', then det $D' =$ det D, where D is the corresponding matrix in A. But det $D = 0$, since $k > r$ and since A is of rank r. The same result holds if D' contains neither the first nor the second row. If D' contains only the first row, we again have det $D' =$ det D. The only remaining case occurs when D' contains the second row but not the first. But then det D' is a linear combination of two determinants of A of order k. Since all determinants of A of order k are zero, we conclude that det $D' = 0$. Thus the determinant of every $k \times k$ submatrix of A' with $k > r$ is zero, and therefore (1) is established for every rank r. The argument just given works for any two matrices which are related to each other by a transformation of type (c). But since A can be obtained from A' in this manner, we conclude that

$$\text{rank } A \leq \text{rank } A'. \tag{2}$$

Combining (1) and (2), we get rank $A =$ rank A'.

We say that two matrices are **equivalent** if and only if it is possible to pass from one to the other by applying a finite number of elementary transformations. We write $A \cong B$ if A and B are equivalent. From Theorem 12 we conclude that *equivalent matrices have the same rank.*

To compute the rank of an $m \times n$ matrix A directly from the definition, we must evaluate the determinant of every square submatrix of A. If m and n are large, this task is laborious, unless A has many zero entries. However, with the aid of elementary transformations and without an undue amount of computation, we can find a matrix equivalent to A whose rank can be determined by inspection. We first illustrate the technique with an example and then establish the appropriate theorem.

EXAMPLE 1. Determine the rank of the 4×5 matrix

$$A = \begin{pmatrix} -2 & -3 & -1 & 1 & 0 \\ 0 & 1 & 7 & 1 & -4 \\ 1 & 2 & 4 & 0 & -2 \\ -2 & -2 & 6 & 2 & -4 \end{pmatrix}$$

Solution. Interchanging the first and third rows, we find that

$$A \cong \begin{pmatrix} 1 & 2 & 4 & 0 & -2 \\ 0 & 1 & 7 & 1 & -4 \\ -2 & -3 & -1 & 1 & 0 \\ -2 & -2 & 6 & 2 & -4 \end{pmatrix}$$

Multiplying the first row of this new matrix by 2 and adding the result to the third and fourth rows, we obtain

$$A \cong \begin{pmatrix} 1 & 2 & 4 & 0 & -2 \\ 0 & 1 & 7 & 1 & -4 \\ 0 & 1 & 7 & 1 & -4 \\ 0 & 2 & 14 & 2 & -8 \end{pmatrix}$$

Multiplying the second row by 1 and 2 and subtracting the results from the third and fourth rows, respectively, we get

$$A \cong \begin{pmatrix} 1 & 2 & 4 & 0 & -2 \\ 0 & 1 & 7 & 1 & 4 \\ 0 & 0 & 0 & 0 & 0 \\ 0 & 0 & 0 & 0 & 0 \end{pmatrix} = A'.$$

By inspection, A' is seen to have rank 2. Therefore A has rank 2.

We now state a theorem which describes the process developed in the above example.

Theorem 13. *By a succession of elementary transformations of type* (a) *and* (c) *operating on rows only, any $m \times n$ matrix A can be reduced to an equivalent matrix A' in which*

$$a'_{ij} = 0 \quad \text{for} \quad j < j_i, \quad i = 1, 2, \ldots, m,$$

where

$$1 \le j_1 < j_2 < \cdots < j_r \le n,$$
$$j_i = n + 1 \quad \text{for} \quad i = r + 1, \ r + 2, \ldots, m. \tag{3}$$

The rank r is the integer equal to the number of rows which do not consist entirely of zeros. A corresponding result holds for columns.

Before proceeding with the proof, we give an example of the content of (3).

$$\begin{array}{l} j_1 = 1 \\ j_2 = 2 \\ j_3 = 4 \\ j_4 = 5 \\ \vdots \\ j_m = n + 1 \end{array} \begin{pmatrix} 3 & 2 & 3 & \cdots & & & \\ 0 & 1 & 2 & 5 & \cdots & & \\ 0 & 0 & 0 & 2 & -1 & 4 & \cdots \\ 0 & 0 & 0 & 0 & 1 & 3 & \cdots \\ \vdots & \vdots & \vdots & & & & \\ 0 & 0 & 0 & 0 & 0 & \cdots & 0 & 0 \end{pmatrix}$$

In other words, by means of elementary transformations we introduce as many zeros as possible in the bottom row, the second largest number of zeros in the second from bottom row, and so on until, in the first row, we introduce the fewest zeros (or perhaps none).

Proof. If A is the zero matrix, then $j_1 = j_2 = \cdots j_m = n + 1$, and A is already in the desired form. The rank is zero. Otherwise we let the j_1st column be the first column which does not consist entirely of zeros. If a nonzero element occurs in the first column, then $j_1 = 1$. We interchange rows, if necessary, so that the element $a_{1,j_1} \neq 0$. Then, by elementary transformations of type (c), we make the remainder of that column all zeros; of course, all the numbers (if any) to the left of the j_1st column are still 0. If all the numbers below the first row are zero, then $j_2 = j_3 = \cdots = j_m = n + 1$, and the matrix is in the desired form; then $r = 1$. Otherwise, we let the j_2nd column (of course $j_2 > j_1$) be the first one which contains a nonzero element below the first row. By interchanging rows and performing elementary transformations of type (c), we arrange that $a_{2,j_2} \neq 0$ but that all the numbers $a_{i,j_2} = 0$ if $i > 2$; of course, all $a_{ij} = 0$ if $i \geq 2$ and $j < j_2$. If all the a_{ij} with $i > 2$ are zero, then $j_3 = \cdots j_m = n + 1$, and the matrix has been reduced to the desired form with $r = 2$. In any case, the process will stop after some step, say the rth, where $r \leq m$; if $r < m$, all the rows below the rth will consist of zeros. If, now, we select the submatrix of A' consisting of the first r rows and the $j_1, \ldots, j_r$ columns, this $r \times r$ submatrix has the form

$$\begin{pmatrix} a_{1j_1} & * & * & * & \cdots & * \\ 0 & a_{2j_2} & * & * & \cdots & * \\ 0 & 0 & & & & \\ 0 & 0 & & & & \\ \vdots & \vdots & & & & \\ 0 & 0 & \cdots & 0 & & a_{rj_r} \end{pmatrix}$$

The diagonal elements are all nonzero and the asterisks stand for numbers which may or may not be zero. The matrix is triangular and nonsingular. The rank of A is r.

The proof for columns is the same.

Corollary 1. *If A is an $n \times n$ matrix, the process described in the proof of Theorem 13 transforms A to a triangular matrix with all zeros below the diagonal.*

Corollary 2. *If A is a nonsingular $n \times n$ matrix, it can be reduced by means of elementary transformations, as in Theorem 13, to a matrix having nonzero elements along the diagonal and zeros elsewhere.*

Proof. By Corollary 1, we make A into the triangular matrix A'. The value of the determinant is then the product of the diagonal elements; hence all diagonal elements are nonzero. Now, starting with the last column and the element in the lower right-hand corner, we use elementary transformation of type (c) to transform to zero all the elements above the bottom one in the last column. Proceeding to the second-from-last column, we transform to zero all the elements above the diagonal element. Continuing, we get a diagonal matrix which is equivalent to A.

Corollary 3. *The basic process as described in the proof of Theorem 13 leads to the simultaneous reduction to a similar form of each of the submatrices A_l, which are obtained by deleting all but the first l columns of A. The rank of A_l is the integer equal to the number of rows in A'_l which have nonzero elements.*

The proof of the theorem shows that, in principle, elementary transformations of type (b) are not needed to carry out the reduction from A to A'. However, if we have matrices with integers or if, in carrying out the reduction, we wish to avoid arithmetical difficulties, transformations of type (b) are helpful. The next example illustrates the use of type (b) transformations in performing the reduction.

EXAMPLE 2. Given the matrix

$$A = \begin{pmatrix} 3 & 2 & -2 & 3 \\ 2 & 3 & -3 & 4 \\ -2 & 4 & 2 & 3 \\ 5 & -2 & 4 & 2 \\ 3 & 4 & 2 & 3 \end{pmatrix}$$

determine the rank of A, as in Example 1. Use transformations of type (b) when convenient.

Solution. We begin by interchanging the first two rows and then multiplying the second, fourth, and fifth rows by 2. In this way we avoid fractions. The reduction then proceeds as follows:

$$A \cong \begin{pmatrix} 2 & 3 & -3 & 4 \\ 6 & 4 & -4 & 6 \\ -2 & 4 & 2 & 3 \\ 10 & -4 & 8 & 4 \\ 6 & 8 & 4 & 6 \end{pmatrix} \cong \begin{pmatrix} 2 & 3 & -3 & 4 \\ 0 & -5 & 5 & -6 \\ 0 & 7 & -1 & 7 \\ 0 & -19 & 23 & -16 \\ 0 & -1 & 13 & -6 \end{pmatrix} \cong \begin{pmatrix} 2 & 3 & -3 & 4 \\ 0 & -1 & 13 & -6 \\ 0 & -5 & 5 & -6 \\ 0 & 7 & -1 & 7 \\ 0 & -19 & 23 & -16 \end{pmatrix}$$

$$\cong \begin{pmatrix} 2 & 3 & -3 & 4 \\ 0 & -1 & 13 & -6 \\ 0 & 0 & -60 & 24 \\ 0 & 0 & 90 & -35 \\ 0 & 0 & -224 & 98 \end{pmatrix} \cong \begin{pmatrix} 2 & 3 & -3 & 4 \\ 0 & -1 & 13 & -6 \\ 0 & 0 & -5 & 2 \\ 0 & 0 & 18 & -7 \\ 0 & 0 & -16 & 7 \end{pmatrix}$$

Multiplying the third row of the last matrix by 3 and combining with the fourth and fifth rows, we get

$$A \cong \begin{pmatrix} 2 & 3 & -3 & 4 \\ 0 & -1 & 13 & -6 \\ 0 & 0 & -5 & 2 \\ 0 & 0 & 3 & -1 \\ 0 & 0 & -1 & 1 \end{pmatrix} \cong \begin{pmatrix} 2 & 3 & -3 & 4 \\ 0 & -1 & 13 & -6 \\ 0 & 0 & -1 & 1 \\ 0 & 0 & 3 & -1 \\ 0 & 0 & -5 & 2 \end{pmatrix} \cong \begin{pmatrix} 2 & 3 & -3 & 4 \\ 0 & -1 & 13 & -6 \\ 0 & 0 & -1 & 1 \\ 0 & 0 & 0 & 2 \\ 0 & 0 & 0 & -3 \end{pmatrix}$$

Finally, we obtain

$$A \cong \begin{pmatrix} 2 & 3 & -3 & 4 \\ 0 & -1 & 13 & -6 \\ 0 & 0 & -1 & 1 \\ 0 & 0 & 0 & 2 \\ 0 & 0 & 0 & 0 \end{pmatrix} = A',$$

and A is of rank 4.

The next example shows how the reduction proceeds when only elementary transformations involving columns are used.

_____

EXAMPLE 3. Given the matrix

$$A = \begin{pmatrix} 2 & 1 & -1 & -2 \\ 4 & 2 & -2 & -4 \\ -5 & -2 & 3 & 2 \\ 1 & -1 & -2 & 8 \\ 8 & 3 & -2 & 1 \end{pmatrix}$$

reduce A to the form described in Theorem 13, using only elementary transformations involving columns. Find the rank of A.

Solution. We begin by interchanging the first two columns; we then introduce zeros in the first row.

$$A \cong \begin{pmatrix} 1 & 2 & -1 & -2 \\ 2 & 4 & -2 & -4 \\ -2 & -5 & 3 & 2 \\ -1 & 1 & -2 & 8 \\ 3 & 8 & -2 & 1 \end{pmatrix} \cong \begin{pmatrix} 1 & 0 & 0 & 0 \\ 2 & 0 & 0 & 0 \\ -2 & -1 & 1 & -2 \\ -1 & 3 & -3 & 6 \\ 3 & 2 & 1 & 7 \end{pmatrix}$$

$$\cong \begin{pmatrix} 1 & 0 & 0 & 0 \\ 2 & 0 & 0 & 0 \\ -2 & -1 & 0 & 0 \\ -1 & 3 & 0 & 0 \\ 3 & 2 & 3 & 3 \end{pmatrix} \cong \begin{pmatrix} 1 & 0 & 0 & 0 \\ 2 & 0 & 0 & 0 \\ -2 & -1 & 0 & 0 \\ -1 & 3 & 0 & 0 \\ 3 & 2 & 3 & 0 \end{pmatrix}.$$

The rank is 3.

EXERCISES

In each of problems 1 through 9, reduce A to a matrix A', using elementary transformations involving rows, in accordance with the procedure of Theorem 13. Determine the rank of A.

1. $\begin{pmatrix} 1 & 4 & 6 & 1 \\ 1 & -1 & 2 & 3 \\ 1 & -11 & -6 & 7 \end{pmatrix}$

2. $\begin{pmatrix} 1 & 4 & -2 & 3 \\ 1 & 5 & 0 & 1 \\ -1 & -1 & 8 & -6 \\ 2 & 10 & 0 & 7 \end{pmatrix}$

3. $\begin{pmatrix} 1 & -1 & 0 & 3 \\ 2 & 1 & 0 & 1 \\ 2 & -2 & 0 & 6 \\ 1 & -2 & 1 & 2 \end{pmatrix}$

4. $\begin{pmatrix} 1 & 0 & -2 & -5 & -1 \\ 0 & -1 & 3 & 1 & 2 \\ 0 & 1 & 0 & 0 & 3 \\ 1 & 1 & -1 & 0 & 4 \end{pmatrix}$

5. $\begin{pmatrix} 1 & 1 & -2 & 0 \\ 2 & 1 & -3 & 0 \\ -4 & 2 & 2 & 0 \\ 6 & -1 & -5 & 0 \\ 7 & -3 & -4 & 1 \end{pmatrix}$

6. $\begin{pmatrix} 1 & 2 & -1 & 3 & 1 \\ -3 & -5 & -1 & 0 & -2 \\ 1 & 1 & 3 & -6 & 0 \\ 4 & 7 & 0 & 3 & 3 \\ 1 & 0 & 7 & -15 & -1 \end{pmatrix}$

7. $\begin{pmatrix} 3 & -3 & 2 & -2 \\ 2 & -1 & 3 & -2 \\ -2 & 3 & -2 & 1 \\ 4 & -2 & 1 & 3 \end{pmatrix}$ 8. $\begin{pmatrix} 2 & 3 & -3 & -2 \\ -3 & 2 & 2 & 3 \\ 4 & -2 & 3 & 2 \\ 5 & 4 & -2 & 2 \end{pmatrix}$

9. $\begin{pmatrix} 3 & \frac{2}{3} & 1 & \frac{1}{2} & 6 \\ 2 & -\frac{1}{4} & 3 & -2 & 1 \\ 6 & -\frac{2}{5} & 3 & -\frac{1}{3} & 5 \end{pmatrix}$

In each of problems 10 through 14, reduce A to a matrix A' as above, but use elementary transformations involving columns only. Determine the rank of A in each case.

10. A is the matrix in problem 3.   11. A is the matrix in problem 4.

12. A is the matrix in problem 7.   13. A is the matrix in problem 8.

14. A is the matrix in problem 9.

In problems 15 through 17, in each case reduce the given matrix to diagonal form as in Corollary 2.

15. $\begin{pmatrix} 1 & -2 & -4 & 2 \\ 1 & -3 & -2 & 1 \\ 2 & -2 & -11 & 4 \\ -1 & 3 & 3 & 0 \end{pmatrix}$ 16. $\begin{pmatrix} 2 & -1 & -2 & 3 \\ 1 & -2 & -1 & 2 \\ 3 & -1 & 2 & 1 \\ -1 & 2 & -3 & 2 \end{pmatrix}$

17. $\begin{pmatrix} 1 & 3 & -1 & 2 & 1 \\ 4 & 1 & 5 & -1 & 2 \\ 3 & 2 & 1 & 6 & 5 \\ 2 & -1 & 4 & -2 & 1 \\ 3 & 1 & -5 & 6 & 3 \end{pmatrix}$

18. Show that every $n \times n$ matrix may be reduced by elementary row transformations to triangular form, with all elements above the diagonal consisting of zeros.

19. Let A be an $m \times n$ matrix with $m \leq n$. If A is of rank r with $r < m$, describe a process using elementary row transformations which reduces A to a form having the first $m - r$ rows consisting entirely of zeros.

20. Show that every $n \times n$ nonsingular matrix A may be reduced by row transformations to a form in which the only nonzero elements are $a_{n1}, a_{n-1,2}, a_{n-2,3}, \ldots,$ $a_{2,n-1}, a_{1n}$.

8. GENERAL LINEAR SYSTEMS

We consider a system of m linear equations in n unknowns in which m may be different from n. In a system such as

$$\left.\begin{array}{c} a_{11}x_1 + a_{12}x_2 + \cdots + a_{1n}x_n = b_1 \\ a_{21}x_1 + a_{22}x_2 + \cdots + a_{2n}x_n = b_2 \\ \vdots \qquad\qquad \vdots \qquad\qquad \vdots \\ a_{m1}x_1 + a_{m2}x_2 + \cdots + a_{mn}x_n = b_m \end{array}\right\} \qquad (1)$$

we call the $m \times n$ matrix

$$A = \begin{pmatrix} a_{11} & a_{12} & \cdots & a_{1n} \\ a_{21} & a_{22} & \cdots & a_{2n} \\ \vdots & \vdots & & \vdots \\ a_{m1} & a_{m2} & \cdots & a_{mn} \end{pmatrix}$$

the **coefficient matrix** of the system (1). The $m \times (n + 1)$ matrix

$$B = \begin{pmatrix} a_{11} & a_{12} & \cdots & a_{1n} & b_1 \\ a_{21} & a_{22} & \cdots & a_{2n} & b_2 \\ \vdots & \vdots & & \vdots & \vdots \\ a_{m1} & a_{m2} & \cdots & a_{mn} & b_m \end{pmatrix}$$

is called the **augmented matrix** of (1).

As an illustration, suppose we wish to solve the system of equations

$$\begin{aligned} 3x + 2y - 2z &= 3, \\ 2x + 3y - 3z &= 4, \\ -2x + 4y + 2z &= 3, \\ 5x - 2y + 4z &= 2, \\ 3x + 4y + 2z &= 3. \end{aligned} \qquad (2)$$

The reader, noticing that there are more equations than there are unknowns, would not expect a solution. However, sometimes such systems do have solutions. Geometrically, the above system represents five planes in three-dimensional space. If it should accidentally happen that the five planes have a point in common, then (2) has a solution. We shall show that the existence of a solution of (2) depends on the behavior of both the coefficient matrix and the augmented matrix.

Suppose now that we proceed to reduce an augmented matrix of a given system such as (1), using the methods of elementary transformations as described in the preceding section. First we note that the coefficient matrix is reduced simultaneously. Second, the interchange of two rows of a matrix corresponds to an interchange of two rows of the system of equations and does not affect the solution. Multiplication of a row by a constant is the same as multiplication of an equation by a constant and also has no effect on the solution. A transformation of type (c) is the same as multiplication of one equation by a constant and addition of the resulting equation to another equation. Again, the solution is unaffected. Thus to each elementary transformation of the augmented matrix involving only rows, there corresponds an operation which we call the **corresponding elementary transformation of the given system of equations.** By now, the following important theorem is evident.

Theorem 14. *Suppose B is the augmented matrix of a certain system of equations and B' is obtained from B by applying a finite number of elementary* *transformations involving only rows. Then B' is the augmented matrix of that —system of equations which is obtained from the original system by performing the corresponding elementary transformations. The coefficient matrix of the transformed system is obtained from B' by omitting the last column. The systems corresponding to B and B' have the same solutions. We say the systems are* **equivalent.**

The augmented matrix of (2) is

$$\begin{pmatrix} 3 & 2 & -2 & 3 \\ 2 & 3 & -3 & 4 \\ -2 & 4 & 2 & 3 \\ 5 & -2 & 4 & 2 \\ 3 & 4 & 2 & 3 \end{pmatrix}$$

We recognize this matrix as the one in Example 2 of the preceding section. The result of reduction shows that

$$\begin{pmatrix} 3 & 2 & -2 & 3 \\ 2 & 3 & -3 & 4 \\ -2 & 4 & 2 & 3 \\ 5 & -2 & 4 & 2 \\ 3 & 4 & 2 & 3 \end{pmatrix} \cong \begin{pmatrix} 2 & 3 & -3 & 4 \\ 0 & -1 & 13 & -6 \\ 0 & 0 & -1 & 1 \\ 0 & 0 & 0 & 2 \\ 0 & 0 & 0 & 0 \end{pmatrix}$$

We now may use Theorem 14 to conclude that the system (2) is equivalent to the system

$$2x + 3y - 3z = 4$$
$$- y + 13z = -6$$
$$-z = 1$$
$$0 = 2$$
$$0 = 0,$$

which has no solution. However, if the column of numbers on the right in (2) is replaced by the column

$$\begin{matrix} -3 \\ -7 \\ -2 \\ 15 \\ 3, \end{matrix}$$

so that the system becomes

$$3x + 2y - 2z = -3$$
$$2x + 3y - 3z = -7$$
$$-2x + 4y + 2z = -2 \qquad\qquad (3)$$
$$5x - 2y + 4z = 15$$
$$3x + 4y + 2z = 3,$$

we see that exactly the same elementary transformations as before bring about an equivalent system. We obtain

$$2x + 3y - 3z = -7$$
$$-y + 13z = 27$$
$$-z = -2$$
$$0 = 0$$
$$0 = 0,$$

which has the solution $x = 1$, $y = -1$, $z = 2$. The student can easily verify that this set of numbers satisfies all equations in (3).

EXAMPLE. Solve, if possible, the following system of equations by using elementary row transformations on the augmented matrix to reduce the system to a simpler equivalent form:

$$x_1 + 2x_2 - 2x_3 + 3x_4 - 4x_5 = -3,$$
$$2x_1 + 4x_2 - 5x_3 + 6x_4 - 5x_5 = -1,$$
$$-x_1 - 2x_2 \qquad\quad - 3x_4 + 11x_5 = 15.$$

Solution. The augmented matrix is

$$B = \begin{pmatrix} 1 & 2 & -2 & 3 & -4 & \vdots & -3 \\ 2 & 4 & -5 & 6 & -5 & \vdots & -1 \\ -1 & -2 & 0 & -3 & 11 & \vdots & 15 \end{pmatrix},$$

and the part of B to the left of the dotted vertical line is the coefficient matrix. We multiply the first row by 2 and subtract from the second row. Then, adding the first row to the third row, we obtain

$$B \cong \begin{pmatrix} 1 & 2 & -2 & 3 & -4 & \vdots & -3 \\ 0 & 0 & -1 & 0 & 3 & \vdots & 5 \\ 0 & 0 & -2 & 0 & 7 & \vdots & 12 \end{pmatrix} \cong \begin{pmatrix} 1 & 2 & -2 & 3 & -4 & \vdots & -3 \\ 0 & 0 & -1 & 0 & 3 & \vdots & 5 \\ 0 & 0 & 0 & 0 & 1 & \vdots & 2 \end{pmatrix}.$$

Therefore, the given system is equivalent to the system

$$x_1 + 2x_2 - 2x_3 + 3x_4 - 4x_5 = -3,$$
$$- x_3 \qquad\qquad + 3x_5 = 5,$$
$$x_5 = 2.$$

From the last two equations we get at once $x_5 = 2$ and $x_3 = 1$. Solving the first equation for x_1, we find that

$$x_1 = 7 - 2x_2 - 3x_4.$$

In other words, regardless of the values we assign to x_2 and x_4, the remaining values of x_1, x_3, and x_5 will satisfy the given system. There is an infinite number of solutions.

We now give a criterion (in terms of the ranks of both the coefficient matrix and the augmented matrix) which determines when a general linear system has a solution. However, the actual solution of a problem is best performed as in the example above.

Theorem 15. (a) *A system of m linear equations in n unknowns has a solution if and only if the rank r of the augmented matrix equals that of the coefficient matrix.*
(b) *If the two matrices have the same rank r and r = n, the solution is unique.*
(c) *If the two matrices have the same rank r and r < n, then at least one set of r of the unknowns can be solved in terms of the remaining n − r unknowns.*

Proof. Let A be the $m \times n$ coefficient matrix and B be the $m \times (n + 1)$ augmented matrix of the given system. Suppose that B is reduced to the matrix B' as in Theorem 13 (using rows only). The coefficient matrix of the equivalent system corresponding to A is the submatrix A' of B' obtained by deleting the last column. If A and B are not of the same rank, neither are A' and B'. Then the last column of B' has at least one nonzero element b' in which all remaining elements of that row are zeros. The corresponding equation of the equivalent system is

$$0 = b',$$

which is clearly impossible.

If A' and B' are of the same rank r and $r = n$, then there are as many equations as unknowns and Cramer's Rule applies. There is a unique solution.

If $r < n$, we select an $r \times r$ submatrix of A' which is nonsingular, and we apply Cramer's Rule to solve for r of the unknowns in terms of the remaining $n - r$ unknowns. Thus part (c) of the theorem is established.

A system of linear equations is said to be **homogeneous** if and only if the numbers on the right [i.e., $b_1, b_2, \ldots, b_m$ in (1)] are all zero. In this case, it is clear that the rank of the augmented matrix equals that of the coefficient matrix. Every homogeneous system has the solution $x_1 = x_2 = \cdots = x_n = 0$. Moreover, if $x_1, x_2, \ldots, x_n$ and $y_1, y_2, \ldots, y_n$ are solutions of a homogeneous system and c is a constant, then

$$x_1 + y_1, x_2 + y_2, \ldots, x_n + y_n \quad \text{and} \quad cx_1, cx_2, \ldots, cx_n$$

are solutions. If the rank of a homogeneous system is n, then the solution is unique and $x_1 = x_2 = \cdots = x_n = 0$ is the only solution. We have just derived the following corollary.

Corollary. *A homogeneous system of m equations in n unknowns has a solution $x_1, x_2, \ldots, x_n$ in which not all the x_j are zero if and only if the rank r of the coefficient matrix is less than n. When r < n, some group of r of the x_j can be expressed in terms of the remaining x_j.*

EXERCISES

Find all the solutions, if any, of the following systems of equations. Begin by reducing the augmented matrix (and thus the coefficient matrix) to the simplified form, using only row transformations. Find the rank r of the coefficient matrix and the rank r^* of the augmented matrix.

1. $\begin{aligned} x_1 - x_2 + 2x_3 &= -2 \\ 3x_1 - 2x_2 + 4x_3 &= -5 \\ 2x_2 - 3x_3 &= 2 \end{aligned}$

2. $\begin{aligned} x_1 + x_2 - 5x_3 &= 26 \\ x_1 + 2x_2 + x_3 &= -4 \\ x_1 + 3x_2 + 7x_3 &= -34 \end{aligned}$

3. $\begin{aligned} 2x_1 + 3x_2 - x_3 &= -15 \\ 3x_1 + 5x_2 + 2x_3 &= 0 \\ x_1 + 3x_2 + 3x_3 &= 11 \\ 7x_1 + 11x_2 &= -30 \end{aligned}$

4. $\begin{aligned} x_1 - x_3 + x_4 &= -2 \\ -x_2 + 2x_3 + x_4 &= 5 \\ x_1 - x_3 + 2x_4 &= 3 \\ 2x_1 + x_2 - x_3 &= -6 \end{aligned}$

5. $\begin{aligned} 3x_1 - x_2 + 2x_3 &= 3 \\ 2x_1 + 2x_2 + x_3 &= 2 \\ x_1 - 3x_2 + x_3 &= 4 \end{aligned}$

6. $\begin{aligned} 4x_1 - 6x_2 + 7x_3 &= 8 \\ x_1 - 2x_2 + 6x_3 &= 4 \\ 8x_1 - 10x_2 - 3x_3 &= 8 \end{aligned}$

7. $\begin{aligned} 2x_1 - x_2 + x_4 &= 2 \\ -3x_1 + x_3 - 2x_4 &= -4 \\ x_1 + x_2 - x_3 + x_4 &= 2 \\ 2x_1 - x_2 + 5x_3 &= 6 \end{aligned}$

8. $\begin{aligned} x_1 + x_2 + 2x_3 - x_4 &= 3 \\ 2x_1 - x_2 + x_3 + x_4 &= 1 \\ x_1 - 5x_2 - 4x_3 + 5x_4 &= -7 \\ 4x_1 - 5x_2 - x_3 + 5x_4 &= -3 \end{aligned}$

9. $\begin{aligned} 2x_1 - 7x_2 - 6x_3 &= 0 \\ 3x_1 + 5x_2 - 2x_3 &= 0 \\ 4x_1 - 2x_2 - 7x_3 &= 0 \end{aligned}$

10. $\begin{aligned} x_1 - x_2 - 5x_3 &= 0 \\ 2x_1 + 3x_2 &= 0 \\ 4x_1 - 5x_2 - 22x_3 &= 0 \end{aligned}$

11. $\begin{aligned} 2x_1 + 3x_2 - x_3 - x_4 &= 0 \\ x_1 - x_2 - 2x_3 - 4x_4 &= 0 \\ 3x_1 + x_2 + 3x_3 - 2x_4 &= 0 \\ 6x_1 + 3x_2 - 7x_4 &= 0 \end{aligned}$

12. $\begin{aligned} x_1 - 2x_2 + 10x_3 - 4x_4 &= 0 \\ 3x_1 - x_2 &= 0 \\ -2x_1 + x_2 + 5x_3 - 2x_4 &= 0 \\ 2x_1 - 3x_2 - 5x_3 + 2x_4 &= 0 \end{aligned}$

13. $\begin{aligned} x_1 + x_2 - 3x_3 + x_4 &= 1 \\ 2x_1 - 4x_2 + 2x_4 &= 2 \\ 3x_1 - 4x_2 - 2x_3 &= 0 \\ x_1 - 2x_3 + 3x_4 &= 3 \end{aligned}$

14. $\begin{aligned} x_1 - 2x_2 + 2x_3 + 3x_4 &= 1 \\ -x_1 + 4x_2 - x_3 - 5x_4 &= 2 \\ 2x_1 - 2x_2 + 5x_3 + 4x_4 &= 5 \\ -x_1 + 6x_2 - x_4 &= 12 \end{aligned}$

15. $\begin{aligned} x_1 + 3x_2 - 2x_3 - 3x_4 + 2x_5 &= 4 \\ -x_1 - 3x_2 + 4x_3 + 4x_4 + 4x_5 &= -1 \\ -x_1 - 3x_2 + 4x_3 + 4x_4 - x_5 &= -2 \\ -2x_1 - 6x_2 + 10x_3 + 9x_4 - 4x_5 &= 1 \end{aligned}$

16. $\begin{aligned} 2x_1 + x_2 - 3x_3 + x_4 + x_5 &= 0 \\ x_1 + 2x_2 - x_3 + 4x_4 + 2x_5 &= 1 \\ 2x_1 - 3x_2 + 2x_3 - x_4 + 3x_5 &= -6 \\ -x_1 + 2x_3 + 3x_4 - x_5 &= 8 \\ 2x_2 + x_3 + 2x_4 + 3x_5 &= -7 \\ 3x_1 - 4x_2 + 5x_3 - 2x_4 + x_5 &= 0 \end{aligned}$

9. NUMERICAL SOLUTIONS BY ITERATIVE METHODS

Suppose we have a system of n linear equations in n unknowns which may be written in the matrix form

$$AX = B, \tag{1}$$

where A is an $n \times n$ matrix and X and B are $n \times 1$ matrices. If A is nonsingular, so that A^{-1} exists, the matrix equation (1) has the unique solution

$$X = A^{-1}B. \tag{2}$$

The individual components of X may be read off from (2), once the inverse of A is found and the product $A^{-1}B$ is computed.

From the numerical point of view, it is difficult to find the inverse of a non-singular $n \times n$ matrix when n is large. For $n = 2, 3,$ or 4, the inverse of a matrix is computable by hand directly from the definition. If n is below 10, it is still possible to find the inverse by using an electric desk calculator. However, if n is 100 or 1000, the situation is clearly hopeless. In recent times, many problems have arisen in engineering, economics, biology, etc., which lead to systems of linear equations with an enormous number of unknowns.

In Section 1 we discussed the solution of a linear system by the method of elimination. If the number of equations is relatively small, such techniques are quite adequate. We shall now take up a method of solving the system (1) which has many computational advantages, especially when the number of equations and unknowns is large and computing machines are available. We shall illustrate the method with an example and then state a theorem which justifies the process.

EXAMPLE 1. Find a solution of the system

$$\begin{aligned}
6x + 2y - 3z &= 5, \\
-x + 8y + 3z &= -10, \\
x + 4y + 12z &= 12.
\end{aligned} \tag{3}$$

Solution. We rewrite the system in the form

$$\begin{aligned}
x &= \frac{5 - 2y + 3z}{6}, \\
y &= \frac{-10 + x - 3z}{8}, \\
z &= \frac{12 - x - 4y}{12},
\end{aligned} \tag{4}$$

in which we solve the first equation in (3) for x in terms of the remaining variables, the second equation for y in terms of x and z, and the third equation for z in terms of x and y. Now we *guess* at the solution. The guess may be good or bad, a fact which we shall discuss later. Suppose the guess is $x = 1, y = 1, z = 1$. We call this guess $x^{(0)} = 1, y^{(0)} = 1, z^{(0)} = 1$ and insert these values in the *right side* of (4). We find, for the left side, the values

$$x = 1, \qquad y = -\tfrac{3}{2}, \qquad z = \tfrac{7}{12}.$$

These values are denoted $x^{(1)} = 1, y^{(1)} = -\tfrac{3}{2}, z^{(1)} = \tfrac{7}{12}$, and we insert these in the right side of (4). The corresponding values on the left, which we call $x^{(2)}, y^{(2)}, z^{(2)}$, are

$$x^{(2)} = \tfrac{13}{8}, \qquad y^{(2)} = -\tfrac{43}{32}, \qquad z^{(2)} = \tfrac{17}{12}.$$

Continuing the process by substituting these values in the right side of (4), we obtain

$$x^{(3)} = \tfrac{191}{96}, \qquad y^{(3)} = -\tfrac{101}{64}, \qquad z^{(3)} = \tfrac{21}{16}.$$

Repeating the iteration once again, we get

$$x^{(4)} = \tfrac{387}{192}, \qquad y^{(4)} = -\tfrac{1147}{768}, \qquad z^{(4)} = \tfrac{1567}{1152}.$$

Of course, a simple calculation shows that the true answer is $x = 2, y = -\tfrac{3}{2}, z = \tfrac{4}{3}$, while the fourth iteration above, in decimal form, is $x^{(4)} = 2.015^{+}, y^{(4)} = -1.493^{+}$, $z^{(4)} = 1.360^{+}$.

If the system (1) is written out in coordinates

$$\begin{aligned}
a_{11}x_1 + a_{12}x_2 + \cdots + a_{1n}x_n &= b_1 \\
a_{21}x_1 + a_{22}x_2 + \cdots + a_{2n}x_n &= b_2 \\
&\;\;\vdots \\
a_{n1}x_1 + a_{n2}x_2 + \cdots + a_{nn}x_n &= b_n
\end{aligned}$$

then the iteration method described in Example 1 has the appearance

$$x_1 = \frac{b_1 - (a_{12}x_2 + \cdots + a_{1n}x_n)}{a_{11}}$$

$$x_2 = \frac{b_2 - (a_{21}x_1 + a_{23}x_3 + \cdots + a_{2n}x_n)}{a_{22}}$$

$$\vdots$$

$$x_n = \frac{b_n - (a_{n1}x_1 + a_{n2}x_2 + \cdots + a_{n,n-1}x_{n-1})}{a_{nn}}.$$

We may write this more compactly as

$$x_i = \frac{1}{a_{ii}}\left[b_i - \sum_{\substack{j=1 \\ j \neq i}}^{n} a_{ij}x_j \right], \qquad i = 1, 2, \ldots, n. \tag{5}$$

The general iteration method begins with the guess

$$x_1^{(0)}, \quad x_2^{(0)}, \quad \ldots, \quad x_n^{(0)},$$

which is inserted in the right side of (5). The values obtained on the left are labeled $x_1^{(1)}, x_2^{(1)}, \ldots, x_n^{(1)}$, and the process is repeated with the new values put into (5) on the right. The general process after k steps is described by the system

$$x_i^{(k+1)} = \frac{1}{a_{ii}}\left[b_i - \sum_{\substack{j=1 \\ j \neq i}}^{n} a_{ij}x_j^{(k)} \right], \qquad i = 1, 2, \ldots, n. \tag{6}$$

It *may* happen that

$$\lim_{k \to \infty} x_i^{(k)} \quad \text{exists for} \quad i = 1, 2, \ldots, n. \tag{7}$$

Calling the limiting values $\bar{x}_1, \bar{x}_2, \ldots, \bar{x}_n$, it is clear that this set of values will satisfy (5) and therefore (1). It is the desired solution.

It is natural to ask when the limit in (7) will exist. At first, we may think it will depend on the initial guess. After all, if by accident we take for

$$x_1^{(0)}, \quad x_2^{(0)}, \quad \ldots, \quad x_n^{(0)}$$

the true solution, the iteration stops after one step. The next theorem shows that under certain circumstances, the initial guess plays no part in the existence of the limit in (7).

The iteration method (5) is actually most useful when the coefficients of the diagonal terms a_{ii} are *much larger* than any of the other coefficients. A sufficient condition for convergence is stated (without proof) in the following theorem.

Theorem 16. *Suppose that the matrix $A = (a_{ij})$ has the property that*

$$|a_{ii}| > \sum_{\substack{j=1 \\ j \neq i}}^{n} |a_{ij}| \quad \text{for} \quad i = 1, 2, \ldots, n. \tag{8}$$

Then the matrix A is nonsingular, and the iteration method (6) converges to the unique solution of (1) no matter what values are selected for the initial guess

$$x_1^{(0)}, \quad x_2^{(0)}, \quad \ldots, \quad x_n^{(0)}.$$

Remarks. Note that in Example 1 the conditions of the Theorem are satisfied. There are many conditions weaker than (8) which are sufficient for the convergence of the iteration method. It is of the utmost importance to observe that a solution is obtained regardless of the initial guess. Suppose that after a certain number of steps, an error is made. Instead of destroying the convergence, this merely means that a new initial guess is at hand and, unless further errors occur, convergence will take place. Of course, the "rate of convergence" is slowed by errors but, unless computing mistakes occur repeatedly, the method must lead to a solution. A good initial guess is helpful in that the number of steps required to obtain a given degree of accuracy is smaller for a good first guess than for a poor one.

The subject of the numerical solution of linear equations is replete with theorems similar to Theorem 16 which guarantee convergence under a variety of hypotheses on A. In order to save as much computation as possible, it is essential to find methods which converge rapidly. The following scheme, called the Gauss-Seidel method, is a refinement of the method (6) above and leads to more rapid convergence than does (6). The Gauss-Seidel iteration method is described by the system

$$x_i^{(k+1)} = \frac{1}{a_{ii}}\left[b_i - \sum_{j=1}^{i-1} a_{ij}x_j^{(k+1)} - \sum_{j=i+1}^{n} a_{ij}x_j^{(k)}\right], \qquad i = 1, 2, \ldots, n, \quad (9)$$

where, if $i = 1$, the first sum on the right is absent while, if $i = n$, the second sum is absent. In essence, the iteration (9) states that at each step we use the latest information available. To illustrate the point, we shall work Example 1, using the iteration scheme (9).

EXAMPLE 2. Find a solution of the system

$$6x + 2y - 3z = 5,$$
$$-x + 8y + 3z = -10,$$
$$x + 4y + 12z = 12,$$

by the Gauss-Seidel method.

Solution. As before, we take $x^{(0)} = y^{(0)} = z^{(0)} = 1$ as the initial guess and find that

$$x^{(1)} = \frac{5 - 2 + 3}{6} = 1,$$

$$y^{(1)} = \frac{-10 + x^{(1)} - 3z^{(0)}}{8} = -\frac{3}{2},$$

$$z^{(1)} = \frac{12 - x^{(1)} - 4y^{(1)}}{12} = \frac{12 - 1 + 6}{12} = \frac{17}{12}.$$

The second iteration is

$$x^{(2)} = \frac{5 - 2y^{(1)} + 3z^{(1)}}{6} = \frac{49}{24},$$

$$y^{(2)} = \frac{-10 + x^{(2)} - 3z^{(1)}}{8} = -\frac{293}{192},$$

$$z^{(2)} = \frac{12 - x^{(2)} - 4y^{(2)}}{12} = \frac{257}{192}.$$

A third iteration yields $x^{(3)} = 2317/1152$, $y^{(3)} = -13,829/9216$, $z^{(3)} = 12,281/9216$ which, in decimal form, is

$$x^{(3)} = 2.012^+, \qquad y^{(3)} = -1.500^+, \qquad z^{(3)} = 1.332^+.$$

We note that the Gauss-Seidel method after three iterations yields values which are closer to the true result than the iteration process (5) does after four iterations.

Remarks. It sometimes happens that a matrix will have a dominant term in each row but that this dominant term is not always the diagonal term. A relabeling of the unknowns or an interchange in the ordering of the equations may bring the matrix A into the appropriate form. It can be shown that convergence of the iteration process takes place if (8) is replaced by

$$|a_{ii}| \geq \sum_{\substack{j=1 \\ j \neq i}}^{n} |a_{ij}| \quad \text{for} \quad i = 1, 2, \ldots, n,$$

and if there is *at least one i* for which the strict inequality holds.

EXERCISES

Solve each of the following systems by the iteration process (5) and by the Gauss-Seidel method (9), repeating the iterations the number of times indicated by the value of n. Use the initial guess as shown. Compare the result with that obtained by elimination.

1. $5x_1 + 2x_2 - x_3 = -2,$
   $2x_1 + 4x_2 + x_3 = -3,$
   $x_1 + 2x_2 + 6x_3 = 15,$ $n = 4.$
   Select $x_1^{(0)} = x_2^{(0)} = x_3^{(0)} = 0.$

2. $4x_1 - x_2 + x_3 = 10,$
   $2x_1 + 5x_2 - 2x_3 = -3,$
   $2x_1 + x_2 + 6x_3 = 9,$ $n = 4.$
   Select $x_1^{(0)} = x_2^{(0)} = x_3^{(0)} = 0.$

3. $7x_1 - 2x_2 + 3x_3 = 19,$
   $x_1 + 8x_2 - 6x_3 = 9,$
   $2x_1 + 4x_2 + 9x_3 = 5,$ $n = 3.$
   Select $x_1^{(0)} = x_2^{(0)} = x_3^{(0)} = 1.$

4. $12x_1 - 2x_2 + 4x_3 = -14,$
   $3x_1 + 8x_2 + 2x_3 = 23,$
   $6x_1 - 3x_2 + 16x_3 = 10,$ $n = 4.$
   Select $x_1^{(0)} = x_2^{(0)} = x_3^{(0)} = 1.$

5. $6x_1 + 2x_2 + 2x_3 \qquad = 20,$
   $x_1 + 5x_2 \qquad + x_4 = -5,$
   $x_1 + 3x_2 + 7x_3 \qquad = -2,$
   $x_2 + 2x_3 - 8x_4 = -10,$ $n = 3.$
   Select $x_1^{(0)} = x_2^{(0)} = x_3^{(0)} = x_4^{(0)} = 1.$

6. $6x_1 + 3x_2 - 2x_3 = 5,$
   $3x_1 + 9x_2 + 2x_3 = 17,$
   $-3x_1 + 3x_2 + 6x_3 = 11,$ $n = 4.$
   Select $x_1^{(0)} = x_2^{(0)} = x_3^{(0)} = 1.$

7. $5x_1 - 2x_2 + 3x_3 = -5,$
   $2x_1 + 3x_2 - 7x_3 = 3,$
   $4x_1 + 5x_2 - x_3 = -17,$ $n = 4.$
   Select $x_1^{(0)} = x_2^{(0)} = x_3^{(0)} = 0.$

CHAPTER 7

Vector Spaces

1. HIGHER DIMENSIONAL SPACES. VECTOR SPACES

The study of analytic geometry in two and three dimensions greatly facilitated our study of functions of two and three variables. The ability to visualize objects in the plane and in space turned out to be particularly useful for the topics on partial differentiation and multiple integration. However, functions of more than three variables often arise in applications, particularly those in economics, engineering, and physics. It would be helpful to have at hand a geometric theory to go with the study of functions of n variables for n larger than 3. While it is obvious that visualization stops abruptly with $n = 3$, it is still possible in such cases to develop much of the formal apparatus which we learned for two- and three-dimensional space and to carry over a great deal of the geometric language.

A coordinate system in the plane consists of sets of ordered number pairs (x_1, x_2). A system in three-space consists of ordered number triples (x_1, x_2, x_3). The feature which characterizes a rectangular system of coordinates is the formula for the distance between two points P_1 and P_2. In the plane this formula is

$$\sqrt{(y_1 - x_1)^2 + (y_2 - x_2)^2}$$

where the coordinates of P_1 and P_2 are (x_1, x_2) and (y_1, y_2), respectively. Similarly, in three-space the formula

$$\sqrt{(y_1 - x_1)^2 + (y_2 - x_2)^2 + (y_3 - x_3)^2}$$

for the distance between the points $P_1(x_1, x_2, x_3)$ and $P_2(y_1, y_2, y_3)$ characterizes a rectangular coordinate system. Notice, however, that this formula does not distinguish a left-handed coordinate system from a right-handed one. We define a **coordinate system** in n-dimensional space as a collection of ordered sets of n-tuples of numbers $(x_1, x_2, \ldots, x_n)$. We say that each such ordered set corresponds to a **point** in n-space. The **distance** d between two points

$$P(x_1, x_2, \ldots, x_n) \quad \text{and} \quad Q(y_1, y_2, \ldots, y_n)$$

may be defined as the natural extension of the distance formulas in the plane and in space; that is

$$d \equiv |PQ| = \sqrt{(y_1 - x_1)^2 + (y_2 - x_2)^2 + \cdots + (y_n - x_n)^2}. \tag{1}$$

Then we say the coordinate system is rectangular. In n-dimensional space, which we denote by R_n, we define the **jth coordinate axis** as the set of all points of the form $(0, \ldots, 0, x_j, 0, \ldots, 0)$. Recalling that in three-space the three coordinate planes are represented by $x = 0$, by $y = 0$, and by $z = 0$, we define a **coordinate hyperplane** in R_n by the equation $x_j = 0$. There are n such hyperplanes.

It is possible to define lines, curves, surfaces, hypersurfaces, and so forth, in R_n by extrapolating from the corresponding definitions in R_2 and R_3. For example, the formula for the midpoint of the line segment joining the points

$$P'(x_1', x_2', \ldots, x_n') \qquad \text{and} \qquad P''(x_1'', x_2'', \ldots, x_n'')$$

is

$$\bar{x}_1 = \frac{x_1' + x_1''}{2}, \qquad \bar{x}_2 = \frac{x_2' + x_2''}{2}, \ldots, \qquad \bar{x}_n = \frac{x_n' + x_n''}{2}.$$

(See Exercises 1 through 7 at the end of this section.)

In Chapter 14 of *First Course* we discussed vectors in the plane, and in Chapter 2 of this volume we described vectors in three-space. It is desirable to define vectors in n-space and we could do so by paralleling the treatment of vectors in two and three dimensions. It is preferable, however, to define directly the general notion of a **vector space**. Such spaces are of paramount importance in higher mathematics, in mathematical physics, and in engineering.

By a **vector space** (or **linear space**) V, we mean a collection of objects, called **vectors**, which satisfy the following axioms:

Axiom A-1. (Closure property.) *To each pair* $(\mathbf{v}, \mathbf{w})$ *in* V, *there corresponds one and only one vector, denoted by* $\mathbf{v} + \mathbf{w}$, *and called their* **sum.**

Axiom A-2. (Commutative law.) $\mathbf{v} + \mathbf{w} = \mathbf{w} + \mathbf{v}$.

Axiom A-3. (Associative law.) $(\mathbf{u} + \mathbf{v}) + \mathbf{w} = \mathbf{u} + (\mathbf{v} + \mathbf{w})$.

Axiom A-4. (Existence of 0, the zero vector.) *There is a unique vector* $\mathbf{0}$ *in* V *such that* $\mathbf{v} + \mathbf{0} = \mathbf{0} + \mathbf{v} = \mathbf{v}$ *for each* $\mathbf{v}$ *in* V.

Axiom A-5. (Inverse of addition.) *If* $\mathbf{v}$ *is any vector in* V, *there is a unique vector, denoted* $-\mathbf{v}$, *such that* $\mathbf{v} + (-\mathbf{v}) = (-\mathbf{v}) + \mathbf{v} = \mathbf{0}$.

Axiom B-1. (Scalar multiplication.) *For every number (scalar)* c *and for each vector* $\mathbf{v}$ *in* V, *there corresponds a vector* $c\mathbf{v}$ *in* V *called the product of* c *and* $\mathbf{v}$.

Axiom B-2. *If* c *and* d *are any numbers and if* $\mathbf{v}$ *and* $\mathbf{w}$ *are any vectors in* V, *scalar multiplication possesses the following properties:*

(i) $c(d\mathbf{v}) = (cd)\mathbf{v}$,

(ii) $1 \cdot \mathbf{v} = \mathbf{v}$,

(iii) $c(\mathbf{v} + \mathbf{w}) = c\mathbf{v} + c\mathbf{w}$,

(iv) $(c + d)\mathbf{v} = c\mathbf{v} + d\mathbf{v}$.

From the axioms it follows easily that

$$0 \cdot \mathbf{v} = \mathbf{0} \qquad \text{and} \qquad (-1)\mathbf{v} = -\mathbf{v}. \tag{2}$$

Letting $c = 1$, $d = 0$ in (iv), we see that $\mathbf{v} = \mathbf{v} + 0\mathbf{v}$ and, using A-4, we conclude that $0 \cdot \mathbf{v} = \mathbf{0}$. Setting $c = 1$ and $d = -1$ in (iv), we have $0 \cdot \mathbf{v} = 1 \cdot \mathbf{v} + (-1) \cdot \mathbf{v}$. We now use (ii) and the fact that $0 \cdot \mathbf{v} = \mathbf{0}$ to write $\mathbf{v} + (-1) \cdot \mathbf{v} = \mathbf{0}$. Since Axiom A-5 states that the additive inverse is unique, we obtain $(-1)\mathbf{v} = -\mathbf{v}$.

It is important to note that these axioms and the simple consequences in (2) are exactly the properties we established for vectors in the plane and in three-space. The totality of vectors in a plane forms a vector space V. The totality of vectors in three-space is another example of a vector space V. There are many examples of vector spaces, a few of which are listed below.

1. **The space** V_n in which vectors are just the points of R_n with addition and scalar multiplication defined by

$$(x_1, x_2, \ldots, x_n) + (y_1, y_2, \ldots, y_n) = (x_1 + y_1, x_2 + y_2, \ldots, x_n + y_n)$$

and

$$c(x_1, x_2, \ldots, x_n) = (cx_1, cx_2, \ldots, cx_n).$$

If $n = 2$ or 3, the above definition states that a vector is a directed line segment in the plane or in three-space, with base at the origin 0 and head at a point P. Addition follows the parallelogram rule and multiplication by a constant c multiplies the length of $\overrightarrow{OP}$ by c. Note that these vectors are *fixed* vectors, while the class of vectors in two and three dimensions which we studied earlier are *free* vectors (equivalence classes of directed line segments).

2. *The totality of $m \times n$ matrices* with m and n fixed and with addition and multiplication by a constant defined as in Chapter 6, Section 2. If $n = 1$ or if $m = 1$, the space is equivalent to V_m or V_n as just described.

3. *The totality of sequences which are convergent*, with addition and scalar multiplication defined in the following way. If

$$\mathbf{a} = (\alpha_1, \alpha_2, \ldots, \alpha_n, \ldots),$$
$$\mathbf{b} = (\beta_1, \beta_2, \ldots, \beta_n, \ldots),$$

then

$$\mathbf{a} + \mathbf{b} = (\alpha_1 + \beta_1, \alpha_2 + \beta_2, \ldots, \alpha_n + \beta_n, \ldots).$$

It is clear that if the sequence $\alpha_1, \alpha_2, \ldots, \alpha_n, \ldots$ converges to L_1 and if the sequence $\beta_1, \beta_2, \ldots, \beta_n, \ldots$ converges to L_2, then $\alpha_1 + \beta_1, \alpha_2 + \beta_2, \ldots, \alpha_n + \beta_n, \ldots$ converges to $L_1 + L_2$. Multiplication by a scalar c is simply defined by

$$c\mathbf{a} = (c\alpha_1, c\alpha_2, \ldots, c\alpha_n, \ldots)$$

and this sequence converges to the value cL_1. The element $\mathbf{0}$ is the sequence $(0, 0, \ldots, 0, \ldots)$.

4. *The totality of functions continuous on a fixed closed interval* $[a, b]$. If $f_1(t)$ and $f_2(t)$ are continuous on $[a, b]$, then addition and scalar multiplication are the functions $f(t)$ and $g(t)$ given by

$$f(t) = f_1(t) + f_2(t), \qquad g(t) = cf_1(t).$$

The functions f and g are continuous whenever f_1 and f_2 are. We observe that the whole function $f_1(t)$ is considered as an element or a point in the space. In other words, a single point is the entire correspondence set up between the domain and the range of values determined by f_1. Spaces in which each point is a function are called *function spaces*.

5. *The totality of polynomials of degree less than or equal to a fixed integer n.* If addition and scalar multiplication are defined as in elementary algebra and if each polynomial is taken to be an element in the space, the axioms for a vector space are easily verified. We have here a second example of a function space, each point of the space being a polynomial.

Suppose that V is a vector space, and suppose that W is a subset (i.e., a part) of the space V. We say that the subset is **closed under addition and multiplication** if the following two conditions are satisfied:

(a) if $\mathbf{u}$ and $\mathbf{v}$ belong to W (we write this $\mathbf{u}, \mathbf{v} \in W$), then the vector $\mathbf{u} + \mathbf{v}$ belongs to W.

(b) for every vector $\mathbf{v} \in W$ and every constant c, the vector $c\mathbf{v} \in W$.

It is a simple matter to verify that the elements of a subset W form a vector space whenever the subset is closed under the operations (a) and (b) above. In this case we say that W is a **subspace** of V. For example, if V is the space of vectors in three dimensions, then the set W of all vectors lying in a plane through the origin forms a subspace. Also the collection of vectors on any line through the origin forms a subspace.

Suppose that $\mathbf{v}_1, \mathbf{v}_2, \ldots, \mathbf{v}_k$ are vectors in a vector space V. If $c_1, c_2, \ldots, c_k$ are any numbers, we say that

$$c_1\mathbf{v}_1 + c_2\mathbf{v}_2 + \cdots + c_k\mathbf{v}_k$$

is a **linear combination** of the vectors $\mathbf{v}_1, \mathbf{v}_2, \ldots, \mathbf{v}_k$. For a given set $\mathbf{v}_1, \mathbf{v}_2, \ldots, \mathbf{v}_k$, the totality of possible linear combinations forms a subset W of V. We can verify that this subset is indeed a subspace, since W is clearly closed under addition and multiplication by scalars.

DEFINITIONS. We say that W, the set of all linear combinations of $\mathbf{v}_1, \mathbf{v}_2, \ldots, \mathbf{v}_k$, is **the space spanned by the k vectors $\mathbf{v}_1, \mathbf{v}_2, \ldots, \mathbf{v}_k$.** We also say that $\mathbf{v}_1, \mathbf{v}_2, \ldots, \mathbf{v}_k$ *span* W. We say that a space is **finite dimensional** if it is spanned by a finite number of vectors. The **dimension** of a finite dimensional vector space V is the smallest number of nonzero vectors of V which span V. If V consists only of the element $\mathbf{0}$, we define the dimension to be zero. If V has dimension $n > 0$, a set $\mathbf{w}_1, \mathbf{w}_2, \ldots, \mathbf{w}_n$ which spans V is called a **basis** for V.

EXAMPLE 1. Denote by P_8 the totality of polynomials of degree ≤ 8. Find a set of polynomials which span this space.

Solution. If f is a polynomial in P_8, it can be represented in the form

$$f(x) = a_0 + a_1x + a_2x^2 + a_3x^3 + a_4x^4 + a_5x^5 + a_6x^6 + a_7x^7 + a_8x^8.$$

Letting $f_k(x) = x^{k-1}$, $k = 1, \ldots, 9$, we see any f in P_8 is given by

$$f = a_0 f_1 + \cdots + a_8 f_9.$$

That is, every f in P_8 is a linear combination of $f_1, \ldots, f_9$.

EXAMPLE 2. Show that the set W of all functions f continuous on $[0, 1]$ with $f(0) = 0$ is a subspace of the space V of all functions continuous on $[0, 1]$.

Solution. The set W is clearly a part of V. If f and g belong to W and c is any scalar, then

$$f(0) + g(0) = 0 + 0 = 0$$

and

$$c f(0) = 0,$$

so that $f + g$ and cf both belong to W. Thus W is closed under addition and scalar multiplication and is therefore a subspace of V.

EXERCISES

In problems 1 through 7, we suppose that V is n-dimensional euclidean space. That is, the rectangular coordinate system is the natural extension of those we have employed in the plane and in three-space, and the formula for the distance between two points is given by (1).

1. Write a formula for the coordinates of the point $\bar{p}$ which divides the line segment from $P'(x_1', x_2', \ldots, x_n')$ to $P''(x_1'', x_2'', \ldots, x_n'')$ in the ratio $p:q$ and verify, by using the distance formula, that $\bar{p}$ satisfies the required condition.

2. (a) Show how to define direction cosines and direction numbers for a line in four-dimensional space.
   (b) Write a formula for the equations of a line through the points

   $$P'(x_1', x_2', x_3', x_4') \qquad \text{and} \qquad P''(x_1'', x_2'', x_3'', x_4'').$$

   (c) Repeat parts (a) and (b) for a line in n-space.

3. (a) A linear equation of the form

   $$A_1 x_1 + A_2 x_2 + A_3 x_3 + A_4 x_4 = B$$

   represents a hyperplane in four-space. State a condition under which four such hyperplanes will have exactly one point of intersection. [*Hint.* Use Cramer's Rule.]
   (b) Repeat part (a) for n hyperplanes in n-space.

4. (a) Given the two hyperplanes

   $$A_1 x_1 + A_2 x_2 + A_3 x_3 + A_4 x_4 = C_1,$$
   $$B_1 x_1 + B_2 x_2 + B_3 x_3 + B_4 x_4 = C_2,$$

   in four-space. State a condition which guarantees that they will have no points in common (i.e., that they will be *parallel*).
   (b) Repeat part (a) for two hyperplanes in n-space.

5. (a) Define the angle between two lines in n-space and, consequently, define perpendicular lines.
(b) State a condition under which a line and a hyperplane in n-space are perpendicular.

6. Write the equation of the locus of all points in n-space which are at a fixed distance r from a given point $P^0(x_1^0, x_2^0, \ldots, x_n^0)$.

7. The equation $x_n^2 = x_1^2 + x_2^2 + \cdots + x_{n-1}^2$ represents a *cone* in n-space. Describe the point sets formed by the intersection of this cone with the hyperplanes $x_n = $ const.

8. Consider all $n \times n$ matrices which have all zero elements on the diagonal. Show that this collection satisfies all the axioms for a vector space.

9. Do all polynomials of even degree (including zeroth degree) up to $n = 10$ form a vector space? Justify your result.

10. Show that the collection of functions continuous on $[0, 1]$ which vanish for $\frac{1}{2} \leq t \leq 1$ forms a subspace of the vector space of all continuous functions on $[0, 1]$.

11. In the space V_4 (as defined on page 309), consider all points (x_1, x_2, x_3, x_4) such that $x_3 = 0$. Does this collection form a subspace of V_4? Justify your result.

12. In the space V_5 (as defined on page 309), consider all points $(x_1, x_2, x_3, x_4, x_5)$ such that $x_2 + x_5 = 0$. Does this collection form a subspace of V_5? Justify your result.

13. In the space V_5 (as defined on page 309), consider all points $(x_1, x_2, x_3, x_4, x_5)$ such that $x_1 + x_4 = 1$. Does this collection form a subspace of V_5? Justify your result.

14. In the space V_4 (as defined on page 309), consider all points (x_1, x_2, x_3, x_4) such that $x_1 x_4 = 0$. Does this collection form a subspace of V_4? Justify your result.

15. In the space V_n show that the vectors $e_i = (0, 0, \ldots, 0, x_i, 0, \ldots, 0)$ where $x_i = 1, i = 1, 2, \ldots, n$, span the space.

16. Do the elements $v_1 = 1$, $v_2 = 3 - x$, $v_3 = x^2$, $v_4 = 1 + 2x - x^2 + 4x^3$ span P_3, the totality of polynomials of degree ≤ 3?

17. Does the collection of $n \times n$ triangular matrices form a subspace of the vector space of all $n \times n$ matrices? Justify your result.

18. Do all $n \times n$ diagonal matrices form a subspace of the vector space of all $n \times n$ matrices? Justify your result.

19. Find a set of 2×2 matrices which spans the vector space of all 2×2 matrices.

20. Consider all $n \times n$ matrices which have zero elements below the diagonal. Is this collection a subspace of the totality of $n \times n$ matrices? Justify your results.

21. Consider all $m \times n$ matrices which have a zero in the upper left-hand corner. Does this collection form a vector space?

22. Consider the collection W of all sequences which converge to zero. Is this a subspace of the space of convergent sequences?

23. Consider the collection of infinite sequences each of which has only a finite number of entries different from zero. Is this collection a subspace of the vector space of convergent sequences?

2. LINEAR DEPENDENCE

A set $\mathbf{v}_1, \mathbf{v}_2, \ldots, \mathbf{v}_k$ of vectors in a vector space V is said to be **linearly dependent** if and only if there are constants $c_1, c_2, \ldots, c_k$, not all zero, such that

$$c_1\mathbf{v}_1 + c_2\mathbf{v}_2 + \cdots + c_k\mathbf{v}_k = \mathbf{0}.$$

If no such constants can be found, the set is said to be **linearly independent**.

For example, in three-space the vectors

$$\mathbf{v}_1 = 2\mathbf{i} + 3\mathbf{k}, \qquad \mathbf{v}_2 = -\mathbf{i} + 2\mathbf{k}, \qquad \mathbf{v}_3 = \mathbf{i} - 4\mathbf{k}$$

are linearly dependent. Taking $c_1 = 2, c_2 = 11, c_3 = 7$, we easily verify that

$$c_1\mathbf{v}_1 + c_2\mathbf{v}_2 + c_3\mathbf{v}_3 = \mathbf{0}.$$

The above three vectors all have representatives in the xz-plane. It can be shown that any three vectors in three-space which have representatives in the same plane must be linearly dependent. Later in this section we shall describe a systematic technique for deciding when a set of vectors is linearly dependent.

Theorem 1. (a) *A set* $\mathbf{v}_1, \mathbf{v}_2, \ldots, \mathbf{v}_k$ *is linearly dependent if and only if at least one of the* $\mathbf{v}_i$ *can be expressed as a linear combination of the others.*

(b) *If* $\mathbf{v}_1, \mathbf{v}_2, \ldots, \mathbf{v}_k$ *are linearly dependent, any finite set containing them is also linearly dependent.*

(c) *If the set* $\mathbf{w}_1, \mathbf{w}_2, \ldots, \mathbf{w}_n$ *is a basis for* V, *the set* $\mathbf{w}_1, \mathbf{w}_2, \ldots, \mathbf{w}_n$ *is linearly independent.*

(d) *If* $\mathbf{v}_1, \mathbf{v}_2, \ldots, \mathbf{v}_k$ *are linearly independent, then no* $\mathbf{v}_j$ *is* $\mathbf{0}$.

(e) *If the set* $\mathbf{v}_1, \mathbf{v}_2, \ldots, \mathbf{v}_k$ *is linearly independent, then so is every (nonempty) subset.*

(f) *If* $\mathbf{v}_1, \mathbf{v}_2, \ldots, \mathbf{v}_k$ *are linearly independent and if*

$$c_1\mathbf{v}_1 + c_2\mathbf{v}_2 + \cdots + c_k\mathbf{v}_k = d_1\mathbf{v}_1 + d_2\mathbf{v}_2 + \cdots + d_k\mathbf{v}_k,$$

then $c_1 = d_1, c_2 = d_2, \ldots, c_k = d_k$.

Proof. (a) If $\mathbf{v}_1, \mathbf{v}_2, \ldots, \mathbf{v}_k$ are linearly dependent then, according to the definition, we can find $c_1, c_2, \ldots, c_k$, not all zero, such that

$$c_1\mathbf{v}_1 + c_2\mathbf{v}_2 + \cdots + c_k\mathbf{v}_k = \mathbf{0}.$$

If $c_j \neq 0$ we can solve this equation for $\mathbf{v}_j$ in terms of the remaining $\mathbf{v}$'s.

(b) Suppose that $\mathbf{v}_1, \mathbf{v}_2, \ldots, \mathbf{v}_k, \mathbf{v}_{k+1}, \mathbf{v}_{k+2}, \ldots, \mathbf{v}_l$ is a set which clearly contains $\mathbf{v}_1, \ldots, \mathbf{v}_k$. Since $\mathbf{v}_1, \mathbf{v}_2, \ldots, \mathbf{v}_k$ are linearly dependent, we select $c_1, c_2, \ldots, c_k$ (not all zero) so that $c_1\mathbf{v}_1 + c_2\mathbf{v}_2 + \cdots + c_k\mathbf{v}_k = 0$. Now we select $c_{k+1} = c_{k+2} = \cdots = c_l = 0$. Therefore

$$\sum_{i=1}^{l} c_i\mathbf{v}_i = \mathbf{0},$$

and not all c_i are zero. Hence the enlarged set is linearly dependent.

(c) If $\mathbf{w}_1, \mathbf{w}_2, \ldots, \mathbf{w}_n$ were linearly dependent then, according to (a), we could express one of them in terms of the remaining $n - 1$ vectors. However, in such a case the remaining $n - 1$ vectors would span V.

(d) If $\mathbf{v}_j = \mathbf{0}$, we select $c_j = 1$ and all $c_i = 0$, $i \neq j$. Then $\sum_{i=1}^{k} c_i \mathbf{v}_i = \mathbf{0}$, and not all c_i are zero. This contradicts the hypothesis.

(e) If a subset of $\mathbf{v}_1, \mathbf{v}_2, \ldots, \mathbf{v}_k$ is linearly dependent, we apply (b) to conclude that the entire set is linearly dependent.

(f) If $\sum_{i=1}^{k} c_i \mathbf{v}_i = \sum_{i=1}^{k} d_i \mathbf{v}_i$, we subtract to obtain

$$(c_1 - d_1)\mathbf{v}_1 + (c_2 - d_2)\mathbf{v}_2 + \cdots + (c_k - d_k)\mathbf{v}_k = \mathbf{0}.$$

If not all the coefficients are zero, the set $\mathbf{v}_1, \mathbf{v}_2, \ldots, \mathbf{v}_k$ would be linearly dependent. Therefore $c_1 = d_1, c_2 = d_2, \ldots, c_k = d_k$.

In the vector space V_n we define the specific vectors

$$\mathbf{e}_1 = (1, 0, \ldots, 0), \qquad \mathbf{e}_2 = (0, 1, 0, \ldots, 0), \ldots, \qquad \mathbf{e}_n = (0, 0, \ldots, 0, 1),$$

in which the vector $\mathbf{e}_i$ has a 1 in the ith place and zeros elsewhere. If, for example, $\mathbf{x} = (x_1, x_2, \ldots, x_n)$ is any vector in V_n, then clearly we can express $\mathbf{x}$ as a linear combination of the vectors $\mathbf{e}_i$. We have

$$\mathbf{x} = \sum_{i=1}^{n} x_i \mathbf{e}_i,$$

so that the vectors $\mathbf{e}_1, \mathbf{e}_2, \ldots, \mathbf{e}_n$ span V. We conclude immediately that the dimension of V_n must be less than or equal to n. Conceivably there could be a basis for V_n with fewer vectors. However, the following theorem, together with the obvious fact that $\mathbf{e}_1, \mathbf{e}_2, \ldots, \mathbf{e}_n$ are linearly independent, allows us to conclude that *the dimension of V_n is exactly n.*

Theorem 2. *Suppose that $\mathbf{v}_1, \mathbf{v}_2, \ldots, \mathbf{v}_k$ are k linearly independent vectors in a vector space V which is spanned by the n vectors $\mathbf{w}_1, \mathbf{w}_2, \ldots, \mathbf{w}_n$. Then $k \leq n$.*

Proof. Let us suppose that $k > n$. We shall reach a contradiction. Since $\mathbf{w}_1, \mathbf{w}_2, \ldots, \mathbf{w}_n$ span V, we may express each $\mathbf{v}_j$ as a linear combination of the $\mathbf{w}_1, \mathbf{w}_2, \ldots, \mathbf{w}_n$. We have

$$\mathbf{v}_j = \sum_{i=1}^{n} a_{ij} \mathbf{w}_i, \qquad j = 1, 2, \ldots, k.$$

The matrix of the coefficients of the $\mathbf{w}_i$ which has n rows and k columns is of rank $\leq n < k$. We now consider the following system of n equations in the k unknowns $c_1, c_2, \ldots, c_k$:

$$a_{11}c_1 + a_{12}c_2 + \cdots + a_{1k}c_k = 0,$$
$$a_{21}c_1 + a_{22}c_2 + \cdots + a_{2k}c_k = 0,$$
$$\vdots \qquad\qquad \vdots \qquad\quad \vdots$$
$$a_{n1}c_1 + a_{n2}c_2 + \cdots + a_{nk}c_k = 0.$$

This homogeneous system with more unknowns than equations always has a non-trivial solution (see the Corollary on page 300). But then

$$\sum_{j=1}^{k} c_j \mathbf{v}_j = \sum_{j=1}^{k} \left(\sum_{i=1}^{n} a_{ij} \mathbf{w}_i \right) c_j = \sum_{i=1}^{n} \left(\sum_{j=1}^{k} a_{ij} c_j \right) \mathbf{w}_i = \sum_{i=1}^{n} 0 \cdot \mathbf{w}_i = \mathbf{0},$$

which contradicts the hypothesis that $\mathbf{v}_1, \mathbf{v}_2, \ldots, \mathbf{v}_k$ are linearly independent.

EXAMPLE 1. Show that the vector space of convergent sequences is not finite dimensional.

Solution. Define the vector $\mathbf{e}_i = (0, 0, \ldots, 0, 1, 0, \ldots)$ in which there is a 1 in the ith place and the remainder of the infinite sequence consists of zeros. Then for every positive integer n, the vectors $\mathbf{e}_1, \mathbf{e}_2, \ldots, \mathbf{e}_n$ are linearly independent. Therefore, according to the above theorem, we cannot possibly have a finite basis.

The same reasoning as in the above example shows that the space of continuous functions on an interval $[a, b]$ cannot be finite dimensional. To see this, we observe that every polynomial is a continuous function on any interval and no n polynomials of different degrees can be linearly dependent.

Suppose that W_1 and W_2 are two subspaces of a vector space V. We define the *intersection of W_1 and W_2* to be the collection of all vectors which are in both W_1 and W_2. Note that the vector $\mathbf{0}$ is in every subspace and, indeed, it may happen that the zero vector is the only one that two subspaces have in common. The following theorem is established by direct verification of the axioms for a vector space. The details are left to the reader.

Theorem 3. *The intersection of any set of subspaces of a vector space V is also a subspace of V.*

The next theorem is important for determining when sets of vectors are linearly independent.

Theorem 4. *Suppose that $\mathbf{w}_1, \mathbf{w}_2, \ldots, \mathbf{w}_n$ form a basis for a vector space V. Let $\mathbf{v}_1, \mathbf{v}_2, \ldots, \mathbf{v}_k$ be k vectors (with $k \leq n$) given by the relations*

$$\begin{aligned}
\mathbf{v}_1 &= a_{11}\mathbf{w}_1 + a_{21}\mathbf{w}_2 + \cdots + a_{n1}\mathbf{w}_n, \\
\mathbf{v}_2 &= a_{12}\mathbf{w}_1 + a_{22}\mathbf{w}_2 + \cdots + a_{n2}\mathbf{w}_n, \\
&\ \ \vdots \qquad \vdots \qquad\qquad\qquad \vdots \\
\mathbf{v}_k &= a_{1k}\mathbf{w}_1 + a_{2k}\mathbf{w}_2 + \cdots + a_{nk}\mathbf{w}_n.
\end{aligned}$$

Then the set $\mathbf{v}_1, \mathbf{v}_2, \ldots, \mathbf{v}_k$ is linearly independent if and only if the $n \times k$ matrix $A = (a_{ij})$ is of rank k.

Proof. The equation

$$\sum_{j=1}^{k} c_j \mathbf{v}_j = \mathbf{0}$$

holds with not all the c_j zero if and only if the homogeneous system

$$\sum_{j=1}^{k} a_{ij} c_j = 0, \qquad i = 1, 2, \ldots, n,$$

has a solution other than the trivial one (i.e., other than the solution $c_1 = c_2 = \cdots = c_k = 0$). According to the Corollary on page 300, the above system has only the trivial solution if and only if A is of rank k. Therefore, if A is of rank less than k, the vectors $\mathbf{v}_1, \mathbf{v}_2, \ldots, \mathbf{v}_k$ are linearly dependent, and if A is of rank k they are linearly independent.

The method we studied for determining the rank of a matrix can now be applied in conjunction with the above theorem to determine when sets of vectors are linearly dependent.

EXAMPLE 2. Suppose that $\mathbf{w}_1, \mathbf{w}_2, \ldots, \mathbf{w}_5$ form a basis for a vector space V. We define the vectors $\mathbf{v}_1, \mathbf{v}_2, \mathbf{v}_3, \mathbf{v}$ by

$$\begin{aligned}
\mathbf{v}_1 &= \mathbf{w}_1 - 2\mathbf{w}_2 + 2\mathbf{w}_3 - \mathbf{w}_4 - 3\mathbf{w}_5, \\
\mathbf{v}_2 &= 2\mathbf{w}_1 + \mathbf{w}_2 - \mathbf{w}_3 + 3\mathbf{w}_4 - 2\mathbf{w}_5, \\
\mathbf{v}_3 &= 3\mathbf{w}_1 - 3\mathbf{w}_2 + \mathbf{w}_3 - 2\mathbf{w}_4 + \mathbf{w}_5, \\
\mathbf{v} &= 2\mathbf{w}_1 - \mathbf{w}_2 + 3\mathbf{w}_3 - 2\mathbf{w}_4 + \mathbf{w}_5.
\end{aligned}$$

Show that $\mathbf{v}_1, \mathbf{v}_2, \mathbf{v}_3$ are linearly independent and determine whether or not $\mathbf{v}$ is in the space spanned by them.

Solution. The vector $\mathbf{v}$ is in the space spanned by $\mathbf{v}_1, \mathbf{v}_2, \mathbf{v}_3$ if and only if $\mathbf{v} = c_1\mathbf{v}_1 + c_2\mathbf{v}_2 + c_3\mathbf{v}_3$. Therefore $\mathbf{v}$ is in the space of $\mathbf{v}_1, \mathbf{v}_2, \mathbf{v}_3$ if we can find a solution to the system

$$\begin{aligned}
c_1 + 2c_2 + 3c_3 &= 2, \\
-2c_1 + c_2 - 3c_3 &= -1, \\
2c_1 - c_2 + c_3 &= 3, \\
-c_1 + 3c_2 - 2c_3 &= -2, \\
-3c_1 - 2c_2 + c_3 &= 1.
\end{aligned} \tag{1}$$

The vectors $\mathbf{v}_1, \mathbf{v}_2, \mathbf{v}_3$ are linearly independent if and only if the system (1) with the right side replaced by zeros has only the zero solution. That is, the three vectors are linearly independent if the matrix of the coefficients in (1) has rank 3. We proceed to reduce the augmented matrix C in (1) and, in doing so, we will automatically reduce the coefficient matrix. We have

$$C = \begin{pmatrix} 1 & 2 & 3 & \vdots & 2 \\ -2 & 1 & -3 & \vdots & -1 \\ 2 & -1 & 1 & \vdots & 3 \\ -1 & 3 & -2 & \vdots & -2 \\ -3 & -2 & 1 & \vdots & 1 \end{pmatrix} \cong \begin{pmatrix} 1 & 2 & 3 & \vdots & 2 \\ 0 & 5 & 3 & \vdots & 3 \\ 0 & -5 & -5 & \vdots & -1 \\ 0 & 5 & 1 & \vdots & 0 \\ 0 & 4 & 10 & \vdots & 7 \end{pmatrix} \cong \begin{pmatrix} 1 & 2 & 3 & \vdots & 2 \\ 0 & 5 & 3 & \vdots & 3 \\ 0 & -5 & -5 & \vdots & -1 \\ 0 & 5 & 1 & \vdots & 0 \\ 0 & 20 & 50 & \vdots & 35 \end{pmatrix}$$

$$C \cong \begin{pmatrix} 1 & 2 & 3 & \vdots & 2 \\ 0 & 5 & 3 & \vdots & 3 \\ 0 & 0 & -2 & \vdots & 2 \\ 0 & 0 & -2 & \vdots & -3 \\ 0 & 0 & 38 & \vdots & 23 \end{pmatrix} \cong \begin{pmatrix} 1 & 2 & 3 & \vdots & 2 \\ 0 & 5 & 3 & \vdots & 3 \\ 0 & 0 & -2 & \vdots & -2 \\ 0 & 0 & 0 & \vdots & -5 \\ 0 & 0 & 0 & \vdots & 61 \end{pmatrix} \cong \begin{pmatrix} 1 & 2 & 3 & \vdots & 2 \\ 0 & 5 & 3 & \vdots & 3 \\ 0 & 0 & -2 & \vdots & 2 \\ 0 & 0 & 0 & \vdots & -5 \\ 0 & 0 & 0 & \vdots & 0 \end{pmatrix}$$

Since the rank of the coefficient matrix is 3 and since the rank of the augmented matrix is 4, we conclude that v_1, v_2, v_3 are linearly independent and that v is not a linear combination of v_1, v_2, and v_3. In fact, the augmented matrix C is the coefficient matrix for v_1, v_2, v_3, and $-v$, so that these four vectors are linearly independent.

Suppose that we are given a set of vectors v_1, v_2, ..., v_k in a vector space V. We have just developed a technique for determining whether or not they are linearly independent. If they happen to be linearly dependent, we seek a way of finding a subset, preferably one as large as possible, which is linearly independent. The next two theorems establish the basis for a method of finding linearly independent subsets.

Theorem 5. *Suppose that* u_1, u_2, ..., u_r *is a sequence of vectors in V such that for each j no u_j is a linear combination of the preceding ones. If* $u_1 \neq 0$, *then* u_1, u_2, ..., u_r *are linearly independent.*

Proof. Suppose that u_1, u_2, ..., u_r are linearly dependent. Then there are constants d_1, d_2, ..., d_r, not all zero, such that

$$d_1 u_1 + d_2 u_2 + \cdots + d_r u_r = 0.$$

Let j be the *largest* integer for which $d_j \neq 0$. Solving the above equation for u_j, we express it as a linear combination of the *preceding* u's. But this contradicts the hypothesis, and therefore u_1, u_2, ..., u_r must be linearly independent.

Theorem 6. (a) *If* v_1, v_2, ..., v_k *span a subspace W which is of dimension $r > 0$ with $r \leq k$, then there is a subset, consisting of r of the v_j, which spans W.*
(b) *If* v_1, v_2, ..., v_k *is a basis for a subspace W of V and if V is of dimension n with $n > k$, then there exist vectors v_{k+1}, v_{k+2}, ..., v_n such that v_1, v_2, ..., v_n form a basis for V.*

Proof. (a) To find the subset which spans W, we begin by denoting the first nonzero vector in the set v_1, v_2, ..., v_k by v_{j_1}. If all the v_j with $j > j_1$ are multiples of v_{j_1}, then v_{j_1} spans W and $r = 1$. Otherwise, denote by v_{j_2} the first v_j in the sequence beyond v_{j_1} which is not a multiple of v_{j_1}. If all the v_j with $j > j_2$ are linear combinations of v_{j_1} and v_{j_2}, then these two vectors span W. Otherwise we select v_{j_3} as the first v_j beyond v_{j_2} which is not a linear combination of v_{j_1} and v_{j_2}. We continue this process, and we finally obtain a set which spans W. By Theorem 5, this set is linearly independent.

(b) Suppose that $\mathbf{w}_1, \mathbf{w}_2, \ldots, \mathbf{w}_n$ is a basis for V. Form the sequence

$$\mathbf{v}_1, \mathbf{v}_2, \ldots, \mathbf{v}_k, \qquad \mathbf{w}_1, \mathbf{w}_2, \ldots, \mathbf{w}_n.$$

Applying the method of part (a), we obtain n vectors, of which the first k are $\mathbf{v}_1, \ldots, \mathbf{v}_k$ and the remaining $n - k$ come from the set $\mathbf{w}_1, \mathbf{w}_2, \ldots, \mathbf{w}_n$.

EXAMPLE 3. Suppose that the set $\mathbf{w}_1, \mathbf{w}_2, \ldots, \mathbf{w}_5$ is a basis for a vector space V, and that the 6 vectors $\mathbf{v}_1, \mathbf{v}_2, \ldots, \mathbf{v}_6$ are given by

$$
\begin{aligned}
\mathbf{v}_1 &= \mathbf{0}, \\
\mathbf{v}_2 &= \mathbf{w}_1 - 2\mathbf{w}_2 - \mathbf{w}_3 + 2\mathbf{w}_4 - 3\mathbf{w}_5, \\
\mathbf{v}_3 &= 2\mathbf{w}_1 - 4\mathbf{w}_2 - 2\mathbf{w}_3 + 4\mathbf{w}_4 - 6\mathbf{w}_5, \\
\mathbf{v}_4 &= 2\mathbf{w}_1 - 3\mathbf{w}_2 - \mathbf{w}_3 + 3\mathbf{w}_4 - 4\mathbf{w}_5, \\
\mathbf{v}_5 &= -\mathbf{w}_1 + 4\mathbf{w}_2 + 3\mathbf{w}_3 - 4\mathbf{w}_4 + 7\mathbf{w}_5, \\
\mathbf{v}_6 &= 2\mathbf{w}_1 - 6\mathbf{w}_2 - 3\mathbf{w}_3 + 7\mathbf{w}_4 - 8\mathbf{w}_5.
\end{aligned}
$$

Find a linearly independent subset of the $\mathbf{v}_j$ by performing the construction described in the proof of Theorem 6(a).

Solution. Since $\mathbf{v}_1 = \mathbf{0}$, we take $j_1 = 2$; i.e., $\mathbf{v}_{j_1} = \mathbf{v}_2$. By inspection, we see that $\mathbf{v}_3 = 2\mathbf{v}_2$ and that $\mathbf{v}_4$ is not a multiple of $\mathbf{v}_2$. We select $j_2 = 4$ and note that $\mathbf{v}_2$ and $\mathbf{v}_4$ are linearly independent. By using the method of Example 2, we determine whether or not $\mathbf{v}_5$ is a linear combination of $\mathbf{v}_2$ and $\mathbf{v}_4$. We omit the details and merely state that $\mathbf{v}_5 = 2\mathbf{v}_4 - 5\mathbf{v}_2$. We proceed to $\mathbf{v}_6$ and see whether or not it is a linear combination of $\mathbf{v}_2$ and $\mathbf{v}_4$. It turns out (by the method of Example 2) that it is not. Therefore $j_3 = 6$ and $\mathbf{v}_2$, $\mathbf{v}_4$, and $\mathbf{v}_6$ form the desired linearly independent set.

Theorem 7. *Suppose that $\mathbf{w}_1, \mathbf{w}_2, \ldots, \mathbf{w}_n$ is a basis for V and that the vectors $\mathbf{v}_1, \mathbf{v}_2, \ldots, \mathbf{v}_k$ are given in terms of the basis vectors by*

$$\mathbf{v}_j = \sum_{i=1}^{n} a_{ij}\mathbf{w}_i, \qquad j = 1, 2, \ldots, k.$$

Then the rank of the matrix $A = (a_{ij})$ equals the dimension of the subspace W spanned by $\mathbf{v}_1, \mathbf{v}_2, \ldots, \mathbf{v}_k$.

Proof. Suppose that the dimension of W is r. From Theorem 6(a), we see that a subset of r linearly independent $\mathbf{v}_j$ spans W. By relabeling the $\mathbf{v}_j$, if necessary, and interchanging the columns of A correspondingly, we get the linearly independent vectors to be $\mathbf{v}_1, \mathbf{v}_2, \ldots, \mathbf{v}_r$. Then $\mathbf{v}_{r+1}, \mathbf{v}_{r+2}, \ldots, \mathbf{v}_k$ are linear combinations of $\mathbf{v}_1, \mathbf{v}_2, \ldots, \mathbf{v}_r$. The rank of the first r columns is r, and the rank of A is r, since the remaining column vectors of A are linear combinations of the first r column vectors.

On the other hand, suppose that the rank of A is r. By following the procedure of Theorem 13, Chapter 6, applied to the *columns* of A, we may reduce it to a form A' in which the column vectors beyond the rth are zero. After this reduction, the first r columns are linearly independent. Since we may pass from A' back to A by elementary column transformations, we see that each of the original column vectors is a linear combination of the first r column vectors in A'. Setting

$$\mathbf{v}'_j = \sum_{i=1}^{n} a'_{ij}\mathbf{w}_i, \qquad j = 1, 2, \ldots, r,$$

where (a'_{ij}) are the elements of A', we conclude that $\mathbf{v}'_1, \mathbf{v}'_2, \ldots, \mathbf{v}'_r$ are linearly independent and that each $\mathbf{v}_j$ is a linear combination of $\mathbf{v}'_1, \mathbf{v}'_2, \ldots, \mathbf{v}'_r$. Consequently, these r vectors span W.

EXAMPLE 4. Consider the vector space V_4 of elements (x_1, x_2, x_3, x_4). Those vectors which satisfy the relations

$$2x_1 - 3x_2 - x_3 + x_4 = 0$$

and

$$x_1 + x_2 + 2x_3 - x_4 = 0$$

can be shown to form a subspace W. Find a basis for W.

Solution. Solving for x_3 and x_4 in terms of x_1 and x_2, we obtain

$$x_3 = -3x_1 + 2x_2, \qquad x_4 = -5x_1 + 5x_2.$$

In other words, every vector (x_1, x_2, x_3, x_4) in W has the form

$$(x_1, x_2, x_3, x_4) = (x_1, x_2, -3x_1 + 2x_2, -5x_1 + 5x_2),$$

which we can write

$$(x_1 + 0, 0 + x_2, -3x_1 + 2x_2, -5x_1 + 5x_2)$$
$$= (x_1, 0, -3x_1, -5x_1) + (0, x_2, 2x_2, 5x_2)$$
$$= x_1(1, 0, -3, -5) + x_2(0, 1, 2, 5).$$

Therefore, we may select $(1, 0, -3, -5)$ and $(0, 1, 2, 5)$ as basis vectors.

EXERCISES

In problems 1 through 10, the vectors $\mathbf{w}_1, \mathbf{w}_2, \ldots, \mathbf{w}_n$ form a basis for V. Determine in each case the dimension r of the subspace W spanned by the $\mathbf{v}_1, \mathbf{v}_2, \ldots, \mathbf{v}_k$, which are defined as indicated. Also determine whether or not $\mathbf{v}$ lies in W; if so, express $\mathbf{v}$ as a linear combination of the vectors $\mathbf{v}_j$.

1. $n = 3$, $\quad k = 2$
   $\mathbf{v}_1 = 2\mathbf{w}_1 - \mathbf{w}_2 - 2\mathbf{w}_3$
   $\mathbf{v}_2 = \mathbf{w}_1 + 2\mathbf{w}_2 - \mathbf{w}_3$
   $\mathbf{v} = 4\mathbf{w}_1 - 7\mathbf{w}_2 - 4\mathbf{w}_3$

2. $n = 3$, $\quad k = 2$
   $\mathbf{v}_1 = \mathbf{w}_1 + 2\mathbf{w}_2 + \mathbf{w}_3$
   $\mathbf{v}_2 = -\mathbf{w}_1 + \mathbf{w}_2 - 2\mathbf{w}_3$
   $\mathbf{v} = 2\mathbf{w}_1 - \mathbf{w}_2 + 3\mathbf{w}_3$

3. $n = 4, \quad k = 2$
   $\mathbf{v}_1 = \mathbf{w}_1 - \mathbf{w}_2 + 2\mathbf{w}_3 + \mathbf{w}_4$
   $\mathbf{v}_2 = 2\mathbf{w}_1 + \mathbf{w}_2 - \mathbf{w}_3 + 2\mathbf{w}_4$
   $\mathbf{v} = \mathbf{w}_1 + 2\mathbf{w}_2 + \mathbf{w}_3 - \mathbf{w}_4$

4. $n = 4, \quad k = 2$
   $\mathbf{v}_1 = 2\mathbf{w}_1 - 3\mathbf{w}_2 - \mathbf{w}_3 + 2\mathbf{w}_4$
   $\mathbf{v}_2 = \mathbf{w}_1 - \mathbf{w}_2 + 2\mathbf{w}_3 - \mathbf{w}_4$
   $\mathbf{v} = \mathbf{w}_1 - 3\mathbf{w}_2 - 8\mathbf{w}_3 + 7\mathbf{w}_4$

5. $n = 5, \quad k = 3$
   $\mathbf{v}_1 = 2\mathbf{w}_1 + \mathbf{w}_2 - \mathbf{w}_3 - 2\mathbf{w}_4 + \mathbf{w}_5$
   $\mathbf{v}_2 = \mathbf{w}_1 - \mathbf{w}_2 + 2\mathbf{w}_3 + \mathbf{w}_4 - \mathbf{w}_5$
   $\mathbf{v}_3 = -\mathbf{w}_1 + 2\mathbf{w}_2 + \mathbf{w}_3 - 2\mathbf{w}_4 - 2\mathbf{w}_5$
   $\mathbf{v} = 7\mathbf{w}_1 - 3\mathbf{w}_2 - 2\mathbf{w}_3 + \mathbf{w}_4 + 5\mathbf{w}_5$

6. $n = 5, \quad k = 3$
   $\mathbf{v}_1 = \mathbf{w}_1 - 2\mathbf{w}_2 + \mathbf{w}_3 + 3\mathbf{w}_4 - \mathbf{w}_5$
   $\mathbf{v}_2 = 2\mathbf{w}_1 + \mathbf{w}_2 - 2\mathbf{w}_3 + 2\mathbf{w}_4 - 3\mathbf{w}_5$
   $\mathbf{v}_3 = -2\mathbf{w}_1 + 2\mathbf{w}_2 + \mathbf{w}_3 - 2\mathbf{w}_4 + \mathbf{w}_5$
   $\mathbf{v} = -9\mathbf{w}_1 + 2\mathbf{w}_2 + 8\mathbf{w}_3 - 7\mathbf{w}_4 + 8\mathbf{w}_5$

7. $n = 5, \quad k = 3$
   $\mathbf{v}_1 = 2\mathbf{w}_1 - \mathbf{w}_2 + \mathbf{w}_3 - 2\mathbf{w}_4 + \mathbf{w}_5$
   $\mathbf{v}_2 = 3\mathbf{w}_1 - 3\mathbf{w}_2 + 4\mathbf{w}_3 - \mathbf{w}_4$
   $\mathbf{v}_3 = \mathbf{w}_1 + \mathbf{w}_2 - 2\mathbf{w}_3 - 3\mathbf{w}_4 + 2\mathbf{w}_5$
   $\mathbf{v} = -\mathbf{w}_1 + 2\mathbf{w}_2 - 7\mathbf{w}_3 + 5\mathbf{w}_4 - 8\mathbf{w}_5$

8. $n = 5, \quad k = 3$
   $\mathbf{v}_1 = 2\mathbf{w}_1 - 3\mathbf{w}_2 + \mathbf{w}_3 - 4\mathbf{w}_4 + \mathbf{w}_5$
   $\mathbf{v}_2 = -2\mathbf{w}_1 + \mathbf{w}_2 + 3\mathbf{w}_3 + 2\mathbf{w}_4 - \mathbf{w}_5$
   $\mathbf{v}_3 = 2\mathbf{w}_1 - 2\mathbf{w}_2 - \mathbf{w}_3 - 3\mathbf{w}_4 + \mathbf{w}_5$
   $\mathbf{v} = -6\mathbf{w}_1 + 4\mathbf{w}_2 + 7\mathbf{w}_3 + 7\mathbf{w}_4 - 3\mathbf{w}_5$

9. $n = 5, \quad k = 3$
   $\mathbf{v}_1 = 2\mathbf{w}_1 + 3\mathbf{w}_2 - 2\mathbf{w}_3 + \mathbf{w}_4 - \mathbf{w}_5$
   $\mathbf{v}_2 = \mathbf{w}_1 + 2\mathbf{w}_2 - \mathbf{w}_3 + 2\mathbf{w}_4 + \mathbf{w}_5$
   $\mathbf{v}_3 = 3\mathbf{w}_1 + 4\mathbf{w}_2 - 3\mathbf{w}_3 - 3\mathbf{w}_5$
   $\mathbf{v} = -\mathbf{w}_2 - 3\mathbf{w}_4 - 3\mathbf{w}_5$

10. $n = 5, \quad k = 3$
    $\mathbf{v}_1 = \mathbf{w}_1 - 2\mathbf{w}_2 + 3\mathbf{w}_3 + 2\mathbf{w}_4 - \mathbf{w}_5$
    $\mathbf{v}_2 = 2\mathbf{w}_1 + \mathbf{w}_2 - \mathbf{w}_3 - 3\mathbf{w}_4 + 2\mathbf{w}_5$
    $\mathbf{v}_3 = -\mathbf{w}_1 + 2\mathbf{w}_2 + 2\mathbf{w}_3 + \mathbf{w}_4 - 3\mathbf{w}_5$
    $\mathbf{v} = -2\mathbf{w}_1 - \mathbf{w}_2 + \mathbf{w}_3 - 2\mathbf{w}_4 - 2\mathbf{w}_5$

In problems 11 through 15, the vectors $\mathbf{w}_1, \mathbf{w}_2, \ldots, \mathbf{w}_n$ form a basis for V. In each case choose a linearly independent subset, as in Example 3.

11. $\mathbf{v}_1 = \mathbf{w}_1 - 2\mathbf{w}_2 + 2\mathbf{w}_3 - \mathbf{w}_4$
    $\mathbf{v}_2 = -2\mathbf{w}_1 + 4\mathbf{w}_2 - 4\mathbf{w}_3 + 2\mathbf{w}_4$
    $\mathbf{v}_3 = 2\mathbf{w}_1 - 3\mathbf{w}_2 + 3\mathbf{w}_3 + \mathbf{w}_4$
    $\mathbf{v}_4 = -\mathbf{w}_1 + \mathbf{w}_2 - 2\mathbf{w}_3 - 2\mathbf{w}_4$

12. $\mathbf{v}_1 = \mathbf{w}_1 + 2\mathbf{w}_2 - \mathbf{w}_3 - 2\mathbf{w}_4$
    $\mathbf{v}_2 = \mathbf{w}_1 + \mathbf{w}_2 - \mathbf{w}_3 - 3\mathbf{w}_4$
    $\mathbf{v}_3 = \mathbf{w}_1 + 3\mathbf{w}_2 - \mathbf{w}_3 - \mathbf{w}_4$
    $\mathbf{v}_4 = 2\mathbf{w}_1 - 3\mathbf{w}_2 + 2\mathbf{w}_3 + \mathbf{w}_4$

13. $\mathbf{v}_1 = \mathbf{w}_1 - 2\mathbf{w}_2 + 2\mathbf{w}_3 + \mathbf{w}_4$
    $\mathbf{v}_2 = 2\mathbf{w}_1 + \mathbf{w}_2 - \mathbf{w}_3 + 3\mathbf{w}_4$
    $\mathbf{v}_3 = -\mathbf{w}_1 + 2\mathbf{w}_2 - 2\mathbf{w}_3 + 2\mathbf{w}_4$
    $\mathbf{v}_4 = \mathbf{w}_1 - \mathbf{w}_2 + \mathbf{w}_3 - \mathbf{w}_4$

14. $\mathbf{v}_1 = \mathbf{w}_1 - 2\mathbf{w}_2 + 2\mathbf{w}_3 - \mathbf{w}_4 - 3\mathbf{w}_5$
    $\mathbf{v}_2 = \mathbf{w}_1 - 3\mathbf{w}_2 + 2\mathbf{w}_3 - \mathbf{w}_4 - 4\mathbf{w}_5$
    $\mathbf{v}_3 = 2\mathbf{w}_1 - 3\mathbf{w}_2 + 4\mathbf{w}_3 - 2\mathbf{w}_4 - 5\mathbf{w}_5$
    $\mathbf{v}_4 = 2\mathbf{w}_1 + 2\mathbf{w}_2 - 3\mathbf{w}_3 + 2\mathbf{w}_4 + 3\mathbf{w}_5$

15. $\mathbf{v}_1 = 2\mathbf{w}_1 - \mathbf{w}_3$
    $\mathbf{v}_2 = 3\mathbf{w}_2 + \mathbf{w}_4$
    $\mathbf{v}_3 = -\mathbf{w}_1 + 2\mathbf{w}_5$
    $\mathbf{v}_4 = \mathbf{w}_1 - 3\mathbf{w}_2 - 2\mathbf{w}_3 - \mathbf{w}_4 + 6\mathbf{w}_5$
    $\mathbf{v}_5 = 6\mathbf{w}_2 + \mathbf{w}_3 + 2\mathbf{w}_4 - 4\mathbf{w}_5$

16. What is the dimension of the vector space of $m \times n$ matrices? Give an argument to support your statement.

17. Show that the totality of vectors (x_1, x_2, x_3, x_4) in V_4 such that the coordinates satisfy the equations

$$a_{11}x_1 + a_{12}x_2 + a_{13}x_3 + a_{14}x_4 = 0,$$
$$a_{21}x_1 + a_{22}x_2 + a_{23}x_3 + a_{24}x_4 = 0,$$

form a subspace of V_4.

18. Show that the totality of vectors $(x_1, x_2, \ldots, x_n)$ in V_n such that the coordinates satisfy a system of m equations of the form

$$\sum_{j=1}^{n} a_{ij}x_j = 0, \qquad i = 1, 2, \ldots, m,$$

form a subspace W of V_n. State a condition on $A = (a_{ij})$ which guarantees that W consists only of $\mathbf{0}$.

In each of problems 19 through 24, choose a set of basis vectors for the locus of each set of equations.

19. $2x_1 - x_2 + 2x_3 = 0$

20. $2x_1 - x_2 + x_3 + 3x_4 = 0$
    $x_1 + x_2 - x_3 - 2x_4 = 0$

21. $x_1 + x_2 - x_3 + 2x_4 - x_5 = 0$
    $-x_1 + 2x_2 + 2x_3 - x_4 + x_5 = 0$

22. $2x_1 + x_2 - x_3 + 2x_4 - 2x_5 = 0$
    $x_1 + 2x_2 + x_3 - x_4 + x_5 = 0$
    $-2x_1 + x_2 + x_3 + 2x_4 - x_5 = 0$

23. $2x_1 + x_2 - x_3 + 2x_4 - x_5 = 0$
    $3x_1 - x_2 - 3x_3 + x_4 + x_5 = 0$
    $x_1 + 3x_2 + x_3 + 3x_4 - 3x_5 = 0$

24. $x_1 + 2x_2 - x_3 - 2x_4 + x_5 = 0$
    $2x_1 - x_2 + x_3 + x_4 - 2x_5 = 0$
    $-x_1 + x_2 + 2x_3 - x_4 + x_5 = 0$

25. Show that the collection of infinite sequences which converge to zero forms an infinite dimensional subspace of the space of convergent sequences.

26. Consider an interval $[a, b]$ and a finite set of points $x_1, x_2, \ldots, x_n$ in this interval. Show that the collection of continuous functions f such that

$$f(x_i) = 0, \qquad i = 1, 2, \ldots, n$$

forms an infinite dimensional subspace of the space of continuous functions.

3. LINEAR TRANSFORMATIONS FROM V_n TO V_n

Let M and N be any two nonempty sets. A **relation** from M to N is a set of ordered pairs (P, Q) in which $P \in M$ and $Q \in N$. The **domain** of the relation consists of all the elements P which occur in any of the pairs, and its **range** consists
of all the elements Q which occur (Fig. 7–1). If the relation is such that no two
of the pairs have the same first element, the relation is called a **transformation**.
We shall use capital letters to denote transformations. For example, a function
of any number of variables, say 3, is a particular case of a transformation in which
the domain M is a collection of ordered number triples and the range N is a portion of the real number system. However, transformations are more general in
that we allow M and N to be arbitrary sets.

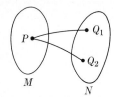
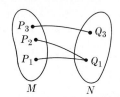
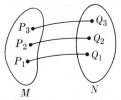

Relation: (P, Q_1) and (P, Q_2) Transformation: No element   One-to-one correspondence
with same first element is   of M has more than one   (c)
possible   image (two elements of M
(a)   may have the same image)
                                  (b)

FIGURE 7–1

If T is a transformation and P is in its domain, we denote the unique corresponding element Q by $T(P)$ and call $T(P)$ the **image of** P under T; we also often
say that "T carries P into Q (or $T(P)$)." If E is a subset of M consisting of points
which are in the domain of T, we call the totality of images $T(P)$ of points P in E
the **image of** E **under** T and denote it by $T(E)$. (Figure 7–2.)

If U is a relation (which may, in particular, be a transformation) from M to N,
we often write

$$U: \quad M \to N.$$

The **inverse relation** of a given relation $U: M \to N$ is defined as the set of all
pairs (Q, P) for which the pair $(P, Q) \in U$. We denote the inverse of U by U^{-1}.
As in the case of functions of one variable, the inverse of a transformation is not
necessarily a transformation. If a transformation U is one to one—that is, if no
two pairs have the same *second* element (as well as not having the same first
element)—then U^{-1} is also a transformation (Fig. 7–3). If U is one to one, then

$$Q = U(P) \quad \text{and} \quad P = U^{-1}(Q)$$

are equivalent transformations. If U carries P into Q, then U^{-1} carries Q back
into P. We can write

$$(U^{-1})^{-1} = U, \quad U^{-1}(U(P)) = P, \quad U(U^{-1}(Q)) = Q$$

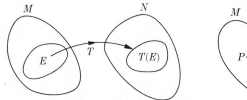

FIGURE 7–2   FIGURE 7–3

for all P in the domain of U and for all Q in the range of U. Note that the range of U is the domain of U^{-1}.

We shall be concerned with transformations in which the domain is the vector space V_n and the range is all or part of the same space V_n. We define a **linear transformation** by the system of equations

$$y_i = \sum_{j=1}^{n} a_{ij}x_j, \qquad i = 1, 2, \ldots, n, \tag{1}$$

in which $\mathbf{x} = (x_1, x_2, \ldots, x_n)$ is an element of the domain and $\mathbf{y} = (y_1, y_2, \ldots, y_n)$ is an element of the range. The transformation is the set of all ordered pairs $(\mathbf{x}, \mathbf{y})$ for which (1) holds. If there were m equations in (1), the system would define a transformation from V_n to V_m. The matrix $A = (a_{ij})$ is called the **matrix of the transformation**. When the transformation (1) is denoted by a single letter, say S, we write

$$S: \quad y_i = \sum_{j=1}^{n} a_{ij}x_j, \qquad i = 1, 2, \ldots, n. \tag{2}$$

EXAMPLE 1. Given the linear transformation

$$T: \quad y_1 = 2x_1 + x_2,$$
$$y_2 = -x_1 + 2x_2.$$

Find $T(1, 2)$, $T(-2, 1)$, and find the equation of the image $T(l)$ consisting of all points $(y_1, y_2) = T(x_1, x_2)$ such that (x_1, x_2) is on the line $l: x_1 - x_2 + 2 = 0$.

Solution. By simple substitution we see that when $x_1 = 1$, $x_2 = 2$, then $y_1 = 4$ and $y_2 = 3$. Therefore $T(1, 2) = (4, 3)$ and, similarly, $T(-2, 1) = (-3, 4)$.

To find the image $T(l)$ of l we first find the inverse to T by solving for x_1, x_2 in terms of y_1, y_2. We get

$$T^{-1}: \quad x_1 = \tfrac{1}{5}(2y_1 - y_2),$$
$$x_2 = \tfrac{1}{5}(y_1 + 2y_2).$$

Then the condition $x_1 - x_2 + 2 = 0$ becomes the condition

$$\tfrac{1}{5}(2y_1 - y_2) - \tfrac{1}{5}(y_1 + 2y_2) + 2 = 0$$

in the image space; that is, $T(l)$ is the set (y_1, y_2) such that $y_1 - 3y_2 + 10 = 0$.

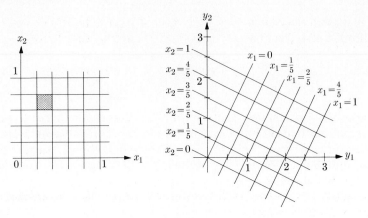

FIGURE 7–4

EXAMPLE 2. Given the transformation T of Example 1, find the equations of the images of the lines $x_1 = i/5$ and $x_2 = j/5$ for $i, j = 0, 1, 2, 3, 4, 5$. Draw and label these images.

Solution. Using the equations for T^{-1} in Example 1, we see that the images of the lines $x_1 = i/5$ and $x_2 = j/5$ have the equations

$$2y_1 - y_2 = i \quad \text{and} \quad y_1 + 2y_2 = j.$$

The given lines in the (x_1, x_2)-plane and their images in the (y_1, y_2)-plane are plotted in Fig. 7–4. The shaded square in the (y_1, y_2) plane is the image of the shaded square in the (x_1, x_2) plane.

Suppose that we have two transformations

$$U: \quad M \to N \quad \text{and} \quad V: \quad N \to R.$$

We define the **product transformation** VU as that transformation having domain D consisting of all P in M such that $U(P)$ is in the domain of V and such that $(VU)(P) = V[U(P)]$ for all P in D. We write

$$VU: \quad M \to R.$$

The transformation VU is also called the **composition** of the transformations U and V.

In particular, if we have two linear transformations,

$$U: \quad y_i = \sum_{j=1}^{n} a_{ij}x_j, \quad i = 1, 2, \ldots, n,$$

$$V: \quad z_i = \sum_{k=1}^{n} b_{ik}y_k, \quad i = 1, 2, \ldots, n,$$

(3)

then we can easily form their product. We observe that since the domain of U

is all of V_n and since the domain of V is all of V_n, the domain of VU is all of V_n. Substitution of y_i from the first system of (3) into the second gives

$$VU: \quad z_i = \sum_{j=1}^{n} \left[\sum_{k=1}^{n} b_{ik} a_{kj} \right] x_j, \quad i = 1, 2, \ldots, n. \tag{4}$$

Letting A and B denote the matrices of the transformations U and V, respectively, we see that the brackets in the system (4) give the formula for the elements in the product of the matrices A and B. (See Chapter 6, Section 2, page 268.) We state the conclusion in the form of a theorem.

Theorem 8. *The matrix of the product VU of two linear transformations is the product BA of their matrices.*

EXAMPLE 3. Verify Theorem 8 in the case that

$$U: \begin{array}{l} y_1 = x_1 - 2x_2, \\ y_2 = 2x_1 + x_2; \end{array} \qquad V: \begin{array}{l} z_1 = y_1 + y_2, \\ z_2 = -y_1 + y_2. \end{array}$$

Solution. We obtain

$$\begin{array}{l} z_1 = 3x_1 - x_2, \\ z_2 = x_1 + 3x_2, \end{array} \quad B = \begin{pmatrix} 1 & 1 \\ -1 & 1 \end{pmatrix}, \quad A = \begin{pmatrix} 1 & -2 \\ 2 & 1 \end{pmatrix},$$

$$BA = \begin{pmatrix} 1 \cdot 1 + 1 \cdot 2 & 1 \cdot (-2) + 1 \cdot 1 \\ -1 \cdot 1 + 1 \cdot 2 & (-1) \cdot (-2) + 1 \cdot 1 \end{pmatrix} = \begin{pmatrix} 3 & -1 \\ 1 & 3 \end{pmatrix}.$$

EXERCISES

In each of problems 1 through 10, find the images of the given points and the equations of the images of the given curves under the transformation T.

1. $T: \begin{array}{l} y_1 = 3x_1 \\ y_2 = 2x_2 \end{array}$ Find images of $(2, -1)$, $x_1 - 2x_2 = 4$, and $x_1^2 + x_2^2 = 1$.

2. $T: \begin{array}{l} y_1 = 3x_2 \\ y_2 = -2x_1 \end{array}$ $(-1, 2)$, $x_1 + x_2 = 3$, and $x_1^2 + x_2^2 = 1$

3. $T: \begin{array}{l} y_1 = x_1 - x_2 \\ y_2 = x_1 + x_2 \end{array}$ $(2, 1)$, $(-1, 2)$, and $x_1^2 + x_2^2 = 1$

4. $T: \begin{array}{l} y_1 = 2x_1 - x_2 \\ y_2 = x_1 - x_2 \end{array}$ Find images of the lines $x_1 = 0, \frac{1}{2}, 1$, and the lines $x_2 = 0, \frac{1}{2}, 1$.

5. $T: \begin{array}{l} y_1 = 2x_1 + x_2 \\ y_2 = x_1 + x_2 \end{array}$ $x_1 + 2x_2 + 3 = 0$ and $x_1^2 + x_2^2 = 4$

6. $T: \begin{array}{l} y_1 = 2x_1 - x_2 \\ y_2 = x_1 + x_2 \end{array}$ Find image of the ellipse $5x_1^2 - 2x_1x_2 + 2x_2^2 = 4$.

7. $T: \begin{array}{l} y_1 = x_1 - 2x_2 \\ y_2 = 2x_1 + x_2 \end{array}$ Lines $x_1 = 0, \frac{1}{2}, 1$, and lines $x_2 = 0, \frac{1}{2}, 1$

8. T:
$$y_1 = 3x_1 - 2x_2$$
$$y_2 = x_1 + 2x_2$$
Hyperbola: $x_1^2 - 2x_1x_2 = 1$

9. T:
$$y_1 = 2x_1 - x_2 + x_3$$
$$y_2 = x_1 + 2x_2 + x_3$$
$$y_3 = x_2 - x_3$$
Find images of the planes
$$x_1 = 0, \quad x_3 = 0, \quad \text{and}$$
$$x_1 + x_2 + x_3 = 1.$$

10. T:
$$y_1 = x_1 + x_2 - x_3$$
$$y_2 = 3x_1 + x_2 - x_3$$
$$y_3 = 2x_1 - x_2 + 4x_3$$
Find images of the line
$$x_1 = 2t, \quad x_2 = t + 1, \quad x_3 = 2t - 1,$$
and the sphere $x_1^2 + x_2^2 + x_3^2 = 1.$

In problems 11 through 14, make a linear change of variables so that the image of the given locus has the equation $y_1^2 + y_2^2 = C_1$ or $y_1^2 + y_2^2 + y_3^2 = C_2$. [*Hint.* Complete the square.]

11. $x_1^2 + 2x_1x_2 + 5x_2^2 = 4$ 12. $3x_1^2 - 6x_1x_2 + 7x_2^2 = 9$

13. $x_1^2 + 5x_2^2 + 6x_3^2 - 2x_1x_2 + 4x_1x_3 - 8x_2x_3 = 4$

14. $6x_1^2 + 2x_2^2 + 2x_3^2 - 2x_1x_2 - 2x_2x_3 = 4$

In problems 15 through 18, verify Theorem 8.

15. U:
$$y_1 = 2x_1 - x_2$$
$$y_2 = x_1 + x_2$$
V:
$$z_1 = y_1 - y_2$$
$$z_2 = y_1 + 2y_2$$

16. U:
$$y_1 = x_1 + x_2$$
$$y_2 = 2x_1 + 3x_2$$
V:
$$z_1 = 2y_1 - y_2$$
$$z_2 = y_1 + y_2$$

17. U:
$$y_1 = x_1 - x_2 - 2x_3$$
$$y_2 = 2x_1 + x_2 - x_3$$
$$y_3 = -x_1 + x_2 + 2x_3$$
V:
$$z_1 = 2y_1 + y_2 - y_3$$
$$z_2 = y_1 - y_2 + 2y_3$$
$$z_3 = -y_1 + y_2 - 2y_3$$

18. U:
$$y_1 = x_1 - x_2 + x_4$$
$$y_2 = x_2 - x_3 + x_4$$
$$y_3 = 2x_1 + 3x_3$$
$$y_4 = x_1 - x_4$$
V:
$$z_1 = 2y_1 + y_2$$
$$z_2 = y_2 - y_3 + y_4$$
$$z_3 = 3y_1 + y_3$$
$$z_4 = y_3 - y_4$$

19. Given the linear transformations

U:
$$y_1 = 2x_1 - x_2 + x_3$$
$$y_2 = x_1 - 2x_2 - 2x_3$$
$$y_3 = x_1 + x_2 - x_3$$
V:
$$z_1 = y_1 - 2y_2 - 3y_3$$
$$z_2 = 2y_1 - y_2 + y_3$$
$$z_3 = y_1 + y_2 - y_3$$

Verify, by computing the appropriate matrices, that $UV \neq VU$.

20. Given the linear transformations

U:
$$y_1 = x_1 - x_2$$
$$y_2 = 2x_1 - x_2$$
V:
$$z_1 = y_1 + 2y_2$$
$$z_2 = 3y_1 - y_2$$
W:
$$w_1 = z_1 + z_2$$
$$w_2 = -z_1 + z_2$$

By computing the appropriate matrices, verify that $W(VU) = (WV)U$.

4. THE INVERSE OF A MATRIX

The **additive inverse** of any real number a is $-a$. It is the number which, when added to a, gives zero. The **multiplicative inverse** of a number a is $1/a$. It is the number which, when multiplied by a, gives 1. We seek the analogous quantities for a matrix A. The **additive inverse** is easy; it is the matrix $-A$. The **multiplicative**

inverse of a matrix A is a much more complicated object. First of all, we shall only consider *square matrices*. Then, recalling that the $n \times n$ matrix I denotes the matrix with ones on the diagonal and zeros elsewhere, we say that the $n \times n$ matrix B is the **multiplicative inverse** (or simply the **inverse**), of the $n \times n$ matrix A if and only if $AB = I$. It is the purpose of this section to determine some properties of such inverse matrices. In order to do so, we first establish a theorem on determinants.

We recall that a square matrix is nonsingular if and only if its determinant is not zero. We begin by proving two lemmas.

Lemma 1. Suppose $a_0, a_1, a_2, \ldots, a_n$ are numbers and h is any positive number. If the polynomial equation in λ

$$a_0 + a_1\lambda + a_2\lambda^2 + \cdots + a_n\lambda^n = 0$$

holds for all $|\lambda| \leq h$, then $a_0 = a_1 = a_2 = \cdots = a_n = 0$.

Proof. Setting $\lambda = 0$, we get $a_0 = 0$. Since all derivatives of the polynomial are zero for $-h \leq \lambda \leq h$, we find by successively evaluating the derivatives with respect to λ at $\lambda = 0$ that $a_1 = 0, a_2 = 0, \ldots, a_n = 0$.

Lemma 2. (a) *The determinant of an $n \times n$ matrix $A = (a_{ij})$ is a continuous function of each a_{ij}.*

(b) *For each $n \times n$ matrix A and each $\epsilon > 0$, there exists a nonsingular matrix $A' = (a'_{ij})$ such that*

$$|a'_{ij} - a_{ij}| < \epsilon \quad for \quad i = 1, 2, \ldots, n, \quad j = 1, 2, \ldots, n. \tag{1}$$

Proof. (a) The result is obvious if $n = 1$. Then, using induction on n and the expansion in cofactors, we obtain the result.

(b) To prove (b), we define

$$A' = A - \lambda I,$$

which implies that

$$a'_{ij} = a_{ij} - \lambda \, \delta_{ij}.$$

Then det A' is a polynomial of degree n in λ, the coefficient of λ^n being $(-1)^n$; hence the polynomial is nonvanishing. If this polynomial were identically zero in an interval $|\lambda| < \epsilon$, then, by Lemma 1, the coefficients would all vanish. Therefore, a sufficiently small λ may be selected so that (1) holds and det $A' \neq 0$.

With the aid of Lemma 2 we can establish the following important theorem.

Theorem 9. *If A and B are $n \times n$ matrices, then*

$$\det AB = (\det A)(\det B).$$

Proof. Suppose, first, that A and B are nonsingular and let $C = AB$. We observe that if A' is obtained from A by an elementary transformation of type (c) (see Chapter 6, Section 7)—i.e., by multiplying a row (or column) by a constant and adding the result to another row (or column)—then det $A' = $ det A. We introduce another type of elementary transformation known as type (d): *interchanging two rows (or columns) and changing the sign of one of them.* If A' is obtained from A by an elementary row transformation of type (d), we again have det $A' = $ det A.

We note that if A' is obtained from A by elementary row transformations of type (c) and (d), then the matrix $C' = A'B$ is obtained from C by the same elementary row transformations. Therefore det $C' = $ det C. Similarly, if B' is obtained from B by elementary column transformations of type (c) and (d), and if $C'' = AB'$, then det $B' = $ det B and det $C'' = $ det C.

Since A and B are nonsingular, we may reduce A to a diagonal matrix A', using only elementary row transformations of type (c) and (d), and B can be reduced to a diagonal matrix B' by such transformations involving only columns. If we define $C''' = A'B'$, we conclude that

$$\det A' = \det A, \qquad \det B' = \det B, \qquad \det C''' = \det C. \qquad (2)$$

But since A', B', and C''' are diagonal, we must have

$$A' = \begin{pmatrix} a'_{11} & & & 0 \\ & a'_{22} & & \\ & & \ddots & \\ 0 & & & a'_{nn} \end{pmatrix}, \qquad B' = \begin{pmatrix} b'_{11} & & & 0 \\ & b'_{22} & & \\ & & \ddots & \\ 0 & & & b'_{nn} \end{pmatrix},$$

$$C''' = \begin{pmatrix} a'_{11}b'_{11} & & & 0 \\ & a'_{22}b'_{22} & & \\ & & \ddots & \\ 0 & & & a'_{nn}b'_{nn} \end{pmatrix},$$

from which we immediately conclude that det $C''' = $ (det A') (det B').

Taking (2) into account, we get the result for nonsingular matrices. Suppose now that A is singular and B is not. According to Lemma 2(b), we can find a nonsingular matrix A' whose elements are as close to those of A as we please. Defining $C' = A'B$, we have det $C' = $ (det A') (det B). Therefore, as $A' \to A$, we conclude from Lemma 1 that det $C' \to $ det C, and so det $C = 0$. When both A and B are singular the proof is similar.

Let U be a linear transformation defined by

$$U: \quad y_i = \sum_{j=1}^{n} a_{ij}x_j, \qquad i = 1, 2, \ldots, n, \qquad (3)$$

with matrix denoted by A. In addition, if U is one to one (so that the range is all of V_n, as is seen below) it has an inverse which we denote by $V = U^{-1}$. Suppose that the matrix of V is B. Then, from the definition of the product of two transformations and the definition of inverse, we have

$$VU = U^{-1}U = I \qquad \text{and} \qquad UV = UU^{-1} = I,$$

where I is the transformation from V_n to V_n which carries every vector into itself. According to Theorem 8, we obtain the matrix equation

$$BA = AB = I,$$

where now I is the $n \times n$ identity matrix. The matrix B corresponding to the inverse transformation $V = U^{-1}$ is called the **inverse matrix** of A, and we denote it by A^{-1}. We write

$$A^{-1}A = AA^{-1} = I.$$

Note that A^{-1} is the multiplicative inverse discussed at the beginning of the section. The term "multiplicative" is usually omitted.

Thinking of vectors in V_n as column vectors (i.e., $n \times 1$ matrices), we can write the transformation (3) in the abbreviated form

$$U: \quad Y = AX,$$

where

$$X = \begin{pmatrix} x_1 \\ x_2 \\ \vdots \\ x_n \end{pmatrix}, \qquad Y = \begin{pmatrix} y_1 \\ y_2 \\ \vdots \\ y_n \end{pmatrix}.$$

If we also have $V: \quad Z = BY$, then the rule for the product of two transformations yields

$$VU: \quad Z = B(AX) = (BA)X,$$

where the associative law for the product of matrices was used to obtain the last equation.

Theorem 10. (a) *The linear transformation U with matrix A has an inverse if and only if A is nonsingular.*
(b) *The inverse of A may be found by solving the system (3) for the x_i in terms of the y_j.*
(c) *If A is nonsingular, then*

$$A^{-1} = \frac{1}{a}(A_{ji}), \qquad where \qquad a = \det A \tag{4}$$

and A_{pq} is the cofactor of a_{pq}; the symbol (A_{ji}) represents the matrix with A_{ji} in the ith row and jth column.

Proof. If U has an inverse, then $AA^{-1} = I$ implies that $(\det A)(\det A^{-1}) = \det I = 1$. Therefore $\det A$ cannot vanish. We can use Cramer's Rule to solve for the x_i in terms of the y_j and get A^{-1}. Assuming, conversely, that $\det A \neq 0$, we choose p between 1 and n and multiply the ith equation in (3) by A_{ip}. We get

$$A_{ip}y_i = \sum_{j=1}^{n} A_{ip}a_{ij}x_j.$$

Now, summing on i from 1 to n, we obtain

$$\sum_{i=1}^{n} A_{ip} y_i = \sum_{i=1}^{n} \sum_{j=1}^{n} A_{ip} a_{ij} x_j = \sum_{j=1}^{n} \left[\sum_{i=1}^{n} A_{ip} a_{ij} \right] x_j.$$

The quantity in the brackets on the right is precisely $a\delta_{pj}$ [see Theorem 10(a) of Chapter 6, page 284]. Therefore

$$\sum_{i=1}^{n} A_{ip} y_i = a \sum_{j=1}^{n} \delta_{pj} x_j = a x_p.$$

The conclusion of part (c) of the theorem and the fact that U has an inverse are both immediate consequences of this last equation.

While formula (4) has theoretical interest and, in fact, will be useful later, it is not practical for finding A^{-1} when A is given. We may use this formula if n is 2 or perhaps 3, but for larger n it is usually faster to solve for the x_i in terms of the y_j and read off A^{-1} as the coefficient matrix of the y_j. Either the method of elimination, or the method of determinants (Cramer's Rule), or the method we learned for reducing augmented matrices may be used to carry out the process.

EXAMPLE 1. Find A^{-1}, given

$$A = \begin{pmatrix} 2 & 1 \\ -1 & 1 \end{pmatrix}.$$

Solution. Since $a = \det A = 3$, we obtain from (4)

$$A^{-1} = \frac{1}{3} \begin{pmatrix} 1 & -1 \\ 1 & 2 \end{pmatrix}.$$

We may check by calculating

$$A^{-1}A = \frac{1}{3} \begin{pmatrix} 1 & -1 \\ 1 & 2 \end{pmatrix} \begin{pmatrix} 2 & 1 \\ -1 & 1 \end{pmatrix} = \frac{1}{3} \begin{pmatrix} 3 & 0 \\ 0 & 3 \end{pmatrix} = \begin{pmatrix} 1 & 0 \\ 0 & 1 \end{pmatrix} = I.$$

EXAMPLE 2. Find A^{-1}, given

$$A = \begin{pmatrix} 1 & -1 & 2 \\ 2 & 1 & -1 \\ 3 & 2 & -2 \end{pmatrix}.$$

Solution. We start with the augmented matrix for the system (3), taking for a_{ij} the elements of A. We get

$$\begin{pmatrix} 1 & -1 & 2 & \vdots & y_1 \\ 2 & 1 & -1 & \vdots & y_2 \\ 3 & 2 & -2 & \vdots & y_3 \end{pmatrix} \cong \begin{pmatrix} 1 & -1 & 2 & \vdots & y_1 \\ 0 & 3 & -5 & \vdots & y_2 - 2y_1 \\ 0 & 5 & -8 & \vdots & y_3 - 3y_1 \end{pmatrix}$$

$$\cong \begin{pmatrix} 1 & -1 & 2 & \vdots & y_1 \\ 0 & 3 & -5 & \vdots & y_2 - 2y_1 \\ 0 & 15 & -24 & \vdots & 3y_3 - 9y_1 \end{pmatrix} \cong \begin{pmatrix} 1 & -1 & 2 & \vdots & y_1 \\ 0 & 3 & -5 & \vdots & y_2 - 2y_1 \\ 0 & 0 & 1 & \vdots & 3y_3 - 5y_2 + y_1 \end{pmatrix}.$$

Therefore, the system with coefficient matrix A is equivalent to the system

$$x_3 = 3y_3 - 5y_2 + y_1,$$
$$3x_2 - 5x_3 = y_2 - 2y_1,$$
$$x_1 - x_2 + 2x_3 = y_1.$$

We easily solve this for the x_i in terms of the y_j, obtaining

$$\begin{aligned} x_1 &= 2y_2 - y_3 \\ x_2 &= y_1 - 8y_2 + 5y_3 \\ x_3 &= y_1 - 5y_2 + 3y_3 \end{aligned} \quad \text{and} \quad A^{-1} = \begin{pmatrix} 0 & 2 & -1 \\ 1 & -8 & 5 \\ 1 & -5 & 3 \end{pmatrix}.$$

Theorem 11. (a) *If A and B are nonsingular matrices, then $(AB)^{-1} = B^{-1}A^{-1}$.*
(b) *If A is nonsingular, then $(A^t)^{-1} = (A^{-1})^t$.*

Proof. (a) We simply form the product $(AB)(B^{-1}A^{-1})$. Using the associative law for matrices, we find $(AB)(B^{-1}A^{-1}) = A(BB^{-1})A^{-1} = AIA^{-1} = AA^{-1} = I$. Therefore $B^{-1}A^{-1}$ must be the inverse of AB. The proof of (b) is left for the student. (See Exercise 15.)

EXERCISES

In problems 1 through 4, verify that det $(AB) = (\det A)(\det B)$.

1. $A = \begin{pmatrix} 2 & 1 \\ 1 & 2 \end{pmatrix},$ $\qquad$ $B = \begin{pmatrix} 1 & -1 \\ -2 & 3 \end{pmatrix}$

2. $A = \begin{pmatrix} 1 & 2 \\ 3 & 4 \end{pmatrix},$ $\qquad$ $B = \begin{pmatrix} 2 & -1 \\ 1 & 2 \end{pmatrix}$

3. $A = \begin{pmatrix} 2 & 1 & -1 \\ 1 & 2 & 1 \\ -1 & -1 & 2 \end{pmatrix},$ $\qquad$ $B = \begin{pmatrix} 1 & -1 & 2 \\ -1 & 2 & 1 \\ 1 & 0 & 2 \end{pmatrix}$

4. $A = \begin{pmatrix} 1 & 0 & 2 & -1 \\ -1 & 1 & 0 & 2 \\ 2 & 0 & 2 & -1 \\ 1 & -1 & 0 & 1 \end{pmatrix},$ $\qquad$ $B = \begin{pmatrix} 2 & 1 & 0 & -1 \\ -1 & 1 & 2 & 0 \\ 0 & -1 & 1 & 2 \\ 1 & -1 & 0 & 2 \end{pmatrix}$

5. If A and B are 2×2 matrices, prove or disprove the statement

$$\det (A + B) \le \det A + \det B.$$

6. (a) Suppose that A and B are $n \times n$ square matrices and C is the $2n \times 2n$ matrix given by

$$C = \left(\begin{array}{c|c} A & 0 \\ \hline 0 & B \end{array} \right),$$

where the zeros indicate that all elements of C except those in A and B are zero. Show that

$$\det C = (\det A)(\det B).$$

[*Hint.* Use elementary transformations.]

(b) Prove the result of part (a) if C has the form

$$C = \begin{pmatrix} A & 0 \\ D & B \end{pmatrix},$$

where D is any matrix.

In problems 7 through 14, find the inverse of each matrix.

7. $\begin{pmatrix} 1 & 1 \\ 1 & 2 \end{pmatrix}$
8. $\begin{pmatrix} 1 & -1 \\ 1 & 2 \end{pmatrix}$
9. $\begin{pmatrix} 2 & -1 \\ 1 & 2 \end{pmatrix}$

10. $\begin{pmatrix} 2 & -1 & 1 \\ -1 & 1 & -2 \\ 2 & -1 & 2 \end{pmatrix}$
11. $\begin{pmatrix} 1 & -1 & 1 \\ -1 & 2 & -1 \\ 2 & -1 & 1 \end{pmatrix}$
12. $\begin{pmatrix} 2 & 0 & -3 \\ 1 & 4 & -1 \\ -2 & 5 & 3 \end{pmatrix}$

13. $\begin{pmatrix} 1 & -1 & 2 & 1 \\ -1 & 2 & -3 & 1 \\ 2 & 0 & 3 & 5 \\ 2 & -1 & 2 & 6 \end{pmatrix}$
14. $\begin{pmatrix} 1 & 2 & -1 & 1 \\ 1 & 3 & -3 & 0 \\ -1 & -1 & 0 & 0 \\ -2 & -2 & -3 & -3 \end{pmatrix}$

15. Prove Theorem 11(b); i.e., show that if A is nonsingular, then $(A^t)^{-1} = (A^{-1})^t$. [*Hint.* Either proceed by induction or use the relation $(AB)^t = B^t A^t$, together with Theorem 11(a).]

16. A square matrix is called **orthogonal** if $A^t = A^{-1}$. Show that if A is orthogonal, $\det A = \pm 1$.

5. LINEAR CHANGES OF VARIABLE IN MULTIPLE INTEGRALS

The matrix of a nonsingular linear transformation has a nonvanishing determinant. We shall now show that the absolute value of such a determinant has an interesting geometric interpretation. In doing so, we obtain an additional tool for the evaluation of double and triple integrals.

The totality of points in the plane which satisfy a linear inequality of the type $ax_1 + bx_2 + c > 0$ forms a half-plane consisting of all points on one side of the line $ax_1 + bx_2 + c = 0$. (See *First Course*, page 57 ff.) Note that we are using x_1 and x_2 for rectangular coordinates in place of x and y. All points (x_1, x_2) in the plane which satisfy a pair of inequalities of the form

$$\alpha \le ax_1 + bx_2 + c \le \beta, \qquad \beta > \alpha,$$

(Fig. 7–5) lie in the strip bounded by the parallel lines

$$ax_1 + bx_2 + c = \alpha \quad \text{and} \quad ax_1 + bx_2 + c = \beta.$$

Since the bounding lines are included in the strip, we say the strip is *closed*. The locus of the four simultaneous inequalities

$$\alpha_1 \le a_1 x_1 + b_1 x_2 + c_1 \le \beta_1, \qquad \alpha_2 \le a_2 x_1 + b_2 x_2 + c_2 \le \beta_2$$

is a closed parallelogram, provided that $a_1 b_2 - a_2 b_1 \ne 0$ (Fig. 7–6).

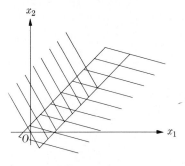

FIGURE 7–5

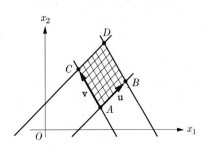

FIGURE 7–6

Now, suppose that $ABDC$ is a parallelogram as shown in Fig. 7–6, and suppose we write

$$\mathbf{u} = \mathbf{u}[\overrightarrow{AB}] = d_1\mathbf{i} + e_1\mathbf{j}, \qquad \mathbf{v} = \mathbf{v}[\overrightarrow{AC}] = d_2\mathbf{i} + e_2\mathbf{j},$$

in which the usual conventions are used for the unit vectors $\mathbf{i}$ and $\mathbf{j}$. Also, $\overrightarrow{AB}$ and $\overrightarrow{AC}$ are representative directed line segments of the vectors $\mathbf{u}$ and $\mathbf{v}$, respectively. If we imagine the (x_1, x_2) plane as part of three-space, we may employ the definition of the cross product of two vectors to write

$$|\mathbf{u} \times \mathbf{v}| = |d_1e_2 - d_2e_1|. \tag{1}$$

It is an exercise in elementary geometry to verify that the quantity (1) is exactly the area of $ABDC$. We obtain the formula

$$\text{Area } ABDC = |d_1e_2 - d_2e_1|.$$

In three-space, the locus of a linear inequality of the form $a_{11}x_1 + a_{12}x_2 + a_{13}x_3 + b_1 > 0$ consists of all points on one side of a plane; it is called a *half-space*. (See Chapter 1, page 39.) In analogy with the above discussion for the relation between linear inequalities and parallelograms, we see that the simultaneous locus of the six inequalities

$$\alpha_i \leq \sum_{j=1}^{3} b_{ij}x_j \leq \beta_i, \qquad \beta_i > \alpha_i, \qquad i = 1, 2, 3$$

is a parallelepiped S as shown in Fig. 7–7. We must assume that the matrix $B = (b_{ij})$ is nonsingular. If $\overrightarrow{AB}, \overrightarrow{AC}, \overrightarrow{AD}$ are three concurrent edges of the parallelepiped S, and if we denote

$$\mathbf{u} = \mathbf{u}[\overrightarrow{AB}] = e_1\mathbf{i} + f_1\mathbf{j} + g_1\mathbf{k}, \qquad \mathbf{v} = \mathbf{v}[\overrightarrow{AC}] = e_2\mathbf{i} + f_2\mathbf{j} + g_2\mathbf{k},$$
$$\mathbf{w} = \mathbf{w}[\overrightarrow{AD}] = e_3\mathbf{i} + f_3\mathbf{j} + g_3\mathbf{k},$$

then we can see that

$$(\mathbf{u} \times \mathbf{v}) \cdot \mathbf{w} = \begin{vmatrix} e_1 & f_1 & g_1 \\ e_2 & f_2 & g_2 \\ e_3 & f_3 & g_3 \end{vmatrix}.$$

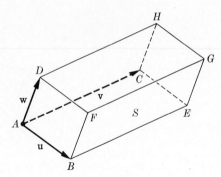

FIGURE 7–7

It can also be shown that the volume $V(S)$ is given by the formula

$$V(S) = |(\mathbf{u} \times \mathbf{v}) \cdot \mathbf{w}| .$$

(See Morrey, *University Calculus*, page 535.)

The basic result of this section depends on the next simple and useful lemma.

Lemma 3. Suppose that

$$U: \quad \begin{aligned} y_1 &= a_{11}x_1 + a_{12}x_2 \\ y_2 &= a_{21}x_1 + a_{22}x_2 \end{aligned} \tag{2}$$

is a nonsingular linear transformation in the plane. If S is a rectangle determined by the inequalities

$$c_1 \leq x_1 \leq d_1, \qquad c_2 \leq x_2 \leq d_2,$$

then its image U(S) is a parallelogram with area

$$|\det A| \cdot \text{area } S,$$

A being the matrix of the transformation.

Proof. Since U is nonsingular, we may solve (2) for x_1, x_2; we obtain

$$\begin{aligned} x_1 &= b_{11}y_1 + b_{12}y_2, \\ x_2 &= b_{21}y_1 + b_{22}y_2, \end{aligned}$$

with $B = A^{-1}$. Therefore the image $U(S)$ consists of all (y_1, y_2) which satisfy the inequalities

$$\begin{aligned} c_1 &\leq b_{11}y_1 + b_{12}y_2 \leq d_1, \\ c_2 &\leq b_{21}y_1 + b_{22}y_2 \leq d_2. \end{aligned}$$

According to the discussion at the beginning of this section, these inequalities determine a parallelogram. This parallelogram has vertices at $U(c_1, c_2)$, $U(d_1, c_2)$, $U(c_1, d_2)$, and $U(d_1, d_2)$, and we denote these by A, B, C, and D, respectively

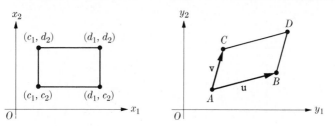

FIGURE 7–8

(Fig. 7–8). By performing the necessary subtractions in (2), we get

$$y_{1B} - y_{1A} = a_{11}(d_1 - c_1), \qquad y_{2B} - y_{2A} = a_{21}(d_1 - c_1),$$
$$y_{1C} - y_{1A} = a_{12}(d_2 - c_2), \qquad y_{2C} - y_{2A} = a_{22}(d_2 - c_2).$$

Therefore

$$\text{area } ABDC = |\mathbf{u}[\overrightarrow{AB}] \times \mathbf{v}[\overrightarrow{AC}]| = |a_{11}a_{22} - a_{12}a_{21}|(d_1 - c_1)(d_2 - c_2)$$
$$= |\det A| A(S).$$

A transformation of the form

$$x_1' = x_1 + h, \qquad y_1' = y_1 + k,$$

consists of a pure translation. Since both areas and shapes of regions are unaffected in such a transformation, we can extend the lemma in the following way.

Corollary to Lemma 3. *The result of Lemma 3 holds for transformations of the form*

$$U: \begin{array}{l} y_1 = a_{11}x_1 + a_{12}x_2 + b_1 \\ y_2 = a_{21}x_1 + a_{22}x_2 + b_2 \end{array}$$

We now establish the theorem to be used in the applications.

Theorem 12. *Suppose that S is any region in the plane with area $A(S)$. If U is the nonsingular transformation*

$$U: \begin{array}{l} y_1 = a_{11}x_1 + a_{12}x_2 + b_1 \\ y_2 = a_{21}x_1 + a_{22}x_2 + b_2 \end{array}$$

then the image $U(S)$ has area $A[U(S)]$ and

$$A[U(S)] = |\det A| \cdot A(S). \tag{3}$$

A denotes the matrix of the transformation.

Proof. Introduce a network of squares in the (x_1, x_2)-plane by drawing lines parallel to the coordinate axes at distance apart $1/2^k$. We get a network for each

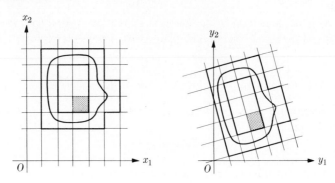

FIGURE 7–9

$k = 0, 1, 2, \ldots$. Let S_k^- denote the sum of all those squares of the kth network such that each one is completely interior to S. Let S_k^+ denote the sum of all those squares such that each one contains one or more points of S. Then $U(S_k^-)$ and $U(S_k^+)$ are each sums of parallelograms which are images under U of the regions S_k^- and S_k^+ (Fig. 7–9). By adding the areas of these parallelograms in the (y_1, y_2)-plane, we obtain

$$A[U(S_k^-)] \le A^-[U(S)] \le A^+[U(S)] \le A[U(S_k^+)].$$

Employing the result of Lemma 3, we have

$$A[U(S_k^-)] = |\det A| A(S_k^-), \qquad A[U(S_k^+)] = |\det A| A(S_k^+).$$

Letting k tend to infinity, we conclude that

$$|\det A| A^-(S) \le A^-[U(S)] \le A^+[U(S)] \le |\det A| A^+(S).$$

Since $A^-(S) = A^+(S) = A(S)$, the result follows.

We state the results in three-space which lead to the formula analogous to (3). The proofs, which follow the same pattern as those in the plane, are left to the reader.

Lemma 4. Suppose that

$$U: \quad y_i = \sum_{j=1}^{3} a_{ij} x_j + b_i, \qquad i = 1, 2, 3, \tag{4}$$

is a nonsingular linear transformation from three-space to three-space. If S is a rectangular parallelepiped of the form $c_i \le x_i \le d_i$, $i = 1, 2, 3$, then its image $U(S)$ is a parallelepiped of volume

$$|\det A| \cdot \text{vol } S.$$

Theorem 13. *Suppose that S is any region in space with volume $V(S)$. If U is the nonsingular transformation (4), then the image $U(S)$ has volume $V[U(S)]$ and*

$$V[U(S)] = |\det A| V(S).$$

Theorem 14. *Suppose that f is a continuous function defined on a closed region G in the xy-plane. Let U be a nonsingular transformation given by*

$$U: \quad \begin{aligned} x &= au + bv + e_1 \\ y &= cu + dv + e_2 \end{aligned}$$

carrying the region F in the uv-plane into G. Then we have the following **change of variable formula:**

$$\iint_G f(x, y) \, dA_{xy} = \iint_F f(au + bv + e_1, cu + dv + e_2)|ad - bc| \, dA_{uv}.$$

The corresponding result holds for triple integrals in three-space.

Proof. Let $\epsilon > 0$ be given and let $G_1, G_2, \ldots, G_n$ be a subdivision of the region G with norm equal to ρ. Then, from the definition of a double integral, we have

$$\left| \sum_{i=1}^{n} f(x_i, y_i)A(G_i) - \iint_G f(x, y) \, dA_{xy} \right| < \frac{\epsilon}{2}, \tag{5}$$

if the norm ρ is sufficiently small and if each (x_i, y_i) is a point in G_i. We make a subdivision $F_1, F_2, \ldots, F_n$ of the region F in the uv-plane with norm so small that the images $G_i = U(F_i)$ have norm less than ρ, and so satisfy (5). For convenience, we introduce the notation

$$h(u, v) = f(au + bv + e_1, cu + dv + e_2),$$

and we further reduce the F-subdivision, if necessary, so that

$$\left| \sum_{i=1}^{n} h(u_i, v_i)|ad - bc|A(F_i) - \iint_F h(u, v)|ad - bc| \, dA_{uv} \right| < \frac{\epsilon}{2} \tag{6}$$

is valid. According to Theorem 12, we have the relation

$$\sum_{i=1}^{n} f(x_i, y_i)A(G_i) = \sum_{i=1}^{n} h(u_i, v_i)|ad - bc|A(F_i). \tag{7}$$

Taking (7) into account, we can write the identity

$$\iint_G f(x, y) \, dA_{xy} - \iint_F h(u, v)|ad - bc| \, dA_{uv}$$

$$= \left\{ \iint_G f(x, y) \, dA_{xy} - \sum_{i=1}^{n} f(x_i, y_i)A(G_i) \right\}$$

$$+ \left\{ \sum_{i=1}^{n} h(u_i, v_i)|ad - bc|A(F_i) - \iint_F h(u, v)|ad - bc| \, dA_{uv} \right\}.$$

Applying (5) and (6), we get

$$\left| \iint\limits_{G} f(x, y)\, dA_{xy} - \iint\limits_{F} h(u, v)|ad - bc|\, dA_{uv} \right| < \epsilon.$$

Since ϵ is arbitrary, the result follows.

EXAMPLE. Evaluate $\iint xy\, dA_{xy}$, where G is the parallelogram bounded by the lines $2x - y = 1$, $2x - y = 3$, $x + y = -2$, and $x + y = 0$. Draw a figure.

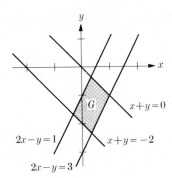

FIGURE 7–10

Solution. See Fig. 7–10. Let $u = 2x - y$, $v = x + y$. Then F is the region determined by the inequalities

$$1 \le u \le 3, \qquad -2 \le v \le 0,$$

and

$$x = \tfrac{1}{3}(u + v), \qquad y = \tfrac{1}{3}(-u + 2v).$$

We compute

$$f(x, y) = xy = \tfrac{1}{9}(-u^2 + uv + 2v^2),$$
$$ad - bc = \tfrac{1}{3}.$$

Therefore

$$\iint\limits_{G} xy\, dA_{xy} = \frac{1}{27} \iint\limits_{F} (-u^2 + uv + 2v^2)\, dA_{uv}$$

$$= \frac{1}{27} \int_{1}^{3} \int_{-2}^{0} (-u^2 + uv + 2v^2)\, dv\, du = -\frac{44}{81}.$$

EXERCISES

In each of problems 1 through 6, evaluate $\iint_{G} x^2\, dA_{xy}$, where G is the parallelogram specified by the given inequalities. Draw a figure in each case.

1. $-1 \le x - y \le 1$
   $0 \le x + y \le 2$

2. $-1 \le 2x + y \le 2$
   $0 \le x + 2y \le 3$

3. $1 \le 2x - y \le 6$
   $-1 \le x + 2y \le 4$

4. $-1 \leq 3x + 2y \leq 3$ 5. $1 \leq 3x + y \leq 6$ 6. $2 \leq 3x - y \leq 9$
     $1 \leq x + 2y \leq 5$ $2 \leq x + 2y \leq 7$ $1 \leq x + 2y \leq 8$

In each of problems 7 through 10, evaluate

$$\iiint_G y^2 \, dV_{xyz}$$

where G is the specified parallelepiped.

7. $-1 \leq x - z \leq 1$ 8. $-1 \leq x - y + z \leq 3$
     $0 \leq y + z \leq 2$ $0 \leq x + y - z \leq 4$
     $1 \leq x + z \leq 3$ $1 \leq -x + y + z \leq 5$

9. $0 \leq 2x - y + z \leq 9$ 10. $-2 \leq 2x + y + z \leq 3$
     $2 \leq x + 2y - z \leq 11$ $-1 \leq x + y - 2z \leq 2$
     $3 \leq -x + y + z \leq 12$ $1 \leq -x + 2y + 2z \leq 6$

In problems 11 and 12, by introducing a linear transformation, find in each case the area bounded by the ellipse.

11. $x^2 + 4xy + 5y^2 = 4$ 12. $2x^2 - 4xy + 3y^2 = 3$

In each of problems 13 and 14, find the volume bounded by the ellipsoid.

13. $x^2 + 8y^2 + 6z^2 + 4xy - 2xz + 4yz = 9$
14. $2x^2 + y^2 + 14z^2 + 2xy + 4xz - 2yz = 4$
15. Prove Lemma 4.
16. Prove Theorem 13.
17. Write out the proof of Theorem 14 for triple integrals.
*18. (a) Give a plausible definition of a rectangular "hyperbox" in four-dimensional euclidean space. (b) Assuming that the volume of regions in four-space can be defined as in two- and three-space, state and prove the analog of Lemma 4. (c) Prove the four-dimensional analog of Theorem 13.
*19. (a) On the basis of problem 18, define a four-fold (quadruple) integral. (b) State and prove the linear change-of-variable theorem in four dimensions.
*20. Repeat problem 18 in n dimensions.
*21. Repeat problem 19 in n dimensions.

6. LINEAR TRANSFORMATIONS FROM V TO V. CHANGE OF BASIS

If V and W are vector spaces, we say that a transformation T from V to W is a **linear transformation** if and only if its domain is the whole of V and if

$$T\left(\sum_{j=1}^{k} c_j \mathbf{u}_j\right) = \sum_{j=1}^{k} c_j T(\mathbf{u}_j)$$

for every set of numbers $c_1, c_2, \ldots, c_k$ and every set of vectors $\mathbf{u}_1, \mathbf{u}_2, \ldots, \mathbf{u}_k$ in V.

We shall be concerned with linear transformations from V to V. If the n vectors $\mathbf{v}_1, \mathbf{v}_2, \ldots, \mathbf{v}_n$ form a basis for V and if T is a linear transformation from V to V, it is possible to associate an $n \times n$ matrix A with this transformation. We define the vectors $\mathbf{w}_1, \mathbf{w}_2, \ldots, \mathbf{w}_n$ by the relations

$$\mathbf{w}_j = T\mathbf{v}_j, \quad j = 1, 2, \ldots, n.$$

Now, since $\mathbf{v}_1, \mathbf{v}_2, \ldots, \mathbf{v}_n$ is a basis, we can represent each $\mathbf{w}_j$ as a linear combination of the basis vectors. Therefore we may write

$$\mathbf{w}_j = \sum_{i=1}^{n} a_{ij} \mathbf{v}_i.$$

The matrix $A = (a_{ij})$ is the one we wish to examine. Letting $\mathbf{v}$ be any vector in V, we may represent it in the form

$$\mathbf{v} = \sum_{j=1}^{n} x_j \mathbf{v}_j.$$

Similarly, the vector $\mathbf{w} = T\mathbf{v}$ has a representation

$$\mathbf{w} = \sum_{i=1}^{n} y_i \mathbf{v}_i.$$

Since T is linear, we have

$$\sum_{i=1}^{n} y_i \mathbf{v}_i = \mathbf{w} = T\mathbf{v} = T\left(\sum_{j=1}^{n} x_j \mathbf{v}_j \right) = \sum_{j=1}^{n} x_j T\mathbf{v}_j$$

$$= \sum_{j=1}^{n} x_j \mathbf{w}_j = \sum_{i=1}^{n} \left[\sum_{j=1}^{n} a_{ij} x_j \right] \mathbf{v}_i.$$

Since the basis vectors are linearly independent, the coefficients on the extreme left and the extreme right above are equal, term by term. In this way, we get the system of equations

$$y_i = \sum_{j=1}^{n} a_{ij} x_j, \quad i = 1, 2, \ldots, n. \tag{1}$$

Moreover, if the basis $\mathbf{v}_1, \mathbf{v}_2, \ldots, \mathbf{v}_n$ is fixed, we get a matrix A for each linear transformation. Conversely, by reversing the steps described above, we see that each $n \times n$ matrix A determines a unique linear transformation T. We are now able to conclude:

Theorem 15. *Suppose that a vector space V has n basis vectors. If T is a linear transformation from V to V, then either* (a), *the inverse of T is also a linear transformation from V to V or* (b), *we have $T\mathbf{v} = \mathbf{0}$ for some $\mathbf{v} \neq \mathbf{0}$.*

Proof. If the matrix A in (1) is nonsingular, then the transformation (1) has an inverse. The transformation corresponding to the matrix A^{-1} is T^{-1}. In this case

(a) holds. If A is singular, then there is a nonzero solution to the system (1) with all the $y_i = 0$. Call the nonzero solution $\bar{x}_1, \bar{x}_2, \ldots, \bar{x}_n$. Defining

$$\mathbf{v} = \bar{x}_1\mathbf{v}_1 + \bar{x}_2\mathbf{v}_2 + \cdots + \bar{x}_n\mathbf{v}_n,$$

we obtain

$$T\mathbf{v} = \sum_{i=1}^{n} y_i\mathbf{v}_i = \mathbf{0}.$$

For any vector space V with a finite fixed basis, there is a unique square matrix A corresponding to a linear transformation T of V to V. Suppose that we employ another basis for the same space V. Then, in general, the matrix B corresponding to T using the second basis will be different from A. We wish to determine the relationship between A and B.

Let the two bases be $\mathbf{v}_1, \mathbf{v}_2, \ldots, \mathbf{v}_n$ and $\mathbf{w}_1, \mathbf{w}_2, \ldots, \mathbf{w}_n$. We have the equations

$$\mathbf{w}_i = \sum_{j=1}^{n} c_{ij}\mathbf{v}_j, \qquad i = 1, 2, \ldots, n,$$

relating the bases in which $C = (c_{ij})$ is a *fixed* nonsingular matrix. Writing $D = C^{-1}$, we also have

$$\mathbf{v}_i = \sum_{j=1}^{n} d_{ij}\mathbf{w}_j, \qquad D = (d_{ij}).$$

Let $\mathbf{v}$ be any vector in V and $\mathbf{w} = T\mathbf{v}$. Both $\mathbf{v}$ and $\mathbf{w}$ may be represented in two ways. We have

$$\mathbf{w} = \sum_{k=1}^{n} y_k\mathbf{v}_k = \sum_{k=1}^{n} y_k \sum_{i=1}^{n} d_{ki}\mathbf{w}_i = \sum_{i=1}^{n} \eta_i\mathbf{w}_i, \tag{2}$$

$$\mathbf{v} = \sum_{l=1}^{n} x_l\mathbf{v}_l = \sum_{l=1}^{n} x_l \sum_{j=1}^{n} d_{lj}\mathbf{w}_j = \sum_{j=1}^{n} \xi_j\mathbf{w}_j. \tag{3}$$

Equating coefficients in the last equation in (2) and in the last equation in (3), we get

$$\eta_i = \sum_{k=1}^{n} d_{ki}y_k, \qquad \xi_j = \sum_{l=1}^{n} d_{lj}x_l. \tag{4}$$

Setting $G = D^t$ and writing $g_{ik} = d_{ki}$, we may rewrite (4) in the more customary form

$$\eta_i = \sum_{k=1}^{n} g_{ik}y_k, \qquad \xi_j = \sum_{l=1}^{n} g_{jl}x_l. \tag{5}$$

We recall that if C is any nonsingular matrix, then (Theorem 11 (b))

$$(C^t)^{-1} = (C^{-1})^t.$$

Therefore, $G = D^t = (C^{-1})^t = (C^t)^{-1}$, and so $G^{-1} = C^t$.

Writing $F = G^{-1} = C^t$, and employing (f_{ij}) for the elements of F, we obtain from (5) the system of equations

$$x_l = \sum_{j=1}^{n} f_{lj}\xi_j. \tag{6}$$

We have not yet used the fact that $\mathbf{w}$ is the image of $\mathbf{v}$ under T. This fact enables us to express η_i in terms of ξ_j and, in turn, we can then obtain a relation between the matrices of T associated with the two bases. Suppose that for the basis $\mathbf{v}_1, \mathbf{v}_2, \ldots, \mathbf{v}_n$, T has the matrix A, so that

$$y_k = \sum_{l=1}^{n} a_{kl}x_l; \tag{7}$$

suppose also that for the basis $\mathbf{w}_1, \mathbf{w}_2, \ldots, \mathbf{w}_n$, T has the matrix B, so that

$$\eta_i = \sum_{j=1}^{n} b_{ij}\xi_j. \tag{8}$$

We write (5), and then we substitute (7), (6), and (8) in turn to get the successive equalities

$$\eta_i = \sum_{k=1}^{n} g_{ik}y_k = \sum_{k=1}^{n} g_{ik} \sum_{l=1}^{n} a_{kl}x_l$$

$$= \sum_{k=1}^{n} \sum_{l=1}^{n} g_{ik}a_{kl} \sum_{j=1}^{n} f_{lj}\xi_j = \sum_{j=1}^{n} b_{ij}\xi_j. \tag{9}$$

The last equality in (9) helps us derive the next theorem.

Theorem 16. *Suppose that T is a linear transformation from V to V, and suppose that $\mathbf{v}_1, \mathbf{v}_2, \ldots, \mathbf{v}_n$ and $\mathbf{w}_1, \mathbf{w}_2, \ldots, \mathbf{w}_n$ are bases for V related by the matrix $C = (c_{ij})$ so that $\mathbf{w}_i = \sum_{j=1}^{n} c_{ij}\mathbf{v}_j$. If the matrix A corresponds to T under the $\mathbf{v}_1, \mathbf{v}_2, \ldots, \mathbf{v}_n$ basis, and if B corresponds to T under the $\mathbf{w}_1, \mathbf{w}_2, \ldots, \mathbf{w}_n$ basis, then*

$$B = (C^t)^{-1}AC^t.$$

Proof. Equating coefficients in the last equality in (9), we obtain the matrix equation

$$B = GAF = D^tAC^t = (C^{-1})^tAC^t = (C^t)^{-1}AC^t.$$

The change of basis itself sets up a linear transformation between the coordinates $(x_1, x_2, \ldots, x_n)$ of a vector $\mathbf{v}$ with respect to the basis $\mathbf{v}_1, \mathbf{v}_2, \ldots, \mathbf{v}_n$ and the coordinates $(\xi_1, \xi_2, \ldots, \xi_n)$ of *the same vector* $\mathbf{v}$ with respect to the basis $\mathbf{w}_1, \mathbf{w}_2, \ldots, \mathbf{w}_n$. This relation is

$$\xi_i = \sum_{j=1}^{n} g_{ij}x_j, \tag{10}$$

and we see that the matrix of the transformation is

$$G = (C^t)^{-1} = (C^{-1})^t.$$

EXAMPLE 1. Suppose that $(\mathbf{v}_1, \mathbf{v}_2)$ and $(\mathbf{w}_1, \mathbf{w}_2)$ are two bases for a vector space V, and suppose that the second set is given in terms of the first by the relations

$$\mathbf{w}_1 = 2\mathbf{v}_1 + \mathbf{v}_2, \qquad \mathbf{w}_2 = \mathbf{v}_1 + \mathbf{v}_2. \tag{11}$$

Solve for the new coordinates (ξ_1, ξ_2) with respect to the old coordinates (x_1, x_2) of the same vector $\mathbf{v}$ in two ways: (a), by using Eqs. (10) and (b), by writing

$$\mathbf{v} = x_1\mathbf{v}_1 + x_2\mathbf{v}_2 = \xi_1\mathbf{w}_1 + \xi_2\mathbf{w}_2,$$

and using the relations (11).

Solution. (a) In the notation of Theorem 16, we have

$$C = \begin{pmatrix} 2 & 1 \\ 1 & 1 \end{pmatrix}, \qquad C^t = \begin{pmatrix} 2 & 1 \\ 1 & 1 \end{pmatrix}, \qquad \det C^t = 1.$$

Therefore

$$G = (C^t)^{-1} = \begin{pmatrix} 1 & -1 \\ -1 & 2 \end{pmatrix}$$

and Eqs. (10) are

$$\xi_1 = x_1 - x_2, \qquad \xi_2 = -x_1 + 2x_2.$$

(b) Solving Eqs. (11) for $\mathbf{v}_1, \mathbf{v}_2$, we find that

$$\mathbf{v}_1 = \mathbf{w}_1 - \mathbf{w}_2, \qquad \mathbf{v}_2 = -\mathbf{w}_1 + 2\mathbf{w}_2.$$

We obtain

$$\begin{aligned} x_1\mathbf{v}_1 + x_2\mathbf{v}_2 &= x_1(\mathbf{w}_1 - \mathbf{w}_2) + x_2(-\mathbf{w}_1 + 2\mathbf{w}_2) \\ &= (x_1 - x_2)\mathbf{w}_1 + (-x_1 + 2x_2)\mathbf{w}_2 \\ &= \xi_1\mathbf{w}_1 + \xi_2\mathbf{w}_2. \end{aligned}$$

Since $\mathbf{w}_1$ and $\mathbf{w}_2$ are linearly independent, we may equate coefficients in the last equation to obtain the result.

EXAMPLE 2. Given the bases $(\mathbf{v}_1, \mathbf{v}_2)$ and $(\mathbf{w}_1, \mathbf{w}_2)$ of the vector space V, related by the equations

$$\mathbf{w}_1 = \mathbf{v}_1 - 2\mathbf{v}_2, \qquad \mathbf{w}_2 = -\mathbf{v}_1 + 3\mathbf{v}_2.$$

If T is a linear transformation from V to V with matrix

$$A = \begin{pmatrix} 1 & 2 \\ -2 & 3 \end{pmatrix}$$

with respect to $(\mathbf{v}_1, \mathbf{v}_2)$, find the matrix B corresponding to T with respect to $(\mathbf{w}_1, \mathbf{w}_2)$.

Solution. We have

$$C = \begin{pmatrix} 1 & -2 \\ -1 & 3 \end{pmatrix}, \qquad C^t = \begin{pmatrix} 1 & -1 \\ -2 & 3 \end{pmatrix}, \qquad \det C^t = 1, \qquad (C^t)^{-1} = \begin{pmatrix} 3 & 1 \\ 2 & 1 \end{pmatrix}.$$

Therefore

$$B = (C^t)^{-1}AC^t = \begin{pmatrix} 3 & 1 \\ 2 & 1 \end{pmatrix}\begin{pmatrix} 1 & 2 \\ -2 & 3 \end{pmatrix}\begin{pmatrix} 1 & -1 \\ -2 & 3 \end{pmatrix} = \begin{pmatrix} -17 & 26 \\ -14 & 21 \end{pmatrix}.$$

EXERCISES

In each of problems 1 through 4, find the relations between the (x_1, x_2) coordinates with respect to the $(\mathbf{v}_1, \mathbf{v}_2)$ basis and the (ξ_1, ξ_2) coordinates with respect to the $(\mathbf{w}_1, \mathbf{w}_2)$ basis by both of the methods in Example 1.

1. $\mathbf{w}_1 = \mathbf{v}_1 - \mathbf{v}_2$
   $\mathbf{w}_2 = \mathbf{v}_1 + \mathbf{v}_2$

2. $\mathbf{w}_1 = 2\mathbf{v}_1 + \mathbf{v}_2$
   $\mathbf{w}_2 = \mathbf{v}_1 + 2\mathbf{v}_2$

3. $\mathbf{w}_1 = 2\mathbf{v}_1 - \mathbf{v}_2$
   $\mathbf{w}_2 = -\mathbf{v}_1 + 2\mathbf{v}_2$

4. $\mathbf{w}_1 = 3\mathbf{v}_1 - 2\mathbf{v}_2$
   $\mathbf{w}_2 = -\mathbf{v}_1 + 2\mathbf{v}_2$

In problems 5 and 6, find $\mathbf{w}_1$ and $\mathbf{w}_2$ in terms of $\mathbf{v}_1$ and $\mathbf{v}_2$.

5. $\xi_1 = 2x_1 - x_2$
   $\xi_2 = -x_1 + x_2$

6. $\xi_1 = x_1 - x_2$
   $\xi_2 = x_1 + x_2$

In problems 7 and 8, find the ξ_i in terms of the x_i.

7. $\mathbf{w}_1 = \mathbf{v}_1 - \mathbf{v}_3$
   $\mathbf{w}_2 = \mathbf{v}_2 + \mathbf{v}_3$
   $\mathbf{w}_3 = \mathbf{v}_1 - \mathbf{v}_2 + \mathbf{v}_3$

8. $\mathbf{w}_1 = \mathbf{v}_1 - \mathbf{v}_2 + 2\mathbf{v}_3$
   $\mathbf{w}_2 = 2\mathbf{v}_1 + \mathbf{v}_2 - \mathbf{v}_3$
   $\mathbf{w}_3 = \mathbf{v}_1 + \mathbf{v}_2 - \mathbf{v}_3$

In problems 9 through 12, find the matrix B.

9. $\mathbf{w}_1 = \mathbf{v}_1 + \mathbf{v}_2,$ $A = \begin{pmatrix} 0 & 2 \\ 2 & 0 \end{pmatrix}$
   $\mathbf{w}_2 = -\mathbf{v}_1 + \mathbf{v}_2$

10. $\mathbf{w}_1 = 2\mathbf{v}_1 + \mathbf{v}_2,$ $A = \begin{pmatrix} 1 & -1 \\ 1 & 1 \end{pmatrix}$
    $\mathbf{w}_2 = -\mathbf{v}_1 + \mathbf{v}_2$

11. $\mathbf{w}_1 = \mathbf{v}_1 - \mathbf{v}_2,$ $A = \begin{pmatrix} 1 & 2 \\ 3 & 4 \end{pmatrix}$
    $\mathbf{w}_2 = 2\mathbf{v}_1 + 3\mathbf{v}_2$

12. $\mathbf{w}_1 = 2\mathbf{v}_1 - \mathbf{v}_2,$ $A = \begin{pmatrix} 3 & -1 \\ 2 & 3 \end{pmatrix}$
    $\mathbf{w}_2 = \mathbf{v}_1 + \mathbf{v}_2$

In problems 13 and 14, find the matrix B by starting with the relation

$$\mathbf{v} = x_1\mathbf{v}_1 + x_2\mathbf{v}_2 = \xi_1\mathbf{w}_1 + \xi_2\mathbf{w}_2$$

and the corresponding ones for $\mathbf{w}$, and solving for the η_i in terms of the ξ_i.

13. $y_1 = 2x_1 - 3x_2,$ $\mathbf{w}_1 = 2\mathbf{v}_1 + \mathbf{v}_2$
    $y_2 = 3x_1 + 2x_2,$ $\mathbf{w}_2 = \mathbf{v}_1 + \mathbf{v}_2$

14. $y_1 = 2x_1 + x_2,$ $\mathbf{w}_1 = \mathbf{v}_1 + 2\mathbf{v}_2$
    $y_2 = x_1 - 2x_2,$ $\mathbf{w}_2 = -\mathbf{v}_1 + \mathbf{v}_2$

7. EUCLIDEAN VECTOR SPACES. ORTHOGONAL BASES

In the study of vectors in the plane and in three-space, we found it useful to define the inner (scalar or dot) product of two vectors. In this way we determined the angle between two vectors, the length of a vector, and various properties of orthogonality and parallelism.

Although an operation called inner product may be defined in many vector spaces, it is not difficult to find examples of vector spaces in which such an operation can never be defined. Vector spaces which possess an inner product are called *Euclidean*.

DEFINITION. A **Euclidean vector space** is a vector space on which can be defined an **inner product** of two vectors, denoted by $\mathbf{u} \cdot \mathbf{v}$. This inner product is to satisfy the following conditions:

(1) $\mathbf{u} \cdot \mathbf{v}$ is defined for all $\mathbf{u}$ and $\mathbf{v}$ in V.

(2) $\mathbf{v} \cdot \mathbf{u} = \mathbf{u} \cdot \mathbf{v}$ for all $\mathbf{u}$ and $\mathbf{v}$.

(3) $(\mathbf{u} + \mathbf{v}) \cdot \mathbf{w} = \mathbf{u} \cdot \mathbf{w} + \mathbf{v} \cdot \mathbf{w}$ for all $\mathbf{u}$ and $\mathbf{v}$.

(4) $(c\mathbf{u}) \cdot \mathbf{v} = c(\mathbf{u} \cdot \mathbf{v})$ for all $\mathbf{u}$, $\mathbf{v}$, and constants c.

(5) $\mathbf{u} \cdot \mathbf{u} > 0$ unless $\mathbf{u} = \mathbf{0}$.

Using laws (2), (3), and (4), we derive easily

(3′) $\mathbf{u} \cdot (\mathbf{v} + \mathbf{w}) = \mathbf{u} \cdot \mathbf{v} + \mathbf{u} \cdot \mathbf{w}$,

(4′) $\mathbf{u} \cdot (c\mathbf{v}) = c(\mathbf{u} \cdot \mathbf{v})$; $\mathbf{u} \cdot \mathbf{v} = 0$ if either $\mathbf{u}$ or $\mathbf{v} = \mathbf{0}$.

Repeated applications of the above laws yield the result

$$\left(\sum_{i=1}^{m} c_i\mathbf{u}_i \right) \cdot \left(\sum_{j=1}^{n} d_j\mathbf{v}_j \right) = \sum_{i=1}^{m} \sum_{j=1}^{n} (c_i \, d_j)(\mathbf{u}_i \cdot \mathbf{v}_j).$$

The reader will recall that these conditions are the same as those satisfied by the inner product in the plane and in three-space. In analogy with the definition of the length of a vector in three-space, we now define the **norm of u**, denoted by $\|\mathbf{u}\|$, by the formula

$$\|\mathbf{u}\| = (\mathbf{u} \cdot \mathbf{u})^{1/2}.$$

Here, as previously, the term "norm" is used as a measure of size.

We give two examples of Euclidean vector spaces:

(1) The space V_n^0 consisting of the elements of V_n with the inner product defined by the relation

$$(x_1, x_2, \ldots, x_n) \cdot (y_1, y_2, \ldots, y_n) = \sum_{i=1}^{n} x_i y_i. \tag{1}$$

For $n = 2$ and 3 we see that this definition coincides with the familiar one we learned.

(2) The totality of functions continuous on an interval $[a, b]$, with inner product defined by

$$f \cdot g = \int_a^b f(x) \, g(x) \, dx.$$

The first four properties of inner product are easily verified for this space. To establish property (5), we note that if $f(x_1) \neq 0$ for some x_1 on $[a, b]$, then $[f(x)]^2 > 0$ for some interval containing x_1 (since f is continuous). Because $[f(x)]^2 \geq 0$ everywhere, it follows that

$$f \cdot f = \int_a^b [f(x)]^2 \, dx > 0 \quad \text{unless} \quad f(x) \equiv 0 \text{ on } [a, b].$$

If W is a subspace of a Euclidean vector space V and if the inner product on W is taken as the same as that on V, then W is a Euclidean vector space. In this way, we see that any subspace of those described in (1) and (2) above is a Euclidean vector space.

The following simple lemma in elementary algebra may be verified by the reader. (See Exercise 15 at the end of this section and the hint given there.)

Lemma 5. Suppose $A > 0$, $C > 0$, and

$$Aa^2 + 2Bab + Cb^2 \geq 0$$

for all a and b. Then $AC - B^2 \geq 0$. The equality $AC - B^2 = 0$ holds if and only if $Aa^2 + 2Bab + Cb^2 = 0$ for some a and b which are not both zero.

We now establish an inequality which, in various forms, has turned out to be one of the most useful in mathematical analysis. It is commonly known as the **Schwarz inequality.**

Theorem 17. (Schwarz Inequality.) *We have*

$$|\mathbf{u} \cdot \mathbf{v}| \leq \|\mathbf{u}\| \, \|\mathbf{v}\| \tag{2}$$

for all $\mathbf{u}$ and $\mathbf{v}$. The equality holds if and only if $\mathbf{u}$ and $\mathbf{v}$ are linearly dependent.

Proof. If either $\mathbf{u}$ or $\mathbf{v}$ is $\mathbf{0}$, both sides of (2) vanish. Suppose both $\mathbf{u}$ and $\mathbf{v}$ are not zero. We define

$$A = \mathbf{u} \cdot \mathbf{u} = \|\mathbf{u}\|^2, \qquad B = \mathbf{u} \cdot \mathbf{v}, \qquad C = \mathbf{v} \cdot \mathbf{v} = \|\mathbf{v}\|^2,$$

and observe that $A > 0$, $C > 0$. Let a and b be any numbers whatsoever; we form the inner product

$$0 \leq (a\mathbf{u} + b\mathbf{v}) \cdot (a\mathbf{u} + b\mathbf{v}) = Aa^2 + 2Bab + Cb^2.$$

Now we apply the Lemma to conclude that

$$AC - B^2 = \|\mathbf{u}\|^2 \cdot \|\mathbf{v}\|^2 - |\mathbf{u} \cdot \mathbf{v}|^2 \geq 0,$$

which is the Schwarz inequality. Equality holds if and only if

$$(a\mathbf{u} + b\mathbf{v}) \cdot (a\mathbf{u} + b\mathbf{v}) = \|a\mathbf{u} + b\mathbf{v}\|^2 = 0$$

for some a and b not both zero. However, the norm of a vector vanishes only if the vector itself vanishes. Therefore

$$a\mathbf{u} + b\mathbf{v} = \mathbf{0},$$

which states that $\mathbf{u}$ and $\mathbf{v}$ are linearly dependent.

The Schwarz inequality enables us to give a consistent definition of the angle between two vectors in a Euclidean vector space.

DEFINITION. If $\mathbf{u}$ and $\mathbf{v}$ are both nonzero vectors, we define the **angle between** $\mathbf{u}$ and $\mathbf{v}$ by the formula

$$\cos \theta = \frac{\mathbf{u} \cdot \mathbf{v}}{\|\mathbf{u}\| \cdot \|\mathbf{v}\|}. \tag{3}$$

If we regard the vectors $\mathbf{u}$ and $\mathbf{v}$ as directed line segments as in Fig. 7–11, the "third side" of the triangle is a representative of $\mathbf{v} - \mathbf{u}$ or $\mathbf{u} - \mathbf{v}$. We have the formula

$$\|\mathbf{u} - \mathbf{v}\|^2 = (\mathbf{u} - \mathbf{v}) \cdot (\mathbf{u} - \mathbf{v})$$
$$= \|\mathbf{u}\|^2 + \|\mathbf{v}\|^2 - 2\mathbf{u} \cdot \mathbf{v}.$$

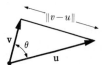

Recalling the Law of Cosines for triangles in the plane, we see that substitution of (3) in the last expression on the right above yields the same Law of Cosines in any vector space with an inner product.

FIGURE 7–11

The next definitions are evident whenever the angle between two vectors can be defined.

DEFINITIONS. Two vectors are **parallel** or **positively proportional** if and only if the angle between them is zero. Two vectors $\mathbf{u}$ and $\mathbf{v}$ are **orthogonal** or **perpendicular** if and only if $\mathbf{u} \cdot \mathbf{v} = 0$. A sequence of nonzero vectors $\mathbf{e}_1, \mathbf{e}_2, \ldots,$ $\mathbf{e}_k$ is called an **orthogonal set** if and only if any two different $\mathbf{e}_i$ are orthogonal. That is,

$$\mathbf{e}_i \cdot \mathbf{e}_j = 0 \quad \text{for } i \neq j, \quad i, j = 1, 2, \ldots, k.$$

We say that an orthogonal set is **normalized**, or that the set is **orthonormal**, if (a) the set is orthogonal and (b) the length of each vector in the set is 1. Note that if $\mathbf{v}_1, \mathbf{v}_2, \ldots, \mathbf{v}_k$ is an orthogonal set, then the set

$$\frac{\mathbf{v}_1}{\|\mathbf{v}_1\|}, \quad \frac{\mathbf{v}_2}{\|\mathbf{v}_2\|}, \quad \ldots, \quad \frac{\mathbf{v}_k}{\|\mathbf{v}_k\|}$$

is an orthonormal set, since each vector in the latter set is of unit length.

We recall that the set $\mathbf{i}, \mathbf{j}$ forms an orthonormal set of vectors in the plane, and that $\mathbf{i}, \mathbf{j}, \mathbf{k}$ form such a set in three-space.

Theorem 18. *Any orthonormal set of vectors is linearly independent.*

Proof. Suppose that $\mathbf{e}_1, \mathbf{e}_2, \ldots, \mathbf{e}_k$ is an orthonormal set and assume that it is linearly dependent. We obtain a contradiction. Linear dependence means that there are constants $c_1, c_2, \ldots, c_k$ not all zero such that $c_1\mathbf{e}_1 + c_2\mathbf{e}_2 + \cdots + c_k\mathbf{e}_k = \mathbf{0}$. We compute

$$0 = \left(\sum_{i=1}^{k} c_i \mathbf{e}_i \right) \cdot \left(\sum_{i=1}^{k} c_i \mathbf{e}_i \right) = c_1^2 + c_2^2 + \cdots + c_k^2,$$

which is clearly impossible.

The Euclidean vector space V_n^0 has the orthonormal basis given by $e_1, e_2, \ldots,$ e_n, where

$$e_i = (0, 0, \overbrace{\ldots, 0, 1, 0}^{i-1}, \ldots, 0);$$

i.e., e_i has a 1 in the ith place and zeros elsewhere. The formula (1) for inner product in V_n^0 establishes at once the orthogonality of this basis.

Theorem 19. *Suppose* $e_1, e_2, \ldots, e_n$ *is an orthonormal basis for a Euclidean vector space* V, *and*

$$\mathbf{u} = \sum_{i=1}^{n} x_i e_i, \qquad \mathbf{v} = \sum_{i=1}^{n} y_i e_i.$$

Then

$$\mathbf{u} \cdot \mathbf{v} = \sum_{i=1}^{n} x_i y_i.$$

Proof. We multiply the terms in the sum according to the rules for multiplying an inner product, and the result is immediate.

EXAMPLE 1. The Euclidean vector space V_5^0 has vectors $\mathbf{u}$ and $\mathbf{v}$ given in terms of an orthonormal basis

$$\mathbf{u} = 2e_1 - e_2 + e_3 - 2e_4 + 3e_5,$$
$$\mathbf{v} = e_1 + e_2 - 2e_3 + e_4 + 2e_5.$$

Find a unit vector orthogonal to $\mathbf{u}$ in the subspace spanned by $\mathbf{u}$ and $\mathbf{v}$.

Solution. (See Fig. 7–12.) The vector $\mathbf{w'} = \mathbf{v} - k\mathbf{u}$, being a linear combination of $\mathbf{u}$ and $\mathbf{v}$, is in the space spanned by them. Then $\mathbf{w'}$ is orthogonal to $\mathbf{u}$ if and only if

$$\mathbf{u} \cdot \mathbf{w'} = \mathbf{u} \cdot \mathbf{v} - k(\mathbf{u} \cdot \mathbf{u}) = 0$$

or, if and only if

$$k = \frac{\mathbf{u} \cdot \mathbf{v}}{(\mathbf{u} \cdot \mathbf{u})}.$$

FIGURE 7–12

We compute: $\mathbf{u} \cdot \mathbf{v} = 3$, $\mathbf{u} \cdot \mathbf{u} = 19$, $k = \frac{3}{19}$; therefore

$$\mathbf{w'} = \mathbf{v} - \tfrac{3}{19}\mathbf{u} = \tfrac{1}{19}(13e_1 + 22e_2 - 41e_3 + 25e_4 + 29e_5).$$

The desired unit vector $\mathbf{w}$ is

$$\mathbf{w} = \frac{\mathbf{w'}}{\|\mathbf{w'}\|} = \frac{1}{10\sqrt{38}}(13e_1 + 22e_2 - 41e_3 + 25e_4 + 29e_5).$$

Note the similarity of this example to Theorem 6 in Chapter 2, Section 2, page 49, which concerns vectors in three-space.

The next theorem and its proof describe a constructive process by which any set of linearly independent vectors may be transformed into an orthogonal set spanning the same space.

Theorem 20. *Suppose that* $\mathbf{v}_1, \mathbf{v}_2, \ldots, \mathbf{v}_k$ *are linearly independent vectors in a Euclidean vector space V. There is an orthonormal set $\mathbf{e}_1, \mathbf{e}_2, \ldots, \mathbf{e}_k$ spanning the same space. Moreover, the $\mathbf{e}_j$ may be chosen so that $\mathbf{e}_1, \mathbf{e}_2, \ldots, \mathbf{e}_j$ spans the same subspace spanned by $\mathbf{v}_1, \mathbf{v}_2, \ldots, \mathbf{v}_j$ for each j. Finally, if for some $r \leq k$ the set $\mathbf{v}_1, \mathbf{v}_2, \ldots, \mathbf{v}_r$ is already orthonormal, the $\mathbf{e}_j$ may be chosen so that $\mathbf{e}_j = \mathbf{v}_j$ for $j = 1, 2, \ldots, r$.*

Proof. We select $\mathbf{e}_1 = \mathbf{v}_1/\|\mathbf{v}_1\|$. Next we define the vector

$$\mathbf{e}_2' = \mathbf{v}_2 - (\mathbf{v}_2 \cdot \mathbf{e}_1) \cdot \mathbf{e}_1.$$

We see that

$$\mathbf{e}_2' \cdot \mathbf{e}_1 = \mathbf{v}_2 \cdot \mathbf{e}_1 - (\mathbf{v}_2 \cdot \mathbf{e}_1)(\mathbf{e}_1 \cdot \mathbf{e}_1) = 0,$$

and so $\mathbf{e}_2'$ and $\mathbf{e}_1$ are orthogonal. We next define $\mathbf{e}_2 = \mathbf{e}_2'/\|\mathbf{e}_2'\|$. Then $\mathbf{e}_1$ and $\mathbf{e}_2$ form an orthonormal set. We proceed by induction. The $(j + 1)$st vector is defined for each j by the relations

$$\mathbf{e}_{j+1}' = \mathbf{v}_{j+1} - \sum_{i=1}^{j} (\mathbf{v}_{j+1} \cdot \mathbf{e}_i)\mathbf{e}_i.$$

We compute for any p between 1 and j:

$$\mathbf{e}_{j+1}' \cdot \mathbf{e}_p = \mathbf{v}_{j+1} \cdot \mathbf{e}_p - \sum_{i=1}^{j} (\mathbf{v}_{j+1} \cdot \mathbf{e}_i)(\mathbf{e}_i \cdot \mathbf{e}_p) = 0.$$

Therefore $\mathbf{e}_{j+1}'$ is orthogonal to all the preceding $\mathbf{e}_p$. We now define

$$\mathbf{e}_{j+1} = \frac{\mathbf{e}_{j+1}'}{\|\mathbf{e}_{j+1}'\|},$$

and the induction is complete. Note that each $\mathbf{e}_{j+1}'$, being a linear combination of the $\mathbf{v}_i$, $i = 1, 2, \ldots, j + 1$, can never be $\mathbf{0}$.

The construction described in the proof of Theorem 20 is quite general. In case we begin with a set $\mathbf{v}_1, \mathbf{v}_2, \ldots, \mathbf{v}_k$ which is not linearly independent, the construction will yield an orthonormal set $\mathbf{e}_1, \mathbf{e}_2, \ldots, \mathbf{e}_l$ for some $l < k$. Those $\mathbf{v}_j$ which are linearly dependent on preceding ones in the sequence are automatically eliminated.

In carrying out the construction, we often find it easier to find an orthogonal basis first and then, when that process is complete, to normalize the vectors in the orthogonal basis. To simplify computations we take into account the following elementary fact: *If $\mathbf{v}_1, \mathbf{v}_2, \ldots, \mathbf{v}_k$ is an orthogonal set, then $c_1\mathbf{v}_1, c_2\mathbf{v}_2, \ldots, c_k\mathbf{v}_k$ is also $(c_j \neq 0)$.*

EXAMPLE 2. Suppose that $e_1, e_2, \ldots, e_5$ is an orthonormal basis for the vectors in a Euclidean five-space. Given that

$$
\begin{aligned}
w_1 &= e_1 - 2e_2 - e_3 + 2e_4 - 3e_5, \\
w_2 &= 2e_1 - 4e_2 - 2e_3 + 4e_4 - 6e_5, \\
w_3 &= 2e_1 - 3e_2 - e_3 + 3e_4 - 4e_5, \\
w_4 &= -e_1 + 4e_2 + 3e_3 - 4e_4 + 7e_5, \\
w_5 &= 2e_1 - 6e_2 - 3e_3 + 7e_4 - 8e_5,
\end{aligned}
$$

find an orthonormal basis for the subspace spanned by these vectors.

Solution. We first find an orthogonal set. We take $v_1' = w_1$. Since $w_2 = 2v_1'$, we omit w_2 and consider w_3. We write

$$
v_2'' = \left[w_3 - \frac{(w_3 \cdot v_1')}{\|v_1'\|^2} v_1' \right],
$$

where

$$
\begin{aligned}
(w_3 \cdot v_1') &= w_3 \cdot w_1 = 2 + 6 + 1 + 6 + 12 = 27, \\
(v_1' \cdot v_1') &= w_1 \cdot w_1 = 1 + 4 + 1 + 4 + 9 = 19.
\end{aligned}
$$

Therefore

$$
v_2'' = w_3 - \tfrac{27}{19}v_1'
$$

is orthogonal to v_1'. But, by the remark above, so is

$$
v_2' = 19v_2'' = 19w_3 - 27v_1' = 11e_1 - 3e_2 + 8e_3 + 3e_4 + 5e_5.
$$

In the attempt to use w_4 to obtain the next orthogonal vector, we get 0; hence we neglect w_4 and write

$$
v_3'' = w_5 - \frac{(w_5 \cdot v_1')}{19} v_1' - \frac{(w_5 \cdot v_2')}{\|v_2'\|^2} v_2'.
$$

Since $w_5 \cdot v_1' = 55$, $w_5 \cdot v_2' = -3$, $\|v_2'\|^2 = 228$, we have $v_3'' = w_5 - \tfrac{55}{19}v_1' + \tfrac{1}{76}v_2'$. The vector

$$
v_3' = 76v_3'' = 19(3e_1 - e_2 + 5e_4 + 3e_5)
$$

is orthogonal to both v_1' and v_2'. The vectors

$$
v_1 = \frac{1}{\sqrt{19}} v_1', \qquad v_2 = \frac{1}{2\sqrt{57}} v_2', \qquad v_3 = \frac{1}{19\sqrt{44}} v_3'
$$

form an orthonormal set which spans the given subspace.

EXERCISES

In each of problems 1 through 4, find the cosine of the angle between u and v if $e_1, e_2, \ldots, e_n$ is a given orthonormal set in V_n^0.

1. $u = 2e_1 + 3e_2 - e_3 + e_4 + 2e_5$
   $v = -e_1 + e_2 - 2e_3 - 3e_4 + e_5$

2. $u = e_2 - e_3 + 2e_4 - e_6$
   $v = e_1 + 2e_2 + e_3 - e_5$

3. $\mathbf{u} = \mathbf{e}_1 + \mathbf{e}_2 - \mathbf{e}_3 + 4\mathbf{e}_4$ 4. $\mathbf{u} = \mathbf{e}_1 + \mathbf{e}_3 - \mathbf{e}_7$
   $\mathbf{v} = 3\mathbf{e}_1 - 4\mathbf{e}_2 + \mathbf{e}_3 - 2\mathbf{e}_4$ $\mathbf{v} = 2\mathbf{e}_1 + \mathbf{e}_5 - 3\mathbf{e}_6$

In each of problems 5 through 7, find orthonormal bases $\mathbf{v}_1, \mathbf{v}_2, \ldots, \mathbf{v}_k$ for the subspaces spanned by the sequences $\mathbf{w}_1, \mathbf{w}_2, \ldots, \mathbf{w}_k$, given that $\mathbf{e}_1, \mathbf{e}_2, \mathbf{e}_3$ constitute an orthonormal basis for V.

5. $\mathbf{w}_1 = \mathbf{e}_1 - \mathbf{e}_2 + \mathbf{e}_3$ 6. $\mathbf{w}_1 = 2\mathbf{e}_1 - 2\mathbf{e}_2 + \mathbf{e}_3$
   $\mathbf{w}_2 = \mathbf{e}_1 + \mathbf{e}_2 - \mathbf{e}_3$ $\mathbf{w}_2 = \mathbf{e}_1 + \mathbf{e}_2 - \mathbf{e}_3$
                                   $\mathbf{w}_3 = 4\mathbf{e}_1 - 8\mathbf{e}_2 + 5\mathbf{e}_3$

7. $\mathbf{w}_1 = 2\mathbf{e}_1 - 2\mathbf{e}_2 + \mathbf{e}_3$
   $\mathbf{w}_2 = 2\mathbf{e}_1 + 2\mathbf{e}_2 + 3\mathbf{e}_3$

In problems 8 through 10, do the same as for 5 through 7, given that $\mathbf{e}_1, \mathbf{e}_2, \mathbf{e}_3, \mathbf{e}_4$ constitute an orthonormal basis for V.

8. $\mathbf{w}_1 = 2\mathbf{e}_1 + \mathbf{e}_2 - \mathbf{e}_3 - 2\mathbf{e}_4$ 9. $\mathbf{w}_1 = 2\mathbf{e}_1 - \mathbf{e}_2 + \mathbf{e}_3 + 3\mathbf{e}_4$
   $\mathbf{w}_2 = \phantom{2\mathbf{e}_1 +} 3\mathbf{e}_2 + 5\mathbf{e}_3 + 4\mathbf{e}_4$ $\mathbf{w}_2 = 3\mathbf{e}_1 - 2\mathbf{e}_2 + 4\mathbf{e}_3 + \mathbf{e}_4$

10. $\mathbf{w}_1 = 2\mathbf{e}_1 - \mathbf{e}_2 \phantom{+ 2\mathbf{e}_3} + \mathbf{e}_4$
    $\mathbf{w}_2 = \mathbf{e}_1 + 2\mathbf{e}_2 + \mathbf{e}_3 - \mathbf{e}_4$
    $\mathbf{w}_3 = \phantom{\mathbf{e}_1 +} \mathbf{e}_2 + 2\mathbf{e}_3 - 2\mathbf{e}_4$
    $\mathbf{w}_4 = \mathbf{e}_1 - 7\mathbf{e}_2 - \mathbf{e}_3 + 3\mathbf{e}_4$

In problems 11 through 14, find orthonormal bases $\mathbf{v}_1, \mathbf{v}_2, \ldots, \mathbf{v}_k$ for the subspaces of V_n^0 spanned by the given $\mathbf{w}_1, \mathbf{w}_2, \ldots, \mathbf{w}_k$.

11. $\mathbf{w}_1 = (1, -1, -1, 1)$ 12. $\mathbf{w}_1 = (2, -1, 1, -2)$
    $\mathbf{w}_2 = (1, -3, -1, 3)$ $\mathbf{w}_2 = (1, -2, 0, -3)$
    $\mathbf{w}_3 = (1, -1, 1, 3)$ $\mathbf{w}_3 = (-1, 4, -6, -1)$

13. $\mathbf{w}_1 = (1, 1, 0, 0, 1, 0)$ 14. $\mathbf{w}_1 = (1, 0, 0, 0, 1)$
    $\mathbf{w}_2 = (2, 1, -1, 0, 0, -1)$ $\mathbf{w}_2 = (2, -1, 0, -1, 1)$
                                             $\mathbf{w}_3 = (0, 1, 0, -1, 0)$
                                             $\mathbf{w}_4 = (2, 1, 1, -1, -1)$

15. Prove Lemma 5. [*Hint.* Write

$$Aa^2 + 2Bab + Cb^2 = A\left(a^2 + \frac{2B}{A}\, ab\right) + Cb^2$$

and complete the square.]

16. Show that if $\mathbf{u}$ and $\mathbf{v}$ are any vectors, then the vectors $\mathbf{w}_1 = \|\mathbf{v}\|\mathbf{u} + \|\mathbf{u}\|\mathbf{v}$ and $\mathbf{w}_2 = \|\mathbf{v}\|\mathbf{u} - \|\mathbf{u}\|\mathbf{v}$ are orthogonal.

17. Given that $\mathbf{e}_1, \mathbf{e}_2, \mathbf{e}_3, \mathbf{e}_4$ is an orthonormal set and that

$$\mathbf{u} = 2\mathbf{e}_1 - \mathbf{e}_2 + \mathbf{e}_3 - \mathbf{e}_4,$$
$$\mathbf{v} = \mathbf{e}_1 + \mathbf{e}_2 - \mathbf{e}_3 - 2\mathbf{e}_4,$$
$$\mathbf{w} = 3\mathbf{e}_1 - 2\mathbf{e}_2 + 2\mathbf{e}_3 - 4\mathbf{e}_4,$$

determine g and h such that $\mathbf{w} - g\mathbf{u} - h\mathbf{v}$ is orthogonal to both $\mathbf{u}$ and $\mathbf{v}$.

18. Use the Schwarz inequality to show that

$$2|(\mathbf{u} \cdot \mathbf{v})| \le \|\mathbf{u}\|^2 + \|\mathbf{v}\|^2.$$

19. If f is continuous on $[a, b]$, show that

$$\left[\int_a^b f(x)\, dx\right]^2 \le (b - a) \int_a^b f^2(x)\, dx.$$

[*Hint.* Write $\int_a^b f(x)\, dx = \int_a^b 1 \cdot f(x)\, dx$ and use the Schwarz inequality.]

20. Let f' be continuous in $[a, b]$, and suppose that $f(a) = 0$.
   (a) Show that for $a \le x \le b$

$$[f(x)]^2 \le (b - a) \int_a^b [f'(x)]^2\, dx.$$

[*Hint.* $f(x) = \int_a^x f'(t)\, dt.$]

*(b) Show that

$$\int_a^b [f(x)]^2\, dx \le \tfrac{1}{2}(b - a)^2 \int_a^b [f'(x)]^2\, dx.$$

8.   ORTHOGONAL MATRICES AND TRANSFORMATIONS

Suppose that $e_1, e_2, \ldots, e_n$ and $e'_1, e'_2, \ldots, e'_n$ are two orthonormal bases for the same vector space V. Then any vector $\mathbf{v}$ has the two representations

$$\mathbf{v} = \sum_{i=1}^n x_i e_i \quad\text{and}\quad \mathbf{v} = \sum_{i=1}^n x'_i e'_i. \tag{1}$$

It is a simple matter to find the relation between the coordinates $x_1, x_2, \ldots, x_n$ and $x'_1, x'_2, \ldots, x'_n$. For any integer p between 1 and n, we have

$$\mathbf{v} \cdot e'_p = x'_p \quad\text{and}\quad \mathbf{v} \cdot e'_p = \sum_{j=1}^n (e_j \cdot e'_p) x_j.$$

Defining the matrix $C = (c_{pj})$ with $c_{pj} = e'_p \cdot e_j$, we get $x'_p = \sum_{j=1}^n c_{pj} x_j$. Taking inner products between $\mathbf{v}$ and e_p, $1 \le p \le n$, we obtain

$$\mathbf{v} \cdot e_p = x_p = \sum_{j=1}^n (e_p \cdot e'_j) x'_j,$$

and so

$$x_p = \sum_{j=1}^n c'_{pj} x'_j \quad\text{where}\quad c'_{pj} = e_p \cdot e'_j = c_{jp}.$$

Since the two bases are orthonormal, it turns out that the matrix C connecting them has certain special properties. To exhibit these, we set $\mathbf{v} = e'_p$ in Eqs. (1) and calculate the coordinates x_i and x'_i. We write

$$e'_p = \sum_{j=1}^n x_j e_j$$

and take the inner product of e'_p with a vector e_q, $1 \le q \le n$. The result is

$$e'_p \cdot e_q = \sum_{j=1}^n x_j e_j \cdot e_q = \sum_{j=1}^n x_j \, \delta_{jq} = x_q.$$

Since, by definition, $\mathbf{e}'_p \cdot \mathbf{e}_q = c_{pq}$, we get $x_q = c_{pq}$; therefore

$$\mathbf{e}'_p = \sum_{j=1}^{n} c_{pj}\mathbf{e}_j. \tag{2}$$

In an entirely similar way, we find

$$\mathbf{e}_q = \sum_{j=1}^{n} c'_{qj}\mathbf{e}'_j. \tag{3}$$

We substitute $\mathbf{e}_q$ from (3) into the right side of (2) and obtain

$$\mathbf{e}'_p = \sum_{j=1}^{n} \sum_{q=1}^{n} c_{pj}c'_{jq}\mathbf{e}'_q = \sum_{q=1}^{n} \left(\sum_{j=1}^{n} c_{pj}c_{qj} \right) \mathbf{e}'_q. \tag{4}$$

Since the $\mathbf{e}'_q$ are linearly independent, the coefficients of $\mathbf{e}'_q$ on the left and right in (4) must agree. That is,

$$\sum_{j=1}^{n} c_{pj}c_{qj} = \delta_{pq}. \tag{5}$$

The left side of (5) is precisely the inner product of two row vectors of the matrix C. Conditions (5) state that two different row vectors of C are orthogonal, while the norm of any row vector is exactly one.

DEFINITION. A matrix C which satisfies relations (5) is called an **orthogonal matrix.**

The basic properties of orthogonal matrices are described in the next theorems.

Theorem 21. (a) *An $n \times n$ square matrix C is orthogonal if and only if C is nonsingular and*

$$C^{-1} = C^t. \tag{6}$$

(b) *If C is orthogonal, then $\det C = \pm 1$.*
(c) *If C is orthogonal, so is C^t, and therefore*

$$\sum_{j=1}^{n} c_{jp}c_{jq} = \delta_{pq}.$$

Proof. (a) The condition $\sum_{j=1}^{n} c_{pj}c_{qj} = \delta_{pq}$ is equivalent to the matrix equation

$$CC^t = I.$$

Multiplying both sides of this equation on the left by C^{-1}, we obtain (6).
   (b) Starting with the relation $CC^t = I$ and taking into account that $\det C^t = \det C$, we get

$$1 = \det I = \det (CC^t) = (\det C)(\det C^t) = (\det C)^2;$$

therefore $\det C = \pm 1$.

To establish (c), we note that

$$(C^t)^{-1} = (C^{-1})^t = (C^t)^t.$$

Therefore C^t satisfies part (a) of the theorem, which is equivalent to the definition of orthogonality.

Remark. Part (c) states that any two column vectors of an orthogonal matrix are orthogonal and that the norm of any column vector is 1.

EXAMPLE. Verify that the matrix

$$A = \begin{pmatrix} \dfrac{1}{2} & \dfrac{1}{2} & -\dfrac{1}{2} & -\dfrac{1}{2} \\[2mm] \dfrac{1}{3\sqrt{2}} & \dfrac{2}{3} & \dfrac{2}{3} & \dfrac{1}{3\sqrt{2}} \\[2mm] \dfrac{1}{2} & -\dfrac{1}{2} & \dfrac{1}{2} & -\dfrac{1}{2} \\[2mm] \dfrac{2}{3} & -\dfrac{1}{3\sqrt{2}} & -\dfrac{1}{3\sqrt{2}} & \dfrac{2}{3} \end{pmatrix}$$

is orthogonal, and find its inverse.

Solution. We verify that (5) is satisfied. Since $\frac{1}{4} + \frac{1}{4} + \frac{1}{4} + \frac{1}{4} = 1$, the first and third rows have unit length. Also, $\frac{1}{18} + \frac{4}{9} + \frac{4}{9} + \frac{1}{18} = 1$ and $\frac{4}{9} + \frac{1}{18} + \frac{1}{18} + \frac{4}{9} = 1$, and so the second and fourth rows also have length 1. To check the orthogonality, we take the inner product of the first two rows:

$$\left(\frac{1}{2}\right)\left(\frac{1}{3\sqrt{2}}\right) + \left(\frac{1}{2}\right)\left(\frac{2}{3}\right) + \left(-\frac{1}{2}\right)\left(\frac{2}{3}\right) + \left(-\frac{1}{2}\right)\left(\frac{1}{3\sqrt{2}}\right) = 0.$$

The inner product of the first and third rows yields $\frac{1}{4} - \frac{1}{4} - \frac{1}{4} + \frac{1}{4} = 0$ and, continuing in this way, we easily verify that A satisfies all the required conditions. To find the inverse, we note that $A^{-1} = A^t$, and so

$$A^{-1} = \begin{pmatrix} \dfrac{1}{2} & \dfrac{1}{3\sqrt{2}} & \dfrac{1}{2} & \dfrac{2}{3} \\[2mm] \dfrac{1}{2} & \dfrac{2}{3} & -\dfrac{1}{2} & -\dfrac{1}{3\sqrt{2}} \\[2mm] -\dfrac{1}{2} & \dfrac{2}{3} & \dfrac{1}{2} & -\dfrac{1}{3\sqrt{2}} \\[2mm] -\dfrac{1}{2} & \dfrac{1}{3\sqrt{2}} & -\dfrac{1}{2} & \dfrac{2}{3} \end{pmatrix}.$$

Theorem 22. (a) *Suppose C is an orthogonal matrix, and suppose* $\mathbf{e}_1, \mathbf{e}_2, \ldots,$ $\mathbf{e}_n$ *is an orthonormal basis for a Euclidean vector space V. If we define*

$$\mathbf{e}'_p = \sum_{j=1}^{n} c_{pj}\mathbf{e}_j, \tag{7}$$

then $\mathbf{e}'_1, \mathbf{e}'_2, \ldots, \mathbf{e}'_n$ *is an orthonormal basis for V. Furthermore, if we define*

$$x'_i = \sum_{j=1}^{n} c_{ij}x_j \quad and \quad y'_i = \sum_{j=1}^{n} c_{ij}y_j,$$

then

$$\sum_{i=1}^{n} x'_i y'_i = \sum_{i=1}^{n} x_i y_i. \tag{8}$$

(b) *The product of any finite number of $n \times n$ orthogonal matrices is orthogonal.*

Proof. To establish that $\mathbf{e}'_1, \mathbf{e}'_2, \ldots, \mathbf{e}'_n$ is an orthonormal basis, we compute

$$\mathbf{e}'_p \cdot \mathbf{e}'_q = \left(\sum_{j=1}^{n} c_{pj}\mathbf{e}_j\right) \cdot \left(\sum_{i=1}^{n} c_{qi}\mathbf{e}_i\right) = \sum_{j=1}^{n} \sum_{i=1}^{n} c_{pj}c_{qi}(\mathbf{e}_j \cdot \mathbf{e}_i)$$

$$= \sum_{j=1}^{n} c_{pj}c_{qj} = \delta_{pq}.$$

To prove relation (8), we proceed directly:

$$\sum_{i=1}^{n} x'_i y'_i = \sum_{i=1}^{n} \left(\sum_{j=1}^{n} c_{ij}x_j\right)\left(\sum_{k=1}^{n} c_{ik}y_k\right)$$

$$= \sum_{j=1}^{n} \sum_{k=1}^{n} \left(\sum_{i=1}^{n} c_{ij}c_{ik}\right) x_j y_k$$

$$= \sum_{j=1}^{n} \sum_{k=1}^{n} \delta_{jk}x_j y_k = \sum_{j=1}^{n} x_j y_j.$$

We obtain the result in (b) for two matrices; the extension to any finite sequence is by induction. Suppose that A and B are orthogonal. We show that AB is also orthogonal. Since $AA^t = BB^t = I$, we have

$$(AB)(AB)^t = (AB)(B^t A^t) = [(AB)B^t]A^t$$
$$= [A(BB^t)]A^t = [AI]A^t = AA^t = I.$$

Suppose that $\mathbf{e}_1, \mathbf{e}_2, \ldots, \mathbf{e}_n$ is a fixed orthonormal basis for a Euclidean vector space V. Let T be a linear transformation from V to V with the property that for any two vectors $\mathbf{v}$ and $\mathbf{w}$ of V,

$$T\mathbf{v} \cdot T\mathbf{w} = \mathbf{v} \cdot \mathbf{w}. \tag{9}$$

If (9) holds, we say that T is an **orthogonal transformation.** Note that if $\mathbf{v}$ and $\mathbf{w}$ are orthogonal, so that $\mathbf{v} \cdot \mathbf{w} = 0$, then the images are orthogonal. Furthermore, selecting $\mathbf{v} = \mathbf{w}$, we see that the image under T of any vector is a vector of the same norm (length). The intimate connection between orthogonal transformations and orthogonal matrices is established in the next theorem.

Theorem 23. *A linear transformation T from a Euclidean space V to V is orthogonal if and only if the matrix corresponding to it with respect to any orthonormal basis is orthogonal.*

Proof. Suppose that $\mathbf{e}_1, \mathbf{e}_2, \ldots, \mathbf{e}_n$ is any orthonormal basis for V. (a) We show first that if the matrix C associated with T is orthogonal, then T is an orthogonal transformation. Any vector $\mathbf{v}$ in V has the representation $\mathbf{v} = \sum_{i=1}^n x_i \mathbf{e}_i$. Therefore

$$T\mathbf{v} = T\left(\sum_{i=1}^n x_i \mathbf{e}_i\right) = \sum_{i=1}^n x_i' \mathbf{e}_i,$$

where

$$x_i' = \sum_{j=1}^n c_{ij} x_j.$$

If $\mathbf{w} = \sum_{i=1}^n y_i \mathbf{e}_i$ is another vector, we compute $T\mathbf{w}$ as above and observe that Eq. (8) is precisely equivalent to (9), the definition of orthogonal transformation.

(b) Now, suppose that T is orthogonal. We show that the associated matrix is orthogonal. If $\mathbf{v}$ and $\mathbf{w}$ are the same as in (a), we have

$$x_i' = \sum_{j=1}^n c_{ij} x_j, \qquad y_i' = \sum_{k=1}^n c_{ik} y_k.$$

Since (9) is assumed to hold for all $\mathbf{v}$ and $\mathbf{w}$ in V, we must have

$$\left(\sum_{j=1}^n c_{ij} x_j\right)\left(\sum_{k=1}^n c_{ik} y_k\right) = \sum_{j=1}^n x_j y_j \tag{10}$$

for all $(x_1, x_2, \ldots, x_n)$ and $(y_1, y_2, \ldots, y_n)$. Setting $x_j = \delta_{jp}$ and $y_k = \delta_{kq}$, we see that Eq. (10) yields

$$\sum_{i=1}^n c_{ip} c_{iq} = \delta_{pq},$$

which is precisely the condition that C^t be orthogonal. From Theorem 21(c), C is orthogonal also.

Remark. If T is an orthogonal transformation such that the determinant of its corresponding matrix is $+1$ with respect to any orthogonal basis, it can then be shown that T is a *rigid motion* of the space V onto itself.

EXERCISES

In problems 1 through 6, verify that the given matrices are orthogonal and find their inverses.

1. $\begin{pmatrix} \frac{1}{2}\sqrt{3} & \frac{1}{2} \\ -\frac{1}{2} & \frac{1}{2}\sqrt{3} \end{pmatrix}$

2. $\begin{pmatrix} \cos\theta & \sin\theta \\ -\sin\theta & \cos\theta \end{pmatrix}$

3. $\begin{pmatrix} \frac{2}{3} & -\frac{2}{3} & \frac{1}{3} \\ \frac{1}{3} & \frac{2}{3} & \frac{2}{3} \\ -\frac{2}{3} & -\frac{1}{3} & \frac{2}{3} \end{pmatrix}$

4. $\begin{pmatrix} \dfrac{2}{\sqrt{6}} & \dfrac{-1}{\sqrt{6}} & \dfrac{1}{\sqrt{6}} \\ \dfrac{1}{\sqrt{3}} & \dfrac{1}{\sqrt{3}} & \dfrac{-1}{\sqrt{3}} \\ 0 & \dfrac{1}{\sqrt{2}} & \dfrac{1}{\sqrt{2}} \end{pmatrix}$

5. $\dfrac{1}{2}\begin{pmatrix} 1 & 1 & -1 & -1 \\ -1 & 1 & 1 & -1 \\ 1 & 1 & 1 & 1 \\ -1 & 1 & -1 & 1 \end{pmatrix}$

6. $\begin{pmatrix} \dfrac{2}{\sqrt{10}} & \dfrac{-1}{\sqrt{10}} & \dfrac{1}{\sqrt{10}} & \dfrac{-2}{\sqrt{10}} \\ \dfrac{1}{2} & \dfrac{1}{2} & \dfrac{1}{2} & \dfrac{1}{2} \\ \dfrac{-1}{\sqrt{10}} & \dfrac{-2}{\sqrt{10}} & \dfrac{2}{\sqrt{10}} & \dfrac{1}{\sqrt{10}} \\ \dfrac{1}{2} & -\dfrac{1}{2} & -\dfrac{1}{2} & \dfrac{1}{2} \end{pmatrix}$

In problems 7 and 8, find an orthonormal basis for the space of vectors (x_1, x_2, x_3, x_4) in V_4^0 whose components satisfy the given systems of equations.

7. $\begin{aligned} 2x_1 + x_2 - x_3 - 2x_4 &= 0 \\ x_1 + 2x_2 - 3x_4 &= 0 \end{aligned}$

8. $\begin{aligned} 3x_1 - 4x_2 + x_3 - 2x_4 &= 0 \\ -3x_1 + 4x_2 - 3x_3 + x_4 &= 0 \end{aligned}$

In problems 9 and 10, find the (x', y', z') equation of the image of the locus whose (x, y, z) equation is given (in V_3^0).

9. $7x^2 + 19y^2 + 10z^2 - 8xy + 28xz + 20yz = 54$
$3x' = 2x - 2y + z, \quad 3y' = x + 2y + 2z, \quad 3z' = -2x - y + 2z$

10. $4x^2 + 7y^2 + 7z^2 + 2xy - 2xz + 4yz = 36$
$x'\sqrt{6} = 2x - y + z, \quad y'\sqrt{3} = x + y - z, \quad z'\sqrt{2} = y + z$

In problems 11 through 13, find the matrix B, given that $\mathbf{v}_1, \mathbf{v}_2, \mathbf{v}_3$ and $\mathbf{w}_1, \mathbf{w}_2, \mathbf{w}_3$ are orthonormal bases for V, and A corresponds to T with respect to the $\mathbf{v}_1, \mathbf{v}_2, \mathbf{v}_3$ basis.

11. $\begin{aligned} 3\mathbf{w}_1 &= 2\mathbf{v}_1 - 2\mathbf{v}_2 + \mathbf{v}_3 \\ 3\mathbf{w}_2 &= \mathbf{v}_1 + 2\mathbf{v}_2 + 2\mathbf{v}_3 \\ 3\mathbf{w}_3 &= -2\mathbf{v}_1 - \mathbf{v}_2 + 2\mathbf{v}_3 \end{aligned}$
$\quad A = \begin{pmatrix} 7 & -4 & 14 \\ -4 & 19 & 10 \\ 14 & 10 & 10 \end{pmatrix}$

12. $\begin{aligned} \sqrt{6}\mathbf{w}_1 &= 2\mathbf{v}_1 - \mathbf{v}_2 + \mathbf{v}_3 \\ \sqrt{3}\mathbf{w}_2 &= \mathbf{v}_1 + \mathbf{v}_2 - \mathbf{v}_3 \\ \sqrt{2}\mathbf{w}_3 &= \mathbf{v}_2 + \mathbf{v}_3 \end{aligned}$
$\quad A = \begin{pmatrix} 4 & 1 & -1 \\ 1 & 7 & 2 \\ -1 & 2 & 7 \end{pmatrix}$

$$7w_1 = 6v_1 + 2v_2 + 3v_3$$
13. $7w_2 = -3v_1 + 6v_2 + 2v_3$ $A = \frac{1}{7}\begin{pmatrix} 88 & 6 & 30 \\ 6 & 51 & 24 \\ 30 & 24 & 29 \end{pmatrix}$
$$7w_3 = -2v_1 - 3v_2 + 6v_3$$

In problems 14 through 17, find in each case an orthogonal matrix whose first two rows are proportional, respectively, to the two rows as given.

14. $\begin{pmatrix} 1 & -1 & 1 \\ 2 & 1 & -1 \end{pmatrix}$ 15. $\begin{pmatrix} 1 & 2 & 3 \\ -1 & -1 & 1 \end{pmatrix}$

16. $\begin{pmatrix} 1 & 2 & -1 & -2 \\ 1 & 1 & 1 & 1 \end{pmatrix}$ 17. $\begin{pmatrix} 1 & 2 & -1 & 1 \\ -1 & 1 & 1 & 0 \end{pmatrix}$

18. If A and B are orthogonal matrices, show that the matrix

$$\begin{pmatrix} A & 0 \\ 0 & B \end{pmatrix}$$

is orthogonal.

19. If A, B, C, D are $n \times n$ orthogonal matrices, show that the $4n \times 4n$ matrix

$$\begin{pmatrix} 0 & 0 & A & 0 \\ 0 & B & 0 & 0 \\ 0 & 0 & 0 & C \\ D & 0 & 0 & 0 \end{pmatrix}$$

is orthogonal.

20. Of the elementary row transformations of type (a), (b), (c), and (d) as described on pages 290 and 328, which ones transform an orthogonal matrix into an orthogonal matrix?

21. A matrix is called *symmetric* if $A = A^t$.
   (a) If A is a symmetric orthogonal matrix, show that $A^2 = I$.
   (b) If A and B are $n \times n$ symmetric orthogonal matrices, and if C is the $2n \times 2n$ matrix

$$C = \begin{pmatrix} A & 0 \\ 0 & B \end{pmatrix},$$

   what can be said about the matrix C^2?
   (c) Suppose that A and B are $n \times n$, symmetric, orthogonal, and $AB = BA$. If

$$D = \begin{pmatrix} 0 & A \\ B & 0 \end{pmatrix},$$

   what form does the matrix D^4 have?

CHAPTER 8

Eigenvalue problems; complex vector spaces

1. EIGENVALUES AND EIGENVECTORS

In this section we shall always assume that V is a finite-dimensional Euclidean vector space. If $\mathbf{v}_1, \mathbf{v}_2, \ldots, \mathbf{v}_n$ is a basis for V and if T is a linear transformation from V to V, then there is an $n \times n$ matrix A which corresponds to T. We recall that if

$$\mathbf{v} = \sum_{i=1}^{n} x_i \mathbf{v}_i, \qquad \mathbf{w} = T\mathbf{v} = \sum_{i=1}^{n} y_i \mathbf{v}_i, \qquad T\mathbf{v}_i = \sum_{j=1}^{n} a_{ji} \mathbf{v}_j,$$

then

$$y_i = \sum_{i=1}^{n} a_{ij} x_j.$$

We are interested in determining solutions, if any, of the vector equation

$$T\mathbf{v} = \lambda \mathbf{v} \tag{1}$$

where λ is a scalar. This vector equation corresponds to the system of linear equations for the components

$$\sum_{j=1}^{n} a_{ij} x_j = \lambda x_i, \qquad i = 1, 2, \ldots, n. \tag{2}$$

The system (2) is a homogeneous system of n equations in the n unknowns $x_1, x_2, \ldots, x_n$, and it may be written in the form

$$\sum_{j=1}^{n} (a_{ij} - \lambda \delta_{ij}) x_j = 0, \qquad i = 1, 2, \ldots, n. \tag{3}$$

According to the Corollary on page 300, a homogeneous system of n equations in n unknowns has a solution different from the trivial one, $x_1 = x_2 = \cdots = x_n = 0$, if and only if the determinant of the coefficient matrix vanishes. The matrix $A - \lambda I$

in (3) has determinant

$$
\begin{vmatrix}
a_{11} - \lambda & a_{12} & \cdots & a_{1n} \\
a_{21} & a_{22} - \lambda & \cdots & a_{2n} \\
\vdots & & & \vdots \\
a_{n1} & a_{n2} & \cdots & a_{nn} - \lambda
\end{vmatrix} = (-1)^n \lambda^n + \cdots, \tag{4}
$$

which is seen to be a polynomial in λ of degree n. The Fundamental Theorem of Algebra states that every polynomial of degree n has exactly n roots (some or all of which may be complex). Therefore there are n values of λ which make the determinant (4) vanish. If λ is a real root of the polynomial equation

$$
\det (A - \lambda I) = 0, \tag{5}
$$

then there is at least one solution $(x_1, x_2, \ldots, x_n)$ of the system (3) which is not identically zero. Consequently, for this same value of λ there is a nonzero solution of the vector equation (1).

DEFINITIONS. Any (real) value of λ for which the equation $T\mathbf{v} = \lambda \mathbf{v}$ has a nonzero solution is called an **eigenvalue** of the transformation T. The terms **characteristic** value and **proper** value are also used for eigenvalue. Any nonzero solution $\mathbf{v}$ of (1) corresponding to an eigenvalue λ is called an **eigenvector** of T.

If A is any $n \times n$ matrix, we may form the polynomial equation (5) and find the n roots. These roots are called the **eigenvalues of the matrix** A, and the polynomial equation

$$
\det (A - \lambda I) = 0
$$

is called the **characteristic equation of the matrix** A. Any real, nonzero solution of the linear system of equations (3) corresponding to an eigenvalue of the matrix A is called an **eigenvector** of A. Of course, given any square matrix A, we may choose a vector space V, a basis $\mathbf{v}_1, \mathbf{v}_2, \ldots, \mathbf{v}_n$, and a linear transformation T so that A corresponds to T for this basis. In this way we have an equivalence between the eigenvalues and eigenvectors of a transformation T in a finite-dimensional vector space and those for $n \times n$ matrices. The problem of determining the eigenvalues of a matrix or of a transformation is called an *eigenvalue problem*. Such problems occur with great frequency in theoretical physics, chemistry, and various branches of engineering. Eigenvalue problems are basic in the study of differential equations and, in particular, those equations which arise in the applications of mathematics to other sciences.

An examination of Eq. (5) shows clearly that the eigenvalues of a matrix do not depend on any associated vector space V or on the basis $\mathbf{v}_1, \mathbf{v}_2, \ldots, \mathbf{v}_n$. We shall now show that if T is a linear transformation from a space V to V, then the eigenvalues of T do not depend on which basis $\mathbf{v}_1, \mathbf{v}_2, \ldots, \mathbf{v}_n$ is selected. However, we saw (page 341) that the matrix A corresponding to T *does* depend on the basis chosen. Therefore, we must show that if A and B are the matrices corresponding to a transformation T with respect to two different bases, then A and B must have the same eigenvalues. According to Theorem 16 of Chapter 7, if A and B correspond to T with two different bases, then there is a nonsingular matrix H with the property that

$$B = H^{-1}AH.$$

DEFINITION. Two $n \times n$ matrices A and B are said to be **similar** if and only if there is a nonsingular $n \times n$ matrix H such that $B = H^{-1}AH$.

We shall show that if A and B are similar matrices, then they have the same characteristic equation. To do so, we must prove that the polynomial equations $\det (B - \lambda I) = 0$ and $\det (A - \lambda I) = 0$ are identical. We have

$$B - \lambda I = H^{-1}AH - \lambda H^{-1}IH = H^{-1}(AH - \lambda IH) = H^{-1}(A - \lambda I)H.$$

Since the determinant of the product of matrices is the product of their determinants (Theorem 9 of Chapter 7), we obtain

$$\det (B - \lambda I) = (\det H^{-1}) (\det (A - \lambda I)) (\det H) = \det (A - \lambda I).$$

We conclude that the *characteristic equations of A and B are identical.* We say, equivalently, that *similar matrices have the same eigenvalues.*

If a square matrix D is diagonal, then it is obvious that the entries along the diagonal are precisely the eigenvalues of D. If we are given a matrix A and can find another matrix B which has the same eigenvalues as A and which is diagonal, then we can read off the eigenvalues of A from those of B. The next theorem describes circumstances under which such a process is possible.

Theorem 1. *Suppose that A is an $n \times n$ matrix and suppose that V, $\mathbf{v}_1$, $\mathbf{v}_2$, ..., $\mathbf{v}_n$, and T are the associated vector space, basis, and linear transformation, respectively. If A has n linearly independent eigenvectors corresponding to real eigenvalues, then a diagonal matrix B can be found which is similar to A and so has the same eigenvalues as A.*

Proof. Let $\mathbf{w}_1$, $\mathbf{w}_2$, ..., $\mathbf{w}_n$ be vectors in V corresponding to the eigenvectors of A; these vectors form a basis for V and so, if $\mathbf{v}$ is any vector and $\mathbf{w} = T\mathbf{v}$, we may write

$$\mathbf{v} = \sum_{i=1}^{n} \xi_i \mathbf{w}_i, \qquad \mathbf{w} = \sum_{i=1}^{n} \eta_i \mathbf{w}_i$$

and, since $T\mathbf{w}_i = \lambda_i \mathbf{w}_i$, $i = 1, 2, \ldots, n$,

$$\sum_{i=1}^{n} \eta_i \mathbf{w}_i = \sum_{i=1}^{n} \xi_i T\mathbf{w}_i = \sum_{i=1}^{n} \lambda_i \xi_i \mathbf{w}_i.$$

We conclude that

$$\eta_i = \lambda_i \xi_i. \qquad (6)$$

If we denote by B the matrix corresponding to T with the basis $\mathbf{w}_1, \mathbf{w}_2, \ldots, \mathbf{w}_n$, then

$$\eta_i = \sum_{j=1}^{n} b_{ij}\xi_j,$$

and we see from (6) that $b_{ij} = 0$, $i \neq j$, $b_{ii} = \lambda_i$. Thus B has the form

$$B = \begin{pmatrix} \lambda_1 & & & 0 \\ & \lambda_2 & & \\ & & \ddots & \\ 0 & & & \lambda_n \end{pmatrix},$$

and $B = H^{-1}AH$ for some nonsingular matrix H.

Remarks. The crucial hypothesis in Theorem 1 is the one which *assumes* that the eigenvectors of A are all linearly independent, a condition not always satisfied. Furthermore, the eigenvalues are assumed to be all real. It can be shown that if the roots of the characteristic equation are all real and distinct (i.e., if all eigenvalues are real and distinct), then the resulting eigenvectors are linearly independent and the reduction to diagonal form is always possible. When there are multiple or complex roots, the reduction may not be possible.

EXAMPLE. Find the eigenvalues and eigenvectors of the matrix

$$A = \begin{pmatrix} 1 & 2 \\ 5 & 4 \end{pmatrix}.$$

Solution. We have

$$A - \lambda I = \begin{pmatrix} 1 - \lambda & 2 \\ 5 & 4 - \lambda \end{pmatrix} \quad \text{and} \quad \det(A - \lambda I) = \lambda^2 - 5\lambda - 6 = (\lambda - 6)(\lambda + 1).$$

The eigenvalues are $\lambda = 6, -1$. If $\lambda = 6$, the system of equations (3) becomes

$$-5x_1 + 2x_2 = 0, \qquad 5x_1 - 2x_2 = 0,$$

which has the solution $x_1 = 2a$, $x_2 = 5a$. The quantity a is arbitrary. If $\lambda = -1$, the equations are

$$2x_1 + 2x_2 = 0, \qquad 5x_1 + 5x_2 = 0,$$

with solution $x_1 = b$, $x_2 = -b$, b arbitrary. The eigenvectors are $(2a, 5a)$ and $(b, -b)$.

Remarks. Suppose that V is a two-dimensional vector space with basis $\mathbf{v}_1, \mathbf{v}_2$, and that T is the transformation on V corresponding to the matrix A of the above example. Since the eigenvectors of A are linearly independent, we may introduce the new basis vectors

$$\mathbf{w}_1 = 2\mathbf{v}_1 + 5\mathbf{v}_2, \qquad \mathbf{w}_2 = \mathbf{v}_1 - \mathbf{v}_2.$$

Then the matrix which governs the transformation from one basis to another, denoted by C (see Section 6 of Chapter 7), is in this case

$$C = \begin{pmatrix} 2 & 5 \\ 1 & -1 \end{pmatrix}, \quad \text{and} \quad C^t = \begin{pmatrix} 2 & 1 \\ 5 & -1 \end{pmatrix}, \quad \det C^t = -7,$$

$$(C^t)^{-1} = \frac{1}{7}\begin{pmatrix} 1 & 1 \\ 5 & -2 \end{pmatrix}.$$

If we write H for C^t, the matrix B corresponding to T with basis vectors $\mathbf{w}_1$, $\mathbf{w}_2$ is $B = H^{-1}AH$. To verify that B is the required diagonal matrix, we calculate

$$B = \frac{1}{7}\begin{pmatrix} 1 & 1 \\ 5 & -2 \end{pmatrix}\begin{pmatrix} 1 & 2 \\ 5 & 4 \end{pmatrix}\begin{pmatrix} 2 & 1 \\ 5 & -1 \end{pmatrix}$$

$$= \frac{1}{7}\begin{pmatrix} 1 & 1 \\ 5 & -2 \end{pmatrix}\begin{pmatrix} 12 & -1 \\ 30 & 1 \end{pmatrix} = \frac{1}{7}\begin{pmatrix} 42 & 0 \\ 0 & -7 \end{pmatrix} = \begin{pmatrix} 6 & 0 \\ 0 & -1 \end{pmatrix}.$$

EXERCISES

In each of problems 1 through 8, find the eigenvalues and eigenvectors of the given matrix A. Assume that V is a vector space with basis $\mathbf{v}_1, \mathbf{v}_2, \ldots, \mathbf{v}_n$ and, by introducing the eigenvectors as new basis vectors, find the matrix H such that $B = H^{-1}AH$ is diagonal.

1. $\begin{pmatrix} 4 & 2 \\ -1 & 1 \end{pmatrix}$

2. $\begin{pmatrix} 3 & -4 \\ -4 & -3 \end{pmatrix}$

3. $\begin{pmatrix} 1 & -2 \\ 8 & 11 \end{pmatrix}$

4. $\begin{pmatrix} 0 & 1 \\ 3 & 2 \end{pmatrix}$

5. $\begin{pmatrix} 1 & 0 & 0 \\ 1 & 2 & 0 \\ 2 & -2 & 3 \end{pmatrix}$

6. $\begin{pmatrix} 3 & 2 & 3 \\ -1 & 0 & -3 \\ 1 & -2 & 1 \end{pmatrix}$

7. $\begin{pmatrix} 13 & -3 & 5 \\ 0 & 4 & 0 \\ -15 & 9 & -7 \end{pmatrix}$

8. $\begin{pmatrix} 5 & -1 & -3 & 3 \\ -1 & 5 & 3 & -3 \\ -3 & 3 & 5 & -1 \\ 3 & -3 & -1 & 5 \end{pmatrix}$

In each of problems 9 through 11, assume that A has eigenvalues $\lambda_1, \lambda_2, \ldots, \lambda_n$.

9. Show that the eigenvalues of $A^2 = AA$ are $\lambda_1^2, \lambda_2^2, \ldots, \lambda_n^2$.

10. Show that the eigenvalues of A^3 are $\lambda_1^3, \lambda_2^3, \ldots, \lambda_n^3$.

*11. Let $P(x) = \sum_{i=0}^{n} a_i x^i$ be a polynomial in x. Show that the eigenvalues of $B = P(A)$ are $P(\lambda_i)$, $i = 1, 2, \ldots, n$. $(A^0 = I.)$

12. Show that there is no nonsingular matrix H such that $H^{-1}AH$ is diagonal, if

$$A = \begin{pmatrix} 3 & 1 \\ -1 & 1 \end{pmatrix}.$$

[*Hint.* Since H may be multiplied by a scalar without changing the result, assume that det $H = 1$, and hence assume that

$$H = \begin{pmatrix} a & b \\ c & d \end{pmatrix}, \qquad H^{-1} = \begin{pmatrix} d & -b \\ -c & a \end{pmatrix}, \qquad ad - bc = 1.$$

Then compute $H^{-1}AH$.]

13. If zero is a root of the characteristic equation of a matrix A, show that the corresponding linear transformation cannot be one to one.

14. Prove that the eigenvalues of any triangular matrix are precisely the elements along the diagonal.

15. Show that if A is an orthogonal matrix and λ is a real eigenvalue, then $\lambda = \pm 1$. [*Hint.* Use the special properties of a transformation T corresponding to an orthogonal matrix.]

2. SYMMETRIC MATRICES. QUADRATIC FORMS

If the eigenvectors of a matrix A are linearly independent and span the space V, the matrix may be reduced to diagonal form. However, if we are given a particular matrix, it is difficult to verify whether or not such a reduction is possible without actually carrying out the entire process. In this section we establish a simple criterion which determines when a matrix may be diagonalized.

A matrix A is said to be **symmetric** if and only if

$$A^t = A; \qquad \text{i.e., } a_{ji} = a_{ij}, \quad i, j = 1, 2, \ldots, n.$$

We simultaneously define a symmetric linear transformation. We say that a linear transformation on a Euclidean vector space is **symmetric** if and only if

$$(T\mathbf{u}) \cdot \mathbf{v} = \mathbf{u} \cdot T\mathbf{v} \qquad \text{for all } \mathbf{u} \text{ and } \mathbf{v}. \tag{1}$$

Theorem 2. *Suppose that* $\mathbf{e}_1, \mathbf{e}_2, \ldots, \mathbf{e}_n$ *is an orthonormal basis for* V. *Then* T *is symmetric if and only if its corresponding matrix* A *is symmetric.*

Proof. Let $\mathbf{u}$ and $\mathbf{v}$ be any vectors and suppose that $\mathbf{s} = T\mathbf{u}, \mathbf{t} = T\mathbf{v}$. We write

$$\mathbf{u} = \sum_{i=1}^{n} u_i \mathbf{e}_i, \quad \mathbf{v} = \sum_{i=1}^{n} v_i \mathbf{e}_i, \quad \mathbf{s} = \sum_{i=1}^{n} s_i \mathbf{e}_i, \quad \mathbf{t} = \sum_{i=1}^{n} t_i \mathbf{e}_i,$$

so that

$$s_i = \sum_{j=1}^{n} a_{ij} u_j, \quad t_i = \sum_{j=1}^{n} a_{ij} v_j.$$

Then, since the basis is orthonormal,

$$(T\mathbf{u}) \cdot \mathbf{v} = \mathbf{s} \cdot \mathbf{v} = \sum_{i=1}^{n} \sum_{j=1}^{n} a_{ij} u_j v_i, \tag{2a}$$

$$\mathbf{u} \cdot T\mathbf{v} = \mathbf{u} \cdot \mathbf{t} = \sum_{i=1}^{n} \sum_{j=1}^{n} a_{ij} u_i v_j = \sum_{i=1}^{n} \sum_{j=1}^{n} a_{ji} u_j v_i. \tag{2b}$$

By making the particular choices $u_j = \delta_{jp}$ and $v_i = \delta_{iq}$, we see that if (1) holds, then the equality of (2a) and (2b) becomes $a_{pq} = a_{qp}$. Since p and q are arbitrary, A must be symmetric. On the other hand, if A is symmetric, then the right sides of (2a) and (2b) are equal and the equality of the left sides is (1) precisely.

Corollary. *If A is a symmetric $n \times n$ matrix and H is an orthogonal $n \times n$ matrix, then $H^{-1}AH$ is symmetric.*

Proof. The basis $\mathbf{e}_1, \mathbf{e}_2, \ldots, \mathbf{e}_n$ in the above theorem is arbitrary. Since any two such orthonormal bases are connected by an orthogonal transformation (Theorem 22 of Chapter 7), the result follows.

DEFINITION. Suppose that W is a subspace of a vector space V. The totality of vectors $\mathbf{v}$ such that $\mathbf{u} \cdot \mathbf{v} = 0$ for all $\mathbf{u}$ in W is **a set called** W **perpendicular and denoted** $W^\perp$.

With this definition the next lemma is easily verified by the reader.

Lemma 1. $W^\perp$ is a subspace having dimension equal to that of V minus that of W.

The next theorem shows that symmetric matrices are always reducible to diagonal form. Since symmetry can frequently be checked by inspection, this criterion is a useful one.

Theorem 3. (a) *If A is a symmetric matrix, the roots of the characteristic equation, $\det (A - \lambda I) = 0$, are all real.*
(b) *If T is a symmetric transformation, there exists an orthonormal basis consisting of eigenvectors of T; any two eigenvectors corresponding to two different eigenvalues are orthogonal.*
(c) *If A is symmetric, there is an orthogonal matrix H such that $H^{-1}AH$ is in diagonal form.*

Proof. The proof of (a) is postponed until Section 6, pages 393–394.

We establish (b) by induction on the dimension n of V. If $n = 1$, the matrix corresponding to T is a 1×1 matrix (a_{11}), and if $\mathbf{e}_1$ and $\mathbf{e}_1'$ are bases for V, we must have $\mathbf{e}_1' = \pm\mathbf{e}_1$. The result is evident.

Now, suppose the theorem has been proved for spaces of all dimensions less than or equal to k, and suppose V has dimension $k + 1$. Let λ_1 be a root of the characteristic equation. By part (a), λ_1 is real; also, it is an eigenvalue of T. Let W denote the totality of vectors $\mathbf{v}$ such that $T\mathbf{v} = \lambda_1\mathbf{v}$. It is easily verified that W is a subspace of V of dimension r, $1 \leq r \leq k + 1$. Let $\mathbf{e}_1, \mathbf{e}_2, \ldots, \mathbf{e}_r$ be an orthonormal basis for W. According to the way we defined W, each $\mathbf{e}_i$, $1 \leq i \leq r$, is an eigenvector for T.

Next consider the space $W^\perp$ as defined above. If $\mathbf{v} \in W^\perp$ then, for each $\mathbf{u}$ in W, we have

$$\mathbf{u} \cdot T\mathbf{v} = (T\mathbf{u}) \cdot \mathbf{v} = \lambda_1\mathbf{u} \cdot \mathbf{v} = 0. \tag{3}$$

In (3), the first equality holds because T is symmetric, the second holds because

u is in W, and the third holds because **u** is in W and **v** is in $W^\perp$. The extreme left and right sides of (3) show that $T\mathbf{v}$ is in $W^\perp$. Now $W^\perp$ is a Euclidean vector space of dimension $k + 1 - r$. If $r = k + 1$, then $\mathbf{e}_1, \mathbf{e}_2, \ldots, \mathbf{e}_r$ is already a basis for V. If $1 \leq r \leq k$, we see that $W^\perp$ is of dimension less than $k + 1$. We define

$$T_1\mathbf{v} = T\mathbf{v} \quad \text{for} \quad \mathbf{v} \in W^\perp.$$

Then T_1 is a symmetric transformation of $W^\perp$ into itself. By the induction hypothesis (dimension of $W^\perp$ is $\leq k$), there is an orthonormal basis $\mathbf{e}_{r+1}, \mathbf{e}_{r+2}, \ldots,$ $\mathbf{e}_{k+1}$ for $W^\perp$ which consists of eigenvectors of T_1. These clearly are also eigenvectors of T. Since every vector in $W^\perp$ is orthogonal to every vector in W, it follows that $\mathbf{e}_1, \mathbf{e}_2, \ldots, \mathbf{e}_{k+1}$ is an orthonormal basis for V. The last statement in (b) follows from the proof in which the selection showed that the eigenvectors are orthogonal.

(c) This result follows from part (b) and the previous discussion concerning the relation of symmetric transformations and symmetric matrices.

EXAMPLE 1. Given the space V_3^0 with the usual orthonormal basis, **i, j, k**. Let W be the subspace consisting of all linear combinations of $\mathbf{u}_1 = 2\mathbf{i} + 3\mathbf{j}, \mathbf{u}_2 = \mathbf{i} - \mathbf{k}$. Find $W^\perp$.

Solution. A vector $\mathbf{v} = a\mathbf{i} + b\mathbf{j} + c\mathbf{k}$ is in $W^\perp$ if

$$\mathbf{v} \cdot \mathbf{u}_1 = 0 \quad \text{and} \quad \mathbf{v} \cdot \mathbf{u}_2 = 0,$$

as then **v** is orthogonal to every linear combination of $\mathbf{u}_1$ and $\mathbf{u}_2$. Therefore

$$(a\mathbf{i} + b\mathbf{j} + c\mathbf{k}) \cdot (2\mathbf{i} + 3\mathbf{j}) = 2a + 3b = 0,$$
$$(a\mathbf{i} + b\mathbf{j} + c\mathbf{k}) \cdot (\mathbf{i} - \mathbf{k}) = a - c = 0.$$

Thus **v** must have the form $a(\mathbf{i} - \frac{2}{3}\mathbf{j} + \mathbf{k})$. The space $W^\perp$ is the one-dimensional subspace determined by these vectors **v**.

A **quadratic form** in n dimensions is a homogeneous quadratic function of the form

$$Q = \sum_{j=1}^{n} \sum_{k=1}^{n} a_{jk} x_j x_k.$$

Without any loss of generality, we may assume that the coefficients in a quadratic form are symmetric. To see this, we suppose that

$$Q' = \sum_{j=1}^{n} \sum_{k=1}^{n} a'_{jk} x_j x_k$$
$$= a'_{11} x_1^2 + a'_{12} x_1 x_2 + a'_{21} x_2 x_1 + a'_{22} x_2^2 + a'_{13} x_1 x_3 + \cdots,$$

in which the coefficients may not be symmetric. Then we combine the coefficients of x_1x_2 and x_2x_1 by defining $a_{12} = a_{21} = \frac{1}{2}(a'_{12} + a'_{21})$; in general, we define $a_{ij} = a_{ji} = \frac{1}{2}(a'_{ij} + a'_{ji})$. Then Q' is unchanged and may be written

$$Q' = \sum_{j=1}^{n} \sum_{k=1}^{n} a_{jk}x_jx_k, \qquad a_{jk} = a_{kj}. \tag{4}$$

The coefficients in (4) may be considered as the elements in a symmetric matrix A. Conversely, the elements of any symmetric matrix may be used as the coefficients in a quadratic form. Therefore, *there is a one-to-one correspondence between quadratic forms and symmetric matrices.* For example, a quadratic form in three variables is written

$$Q = a_{11}x_1^2 + a_{22}x_2^2 + a_{33}x_3^2 + 2a_{12}x_1x_2 + 2a_{13}x_1x_3 + 2a_{23}x_2x_3.$$

Note the factor 2 which appears in the x_ix_j terms, $i \neq j$.

We can apply Theorem 3 to find a change of coordinates which reduces a quadratic form to a sum of squares. The student has already encountered this process in the study of conics. A rotation of coordinates was employed to eliminate the xy-term in quadratic expressions of the form $Ax^2 + 2Bxy + Cy^2$. (See *First Course*, page 324.)

Suppose the new coordinates are $x'_1, x'_2, \ldots, x'_n$, and the x_i are related to the x'_i by an orthogonal transformation. That is, H is an orthogonal matrix, and

$$x_i = \sum_{r=1}^{n} h_{ir}x'_r.$$

The quadratic form (4) becomes

$$Q = \sum_{r=1}^{n} \sum_{s=1}^{n} a'_{rs}x'_rx'_s, \qquad a'_{rs} = \sum_{j=1}^{n} \sum_{k=1}^{n} h_{jr}a_{jk}h_{ks}. \tag{5}$$

The matrix equation corresponding to (5) is

$$A' = H^tAH = H^{-1}AH, \tag{6}$$

since H is orthogonal. However, (6) is just the formula for a change of basis. We conclude that the procedure for changing coordinate axes to reduce a quadratic form to a sum of the form $\sum_{i=1}^{n} a'_{ii}(x'_i)^2$ is: (a) find the roots of the characteristic equation of the symmetric matrix A, and (b) find a complete set of normalized eigenvectors. We illustrate with an example.

EXAMPLE 2. Let Q be the quadratic form

$$Q \equiv 5x_1^2 + 5x_2^2 + 2x_3^2 + 8x_1x_2 + 4x_1x_3 + 4x_2x_3.$$

By using an orthogonal change of variables, reduce Q to a form without the "cross terms," i.e., without the terms $a_{ij}x_ix_j$, $i \neq j$.

Solution. The matrices A and $A - \lambda I$ are

$$A = \begin{pmatrix} 5 & 4 & 2 \\ 4 & 5 & 2 \\ 2 & 2 & 2 \end{pmatrix}, \qquad A - \lambda I = \begin{pmatrix} 5 - \lambda & 4 & 2 \\ 4 & 5 - \lambda & 2 \\ 2 & 2 & 2 - \lambda \end{pmatrix}.$$

After a computation, we see that the characteristic equation is

$$\det (A - \lambda I) = (\lambda - 1)^2 (10 - \lambda).$$

If $\lambda = 10$, the corresponding system of linear equations is

$$\begin{pmatrix} -5 & 4 & 2 \\ 4 & -5 & 2 \\ 2 & 2 & -8 \end{pmatrix} \begin{pmatrix} x_1 \\ x_2 \\ x_3 \end{pmatrix} = \begin{pmatrix} 0 \\ 0 \\ 0 \end{pmatrix}$$

or

$$\begin{aligned} -5x_1 + 4x_2 + 2x_3 &= 0, \\ 4x_1 - 5x_2 + 2x_3 &= 0, \\ 2x_1 + 2x_2 - 8x_3 &= 0. \end{aligned}$$

A solution is $x_1 = 2a$, $x_2 = 2a$, $x_3 = a$, with a arbitrary. Hence, for $\lambda = 10$, we may take

$$e_1' = \tfrac{1}{3}(2e_1 + 2e_2 + e_3),$$

assuming that e_1, e_2, e_3 is the basis yielding the x-coordinates. The factor $\tfrac{1}{3}$ is inserted so that e_1' is of unit length. If $\lambda = 1$, the linear system is

$$\begin{pmatrix} 4 & 4 & 2 \\ 4 & 4 & 2 \\ 2 & 2 & 1 \end{pmatrix} \begin{pmatrix} x_1 \\ x_2 \\ x_3 \end{pmatrix} = \begin{pmatrix} 0 \\ 0 \\ 0 \end{pmatrix}$$

or

$$\begin{aligned} 4x_1 + 4x_2 + 2x_3 &= 0, & x_1 \text{ arbitrary} &= b, \\ 4x_1 + 4x_2 + 2x_3 &= 0, & x_2 \text{ arbitrary} &= c, \\ 2x_1 + 2x_2 + x_3 &= 0, & x_3 = -2x_1 - 2x_2 &= -2b - 2c. \end{aligned}$$

Corresponding to the double root $\lambda = 1$, we have a two-dimensional vector space, every vector of which is orthogonal to e_1'. We merely have to choose a basis for this space. Taking any convenient values for b and c, say $b = 1$, $c = 0$, we get $\mathbf{v} = e_1 - 2e_3$. To get a second basis vector $\mathbf{w}$, we take any value of b, say $b = 2$, and select c so that $\mathbf{v} \cdot \mathbf{w} = 0$. We find

$$(e_1 - 2e_3) \cdot (2e_1 + ce_2 - (4 + 2c)e_3) = 0;$$

then $2 + 8 + 4c = 0$ and $c = -\tfrac{5}{2}$. Therefore

$$\mathbf{v} = e_1 - 2e_3, \qquad \mathbf{w} = 2e_1 - \tfrac{5}{2}e_2 + e_3.$$

Normalizing, we get

$$e_2' = \frac{1}{\sqrt{5}}(e_1 - 2e_3), \qquad e_3' = \frac{1}{3\sqrt{5}}(4e_1 - 5e_2 + 2e_3).$$

The new matrix A' and the orthogonal matrix $C = H^t$ are

$$A' = \begin{pmatrix} 10 & 0 & 0 \\ 0 & 1 & 0 \\ 0 & 0 & 1 \end{pmatrix}, \qquad C = \begin{pmatrix} \dfrac{2}{3} & \dfrac{2}{3} & \dfrac{1}{3} \\[2mm] \dfrac{1}{\sqrt{5}} & 0 & \dfrac{-2}{\sqrt{5}} \\[2mm] \dfrac{4}{3\sqrt{5}} & \dfrac{-5}{3\sqrt{5}} & \dfrac{2}{3\sqrt{5}} \end{pmatrix}.$$

The new equation is

$$Q = 10(x'_1)^2 + x'^2_2 + x'^2_3,$$

and the new variables are introduced by the equations

$$x_1 = \frac{2}{3}x'_1 + \frac{1}{3\sqrt{5}}(3x'_2 + 4x'_3), \qquad x_2 = \frac{2}{3}x'_1 - \frac{5}{3\sqrt{5}}x'_3,$$

$$x_3 = \frac{1}{3}x'_1 + \frac{1}{3\sqrt{5}}(-6x'_2 + 2x'_3).$$

EXERCISES

In problems 1 through 6, the vectors $e_1, e_2, \ldots, e_n$ form an orthonormal basis for V^0_n. The subspace W is the space spanned by the given vectors. In each case find a basis for $W^\perp$.

1. W spanned by $v_1 = e_1 + 2e_2 - e_3$, $\qquad v_2 = e_1 - 2e_3$

2. W spanned by $v_1 = e_1 + e_2$, $\qquad v_2 = -e_1 + 2e_3$

3. W spanned by $v_1 = e_1 - 2e_3 + e_4$, $\qquad v_2 = -e_1 + e_2 - e_4$

4. W spanned by $v_1 = e_1 + 2e_2 - e_3 + e_4$

5. W spanned by $v_1 = e_1 + 2e_3 - e_5$, $\qquad v_2 = e_1 + e_2 - e_3$, $\qquad v_3 = e_2 - e_4$, $v_4 = e_5$

6. W spanned by $v_1 = e_1 + e_4$, $\qquad v_2 = e_1 + e_2 - e_5$

In each of problems 7 through 13, find the symmetric matrix A of the given quadratic form and find its eigenvalues. Also find an orthogonal matrix H which transforms Q to a sum of squares, write the change of variables, and find the expression for Q in the new variables.

7. $Q = 2x_1^2 + x_2^2 - 4x_1x_2 - 4x_2x_3$

8. $Q = 4x_1^2 + x_2^2 - 8x_3^2 + 4x_1x_2 - 4x_1x_3 + 8x_2x_3$

9. $Q = 3x_1^2 + x_2^2 + x_3^2 - 2x_1x_2 + 2x_1x_3 - 2x_2x_3$

10. $Q = 5x_1^2 + 5x_2^2 + 3x_3^2 - 2x_1x_2 + 2x_1x_3 + 2x_2x_3$

11. $Q = x_1^2 + 2x_1x_2 - 2x_1x_3 - 4x_2x_3$

12. $Q = 2x_1^2 + 2x_2^2 - x_3^2 + 8x_1x_2 - 4x_1x_3 - 4x_2x_3$

13. $Q = 4x_1^2 + 6x_2^2 + 4x_3^2 - 4x_1x_3$

14. Let A be the symmetric matrix corresponding to a quadratic form Q in three variables. If the eigenvalues $\lambda_1, \lambda_2, \lambda_3$ of A are always ones, minus ones, or zeros, how many different types of sums of squares can the reduced Q have? For example, one possibility is

$$Q = (x_1')^2 + (x_2')^2 + (x_3')^2;$$

another is

$$Q = (x_1')^2 - (x_2')^2 - (x_3')^2.$$

List all remaining possibilities.

15. (a) A quadratic form Q is **positive definite** if and only if $Q > 0$ except when $x_1 = x_2 = \cdots = x_n = 0$. Show that a quadratic form Q is positive definite if and only if all the eigenvalues of the associated matrix A are positive. What happens if one or more of the eigenvalues are zero and the remainder are positive?

(b) A quadratic form Q is **positive semidefinite** if $Q \geq 0$, always. State and prove a theorem relating the eigenvalues of A to semidefinite forms.

3.   COMPLEX NUMBERS

In studying the eigenvalues of matrices we did not consider the cases in which the characteristic equation has complex roots. For example, the characteristic equation of the simple matrix

$$\begin{pmatrix} 1 & 1 \\ -1 & 1 \end{pmatrix} \quad \text{is} \quad \lambda^2 - 2\lambda + 2 = 0,$$

and the eigenvalues are the complex numbers $1 + i, 1 - i$. Many of the results of Sections 1 and 2 are not applicable for such matrices. In order to develop a theory which encompasses matrices having complex eigenvalues, we must first establish a theory of linear vector spaces with complex scalars. In this section we shall derive some of the elementary properties of complex numbers, many of which the student has encountered in high-school algebra. We shall then take up the study of complex vector spaces in the next section.

We define a **complex number** as an ordered pair of real numbers (a, b) which we write in the convenient and familiar form

$$a + bi.$$

The **sum** and **product** of two complex numbers are defined by the formulas

$$(a + bi) + (c + di) = (a + c) + (b + d)i, \tag{1}$$

$$(a + bi) \cdot (c + di) = (ac - bd) + (ad + bc)i. \tag{2}$$

The complex numbers 0, 1, and i are given by

$$0 = 0 + 0i, \quad 1 = 1 + 0i, \quad i = 0 + 1i,$$

and, in fact, any real number a is identified with the complex number $a + 0i$. A special case of the multiplication rule (2) tells us that*

$$i^2 = -1.$$

Since *all* the rules for addition, subtraction, multiplication, and division of real numbers hold also for complex numbers, it follows that *all the laws of algebra which involve only these elementary operations of arithmetic hold also for complex numbers.* We observe that if complex numbers are allowed, the problem of obtaining solutions of polynomial and other nonlinear equations is radically different. For example, the simple equation $x^2 + 1 = 0$ has no solution when we restrict ourselves to real numbers, but it has the complex solutions $x = \pm i$.

The rules of inequality for real numbers, which we used to such a great extent in the development of calculus, *do not apply for complex numbers.* There is no way to arrange the complex numbers in a simple ordering according to size. However, we define the **absolute value** or **modulus** of a complex number $a + bi$, denoted $|a + bi|$, by

$$|a + bi| = \sqrt{a^2 + b^2}.$$

The absolute value is a real number and, of course, the rules of inequality hold for it.

We shall find it useful to define the **conjugate** of $a + bi$, which we denote by

$$\overline{(a + bi)} = a - bi.$$

Note that the conjugate of a real number a is a itself. The conjugate of the number bi is $-bi$. We shall frequently denote complex numbers by single letters; in particular, we shall use z and w with various subscripts. If $z = x + yi$, where x and y are real numbers, we call x the **real part** of z and denote it by Re z; the quantity y is called the **imaginary part** of z and is denoted by Im z. Finally, it is important to note that two complex numbers $z = x + yi$ and $w = u + vi$ (with x, y, u, v all real) are *equal* if and only if

$$x = u \qquad \text{and} \qquad y = v.$$

That is, two complex numbers are the same if *both* the real parts and the imaginary parts coincide.

The following example recalls the procedures for multiplying and dividing complex numbers.

---

* If we retain the ordered-pair notation we see that we have defined

$$(a, b) + (c, d) = (a + c, b + d),$$
$$(a, b) \cdot (c, d) = (ac - bd, ad + bc).$$

The complex numbers 0, 1, and i correspond to the respective ordered pairs

$$(0, 0), \quad (1, 0), \quad \text{and} \quad (0, 1).$$

The rule of multiplication gives $(0, 1) \cdot (0, 1) = (-1, 0)$.

EXAMPLE 1. Write the complex quantity

$$\frac{3 + 2i}{2 - i}$$

in the form $a + bi$.

Solution. We have

$$\frac{3 + 2i}{2 - i} = \frac{(3 + 2i)(2 + i)}{(2 - i)(2 + i)} = \frac{4 + 7i}{5} = \frac{4}{5} + \frac{7}{5}i.$$

Remark. Only the laws of fractions were used in the solution. Note that we multiplied numerator and denominator by the conjugate of the denominator.

The next example shows that matrices and determinants with complex elements may be treated in the expected manner.

EXAMPLE 2. Evaluate the determinant

$$D = \begin{vmatrix} 2 - i & 1 + i & -1 + i \\ 2 + i & 2i & 1 \\ 1 - 2i & 3 - i & i \end{vmatrix}.$$

Solution. We multiply the second row by $1 - i$ and add to the first row; then we multiply the second row by i and subtract from the third row. The result is

$$D = \begin{vmatrix} 5 - 2i & 3 + 3i & 0 \\ 2 + i & 2i & 1 \\ 2 - 4i & 5 - i & 0 \end{vmatrix} = - \begin{vmatrix} 5 - 2i & 3 + 3i \\ 2 - 4i & 5 - i \end{vmatrix}$$

$$= -[(5 - 2i)(5 - i) - (3 + 3i)(2 - 4i)] = -5 + 9i.$$

Since a complex number is an ordered number pair, it may be represented as a point in the plane. This representation is known as the **Argand diagram.** If $z = x + yi$, then z is the point (x, y) in the plane (Fig. 8–1). The absolute value of z is

$$|z| = \sqrt{x^2 + y^2},$$

which is the distance of z from the origin. The conjugate, $\bar{z}$, is the reflection of z in the x-axis. The x-axis is called the **axis of reals** or the **real axis,** while the y-axis is frequently called the **imaginary axis** or the **axis of imaginaries.**

The Argand diagram suggests that we may introduce polar coordinates (r, θ) with the selection

$$r = |z| = +\sqrt{x^2 + y^2}$$

as one coordinate. We recall that it is always possible to represent points in polar

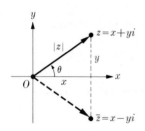

FIGURE 8–1

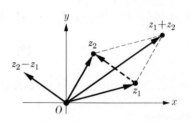

FIGURE 8–2

coordinates with $r \geq 0$. The number θ, which satisfies the equations

$$x = r \cos \theta, \qquad y = r \sin \theta,$$

is called an **argument** of z and is denoted by arg z. Clearly, if θ is an argument of z, then $\theta \pm 2n\pi$, $n = 1, 2, \ldots$, is also one; hence every complex number has infinitely many arguments. The particular value of θ for which $-\pi < \theta \leq \pi$ is called the **principal argument** and is denoted by **Arg** z. If we write two complex numbers z_1 and z_2 in the form

$$z_1 = r_1 \cos \theta_1 + (r_1 \sin \theta_1)i, \qquad z_2 = r_2 \cos \theta_2 + (r_2 \sin \theta_2)i,$$

then we see that $z_1 = z_2$ if and only if

$$r_1 = r_2 \qquad and \qquad \theta_2 - \theta_1 = 2k\pi \text{ for some integer } k.$$

The representation $r (\cos \theta + i \sin \theta)$ for a complex number z is called the **polar form** or **polar representation** of the number.

With each complex number z we can associate a vector **v** by selecting as a representative the directed line segment from the origin to the point z in the Argand diagram. We write $\mathbf{v} = \mathbf{v}(0z)$. Formula (1) for the sum $z_1 + z_2$ of two complex numbers z_1 and z_2 shows that the rule is the same as that for the addition of vectors (Fig. 8–2). Since we may write $z_2 - z_1 = z_2 + (-z_1)$, the subtraction of complex numbers follows the rule for vector subtraction.

Theorem 4. *If z_1, z_2 have the polar representations*

$$z_1 = r_1 (\cos \theta_1 + i \sin \theta_1),$$
$$z_2 = r_2 (\cos \theta_2 + i \sin \theta_2), \qquad (3)$$

then

$$z_1 z_2 = r_1 r_2 [\cos (\theta_1 + \theta_2) + i \sin (\theta_1 + \theta_2)].$$

Proof. We multiply to obtain (Fig. 8–3)

$$z_1 z_2 = r_1 r_2 (\cos \theta_1 \cos \theta_2 - \sin \theta_1 \sin \theta_2)$$
$$+ i r_1 r_2 (\sin \theta_1 \cos \theta_2 + \cos \theta_1 \sin \theta_2)$$
$$= r_1 r_2 \cos (\theta_1 + \theta_2) + i r_1 r_2 \sin (\theta_1 + \theta_2).$$

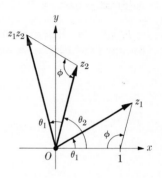

FIGURE 8–3

Corollary 1. *If z_1 and z_2 are defined by (3) and if $z_2 \neq 0$, then*

$$\frac{z_1}{z_2} = \frac{r_1}{r_2} \cos(\theta_1 - \theta_2) + i\frac{r_1}{r_2} \sin(\theta_1 - \theta_2).$$

Proof. We write $z_1 = (z_2)(z_1/z_2)$ and apply Theorem 4.

Corollary 2.

$$|z_1 \cdot z_2| = |z_1| \cdot |z_2|; \qquad \left|\frac{z_1}{z_2}\right| = \frac{|z_1|}{|z_2|} \qquad if \ z_2 \neq 0.$$

The proof is left to the reader. (See Exercise 36 at the end of this section.)

Corollary 3. (DeMoivre's Theorem.) *If*

$$z = r(\cos \theta + i \sin \theta)$$

and n is a positive integer, then

$$z^n = r^n(\cos n\theta + i \sin n\theta). \tag{4}$$

If $z \neq 0$, the result holds if n is a negative integer.

Proof. If $n = 2$, we see that Theorem 4 applies with $\theta_1 = \theta_2$. The result then follows by induction on n. If we select $z_1 = 1$, $z_2 = z^n$ in Corollary 1 and apply formula (4), we obtain DeMoivre's Theorem for negative exponents.

The results of the next theorem are used constantly in the algebra of complex numbers.

Theorem 5.

(a) $|z_1 + z_2| \leq |z_1| + |z_2|$;   (b) $\overline{(z_1 + z_2)} = \bar{z}_1 + \bar{z}_2$;

(c) $\overline{z_1 z_2} = \bar{z}_1 \bar{z}_2$;   (d) $\overline{(z_1/z_2)} = (\bar{z}_1/\bar{z}_2)$;

(e) $z\bar{z} = |z|^2 = |\bar{z}|^2$;   (f) $|-z| = |z|$.

Proof. (a) Referring to Fig. 8–2 and recalling that $|\mathbf{v}_1 + \mathbf{v}_2| \leq |\mathbf{v}_1| + |\mathbf{v}_2|$ for any two vectors, we obtain the result. To prove (b) we write $z_1 = x_1 + y_1 i$, $z_2 = x_2 + y_2 i$. Then

$$\overline{z_1 + z_2} = \overline{x_1 + x_2 + (y_1 + y_2)i} = x_1 + x_2 - (y_1 + y_2)i.$$

Also,

$$\bar{z}_1 = x_1 - y_1 i, \qquad \bar{z}_2 = x_2 - y_2 i$$

and

$$\bar{z}_1 + \bar{z}_2 = x_1 + x_2 - (y_1 + y_2)i.$$

The proofs of (c) through (f) are left to the reader. (See Exercise 38.)

Matrices with complex elements obey the same rules of addition and multiplication as do those with real elements. If $A = (a_{ij})$ is a matrix of complex numbers, we define its **conjugate matrix** $\bar{A} = (\bar{a}_{ij})$. Note that if the elements of A happen to be real, then $\bar{A} = A$.

Theorem 6. *Suppose A and B are matrices which can be multiplied. We have:* (a) $\overline{AB} = \bar{A}\,\bar{B}$; (b) *if A is nonsingular,* $\bar{A}$ *is also, and* $\overline{(A^{-1})} = (\bar{A})^{-1}$; (c) $\det \bar{A} = \overline{\det A}$.

The proofs of these facts are left to the reader. (See Exercises 39 and 40.)

The next example illustrates an important application of DeMoivre's Theorem.

EXAMPLE 3. Find the cube roots of $2 + 2i$.

Solution. Writing $2 + 2i$ in polar form, we have

$$2 + 2i = 2^{3/2}\left(\cos\frac{\pi}{4} + i\sin\frac{\pi}{4}\right).$$

We wish to solve for z in the equation $z^3 = 2 + 2i$. Writing $z = r(\cos\theta + i\sin\theta)$, this equation becomes (using DeMoivre's Theorem)

$$z^3 = r^3(\cos 3\theta + i\sin 3\theta) = 2^{3/2}\left(\cos\frac{\pi}{4} + i\sin\frac{\pi}{4}\right) = 2 + 2i.$$

One solution is seen to be

$$z_1 = 2^{1/2}\left(\cos\frac{\pi}{12} + i\sin\frac{\pi}{12}\right).$$

Recalling that two numbers in polar form are equal if their moduli are equal and if their arguments differ by a multiple of 2π, we can get additional solutions. In fact, we find

$$3\theta = \frac{\pi}{4} + 2k\pi \qquad \text{or} \qquad \theta = \frac{\pi}{12} + \frac{2k\pi}{3},$$

so that

$$z = 2^{1/2}\left[\cos\left(\frac{\pi}{12} + \frac{2k\pi}{3}\right) + i\sin\left(\frac{\pi}{12} + \frac{2k\pi}{3}\right)\right]$$

represents the totality of solutions if $k = 0, \pm1, \pm2, \ldots$. In setting $k = 1$ and 2, we obtain

$$z_2 = 2^{1/2}\left(\cos\frac{3\pi}{4} + i\sin\frac{3\pi}{4}\right)$$

and

$$z_3 = 2^{1/2}\left(\cos\frac{17\pi}{12} + i\sin\frac{17\pi}{12}\right).$$

All other values of k yield complex numbers which are identical with z_1, z_2, or z_3. Thus these three numbers are all the cube roots of $2 + 2i$.

Remark. In general, if z_0 is given, then solving the equation $z^n = z_0$ for z is easily done by writing

$$r^n (\cos n\theta + i \sin n\theta) = r_0 (\cos \theta_0 + i \sin \theta_0),$$

and noting that the solutions are determined by

$$r = r_0^{1/n}, \qquad \theta = \frac{\theta_0}{n} + \frac{2k\pi}{n}, \qquad k = 0, 1, 2, \ldots, n - 1.$$

EXERCISES

1. Write in the form $a + bi$

    (a) $(2 - i)(1 + i)$ (b) $\dfrac{1 + 2i}{3 - i}$ (c) $\dfrac{2 - 3i}{3 + 2i}$ (d) $\dfrac{(3 - i)^2}{(2 + i)^3}$.

In each of problems 2 through 6, the numbers z_1 and z_2 are given. Sketch the directed line segment from the origin to z_1 and to z_2. Calculate $z_1 + z_2$, $z_1 - z_2$, $z_1 z_2$, z_1/z_2, and draw the directed line segments to each of these.

2. $z_1 = 3 + 4i$, $z_2 = 2 + i$ 3. $z_1 = -1 + 2i$, $z_2 = 2 + 2i$

4. $z_1 = 2 - i$, $z_2 = -3 + i$ 5. $z_1 = 3 - i$, $z_2 = -2 - 2i$

6. $z_1 = -2 + i$, $z_2 = -3 - 4i$

In each of problems 7 and 8, evaluate D in the form $a + bi$.

7. $D = \begin{vmatrix} 1 + i & 2 + i & 1 - i \\ 1 & i & -1 - i \\ 2 - i & 2 & 2 - i \end{vmatrix}$ 8. $D = \begin{vmatrix} 2 - i & 1 + i & 1 \\ 3 + i & 1 - i & -i \\ 3 - i & 2i & 2 + i \end{vmatrix}$

9. Verify that $\overline{AB} = \bar{A}\,\bar{B}$ if

$$A = \begin{pmatrix} 1 + i & 2 & 2 - i \\ i & 0 & 3 + i \\ -2 & 1 & -1 + 2i \end{pmatrix}, \qquad B = \begin{pmatrix} 2 - 2i & -2i & 3 \\ 1 & 0 & 2i \\ -1 + i & 2 + i & 3 - i \end{pmatrix}.$$

In problems 10 through 15, use the methods of Section 8, Chapter 6 to find the solutions, if any, to each of the following systems of linear equations.

10. $x - (1 + i)y = 2 - i$ 11. $2ix - (1 - i)y = 1 + i$
      $(1 - i)x \qquad\;\; + 2iy = i$ $(1 - i)x \qquad\;\; + y = 1 - i$

12. $(1 + i)x - iy = 2 - i$ 13. $(2 - i)x + (1 + i)y = 2 + i$
    $(1 - i)x - y = -1 - 2i$ $(1 + 2i)x - (1 - i)y = i$

14. $x + (1 + i)y \qquad\;\; + iz = 2 + i$
      $(1 - i)x \qquad + iy \qquad\;\; - z = 3$
      $2x + (2 + 3i)y + (1 + 2i)z = 5 + 4i$

15. $x - (1 - i)y + (2 + i)z = 2 + i$
    $(1 + i)x - (2 - i)y + (2 - i)z = 2 + i$
      $ix - (2 + i)y + (3 + 3i)z = 1 - i$

In problems 16 and 17, find the moduli and principal arguments of the given numbers.

16. (a) $1 - i\sqrt{3}$ (b) $-2 + 2i$ (c) -2 (d) i^3

17. (a) $\dfrac{1}{1 - i}$ (b) $\dfrac{1 + i}{1 - i}$ (c) $\dfrac{2}{(\sqrt{3} - i)^2}$ (d) $(1 + i)^4$

18. Given that $z = x + yi$, write each of the following in the form $a + bi$:

(a) $\dfrac{1}{z}$ (b) $\dfrac{1}{1 + z}$ (c) $z^2 + 2z - 3$ (d) $\dfrac{1}{z - i}$.

In problems 19 through 22, we are given that $z = x + yi$. Describe and sketch each locus.

19. (a) $|z| = 1$ (b) $|z| < 2$ (c) $|z| > 1$
20. (a) $\text{Re}\,(z) > 1$ (b) $\text{Im}\,(z) > -1$ (c) $\text{Im}\,(z^2) > 2$
21. (a) $|z - 1| < 1$ (b) $|(z - 2)/(z + 1)| = 2$
22. (a) $|z - 2| > 3$ (b) the common part of $|z - 2| < 3$ and $|z + 1| < 1$

In each of problems 23 through 34, find all the solutions and represent them graphically.

23. $z^2 = -2 + 2i\sqrt{3}$ 24. $z^2 = 2i$
25. $z^3 = i$ 26. $z^3 = 2 - 2i$
27. $z^3 = -8$ 28. $z^3 = -2 - 2i$
29. $z^4 = -1$ 30. $z^4 = -8 - 8\sqrt{3}\,i$
31. $z^5 = 1$ 32. $z^5 = -4 - 4i$
33. $z^6 = 1$ 34. $z^6 = -8$

35. Prove the quadratic formula when the coefficients are complex.
36. Prove Corollary 2 on page 374.
37. Prove Theorem 5(a) without reference to the result for vectors.
38. Prove Theorem 5, parts (c), (d), (e), and (f).
39. Prove Theorem 6, parts (a) and (b).
40. Prove Theorem 6, part (c).

4. COMPLEX VECTOR SPACES

A **complex vector space** is defined exactly as was a (real) vector space in Section 1 of Chapter 7, *except that the scalars in Axioms B-1 and B-2 are allowed to be complex numbers.* Some examples of complex vector spaces are:

(i) The space C_n of sequences $(z_1, z_2, \ldots, z_n)$ of complex numbers with addition and scalar multiplication defined by

$$(z_1, z_2, \ldots, z_n) + (w_1, w_2, \ldots, w_n) = (z_1 + w_1, z_2 + w_2, \ldots, z_n + w_n),$$
$$c(z_1, z_2, \ldots, z_n) = (cz_1, cz_2, \ldots, cz_n), \quad c \text{ complex.}$$

(ii) The totality of polynomials with complex coefficients with the usual rules for addition of polynomials and multiplication of a polynomial by a constant.

(iii) The totality of complex-valued continuous functions defined on an interval $[a, b]$—that is, all functions of the form $f(x) = f_1(x) + if_2(x)$,where f_1 and f_2 are real-valued, continuous functions on $[a, b]$.

A **complex vector subspace** W of a complex vector space V is defined as before, but with real scalars replaced by complex ones. The definitions of **linear dependence** and **linear independence** of vectors are unchanged, the scalars being complex. We say that the vectors $\mathbf{v}_1, \mathbf{v}_2, \ldots, \mathbf{v}_k$ **span** the subspace W (or that W is spanned by $\mathbf{v}_1, \mathbf{v}_2, \ldots, \mathbf{v}_k$) provided that every vector in W is a linear combination of $\mathbf{v}_1, \mathbf{v}_2, \ldots, \mathbf{v}_k$ with complex coefficients. Moreover, if $\mathbf{v}_1, \mathbf{v}_2, \ldots, \mathbf{v}_k$ are vectors in a complex vector space V, the totality of linear combinations of these vectors (with complex coefficients) forms the **subspace W which is spanned by** $\mathbf{v}_1, \mathbf{v}_2, \ldots, \mathbf{v}_k$.

The definition of **finite dimensional space** and the definition of the **complex dimension** of a complex vector space are exactly the same as the definitions for real vector spaces. In discussing complex dimension, we must adjust our intuitive picture appropriately. For example, a one-dimensional complex vector space consists of all complex multiples of a single nonzero vector. This collection can be put into one-to-one correspondence with all the complex numbers, i.e., with all points in the complex plane (Argand diagram). Since the elements of a two-dimensional real vector space can be put into one-to-one correspondence with the points in the xy-plane, we see that one complex dimension corresponds to two dimensions in real vector spaces. Similarly, a two-complex-dimensional space is the same, dimensionally, as a four-dimensional real vector space. Theorems 1 and 2 of Chapter 7 concerning linear independence and bases, and Theorems 3 through 7 of Chapter 7 carry over in statement and proof, simply with the understanding that the scalars are complex.

Linear transformations from C_n to C_n and the various properties of such transformations are the same as in Section 3, when all the numbers involved are complex. Although the matrix of a linear transformation now has complex elements, it otherwise exhibits the same behavior as before. If V is a complex vector space, a transformation T from V to V is said to be **linear** if and only if its domain is all of V and

$$T\left(\sum_{i=1}^{k} c_i\mathbf{v}_i\right) = \sum_{i=1}^{k} c_i T\mathbf{v}_i.$$

for all complex numbers $c_1, c_2, \ldots, c_k$ and all vectors $\mathbf{v}_1, \mathbf{v}_2, \ldots, \mathbf{v}_k$ in V. If $\mathbf{v}_1, \mathbf{v}_2, \ldots, \mathbf{v}_n$ is a basis for V and T is a linear transformation from V to V, we associate a matrix of complex numbers A and a linear transformation from C_n to C_n, as in Section 6 of Chapter 7. That is, if we define

$$\mathbf{w}_i = T\mathbf{v}_i \quad \text{and} \quad \mathbf{w}_j = \sum_{j=1}^{n} a_{ij}\mathbf{v}_i,$$

then the complex matrix $A = (a_{ij})$ is the matrix associated with T for the given basis $v_1, v_2, \ldots, v_n$. The matrices and formulas associated with a change of basis carry over without change. The next example illustrates this point.

EXAMPLE 1. Suppose the set v_1, v_2 is a basis for V and T is the linear transformation defined by

$$w = Tv, \qquad v = x_1v_1 + x_2v_2, \qquad w = y_1v_1 + y_2v_2;$$

$$T: \begin{array}{l} y_1 = 3x_1 + (1 + i)x_2, \\ y_2 = -(1 - i)x_1 + 2x_2. \end{array} \tag{1}$$

Suppose that a second basis w_1, w_2 is introduced, related to the basis v_1, v_2 by

$$w_1 = 2v_1 + iv_2, \qquad w_2 = -iv_1 + v_2. \tag{2}$$

Find the equations corresponding to (1) connecting $\xi_1, \xi_2, \eta_1, \eta_2$ in the new basis.

Solution. With the original basis, the matrix A of the transformation T is

$$A = \begin{pmatrix} 3 & 1 + i \\ -(1 - i) & 2 \end{pmatrix}.$$

The matrix C connecting the bases is

$$C = \begin{pmatrix} 2 & i \\ -i & 1 \end{pmatrix},$$

and the matrix $B = H^{-1}AH$ with $H = C^t$ corresponds to T in the new basis. To find H^{-1} we find C^{-1} by solving for v_1, v_2 in terms of w_1, w_2 in Eqs. (2) and then taking the transpose. We get

$$\begin{array}{l} v_1 = w_1 - iw_2 \\ v_2 = iw_1 + 2w_2 \end{array}; \qquad H^{-1} = \begin{pmatrix} 1 & i \\ -i & 2 \end{pmatrix}.$$

Therefore

$$B = H^{-1}AH = \begin{pmatrix} 1 & i \\ -i & 2 \end{pmatrix} \begin{pmatrix} 3 & 1 + i \\ -(1 - i) & 2 \end{pmatrix} \begin{pmatrix} 2 & -i \\ i & 1 \end{pmatrix}$$

$$= \begin{pmatrix} 2 - i & 1 + 3i \\ -2 - i & 5 - i \end{pmatrix} \begin{pmatrix} 2 & -i \\ i & 1 \end{pmatrix} = \begin{pmatrix} 1 - i & i \\ -3 + 3i & 4 + i \end{pmatrix}.$$

The final result is

$$\eta_1 = (1 - i)\xi_1 + i\xi_2, \qquad \eta_2 = (-3 + 3i)\xi_1 + (4 + i)\xi_2.$$

We recall that a Euclidean vector space of finite dimension is one in which an inner product may be defined. It is in the definition of inner product that the first basic distinction occurs between real and complex vector spaces. The **inner product** in a complex vector space is denoted by (u, v) and is supposed to satisfy the following conditions:

(1) (u, v) *is a complex number defined for all* u *and* v *in the space.*
(2) $(v, u) = \overline{(u, v)}$ *for all* u, v.
(3) $(cu + dv, w) = c(u, w) + d(v, w)$ *for all* u, v, w *and all complex numbers* c, d.
(4) $(u, u) > 0$ *unless* $u = 0$. *Note that, by* (2), (u, u) *is always real.*

The inner product in a complex vector space is complex-valued, and the symmetry condition for inner product in real vector spaces is replaced by the "conjugate symmetry" condition (2), above. A finite-dimensional complex vector space in which an inner product is defined is called a **unitary space.** Unitary spaces are the appropriate generalization of real Euclidean vector spaces of finite dimension.

A complex vector space which is not necessarily of finite dimension, but on which an inner product is defined, is called a **pre-Hilbert space.** If a certain additional completeness condition is satisfied, we obtain a **Hilbert space.** Hilbert spaces are valuable tools for many problems in theoretical physics and applied mathematics. A detailed treatment of these spaces occurs in more advanced mathematics courses.

The space of complex-valued continuous functions on $[a, b]$ becomes a pre-Hilbert space if we define

$$(f, g) = \int_a^b f(x)\overline{g(x)} \, dx.$$

Items (1) through (4) in the definition of inner product are easily verified in this case.

The space C_n is a unitary space if we define

$$(\mathbf{z}, \mathbf{w}) = \sum_{j=1}^{n} z_j \overline{w}_j,$$

where $\mathbf{z} = (z_1, z_2, \ldots, z_n)$ and $\mathbf{w} = (w_1, w_2, \ldots, w_n)$. For $\mathbf{z} \neq \mathbf{0}$, we easily verify that $(\mathbf{z}, \mathbf{z})$ is real and positive. We have

$$(\mathbf{z}, \mathbf{z}) = \sum_{j=1}^{n} z_j \overline{z}_j = \sum_{j=1}^{n} |z_j|^2 = \sum_{j=1}^{n} (x_j^2 + y_j^2) \qquad \text{if } z_j = x_j + y_j i.$$

The quantity $(\mathbf{z}, \mathbf{z})$ is the square of the length of the vector in V_{2n}^0 with components $(x_1, x_2, \ldots, x_n, y_1, y_2, \ldots, y_n)$. We can frequently translate statements and results about unitary spaces into statements and results about Euclidean vector spaces with twice the dimension. In any unitary or pre-Hilbert space we define the **norm** $\|\mathbf{z}\|$ by

$$\|\mathbf{z}\| = (\mathbf{z}, \mathbf{z})^{1/2}.$$

Combining items (2) and (3) in the definition of inner product, we easily obtain

$$(\mathbf{u}, c\mathbf{v}) = \overline{(c\mathbf{v}, \mathbf{u})} = \overline{c(\mathbf{v}, \mathbf{u})} = \overline{c}\,\overline{(\mathbf{v}, \mathbf{u})} = \overline{c}\,(\mathbf{u}, \mathbf{v})$$

or, more generally,

$$(\mathbf{u}, c\mathbf{v} + d\mathbf{w}) = \overline{c}(\mathbf{u}, \mathbf{v}) + \overline{d}(\mathbf{u}, \mathbf{w});$$

also, we find

$$(\mathbf{u}, \mathbf{0}) = (\mathbf{0}, \mathbf{u}) = 0.$$

If c is any complex number, we have

$$\|c\mathbf{u}\|^2 = (c\mathbf{u}, c\mathbf{u}) = c(\mathbf{u}, c\mathbf{u}) = c\overline{c}(\mathbf{u}, \mathbf{u}) = |c|^2(\mathbf{u}, \mathbf{u}),$$

and therefore

$$\|c\mathbf{u}\| = |c| \|\mathbf{u}\|.$$

We recall that for real vector spaces the Schwarz inequality states that

$$|(\mathbf{x} \cdot \mathbf{y})| \leq |\mathbf{x}| \cdot |\mathbf{y}|.$$

The proof of this result for complex spaces, which we now establish, is somewhat more complicated than the proof for real vector spaces.

Theorem 7. (The Schwarz Inequality.) *For any vectors* $\mathbf{u}$ *and* $\mathbf{v}$ *in a complex vector space with an inner product, we have*

$$|(\mathbf{u}, \mathbf{v})| \leq \|\mathbf{u}\| \|\mathbf{v}\|.$$

Proof. If either $\mathbf{u}$ or $\mathbf{v} = \mathbf{0}$, the result is evident, so we assume that neither $\mathbf{u}$ nor $\mathbf{v}$ is $\mathbf{0}$. Let a and b be arbitrary *real* numbers. Then we have

$$0 \leq (a\mathbf{u} + b\mathbf{v}, a\mathbf{u} + b\mathbf{v}) = a^2(\mathbf{u}, \mathbf{u}) + ab[(\mathbf{u}, \mathbf{v}) + (\mathbf{v}, \mathbf{u})]$$
$$+ b^2(\mathbf{v}, \mathbf{v}). \tag{3}$$

Observe that $(\mathbf{u}, \mathbf{v}) + (\mathbf{v}, \mathbf{u}) = (\mathbf{u}, \mathbf{v}) + \overline{(\mathbf{u}, \mathbf{v})} = 2\,\mathrm{Re}\,(\mathbf{u}, \mathbf{v})$, since any complex number plus its conjugate is twice its real part. We define

$$A = (\mathbf{u}, \mathbf{u}), \qquad B = \mathrm{Re}\,(\mathbf{u}, \mathbf{v}), \qquad C = (\mathbf{v}, \mathbf{v}),$$

and the inequality (3) becomes

$$0 \leq Aa^2 + 2Bab + Cb^2$$

for arbitrary real numbers a and b. We now apply Lemma 5 to conclude that $B^2 \leq AC$ or

$$|\mathrm{Re}\,(\mathbf{u}, \mathbf{v})| \leq \|\mathbf{u}\| \|\mathbf{v}\|. \tag{4}$$

We now write $(\mathbf{u}, \mathbf{v})$ in the polar form

$$(\mathbf{u}, \mathbf{v}) = r\,(\cos \phi + i \sin \phi), \qquad r = |(\mathbf{u}, \mathbf{v})| \geq 0.$$

Defining $\lambda = \cos \phi - i \sin \phi$, we note that $|\lambda| = 1$ and that

$$(\lambda\mathbf{u}, \mathbf{v}) = \lambda(\mathbf{u}, \mathbf{v}) = r.$$

Since (4) holds for *any* two vectors $\mathbf{u}$ and $\mathbf{v}$, we apply it for $\lambda\mathbf{u}$ and $\mathbf{v}$. The result is

$$|\mathrm{Re}\,(\lambda\mathbf{u}, \mathbf{v})| \leq \|\lambda\mathbf{u}\| \|\mathbf{v}\| = |\lambda| \|\mathbf{u}\| \|\mathbf{v}\| = \|\mathbf{u}\| \|\mathbf{v}\|.$$

On the other hand, $(\lambda\mathbf{u}, \mathbf{v})$ is already real and nonnegative, being equal to r. Therefore

$$|\mathrm{Re}\,(\lambda\mathbf{u}, \mathbf{v})| = |\lambda(\mathbf{u}, \mathbf{v})| = |(\mathbf{u}, \mathbf{v})| \leq \|\mathbf{u}\| \|\mathbf{v}\|.$$

Corollary. $\|\mathbf{u} + \mathbf{v}\| \le \|\mathbf{u}\| + \|\mathbf{v}\|$.

The proof is left to the reader. (See Exercise 29 at the end of this section.)

We say that $\mathbf{u}$ and $\mathbf{v}$ are **complex-orthogonal** or simply **orthogonal** if and only if $(\mathbf{u}, \mathbf{v}) = 0$. Two (complex) *subspaces* of a given unitary or pre-Hilbert space are **orthogonal** if and only if each vector in one is orthogonal to each vector in the other. If E is a subspace, we denote by $E^\perp$ (read "E **perpendicular**") the set of all vectors orthogonal to every vector in E. A set $\mathbf{e}_1, \mathbf{e}_2, \ldots, \mathbf{e}_n$ forms an **orthonormal basis** for a unitary space if and only if

$$(\mathbf{e}_i, \mathbf{e}_j) = \delta_{ij}, \qquad i, j = 1, 2, \ldots, n.$$

Theorem 8. *Suppose that* $\mathbf{e}_1, \mathbf{e}_2, \ldots, \mathbf{e}_n$ *form an orthonormal set in a complex vector space* V. *If*

$$\mathbf{u} = \sum_{i=1}^{n} x_i \mathbf{e}_i, \qquad \mathbf{v} = \sum_{i=1}^{n} y_i \mathbf{e}_i,$$

then

$$x_i = (\mathbf{u}, \mathbf{e}_i), \qquad (\mathbf{u}, \mathbf{v}) = \sum_{i=1}^{n} x_i \bar{y}_i, \qquad (\mathbf{u}, \mathbf{u}) = \sum_{i=1}^{n} |x_i|^2. \qquad (5)$$

Consequently, the set $\mathbf{e}_1, \mathbf{e}_2, \ldots, \mathbf{e}_n$ *is linearly independent. If* E *is any subspace of* V, *then* $E^\perp$ *is a subspace also.*

Proof. We calculate

$$(\mathbf{u}, \mathbf{e}_j) = \sum_{i=1}^{n} x_i(\mathbf{e}_i, \mathbf{e}_j) = \sum_{i=1}^{n} x_i \, \delta_{ij} = x_j.$$

The proof of the remaining formulas is like that of Theorem 19 (Chapter 7) for real inner products. To show that $E^\perp$ is a subspace, we observe that if

$$(\mathbf{u}, \mathbf{w}) = (\mathbf{v}, \mathbf{w}) = 0 \qquad \text{for every } \mathbf{w} \text{ in } E,$$

then

$$(c\mathbf{u} + d\mathbf{v}, \mathbf{w}) = c(\mathbf{u}, \mathbf{w}) + d(\mathbf{v}, \mathbf{w}) = 0 \qquad \text{for all } \mathbf{w} \text{ in } E.$$

Hence, if $\mathbf{u}, \mathbf{v}$ belong to $E^\perp$, so does $c\mathbf{u} + d\mathbf{v}$; i.e., $E^\perp$ is a subspace.

EXAMPLE 2. Let $\mathbf{e}_1, \mathbf{e}_2, \mathbf{e}_3, \mathbf{e}_4$ be an orthonormal basis for a complex vector space V. Let E be the subspace spanned by

$$\mathbf{v}_1 = 2\mathbf{e}_1 + (3 - i)\mathbf{e}_3, \qquad \mathbf{v}_2 = i\mathbf{e}_1 + 2\mathbf{e}_2 - (1 + i)\mathbf{e}_4.$$

Find an orthogonal basis for the subspace $E^\perp$.

Solution. A vector $\mathbf{w}_1 = a_1\mathbf{e}_1 + b_1\mathbf{e}_2 + c_1\mathbf{e}_3 + d_1\mathbf{e}_4$ is in $E^\perp$ if

$$(\mathbf{w}_1, \mathbf{v}_1) = (\mathbf{w}_1, \mathbf{v}_2) = 0.$$

We have

$$(\mathbf{w}_1, \mathbf{v}_1) = (a_1\mathbf{e}_1, 2\mathbf{e}_1) + (c_1\mathbf{e}_3, (3 - i)\mathbf{e}_3) = 0$$

$$= 2a_1 + c_1 \overline{(3 - i)} = 0;$$

$$(\mathbf{w}_1, \mathbf{v}_2) = (a_1\mathbf{e}_1, i\mathbf{e}_1) + (b_1\mathbf{e}_2, 2\mathbf{e}_2) + (d_1\mathbf{e}_4, -(1 + i)\mathbf{e}_4) = 0$$

$$= -ia_1 + 2b_1 + (-1 + i)d_1 = 0.$$

We select $c_1 = 2$, $d_1 = 0$ and find that $a_1 = -(3 + i)$, $b_1 = \frac{1}{2}(1 - 3i)$. A second vector $\mathbf{w}_2 = a_2\mathbf{e}_1 + b_2\mathbf{e}_2 + c_2\mathbf{e}_3 + d_2\mathbf{e}_4$ is in $E^\perp$ and is orthogonal to $\mathbf{w}_1$ if

$$(\mathbf{w}_2, \mathbf{v}_1) = 2a_2 + (3 + i)c_2 = 0,$$

$$(\mathbf{w}_2, \mathbf{v}_2) = -ia_2 + 2b_2 + (-1 + i)d_2 = 0,$$

$$(\mathbf{w}_2, \mathbf{w}_1) = -(3 - i)a_2 + \frac{1}{2}(1 + 3i)b_2 + 2c_2 = 0.$$

We have three homogeneous equations in four unknowns. The rank of the coefficient matrix is 3, and a solution can be obtained. We find that

$$a_2 = 1, \ b_2 = -\tfrac{14}{5}i, \qquad c_2 = -\tfrac{1}{5}(3 - i), \qquad d_2 = \tfrac{33}{10}(1 - i).$$

The orthogonal vectors

$$\mathbf{w}_1 = -(3 + i)\mathbf{e}_1 + \tfrac{1}{2}(1 - 3i)\mathbf{e}_2 + 2\mathbf{e}_3,$$

$$\mathbf{w}_2 = \mathbf{e}_1 - \tfrac{14}{5}i\mathbf{e}_2 - \tfrac{1}{5}(3 - i)\mathbf{e}_3 + \tfrac{33}{10}(1 - i)\mathbf{e}_4,$$

span the subspace $E^\perp$.

Orthogonal vectors $\mathbf{w}_1$ and $\mathbf{w}_2$ spanning $E^\perp$ could also have been found by first finding *any* basis vectors $\mathbf{u}_1$ and $\mathbf{u}_2$ for $E^\perp$ and then setting $\mathbf{w}_1 = \mathbf{u}_1$ and $\mathbf{w}_2 = \mathbf{u}_2 - k\mathbf{u}_1$ where k is determined so that $(\mathbf{w}_2, \mathbf{w}_1) = 0$.

EXERCISES

In each of problems 1 through 5, find the inverse of the given matrix.

1. $\begin{pmatrix} 1 + i & 2 + i \\ 3 - i & 2 - i \end{pmatrix}$

2. $\begin{pmatrix} 1 - i & 2 \\ 2 + i & 3 - i \end{pmatrix}$

3. $\begin{pmatrix} 2 & i & 3 \\ 1 + i & 0 & 1 - i \\ 2 & 1 & 2 + i \end{pmatrix}$

4. $\begin{pmatrix} 1 & 1 - i & 2 + i \\ 2 - i & 2 - 3i & 5 + i \\ 1 + i & 4 - i & 1 + 3i \end{pmatrix}$

5. $\begin{pmatrix} 1 & 2 + i & i \\ 2 - i & 6 & 2 + 3i \\ 1 + i & 2 + 2i & 3 + i \end{pmatrix}$

In problems 6 through 9, assume that $\mathbf{u}_1, \mathbf{u}_2, \mathbf{u}_3$ form a basis for the complex vector space V and determine whether or not $\mathbf{v}$ is in the space spanned by $\mathbf{v}_1$ and $\mathbf{v}_2$; if it is, express $\mathbf{v}$ in terms of $\mathbf{v}_1$ and $\mathbf{v}_2$. If $\mathbf{v}_1$ and $\mathbf{v}_2$ are linearly dependent, note that fact.

6. $\mathbf{v}_1 = \mathbf{u}_1 + (1 + i)\mathbf{u}_2 - i\mathbf{u}_3$

$\mathbf{v}_2 = i\mathbf{u}_1 + i\mathbf{u}_2 + (2 - i)\mathbf{u}_3$

$\mathbf{v} = (2 + 2i)\mathbf{u}_1 + (1 + 3i)\mathbf{u}_2 + (2 - 4i)\mathbf{u}_3$

7. $v_1 = u_1 + (1 - i)u_2 + 2iu_3$
   $v_2 = (1 + i)u_1 + 3u_2 + 3iu_3$
   $v = (2 + i)u_1 + (4 - 2i)u_2 + (2 + 3i)u_3$

8. $v_1 = u_1 + (2 - i)u_2 + (1 + 2i)u_3$
   $v_2 = (1 + i)u_1 + (3 + i)u_2 - (1 - 3i)u_3$
   $v = (1 - i)u_1 + (2 + i)u_2 + iu_3$

9. $v_1 = u_1 + (2 + i)u_2 + (1 - i)u_3$
   $v_2 = (2 + i)u_1 + (4 + 4i)u_2 + (2 - i)u_3$
   $v = -(1 - 2i)u_1 - (5 - 4i)u_2 + (2 + 2i)u_3$

In problems 10 through 12, the vectors u_1, u_2 form a basis. A new basis v_1, v_2 is given in terms of the original one. Find in each case the new coordinates (ξ_1, ξ_2) in terms of the old coordinates (x_1, x_2).

10. $v_1 = (2 + i)u_1 + (1 + i)u_2$
    $v_2 = (1 + i)u_1 + iu_2$

11. $v_1 = (1 - i)u_1 + (2 + i)u_2$
    $v_2 = (1 + i)u_1 + (1 + 2i)u_2$

12. $v_1 = 2iu_1 - (2 - i)u_2$
    $v_2 = (1 - 2i)u_1 + 2u_2$

In problems 13 through 16, e_1, e_2, e_3 form an orthonormal basis. Verify that v_1 and v_2 are orthogonal and find a vector v_3 orthogonal to both.

13. $v_1 = e_1 + ie_2$
    $v_2 = e_1 - ie_2 + (1 + i)e_3$

14. $v_1 = e_1 + (1 + i)e_2 - (3 + 2i)e_3$
    $v_2 = (1 - 6i)e_1 - (1 - 3i)e_2 + e_3$

15. $v_1 = e_1 + (4 + 6i)e_2 - (1 + 3i)e_3$
    $v_2 = (3 - 5i)e_1 - 2ie_2 + (3 - 4i)e_3$

16. $v_1 = e_1 - (9 + 21i)e_2 - (4 + 14i)e_3$
    $v_2 = (5 - 13i)e_1 - (3 - 12i)e_2 + (8 - 18i)e_3$

In problems 17 through 19, $v_1, v_2, \ldots, v_k$ is a basis and T is a linear transformation as shown by the given matrix. If $w_1, w_2, \ldots, w_k$ is a second basis, find in each case the matrix corresponding to T with respect to the second basis.

17. $T: \begin{pmatrix} 2i & 1 \\ i & -i \end{pmatrix}$
    $w_1 = iv_1 + (1 + i)v_2$
    $w_2 = 2iv_1 + (1 - i)v_2$

18. $T: \begin{pmatrix} 1 + i & 1 - i \\ 2 - i & 2 + i \end{pmatrix}$
    $w_1 = 3iv_1 + (-1 + 2i)v_2$
    $w_2 = -2iv_1 + (1 - 3i)v_2$

19. $T: \begin{pmatrix} 1 + i & 2 & i \\ 0 & 3 - i & 1 \\ -1 & 1 - i & 0 \end{pmatrix}$
    $w_1 = 2v_1 - iv_2 + v_3$
    $w_2 = v_1 - v_2 + (1 - i)v_3$
    $w_3 = v_1 \qquad + iv_3$

In problems 20 through 25, the vectors $e_1, e_2, \ldots, e_n$ form an orthonormal basis of V. The vectors $v_1, v_2, \ldots, v_k$ span a subspace E. Find an orthogonal basis for $E^\perp$.

20. E is spanned by $v_1 = 2e_1 + 3e_2$, $v_2 = ie_1 + 2e_3$; $\quad n = 3$.

21. E is spanned by $v_1 = (1 + i)e_1 + 2e_2 + (1 - i)e_3$; $\quad n = 3$.

22. E is spanned by $v_1 = ie_1 - 2ie_2 + 3e_3 - e_4$,
    $v_2 = e_1 - ie_3$, $v_3 = e_1 + 2e_2 - ie_3$; $\quad n = 4$.

23. E is spanned by $\mathbf{v}_1 = (2 - i)\mathbf{e}_1 + \mathbf{e}_2$; $n = 4$.

24. E is spanned by $\mathbf{v}_1 = (1 - i)\mathbf{e}_1 + \mathbf{e}_3 - 2i\mathbf{e}_4$,
    $\mathbf{v}_2 = (1 + i)\mathbf{e}_1 + \mathbf{e}_2 - 3i\mathbf{e}_3 + \mathbf{e}_4$; $n = 4$.

25. If E_1 and E_2 are subspaces of a complex vector space V, we define the set E as all vectors of the form $\mathbf{v} = \mathbf{v}_1 + \mathbf{v}_2$ where $\mathbf{v}_1 \in E_1$, $\mathbf{v}_2 \in E_2$. We write $E = E_1 \oplus E_2$. Show that E is a subspace of V.

26. In the notation of problem 25, we define the set $F = E_1 \cap E_2$ as the set of vectors $\mathbf{v}$ which are in both E_1 and E_2. Show that F is a subspace.

27. In the notation of problems 25 and 26, show that $(E_1 \oplus E_2)^\perp = E_1^\perp \cap E_2^\perp$.

28. In the notation of problems 25 and 26, show that $E_1^\perp \oplus E_2^\perp = (E_1 \cap E_2)^\perp$.

29. Prove the corollary to Theorem 7.

5. UNITARY MATRICES AND UNITARY TRANSFORMATIONS

The theorems on orthonormal bases established in Sections 7 and 8 of Chapter 7 have analogs in complex vector spaces. The reader can now prove the next theorem and corollary by methods similar to those used for real vector spaces.

Theorem 9. *Suppose that* $\mathbf{v}_1, \mathbf{v}_2, \ldots, \mathbf{v}_k$ *are linearly independent vectors in a unitary space* V. *Then there is an orthonormal sequence* $\mathbf{e}_1, \mathbf{e}_2, \ldots, \mathbf{e}_k$ *spanning the same space. Moreover, the* $\mathbf{e}_j$ *may be chosen so that for each* j *the set* $\mathbf{e}_1, \mathbf{e}_2, \ldots, \mathbf{e}_j$ *spans the space spanned by* $\mathbf{v}_1, \mathbf{v}_2, \ldots, \mathbf{v}_j$. *If for some* $r \leq k$, *the set* $\mathbf{v}_1, \mathbf{v}_2, \ldots, \mathbf{v}_r$ *is already orthonormal, the* $\mathbf{e}_j$ *may be chosen so that* $\mathbf{e}_j = \mathbf{v}_j$ *for* $j = 1, 2, \ldots, r$.

Corollary. *If* W *is a* k-*dimensional subspace of a unitary space* V, *then there exists an orthonormal basis* $\mathbf{e}_1, \mathbf{e}_2, \ldots, \mathbf{e}_n$ *for* V *in which* $\mathbf{e}_1, \mathbf{e}_2, \ldots, \mathbf{e}_k$ *is a basis for* W. *In fact,* $\mathbf{e}_1, \mathbf{e}_2, \ldots, \mathbf{e}_k$ *may be prescribed.*

EXAMPLE. A complex vector space V has the orthonormal basis $\mathbf{e}_1, \mathbf{e}_2, \mathbf{e}_3$. Find an orthonormal basis for the subspace W spanned by $\mathbf{v}_1 = \mathbf{e}_1 + i\mathbf{e}_2 - (1 + i)\mathbf{e}_3$ and $\mathbf{v}_2 = i\mathbf{e}_1 + (2 + i)\mathbf{e}_2 + (1 - 2i)\mathbf{e}_3$.

Solution. We first find a vector $\mathbf{w}$ in W which is orthogonal to $\mathbf{v}_1$. Any vector $\mathbf{w} = \mathbf{v}_1 - k\mathbf{v}_2$ is in W, and we select k so that $\mathbf{w}$ and $\mathbf{v}_1$ are orthogonal. We have

$$0 = (\mathbf{v}_1, \mathbf{w}) = (\mathbf{v}_1, \mathbf{v}_1 - k\mathbf{v}_2) = (\mathbf{v}_1, \mathbf{v}_1) - (\mathbf{v}_1, k\mathbf{v}_2)$$
$$= (\mathbf{v}_1, \mathbf{v}_1) - \overline{k}(\mathbf{v}_1, \mathbf{v}_2).$$

We compute

$$\|\mathbf{v}_1\|^2 = (\mathbf{v}_1, \mathbf{v}_1) = 1 + 1 + 2 = 4,$$
$$(\mathbf{v}_1, \mathbf{v}_2) = -i + i(2 - i) - (1 + i)(1 + 2i) = 2 - 2i,$$

and therefore

$$\bar{k} = \frac{4}{2 - 2i} = \frac{4(2 + 2i)}{8} = 1 + i.$$

The vector $\mathbf{w}$ is given by

$$\mathbf{w} = \mathbf{v}_1 - (1 - i)\mathbf{v}_2 = -i\mathbf{e}_1 + (-3 + 2i)\mathbf{e}_2 + 2i\mathbf{e}_3.$$

The norm of $\mathbf{w}$ is $\|\mathbf{w}\| = \sqrt{1 + 13 + 4} = \sqrt{18}$. Hence we may take

$$\mathbf{e}_1' = \frac{\mathbf{v}_1}{\|\mathbf{v}_1\|} = \frac{1}{2}[\mathbf{e}_1 + i\mathbf{e}_2 - (1 + i)\mathbf{e}_3],$$

$$\mathbf{e}_2' = \frac{\mathbf{w}}{\|\mathbf{w}\|} = \frac{1}{3\sqrt{2}}[-i\mathbf{e}_1 - (3 - 2i)\mathbf{e}_2 + 2i\mathbf{e}_3],$$

and $\mathbf{e}_1'$, $\mathbf{e}_2'$ form the desired orthonormal basis.

Suppose that $\mathbf{e}_1, \mathbf{e}_2, \ldots, \mathbf{e}_n$ and $\mathbf{e}_1', \mathbf{e}_2', \ldots, \mathbf{e}_n'$ are two orthonormal bases for a unitary space V. Any vector $\mathbf{v}$ has the two representations

$$\mathbf{v} = \sum_{i=1}^{n} x_i \mathbf{e}_i, \qquad \mathbf{v} = \sum_{i=1}^{n} x_i' \mathbf{e}_i'. \tag{1}$$

By following the same procedure as with real vector spaces, we can find the relation between these representations. We take the inner product *on the right* between $\mathbf{v}$ and the basis vectors. We find

$$(\mathbf{v}, \mathbf{e}_p') = \sum_{i=1}^{n} x_i(\mathbf{e}_i, \mathbf{e}_p'),$$

$$(\mathbf{v}, \mathbf{e}_p') = \sum_{i=1}^{n} x_i'(\mathbf{e}_i', \mathbf{e}_p') = x_p',$$

and so

$$x_p' = \sum_{i=1}^{n} x_i(\mathbf{e}_i, \mathbf{e}_p'). \tag{2}$$

Similarly, taking inner products between $\mathbf{v}$ and $\mathbf{e}_p$, we get

$$x_p = \sum_{i=1}^{n} x_i'(\mathbf{e}_i', \mathbf{e}_p). \tag{3}$$

Defining the constants c_{pj} and c_{pj}' by the relations

$$c_{pj} = (\mathbf{e}_j, \mathbf{e}_p'), \qquad c_{pj}' = (\mathbf{e}_j', \mathbf{e}_p) = \overline{(\mathbf{e}_p, \mathbf{e}_j')} = \bar{c}_{pj},$$

we may write (2) and (3) in the form

$$x_p' = \sum_{j=1}^{n} c_{pj} x_j, \qquad x_p = \sum_{j=1}^{n} c_{pj}' x_j'. \tag{4}$$

In Eq. (1) we make the special choice of $\mathbf{e}'_p$ for $\mathbf{v}$. That is, we choose $x'_i = \delta_{ip}$ and write [using (4)]

$$\mathbf{e}'_p = \sum_{j=1}^n x_j \mathbf{e}_j = \sum_{j=1}^n \left(\sum_{q=1}^n c'_{jq} x'_q \right) \mathbf{e}_j = \sum_{j=1}^n \sum_{q=1}^n c'_{jq} \delta_{qp} \mathbf{e}_j.$$

Therefore

$$\mathbf{e}'_p = \sum_{j=1}^n c'_{jp} \mathbf{e}_j = \sum_{j=1}^n \overline{c}_{pj} \mathbf{e}_j. \tag{5}$$

Selecting $\mathbf{e}_p$ for $\mathbf{v}$ in (1) and proceeding similarly, we find

$$\mathbf{e}_p = \sum_{j=1}^n c_{jp} \mathbf{e}'_j. \tag{6}$$

Substituting (6) into the right side of (5) and taking the inner product $(\mathbf{e}'_p, \mathbf{e}'_q) = \delta_{pq}$, we obtain

$$\sum_{j=1}^n c_{pj} \overline{c}_{qj} = \delta_{pq}, \qquad p, q = 1, 2, \ldots, n. \tag{7}$$

An $n \times n$ matrix C which satisfies conditions (7) is called a **unitary matrix.** If the elements of C happen to be real, then a unitary matrix is an orthogonal matrix. The next theorem is the appropriate generalization to complex vector spaces of Theorem 21 in Chapter 7.

Theorem 10. (a) *An $n \times n$ matrix C is unitary if and only if C is nonsingular and*

$$C^{-1} = \overline{C}^t. \tag{8}$$

That is, if we denote $D = C^{-1}$, then (8) may be written:

$$d_{pq} = \overline{c}_{qp}, \qquad p, q = 1, 2, \ldots, n.$$

(b) *If C is unitary, then $|\det C| = 1$.*
(c) *If C is unitary, then $\overline{C}$, C^t, and $C^{-1} = \overline{C}^t$ are unitary, and*

$$\sum_{j=1}^n c_{jp} \overline{c}_{jq} = \delta_{pq}, \qquad p, q = 1, 2, \ldots, n. \tag{9}$$

Proof. (a) Equation (7) in matrix form states that

$$C \overline{C}^t = I, \tag{10}$$

which is equivalent to (8).
(b) This result follows from the rules for the product of determinants and (8).
(c) Taking conjugates in (10), we get

$$\overline{C \overline{C}^t} = \overline{C} C^t = \overline{I} = I.$$

Therefore $\overline{C}$ is unitary. By taking the transpose we find C^t is unitary; similarly, C^{-1} is.

Theorem 11. (a) *Suppose that C is a unitary matrix and V is a unitary space; let* $\mathbf{e}_1, \mathbf{e}_2, \ldots, \mathbf{e}_n$ *be an orthonormal basis for V and define*

$$\mathbf{e}'_p = \sum_{j=1}^n c_{pj}\mathbf{e}_j.$$

Then $\mathbf{e}'_1, \mathbf{e}'_2, \ldots, \mathbf{e}'_n$ *is an orthonormal basis for V.*

(b) *The product of any finite sequence of unitary matrices (each* $n \times n$ *) is unitary.*

(c) *If the hypotheses of* (a) *hold and if we define*

$$x'_i = \sum_{j=1}^n c_{ij}x_j \qquad and \qquad y'_i = \sum_{j=1}^n c_{ij}y_j,$$

then

$$\sum_{i=1}^n x'_i\bar{y}'_i = \sum_{i=1}^n x_i\bar{y}_i. \tag{11}$$

Proof. Parts (a) and (c) result from straight computation from the definition of a unitary matrix. Part (b) follows from part (a). (See Exercise 13 at the end of this section.)

Remarks. Condition (7), in the definition of unitary matrix, states that the row vectors form an orthonormal set. Condition (8) is equivalent to (7) and is sometimes used as a definition of a unitary matrix; the relations (9) assert that the column vectors of a unitary matrix form an orthonormal set.

A linear transformation T from a unitary space V into itself is called a **unitary transformation** if and only if

$$(T\mathbf{u}, T\mathbf{v}) = (\mathbf{u}, \mathbf{v}) \quad \text{for all} \quad \mathbf{u} \in V, \quad \mathbf{v} \in V. \tag{12}$$

Theorem 12. *A linear transformation T from a unitary space into itself is unitary if and only if the matrix corresponding to it with respect to any orthonormal basis is unitary.*

The proof is similar to the proof of the corresponding theorem for orthogonal transformations (Theorem 23 of Chapter 7) and is left to the reader. (See Exercise 14 at the end of this section.)

A unitary transformation leaves the norm of any element unchanged, as is shown by setting $\mathbf{u} = \mathbf{v}$ in (12). The corresponding statement in terms of coordinates is reflected in relation (11) for unitary matrices.

There are occasions when, in considering transformations from a *real* vector space V to V, we encounter a complex root of the characteristic equation. We should like to draw conclusions concerning the real transformations even when there is such a complex eigenvalue. This can be done by employing a device which "complexifies" the space V. One of the applications of complexification, given in the next section, is a proof of the theorem that all the eigenvalues of a symmetric matrix with real elements are real.

If we think about the way in which the complex numbers are obtained from the reals, we would expect that a complexification of a real vector space V would consist of all vectors of the form

$$\mathbf{u} + i\mathbf{v},$$

with $\mathbf{u}$ and $\mathbf{v}$ in V. With the appropriate interpretation of the various rules, this is indeed the case.

Given any real vector space V, we define the **complexification** V^C as the collection of all ordered pairs $[\mathbf{u}, \mathbf{v}]$ with $\mathbf{u}$ and $\mathbf{v}$ in V. So far we have merely a set of elements V^C. We must establish rules for addition and multiplication by scalars and then prove that what we have obtained is a complex vector space. In analogy with the method used for obtaining complex numbers, we proceed by writing an ordered pair in V^C in the form $\mathbf{u} \dotplus i\mathbf{v}$. The sum of two elements $\mathbf{u}_1 \dotplus i\mathbf{v}_1$ and $\mathbf{u}_2 \dotplus i\mathbf{v}_2$ in V^C is defined by

$$(\mathbf{u}_1 \dotplus i\mathbf{v}_1) + (\mathbf{u}_2 \dotplus i\mathbf{v}_2) = (\mathbf{u}_1 + \mathbf{u}_2) \dotplus i(\mathbf{v}_1 + \mathbf{v}_2). \tag{13}$$

The symbol $\dotplus$ indicates an element in V^C, while the ordinary plus sign between vectors indicates addition in V. Multiplication by scalars is defined by

$$(c + id)(\mathbf{u} \dotplus i\mathbf{v}) = c\mathbf{u} + (-d)\mathbf{v} \dotplus i(d\mathbf{u} + c\mathbf{v}). \tag{14}$$

To check that V^C is a complex vector space, we must verify axioms A-1 through A-5 and B-1 through B-2 (in Section 1 of Chapter 7) with regard to the operations (13) and (14). This verification is left to the reader. (See Exercise 15 at the end of this section.)

If the real vector space V has an inner product (so that it is Euclidean), we are able to define a scalar product in V^C. This scalar product is defined by the formula

$$(\mathbf{u}_1 \dotplus i\mathbf{v}_1, \mathbf{u}_2 \dotplus i\mathbf{v}_2) = \mathbf{u}_1 \cdot \mathbf{u}_2 + \mathbf{v}_1 \cdot \mathbf{v}_2 - i(\mathbf{u}_1 \cdot \mathbf{v}_2 - \mathbf{u}_2 \cdot \mathbf{v}_1). \tag{15}$$

It may be verified (see Exercise 16) that if V is Euclidean, the above inner product makes V^C a unitary or pre-Hilbert space. In such a case we say that we have a **unitary complexification** of V and denote it by V_0^C. Note that the rule for the scalar product is suggested by the usual distributive law for scalar product plus the rules for dealing with complex numbers. We use the customary dot product for real vectors. The right side of (15) is an ordinary complex number.

Theorem 13. (a) *If* $\mathbf{v}_1, \mathbf{v}_2, \ldots, \mathbf{v}_n$ *is a basis for a real vector space* V, *then* $\mathbf{v}_1 \dotplus i\mathbf{0}, \mathbf{v}_2 \dotplus i\mathbf{0}, \ldots, \mathbf{v}_n \dotplus i\mathbf{0}$ *is a basis for* V^C.
(b) *Suppose that* V *is also Euclidean; then* V_0^C *is unitary or pre-Hilbert. If* $\mathbf{e}_1,$ $\mathbf{e}_2, \ldots, \mathbf{e}_n$ *is an orthonormal basis for* V, $\mathbf{e}_1 \dotplus i\mathbf{0}, \mathbf{e}_2 \dotplus i\mathbf{0}, \ldots, \mathbf{e}_n \dotplus i\mathbf{0}$ *is an orthonormal basis for* V_0^C.

Proof. To prove part (a), we let

$$\mathbf{u} = \sum_{j=1}^{n} c_j \mathbf{v}_j, \qquad \mathbf{v} = \sum_{j=1}^{n} d_j \mathbf{v}_j, \qquad c_j, d_j \text{ real.}$$

Then, according to the definition of V^C, we obtain

$$\mathbf{u} + i\mathbf{v} = \left(\sum_{j=1}^{n} c_j \mathbf{v}_j\right) + i\left(\sum_{j=1}^{n} d_j \mathbf{v}_j\right) = \sum_{j=1}^{n} (c_j + i\, d_j)(\mathbf{v}_j + i\mathbf{0}).$$

The proof of part (b) is similar. (See Exercise 17.)

Recalling the definition of C_n, we see that it is essentially the complexification of V_n. The space of complex-valued continuous functions on an interval $[a, b]$ is the complexification of the space of real-valued continuous functions on $[a, b]$. Actually, it can be shown that any finite-dimensional complex vector space can be obtained as the complexification of a finite dimensional real vector space. In particular, any unitary space can be obtained as the unitary complexification of a Euclidean vector space.

EXERCISES

In each of problems 1 through 5, it is assumed that $\mathbf{e}_1, \mathbf{e}_2, \ldots, \mathbf{e}_n$ is an orthonormal basis for the complex vector space V. If $\mathbf{u}$ and $\mathbf{v}$ are as given, determine k such that the vector $\mathbf{w} = \mathbf{u} - k\mathbf{v}$ is orthogonal to $\mathbf{u}$.

1. $\mathbf{u} = i\mathbf{e}_1 + (2 - i)\mathbf{e}_2,$ $\mathbf{v} = \mathbf{e}_1 - i\mathbf{e}_2$
2. $\mathbf{u} = \mathbf{e}_1 + (2 + i)\mathbf{e}_2 - \mathbf{e}_3,$ $\mathbf{v} = (1 - i)\mathbf{e}_1 + i\mathbf{e}_2$
3. $\mathbf{u} = \mathbf{e}_1 - \mathbf{e}_2 + 2i\mathbf{e}_3,$ $\mathbf{v} = -\mathbf{e}_1 + i\mathbf{e}_2 + (1 - 2i)\mathbf{e}_3$
4. $\mathbf{u} = 2\mathbf{e}_1 + i\mathbf{e}_2 - (1 - i)\mathbf{e}_3,$ $\mathbf{v} = 2\mathbf{e}_1 + 3\mathbf{e}_2 - (1 + 3i)\mathbf{e}_3$
5. $\mathbf{u} = \mathbf{e}_1 - \mathbf{e}_2 + i\mathbf{e}_3 - 2\mathbf{e}_4,$ $\mathbf{v} = (1 + i)\mathbf{e}_1 - i\mathbf{e}_2 + \mathbf{e}_4$
6. If $\mathbf{u}$ and $\mathbf{v}$ are any vectors in a unitary space, show that

$$\mathrm{Re}\,(\|\mathbf{v}\|\mathbf{u} + \|\mathbf{u}\|\mathbf{v}, \|\mathbf{v}\|\mathbf{u} - \|\mathbf{u}\|\mathbf{v}) = 0.$$

In each of problems 7 through 10, assume that $\mathbf{e}_1, \mathbf{e}_2, \mathbf{e}_3$ form an orthonormal basis for the complex vector space V. Find an orthonormal basis for the subspace W spanned by the given vectors $\mathbf{v}_1, \mathbf{v}_2$.

7. $\mathbf{v}_1 = \mathbf{e}_1 + i\mathbf{e}_2$
   $\mathbf{v}_2 = \mathbf{e}_1 - i\mathbf{e}_2 + (1 + i)\mathbf{e}_3$
8. $\mathbf{v}_1 = \mathbf{e}_1 - \mathbf{e}_2 - i\mathbf{e}_3$
   $\mathbf{v}_2 = (2 + 2i)\mathbf{e}_1 - (1 - i)\mathbf{e}_2 + (2 - 3i)\mathbf{e}_3$
9. $\mathbf{v}_1 = \mathbf{e}_1 + (1 + i)\mathbf{e}_2 - (2 - i)\mathbf{e}_3$
   $\mathbf{v}_2 = (2 + i)\mathbf{e}_1 - (2 - 2i)\mathbf{e}_2 - 3\mathbf{e}_3$
10. $\mathbf{v}_1 = (1 + i)\mathbf{e}_1 - i\mathbf{e}_2 - (1 - i)\mathbf{e}_3$
    $\mathbf{v}_2 = (2 + i)\mathbf{e}_1 + (2 - 2i)\mathbf{e}_2 + 3(1 + i)\mathbf{e}_3$

In problems 11 and 12, assume that $\mathbf{e}_1, \mathbf{e}_2, \mathbf{e}_3$ form an orthonormal basis for the complex space V. Determine k_1 and k_2 so that the vector $\mathbf{w} - k_1\mathbf{u} - k_2\mathbf{v}$ is orthogonal to both $\mathbf{u}$ and $\mathbf{v}$.

11. $\mathbf{u} = \mathbf{e}_1 + i\mathbf{e}_2 - \mathbf{e}_3,$ $\mathbf{v} = \mathbf{e}_1 + (1 - i)\mathbf{e}_2,$ $\mathbf{w} = \mathbf{e}_2 - i\mathbf{e}_3$

12. $\mathbf{u} = \mathbf{e}_1 - 2\mathbf{e}_3,$ $\mathbf{v} = i\mathbf{e}_1 + (1 - i)\mathbf{e}_2,$ $\mathbf{w} = 2i\mathbf{e}_2 - (2 + i)\mathbf{e}_3$

13. Write out the details of the proof of Theorem 11(b).

14. Prove Theorem 12.

15. Prove that the set V^C with rules (13) and (14) satisfies all the axioms for a vector space.

16. Prove that the inner product defined by formula (15) makes the complexification of a Euclidean space into a unitary space.

17. Prove Theorem 13(b).

18. Let T be a linear transformation on a real vector space V. Form the complexification V^C of V. If $\mathbf{u} \dot{+} i\mathbf{v}$ is an element of $V^C(\mathbf{u}, \mathbf{v} \in V)$, show that the transformation T^C defined for all $\mathbf{u}, \mathbf{v}$ by

$$T^C(\mathbf{u} \dot{+} i\mathbf{v}) = T\mathbf{u} \dot{+} iT\mathbf{v}$$

is a linear transformation on V^C.

19. Is anything new obtained if a complex vector space is complexified?

20. Describe the complexification of the linear vector space of all convergent sequences of real numbers.

6. EIGENVALUES AND EIGENVECTORS IN COMPLEX VECTOR SPACES

Let V be a complex vector space with the basis $\mathbf{v}_1, \mathbf{v}_2, \ldots, \mathbf{v}_n$. Suppose that T is a linear transformation from V to V and that A is the $n \times n$ matrix of complex numbers corresponding to T with respect to the basis $\mathbf{v}_1, \mathbf{v}_2, \ldots, \mathbf{v}_n$. The vector equation

$$T\mathbf{v} = \lambda\mathbf{v}$$

corresponds to the homogeneous system of linear equations

$$\sum_{j=1}^{n} (a_{ij} - \lambda\delta_{ij})x_j = 0, \qquad i = 1, 2, \ldots, n, \tag{1}$$

which is, in matrix notation,

$$(A - \lambda I)X = 0.$$

The system (1) has a nonzero solution if and only if the characteristic equation

$$\det (A - \lambda I) = 0 \tag{2}$$

has a solution. According to the Fundamental Theorem of Algebra, however, the polynomial equation (2), which is of the nth degree, has exactly n roots if multiplicities are counted. Of course, there may be only one root with multiplicity n, but we know that (2) always has at least one root. This root is an eigenvalue, and corresponding to it there is at least one eigenvector. We have just established the next theorem.

Theorem 14. *If V is a finite-dimensional complex vector space and if T is a linear transformation from V to V, then there exists at least one complex number λ for which the equation $T\mathbf{v} = \lambda\mathbf{v}$ has a nonzero solution $\mathbf{v}$. For each such λ, the totality of such solutions forms a subspace of dimension ≥ 1.*

The last sentence of the theorem follows at once from the preceding statements. Recall that for real vector spaces this result is false, since a characteristic equation (2) with real coefficients does not necessarily have real roots.

It can be shown (although we shall not do so) that if $\lambda_1, \lambda_2, \ldots, \lambda_k$ are *distinct* roots and if $\mathbf{u}_1, \mathbf{u}_2, \ldots, \mathbf{u}_k$ are the corresponding eigenvectors, then these eigenvectors are linearly independent. Therefore, if the n roots of the characteristic equation (2) are distinct, we may choose a set of eigenvectors as a basis. In terms of this basis the matrix corresponding to T is the diagonal matrix B. Unfortunately, if there are multiple roots, the reduction to diagonal form is not always possible. A change of basis is of no help, since we recall that the characteristic equation (2) does not depend on the basis chosen, even though the matrix A does.

EXAMPLE 1. Find the eigenvalues and eigenvectors of the matrix

$$A = \begin{pmatrix} 1 & 1 \\ -1 & 1 \end{pmatrix}.$$

Solution. We have

$$\det (A - \lambda I) = \begin{vmatrix} 1 - \lambda & 1 \\ -1 & 1 - \lambda \end{vmatrix} = (\lambda - 1)^2 + 1 = 0,$$

and $\lambda = 1 + i, 1 - i$. If $\lambda = 1 + i$, the linear system (1) becomes

$$\begin{pmatrix} -i & 1 \\ -1 & -i \end{pmatrix} \begin{pmatrix} x_1 \\ x_2 \end{pmatrix} = \begin{pmatrix} -ix_1 + x_2 \\ -x_1 - ix_2 \end{pmatrix} = \begin{pmatrix} 0 \\ 0 \end{pmatrix},$$

and we obtain the relation $x_2 = ix_1$. Any vector of the form $a(1, i)$ with a complex is an eigenvector. In a similar way we find that for $\lambda = 1 - i$, any vector of the form $b(1, -i)$ with b complex is an eigenvector. The matrix $H = C^t$ defined in the change of basis is obtained by selecting the eigenvectors as columns of H. We take

$$H = \begin{pmatrix} 1 & 1 \\ i & -i \end{pmatrix}, \qquad H^{-1} = \frac{i}{2} \begin{pmatrix} -i & -1 \\ -i & 1 \end{pmatrix},$$

and we find

$$H^{-1}AH = \begin{pmatrix} 1 + i & 0 \\ 0 & 1 - i \end{pmatrix},$$

as it should be.

There is an important class of matrices for which the reduction to diagonal form is possible. A matrix A is called **Hermitian** if and only if

$$A^t = \overline{A}, \qquad \text{i.e.,} \qquad a_{ji} = \overline{a}_{ij}, \quad i, j = 1, 2, \ldots, n.$$

If $i = j$ we see that $a_{ii} = \bar{a}_{ii}$, and so the diagonal elements of any Hermitian matrix are always real. We also note that if A happens to be real (i.e., if all a_{ij} are real numbers) and if A is symmetric as well, then A is automatically Hermitian.

We now assume for the remainder of this section that E is a unitary space. Let T be a linear transformation from E to E. We say that T is a **Hermitian transformation** if and only if

$$(T\mathbf{u}, \mathbf{v}) = (\mathbf{u}, T\mathbf{v}) \qquad \text{for all } \mathbf{u}, \mathbf{v} \text{ in } E. \tag{3}$$

Theorem 15. *Suppose that* $\mathbf{e}_1, \mathbf{e}_2, \ldots, \mathbf{e}_n$ *is an orthonormal basis for E. Then T is a Hermitian transformation if and only if its corresponding matrix is a Hermitian matrix.*

Proof. Let the matrix A correspond to T. We write

$$\mathbf{u} = \sum_{i=1}^{n} u_i \mathbf{e}_i, \quad \mathbf{v} = \sum_{i=1}^{n} v_i \mathbf{e}_i, \quad \mathbf{s} = \sum_{i=1}^{n} s_i \mathbf{e}_i, \quad \mathbf{t} = \sum_{i=1}^{n} t_i \mathbf{e}_i.$$

If $\mathbf{s} = T\mathbf{u}$ and $\mathbf{t} = T\mathbf{v}$, then

$$s_i = \sum_{j=1}^{n} a_{ij} u_j, \quad t_i = \sum_{j=1}^{n} a_{ij} v_j.$$

Therefore

$$(T\mathbf{u}, \mathbf{v}) = (\mathbf{s}, \mathbf{v}) = \sum_{i=1}^{n} s_i \bar{v}_i = \sum_{i=1}^{n} \sum_{j=1}^{n} a_{ij} u_j \bar{v}_i,$$

$$(\mathbf{u}, T\mathbf{v}) = (\mathbf{u}, \mathbf{t}) = \sum_{i=1}^{n} u_i \bar{t}_i = \sum_{i=1}^{n} \sum_{j=1}^{n} \bar{a}_{ij} u_i \bar{v}_j$$

$$= \sum_{i=1}^{n} \sum_{j=1}^{n} \bar{a}_{ji} u_j \bar{v}_i.$$

By choosing $\mathbf{u} = \mathbf{e}_p$ and $\mathbf{v} = \mathbf{e}_q$, which is the same as taking $u_j = \delta_{jp}$, $v_j = \delta_{jq}$, we see that (3) holds if and only if $a_{ij} = \bar{a}_{ji}$ for all i, j.

Note the similarity of this proof to the proof of Theorem 2.

Corollary. *If A is Hermitian and H is unitary, then* $H^{-1}AH$ *is Hermitian.*

Proof. Theorem 15 holds for any orthonormal basis. If A corresponds to T under one such basis, then T will correspond to $H^{-1}AH$ under another orthonormal basis, H being an appropriate unitary matrix.

Theorem 16. (a) *The eigenvalues of a Hermitian matrix are real.*
(b) *The eigenvalues of a Hermitian transformation are real.*

Proof. Because of Theorems 14 and 15 it is sufficient to prove only (b). Suppose that λ is an eigenvalue and $\mathbf{u} \neq \mathbf{0}$ is the corresponding eigenvector. Since T is

Hermitian, we have

$$(T\mathbf{u}, \mathbf{u}) = \lambda(\mathbf{u}, \mathbf{u}) = (\mathbf{u}, T\mathbf{u}).$$

But, from the definition of inner product, we have

$$(T\mathbf{u}, \mathbf{u}) = \overline{(\mathbf{u}, T\mathbf{u})},$$

and so $(T\mathbf{u}, \mathbf{u}) = \lambda(\mathbf{u}, \mathbf{u})$ must be real. Since $(\mathbf{u}, \mathbf{u}) > 0$, we conclude that λ is real.

Corollary. *If A is a real and symmetric matrix, then all the eigenvalues of A are real. That is, Theorem 3(a) holds.*

The corollary follows since a real, symmetric matrix is Hermitian.
The next theorem shows that Hermitian matrices may be diagonalized.

Theorem 17. (a) *If T is a Hermitian transformation, there is an orthonormal basis for E consisting of eigenvectors of T.*
(b) *If A is a Hermitian matrix, there is a unitary matrix H such that $H^{-1}AH$ is a diagonal matrix.*

Proof. Part (b) follows from both the discussion above and part (a). Part (a) may be established by induction on the dimension n of E. Since the proof is much the same as the proof of Theorem 3(b), it will be left to the reader. (See Exercise 13.)

EXAMPLE 2. Find the eigenvalues and eigenvectors of the matrix

$$A = \begin{pmatrix} 0 & 2+i \\ 2-i & 4 \end{pmatrix}.$$

Is A Hermitian? Verify that the eigenvectors are orthogonal.

Solution. Since $a_{21} = \bar{a}_{12} = 2 - i = \overline{2+i}$ and since a_{11}, a_{22} are real, A is Hermitian. We have

$$\det (A - \lambda I) = \begin{vmatrix} -\lambda & 2+i \\ 2-i & 4-\lambda \end{vmatrix} = \lambda^2 - 4\lambda - 5 = (\lambda + 1)(\lambda - 5).$$

Setting $\lambda = -1$, we find

$$\begin{pmatrix} 1 & 2+i \\ 2-i & 5 \end{pmatrix}\begin{pmatrix} x_1 \\ x_2 \end{pmatrix} = \begin{pmatrix} x_1 + (2+i)x_2 \\ (2-i)x_1 + 5x_2 \end{pmatrix} = \begin{pmatrix} 0 \\ 0 \end{pmatrix},$$

and therefore $x_1 = -(2+i)x_2$. Any vector of the form $a[-(2+i), 1]$ with a complex is an eigenvector. Setting $\lambda = 5$, we find that any vector of the form $b[1, (2-i)]$ is an eigenvector. The inner product of the eigenvectors is

$$(a[-(2+i), 1], b[1, (2-i)]) = a\bar{b}[-(2+i)\cdot 1 + 1 \cdot \overline{(2-i)}]$$
$$= a\bar{b}[-2 - i + 2 + i] = 0.$$

EXERCISES

In each of problems 1 through 6, find the eigenvalues and eigenvectors of the given matrix and form H.

1. $\begin{pmatrix} 1 & i \\ -i & 1 \end{pmatrix}$

2. $\begin{pmatrix} 1 + 2i & -i \\ 3i & 1 - 2i \end{pmatrix}$

3. $\begin{pmatrix} 4 - i & 1 + i \\ 2(1 + i) & 2 + i \end{pmatrix}$

4. $\begin{pmatrix} -1 + i & 2 + 2i \\ 4 + 8i & 5 - i \end{pmatrix}$

5. $\begin{pmatrix} 3 + 4i & -1 + i \\ -1 + i & 1 + 2i \end{pmatrix}$

6. $\begin{pmatrix} -2 + i & 2 - i \\ -2 - 4i & 4 + 3i \end{pmatrix}$

In each of problems 7 through 12, find the eigenvalues and eigenvectors of the given matrix. Verify that eigenvectors corresponding to different eigenvalues are orthogonal, and choose a unitary matrix H.

7. $\begin{pmatrix} 2 & i \\ -i & 2 \end{pmatrix}$

8. $\begin{pmatrix} 1 & 1 - i \\ 1 + i & 2 \end{pmatrix}$

9. $\begin{pmatrix} -1 & 2 + i \\ 2 - i & 3 \end{pmatrix}$

10. $\begin{pmatrix} 2 & 1 - 2i \\ 1 + 2i & 6 \end{pmatrix}$

11. $\begin{pmatrix} 3 & -3i & 2 - 2i \\ 3i & 3 & -(2 + 2i) \\ 2 + 2i & -(2 - 2i) & 0 \end{pmatrix}$

12. $\begin{pmatrix} 3 & 0 & 0 \\ 0 & 2 & i \\ 0 & -i & 2 \end{pmatrix}$

13. Write out a proof of Theorem 17(a).

14. A matrix A is called *normal* if

$$A\bar{A}^t = \bar{A}^t A.$$

Suppose that $A = B + iC$, where B and C are real matrices. Show that A is normal if and only if $BC = CB$.

15. (a) Show that any unitary matrix is normal (see Exercise 14). (b) Show that any Hermitian matrix is normal. (c) If A is Hermitian and c is a complex number, show that the matrix $A_1 = cA$ is normal. (d) Find a 2×2 normal matrix which is neither symmetric, nor Hermitian, nor unitary, nor of the form cA where A is one of these.

CHAPTER 9

Advanced Topics in Infinite Series

1. UNIFORM CONVERGENCE. SEQUENCES OF FUNCTIONS

We studied the elementary properties of infinite sequences and infinite series in Chapter 3. We learned that *an infinite sequence*

$$a_1, a_2, \ldots, a_n, \ldots$$

has the limit c if and only if for each $\epsilon > 0$ there is a positive integer N (the size of N depending on ϵ) such that

$$|a_n - c| < \epsilon \qquad \text{for all } n > N.$$

Frequently the dependence of N upon ϵ is indicated by writing $N = N(\epsilon)$.

Consider a sequence of functions

$$f_1(x), f_2(x), \ldots, f_n(x), \ldots \tag{1}$$

with each function in the sequence defined for x in an interval $[a, b]$. For any particular value of x, the sequence (1) is a sequence of numbers which may or may not have a limit. If the limit does exist for some values of x, the limiting values form a function of x which we may denote by $f(x)$. We write

$$\lim_{n \to \infty} f_n(x) = f(x)$$

where it is understood that f is defined only for those x for which the sequence converges.

For example, the collection of functions $f_n(x) = x^n$, $n = 1, 2, \ldots$, which we may write

$$x, x^2, x^3, \ldots, x^n, \ldots,$$

is a sequence of functions which converges if x is in the half-open interval $(-1, 1]$ but does not converge otherwise. The student may easily verify that the limit function f is

$$f(x) = \begin{cases} 0, & -1 < x < 1, \\ 1, & x = 1. \end{cases}$$

Among the most important limiting processes studied in the calculus are those of differentiation and integration of elementary functions and those concerned with the convergence of infinite sequences and series. In the applications of mathematics to various branches of technology as well as in the development of mathematical theory, it frequently happens that two, three, or even more limiting processes have to be applied successively. Does it matter in which order these limiting processes are performed? The answer is emphatically yes! The computation of two successive limits in one order usually yields an answer different from that obtained by computing them in the reverse order. Therefore it is of the utmost importance to know exactly when reversing the order in which two limits are computed does not change the answer. Mathematical literature abounds with erroneous results caused by the invalid interchange of two successive limiting processes.

Topics concerning sequences of functions frequently involve several successive limiting processes. A basic tool in the study of the evaluation of such limiting processes is the notion of the uniform convergence of a sequence.

DEFINITION. We say that a sequence

$$f_1(x), f_2(x), \ldots, f_n(x), \ldots$$

of functions **converges uniformly on the interval** I to the function $f(x)$ if and only if for each $\epsilon > 0$ there is an integer N **independent of** x such that

$$|f_n(x) - f(x)| < \epsilon \qquad \text{for all } x \text{ on } I \text{ and all } n > N. \tag{2}$$

The important fact which makes uniform convergence differ from ordinary convergence is that N does not depend on x, although it naturally depends on ϵ.

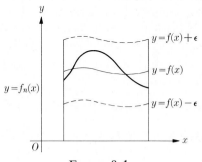

FIGURE 9–1

The geometric meaning of uniform convergence is illustrated in Fig. 9–1. Condition (2) in the definition states that if ϵ is any positive number, then the graph of $f_n(x)$ lies below the graph of $f(x) + \epsilon$ and above the graph of $f(x) - \epsilon$. The *uniformity* condition states that the graph of f_n must lie in this band of width 2ϵ not only for all n larger than N but also for all x in the *entire* length of the interval I.

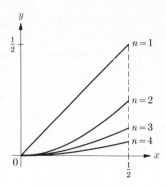

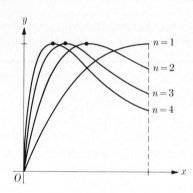

FIGURE 9–2   FIGURE 9–3

An example of uniform convergence is given by the sequence of functions $f_n(x) = x^n$, $0 \le x \le \frac{1}{2}$, $n = 1, 2, \ldots$ (Fig. 9–2). We see that the limit function is zero and that the largest value of $f_n(x)$ occurs at $x = \frac{1}{2}$ with the value $1/2^n$. If we are given an $\epsilon > 0$, we select N so that $1/2^N < \epsilon$, and then $f_n(x)$ will be in the desired band for all $n > N$ and *all* x *in* the interval $[0, \frac{1}{2}]$. Note that the selection of N depended only on ϵ and not on the various values of x in the interval $[0, \frac{1}{2}]$.

It is important to observe that a sequence may converge uniformly in one interval and not in another. The same sequence, $f_n(x) = x^n$, $n = 1, 2, \ldots$, does not converge uniformly in the interval $[0, 1]$, although it does converge at every point of this interval. (See Exercise 15.) Of course, the limit function is discontinuous at $x = 1$.

It can happen that a sequence of continuous functions converges to a limit $f(x)$ for each x in an interval, that the limit function $f(x)$ is continuous, and that the convergence is not uniform. We shall show that such is the case for the sequence

$$f_n(x) = \frac{2nx}{1 + n^2x^2}, \qquad I: \ 0 \le x \le 1, \quad n = 1, 2, \ldots.$$

The graphs for the first few values of n are shown in Fig. 9–3. If $x \ne 0$, we write

$$f_n(x) = \frac{2x/n}{x^2 + (1/n^2)},$$

and $\lim_{n \to \infty} f_n(x) = 0$ for $0 < x \le 1$. Furthermore, $f_n(0) = 0$ for all n, and we conclude that $f_n(x) \to 0$ for all x in the closed interval $0 \le x \le 1$. The graph of $f_n(x)$ can be plotted accurately if we locate the maximum of this function. Taking the derivative,

$$f_n'(x) = \frac{2n(1 - n^2x^2)}{(1 + n^2x^2)^2},$$

we see that f_n has a maximum at $x = 1/n$ with the value $f_n(1/n) = 1$. Thus, as we proceed from right to left, every function in the sequence rises to the value 1 and then drops off to zero at $x = 0$. If we select $\epsilon = \frac{1}{2}$, the condition of uniform convergence requires that

$$f(x) - \tfrac{1}{2} \leq f_n(x) \leq f(x) + \tfrac{1}{2}.$$

Since, in our example, $f(x) = 0$, $0 \leq x \leq 1$, the above condition becomes $-\frac{1}{2} \leq f_n(x) \leq \frac{1}{2}$. But *every* function $f_n(x)$ has the value 1 somewhere in the interval, and so the sequence cannot converge uniformly.

Remark. In calculating the limit of $f_n(x)$, it was necessary to consider the case $x = 0$ separately since, for $x = 0$, the limit of the denominators $x^2 + (1/n^2)$ is zero.

In most cases it is not possible to verify directly from the definition whether or not a sequence converges uniformly. Therefore it is important to develop simple criteria which guarantee that a given sequence converges uniformly. We now derive a rule which is sometimes useful.

Theorem 1. *Suppose that $\{f_n(x)\}$, $n = 1, 2, \ldots$, and $f(x)$ are continuous on the closed interval I: $a \leq x \leq b$. Then the sequence $\{f_n(x)\}$ converges uniformly to $f(x)$ on I if and only if the maximum value ϵ_n of $|f_n(x) - f(x)|$ converges to zero.*

Proof. We must show (a), that if the convergence is uniform, then $\epsilon_n \to 0$, and (b), that if $\epsilon_n \to 0$, the convergence is uniform. To prove (a), we let

$$\epsilon_n = \max |f_n(x) - f(x)|$$

for x on I. Then, since $f_n(x) - f(x)$ is continuous on I, for each n there is a value x_n such that $\epsilon_n = |f_n(x_n) - f(x_n)|$. From the definition of uniform convergence, we know that for any $\epsilon > 0$ there is an N such that $|f_n(x) - f(x)| < \epsilon$ for all $n > N$ and all x on I. Thus, for $n > N$, we must have

$$\epsilon_n = |f_n(x_n) - f(x_n)| < \epsilon.$$

Therefore $\epsilon_n \to 0$ as $n \to \infty$.

(b) Suppose that $\lim_{n \to \infty} \epsilon_n = 0$. Let $\epsilon > 0$ be given. There is an N such that $\epsilon_n < \epsilon$ for $n > N$. But then, since $\epsilon_n = \max |f_n(x) - f(x)|$, we must have

$$0 \leq |f_n(x) - f(x)| \leq \epsilon_n < \epsilon$$

for all x on I and all $n > N$.

EXAMPLE 1. Given the sequence

$$f_n(x) = \frac{n^2 x}{1 + n^3 x^2},$$

show that $f_n(x) \to 0$ for each x on $[0, 1]$, and determine whether or not the convergence is uniform.

Solution. Since $f_n(0) = 0$ for every n, we have $f_n(0) \to 0$ as $n \to \infty$. For $0 < x \leq 1$, we see that

$$\lim_{n \to \infty} f_n(x) = \lim_{n \to \infty} \frac{(x/n)}{x^2 + (1/n^3)} = \frac{0}{x^2} = 0.$$

The limit function $f(x)$ vanishes for $0 \leq x \leq 1$. Therefore $\epsilon_n = \max |f_n(x) - f(x)| = \max f_n(x)$. To find ϵ_n, we differentiate $f_n(x)$, getting

$$f_n'(x) = \frac{(1 + n^3 x^2) \cdot n^2 - n^2 x \cdot 2n^3 x}{(1 + n^3 x^2)^2} = \frac{n^2(1 - n^3 x^2)}{(1 + n^3 x^2)^2}.$$

Setting this derivative equal to zero, we obtain

$$x = x_n = n^{-3/2}; \qquad \epsilon_n = f_n(x_n) = \frac{n^2 \cdot n^{-3/2}}{1 + n^3 \cdot n^{-3}} = \frac{\sqrt{n}}{2} \to \infty \text{ as } n \to \infty.$$

Since ϵ_n does not tend to 0, the convergence is not uniform (Fig. 9–4).

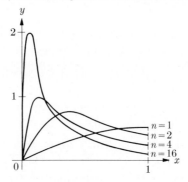

FIGURE 9–4

Although Theorem 1 is useful in illustrating the idea of uniform convergence, it cannot be applied unless an explicit expression for the limit function $f(x)$ is known.

The next result shows that if a sequence of continuous functions converges uniformly on some interval, the limit function must be continuous on this interval.

Theorem 2. *Suppose that the sequence $\{f_n(x)\}$, $n = 1, 2, \ldots$, converges uniformly to $f(x)$ on the interval I, with each $f_n(x)$ continuous on I. Then $f(x)$ is continuous on I.*

Proof. Suppose that I is the open interval $a < x < b$. Let x_0 be any point of I and suppose that ϵ is any positive number. Then, by the definition of uniform convergence, we can select N so large that

$$|f_n(x) - f(x)| < \frac{\epsilon}{3} \qquad \text{for all } n > N \text{ and all } x \text{ on } I. \tag{3}$$

Since f_{N+1} is continuous on I, we apply the definition of continuity to state that for any $\epsilon > 0$ there is a $\delta > 0$ such that

$$|f_{N+1}(x) - f_{N+1}(x_0)| < \frac{\epsilon}{3} \qquad \text{for all } x \text{ in the interval } |x - x_0| < \delta. \quad (4)$$

At this point, the trick is to express $f(x) - f(x_0)$ in a more complicated way. We write

$$f(x) - f(x_0) = f(x) - f_{N+1}(x) + f_{N+1}(x) - f_{N+1}(x_0) + f_{N+1}(x_0) - f(x_0)$$

and, using our knowledge of absolute values,

$$|f(x) - f(x_0)| \leq |f(x) - f_{N+1}(x)| + |f_{N+1}(x) - f_{N+1}(x_0)| \\ + |f_{N+1}(x_0) - f(x_0)|.$$

From (3), the first and third terms on the right are less than $\epsilon/3$ while, according to (4), the second term on the right is less than $\epsilon/3$. Hence

$$|f(x) - f(x_0)| \leq \frac{\epsilon}{3} + \frac{\epsilon}{3} + \frac{\epsilon}{3} = \epsilon. \quad (5)$$

Since (5) holds for $|x - x_0| < \delta$, we conclude that f is continuous at each point x_0 on I.

If I includes an endpoint and if each f_n is continuous at this endpoint, the proof above shows that f is continuous at the endpoint also.

Suppose that $\{f_n(x)\}$ is a uniformly convergent sequence of continuous functions on an interval I. If c is any point on I, we may form the new sequence $\{F_n(x)\}$, where for each n,

$$F_n(x) = \int_c^x f_n(t)\, dt.$$

The next theorem establishes the convergence of such an integrated sequence.

Theorem 3. *Suppose that the sequence $\{f_n(x)\}$ converges uniformly to $f(x)$ on the bounded interval I, and suppose that each f_n is continuous on I. Then the sequence $\{F_n(x)\}$ defined by*

$$F_n(x) = \int_c^x f_n(t)\, dt, \qquad n = 1, 2, \ldots$$

converges uniformly to $F(x) = \int_c^x f(t)\, dt$ on I.

Proof. First of all, by Theorem 2, $f(x)$ is continuous and so $F(x)$ may be defined. Let L be the length of I. For any $\epsilon > 0$ it follows from the uniform convergence that there is an N such that

$$|f_n(t) - f(t)| < \frac{\epsilon}{L} \qquad \text{for all } n > N \text{ and all } t \text{ on } I.$$

The basic rule for estimating integrals may be found in *First Course*, page 230.

We use it to conclude that

$$|F_n(x) - F(x)| = \left| \int_c^x \{f_n(t) - f(t)\} \, dt \right| \le \left| \int_c^x |f_n(t) - f(t)| \, dt \right|$$

$$\le \frac{\epsilon}{L} |x - c| \le \epsilon \qquad \text{for } n > N \text{ and all } x \text{ on } I.$$

Hence $\{F_n(x)\}$ converges uniformly to $F(x)$.

EXAMPLE 2. Show that the sequence

$$F_n(x) = \frac{\ln (1 + n^3 x^2)}{n^2}, \qquad n = 1, 2, \ldots,$$

converges uniformly on the interval $0 \le x \le 1$.

Solution. The sequence

$$F_n'(x) = \frac{2nx}{1 + n^3 x^2} \equiv f_n(x)$$

is easily shown to converge to zero uniformly. In fact, we have $\lim_{n \to \infty} f_n(x) = 0$ for $0 \le x \le 1$, and we can calculate ϵ_n in order to apply Theorem 1. The result is $\epsilon_n = f_n(x_n) = 1/\sqrt{n} \to 0$ and $n \to \infty$. Therefore $\{f_n(x)\}$ converges uniformly, and we now apply Theorem 3 to conclude that $\{F_n(x)\}$ converges uniformly.

The next theorem allows us to draw conclusions about sequences which are differentiated term by term.

Theorem 4. *Suppose that $\{f_n(x)\}$ is a sequence of functions each having a continuous derivative on a bounded interval I. Suppose that $f_n(x)$ converges to $f(x)$ on I and that the derived sequence $\{f_n'(x)\}$ converges uniformly to $g(x)$ on I. Then $g(x) = f'(x)$ on I.*

Proof. According to Theorem 2, $g(x)$ is continuous on I. Let c be any point on I. For each n we write

$$\int_c^x f_n'(t) \, dt = f_n(x) - f_n(c),$$

and we observe that the hypotheses of Theorem 3 are satisfied for this sequence of integrals. Therefore

$$\int_c^x g(t) \, dt = \lim_{n \to \infty} \int_c^x f_n'(t) \, dt = \lim_{n \to \infty} [f_n(x) - f_n(c)].$$

However, $\{f_n(x)\}$ converges to $f(x)$ for each x, and so

$$\int_c^x g(t) \, dt = f(x) - f(c).$$

The left side may be differentiated with respect to x and, applying the Fundamental Theorem of Calculus, we find $g(x) = f'(x)$.

It is natural to ask whether or not the various hypotheses of the above theorems are absolutely essential. For example, suppose that $\{f_n(x)\}$ is a sequence of continuous functions which converges on an interval I to a continuous function $f(x)$. Is it true that

$$\int_c^x f_n(x)\,dx \to \int_c^x f(x)\,dx \quad \text{as} \quad n \to \infty ?$$

The answer is no! The hypothesis of uniform convergence (or some similar hypothesis) is required. To illustrate this point, we form the sequence

$$f_n(x) = \begin{cases} n^2 x, & 0 \le x \le \dfrac{1}{n}, \\[2mm] -n^2 x + 2n, & \dfrac{1}{n} \le x \le \dfrac{2}{n}, \\[2mm] 0, & \dfrac{2}{n} \le x \le 1. \end{cases}$$

The graph of f_n is shown in Fig. 9–5. As n increases, the triangle becomes narrower and taller. We see easily that

$$f_n(x) \to f(x) \equiv 0 \qquad \text{for all } x \text{ on } [0, 1].$$

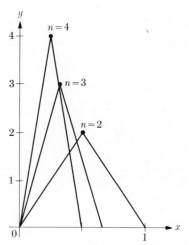

FIGURE 9–5

On the other hand, $\int_0^1 f_n(x)\,dx = 1$ for every n, since the integral is exactly the area of the triangle. This area is the same for every function. But $\int_0^1 f(x)\,dx = 0$. Hence

$$1 = \int_0^1 f_n(x)\,dx \nrightarrow \int_0^1 f(x)\,dx = 0.$$

(See also Exercise 16.)

EXERCISES

In problems 1 through 14, show that $f_n(x)$ converges to $f(x)$ for each x on I, and determine whether or not the convergence is uniform.

1. $f_n(x) = \dfrac{x}{1 + nx}$, $f(x) = 0$, $I = [0, 1]$

2. $f_n(x) = \dfrac{\sin nx}{\sqrt{n}}$, $f(x) = 0$, $I = [0, 1]$

3. $f_n(x) = \dfrac{nx^2}{1 + nx}$, $f(x) = x$, $I = [0, 1]$

4. $f_n(x) = \dfrac{nx}{1 + n^2 x^2}$, $f(x) = 0$, $I = [1, 2]$

5. $f_n(x) = \dfrac{n^2 x}{1 + n^3 x^2}$, $f(x) = 0$, $I = [a, \infty)$, $a > 0$

6. $f_n(x) = nxe^{-nx^2}$, $f(x) = 0$, $I = [0, 1]$

7. $f_n(x) = \dfrac{1}{x} + \dfrac{1}{n}\sin\dfrac{1}{nx}$, $f(x) = \dfrac{1}{x}$, $I = (0, 1]$

8. $f_n(x) = \dfrac{\sin nx}{nx}$, $f(x) = 0$, $I = (0, \infty)$

9. $f_n(x) = (x^n - x^{n+1})$, $f(x) = 0$, $I = [0, 1]$

10. $f_n(x) = \sqrt{n}\,(x^n - x^{n+1})$, $f(x) = 0$, $I = [0, 1]$

11. $f_n(x) = \dfrac{1 - x^{n+1}}{1 - x}$, $f(x) = \dfrac{1}{1 - x}$, $I = [-\frac{1}{2}, \frac{1}{2}]$

12. $f_n(x) = \dfrac{1 - x^{n+1}}{1 - x}$, $f(x) = \dfrac{1}{1 - x}$, $I = (-1, 1)$

13. $f_n(x) = n^2 x^n(1 - x)$, $f(x) = 0$, $I = [0, 1]$

14. $f_n(x) = n^3 x(1 - x)^n$, $f(x) = 0$, $I = [0, 1]$

15. Show that the sequence $f_n(x) = x^n$ does not converge uniformly on $[0, 1]$.

16. Given the sequence $f_n(x) = (n + 1)(n + 2)x^n(1 - x)$, show that $f_n(x) \to f(x) \equiv 0$ for x on $[0, 1]$. Decide whether or not

$$\int_0^1 f_n(x)\, dx \to \int_0^1 f(x)\, dx.$$

What can you conclude about the uniformity of the convergence of $f_n(x)$?

17. Given $f_n(x) = 2n^\alpha x/(1 + n^\beta x^2)$ with $\beta > \alpha \geq 0$. Find the values of α and β for which this sequence converges uniformly on $[0, 1]$.

18. (a) If $f_n(x)$ and $g_n(x)$ converge uniformly on I, show that $f_n(x) + g_n(x)$ converges uniformly on I. (b) If $f_n(x)$ converges uniformly on I and if c is a constant, show that $g_n(x) = cf_n(x)$ converges uniformly on I. (c) Let V be the set of elements $\mathbf{v} = \{f_n\}$, where f_n converges uniformly on I. Show that V is a linear vector space. (d) Describe the space which is the complexification of V.

19. Show that the sequence

$$f_n(x) = x - \frac{x^n}{n}$$

converges uniformly on $[0, 1]$. Show that the sequence

$$f_n'(x) = 1 - x^{n-1}$$

does not converge uniformly on $[0, 1]$.

In each of problems 20 through 23, a sequence $f_n(x)$ is given. Decide whether or not $f_n'(x)$ converges uniformly. Also decide whether or not $F_n(x) = \int_0^x f_n(t)\, dt$ converges uniformly.

20. $f_n(x) = \dfrac{x}{1 + n^2 x}$, $I = [0, 1]$ 21. $f_n(x) = \dfrac{n^2 x}{1 + n^3 x^2}$, $I = [0, 1]$

22. $f_n(x) = \dfrac{2 + nx^2}{2 + nx}$, $I = [0, 1]$ 23. $f_n(x) = nxe^{-nx^2}$, $I = [0, 1]$

24. Show that the sequence $f_n(x) = (\sin x)^{1/n}$ converges, but not uniformly, on $I = [0, \pi]$.

25. Show that the sequence

$$f_n(x) = \left(\frac{\sin x}{x}\right)^{1/n}$$

converges, but not uniformly, on $I = (0, \pi)$.

26. Suppose that $f_n(x)$ and $g_n(x)$ are sequences of continuous functions which converge uniformly on a closed, bounded interval I, to the functions $f(x)$ and $g(x)$, respectively. Show that $h_n(x) = f_n(x)g_n(x)$ converges uniformly on I to $h(x) = f(x)g(x)$.

2. UNIFORM CONVERGENCE OF SERIES

An infinite series

$$\sum_{k=1}^{\infty} a_k = a_1 + a_2 + \cdots + a_n + \cdots \tag{1}$$

has the sequence of partial sums

$$s_1, s_2, \ldots, s_n, \ldots \tag{2}$$

defined by

$$s_n = \sum_{k=1}^{n} a_k.$$

The convergence of the infinite series (1) is equivalent, by definition, to the convergence of the sequence of partial sums (2). The infinite series of functions

$$\sum_{k=1}^{\infty} u_k(x)$$

converges uniformly on an interval I if and only if the sequence $\{s_n(x)\}$ of partial sums converges uniformly on this interval. We observe that all the results of Section 1 on the uniform convergence of sequences translates at once into the corresponding results for series. In this way we obtain the next three theorems.

Theorem 2'. *Suppose that the series $\sum_{k=1}^{\infty} u_k(x)$ converges uniformly to $s(x)$ on the interval I. If each $u_k(x)$ is continuous on I, then $s(x)$ is continuous on I.*

Theorem 3'. *Suppose that the series $\sum_{k=1}^{n} u_k(x)$ converges uniformly to $s(x)$ on I and that each $u_k(x)$ is continuous on I. If we define $U_k(x) = \int_c^x u_k(t)\, dt$, $S(x) = \int_c^x s(t)\, dt$, then $\sum_{k=1}^{\infty} U_k(x)$ converges uniformly to $S(x)$ on I.*

Theorem 4'. *Suppose that $\sum_{k=1}^{\infty} u_k(x)$ converges to $s(x)$ on I and that $\sum_{k=1}^{\infty} u_k'(x)$ converges uniformly to $t(x)$ on I. If each $u_k'(x)$ is continuous on I, then $t(x) = s'(x)$.*

The following theorem gives a simple and useful *indirect* test for uniform convergence of series. One of its virtues is that we may apply the test without finding the sum of the series—i.e., without obtaining the limit of the sequence of partial sums.

Theorem 5. (Weierstrass *M*-test.)   *Suppose that $|u_k(x)| \le M_k$ for all x on I where, for each k, M_k is a positive constant. If the series $\sum_{k=1}^{\infty} M_k$ converges, then the series $\sum_{k=1}^{\infty} u_k(x)$ converges uniformly on I.*

Proof. For each x on I, the series $\sum_{k=1}^{\infty} |u_k(x)|$ converges by the Comparison Test (Chapter 3, Theorem 9). Therefore $\sum_{k=1}^{\infty} u_k(x)$ converges, and we call the limit $s(x)$. We define

$$S = \sum_{j=1}^{\infty} M_j, \qquad S_n = \sum_{j=1}^{n} M_j,$$

and we note that $\lim_{n \to \infty} (S - S_n) = 0$. Since $s_n(x) = \sum_{j=1}^{n} u_j(x)$, we have

$$|s(x) - s_n(x)| = \left| \sum_{j=n+1}^{\infty} u_j(x) \right| \le \sum_{j=n+1}^{\infty} |u_j(x)| \le \sum_{j=n+1}^{\infty} M_j = S - S_n \to 0.$$

Since the numbers $S - S_n$ are independent of x, the convergence is uniform.

Remarks. If the interval I is closed and if each $u_n(x)$ is continuous, then the theorem requires that M_n be either the *maximum* of $|u_n(x)|$ on I or some conveniently chosen number larger than this maximum. If I is open or half-open and $u_n(x)$ tends to a limit at each open endpoint, then M_n must be larger than the maximum of the function extended to the closed interval. We shall see that the quantities M_n may frequently be found by inspection. It is important also that we remember many of the convergent series of positive constants studied in Chapter 3.

EXAMPLE 1. Show that the series

$$\sum_{n=1}^{\infty} \frac{\cos nx}{n^2}$$

converges uniformly for all x.

Solution. We have

$$\left| \frac{\cos nx}{n^2} \right| = \frac{|\cos nx|}{n^2} \le \frac{1}{n^2}.$$

Since $\sum_{n=1}^{\infty} 1/n^2$ is a known convergent series, the result follows.

The Weierstrass *M*-test may frequently be combined with the ratio test for convergence of constants to yield results on uniform convergence. We illustrate with an example.

EXAMPLE 2. Given the series

$$\sum_{n=0}^{\infty} (n+1)x^n,$$

determine an interval of the form $|x| \le h$ on which the series is uniformly convergent.

Solution. If $|x| \le h$, then $|(n+1)x^n| \le (n+1)h^n = M_n$. We apply the Ratio Test to determine when $\sum_{n=1}^{\infty} M_n$ converges. We have

$$\frac{M_{n+1}}{M_n} = \frac{(n+2)h^{n+1}}{(n+1)h^n} = h\frac{1+(2/n)}{1+(1/n)} \to h \text{ as } n \to \infty.$$

Accordingly, the series $\sum_{n=1}^{\infty} M_n$ converges if $h < 1$ and does not converge if $h \ge 1$. The original series converges uniformly on any interval of the form $|x| \le h$ with $h < 1$.

EXAMPLE 3. Discuss for uniform convergence the series

$$\sum_{n=1}^{\infty} \frac{(-1)^{n-1}x^n}{n^2}.$$

Solution. If $|x| \le h$, then

$$\left|\frac{(-1)^{n-1}x^n}{n^2}\right| \le \frac{h^n}{n^2} = M_n.$$

The series $\sum_{n=1}^{\infty} M_n$ converges for $h \le 1$. The given series converges uniformly on any interval of the form $|x| \le h$ with $h \le 1$.

EXAMPLE 4. Given the relation

$$\frac{1}{1-x^4} = \sum_{n=0}^{\infty} x^{4n}, \quad -1 < x < 1,$$

show that

$$\frac{1}{(1-x^4)^2} = \sum_{n=1}^{\infty} nx^{4(n-1)}, \quad -1 < x < 1.$$

Solution. Setting $u_n(x) = x^{4n}$, we see that the series $\sum_{n=1}^{\infty} u_n'(x) = \sum_{n=1}^{\infty} 4nx^{4n-1}$ converges uniformly for $|x| \le h$ with $h < 1$. Therefore, applying Theorem 4', we find that the sum of the derived series is

$$\frac{d}{dx}[(1-x^4)^{-1}] = 4x^3(1-x^4)^{-2} = \sum_{n=1}^{\infty} 4nx^{4n-1}.$$

Dividing by $4x^3$, we get the desired expansion.

EXERCISES

In each of problems 1 through 14, determine h such that the given series converges uniformly on I.

1. $\sum_{n=0}^{\infty} x^n,$ $I:\ |x| \le h$

2. $\sum_{n=0}^{\infty} \frac{(-1)^n x^n}{n^2},$ $I:\ |x| \le h$

3. $\displaystyle\sum_{n=0}^{\infty} n^2 x^n$, $I: \ |x| \leq h$

4. $\displaystyle\sum_{n=0}^{\infty} \frac{x^n}{n!}$, $I: \ |x| \leq h$

5. $\displaystyle\sum_{n=0}^{\infty} \frac{(10x)^n}{n!}$, $I: \ |x| \leq h$

6. $\displaystyle\sum_{n=0}^{\infty} \frac{(-1)^n 3^n x^n}{(n+1)2^n}$, $I: \ |x| \leq h$

7. $\displaystyle\sum_{n=1}^{\infty} \frac{n!(x-3)^n}{1 \cdot 3 \cdot 5 \cdots (2n-1)}$, $I: \ |x-3| \leq h$

8. $\displaystyle\sum_{n=1}^{\infty} \frac{(n!)^2 (x+1)^n}{(2n)!}$, $I: \ |x+1| \leq h$

9. $\displaystyle\sum_{n=1}^{\infty} \frac{(-1)^n x^n}{(n+1)\ln(n+1)}$, $I: \ |x| \leq h$

10. $\displaystyle\sum_{n=1}^{\infty} \frac{(\ln n)2^n x^n}{3^n n^2}$, $I: \ |x| \leq h$

11. $\displaystyle\sum_{n=1}^{\infty} \frac{(1-x^{2n})^{1/2}}{3^n}$, $I: \ |x| \leq h$

12. $\displaystyle\sum_{n=1}^{\infty} x(1-x)^n$, $I: \ |x| \leq h$

13. $\displaystyle\sum_{n=1}^{\infty} \frac{x^2}{n(4+nx^2)}$, $I: \ |x| \leq h$

14. $\displaystyle\sum_{n=1}^{\infty} \frac{1}{n^x}$, $I: \ \infty > x \geq h$

15. Given that $\sum_{n=1}^{\infty} |b_n|$ converges, show that $\sum_{n=1}^{\infty} b_n \sin nx$ converges uniformly for all x.

16. Given that $\sum_{n=1}^{\infty} n|b_n|$ converges and that $f(x) = \sum_{n=1}^{\infty} b_n \sin nx$, show that $f'(x) = \sum_{n=1}^{\infty} n b_n \cos nx$, and that both series converge uniformly for all x.

17. Find those values of h for which $\sum_{n=1}^{\infty} (x \ln x)^n$ converges uniformly for $0 < x \leq h$.

18. Find the values of h such that $\sum_{n=1}^{\infty} (\sin x)^n$ converges uniformly for $|x| \leq h$.

19. Show that $\sum_{n=0}^{\infty} (1 + x)x^n$ converges uniformly for $-1 \leq x \leq 0$.

20. Given the series expansions

$$A(x, \lambda) = \sum_{n=0}^{\infty} \frac{(-1)^n \lambda^{2n+1} x^{2n+1}}{(2n+1)!},$$

$$B(x, \lambda) = \sum_{n=0}^{\infty} \frac{(-1)^n \lambda^{2n} x^{2n}}{(2n)!},$$

use the theorems of this section to show that (a) $dA/dx = \lambda B$, (b) $dB/dx = -\lambda A$, (c) $d^2A/dx^2 + \lambda^2 A = 0$, (d) $d^2B/dx^2 + \lambda^2 B = 0$. Do not use the fact that $A(x, \lambda) = \sin \lambda x$, $B(x, \lambda) = \cos \lambda x$.

21. Show, by successive differentiation of the series $(1 - x)^{-1} = \sum_{n=0}^{\infty} x^n$ that

$$\frac{1}{(1 - x)^k} = \sum_{n=0}^{\infty} \frac{(n + 1)(n + 2) \cdots (n + k - 1)}{(k - 1)!} x^n, \quad k \geq 2$$

and that the convergence is uniform for $|x| \leq h$, with $h < 1$.

22. Let A be an $n \times n$ matrix, and consider the series

$$I + \frac{1}{1!} A + \frac{1}{2!} A^2 + \cdots + \frac{1}{k!} A^k + \cdots.$$

(a) Define convergence for such a series in terms of the limit of the matrices forming the partial sums.

(b) Show that the above series converges if

$$A = \begin{pmatrix} 1 & 1 \\ 0 & 1 \end{pmatrix}.$$

3. INTEGRATION AND DIFFERENTIATION OF POWER SERIES

The elementary properties of power series were discussed in Chapter 3, beginning with Section 5. In this section we shall show how the notion of uniform convergence can be used to extend and amplify many of the earlier results. In addition, we shall supply the proof of one theorem in Chapter 3 which was stated there without proof.

Theorem 6. *Suppose that the power series $\sum_{n=0}^{\infty} a_n x^n$ converges for some $x_1 \neq 0$. Then (a), the series converges absolutely for all x with $|x| < |x_1|$ and (b), it converges uniformly on any interval $|x| \leq h$ with $h < |x_1|$.*

Proof. We recognize part (a) as Theorem 15 of Chapter 3 (page 93). As for (b), the proof given in Chapter 3 shows that the convergence is uniform (although we did not say so at the time). Referring to the proof there, we see that the series

$$\sum_{n=0}^{\infty} M \left| \frac{x}{x_1} \right|^n \leq \sum_{n=0}^{\infty} M \left(\frac{h}{|x_1|} \right)^n = \sum_{n=0}^{\infty} M_n.$$

The series $\sum_{n=0}^{\infty} M_n$ is a convergent geometric series, and the Weierstrass M-test may be applied to yield uniform convergence.

Remark. Theorem 6 also holds for the series $\sum_{n=0}^{\infty} a_n(x - c)^n$ with $|x - c| \leq h$.

Corollary. *If $\sum_{n=0}^{\infty} a_n x_1^n$ diverges, then $\sum_{n=0}^{\infty} a_n x^n$ diverges for all x with $|x| > |x_1|$.*

Proof. If $\sum_{n=0}^{\infty} a_n x_2^n$ were convergent for $|x_2| > |x_1|$, then we could apply the above theorem to conclude that $\sum_{n=0}^{\infty} a_n x_1^n$ converges, a contradiction.

The next theorem shows the pattern for the convergence properties of power series in general. The result which we now derive was stated without proof in Chapter 3 (Theorem 16, page 94).

Theorem 7. *Let $\sum_{n=0}^{\infty} a_n x^n$ be any power series. Then either*
(i)   *the series converges only for $x = 0$; or*
(ii)   *the series converges for all x; or*
(iii)   *there is a number R such that the series converges for all x with $|x| < R$ and diverges for all x with $|x| > R$.*

Proof. We have already encountered examples of series where (i) and (ii) hold. The series $\sum_{n=0}^{\infty} n! x^n$ converges only for $x = 0$, and the series $\sum_{n=0}^{\infty} x^n/n!$ converges for all x. Therefore we need consider only alternative (iii). Then there is an $x_1 \neq 0$ where the series converges and an X_1 where the series diverges. We let

$$r_1 = |x_1| \qquad \text{and} \qquad R_1 = |X_1|$$

and note that $R_1 \geq r_1 > 0$. If $R_1 = r_1$, we select this value for R and (iii) is established. So we suppose that $R_1 > r_1$. We define an increasing sequence r_1, $r_2, \ldots, r_n, \ldots$ and a decreasing sequence $R_1, R_2, \ldots, R_n, \ldots$ in the following way. If the series converges for $\frac{1}{2}(r_1 + R_1)$, we define

$$r_2 = \frac{r_1 + R_1}{2} \qquad \text{and} \qquad R_2 = R_1.$$

If the series diverges for $\frac{1}{2}(r_1 + R_1)$, we define

$$r_2 = r_1 \qquad \text{and} \qquad R_2 = \frac{r_1 + R_1}{2}.$$

We continue the process inductively. If the series converges for $\frac{1}{2}(r_k + R_k)$, we define

$$r_{k+1} = \frac{r_k + R_k}{2} \qquad \text{and} \qquad R_{k+1} = R_k.$$

If the series diverges for $\frac{1}{2}(r_k + R_k)$, we define

$$r_{k+1} = r_k \qquad \text{and} \qquad R_{k+1} = \frac{r_k + R_k}{2}.$$

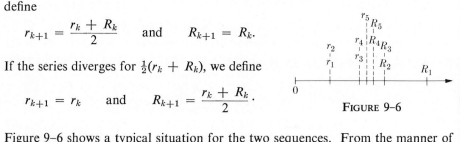

FIGURE 9–6

Figure 9–6 shows a typical situation for the two sequences. From the manner of construction we see that
(a) $r_k \leq r_{k+1} < R_{k+1} \leq R_k$, $k = 1, 2, \ldots$.
(b) $R_k - r_k = (R_1 - r_1)/2^{k-1}$, $k = 1, 2, \ldots$.
(c) The series converges for all x with $|x| < r_k$ and diverges for all x with $|x| > R_k$, $k = 1, 2, \ldots$. Now, from the Axiom of Continuity (*First Course*, page 122), (a) and (b) together imply that there is a number R such that $R_k \to R$ and $r_k \to R$. Part (c) shows that the series converges if $|x| < R$ and diverges if $|x| > R$.

DEFINITIONS. The number R in Theorem 7 is called the **radius of convergence** of the series. We define $R = 0$ when (i) holds and $R = +\infty$ when (ii) holds.

Corollary. *If a power series has a positive radius of convergence R, it converges uniformly on any interval $|x| < h$ where $h < R$. In fact, the series of absolute values $\sum_{n=0}^{\infty} |a_n x^n|$ also converges uniformly.*

Proof. Given h, we select $x_1 = \frac{1}{2}(h + R)$ and observe that the series $\sum_{n=0}^{\infty} a_n x_1^n$ converges. Now we apply Theorem 6.

The theorems on term-by-term differentiation and term-by-term integration of power series which were proved in Chapter 3 (Theorems 19 and 21) may now be established as simple consequences of the uniform convergence properties of power series. We state the results, leaving as exercises for the student those proofs which use the more advanced methods. (See Exercise 12.)

Theorem 8. *Suppose that $R > 0$ and that the power series $\sum_{n=0}^{\infty} a_n x^n$ converges for $|x| < R$. Then the series obtained by differentiating it term by term also converges for $|x| < R$.*

(The above theorem is proved in Chapter 3, page 105.)

Theorem 9. *Suppose that $R > 0$ and that*

$$f(x) = \sum_{n=0}^{\infty} a_n (x - c)^n$$

converges for $|x - c| < R$. Then f is continuous and has continuous derivatives of all orders, which are given on $|x - c| < R$ by differentiating the series the appropriate number of times. Moreover, if

$$F(x) = \int_c^x f(t)\, dt,$$

then $F(x)$ is given on the same interval by the series obtained by term-by-term integration of the series for f.

The fact that f is continuous is the content of Theorem 20 in Chapter 3. However, since $u_n(x) = a_n(x - c)^n$ is continuous for each n, Theorem 2' yields the continuity of f at once.

Corollary. *If $R > 0$ and*

$$f(x) = \sum_{n=0}^{\infty} a_n (x - c)^n \qquad \text{for } |x - c| < R,$$

then

$$a_n = \frac{f^{(n)}(c)}{n!}.$$

The reader may verify the formula for a_n by repeated differentiation of the series for f, after which the value $x = c$ is inserted.

Remark. In Theorems 8 and 9 nothing has been said about convergence at the endpoints. If a power series with radius of convergence $R > 0$ converges uniformly for $|x - c| \leq R$, then the *integrated* series will also converge uniformly for $|x - c| \leq R$ (Theorem 2'). On the other hand, the differentiated series may not converge at all at the endpoints. For example, the series

$$f(x) = \sum_{n=2}^{\infty} \frac{x^n}{n(n - 1)}$$

converges uniformly for $|x| \leq 1$. But the series

$$f''(x) = \sum_{n=0}^{\infty} x^n, \qquad |x| < 1,$$

does not converge at either endpoint.

Once a power series expansion for a function is known (say by a Taylor or Maclaurin expansion), the series expansions for related functions may be found by differentiation and integration. Further results may be obtained by using these in connection with the next theorem, on substitution.

Theorem 10. (Simple Substitutions.) (a) *If* $f(u) = \sum_{n=0}^{\infty} a_n(u - c_0)^n$ *for* $|u - c_0| < R$, *and if* $c_0 = bc + d$ *with* $b \neq 0$, *then*

$$f(bx + d) = \sum_{n=0}^{\infty} a_n b^n (x - c)^n \qquad \text{for } |x - c| < \frac{R}{|b|}.$$

(b) *If* $f(u) = \sum_{n=0}^{\infty} a_n u^n$ *for* $|u| < R$, *then for any positive integer* k,

$$f(x^k) = \sum_{n=0}^{\infty} a_n x^{kn} \qquad \text{for } |x| < R^{1/k}.$$

Proof. (a) If $u = bx + d$, then $u - c_0 = b(x - c)$ and

$$a_n(u - c_0)^n = a_n b^n (x - c)^n$$

for each n. Therefore the following inequalities are equivalent:

$$|u - c_0| < R \leftrightarrow |b| \, |x - c| < R \leftrightarrow |x - c| < \frac{R}{|b|}.$$

The proof of part (b) is similar.

EXAMPLE 1. Use simple substitution to obtain the expansion

$$(1 - x^8)^{-1} = \sum_{n=0}^{\infty} x^{8n}, \qquad |x| < 1.$$

Solution. We have

$$(1 - u)^{-1} = \sum_{n=0}^{\infty} u^n,$$

and we use Theorem 10(b) with $u = x^8$ to get the result. Note that since $R = 1$, $R^{1/8} = 1$.

EXAMPLE 2. Find the Maclaurin expansion for Arctan (x^2).

Solution. We have the expansion

$$\frac{1}{1 + u} = \sum_{n=0}^{\infty} (-1)^n u^n, \qquad |u| < 1.$$

Therefore, letting $u = v^2$, we get

$$\frac{1}{1 + v^2} = \sum_{n=0}^{\infty} (-1)^n v^{2n}, \qquad |v| < 1.$$

Now we integrate term by term, obtaining

$$\text{Arctan } v = \sum_{n=0}^{\infty} \frac{(-1)^n v^{2n+1}}{2n + 1}, \qquad |v| < 1.$$

Setting $v = x^2$, we conclude that

$$\text{Arctan } (x^2) = \sum_{n=0}^{\infty} \frac{(-1)^n x^{4n+2}}{2n + 1}, \qquad |x| < 1.$$

EXERCISES

In each of problems 1 through 10, find the Maclaurin expansion for the function f, using differentiation, integration, and simple substitution, whenever necessary.

1. $f(x) = \dfrac{1}{(1 - x^2)^2}$

2. $f(x) = \ln (1 + x^3)$

3. $f(x) = e^{(1+x^2)}$

4. $f(x) = \arctan x^3$

5. $f(x) = [1 + (x/3)^3]^{-3}$

6. $f(x) = \arcsin (x^2)$

7. $f(x) = \sin \left(\dfrac{3x^2}{\pi} \right)$

8. $f(x) = \text{argtanh } (x^3)$

9. $f(x) = \begin{cases} (e^{x^2} - 1)/x^2, & x \neq 0 \\ 1, & x = 0 \end{cases}$

10. $f(x) = (1 + x^2)^{-3/2}$

11. Write a complete proof of Theorem 6(b).

12. Use the results on uniform convergence to prove Theorem 8.

13. Use the results on uniform convergence to prove Theorem 9.

14. Write a complete proof of the Corollary to Theorem 9.

15. Prove Theorem 10(b).

16. Show that if a power series converges absolutely at an endpoint, then the series obtained by term-by-term integration converges at the same point.

17. Let k be any positive integer. Give an example of a power series for a function f such that all the series obtained by term-by-term differentiation k times converge at the endpoint, while the series obtained by differentiating $k + 1$ times diverges at the endpoint.

4. DOUBLE SERIES

Finite double sequences and finite double sums were discussed in Chapter 6, Section 3, in connection with matrices. We obtain a simple way of looking at a double sequence by writing a rectangular array, as shown in Fig. 9–7. If all the rows and columns of such an array terminate, then the sequence is called **finite**; if the rows and columns continue indefinitely to the right and downward, the double sequence is called **infinite**.

$$
\begin{array}{ccccc}
u_{11} & u_{12} & \cdots & u_{1n} & \cdots \\
u_{21} & u_{22} & \cdots & u_{2n} & \cdots \\
\vdots & \vdots & \cdots & \vdots & \cdots \\
u_{m1} & u_{m2} & \cdots & u_{mn} & \cdots \\
\vdots & \vdots & \cdots & \vdots & \cdots
\end{array}
$$

FIGURE 9–7

DEFINITION. We say that **the double sequence** $\{u_{mn}\}$, $m, n = 1, 2, \ldots,$ **tends to a limit L as** $(m, n) \to \infty$ if and only if for each $\epsilon > 0$ there is a positive integer N such that

$$|u_{mn} - L| < \epsilon \qquad \text{whenever } both\ m > N \text{ and } n > N.$$

We observe the resemblance of this definition to that of the limit of a function of two variables (Chapter 4, page 121). It can be shown that if such a number L exists, then it must be unique. Also, the customary theorems on limits which we established for ordinary sequences are easily extended to double sequences. We shall use the symbols

$$\lim_{(m,n)\to\infty} u_{mn} = L \qquad \text{and} \qquad u_{mn} \to L \text{ as } (m, n) \to \infty$$

for the double limit of a sequence.

Given a double sequence $\{a_{mn}\}$, $m, n = 1, 2, \ldots,$ we define its **partial sum** s_{mn} by the formula

$$s_{mn} = \sum_{j=1}^{m} \sum_{k=1}^{n} a_{jk}. \tag{1}$$

Pictorially, s_{mn} denotes the sum of all the terms in the rectangular array indicated in Fig. 9–8. In terms of matrices, s_{mn} is the sum of all the elements of the $m \times n$ matrix obtained by selecting the m rows and n columns in the upper left-hand corner of the double sequence $\{a_{mn}\}$.

$$
\begin{array}{cccc|cc}
a_{11} & a_{12} & \cdots & a_{1n} & a_{1,n+1} & \cdots \\
a_{21} & a_{22} & \cdots & a_{2n} & a_{2,n+1} & \cdots \\
\vdots & \vdots & \cdots & \vdots & & \cdots \\
a_{m1} & a_{m2} & \cdots & a_{mn} & a_{m,n+1} & \cdots \\
\hline
a_{m+1,1} & a_{m+1,2} & \cdots & & &
\end{array}
$$

FIGURE 9–8

DEFINITION. The **sum of a double sequence** $\{a_{mn}\}$, $m, n = 1, 2, \ldots$, is defined as

$$
\lim_{m,n\to\infty} s_{mn} \tag{2}
$$

where s_{mn} is given by (1). The limit may or may not exist. We write the expression

$$
\sum_{m,n=1}^{\infty} a_{mn}
$$

instead of (2) and call this the **infinite double series whose terms are** a_{mn}. If the limit in (2) exists, we say that the double series **converges**; otherwise we say it **diverges**.

Many of the theorems on double series are direct extensions of those for single series. The next theorem follows immediately from the theory of limits.

Theorem 11. *Suppose that $\sum_{m,n=1}^{\infty} a_{mn}$ and $\sum_{m,n=1}^{\infty} b_{mn}$ are convergent double series and suppose that c and d are constants. Then $\sum_{m,n=1}^{\infty} (ca_{mn} + db_{mn})$ is convergent and*

$$
\sum_{m,n=1}^{\infty} (ca_{mn} + db_{mn}) = c \sum_{m,n=1}^{\infty} a_{mn} + d \sum_{m,n=1}^{\infty} b_{mn}.
$$

The following theorem on double series with positive terms is a direct analog of the corresponding theorem for single series. Note that part (ii) of the theorem is a comparison test.

Theorem 12. *Suppose that $a_{mn} \geq 0$ for all m and n.*
(i) *If there is a number M such that the partial sums $s_{mn} \leq M$, then $\sum_{m,n=1}^{\infty} a_{mn}$ converges to a number $s \leq M$, and each $s_{mn} \leq s$.*
(ii) *(Comparison Test.) If $0 \leq a_{mn} \leq A_{mn}$ and $\sum_{m,n=1}^{\infty} A_{mn}$ converges, then $\sum_{m,n=1}^{\infty} a_{mn}$ does also, and $\sum_{m,n=1}^{\infty} a_{mn} \leq \sum_{m,n=1}^{\infty} A_{mn}$.*

Proof. (i) It is clear that the *special* partial sums s_{nn} are nondecreasing and $s_{nn} \leq M$ for all n. By Axiom C (*First Course*, page 122) we know there is a number $s \leq M$ such that $s_{nn} \to s$ and $s_{nn} \leq s$ for each n. Therefore, for any $\epsilon > 0$, there is an N such that

$$
s - \epsilon \leq s_{nn} \leq s \qquad \text{for all } n > N.
$$

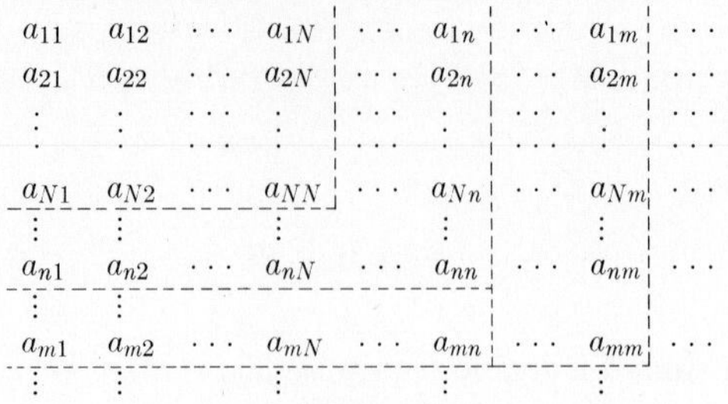

$$
\begin{array}{cccccccccc}
a_{11} & a_{12} & \cdots & a_{1N} & \cdots & a_{1n} & \cdots & a_{1m} & \cdots \\
a_{21} & a_{22} & \cdots & a_{2N} & \cdots & a_{2n} & \cdots & a_{2m} & \cdots \\
\vdots & \vdots & \cdots & \vdots & \cdots & \vdots & \cdots & \vdots & \cdots \\
 & & \cdots & & \cdots & & \cdots & & \cdots \\
a_{N1} & a_{N2} & \cdots & a_{NN} & \cdots & a_{Nn} & \cdots & a_{Nm} & \cdots \\
\vdots & & & & & \vdots & & \vdots & \\
a_{n1} & a_{n2} & \cdots & a_{nN} & \cdots & a_{nn} & \cdots & a_{nm} & \cdots \\
\vdots & & & & & & & \vdots & \\
a_{m1} & a_{m2} & \cdots & a_{mN} & \cdots & a_{mn} & \cdots & a_{mm} & \cdots \\
\vdots & & & \vdots & & \vdots & & \vdots &
\end{array}
$$

FIGURE 9–9

Let m, n be any two numbers larger than N. For convenience, suppose that $m \geq n$. It follows that

$$
s - \epsilon \leq s_{nn} \leq s_{mn} \leq s_{mm} \leq s.
$$

Figure 9–9 shows the various rectangular blocks of a_{ij} which make up the partial sums. Hence $s_{mn} \to s$ as $(m, n) \to \infty$ and each $s_{mn} \leq s$. Part (ii) follows directly from (i).

EXAMPLE 1. Show that the double series

$$
\sum_{p,q=1}^{\infty} \frac{1}{p^2 q^2}
$$

is convergent.

Solution. We have

$$
\sum_{p,q=1}^{m,n} \frac{1}{p^2 q^2} = s_{mn},
$$

and it is easy to see that

$$
s_{mn} = \left(\sum_{p=1}^{m} \frac{1}{p^2} \right) \left(\sum_{q=1}^{n} \frac{1}{q^2} \right).
$$

Since $\sum_{k=1}^{\infty} 1/k^2 = M < \infty$, it follows that $s_{mn} < M^2$ for all m, n. Therefore the double series is convergent.

EXAMPLE 2. Show that the series

$$
\sum_{m,n=1}^{\infty} \frac{1}{m^4 + n^4}
$$

is convergent.

Solution. Since, by Example 1,

$$
\frac{1}{2} \sum_{p,q=1}^{\infty} \frac{1}{p^2 q^2} = \sum_{p,q=1}^{\infty} \frac{1}{2p^2 q^2}
$$

converges and since, for any numbers p and q, $2p^2 q^2 \leq p^4 + q^4$, the comparison test may be used to yield the result.

For single series with positive and negative terms the notion of absolute convergence plays an important part. We now establish the basic theorems for absolute convergence of double series.

Theorem 13. *Suppose that $\sum_{m,n=1}^{\infty} |a_{mn}|$ converges. If we define*

$$
b_{mn} = \begin{cases} a_{mn} & \text{whenever} \quad a_{mn} \geq 0, \\ 0 & \text{whenever} \quad a_{mn} < 0, \end{cases}
$$

$$
c_{mn} = \begin{cases} 0 & \text{whenever} \quad a_{mn} \geq 0, \\ -a_{mn} & \text{whenever} \quad a_{mn} < 0, \end{cases}
$$

then $b_{mn} + c_{mn} = |a_{mn}|$, $b_{mn} - c_{mn} = a_{mn}$, and $\sum_{m,n=1}^{\infty} b_{mn}$, $\sum_{m,n=1}^{\infty} c_{mn}$ converge. Calling the sum of these last series B and C, respectively, we have

$$
\sum_{m,n=1}^{\infty} a_{mn} = B - C, \qquad \sum_{m,n=1}^{\infty} |a_{mn}| = B + C.
$$

Furthermore,

$$
\left| \sum_{m,n=1}^{\infty} a_{mn} \right| \leq \sum_{m,n=1}^{\infty} |a_{mn}|.
$$

Proof. Since $b_{mn} \geq 0$, $c_{mn} \geq 0$, and $b_{mn} + c_{mn} = |a_{mn}|$, the Comparison Test (Theorem 12 (ii)) shows that $\sum_{m,n=1}^{\infty} b_{mn}$ and $\sum_{m,n=1}^{\infty} c_{mn}$ converge. The remaining results follow from Theorem 11 and from the fact that $|B - C| \leq B + C$, when B and C are any nonnegative numbers.

DEFINITION. A series $\sum_{m,n=1}^{\infty} a_{mn}$ is called **absolutely convergent** if the series $\sum_{m,n=1}^{\infty} |a_{mn}|$ converges.

Theorem 13 states that *a series which is absolutely convergent is itself convergent.* The next theorem establishes the basic relation between double series, single series, and repeated or iterated series.

Theorem 14. *Suppose that $\sum_{m,n=1}^{\infty} a_{mn}$ is absolutely convergent. Then*
(i) *$\sum_{n=1}^{\infty} a_{mn}$ is absolutely convergent. (Each row of the rectangular array, considered as a single series, is absolutely convergent.)*
(ii) *$\sum_{m=1}^{\infty} \left[\sum_{n=1}^{\infty} a_{mn} \right]$ is absolutely convergent. (The iterated sum, taking rows first, is absolutely convergent.)*
(iii) *$\sum_{m=1}^{\infty} \left[\sum_{n=1}^{\infty} a_{mn} \right] = \sum_{m,n=1}^{\infty} a_{mn}$. (The iterated sum is equal to the double sum.) The same results hold if the roles of m and n are interchanged.*
(iv) *$\sum_{p=2}^{\infty} \left[\sum_{m+n=p} a_{mn} \right]$ converges absolutely and equals the double sum.*

Proof. Because of Theorem 13, it suffices to prove the result when $a_{mn} \geq 0$.

Let $s = \sum_{m,n=1}^{\infty} a_{mn}$. Then it follows that $\sum_{n=1}^{N} a_{mn} \leq s$ for every N and every m. Thus (i) holds. Therefore we may write

$$A_m = \sum_{n=1}^{\infty} a_{mn} = \lim_{N \to \infty} \sum_{n=1}^{N} a_{mn}.$$

Also, since $s_{MN} \leq s$ always, we see that

$$\sum_{m=1}^{M} A_m = \lim_{N \to \infty} s_{MN} \leq s \qquad \text{for each } M.$$

We conclude that $\sum_{m=1}^{\infty} A_m$ is a convergent series, and so (ii) holds. Now let $\epsilon > 0$ be given. There is an N_0 such that

$$s - \epsilon < s_{MN} \leq \sum_{m=1}^{M} A_m \leq s \qquad \text{if } M > N_0 \text{ and } N > N_0.$$

This last statement establishes (iii).

To prove (iv), we first consider the meaning of the sum

$$\sum_{p=2}^{\infty} \sum_{m+n=p} a_{mn}.$$

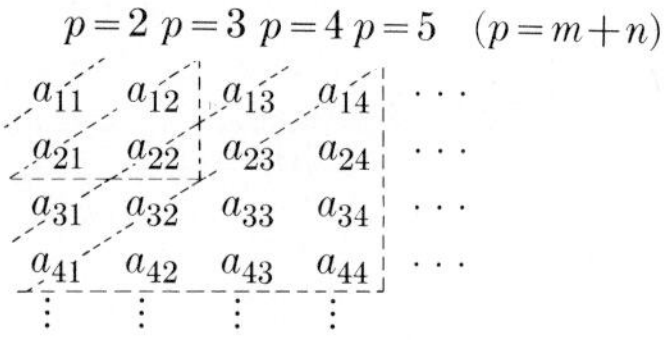

$$p=2 \; p=3 \; p=4 \; p=5 \quad (p=m+n)$$

$$
\begin{array}{llll}
a_{11} & a_{12} & a_{13} & a_{14} & \cdots \\
a_{21} & a_{22} & a_{23} & a_{24} & \cdots \\
a_{31} & a_{32} & a_{33} & a_{34} & \cdots \\
a_{41} & a_{42} & a_{43} & a_{44} & \cdots \\
\vdots & \vdots & \vdots & \vdots &
\end{array}
$$

FIGURE 9–10

The inner sum adds the elements of a diagonal, as shown in Fig. 9–10, and the outer sum adds all the diagonals. For any $\epsilon > 0$, there is an N_0 such that $s - \epsilon < s_{MN} \leq s$ whenever $M > N_0$ and $N > N_0$. Suppose that $P = 2N_0 + 2$. Then the triangular set (Fig. 9–10) of all (m, n) such that $m + n = p \leq P$ contains the set of all (m, n) such that $m \leq N_0 + 1$ and $n \leq N_0 + 1$. This set is also contained in the set of all (m, n) such that $m \leq P$ and $n \leq P$. Therefore

$$s - \epsilon < s_{N_0+1, N_0+1} \leq \sum_{p=0}^{P} \left[\sum_{m+n=p} a_{mn} \right] \leq s_{PP} \leq s,$$

which implies statement (iv).

Theorem 15. *Suppose that $\sum_{n=0}^{\infty} a_{mn}$ converges absolutely for each m and that*

$$\sum_{m=1}^{\infty} \left[\sum_{n=1}^{\infty} |a_{mn}| \right] \tag{3}$$

converges. Then the double series converges absolutely. The same result holds with the roles of m and n interchanged.

Proof. If s is the sum of the iterated series (3), we see immediately that

$$s_{MN} = \sum_{m=1}^{M} \left[\sum_{n=1}^{N} |a_{mn}| \right] \leq \sum_{m=1}^{M} \left[\sum_{n=1}^{\infty} |a_{mn}| \right] \leq s.$$

The result follows from Theorems 12 and 14.

Remark. The above theorem states that *if each row of a double series is absolutely convergent and if the iterated series of absolute values converges, then the double series converges.*

A similar statement is true for columns.

EXAMPLE 3. Show that the double series

$$\sum_{m,n=1}^{\infty} \frac{1}{m^2(1 + n^{3/2})}$$

converges.

Solution. For fixed m, the series

$$\sum_{n=1}^{\infty} \frac{1}{m^2(1 + n^{3/2})} < \sum_{n=1}^{\infty} \frac{1}{n^{3/2}},$$

which is a convergent p-series with $p = 3/2$. Denoting

$$A = \sum_{n=1}^{\infty} \frac{1}{n^{3/2}},$$

we see that

$$\sum_{m=1}^{\infty} \left[\sum_{n=1}^{\infty} \frac{1}{m^2(1 + n^{3/2})} \right] \leq \sum_{m=1}^{\infty} \frac{1}{m^2} A,$$

which converges. The hypotheses of Theorem 15 are satisfied and the double series converges.

Theorem 16. *Suppose that $\sum_{m=1}^{\infty} a_m$ and $\sum_{n=1}^{\infty} b_n$ are each absolutely convergent. Then the double series $\sum_{m,n=1}^{\infty} a_m b_n$ is absolutely convergent, and*

$$\sum_{m,n=1}^{\infty} a_m b_n = \left[\sum_{m=1}^{\infty} a_m \right] \left[\sum_{n=1}^{\infty} b_n \right]. \tag{4}$$

Proof. Let

$$A = \sum_{m=1}^{\infty} |a_m|, \qquad B = \sum_{n=1}^{\infty} |b_n|,$$

$$s_{mn} = \sum_{j=1}^{m} \sum_{k=1}^{n} a_j b_k, \qquad \text{and} \qquad S_{mn} = \sum_{j=1}^{m} \sum_{k=1}^{n} |a_j| |b_k|.$$

Then we see that for *every* m and n, $S_{mn} \le AB$. Hence the double series is absolutely convergent. Also

$$S_{mn} = \left[\sum_{j=1}^{m} a_j\right]\left[\sum_{k=1}^{n} b_k\right]. \tag{5}$$

The formula (4) results from passing to the limit in (5).

In Chapter 3, page 117, we stated without proof the following theorem concerning the Cauchy product of two power series.

Theorem 17. *Suppose that*

$$f(x) = \sum_{m=0}^{\infty} a_m x^m \qquad \text{for } |x| < R,$$

$$g(x) = \sum_{n=0}^{\infty} b_n x^n \qquad \text{for } |x| < R,$$

where $R > 0$. Then

$$f(x)g(x) = \sum_{p=0}^{\infty}\left[\sum_{m+n=p} a_m b_n\right] x^p \qquad \text{for } |x| < R.$$

Proof. Since the series for $f(x)$ and $g(x)$ both converge absolutely for $|x| < R$ then, by Theorem 16, their product

$$\sum_{m,n=0}^{\infty} a_m b_n x^{m+n}$$

converges absolutely as a double series (Fig. 9–11). The convergence of the Cauchy product series to $f(x)g(x)$ follows from Theorem 14 (iv).

$$
\begin{array}{llll}
a_0 b_0 & + a_0 b_1 x & + \cdots + a_0 b_n x^n & + \cdots \\
a_1 b_0 x & + a_1 b_1 x^2 & + \cdots + a_1 b_n x^{n+1} & + \cdots \\
\vdots & \vdots & & \vdots \\
a_m b_0 x^m & + a_m b_1 x^{m+1} & + \cdots + a_m b_n x^{m+n} & + \cdots \\
\vdots & \vdots & & \vdots
\end{array}
$$

FIGURE 9–11

Remark. Theorem 17 applies equally well for power series of the form

$$\sum_{m=0}^{\infty} a_m(x - c)^m \qquad \text{and} \qquad \sum_{n=0}^{\infty} b_n(x - c)^n.$$

EXAMPLE 4. Use Theorem 17 to find the power-series expansion of all terms up to x^5 of the function $\sqrt{1 + x}\cos x$.

Solution. We have

$$(1 + x)^{1/2} = 1 + \frac{1}{2}x - \frac{1}{8}x^2 + \frac{1}{16}x^3 - \frac{5}{128}x^4 + \frac{7}{256}x^5 - \cdots,$$

$$\cos x = 1 - \frac{x^2}{2!} + \frac{x^4}{4!} - \cdots.$$

Taking the Cauchy product, we find

$$\sqrt{1 + x} \cos x \approx \sum_{p=0}^{5} \left[\sum_{m+n=p} a_m b_n \right] x^p = 1 + \frac{1}{2} x - \frac{5}{8} x^2 - \frac{3}{16} x^3$$

$$+ \frac{25}{384} x^4 + \frac{13}{768} x^5.$$

Double sequences and series in which the elements are functions are defined in the same way as are single sequences and series of functions. The sequence $s_{mn}(x)$ is said to **converge uniformly to** $s(x)$ for x on some interval I if and only if for each $\epsilon > 0$ there is an N *independent of* x such that

$$|s_{mn}(x) - s(x)| < \epsilon \qquad \text{for all } m > N \text{ and } n > N.$$

The **uniform convergence of a double series** is equivalent to the uniform convergence of the sequence of its partial sums. The individual terms in a sequence or series may consist of functions of several variables. We define uniform convergence in the natural way; a sequence such as $\{s_{mn}(x_1, x_2, x_3)\}$ is said to converge uniformly for (x_1, x_2, x_3) in some region R if the index N in the definition above does not depend on the location of the point (x_1, x_2, x_3) in R.

We can now extend the Weierstrass M-test to double series.

Theorem 18. *Suppose that* $|u_{mn}(x, y)| < M_{mn}$ *for all* (x, y) *in some region* R *of the plane. If the double series* $\sum_{m,n=0}^{\infty} M_{mn}$ *converges, then* $\sum_{m,n=0}^{\infty} u_{mn}(x, y)$ *converges uniformly on* R.

Proof. By Theorem 12, $\sum_{m,n=0}^{\infty} u_{mn}(x, y)$ converges absolutely for each fixed (x, y) in R. Let $s(x, y)$ be the sum of the series and $s_{mn}(x, y)$ its partial sum. Denote

$$S = \sum_{m,n=0}^{\infty} M_{mn} \qquad \text{and} \qquad S_{mn} = \sum_{j,k=0}^{m,n} M_{jk}.$$

Let $\epsilon > 0$ be given. Then there is an N such that

$$|S - S_{mn}| < \epsilon \quad \text{for all} \quad (m, n), \qquad \text{with} \quad m > N, \qquad n > N.$$

Therefore, for each (x, y) in R,

$$|s(x, y) - s_{mn}(x, y)| \le |S - S_{mn}| < \epsilon, \qquad \text{whenever } m > N \text{ and } n > N.$$

Since N was chosen without regard to (x, y), the convergence is uniform.

Theorems on the continuity of the uniform limit of double series of continuous functions read the same as for single series. Similarly the results on term-by-term integration and differentiation (partial differentiation for functions of several variables) are all quite analogous to those obtained in Section 1. (See Exercises 24, 25, and 26.) We shall restrict ourselves to the statement of some results for double power series of the form

$$\sum_{m,n=0}^{\infty} a_{mn}x^m y^n. \tag{6}$$

Theorem 19. *If the double power series (6) converges absolutely for some $x_0 \neq 0$ and $y_0 \neq 0$, then the series (6) and the series of absolute values converge uniformly for $|x| \leq |x_0|$ and $|y| \leq |y_0|$.*

Proof. Since $|a_{mn}x^m y^n| \leq |a_{mn}x_0^m y_0^n| \equiv M_{mn}$, we may apply Theorem 12, and the result follows.

Theorem 20. *If the double power series (6) converges absolutely for $x_0 \neq 0$ and $y_0 \neq 0$, then all the series obtained by differentiating term by term with respect to x and y converge for all (x, y) in the rectangle $|x| < |x_0|, |y| < |y_0|$. The convergence is uniform on any rectangle $|x| \leq h, |y| \leq k$, where $h < |x_0|$, $k < |y_0|$.*

The proof parallels that for single power series. (See Exercise 27.)

Theorem 21. *If the double power series (6) converges absolutely for $x_0 \neq 0$ and $y_0 \neq 0$, and if f is defined by the series, so that*

$$f(x, y) = \sum_{m,n=0}^{\infty} a_{mn}x^m y^n \quad \text{for } |x| < |x_0|, \quad |y| < |y_0|, \tag{7}$$

then f is continuous and has partial derivatives of all orders which are given in the rectangle $|x| < |x_0|, |y| < |y_0|$ by the appropriate series obtained by term-by-term differentiation.

Corollary. *Under the assumptions of Theorem 21, we have*

$$a_{mn} = \frac{1}{m!n!} \frac{\partial^{m+n} f(0, 0)}{\partial x^m \partial y^n}. \tag{8}$$

Remark. All the results on double power series are valid for series of the form $\sum a_{mn}(x - c)^m(y - d)^n$, with the usual modifications; e.g., the evaluation in (8) is at (c, d) instead of $(0, 0)$.

EXAMPLE 5. Find the first six nonvanishing terms of the double power series expansion of e^{xy} about the point $(1, 0)$. Assume that the series is convergent in a rectangle containing $(1, 0)$.

Solution. We have $f(1, 0) = 1$. Evaluating all partial derivatives at $(1, 0)$, we find

$$\frac{\partial f}{\partial x} = 0, \quad \frac{\partial f}{\partial y} = 1, \quad \frac{\partial^2 f}{\partial x^2} = 0, \quad \frac{\partial^2 f}{\partial x \partial y} = 1, \quad \frac{\partial^2 f}{\partial y^2} = 1,$$

$$\frac{\partial^3 f}{\partial x^3} = \frac{\partial^3 f}{\partial x^2 \partial y} = 0, \quad \frac{\partial^3 f}{\partial x \partial y^2} = 2, \quad \frac{\partial^3 f}{\partial y^3} = 1.$$

Therefore

$$f = 1 + y + (x - 1)y + \frac{1}{2!}y^2 + (x - 1)y^2 + \frac{1}{3!}y^3 + \cdots .$$

EXAMPLE 6. Using power-series expansions, estimate the error made in computing $(e^{0.2} - 1)^2$ from the terms of the series for $(e^x - 1)^2$ out to and including the terms in x^5.

Solution. We use the Cauchy product of the series $e^x - 1$ with itself. We write

$$e^x - 1 = x + \frac{x^2}{2} + \frac{x^3}{6} + \frac{x^4}{24} + \frac{x^5}{120} + \cdots$$

$$e^x - 1 = x + \frac{x^2}{2} + \frac{x^3}{6} + \frac{x^4}{24} + \frac{x^5}{120} + \cdots$$

$$x^2 + \frac{x^3}{2} + \frac{x^4}{6} + \frac{x^5}{24} + \frac{x^6}{120}\left(1 + \frac{x}{6} + \frac{x^2}{6 \cdot 7} + \cdots\right)$$

$$+ \frac{x^3}{2} + \frac{x^4}{4} + \frac{x^5}{12} + \frac{x^6}{48}\left(1 + \frac{x}{5} + \frac{x^2}{5 \cdot 6} + \cdots\right)$$

$$+ \frac{x^4}{6} + \frac{x^5}{12} + \frac{x^6}{36}\left(1 + \frac{x}{4} + \frac{x^2}{4 \cdot 5} + \cdots\right)$$

$$+ \frac{x^5}{24} + \frac{x^6}{48}\left(1 + \frac{x}{3} + \frac{x^2}{3 \cdot 4} + \cdots\right)$$

$$+ \frac{x^6}{120}\left(1 + \frac{x}{6} + \frac{x^2}{6 \cdot 7} + \cdots\right)\left(1 + \frac{x}{2} + \frac{x^2}{2 \cdot 3} + \cdots\right).$$

Therefore

$$(e^x - 1)^2 = x^2 + x^3 + \frac{7x^4}{12} + \frac{x^5}{4} + \epsilon$$

where, by replacing each series in parentheses by the geometric series with the same first two terms, we obtain

$$\epsilon < \frac{x^6}{120(1 - x/6)} + \frac{x^6}{48(1 - x/5)} + \frac{x^6}{36(1 - x/4)} + \frac{x^6}{48(1 - x/3)}$$

$$+ \frac{x^6}{120(1 - x/6)(1 - x/2)}$$

$$< \frac{(0.2)^6}{12}\left(\frac{30}{290} + \frac{25}{96} + \frac{20}{19 \cdot 3} + \frac{15}{56} + \frac{30 \cdot 10}{10 \cdot 29 \cdot 9}\right)$$

$$< \frac{1}{12}(0.000064)(0.104 + 0.261 + 0.351 + 0.268 + 0.115)$$

$$< 59 \times 10^{-7}.$$

In an actual computation the rounding error would have to be added to obtain an estimate of the total error committed. Thus

$$(e^{0.2} - 1)^2 = 0.049013 \pm 0.000006.$$

EXERCISES

1. Show that the double series

$$\sum_{m,n=0}^{\infty} \frac{1}{(m+n)!}$$

is convergent. [*Hint.* Show that $(m+n)! \geq m!n!$.]

2. Show that

$$\sum_{m,n=1}^{\infty} \frac{1}{mn}$$

is divergent.

3. Test for convergence:

$$\sum_{m,n=0}^{\infty} \frac{2^{m+n}}{m!n!}.$$

4. Show that

$$\sum_{m,n=1}^{\infty} \frac{1}{(m^2+n^2)^p}$$

converges if $p > 1$. [*Hint.* Show first that $m^2 + n^2 \geq 2mn$.]

5. Show that

$$\sum_{m,n=1}^{\infty} \frac{1}{m^2+n^2}$$

is divergent.   *Hint.*   Note that

$$\sum_{m,n=1}^{N} \frac{1}{(m^2+n^2)} > \sum_{m=1}^{N} \left[\sum_{n=1}^{m} \frac{1}{(m^2+n^2)} \right] \geq \sum_{m=1}^{N} \frac{1}{2m}.$$

6. Show that

$$\sum_{m,n=0}^{\infty} \frac{(m+n)^5}{m!n!}$$

is convergent. [*Hint.* Let

$$a_m = \sum_{n=0}^{\infty} \frac{(m+n)^5}{m!n!} = \sum_{n=0}^{m} \frac{(m+n)^5}{m!n!} + \sum_{n=m+1}^{\infty} \frac{(m+n)^5}{m!n!}, \qquad m \geq 0,$$

and show that

$$a_m \leq e \cdot \frac{(2m)^5}{m!} + \frac{1}{m!} \sum_{n=0}^{\infty} \frac{(2n)^5}{n!} = A_m.$$

Then show that $\sum_{m=0}^{\infty} A_m$ converges and use Theorem 15.]

7. Test for convergence:

$$\sum_{\substack{m,n=1 \\ m \neq n}}^{\infty} \frac{1}{m^2-n^2}.$$

In each of problems 8 through 12, use Maclaurin expansions for the separate functions, and the formula for the Cauchy product to obtain the Maclaurin expansions for the given functions. Carry the process out to the term ax^n, where n is given.

8. $e^{2x} \sin 3x$, $n = 4$
9. $(1 + x)^{-2} \cos x$, $n = 5$
10. $(1 + x)^{-1/2} e^x$, $n = 4$
11. $e^x \ln (1 + x)$, $n = 5$
12. $(\cos x) \ln (1 + x)$, $n = 5$.

In each of problems 13 through 18, find the terms in the double (Maclaurin) series of the given function up to and including terms of degree three.

13. $e^x \cos y$

14. $\dfrac{1}{1 - x - 2y + x^2}$

15. $\dfrac{1}{e^x \cos y}$

16. $e^{-2x} \ln (1 + y)$

17. $\cos xy$

18. $(1 + x + y)^{-1/2}$

In each of problems 19 through 22, estimate the error made in computing each function from its series, as in Example 6 above.

19. $\dfrac{e^x}{1 - x}$, $x = 0.2$, $n = 4$
20. $\dfrac{\cosh x}{1 - x^2}$, $x = 0.2$, $n = 4$

21. $\dfrac{e^{-x}}{1 + x}$, $x = 0.2$, $n = 4$
22. $\dfrac{\sinh x}{1 - x^2}$, $x = 0.2$, $n = 5$

23. Prove Theorem 11.

24. State and prove a theorem on the limit of a uniformly convergent double series of continuous functions of n variables.

25. State and prove a theorem on the term-by-term differentiation of a uniformly convergent double series of functions of n variables.

26. State and prove a theorem on the term-by-term integration of a uniformly convergent double series of functions of three variables.

27. Write out the proof of Theorem 20.

28. Write out the proof of Theorem 21.

29. Prove the Corollary to Theorem 21.

5. COMPLEX FUNCTIONS. COMPLEX SERIES

Although functions of one or more complex variables are defined formally in the same manner as are functions of real variables, we shall see that the implications of the definition are quite different in the complex case. Denoting the collection of all complex numbers $a + bi$ by C_1, we define a **relation from C_1 to C_1** as a collection of ordered pairs (z, w) in which z is a complex number and w is a complex number. A **relation from C_n to C_1** is a set of ordered pairs $[(z_1, z_2, \ldots, z_n), w]$ in which $(z_1, z_2, \ldots, z_n)$ is an ordered array of n complex numbers and w is a complex number. A relation from C_1 to C_1 is a **function on C_1** if and only if no two distinct pairs have the same first element. We use the usual functional notation and write $w = f(z)$ for a function of one variable. Functions of more than one variable are defined similarly, and we write $w = f(z_1, z_2, \ldots, z_n)$ for a function on C_n. The **domain** and **range** of a function are defined precisely as in the case of a real variable (*First Course*, pages 17 and 184).

For problems concerning a function of a real variable, we found it helpful to interpret $y = f(x)$ as a set of points (frequently a curve or an arc) in the plane of analytic geometry. If $w = f(z)$ is a complex-valued function of a complex variable z, then both the domain and the range are sets of complex numbers. An aid to visualization is obtained by drawing *two* complex planes side by side, denoting one of them the z-plane and the other the w-plane (Fig. 9–12). The domain of a function f is a set of points S in the z-plane, and the range is a set of points T in the w-plane. The function f assigns a value Q in T to each point P in S. The functions we shall consider will usually have domains which are either a region in the z-plane or the entire z-plane.

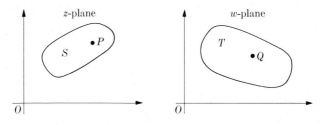

FIGURE 9–12

DEFINITION.   Suppose that f is a function on C_1 and c and L are complex numbers. We say that $f(z)$ *has the limit L as z tends to c* if and only if for each $\epsilon > 0$ there is a $\delta > 0$ such that

$$|f(z) - L| < \epsilon \qquad \text{whenever} \qquad 0 < |z - c| < \delta.$$

We also write $f(z) \to L$ as $z \to c$ and $\lim_{z \to c} f(z) = L$.

Remarks.   (i) Since absolute values were defined for complex numbers, the above definition makes sense when the function is complex-valued as well as when it is real-valued.

(ii) The definition has a simple interpretation in terms of two complex planes. The set of points $|z - c| < \delta$ consists of all the points z which are not farther away from c than δ. That is, the set consists of all points in a circle of radius δ with center at c (Fig. 9–13). The inequality $0 < |z - c|$ means that z is not allowed to be equal to c itself. The points $w = f(z)$ which satisfy the inequality $|f(z) - L| < \epsilon$ must lie in the circle of radius ϵ with center at L (Fig. 9–14).

(iii) In order for the definition to make sense, the circle $0 < |z - c| < \delta$ must lie in the domain S of f. If S consists of only part of this circle, then it is automatically understood that z is restricted to be a point of S.

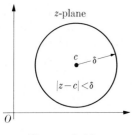

FIGURE 9–13

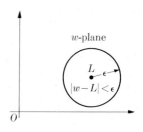

FIGURE 9–14

(iv) The definition of limit for a function of several complex variables is analogous to that for real variables. However, since a geometric interpretation is no longer readily available, we must lean more heavily on the analytic statements.

(v) The usual theorems on limits, such as those for limits of sums, products, and quotients, have the identical statements and proofs given for real variables and need not be repeated.

If we set $z = x + iy$ and write $w = u + iv$, then a function $w = f(z)$ may be written

$$f(z) = u(x, y) + iv(x, y),$$

in which u and v are each a function of the *two real variables* x and y. In other words, one complex-valued function of one complex variable may be considered as two real-valued functions of two real variables. If a, b, M, N are real numbers, then we see that the statement

$$f(z) \to M + iN \qquad \text{as} \qquad z \to a + ib$$

is equivalent to

$$u(x, y) \to M, \qquad v(x, y) \to N \qquad \text{as} \qquad (x, y) \to (a, b).$$

Thus notions of limits and continuity for a complex function may be reduced to the corresponding statements for pairs of functions of two real variables. We say that f **is continuous at** z_0 if and only if

$$\lim_{z \to z_0} f(z) = f(z_0).$$

If u and v are continuous at (x_0, y_0), then f is continuous at $z_0 = x_0 + iy_0$, and conversely.

The processes of differentiation and integration for complex functions are substantially *different* from those for real-valued functions. There is a wealth of material in complex analysis, and the interested student will find many texts devoted entirely to this subject. Most universities and colleges offer one or more advanced courses in the theory of functions of a complex variable.

We shall be concerned here mostly with sequences and series of complex numbers. A sequence

$$s_1, s_2, \ldots, s_n, \ldots$$

of complex numbers may be written in the form

$$r_1 + it_1, r_2 + it_2, \ldots, r_n + it_n, \ldots,$$

where r_k and t_k are the real and imaginary parts of s_k, respectively. We say that the sequence $\{s_n\}$ **is convergent** if and only if the two sequences $\{r_n\}$, $\{t_n\}$ of real numbers are convergent. The **infinite series**

$$\sum_{n=1}^{\infty} b_n \tag{1}$$

of complex numbers is convergent if and only if the sequence

$$s_n = \sum_{k=1}^{n} b_k$$

of partial sums is convergent. Many of the theorems established for sequences and series of real numbers carry over to complex sequences and series without change in statement or proof. A series of complex numbers such as (1) is said to be **absolutely convergent** if the real series

$$\sum_{n=1}^{\infty} |b_n|$$

is convergent. As in the case of Theorem 12 of Chapter 3, we can easily show that an absolutely convergent series of complex numbers is convergent.

A power series with complex terms is one of the form

$$\sum_{n=0}^{\infty} a_n z^n$$

in which the coefficients $a_0, a_1, \ldots$ are complex numbers. The Ratio Test (Chapter 3, Theorem 14, page 86) is valid without change for complex numbers as well as for real numbers. We now state the basic theorem on the convergence of complex power series.

Theorem 22. *Let $\sum_{n=0}^{\infty} a_n z^n$ be any power series. Then either*
(i) *the series converges only for $z = 0$; or*
(ii) *the series converges for all z; or*
(iii) *there is a number R such that the series converges for all z with $|z| < R$ and diverges for all z with $|z| > R$.*

The statement and proof of this theorem are the same as those of Theorem 7 in Section 3. However, we now see that the interpretation of the set $|z| < R$ of points of convergence is a circle of radius R with center at the origin (Fig. 9–15). We also see why the quantity R is called the *radius of convergence*. For series of the form $\sum_{n=0}^{\infty} a_n(z - c)^n$, the circle of convergence has radius R with its center at the number c (Fig. 9–15).

The elementary functions of algebra, trigonometry, and calculus may be defined as functions of a complex variable. A certain amount of care is necessary, since a function such as $\sin \omega$, which has a perfectly good meaning in terms of angles when ω is real, has no geometric interpretation or definition when ω is complex. We solve the problem by defining functions in terms of their power series expansions. We recall that e^x with x real has the Taylor expansion

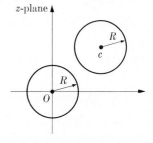

$$e^x = \sum_{n=0}^{\infty} \frac{x^n}{n!}, \tag{2}$$

which is convergent for *all* x. We *define* e^z by the series

$$e^z = \sum_{n=0}^{\infty} \frac{z^n}{n!} \quad \text{for all complex } z. \tag{3}$$

FIGURE 9–15

It is a simple exercise (Ratio Test) to show that the series (3) is absolutely convergent for all z. If z is real, then (3) becomes (2) and the definition is consistent. The next theorem gives the basic properties of e^z when z is complex.

Theorem 23. (a) $e^z \cdot e^w = e^{z+w}$ *for all complex z and w.*

(b) $e^{x+yi} = e^x (\cos y + i \sin y)$ *for all real x and y.*

(c) *If* $f(x) = e^{(a+bi)x}$, x *real, then* $f'(x) = (a + bi)e^{(a+bi)x}$.

Proof. (a) We write

$$e^z \cdot e^w = \left(\sum_{m=0}^{\infty} \frac{z^m}{m!} \right) \left(\sum_{n=0}^{\infty} \frac{w^n}{n!} \right)$$

and apply Theorems 16 and 14 of Section 4. We obtain

$$e^z \cdot e^w = \sum_{p=0}^{\infty} \left[\frac{1}{p!} \sum_{m+n=p} \frac{(m+n)!}{m!n!} z^m w^n \right] = \sum_{p=0}^{\infty} \frac{(z+w)^p}{p!} = e^{z+w}.$$

(b) From (a) we find

$$e^{x+yi} = e^x \cdot e^{yi} = e^x \sum_{n=0}^{\infty} \frac{(yi)^n}{n!}$$

$$= e^x \left\{ \sum_{k=0}^{\infty} \frac{(-1)^k y^{2k}}{(2k)!} + i \sum_{k=0}^{\infty} \frac{(-1)^k y^{2k+1}}{(2k+1)!} \right\} = e^x (\cos y + i \sin y).$$

To prove (c), we have

$$f(x) = e^{ax} (\cos bx + i \sin bx)$$

and

$$f'(x) = e^{ax}(-b \sin bx + ib \cos bx) + ae^{ax} (\cos bx + i \sin bx)$$
$$= (a + ib)e^{ax} (\cos bx + i \sin bx) = (a + ib)f(x).$$

The trigonometric and hyperbolic functions of a complex variable are *defined* by the formulas

$$\sin z = \frac{e^{iz} - e^{-iz}}{2i}, \qquad \cos z = \frac{e^{iz} + e^{-iz}}{2},$$

$$\sinh z = \frac{e^{z} - e^{-z}}{2}, \qquad \cosh z = \frac{e^{z} + e^{-z}}{2},$$

for all complex z. The remaining trigonometric and hyperbolic functions are defined in the usual way. That is,

$$\tan z = \frac{\sin z}{\cos z}, \qquad \sec z = \frac{1}{\cos z},$$

and so forth.

The power series expansion for e^z and the definitions of the trigonometric functions may be used to get the familiar expansions

$$\sin z = \sum_{k=0}^{\infty} \frac{(-1)^k z^{2k+1}}{(2k+1)!}, \qquad \cos z = \sum_{k=0}^{\infty} \frac{(-1)^k z^{2k}}{(2k)!}, \qquad (4)$$

valid for all complex z. Expansions for the remaining functions are obtained similarly.

Theorem 24. (a) *The addition theorems and double-angle formulas for* $\sin z$, $\cos z$, $\sinh z$ *and* $\cosh z$ *hold for all complex numbers.*
(b) *We have for all z:*

$$\cos iz = \cosh z, \qquad \cosh iz = \cos z,$$
$$\sin iz = i \sinh z, \qquad \sinh iz = i \sin z,$$
$$\sin (x + iy) = \sin x \cosh y + i \cos x \sinh y,$$
$$\cos (x + iy) = \cos x \cosh y - i \sin x \sinh y.$$

Proof. (a) We show that

$$\sin (z + w) = \sin z \cos w + \cos z \sin w.$$

From the definition of the trigonometric functions, we may write

$$\sin z \cos w + \cos z \sin w$$

$$= \frac{1}{4i} [(e^{iz} - e^{-iz})(e^{iw} + e^{-iw}) + (e^{iz} + e^{-iz})(e^{iw} - e^{-iw})]$$

$$= \frac{1}{4i} [e^{i(z+w)} + e^{i(z-w)} - e^{-i(z-w)} - e^{-i(z+w)} + e^{i(z+w)} - e^{-i(z+w)}]$$

$$= \frac{1}{2i} [e^{i(z+w)} - e^{-i(z+w)}] = \sin (z + w).$$

The remaining portions of the theorem are proved in a similar manner.

EXAMPLE 1. Write $\sin (1 + i)$ in the form $a + ib$.

Solution. $\sin (1 + i) = \sin 1 \cdot \cosh 1 + i \cos 1 \cdot \sinh 1$. Therefore $a = \sin 1 \cosh 1$, $b = \cos 1 \sinh 1$.

The student may verify the following formulas:

$$\tan z = z + \frac{z^3}{3} + \frac{2z^5}{15} + \cdots, \tag{5}$$

$$\sec z = 1 + \frac{z^2}{2} + \frac{5z^4}{24} + \frac{61z^6}{720} + \cdots, \tag{6}$$

$$\tanh z = z - \frac{z^3}{3} + \frac{2z^5}{15} - \cdots, \tag{7}$$

$$\operatorname{sech} z = 1 - \frac{z^2}{2} + \frac{5z^4}{24} + \frac{61z^6}{720} + \cdots. \tag{8}$$

EXAMPLE 2. Solve for z: $\cos z = 2$.

Solution. We have

$$\cos z = \cos (x + iy) = \cos x \cosh y - i \sin x \sinh y = 2.$$

Therefore

$$\cos x \cosh y = 2 \quad \text{and} \quad \sin x \sinh y = 0.$$

Taking $\sin x \sinh y = 0$ first, we see that $\sinh y = 0$ if and only if $y = 0$. But then from the first equation $\cosh y = 1$ and $\cos x$ must be 2, which is impossible if x is *real*. So $y = 0$ is excluded. The equation $\sin x \sinh y = 0$ can also hold if $\sin x = 0$, which occurs if $x = \pm n\pi$. Then $\cos x = \pm 1$ and, ruling out the negative values, we get $x = \pm 2n\pi$ and $\cos x = +1$. The first equation then implies that $\cosh y = 2$ or $y = \operatorname{argcosh} 2 = \ln(2 + \sqrt{3})$. The answer is

$$z = x + iy = \pm 2n\pi + i \ln (2 + \sqrt{3}).$$

Remark. Example 2 shows that the rules about the range of the various trigonometric functions, which we learned for the case when the domain is real, no longer hold when the domain is complex. For example, the functions $\sin z$ and $\cos z$ may have arbitrarily large values if z is complex.

When x is real, the function e^x is the inverse of the logarithm function. For complex functions we must proceed quite differently. With the observation that

$$e^{2\pi i} = (\cos 2\pi + i \sin 2\pi) = 1,$$

we conclude

$$e^{z+2\pi i} = e^z \cdot e^{2\pi i} = e^z,$$

and so e^z *is a periodic function with period* $2\pi i$. In attempting to define the logarithm as the inverse of $w = e^z$ we are stymied, because to each value of w there corresponds the infinite collection of values $z \pm 2n\pi i$, $n = 1, 2, \ldots$. The inverse relation is not a function. However, we can proceed by writing w in polar form. That is, if $w = r(\cos \theta + i \sin \theta)$, then

$$w = e^z = e^{x+iy} = e^x (\cos y + i \sin y) = r(\cos \theta + i \sin \theta).$$

The last equality yields $r = e^x$ and $y = \theta \pm 2n\pi$. This suggests defining the *principal inverse function* by

$$\text{Ln } w = \ln r + i\theta \quad \text{where} \quad w = re^{i\theta}, \quad -\pi < \theta \le \pi.$$

Thus we find that

$$e^z = w \text{ is equivalent to } z = \text{Ln } w \pm 2n\pi i.$$

We observe that if w is real and positive, then $\theta = 0$ and $\text{Ln } w$ is just the ordinary natural logarithm of w.

EXAMPLE 3. Find the value of $\text{Ln}(-3)$.

Solution. We have $-3 = 3e^{\pi i}$, since $e^{\pi i} = \cos \pi + i \sin \pi = -1$. Therefore $\text{Ln}(-3) = \ln 3 + \pi i$.

Remarks. The trick in Example 3 is to write the number -3 in the form $re^{i\theta}$ with $r \ge 0$ and with θ always in the interval $-\pi < \theta \le \pi$. *Every* complex number can be so written. Note that the myth prevalent in elementary trigonometry courses concerning the nonexistence of logarithms of negative numbers evaporates as we enter the complex domain.

The inverses of the trigonometric and hyperbolic functions are also multiple-valued relations and so are not functions. The definitions of principal inverses of these functions are usually given in texts on complex function theory and will be omitted here.

EXAMPLE 4. Express all the solutions of $\sinh w = z$ in terms of the Ln function.

Solution. From the equation $\sinh w = z$, we have

$$e^w - e^{-w} = 2z \quad \text{or} \quad e^{2w} - 2ze^w - 1 = 0.$$

This is a quadratic equation in e^w. Therefore

$$e^w = z \pm \sqrt{z^2 + 1}$$

and

$$w = \text{Ln}(z \pm \sqrt{z^2 + 1}) \pm 2n\pi i.$$

Remark. Since z is complex, it is not clear what meaning should be attached to the expression $\sqrt{z^2 + 1}$. If $\zeta = \rho e^{i\phi}$, $-\pi < \phi \le \pi$ is any complex number, the two square roots of ζ are

$$\sqrt{\rho}\, e^{i\phi/2} \quad \text{and} \quad -\sqrt{\rho}\, e^{i\phi/2}, \quad \text{with} \quad -\pi < \phi \leq \pi.$$

The first of these numbers is in the right-hand portion of the z-plane (has positive real part), while the second is in the left-hand portion. We call the first one the *positive* square root of ζ and the second the negative square root (except when $\phi = \pi$ and both square roots are on the imaginary axis).

EXERCISES

1. Show, by the Ratio Test, that

$$\sum_{n=0}^{\infty} \frac{z^n}{n!}$$

converges for all complex z.

2. Prove that $\sin z$ and $\cos z$ are given by their series expansions (4).

3. Prove that the functions $\sinh z$ and $\cosh z$ are given, respectively, by the series

$$\sum_{n=0}^{\infty} \frac{z^{2n+1}}{(2n + 1)!}, \qquad \sum_{n=0}^{\infty} \frac{z^{2n}}{(2n)!},$$

which are valid for all z.

4. Prove that for all complex z, w:

$$\cos (z + w) = \cos z \cos w - \sin z \sin w.$$

5. Prove that for all complex z, w:

$$\sinh (z + w) = \sinh z \cosh w + \cosh z \sinh w.$$

6. Prove that for all complex z, w:

$$\cosh (z + w) = \cosh z \cosh w + \sinh z \sinh w.$$

7. Derive formulas for $\sin 2z$, $\cos 2z$, $\sinh 2z$, $\cosh 2z$, z complex.

8. Prove the validity of the formulas in Theorem 24(b).

9. Verify the formula for $\tan z$ in (5).

10. Verify the formula for $\sec z$ in (6).

11. Verify the formula for $\tanh z$ in (7).

12. Verify the formula for $\operatorname{sech} z$ in (8).

13. Write in the form $a + ib$:

(a) e^{1+i},   (b) $\sin \left(\dfrac{4\pi}{3} + i\right)$,   (c) $\cosh \left(2 + \dfrac{i\pi}{3}\right)$

14. Write in the form $a + ib$:
    (a) $\sinh (1 - i)$,   (b) $\tanh (2 + 3i)$,   (c) $e^{\pi + 2i}$

15. Write in the form $a + ib$:
    (a) Ln (-4),   (b) Ln $(1 + i)$,   (c) Ln $(-i)$

16. Write in the form $a + ib$:
    (a) $e^{\sin[(\pi+i)/2]}$,   (b) Ln $(e^{(1+\pi i)/4})$,   (c) $\sin(\sqrt{1 - i})$

17. Write in the form $u(x, y) + iv(x, y)$:
    (a) $\sin^2 z$,   (b) $\tan z$

18. Write in the form $u(x, y) + iv(x, y)$:
    (a) e^{z^2},   (b) $e^{(1/z)}$

19. Write in the form $u(x, y) + iv(x, y)$:
    (a) Ln $(3 + 3i)$,   (b) $e^{\sin(z^2)}$

20. Find the circle of convergence of the series

$$\text{Ln } (1 + z) = z - \frac{z^2}{2} + \frac{z^3}{3} - \frac{z^4}{4} + \cdots (-1)^{n+1}\frac{z^n}{n} + \cdots.$$

21. Find the Maclaurin expansion for $(1 + x)^{1/2}$, replace each x by z and find the circle of convergence of the resulting complex series.

22. Repeat problem 21 for $(1 - x^2)^{-1/2}$.

23. Show that $\tanh w = z$ if and only if

$$w = \frac{1}{2} \text{Ln } \frac{1 + z}{1 - z} \pm in\pi.$$

24. Show that $\tan w = z$ if and only if

$$w = \pm n\pi + \frac{i}{2} \text{Ln } \frac{i + z}{i - z}.$$

25. Show that $\sin w = z$ if and only if

$$w = \frac{1}{i} \text{Ln } (iz \pm \sqrt{1 - z^2}) \pm 2n\pi.$$

Are there any complex numbers z for which there is no solution w? Note that Ln z is defined for all $z \neq 0$.

26. Show that $\cos w = z$ if and only if

$$w = \frac{1}{i} \text{Ln } (z \pm \sqrt{z^2 - 1}) \pm 2n\pi.$$

Are there any complex numbers z for which there is no solution w?

CHAPTER 10

Fourier Series

1. FOURIER SERIES

In the study of infinite series, the functions

$$1, x, x^2, \ldots, x^n, \ldots$$

play a central role. Most of the elementary functions of algebra, trigonometry, and calculus may be expanded in series which are sums of powers of x—i.e., in power series. The coefficients in such a Taylor or Maclaurin series are the successive derivatives of the given function evaluated at a point.

The simple function $f(x) = |x|$ cannot be expanded in a Maclaurin series. Since f does not have a derivative at $x = 0$ (Fig. 10–1), there is no way to compute the coefficient of x^n, $n \geq 1$, in such an expansion.

The collection of functions

$$1, \cos x, \cos 2x, \ldots, \cos nx, \ldots,$$
$$\sin x, \sin 2x, \ldots, \sin nx, \ldots$$

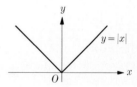

FIGURE 10–1

all have period 2π. We consider a function $f(x)$ which is periodic with period 2π and try to represent it in a series of the form

$$f(x) = \frac{a_0}{2} + \sum_{n=1}^{\infty} (a_n \cos nx + b_n \sin nx), \tag{1}$$

where all the a_n and b_n, $n = 0, 1, 2, \ldots$, are constants. Suppose the above series (1) converges uniformly to $f(x)$ on the interval $-\pi \leq x \leq \pi$. Then, since $u_n(x) = a_n \cos nx + b_n \sin nx$ is continuous, we know (Theorem 2 in Chapter 9) that $f(x)$ is continuous on $-\pi \leq x \leq \pi$. Furthermore, we may integrate term by term and perform various manipulations with uniformly convergent series. Proceeding formally for the moment, we let m be a *fixed* integer and multiply the series (1) by $\cos mx$. We get

$$f(x) \cos mx = \frac{a_0}{2} \cos mx + \sum_{n=1}^{\infty} (a_n \cos nx \cos mx + b_n \sin nx \cos mx).$$

Now we integrate this series on the interval $-\pi$ to π, obtaining

$$\int_{-\pi}^{\pi} f(x) \cos mx \, dx = \frac{a_0}{2} \int_{-\pi}^{\pi} \cos mx \, dx$$

$$+ \sum_{n=1}^{\infty} \left[a_n \int_{-\pi}^{\pi} \cos nx \cos mx \, dx + b_n \int_{-\pi}^{\pi} \sin nx \cos mx \, dx \right]. \qquad (2)$$

All the integrals on the right in the above expression may be calculated by elementary means. The student can easily verify that

$$\int_{-\pi}^{\pi} \cos nx \, dx = \int_{-\pi}^{\pi} \sin nx \, dx = 0 \qquad \text{for } n = 1, 2, \ldots.$$

Also, using trigonometric relations such as

$$\cos mx \cos nx = \tfrac{1}{2} [\cos (m + n)x + \cos (m - n)x],$$

we find it a simple matter to verify (see Exercise 16) that

$$\int_{-\pi}^{\pi} \cos mx \cos nx \, dx = \int_{-\pi}^{\pi} \sin mx \sin nx \, dx = \begin{cases} \pi & \text{if } m = n, \\ 0 & \text{if } m \neq n, \end{cases}$$

and

$$\int_{-\pi}^{\pi} \cos mx \sin nx \, dx = 0$$

for $m, n = 1, 2, \ldots.$ In Eq. (2), all the integrals on the right have the value zero except the term in which $n = m$. We conclude (on the basis of the formal manipulations) that

$$\int_{-\pi}^{\pi} f(x) \cos mx \, dx = \pi a_m, \qquad m = 0, 1, 2, \ldots.$$

Multiplying the series (1) by $\sin mx$ with m fixed and then integrating term by term, we obtain the corresponding formulas for b_m. They are

$$\int_{-\pi}^{\pi} f(x) \sin mx \, dx = \pi b_m, \qquad m = 1, 2, \ldots.$$

The above discussion leads to the following theorem:

Theorem 1. *Suppose that f is continuous for all x and periodic with period 2π. Suppose that the series*

$$f(x) = \frac{a_0}{2} + \sum_{n=1}^{\infty} (a_n \cos nx + b_n \sin nx) \qquad (3)$$

converges uniformly to $f(x)$ for all x. Then

$$a_n = \frac{1}{\pi} \int_{-\pi}^{\pi} f(t) \cos nt \, dt, \quad b_n = \frac{1}{\pi} \int_{-\pi}^{\pi} f(t) \sin nt \, dt, \quad n = 0, 1, 2 \ldots. \qquad (4)$$

Proof. Let

$$s_p(x) = \frac{a_0}{2} + \sum_{k=1}^{p} (a_k \cos kx + b_k \sin kx)$$

be the pth partial sum. Since the sequence $s_p(x)$ converges uniformly to $f(x)$, it follows that $s_p(x) \cos nx$ converges uniformly to $f(x) \cos nx$ for each fixed n. In fact,

$$|s_p(x) \cos nx - f(x) \cos nx| = |s_p(x) - f(x)| \, |\cos nx| \le |s_p(x) - f(x)|,$$

and the last expression on the right tends to zero uniformly as $p \to \infty$. Similarly, $s_p(x) \sin nx$ converges uniformly to $f(x) \sin nx$. Therefore, the series for $s_p(x) \cos nx$ and for $s_p(x) \sin nx$ may be integrated term by term (Theorem 2 in Chapter 9). Performing this process as outlined in the discussion preceding the theorem, we obtain the formulas for a_n and b_n.

DEFINITIONS. The series (3) is called the **Fourier series** of the function f, and the numbers a_n and b_n as given by (4) are called the **Fourier coefficients of** f.

The fact that Theorem 1 is valid only for functions which are periodic with period 2π may be considered an unsatisfactory feature of a Fourier series. Functions such as e^x, $\ln (1 + x)$, etc., which have Taylor expansions, are not periodic at all. Later we shall see how to extend the study of Fourier series so that this problem may be partially overcome. (See Theorem 2 on page 439.)

A more serious objection to Theorem 1 is the requirement that the series (3) converge uniformly. If f is any continuous function, the coefficients a_n and b_n may be calculated by (4). It turns out that there are continuous functions f such that the resulting Fourier series does not converge to the function. On the other hand, there are *discontinuous* functions f such that the series *does* converge to the function. *If f is discontinuous, the convergence cannot be uniform* because of Theorem 2 of Chapter 9. In order to obtain a useful convergence theorem—one which includes convergence for discontinuous functions—we introduce the class of functions described in the next paragraph.

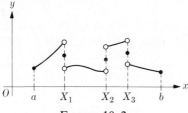

FIGURE 10–2

A function f is said to be **piecewise continuous** on $[a, b]$ if and only if there is a finite subdivision $a = X_0 < X_1 < \cdots < X_{k-1} < X_k = b$ such that the function f is continuous on each subinterval (X_{i-1}, X_i). Furthermore, the one-sided limits of f at each of the subdivision points must exist (Fig. 10–2). At each subdivision point X_i the function may be discontinuous, and we call the difference

$f(X_i +) - f(X_i -)$ the **jump** of f at X_i. The function f may have any value at X_i. In other words, a piecewise continuous function on $[a, b]$ is one which is continuous except at a finite number of points where it has jumps. A function f is **piecewise smooth** on $[a, b]$ if and only if f is piecewise continuous and f' is piecewise continuous, with the jumps of f' occurring at $X_0, X_1, \ldots, X_k$. The value of a piecewise continuous function at one of the points of discontinuity plays an important part in Fourier analysis. We say that a piecewise continuous function f is **normalized** if and only if its value at a jump point X_i is given by

$$f(X_i) = \tfrac{1}{2}[f(X_i -) + f(X_i +)], \qquad i = 1, 2, \ldots, k - 1.$$

That is, the function value is halfway between the limit values from the left and right (Fig. 10–2). We say that a function f is **smooth** on $[a, b]$ if and only if f and f' are continuous throughout $[a, b]$.

The integral of a function f, thus far defined only for continuous functions, may now easily be defined for piecewise continuous functions. If f is continuous on $[a, b]$ except at $X_1, X_2, \ldots, X_{k-1}$ where it has jumps, we define

$$\int_a^b f(x)\, dx = \sum_{i=1}^k \int_{X_{i-1}}^{X_i} f(x)\, dx. \tag{5}$$

The values of the integrals on the right in (5) are not influenced by the value of f at X_i. Therefore, normalizing a piecewise continuous function does not affect its integral. Since any finite linear combination of piecewise continuous functions is piecewise continuous, the law

$$\int_a^b [cf(x) + dg(x)]\, dx = c\int_a^b f(x)\, dx + d\int_a^b g(x)\, dx$$

still holds. Furthermore, for piecewise continuous f,

$$m(b - a) \le \int_a^b f(x)\, dx \le M(b - a) \qquad \text{if } m \le f(x) \le M,$$

and

$$\int_a^b f(x)\, dx = \int_a^c f(x)\, dx + \int_c^b f(x)\, dx,$$

and so forth. In addition, if f is piecewise continuous on $[a, b]$, and if F is given by

$$F(x) = \int_a^x f(t)\, dt,$$

then F and F' are continuous except at $X_1, X_2, \ldots, X_{k-1}$. Finally, F *is continuous on* $[a, b]$. To see this, we note that

$$|F(x_2) - F(x_1)| = \left| \int_{x_1}^{x_2} f(t)\, dt \right| \le M|x_2 - x_1|,$$

and so F satisfies the definition of continuity. Also, F is piecewise smooth on $[a, b]$.

The student may verify that the product fg of a piecewise continuous function f and a continuous function g is piecewise continuous. Therefore $f(x) \cos nx$ and $f(x) \sin nx$ are piecewise continuous on $[-\pi, \pi]$ whenever f is. The coefficients a_n and b_n may be defined according to (4) for any piecewise continuous f, with the interpretation of integrals as described above. We now state a remarkable convergence theorem which, while not the most general one, is sufficiently broad for most applications.

Theorem 2. *Suppose that f is piecewise smooth and normalized on any finite interval and f is periodic with period 2π. Then its Fourier series converges to $f(x)$ for each x. Furthermore, if f is smooth on the closed interval $[c, d]$, then the convergence is uniform on $[c, d]$.*

We shall prove this theorem in Section 6.

Suppose that a piecewise smooth function f is defined on the interval $[-\pi, \pi]$ and we wish to expand it in a Fourier series. First we define the **periodic extension** of f as the function $f_0(x)$ defined for all x by the relation

$$f_0(x) = \begin{cases} f(x) & \text{for } -\pi < x < \pi, \\ f_0(x - 2\pi) & \text{for all other } x. \end{cases}$$

Then we *normalize* f_0, if necessary (Fig. 10–3). No matter how smooth the function f may be, the periodic extension f_0 will introduce a discontinuity at π and $-\pi$ whenever $f(-\pi) \neq f(\pi)$. The study of Fourier series would have limited value if we were restricted from the beginning to functions which are periodic. The fact that the basic theory enables us to handle functions with jumps means that we may start with any integrable function defined on $[-\pi, \pi]$, form its periodic extension, and apply the theory. We illustrate this point with examples.

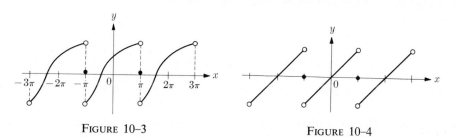

FIGURE 10–3   FIGURE 10–4

EXAMPLE 1. Find the Fourier series for the function

$$f(x) = x \qquad \text{on } (-\pi, \pi).$$

Solution. We form the periodic extension of f and normalize it (Fig. 10–4). The normalization yields $f_0(-\pi) = f_0(\pi) = 0$. We compute a_n and b_n, using the formulas (4) for the coefficients:

$$a_n = \frac{1}{\pi} \int_{-\pi}^{\pi} x \cos nx \, dx, \qquad b_n = \frac{1}{\pi} \int_{-\pi}^{\pi} x \sin nx \, dx$$

Integrating by parts, we find

$$a_n = \frac{1}{\pi} \left[\frac{x \sin nx}{n} \right]_{-\pi}^{\pi} - \frac{1}{n\pi} \int_{-\pi}^{\pi} \sin nx \, dx = 0, \qquad n = 1, 2, \ldots,$$

$$b_n = \frac{1}{\pi} \left[-\frac{x \cos nx}{n} \right]_{-\pi}^{\pi} + \frac{1}{n\pi} \int_{-\pi}^{\pi} \cos nx \, dx = -\frac{2}{n} \cos n\pi = (-1)^{n-1} \frac{2}{n}.$$

Since $a_0 = 0$, the Fourier series for $f(x) = x$ is

$$x = 2 \left[\sin x - \frac{\sin 2x}{2} + \frac{\sin 3x}{3} - \frac{\sin 4x}{4} + \cdots \right], \qquad -\pi < x < \pi.$$

For $x = \pm\pi$, the series converges not to f but to the normalized value of f_0, the extension of f. In this case we notice that at $-\pi$ and π all the terms in the series vanish, and we verify that the series converges to $f_0(-\pi) = f_0(\pi) = 0$. A graph of the first few terms of the series is shown in Fig. 10–5.

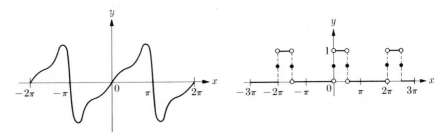

FIGURE 10–5   FIGURE 10–6

EXAMPLE 2. Find the Fourier series of the function

$$f(x) = \begin{cases} 0, & -\pi \le x \le 0, \\ 1, & 0 < x < \dfrac{\pi}{2}, \\ 0, & \dfrac{\pi}{2} \le x \le \pi. \end{cases}$$

Solution. We first extend f periodically and normalize it. The result is shown in Fig. 10–6. The dots in the figure show the normalized values at the jumps. We compute the coefficients:

$$a_0 = \frac{1}{\pi} \int_{-\pi}^{\pi} f(x) \, dx = \frac{1}{\pi} \int_{0}^{\pi/2} 1 \, dx = \frac{1}{2},$$

$$a_n = \frac{1}{\pi} \int_{-\pi}^{\pi} f(x) \cos nx \, dx = \frac{1}{\pi} \int_{0}^{\pi/2} \cos nx \, dx = \frac{1}{n\pi} \sin nx \Big]_{0}^{\pi/2}.$$

Therefore

$$
a_n = \begin{cases} 0 & \text{if } n \text{ is even} \\[2mm] \dfrac{(-1)^k}{(2k+1)\pi} & \text{if } n = 2k+1, \quad k = 0, 1, 2, \dots. \end{cases}
$$

To find b_n, we write

$$
b_n = \frac{1}{\pi} \int_0^{\pi/2} \sin nx \, dx = -\frac{\cos nx}{n\pi} \Big]_0^{\pi/2} = \begin{cases} \dfrac{1}{(2k+1)\pi} & \text{if } n = 2k+1, \\[2mm] \dfrac{1-(-1)^k}{2k\pi} & \text{if } n = 2k. \end{cases}
$$

The values of $\sin nx$ and $\cos nx$ at odd and even multiples of $\pi/2$ occur often in Fourier series; the reader should study carefully the evaluations above to be sure he understands how they were obtained. The desired Fourier series for f is

$$
f(x) = \frac{1}{4} + \frac{1}{\pi}\bigg[\cos x + \sin x + \sin 2x - \frac{\cos 3x}{3} + \frac{\sin 3x}{3}
$$
$$
+ \frac{\cos 5x}{5} + \frac{\sin 5x}{5} + \frac{\sin 6x}{3} + \cdots \bigg].
$$

Note that at $x = 0, \pi/2, 2\pi, 5\pi/2, \dots$, the series must converge to the value $1/2$.

In computing the Fourier coefficients we may frequently save a great deal of labor by using certain properties of even and odd functions. We recall that a function $f(x)$ is **even** if

$$
f(-x) = f(x)
$$

for all x. A function $g(x)$ is **odd** if

$$
g(-x) = -g(x)
$$

for all x. Using the definition of an integral, we observe that if f is even and g is odd then, for any value a,

$$
\int_{-a}^{a} f(x) \, dx = 2 \int_0^a f(x) \, dx, \qquad \int_{-a}^{a} g(x) \, dx = 0.
$$

The product of two even functions is even, the product of two odd functions is even, and the product of an even and an odd function is odd. For every integer n, the function $\cos nx$ is even and the function $\sin nx$ is odd. If we observe in Example 1 that $f(x) = x$ is *odd*, then clearly $f(x) \cos nx$ is odd for every n; we can thus conclude without any computation at all that $a_n = 0$, for $n = 0, 1, 2, \dots$. In Example 2, since the function f is neither even nor odd, no simplification can be made on this basis. The preceding discussion is now stated in the form of a theorem.

Theorem 3. *Suppose that f is periodic with period 2π and is piecewise continuous and normalized. Then*

(a) *if f is odd on $(-\pi, \pi)$, we have $a_n = 0$ for $n = 0, 1, 2, \ldots$, and*

$$b_n = \frac{2}{\pi} \int_0^\pi f(x) \sin nx \, dx;$$

(b) *if f is even on $(-\pi, \pi)$, we have $b_n = 0$ for $n = 1, 2, \ldots$, and*

$$a_n = \frac{2}{\pi} \int_0^\pi f(x) \cos nx \, dx.$$

EXAMPLE 3. Find the Fourier series of the function

$$f(x) = |x|, \qquad -\pi < x < \pi.$$

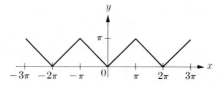

FIGURE 10–7

Solution. We form the periodic extension of f as shown in Fig. 10–7. Note that the extended function happens to be continuous; normalization is therefore unnecessary. Because f is even, we conclude at once that $b_n = 0$ for all n, that $a_0 = \pi$, and that

$$a_n = \frac{2}{\pi} \int_0^\pi f(x) \cos nx \, dx.$$

Since $f(x) = x$ for $x \geq 0$ we obtain, upon integrating by parts,

$$a_n = \frac{2}{\pi} \left[\frac{x \sin nx}{n} \right]_0^\pi - \frac{2}{n\pi} \int_0^\pi \sin nx \, dx = \frac{2}{n^2\pi} \cos nx \Big]_0^\pi$$

$$= \frac{2}{n^2\pi} [\cos n\pi - 1] = \frac{2}{n^2\pi} [(-1)^n - 1]$$

$$= \begin{cases} \dfrac{-4}{(2k+1)^2\pi} & \text{for } n = 2k+1 \\ 0 & \text{for } n = 2k \end{cases}, \qquad k = 0, 1, 2, \ldots.$$

The desired series is

$$f(x) = \frac{\pi}{2} - \frac{4}{\pi} \left[\frac{\cos x}{1^2} + \frac{\cos 3x}{3^2} + \cdots + \frac{\cos (2k+1)x}{(2k+1)^2} + \cdots \right].$$

Setting $x = 0$ and noting that $f(0) = 0$, we obtain the remarkable formula

$$\pi^2 = 8(1 + \tfrac{1}{9} + \tfrac{1}{25} + \tfrac{1}{49} + \cdots).$$

EXERCISES

In each of problems 1 through 15, find the Fourier series for the given function. Draw a graph of the periodic, normalized extension f_0 in the interval $[-3\pi, 3\pi]$.

1. $f(x) = \begin{cases} \pi/4 & \text{for } 0 < x < \pi \\ -\pi/4 & \text{for } -\pi < x < 0 \end{cases}$

2. $f(x) = \begin{cases} 0 & \text{for } -\pi < x < 0 \\ 1 & \text{for } 0 < x < \pi \end{cases}$

3. $f(x) = x^2 \quad \text{for } -\pi \le x \le \pi$

4. $f(x) = \begin{cases} 0 & \text{for } -\pi < x < \pi/2 \\ 1 & \text{for } \pi/2 < x < \pi \end{cases}$

5. $f(x) = \begin{cases} -1 & \text{for } -\pi < x < -\pi/2 \\ 0 & \text{for } -\pi/2 < x < \pi/2 \\ 1 & \text{for } \pi/2 < x < \pi \end{cases}$

6. $f(x) = \begin{cases} 0 & \text{for } -\pi < x < 0 \\ x & \text{for } 0 \le x < \pi \end{cases}$

7. $f(x) = \begin{cases} 1 & \text{for } -\pi < x < -\pi/2 \\ 0 & \text{for } -\pi/2 < x < \pi/2 \\ 1 & \text{for } \pi/2 < x < \pi \end{cases}$

8. $f(x) = |\sin x| \quad \text{for } -\pi \le x \le \pi$

9. $f(x) = |\cos x| \quad \text{for } -\pi \le x \le \pi$

10. $f(x) = x^3 \quad \text{for } -\pi < x < \pi$

11. $f(x) = e^x \text{ for } -\pi < x < \pi$

12. $f(x) = \begin{cases} 0 & \text{for } -\pi < x < 0 \\ \sin x & \text{for } 0 < x < \pi \end{cases}$

13. $f(x) = \sin^2 x \quad \text{for } -\pi < x < \pi$

14. $f(x) = x \sin x \quad \text{for } -\pi < x < \pi$

15. $f(x) = \begin{cases} -\pi & \text{for } -\pi < x < 0 \\ x & \text{for } 0 < x < \pi \end{cases}$

16. Verify the formulas

$$\int_{-\pi}^{\pi} \cos mx \cos nx \, dx = \int_{-\pi}^{\pi} \sin mx \sin nx \, dx = \begin{cases} \pi & \text{if } m = n, \\ 0 & \text{if } m \ne n. \end{cases}$$

Also show that $\int_{-\pi}^{\pi} \cos mx \sin nx \, dx = 0 \quad$ for $m, n = 1, 2, \ldots$.

17. For the series in problem 15, show that for $x = 0$, we get the relation

$$\frac{\pi^2}{8} = \frac{1}{1^2} + \frac{1}{3^2} + \frac{1}{5^2} + \cdots.$$

18. Find an expansion for π^2 in Example 3 by evaluation of the series at $x = \pi/4$. Can the error after n terms be estimated?

19. Given that $f(x) = x + x^2$, $-\pi < x < \pi$, find the Fourier expansion of f. Show that

$$\frac{\pi^2}{6} = \sum_{n=1}^{\infty} \frac{1}{n^2}.$$

20. Using the series expansion in problem 1, show that

(a) $\dfrac{\pi}{4} = 1 - \dfrac{1}{3} + \dfrac{1}{5} - \dfrac{1}{7} + \cdots$

(b) $\dfrac{\pi}{3} = 1 + \dfrac{1}{5} - \dfrac{1}{7} - \dfrac{1}{11} + \dfrac{1}{13} + \dfrac{1}{17} - \cdots$

(c) $\dfrac{\sqrt{3}}{6}\pi = 1 - \dfrac{1}{5} + \dfrac{1}{7} - \dfrac{1}{11} + \dfrac{1}{13} - \dfrac{1}{17} + \cdots$

2.   HALF-RANGE EXPANSIONS

Suppose that a function f is defined on $0 \leq x \leq \pi$ and we wish to expand it in a Fourier series. Since the coefficients a_n and b_n involve integrals from $-\pi$ to π, we must somehow extend the definition of f to the interval $(-\pi, \pi)$. We can do this in many ways at our own convenience. One way is to extend f so that it is an *even* function on the interval $(-\pi, \pi)$ (Fig. 10–8). Since an even function has $b_n = 0$ for $n = 1, 2, \ldots$, the Fourier series has only cosine terms. We call such a series a **cosine series** and, as the original function is represented on $(0, \pi)$, the expansion is called a **half-range expansion.**

A function f defined on $(0, \pi)$ may also be extended to the interval $(-\pi, \pi)$ as an *odd* function. Figure 10–9 shows such an extension, and we notice that discontinuities are introduced at $-\pi$, 0, and π, unless $f(-\pi) = f(0) = f(\pi) = 0$. We are not disturbed by this fact, since the convergence theorem for Fourier series is valid for piecewise continuous functions. Since the Fourier series of an odd function has $a_n = 0$, $n = 0, 1, 2, \ldots$, the resulting series is called a **sine series.** Examples illustrate the process for obtaining half-range expansions.

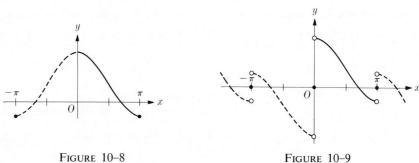

FIGURE 10–8   FIGURE 10–9

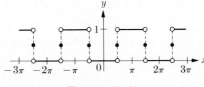

FIGURE 10–10

EXAMPLE 1.   Given the function

$$f(x) = \begin{cases} 0' & \text{for } 0 < x < \pi/2, \\ 1 & \text{for } \pi/2 < x < \pi, \end{cases}$$

expand f in a cosine series and draw a graph of the extended function on the interval $[-3\pi, 3\pi]$.

Solution. The graph of the even function is shown in Fig. 10–10. Since f is extended to be even, we have $b_n = 0$, $n = 1, 2, \ldots$, and

$$a_n = \frac{2}{\pi} \int_0^\pi f(x) \cos nx \, dx, \qquad n = 0, 1, 2, \ldots. \tag{1}$$

A simple calculation shows that

$$a_0 = \frac{2}{\pi} \int_{\pi/2}^{\pi} dx = 1 \quad \text{and} \quad a_n = \frac{2}{\pi} \int_{\pi/2}^{\pi} \cos nx \, dx = \left[\frac{2}{n\pi} \sin nx \right]_{\pi/2}^{\pi}$$

with

$$a_n = \begin{cases} 0 & \text{if } n \text{ is even} \\ -\dfrac{2(-1)^{k-1}}{(2k-1)\pi} & \text{if } n = 2k-1 \end{cases}, \quad k = 1, 2, \ldots.$$

Therefore

$$f(x) = \frac{1}{2} - \frac{2}{\pi} \left[\frac{\cos x}{1} - \frac{\cos 3x}{3} + \frac{\cos 5x}{5} - \cdots \right], \quad 0 < x < \pi.$$

Remark. Setting $x = 0$ in this result, we obtain the interesting formula

$$\frac{\pi}{4} = 1 - \frac{1}{3} + \frac{1}{5} - \frac{1}{7} + \cdots.$$

(See also Exercise 20 of Section 1.)

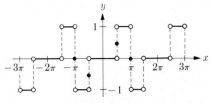

FIGURE 10–11

EXAMPLE 2. Expand the function f of Example 1 in a sine series and draw a graph of the extended function in the interval $[-3\pi, 3\pi]$.

Solution. The graph of the odd function is shown in Fig. 10–11. We have $a_n = 0$, $n = 0, 1, 2, \ldots$, and

$$b_n = \frac{2}{\pi} \int_0^{\pi} f(x) \sin nx \, dx, \quad n = 1, 2, \ldots. \tag{2}$$

Therefore

$$b_n = \frac{2}{\pi} \int_{\pi/2}^{\pi} \sin nx \, dx = \left[-\frac{2}{n\pi} \cos nx \right]_{\pi/2}^{\pi} = \begin{cases} \dfrac{2}{n\pi} & \text{if } n \text{ is odd,} \\ \dfrac{2}{2k\pi}[(-1)^k - 1] & \text{if } n = 2k. \end{cases}$$

We conclude that

$$f(x) = \frac{2}{\pi} \left[\frac{\sin x}{1} - \frac{2 \sin 2x}{2} + \frac{\sin 3x}{3} + \frac{0 \cdot \sin 4x}{4} + \frac{\sin 5x}{5} \right.$$
$$\left. - \frac{2 \sin 6x}{6} + \frac{\sin 7x}{7} + \frac{0 \cdot \sin 8x}{8} + \cdots \right], \quad 0 < x < \pi.$$

Remarks. Equation (1) shows that a cosine series may be obtained by using only the definition of f on $(0, \pi)$. The extension of f as an even function is a mental convenience which is used to set $b_n = 0$. The evaluation of a_n as in (1) does not use the extended function at all. Similarly, the evaluation of b_n in a sine series shows that the extended function plays no part except to help us set $a_n = 0$ and to permit us to use formula (2). Of course, for both cosine and sine series the expansion represents the original function on the interval $(0, \pi)$ only.

EXERCISES

In each of problems 1 through 8, expand each function in a cosine series and draw the graph of the extended function on $[-3\pi, 3\pi]$.

1. $f(x) = \begin{cases} 1 & \text{for } 0 < x < \pi/2 \\ 0 & \text{for } \pi/2 < x < \pi \end{cases}$ 2. $f(x) = \begin{cases} x & \text{for } 0 < x < \pi/2 \\ \pi - x & \text{for } \pi/2 < x < \pi \end{cases}$

3. $f(x) = \sin x$ for $0 \le x \le \pi$ 4. $f(x) = |\cos x|$ for $0 \le x \le \pi$

5. $f(x) = x$ for $0 \le x \le \pi$ 6. $f(x) = x^2$ for $0 \le x \le \pi$

7. $f(x) = x^3$ for $0 \le x \le \pi$ 8. $f(x) = e^x$ for $0 \le x \le \pi$

In problems 9 through 16, expand each function in a sine series and draw the graph of the extended function on $[-3\pi, 3\pi]$.

9. f is the function in problem 1.   10. f is the function in problem 2.

11. $f(x) = \cos x$ for $0 < x < \pi$ 12. $f(x) = |\cos x|$ for $0 < x < \pi$

13. $f(x) = x$ for $0 < x < \pi$ 14. $f(x) = x^2$ for $0 < x < \pi$

15. $f(x) = x^3$ for $0 < x < \pi$

16. $f(x) = e^{ax}$ for $0 < x < \pi$, a constant.

3. EXPANSIONS ON OTHER INTERVALS

If f is a piecewise smooth function defined on the interval $[c - \pi, c + \pi]$, we form the periodic extension f_0 precisely as we did before. According to Theorem 2, the Fourier coefficients of the extended function, which we denote $\{a_n^0\}$ and $\{b_n^0\}$, are given by the formulas

$$a_n^0 = \frac{1}{\pi} \int_{-\pi}^{\pi} f_0(x) \cos nx \, dx, \qquad b_n^0 = \frac{1}{\pi} \int_{-\pi}^{\pi} f_0(x) \sin nx \, dx. \tag{1}$$

But, since f_0, $\cos nx$, and $\sin nx$ all have period 2π, we may replace the interval $[-\pi, \pi]$ in (1) by any other interval of length 2π. In particular, we may replace $[-\pi, \pi]$ by the interval $[c - \pi, c + \pi]$, on which f_0 coincides with f. Denoting by $\{a_n\}$ and $\{b_n\}$ the quantities

$$a_n = \frac{1}{\pi} \int_{c-\pi}^{c+\pi} f(x) \cos nx \, dx, \qquad b_n = \frac{1}{\pi} \int_{c-\pi}^{c+\pi} f(x) \sin nx \, dx, \tag{2}$$

we see that $a_n = a_n^0$ and $b_n = b_n^0$. The Fourier series formed with the coefficients in (2) yield an expansion for f which is valid in the interval $(c - \pi) < x < (c + \pi)$.

EXAMPLE 1. If $f(x) = x$ for $0 < x < 2\pi$, expand f in a Fourier series on the interval $(0, 2\pi)$. Draw a graph of f_0, the periodic extension of f, on the interval $[-2\pi, 4\pi]$.

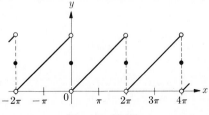

FIGURE 10–12

Solution. The graph of f_0 is drawn in Fig. 10–12. We compute

$$a_0 = \frac{1}{\pi} \int_0^{2\pi} x \, dx = 2\pi,$$

$$a_n = \frac{1}{\pi} \int_0^{2\pi} x \cos nx \, dx = \frac{1}{\pi} \left[\frac{x \sin nx}{n} + \frac{\cos nx}{n^2} \right]_0^{2\pi} = 0,$$

$$b_n = \frac{1}{\pi} \int_0^{2\pi} x \sin nx \, dx = \frac{1}{\pi} \left[-\frac{x \cos nx}{n} + \frac{\sin nx}{n^2} \right]_0^{2\pi} = -\frac{2}{n}.$$

Therefore

$$x = \pi - 2 \sum_{n=1}^{\infty} \frac{\sin nx}{n} \qquad \text{for } 0 < x < 2\pi.$$

Remark. Of course, we could make the computations on the interval $[-\pi, \pi]$ using the extended function f_0. However, in that case the integrals which determine $\{a_n^0\}$ and $\{b_n^0\}$ would have to be separated into two parts, even though the final result would be the same.

A function f which is piecewise smooth and normalized on an interval $[-L, L]$ can be expanded in a **modified Fourier series.** If $f(x)$ is given on $[-L, L]$, we introduce a change of variable and define y and $g(y)$ by the relations

$$y = \frac{\pi x}{L}, \qquad f(x) = f\left(\frac{Ly}{\pi}\right) = g(y) = g\left(\frac{\pi x}{L}\right).$$

This transformation maps $[-L, L]$ onto $[-\pi, \pi]$. We see that $g(y)$ is piecewise smooth and normalized on $[-\pi, \pi]$ and so can be expanded in a Fourier series

$$g(y) = \frac{a_0}{2} + \sum_{n=1}^{\infty} (a_n \cos ny + b_n \sin ny), \qquad -\pi < y < \pi, \qquad (3)$$

with

$$a_n = \frac{1}{\pi} \int_{-\pi}^{\pi} g(y) \cos ny \, dy, \qquad b_n = \frac{1}{\pi} \int_{-\pi}^{\pi} g(y) \sin ny \, dy. \qquad (4)$$

Using the change of variable $y = \pi x/L, dy = (\pi/L) \, dx$, we may write (4) in the form

$$a_n = \frac{1}{L} \int_{-L}^{L} f(x) \cos \frac{n\pi x}{L} \, dx, \qquad b_n = \frac{1}{L} \int_{-L}^{L} f(x) \sin \frac{n\pi x}{L} \, dx. \qquad (5)$$

The series (3) becomes

$$f(x) = \frac{a_0}{2} + \sum_{n=1}^{\infty} \left(a_n \cos \frac{n\pi x}{L} + b_n \sin \frac{n\pi x}{L} \right), \qquad -L < x < L. \qquad (6)$$

The modified Fourier series (6), with formulas (5) for the coefficients, shows that a function f defined on any interval of *finite* length may be expanded in a trigonometric series of the type we have been discussing. Hence such functions as e^x, $1/(1 + x^2)$, etc., have Fourier expansions on intervals of any finite size. However, we recall that there are some power series such as the one for e^x which converge for $-\infty < x < \infty$. The corresponding result does not exist for Fourier series. There are no Fourier series which are valid on an infinite interval if the function is nonperiodic. The situation is not completely hopeless because an extension of the idea of Fourier series can be employed. This special representation, called the Fourier Integral, exhibits many properties for nonperiodic functions which Fourier series exhibit for periodic ones. Furthermore, the Fourier Integral is a useful tool in both theoretical and applied investigations of differential equations.

Theorem 3, for the expansions of even and odd functions, is applicable to modified Fourier series on an interval $(0, L)$. For example, a function f defined on $(0, L)$ has the sine series

$$f(x) = \sum_{n=1}^{\infty} b_n \sin \frac{n\pi x}{L}, \qquad b_n = \frac{2}{L} \int_{0}^{L} f(x) \sin \frac{n\pi x}{L} \, dx.$$

Half-range expansions may be carried out on any interval of length L, either by making a change of variable or by forming the periodic extension and computing all the formulas for the interval $(0, L)$.

EXAMPLE 2. Given

$$f(x) = \begin{cases} x + 1 & \text{for } -1 < x < 0, \\ x - 1 & \text{for } 0 < x < 1, \end{cases}$$

expand f into a Fourier series on $(-1, 1)$ and draw the graph of its periodic normalized extension on $[-3, 3]$.

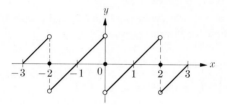

FIGURE 10–13

Solution. The graph of the periodic extension f_0 is drawn in Fig. 10–13. The function f_0 is odd, so all $a_n = 0$. Then

$$b_n = 2\int_0^1 (x - 1) \sin n\pi x \, dx = 2\left[-\frac{(x - 1)\cos n\pi x}{n\pi} + \frac{\sin n\pi x}{n^2\pi^2} \right]_0^1 = -\frac{2}{n\pi}.$$

We obtain

$$f(x) = -\frac{2}{\pi}\sum_{n=1}^{\infty} \frac{\sin n\pi x}{n}, \qquad -1 < x < 1.$$

EXERCISES

In problems 1 through 11, expand each function as indicated. Draw the appropriate extension f_0 on an interval of length three full periods.

1. $f(x) = \begin{cases} 1 & \text{for } 0 < x < \pi, \\ -1 & \text{for } \pi < x < 2\pi, \end{cases}$ full series on $[0, 2\pi]$

2. $f(x) = \begin{cases} 1 & \text{for } 0 < x < \pi/2, \\ 0 & \text{for } \pi/2 < x < 2\pi, \end{cases}$ full series on $[0, 2\pi]$

3. $f(x) = \sin(x/2)$,   full series on $[0, 2\pi]$

4. $f(x) = x - \pi$,   full series on $[\pi, 3\pi]$

5. $f(x) = \begin{cases} 0 & \text{for } -2 < x < 0, \\ x & \text{for } 0 \le x < 2, \end{cases}$ full series on $[-2, 2]$

6. $f(x) = \sin x$,   full series on $[-\pi/2, \pi/2]$

7. $f(x) = 1 - |x|$,   full series on $[-1, 1]$

8. $f(x) = \begin{cases} 1 & \text{for } 0 < x < \pi, \\ 0 & \text{for } \pi < x < 2\pi, \end{cases}$ cosine series on $[0, 2\pi]$

9. $f(x) = x^2$ for $0 < x < 2$,   sine series on $[0, 2]$

10. $f(x) = x^2$ for $0 < x < 1$,   cosine series on $[0, 1]$

11. $f(x) = 1 - 2x$ for $0 < x < 1$,   sine series on $[0, 1]$

12. Denote the Fourier series of a function f by the symbol **F**, and that of a function g by the symbol **G**. Show that if c is a constant, the Fourier series of the function cf is $c\mathbf{F}$. If f and g are periodic with the same period, show that the Fourier series of $f + g$ is $\mathbf{F} + \mathbf{G}$. What collection of Fourier series forms a linear vector space?

13. By combining the series in problems 1 and 2, and taking into account the results of problem 12, find the Fourier series of

$$f(x) = \begin{cases} 2 & \text{for } 0 < x < \pi/2, \\ 1 & \text{for } \pi/2 < x < \pi, \\ -1 & \text{for } \pi < x < 2\pi, \end{cases} \qquad \text{full series on } [0, 2\pi].$$

14. Combine the results of problems 2 and 3, and take the results of problem 12 into account, to get an expansion for

$$f(x) = \begin{cases} 3 - 2 \sin (x/2) \text{ for } 0 < x < \pi/2, \\ - 2 \sin (x/2) \text{ for } \pi/2 < x < 2\pi, \end{cases} \qquad \text{full series on } [0, 2\pi].$$

4. ORTHOGONAL EXPANSIONS

The theory of Fourier series is part of a more general structure which we now investigate. When we select the collection of sines and cosines for the expansion of piecewise smooth, periodic functions, it is natural for the student to ask whether or not there are classes of functions other than sines and cosines which would do just as well. The answer is that there are. However, instead of selecting one class of functions after another and going through the procedure of Sections 1 through 3, we shall extract those properties which are common to all such expansions and describe them at once. It is on such occasions that the study of linear vector spaces is put to remarkably good use.

The student may easily verify that the collection of piecewise continuous, normalized functions on an interval $[a, b]$ form a vector space V. (See Chapter 7, page 308, for the statement of the axioms.) Furthermore we may introduce an *inner product* (see Chapter 7, Section 7) by defining for any f and g in V,

$$(f, g) = \int_a^b f(x)g(x)\, dx.$$

Defining the *norm* of a function f by

$$\|f\| = (f, f)^{1/2}, \tag{1}$$

we show that the requirements of a norm are satisfied for functions in V. First of all, if $f \equiv 0$, it is clear that $\|f\| = 0$. So suppose that $f(x) \not\equiv 0$. Then there is a value x_0 with $f(x_0) \neq 0$. If f is continuous at x_0, then $f(x) \neq 0$ in an interval containing x_0. If f has a jump at x_0, it follows from the normalization that either $f(x_0-) \neq 0$ or $f(x_0+) \neq 0$ (or both). Once again, $f(x) \neq 0$ in an interval. Since $f^2(x) \geq 0$ everywhere and $f^2(x) > 0$ at least over part of the interval $[a, b]$, we conclude that

$$\|f\|^2 = \int_a^b [f(x)]^2\, dx > 0.$$

Therefore the definition given by (1) satisfies the conditions for a norm. With the inner product and norm as defined, the vector space V of piecewise continuous, normalized functions on $[a, b]$ is a *Euclidean* space.

We select $[-\pi, \pi]$ for the interval $[a, b]$, and consider the functions

$$1, \quad \cos x, \quad \sin x, \quad \cos 2x, \quad \sin 2x, \quad \ldots, \quad \cos nx, \quad \sin nx, \quad \ldots, \tag{2}$$

which clearly are in V. Then the relationships

$$\int_{-\pi}^{\pi} \cos nx \, dx = \int_{-\pi}^{\pi} \sin nx \, dx = \int_{-\pi}^{\pi} \sin nx \cos mx \, dx = 0 \quad \text{for all } m, n,$$

$$\int_{-\pi}^{\pi} \cos mx \cos nx \, dx = \int_{-\pi}^{\pi} \sin mx \sin nx \, dx = 0 \quad \text{if } m \neq n,$$

are equivalent to the statement that the inner product of any two functions in (2) is zero. In vector space terminology, we say that if f and g are elements of (2), then f and g are *orthogonal*, or we say that the elements of (2) form an *orthogonal set*. (See Chapter 7, Section 7.)

EXAMPLE 1. Show that if $[0, \pi]$ is selected for $[a, b]$, and V is the space as described above, then the collection

$$1, \cos x, \cos 2x, \ldots, \cos nx, \ldots \tag{3}$$

is an orthogonal set on V.

Solution. We must show that

$$\int_0^{\pi} 1 \cdot \cos nx \, dx = \int_0^{\pi} \cos nx \cdot \cos mx \, dx = 0 \qquad \text{for all } m \neq n.$$

We have

$$\int_0^{\pi} \cos nx \, dx = \left[\frac{1}{n} \sin nx\right]_0^{\pi} = 0$$

and

$$\int_0^{\pi} \cos nx \cos mx \, dx = \frac{1}{2} \int_0^{\pi} [\cos (m + n)x + \cos (m - n)x] \, dx$$

$$= \left[\frac{1}{2(m + n)} \sin (m + n)x + \frac{1}{2(m - n)} \sin (m - n)x\right]_0^{\pi} = 0 \qquad \text{if } m \neq n.$$

In a similar way it may be shown that the collection

$$\sin x, \sin 2x, \ldots, \sin nx, \ldots \tag{4}$$

is an orthogonal set if $[0, \pi]$ is selected for $[a, b]$. (See Exercise 2.)

An orthogonal collection may be made orthonormal by dividing each element by its norm. The collection (2) becomes the orthonormal set

$$\frac{1}{\sqrt{2\pi}}, \quad \frac{\cos x}{\sqrt{\pi}}, \quad \frac{\sin x}{\sqrt{\pi}}, \ldots, \quad \frac{\cos nx}{\sqrt{\pi}}, \quad \frac{\sin nx}{\sqrt{\pi}}, \ldots.$$

The set in Example 1 leads to the orthonormal set

$$\frac{1}{\sqrt{\pi}}, \quad \sqrt{\frac{2}{\pi}} \cos x, \quad \ldots, \quad \sqrt{\frac{2}{\pi}} \cos nx, \quad \ldots$$

and, corresponding to (4), we obtain the orthonormal set

$$\sqrt{\frac{2}{\pi}} \sin x, \quad \sqrt{\frac{2}{\pi}} \sin 2x, \quad \ldots, \quad \sqrt{\frac{2}{\pi}} \sin nx, \quad \ldots .$$

Because the space V has (for any $[a, b]$) an infinite collection of orthogonal vectors, it is clear that the space is not finite dimensional.

As in the case of finite-dimensional Euclidean vector spaces, there are many ways of choosing orthogonal and orthonormal sets. The collections (3) and (4) each form orthogonal sets on $V[0, \pi]$. There are many examples of orthonormal sets other than the trigonometric classes mentioned above. One other set is the collection of Legendre polynomials, which is described in the next section. In the remainder of this section we shall restrict ourselves to a discussion of general orthogonal expansions.

For a finite-dimensional vector space V, we recall that a *basis* for V is a finite set of elements with the property that *every other element in V* is a linear combination of elements of the basis. The question we raise now is how to define a "basis" for a space V when it is not finite dimensional. In view of our experience with Fourier series, we are led to the generalization that a collection

$$\phi_1, \phi_2, \ldots, \phi_n, \ldots$$

of elements of V forms a basis for V if every function f in V can be expanded in a series of the form

$$f = \sum_{j=1}^{\infty} c_j \phi_j, \tag{5}$$

with the $\{c_j\}$ constants and, of course, with the series convergent. The crux of the matter comes in the meaning of the word "convergent." The function f has two interpretations: (1), the usual meaning of $f(x)$ as a function defined for each x on an interval $[a, b]$ and (2), an element or point in the space V. Convergence of the series (5) in the above discussion means convergence to an element in V and not necessarily convergence for any particular value of x which may be inserted in each ϕ_j and f. To be more precise, if we define the partial sum

$$s_n = \sum_{j=1}^{n} c_j \phi_j,$$

then convergence of the series (5) means

$$\lim_{n \to \infty} \|f - s_n\| = 0. \tag{6}$$

The convergence in (6) is called **mean-square convergence** and, taking into account the definition of norm, we may write

$$\lim_{n \to \infty} \int_a^b [f(x) - s_n(x)]^2 \, dx = 0. \tag{7}$$

It is clear that if $s_n(x)$ converges uniformly to f then, from the very definition of uniform convergence, we conclude that (7) holds. On the other hand, it can be shown that (7) may hold without $s_n(x)$ converging to $f(x)$ at *any* point x. (In this connection see the example below, and Exercise 4 at the end of this section.) Nevertheless, mean-square convergence is useful in many connections.

EXAMPLE 2. Given the sequence of functions $\{s_n(x)\}$ defined by

$$s_n(x) = \begin{cases} 0, & -1 \le x \le -\dfrac{1}{2^n}, \\[2mm] 2^n x + 1, & -\dfrac{1}{2^n} \le x \le 0, \\[2mm] -2^n x + 1, & 0 \le x \le \dfrac{1}{2^n}, \\[2mm] 0, & \dfrac{1}{2^n} \le x \le 1. \end{cases} \quad , n = 1, 2, \ldots$$

A graph of $s_n(x)$ is shown in Fig. 10–14. Let $f(x) \equiv 0$, $-1 \le x \le 1$. Show that $s_n(x) \to f(x)$ in the mean-square sense, that $s_n(x) \to f(x)$, $x \ne 0$, and that $s_n(0) \nrightarrow f(0) = 0$.

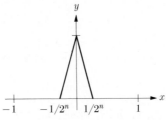

FIGURE 10–14

Solution. We have

$$\|f - s_n\|^2 = \int_a^b |0 - s_n(x)|^2 \, dx = \int_{-1/2^n}^{1/2^n} |s_n(x)|^2 \, dx = 2 \int_0^{1/2^n} |s_n(x)|^2 \, dx$$

$$= \int_0^{1/2^n} (1 - 2^n x)^2 \, dx = \left[x - 2^n x^2 + \frac{2^{2n} x^3}{3} \right]_0^{1/2^n} = \frac{1}{3 \cdot 2^n} \to 0 \quad \text{as} \quad n \to \infty.$$

Geometrically, we see that the areas of the terms $\{s_n^2(x)\}$ must tend to zero as $n \to \infty$.

If $x \ne 0$ then, for N sufficiently large, $s_n(x) = 0$ for all $n > N$. Therefore $s_n(x) \to f(x) = 0$. On the other hand, $s_n(0) = 1$ for all n, and so $\lim_{n \to \infty} s(0) = 1 \ne f(0) = 0$.

DEFINITION. An orthonormal set $\{\phi_j\}$ in a Euclidean vector space V is said to be **complete** if and only if every element f of V is the limit in the sense of mean-square convergence of a sequence $\{f_n\}$, with each f_n consisting of a finite linear combination of the ϕ_j. We use the term **complete orthonormal set** instead of "orthonormal basis" when working in a space with infinitely many orthogonal elements. Any space with infinitely many orthogonal elements is called an **infinite dimensional space.**

Subspaces of the space V of piecewise continuous, normalized functions on $[-\pi, \pi]$ may be formed in an endless variety of ways. All the linear combinations of a fixed finite number of sines and cosines form a finite-dimensional subspace. All finite linear combinations of the functions

$$\cos 2x, \cos 4x, \cos 6x, \ldots, \cos 2nx, \ldots$$

form an infinite-dimensional subspace of V. Another subspace, and one of the most interesting, is that formed by the totality of *finite* linear combinations of sines and cosines. This subspace, denoted W, is infinite dimensional but is not the whole of V. It bears the same relation to V that the rational numbers bear to the entire collection of real numbers.

The next theorem states, in a general form, the formula for the coefficients of a Fourier series. We recognize the result when we bear in mind the definitions of orthogonal set and inner product in a vector space, and then we apply them to the trigonometric functions.

Theorem 4. *Suppose $\{\phi_j\}$ is a finite or infinite orthonormal set, and suppose f is given by*

$$f = \sum_{j=1}^{\infty} c_j \phi_j,$$

with convergence of the series in the mean-square sense [as in (6)]. Then

$$c_p = (f, \phi_p).$$

Proof. Of course, if $\{\phi_j\}$ is finite, then taking the inner product (f, ϕ_p) yields the result at once. In general,

$$|(f, \phi_p) - (s_n, \phi_p)| = |(f - s_n, \phi_p)|,$$

and we apply the Schwarz Inequality to the term on the right. Therefore, since $\|\phi_p\| = 1$ for all p,

$$|(f, \phi_p) - (s_n, \phi_p)| \leq \|f - s_n\| \cdot \|\phi_p\| = \|f - s_n\|.$$

Now,

$$(s_n, \phi_p) = \sum_{j=1}^{n} c_j(\phi_j, \phi_p) = c_p \qquad \text{for all } n \geq p;$$

and so, if n is large enough, we have

$$|(f, \phi_p) - c_p| \leq \|f - s_n\|.$$

The left side is independent of n and the right side tends to zero by hypothesis. Therefore $c_p = (f, \phi_p)$.

The student should compare this theorem with Theorem 1. The proof above shows that the uniform convergence in Theorem 1 is not necessary to obtain the formulas for the Fourier coefficients.

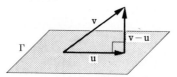

FIGURE 10–15

Returning for the moment to vectors in three-space, we consider a vector $\mathbf{v}$ and a plane Γ which does not contain $\mathbf{v}$. Draw a representative of $\mathbf{v}$ which has its base resting in the plane Γ as shown in Fig. 10–15. The projection of $\mathbf{v}$ on this plane is a vector $\mathbf{u}$, and the vector $\mathbf{v} - \mathbf{u}$ is clearly orthogonal to $\mathbf{u}$. Referring to the diagram, we obtain at once

$$(\mathbf{v} - \mathbf{u}) \cdot \mathbf{u} = 0 \qquad \text{and} \qquad |\mathbf{u}|^2 + |\mathbf{v} - \mathbf{u}|^2 = |\mathbf{v}|^2. \tag{8}$$

However, we may observe two more facts: (1) if $\mathbf{u}'$ is *any other* vector in Γ, then the vector $\mathbf{v} - \mathbf{u}'$ is longer than $\mathbf{v} - \mathbf{u}$ and (2), if $\mathbf{i}$ and $\mathbf{j}$ are unit orthogonal vectors in the plane Γ (if, for example, Γ is the xy-plane) then

$$\mathbf{u} = (\mathbf{v} \cdot \mathbf{i})\mathbf{i} + (\mathbf{v} \cdot \mathbf{j})\mathbf{j}. \tag{9}$$

The reader may verify (9) by the methods of Chapter 2.

The next theorem shows that the above discussion for vectors in three-space can be extended to more general vector spaces.

Theorem 5. *Suppose that $\phi_1, \phi_2, \ldots, \phi_n$ is a finite orthogonal set and W is the space spanned by the ϕ_j. Then for any f in V there is a g in W with the following properties:*

(i) $(f - g, g) = 0.$
(ii) $\|g\|^2 + \|f - g\|^2 = \|f\|^2.$
(iii) *If g' is any other element of W, then $\|f - g\|$ is smaller than $\|f - g'\|$.*
(iv) *The element g is given by*

$$g = \sum_{j=1}^{n} c_j \phi_j, \qquad \text{with } c_j = (f, \phi_j).$$

Remark. We recognize the extension of properties (8) and (9) in the statement of the theorem; the proof will also establish the result for vectors in three-space

Proof. Let $g' = \sum_{j=1}^{n} a_j\phi_j$ be any element in W. Then

$$\|f - g'\|^2 = (f - g', f - g') = (f, f) - 2(f, g') + (g', g')$$

$$= (f, f) - 2\sum_{j=1}^{n} (f, a_j\phi_j) + \left(\sum_{j=1}^{n} a_j\phi_j, \sum_{j=1}^{n} a_j\phi_j\right).$$

Since the ϕ_j are orthonormal and $c_j = (f, \phi_j)$ by definition, we get

$$\|f - g'\|^2 = (f, f) - 2\sum_{j=1}^{n} a_j c_j + \sum_{j=1}^{n} a_j^2$$

$$= (f, f) + \sum_{j=1}^{n} (a_j - c_j)^2 - \sum_{j=1}^{n} c_j^2. \tag{10}$$

This last expression has its minimum value when $a_j = c_j$, i.e., when $g' = g$. Thus (iii) and (iv) are established. Since $\|g\|^2 = \sum_{j=1}^{n} c_j^2$, we may set $a_j = c_j$ in (10) to get (ii) in the theorem. Now, since

$$(f - g, g) = (f, g) - (g, g) = \left(f, \sum_{j=1}^{n} c_j\phi_j\right) - \sum_{j=1}^{n} c_j^2$$

$$= \sum_{j=1}^{n} c_j^2 - \sum_{j=1}^{n} c_j^2 = 0,$$

we see that (i) holds.

DEFINITION. The function g determined in Theorem 5 is called the **best mean-square approximation** to f in the space W. In terms of the norm in V, g is that element of W which is "closest" to f.

EXAMPLE 3. Let $f = x + x^2$ be an element in V on $[-\pi, \pi]$. Let W be the subspace spanned by the orthonormal elements

$$\phi_1 = \frac{1}{\sqrt{\pi}} \cos x, \qquad \phi_2 = \frac{1}{\sqrt{\pi}} \sin 2x, \qquad \phi_3 = \frac{1}{\sqrt{\pi}} \cos 3x.$$

Find the best mean-square approximation to f in W.

Solution. The required element is

$$g = c_1 \frac{1}{\sqrt{\pi}} \cos x + c_2 \frac{1}{\sqrt{\pi}} \sin 2x + c_3 \frac{1}{\sqrt{\pi}} \cos 3x,$$

with

$$c_1 = \frac{1}{\sqrt{\pi}} \int_{-\pi}^{\pi} (x + x^2) \cos x \, dx,$$

$$c_2 = \frac{1}{\sqrt{\pi}} \int_{-\pi}^{\pi} (x + x^2) \sin 2x \, dx, \qquad c_3 = \frac{1}{\sqrt{\pi}} \int_{-\pi}^{\pi} (x + x^2) \cos 3x \, dx.$$

A computation yields $c_1 = -4\sqrt{\pi}$, $c_2 = -\sqrt{\pi}$, $c_3 = -\frac{4}{9}\sqrt{\pi}$. The result is

$$g = -4\cos x - \sin 2x - \tfrac{4}{9}\cos 3x.$$

DEFINITION. *Let* $\phi_1, \phi_2, \ldots, \phi_n, \ldots$ *be an infinite set of orthonormal elements in* V. *For each* n, *we define* **the subspace** W_n *as the space spanned by* $\phi_1, \phi_2, \ldots, \phi_n$.

Corollary. *Let* $\phi_1, \phi_2, \ldots, \phi_n, \ldots$ *be an infinite orthonormal set, and let* f *be an element of* V *with Fourier coefficients:* $c_j = (f, \phi_j)$, $j = 1, 2, \ldots$. *If* s_n *is the* nth *partial sum* $s_n = \sum_{j=1}^{n} c_j\phi_j$, *then*

(i) *for each* n, s_n *is the best mean-square approximation to* f *in the space* W_n;
(ii) *for each* n, *we have the inequality*

$$\sum_{j=1}^{n} c_j^2 \leq \|f\|^2; \qquad \textit{Bessel's} \qquad (11)$$

(iii) *the series* $\sum_{j=1}^{\infty} c_j^2$ *converges;*
(iv) $\lim_{n\to\infty} c_n = 0$.

Proof. (i) follows, since s_n is the element g in Theorem 5. As for (ii), we have

$$\sum_{j=1}^{n} c_j^2 + \|f - s_n\|^2 = \|f\|^2,$$

and the inequality follows by discarding $\|f - s_n\|^2$. Since $\|f\|$ is independent of n, we may let $n \to \infty$ in (11), and (iii) is the result. Since the nth term of a convergent series must tend to zero, we conclude that *the Fourier coefficients of any square integrable function must tend to zero;* i.e., (iv) holds.

In the case of Fourier series of trigonometric functions, we obtain **Bessel's Inequalities:**

$$\pi\left[\frac{a_0^2}{2} + \sum_{n=1}^{\infty} (a_n^2 + b_n^2)\right] \leq \int_{-\pi}^{\pi} [f(x)]^2 \, dx \qquad \text{for any } f \in V[-\pi, \pi]$$

and

$$\frac{\pi}{2}\left[\frac{a_0^2}{2} + \sum_{n=1}^{\infty} a_n^2\right] \leq \int_{0}^{\pi} [f(x)]^2 \, dx, \qquad \text{cosine series,}$$

$$\frac{\pi}{2}\left[\sum_{n=1}^{\infty} b_n^2\right] \leq \int_{0}^{\pi} [f(x)]^2 \, dx, \qquad \text{sine series.}$$

Furthermore, $\lim_{n\to\infty} a_n = \lim_{n\to\infty} b_n = 0$.

The next theorem is the principal result concerning general orthogonal expansions.

Theorem 6. *Suppose that* $\phi_1, \phi_2, \ldots, \phi_n, \ldots$ *is a complete orthonormal set, that* $f \in V$ *and that* $c_j = (f, \phi_j)$ *are the Fourier coefficients of* f. *Then the series*

$$\sum_{n=1}^{\infty} c_n \phi_n$$

converges to f *in the mean-square sense and*

$$\sum_{n=1}^{\infty} c_n^2 = \|f\|^2, \quad \textit{Parseval's Formula.}$$

Proof. Let W_n be defined as on page 457, and denote by W the subspace consisting of elements which are a linear combination of a finite number of the ϕ_j. Denoting $s_n = \sum_{j=1}^{n} c_j \phi_j$, we have (using Theorem 5)

$$\|s_n\|^2 = \sum_{j=1}^{n} c_j^2, \qquad \|f\|^2 = \|s_n\|^2 + \|f - s_n\|^2.$$

Since the set $\{\phi_j\}$ is complete, there is a sequence

$$f_1, f_2, \ldots, f_p, \ldots$$

of elements in W such that $\|f - f_p\| \to 0$ as $p \to \infty$. Since each f_p is a *finite* linear combination of the ϕ_j, each f_p must lie in some W_n space which we denote W_{n_p}. We observe that the W_n spaces are increasing in size; that is, W_r is part of W_s if $r < s$. We may assume that the spaces $W_{n_1}, W_{n_2}, \ldots, W_{n_p}, \ldots$ are increasing as $p \to \infty$.

Let $\epsilon > 0$ be given. Then there is an integer P such that

$$\|f - f_p\|^2 < \epsilon \qquad \text{if } p > P.$$

Since each s_n gives the *best* mean-square approximation to f in W_n, we conclude that

$$\|f - s_n\|^2 \le \|f - s_{n_{p+1}}\|^2 \le \|f - f_{p+1}\|^2 < \epsilon \qquad \text{for all } n > n_{p+1}.$$

Hence $\|f - s_n\|^2 \to 0$ and the theorem is proved.

Remark. The orthonormal set

$$\frac{1}{\sqrt{2\pi}}, \quad \frac{1}{\sqrt{\pi}} \cos nx, \quad \frac{1}{\sqrt{\pi}} \sin nx, \quad n = 1, 2, \ldots,$$

can be shown to be complete in the space V of piecewise smooth, normalized, periodic functions of period 2π. (See Appendix 1.)

EXERCISES

1. Verify that the collection of piecewise continuous, normalized functions on an interval $[a, b]$ form a vector space.

2. Show that the collection

$$\sin x, \sin 2x, \ldots, \sin nx, \ldots$$

is an orthogonal set on $[0, \pi]$.

3. Find an infinite collection of sines and cosines which are orthonormal on an interval $[a, b]$.

4. (a) Construct a set of functions $\{s_n(x)\}$ as in Example 2 with the property that $\|s_n - f\| \to 0$ on $[-1, 1]$, and such that $s_n(x)$ does not converge to $f(x)$ for *two* values of x.

(b) Repeat part (a), with $s_n(x)$ not converging to $f(x)$ for k values of x, where k is any positive integer.

*(c) By performing (b) for every integer k, show how a sequence $\{s_n\}$ may be obtained such that $\|s_n - f\| \to 0$ and $s_n(x) \nrightarrow f(x)$ at an infinite collection of values of x.

5. If $\mathbf{v}$ is a vector in three-space and $\mathbf{u}$ is the projection of $\mathbf{v}$ on the xz-plane, use the methods of Chapter 2 to show that $\mathbf{u} = (\mathbf{v} \cdot \mathbf{i})\mathbf{i} + (\mathbf{v} \cdot \mathbf{k})\mathbf{k}$.

6. Given $\mathbf{v} = 4\mathbf{i} + 2\mathbf{j} + 3\mathbf{k}$ and the plane Γ determined by linear combinations of $\mathbf{a}$ and $\mathbf{b}$, where

$$\mathbf{a} = \mathbf{i} - 2\mathbf{j} + \mathbf{k}, \qquad \mathbf{b} = 2\mathbf{i} - \mathbf{j} - 4\mathbf{k}.$$

Find the vector $\mathbf{u}$ in Γ which is "closest" to $\mathbf{v}$ in the mean-square sense.

In each of problems 7 through 10, a function f in $V[-\pi, \pi]$ and a subspace W spanned by the given orthonormal elements $\phi_1, \phi_2, \ldots, \phi_n$ are given. Find the element g in W which is the best approximation to f in the mean-square sense.

7. $f = 1 + 2x, \qquad \phi_1 = \dfrac{1}{\sqrt{2\pi}}, \qquad \phi_2 = \dfrac{\sin 3x}{\sqrt{\pi}}, \qquad \phi_3 = \dfrac{\cos 2x}{\sqrt{\pi}}$

8. $f(x) = \sin \tfrac{1}{2}x, \qquad \phi_1 = \dfrac{1}{\sqrt{2\pi}}, \qquad \phi_2 = \dfrac{\sin x}{\sqrt{\pi}}$

9. $f(x) = e^x, \qquad \phi_1 = \dfrac{\sin 2x}{\sqrt{\pi}}, \qquad \phi_2 = \dfrac{\cos 4x}{\sqrt{\pi}}, \qquad \phi_3 = \dfrac{\sin 5x}{\sqrt{\pi}}$

10. $f(x) = \begin{cases} -1, & -\pi < x < 0, \\ 1, & 0 < x \le \pi/2, \\ 0, & \pi/2 < x < \pi, \end{cases} \qquad \begin{aligned} \phi_1 &= \dfrac{1}{\sqrt{2\pi}}, & \phi_2 &= \dfrac{\sin x}{\sqrt{\pi}}, \\[2mm] \phi_3 &= \dfrac{\sin 4x}{\sqrt{\pi}}, & \phi_4 &= \dfrac{\cos 5x}{\sqrt{\pi}} \end{aligned}$

In each of problems 11 through 14, assume that the trigonometric set corresponding to $[-\pi, \pi]$ is complete. Then from Theorem 6, it follows that

$$\int_{-\pi}^{\pi} [f(x)]^2 \, dx = \pi \left[\frac{a_0^2}{2} + \sum_{n=1}^{\infty} (a_n^2 + b_n^2) \right], \qquad \text{(Parseval's Formula)}.$$

11. Use the result of Example 1, Section 1, to deduce that

$$\sum_{n=1}^{\infty} \frac{1}{n^2} = \frac{\pi^2}{6}.$$

12. Use the result of Example 1, Section 2, to show that

$$\frac{1}{1^2} + \frac{1}{3^2} + \frac{1}{5^2} + \cdots + \frac{1}{(2n+1)^2} + \cdots = \frac{\pi^2}{8}.$$

13. Find the Fourier series on $[-\pi, \pi]$ for $f(x) = |x| - (\pi/2)$, and use the result to show that

$$\frac{1}{1^4} + \frac{1}{3^4} + \frac{1}{5^4} + \cdots = \frac{\pi^4}{96}.$$

14. Find the Fourier series for $x^2 - (\pi^2/3)$ on $[-\pi, \pi]$, and deduce that

$$\sum_{n=1}^{\infty} \frac{1}{n^4} = \frac{\pi^4}{90}.$$

The functions

$$\phi_1 = \frac{1}{\sqrt{2}}, \qquad \phi_2 = \sqrt{\frac{3}{2}}\, x, \qquad \phi_3 = \frac{\sqrt{10}}{4}(3x^2 - 1)$$

form an orthonormal set in the space $V[-1, 1]$. In each of problems 15 through 18, find the function g in the subspace spanned by ϕ_1, ϕ_2, ϕ_3 which best approximates the given function f in the mean-square sense.

15. $f(x)$ $\begin{cases} -1 & \text{for } -1 < x < 0 \\ 1 & \text{for } 0 < x < 1 \end{cases}$ 16. $f(x) = |x|, \quad -1 \le x \le 1$

17. $f(x) = \sin(\pi x/2), \quad -1 \le x \le 1$ 18. $f(x) = \cos(\pi x/2), \quad -1 \le x \le 1$

5. LEGENDRE POLYNOMIALS

Let V be the linear space of piecewise continuous functions on $[-1, 1]$. Since no polynomial vanishes everywhere on $[-1, 1]$ unless all its coefficients are zero, it follows that for each positive integer n the functions

$$1, x, x^2, \ldots, x^n$$

are linearly independent on $[-1, 1]$. By following the procedure of Chapter 7, Section 7, for transforming a linearly independent set into an orthonormal set, we can obtain an infinite orthonormal set of polynomials $g_0, g_1, g_2, \ldots, g_n, \ldots$ in V, with each g_n being of degree n.

Using the notation $f_n(x) = x^n$ for $n \ge 0$, we shall calculate the first few g_i. Since

$$\|f_0\|^2 = \int_{-1}^{1} f_0^2(x)\, dx = \int_{-1}^{1} 1\, dx = 2,$$

we define

$$g_0 = \frac{f_0}{\|f_0\|} = \frac{1}{\sqrt{2}}.$$

Proceeding to g_1, we define

$$g_1' = f_1 - (f_1, g_0)g_0.$$

Since $(f_1, g_0) = \int_{-1}^{1} x\, dx = 0$, we see that $g_1' = f_1 = x$. The norm of g_1' is given by

$$\|g_1'\|^2 = \int_{-1}^{1} x^2\, dx = \frac{2}{3}.$$

The element g_1 is

$$g_1 = \frac{g_1'}{\|g_1'\|} = \frac{x}{\sqrt{2/3}} = \sqrt{\frac{3}{2}}\, x.$$

Next we define

$$g_2' = f_2 - (f_2, g_0)g_0 - (f_2, g_1)g_1.$$

A simple calculation shows that

$$(f_2, g_0) = \int_{-1}^{1} x^2 \cdot \frac{1}{\sqrt{2}}\, dx = \frac{1}{3}\sqrt{2}$$

and

$$(f_2, g_1) = \sqrt{\frac{3}{2}} \int_{-1}^{1} x^3\, dx = 0.$$

Therefore $g_2' = x^2 - \frac{1}{3}$, $\|g_2'\|^2 = \int_{-1}^{1} (x^2 - \frac{1}{3})^2\, dx = \frac{8}{45}$, and so

$$g_2 = \frac{g_2'}{\|g_2'\|} = \frac{\sqrt{10}}{4}(3x^2 - 1).$$

We define the **Legendre polynomial** according to the formula of Rodrigues:

$$P_0(x) = 1, \qquad P_n(x) = \frac{1}{2^n n!}\frac{d^n}{dx^n}[(x^2 - 1)^n], \qquad n = 1, 2, \ldots.$$

Calculating the first few such polynomials, we obtain

$$P_1(x) = \frac{1}{2}\frac{d}{dx}(x^2 - 1) = x,$$

$$P_2(x) = \frac{1}{2^2 \cdot 2!}\frac{d^2}{dx^2}[(x^2 - 1)^2] = \frac{1}{2}(3x^2 - 1),$$

$$P_3(x) = \frac{1}{2^3 \cdot 3!}\frac{d^3}{dx^3}(x^6 - 3x^4 + 3x^2 - 1) = \frac{5}{2}x^3 - \frac{3}{2}x,$$

$$P_4(x) = \frac{1}{2^4 \cdot 4!}\frac{d^4}{dx^4}[(x^2 - 1)^4] = \frac{1}{8}(35x^4 - 30x^2 + 3).$$

(1)

From the formula of Rodrigues, we can prove the following theorem concerning these polynomials:

Theorem 7. (a) *For each $n \geq 1$, P_n is a polynomial of degree exactly n.*
(b) *The P_n form an orthogonal set.*

Proof. (a) follows since

$$P_n(x) = \frac{1}{2^n n!} \frac{d^n}{dx^n} (x^{2n} + \cdots)$$

$$= \frac{2n \cdot (2n - 1) \cdot \ldots \cdot (n + 1)}{2^n n!} x^n + \cdots .$$

To prove (b), we note first that

$$\int_{-1}^{1} P_0(x) P_n(x)\, dx = C_n \int_{-1}^{1} \frac{d^n}{dx^n} [(x^2 - 1)^n]\, dx$$

$$= \left[C_n \frac{d^{n-1}}{dx^{n-1}} (x^2 - 1)^n \right]_{-1}^{1} = 0, \qquad \text{where } C_n = \frac{1}{2^n n!},$$

since the quantity in the bracket is zero for $x = \pm 1$. Next, let us suppose that $1 \leq m < n$. Then, using the symbol D^k for d^k/dx^k,

$$\int_{-1}^{1} P_m(x) P_n(x)\, dx = C_m C_n \int_{-1}^{1} [D^m(x^2 - 1)^m][D^n(x^2 - 1)^n]\, dx.$$

We integrate by parts, differentiating the first term and integrating the second, to obtain

$$\int_{-1}^{1} P_m(x) P_n(x)\, dx = -C_m C_n \int_{-1}^{1} [D^{m+1}(x^2 - 1)^m][D^{n-1}(x^2 - 1)^n]\, dx$$

$$+ \{C_m C_n [D^m(x^2 - 1)^m][D^{n-1}(x^2 - 1)^n]\}_{-1}^{1}.$$

The second term vanishes as before. Continuing this process n times, we find that

$$\int_{-1}^{1} P_m(x) P_n(x)\, dx = (-1)^n C_m C_n \int_{-1}^{1} [D^{m+n}(x^2 - 1)^m](x^2 - 1)^n\, dx = 0,$$

since $m + n > 2m$ and $(x^2 - 1)^m$ is of degree only $2m$.

Remarks. From part (a), it follows that the space spanned by $P_0, \ldots, P_n$ is, for each n, the same as that spanned by $1, x, \ldots, x^n$. Accordingly, by induction, it follows that we must have

$$g_n = \frac{P_n}{\|P_n\|} \qquad \text{for each } n.$$

It can be shown that the collection of all g_n is complete.

EXERCISES

1. Find the function g_3 by setting $g_3' = f_3 - (f_3, g_0)g_0 - (f_3, g_1)g_1 - (f_3, g_2)g_2$ and then defining $g_3 = g_3'/\|g_3'\|$. Check the result by normalizing $P_3(x)$ as given in (1).

2. Find g_4 by continuing the procedure of problem 1.

3. By using oddness and evenness properties of the functions f_n on $[-1, 1]$, show that g_{2n} must contain only terms of even degree and g_{2n+1} must contain only terms of odd degree. [*Hint.* Use induction.]

4. Compute $P_5(x)$, using Rodrigues' Formula.

5. Using the *Recurrence Formula* for $P_n(x)$,

$$(n + 1)P_{n+1}(x) - (2n + 1)xP_n(x) + nP_{n-1}(x) = 0,$$

and the fact that $P_0(x) = 1$, $P_1(x) = x$, calculate $P_2(x)$, $P_3(x)$, $P_4(x)$, and $P_5(x)$. Check your result with the results using Rodrigues' Formula and with the result in problem 4.

6. Verify that the *Recurrence Formula* for $P_n(x)$,

$$\frac{d}{dx} P_{n+1}(x) - x \frac{d}{dx} P_n(x) = (n + 1)P_n(x),$$

is satisfied for $n = 2$ and 3.

7. For $n = 2$, 3, and 4 verify that $P_n(x)$ satisfies the *Legendre differential equation*

$$(1 - x^2)\frac{d^2P_n}{dx^2} - 2x\frac{dP_n}{dx} + n(n + 1)P_n = 0.$$

8. For any n, each Legendre polynomial $P_n(x)$ is given by *Laplace's Integral*,

$$P_n(x) = \frac{1}{\pi} \int_0^\pi [x + \sqrt{x^2 - 1} \cos \phi]^n \, d\phi.$$

Verify this formula for $n = 2$ and 3. Note that x is kept constant in performing the integration.

9. On the interval $[0, 1]$, let $f_0 = 1$, $f_1 = x$, and $f_2 = x^2$. Find the orthonormal set g_0, g_1, g_2 by following the procedure described in this section.

10. Continue the work as in problem 9, by letting $f_3 = x^3$ and finding g_3.

11. Proceed as in problem 9, with the interval $[0, \pi]$, given that $f_0 = 1$, $f_1 = \cos x$, $f_2 = \sin x$.

6. CONVERGENCE THEOREMS. DIFFERENTIATION AND INTEGRATION OF FOURIER SERIES

We return to the study of Fourier series of trigonometric functions with the principal intention of establishing Theorem 2 on the convergence of a Fourier series to the function it represents. Before repeating the statement of the theorem and deriving the main result, we shall prove three simple lemmas.

Lemma 1. *For every positive integer n and for all x we have the identity*

$$\frac{\sin (n + \frac{1}{2})x}{2 \sin \frac{1}{2}x} = \frac{1}{2} + \sum_{k=1}^{n} \cos kx, \qquad x \neq 0.$$

Proof. This is the type of trigonometric identity which occurs frequently in elementary trigonometry courses. It is sufficient to show that

$$\sin (n + \tfrac{1}{2})x = \sin \tfrac{1}{2}x + \sum_{k=1}^{n} 2 \sin \tfrac{1}{2}x \cos kx.$$

We employ the formula

$$2 \cos A \sin B = \sin (A + B) - \sin (A - B),$$

with $A = kx$, $B = \frac{1}{2}x$, to get

$$\sin (n + \tfrac{1}{2})x = \sin \tfrac{1}{2}x + \sum_{k=1}^{n} [\sin (k + \tfrac{1}{2})x - \sin (k - \tfrac{1}{2})x].$$

All the terms on the right "telescope" except for the last, yielding

$$\sin (n + \tfrac{1}{2})x = \sin (n + \tfrac{1}{2})x.$$

Since each step is reversible, the identity is established.

DEFINITION. We call the quantity

$$D_n(x) = \frac{\sin (n + \frac{1}{2})x}{2 \sin \frac{1}{2}x} \equiv \frac{1}{2} + \sum_{k=1}^{n} \cos kx$$

the **Dirichlet kernel.** It is an expression which occurs frequently in the study of Fourier series and plays a central role in many proofs. Three properties of $D_n(x)$ which we shall use are the following:

(i) $D_n(x)$ is an *even* function of x. This fact is readily seen when we note that $\cos kx$ is even for every k;

(ii)   for every n, we have

$$\frac{1}{\pi} \int_0^{\pi} D_n(x) \, dx = \frac{1}{2}. \tag{1}$$

This result is obtained directly by integrating $\frac{1}{2} + \sum_{k=1}^{n} \cos kx$;

(iii) $D_n(x)$ is periodic with period 2π. This result is evident from the formula defining $D_n(x)$.

Lemma 2. *For all x such that $0 < x \leq \pi/2$, we have*

$$1 < \frac{x}{\sin x} \leq \frac{\pi}{2}, \qquad \textit{Jordan's Inequality.}$$

Proof. We note that $h(x) = x/\sin x$ tends to 1 when $x \to 0$ and has the value $\pi/2$ when $x = \pi/2$. If we can show that h is an increasing function in this interval, the result is established (Fig. 10–16). It suffices to show that $h'(x) \geq 0$ on $(0, \pi/2)$. We leave this verification for the student. (See Exercise 1.)

Let f be a piecewise smooth, periodic function with period 2π, and form the Fourier series

$$\frac{a_0}{2} + \sum_{k=1}^{\infty} (a_k \cos kx + b_k \sin kx), \qquad (2)$$

where

$$a_k = \frac{1}{\pi} \int_{-\pi}^{\pi} f(t) \cos kt \, dt,$$

$$b_k = \frac{1}{\pi} \int_{-\pi}^{\pi} f(t) \sin kt \, dt. \qquad (3)$$

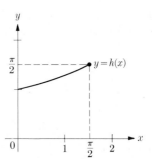

FIGURE 10–16

The partial sum $s_n(x)$ is the sum of the first n terms in (2).

The next lemma establishes a basic formula for $s_n(x)$.

Lemma 3. If $s_n(x)$ is the nth partial sum of the Fourier series of f, and if $D_n(x)$ is the Dirichlet kernel, then

$$s_n(x) - f(x) = \frac{1}{\pi} \int_{0}^{\pi} [f(x+u) - f(x+)] D_n(u) \, du$$

$$+ \frac{1}{\pi} \int_{0}^{\pi} [f(x-u) - f(x-)] D_n(u) \, du. \qquad (4)$$

Proof. Since

$$s_n(x) = \frac{1}{2} + \sum_{k=1}^{n} (a_k \cos kx + b_k \sin kx),$$

we may insert the formulas for a_k and b_k from (3). We obtain

$$s_n(x) = \frac{1}{\pi} \int_{-\pi}^{\pi} f(t) \left[\frac{1}{2} + \sum_{k=1}^{n} (\cos kt \cos kx + \sin kt \sin kx) \right] dt.$$

From trigonometry we know that

$$\cos (kt - kx) = \cos kt \cos kx + \sin kt \sin kx,$$

and so

$$s_n(x) = \frac{1}{\pi} \int_{-\pi}^{\pi} f(t) \left[\frac{1}{2} + \sum_{k=1}^{n} \cos k(t - x) \right] dt. \qquad (5)$$

Now, in (5), we hold x fixed and change the variable of integration by setting $t = x + u$. Then (5) becomes

$$s_n(x) = \frac{1}{\pi} \int_{-\pi-x}^{\pi-x} f(x + u) \left[\frac{1}{2} + \sum_{k=1}^{n} \cos ku \right] du.$$

Since both f and the quantity in the bracket are periodic with period 2π, we may adjust the interval of integration to $[-\pi, \pi]$. We also recognize the Dirichlet kernel in the bracket. Therefore

$$s_n(x) = \frac{1}{\pi} \int_{-\pi}^{\pi} f(x + u) \, D_n(u) \, du$$

$$= \frac{1}{\pi} \left[\int_{-\pi}^{0} f(x + v) \, D_n(v) \, dv + \int_{0}^{\pi} f(x + u) \, D_n(u) \, du \right].$$

Since $D_n(v)$ is even, we may replace v by $-u$ in the first integral on the right above and get

$$s_n(x) = \frac{1}{\pi} \int_{0}^{\pi} [f(x + u) + f(x - u)] \, D_n(u) \, du. \tag{6}$$

We have almost finished. We write $f(x)$ in a tricky form (a simple device used frequently in Fourier series),

$$f(x) = 2 \cdot \frac{1}{\pi} \int_{0}^{\pi} f(x) \, D_n(u) \, du,$$

which we can do because of (1). Note that $f(x)$ is constant so far as the integration is concerned. Since $f(x) = \frac{1}{2}[f(x +) + f(x -)]$, we have

$$f(x) = \frac{1}{\pi} \int_{0}^{\pi} [f(x+) + f(x-)] \, D_n(u) \, du.$$

Subtracting this expression from (6), we obtain (4) precisely.

We are ready to prove the convergence theorem of Section 1, which we restate.

Theorem 2. *Suppose that f is piecewise smooth and normalized on any finite interval and that f is periodic with period 2π. Then its Fourier series converges to $f(x)$ for each x. Furthermore, if f is smooth on the closed interval $[c, d]$, then the convergence is uniform on $[c, d]$.*

Proof. To show that the Fourier series converges at a fixed value of x, we shall prove that each integral on the right in (4) tends to zero as n tends to infinity. Writing

$$T_n(x) = \frac{1}{\pi} \int_{0}^{\pi} [f(x - u) - f(x-)] \, D_n(u) \, du,$$

we employ Lemma 1 to obtain

$$T_n(x) = \frac{1}{\pi} \int_{0}^{\pi} \frac{f(x - u) - f(x-)}{2 \sin \frac{1}{2}u} \sin \left(n + \frac{1}{2} \right) u \, du$$

$$= \frac{1}{\pi} \int_{0}^{\pi} \frac{f(x - u) - f(x-)}{2 \sin \frac{1}{2}u} \left[\sin nu \cos \frac{1}{2} u + \cos nu \sin \frac{1}{2} u \right] du.$$

We define

$$\phi(x, u) = \frac{f(x - u) - f(x-)}{2 \sin \frac{1}{2}u}, \qquad \phi_1(x, u) = \phi(x, u) \sin \frac{1}{2} u,$$

$$\phi_2(x, u) = \phi(x, u) \cos \frac{1}{2}u.$$

The functions ϕ, ϕ_1, and ϕ_2 are all piecewise smooth whenever f is, except possibly at $u = 0$. Applying L'Hôpital's Rule to ϕ at $u = 0$, we find that

$$\phi(x, 0 +) = -f'(x -)$$

and conclude that ϕ, ϕ_1, and ϕ_2 are piecewise continuous everywhere. Therefore we find that

$$T_n(x) = \frac{1}{\pi} \int_0^\pi [\phi_1(x, u) \cos nu + \phi_2(x, u) \sin nu] \, du.$$

For *fixed* x this expression for $T_n(x)$ is the sum of the nth Fourier cosine and sine coefficient for $\frac{1}{2}\phi_1$ and $\frac{1}{2}\phi_2$, respectively. The conditions for the Bessel Inequality (Section 4, page 457) apply, and we conclude that $T_n(x) \to 0$ as $n \to \infty$. The proof for the first integral in (4) is identical.

To establish uniform convergence, we again consider $T_n(x)$, the discussion for the first integral in (4) being the same. We first assume that f is smooth throughout $[-\pi, \pi]$. Writing the abbreviation $\phi(u)$ instead of $\phi(x, u)$, we have

$$T_n(x) = \int_0^{h/n} \phi(u) \sin \left(n + \frac{1}{2} \right) u \, du + \int_{h/n}^\pi \phi(u) \sin \left(n + \frac{1}{2} \right) u \, du, \qquad (7)$$

where h is any *fixed* positive number such that $0 < h < \pi$. Since f and f' are bounded on $[-\pi, \pi]$, we write

$$|f| \le M, \qquad |f'| \le L.$$

We apply the Theorem of the Mean and Jordan's Inequality to the expression

$$\phi(u) = \frac{f(x - u) - f(x)}{u} \cdot \frac{u/2}{\sin (u/2)}$$

and get

$$|\phi(u)| \le L \cdot \frac{\pi}{2} \cdot$$

Furthermore,

$$\phi'(u) = -\frac{f'(x - u)}{u} \frac{u}{2 \sin (u/2)} - \left[\frac{u}{2 \sin (u/2)} \right]^2 \frac{f(x - u) - f(x)}{u^2},$$

and thus it is easy to obtain the inequality

$$|\phi'(u)| \le \frac{\pi}{2} \cdot \frac{L}{u} + \frac{\pi^2}{4} \cdot \frac{L}{u} \cdot$$

We estimate the first integral in Eq. (7) directly, and we integrate by parts in the second. We find

$$T_n(x) = \int_0^{h/n} \phi(u) \sin\left(n + \frac{1}{2}\right) u \, du + \left[-\frac{\phi(u) \cos\left(n + \frac{1}{2}\right)u}{n + \frac{1}{2}}\right]_{h/n}^{\pi}$$

$$+ \frac{1}{n + \frac{1}{2}} \int_{h/n}^{\pi} \phi'(u) \cos\left(n + \frac{1}{2}\right) u \, du,$$

and

$$|T_n(x)| \le \frac{Mh}{n} + \frac{2M}{n + \frac{1}{2}} + \frac{1}{n + \frac{1}{2}} \left(\frac{\pi}{2} + \frac{\pi^2}{4}\right) L \int_{h/n}^{\pi} \frac{du}{u}.$$

Therefore, since h is fixed, we obtain

$$|T_n(x)| \le \frac{M(2 + h)}{n} + \frac{\pi L}{4n}(2 + \pi)[\ln \pi - \ln h + \ln n].$$

According to l'Hôpital's Rule,

$$\frac{\ln n}{n} \to 0 \quad \text{as} \quad n \to \infty,$$

and so $|T_n(x)| \to 0$ as $n \to \infty$, independently of x. That is, the convergence is uniform.

If f is piecewise smooth in $[-\pi, \pi]$, we wish to establish uniform convergence on an interval $[c, d]$ where, for some $h > 0$,

$$X_{i-1} + h \le c < d \le X_i - h. \tag{8}$$

The quantities $X_0, X_1, \ldots, X_m$ are those values of x at which jumps in f occur. The technique we employ is similar to that used in the above proof for smooth f. Since we cannot integrate by parts directly for piecewise smooth functions, we subdivide the integral for $T_n(x)$ into a finite number of integrals extending from one jump of the integrand to the next. In each of these integrals an integration by parts is possible.

Selecting h to be the quantity in (8), we can obtain the bounds

$$|\phi(u)| \le \frac{\pi L}{2} \quad \text{and} \quad |\phi'(u)| \le \frac{1}{u}\left(\frac{\pi}{2} + \frac{\pi^2}{4}\right) L \qquad \text{for } 0 \le u \le h,$$

$$|\phi(u)| \le \frac{\pi L}{h} \quad \text{and} \quad |\phi'(u)| \le \frac{1}{h}\left(\frac{\pi}{2} + \frac{\pi^2}{4}\right) L \qquad \text{for } h \le u \le \pi. \tag{9}$$

Since the integrand in $T_n(x)$ depends on x, the jumps of this integrand, labeled $a_1, a_2, \ldots, a_m = \pi$, will also depend on x. We now write (noting by (8) that $a_1 \ge h$)

$$T_n(x) = \int_0^{h/n} \phi(u) \sin\left(n + \frac{1}{2}\right) u \, du + \int_{h/n}^{a_1} \phi(u) \sin\left(n + \frac{1}{2}\right) u \, du$$

$$+ \sum_{j=2}^{m} \int_{a_{j-1}}^{a_j} \phi(u) \sin\left(n + \frac{1}{2}\right) u \, du. \tag{10}$$

Integrating by parts in all the integrals in (10) except the first and taking into account the estimates (9), we conclude that $T_n(x) \to 0$ uniformly on $[c, d]$.

If a function f is given by a power series with a positive radius of convergence, we saw (in Section 3 of Chapter 9) that f possesses derivatives of all orders. These derivatives are obtained by differentiating the series for f the appropriate number of times, and the various derived series will have the same radius of convergence that f does. Moreover, if $F'(x) = f(x)$, then F is represented by integrating the series for f term by term. The situation is completely different in the case of Fourier series. We showed that any piecewise smooth function is representable by its Fourier series. Therefore it is intuitively clear that such a series cannot be differentiated at will. While it is true that if f is piecewise smooth, then f' is piecewise continuous, we observe that if f is not continuous everywhere, then it is not true necessarily that

$$f(\beta) - f(\alpha) = \int_\alpha^\beta f'(x) \, dx.$$

In general, if f is piecewise smooth but not continuous, the series obtained by term-by-term differentiation of the Fourier series for f does not converge to f'; in fact, it ordinarily does not converge at all.

An example to illustrate this point is given by

$$f(x) = \begin{cases} -1 & \text{for} \quad -\pi < x < 0, \\ 1 & \text{for} \quad 0 < x < \pi, \end{cases}$$

with the Fourier series

$$f(x) = \frac{4}{\pi}\left(\sin x - \frac{\sin 3x}{3} + \frac{\sin 5x}{5} - \cdots\right).$$

The derivative series is

$$\frac{4}{\pi}(\cos x - \cos 3x + \cos 5x - \cos 7x + \cdots),$$

which can be shown to diverge for every x.

Since integration is a "smoothing" process while differentiation is a "scrambling" process, we can expect the behavior for term-by-term integration of Fourier series to be quite different from that for term-by-term differentiation. The main results in this direction are established in the following lemma and in the next two theorems.

Lemma 4. Suppose that f is periodic with period 2π and piecewise smooth on every finite interval. Then its Fourier coefficients satisfy the bounds

$$|a_n| \le \frac{C}{n}, \qquad |b_n| \le \frac{C}{n}, \qquad n = 1, 2, \ldots,$$

where C is a constant which depends on f but not on n.

Proof. If the jumps of f occur at $X_1, X_2, \ldots, X_{L-1}$, we have

$$a_n = \frac{1}{\pi} \int_{-\pi}^{\pi} f(t) \cos nt \, dt = \frac{1}{\pi} \sum_{i=1}^{L} \int_{X_{i-1}}^{X_i} f(t) \cos nt \, dt,$$

where we set $-\pi = X_0$ and $\pi = X_L$. We integrate by parts in each of the integrals and use the fact that $|f(x)| \leq M$, $|f'(x)| \leq N$ for $[-\pi, \pi]$. The details are left to the reader. (See Exercise 9.)

Theorem 8. *Suppose that F is continuous for all x and is periodic with period 2π, and suppose that F' is piecewise smooth on any finite interval. Then*

(i)   *the series obtained by differentiating the Fourier series for F converges at each point x to $\frac{1}{2}[F'(x +) + F'(x -)]$;*

(ii) *the Fourier series for F converges uniformly to $F(x)$ for all x;*

(iii) *the Fourier coefficients A_n, B_n of F satisfy the inequalities*

$$|A_n| \leq \frac{C}{n^2}, \qquad |B_n| \leq \frac{C}{n^2}, \qquad n = 1, 2, \ldots, \tag{11}$$

where C is a constant which depends on f but not on n.

Proof. Let the jumps of $F'(x)$ occur at $X_0, X_1, \ldots, X_L$. If we define

$$G(x) = \int_{-\pi}^{x} F'(t) \, dt,$$

we see that $G(x)$ is continuous and that $G'(x) - F'(x) = 0$ on each (X_{i-1}, X_i). By the Theorem of the Mean, the function $G(x) - F(x)$ is constant on each (X_{i-1}, X_i), and therefore for all x, since G and F are both continuous. Denote the Fourier coefficients of F' by a_n, b_n. We have

$$A_n = \frac{1}{\pi} \int_{-\pi}^{\pi} F(x) \cos nx \, dx = \frac{1}{\pi} \sum_{i=1}^{L} \int_{X_{i-1}}^{X_i} F(x) \cos nx \, dx$$

and, upon integration by parts,

$$A_n = \frac{1}{\pi} \sum_{i=1}^{k} \frac{F(X_i) \sin nX_i - F(X_{i-1}) \sin nX_{i-1}}{n} + \frac{1}{n} \int_{X_{i-1}}^{X_i} F'(x) \sin nx \, dx.$$

The terms in the first sum "telescope" except for the first and last ones, and the periodicity of F and $\sin nx$ cancel these two. Therefore

$$A_n = \frac{1}{n\pi} \int_{-\pi}^{\pi} F'(x) \sin nx \, dx = \frac{b_n}{n}. \tag{12}$$

Similarly,

$$B_n = -\frac{a_n}{n}. \tag{13}$$

Since F' is piecewise smooth, its series converges to $F'(x)$ for each x (Theorem 2). The bounds (11), the uniform convergence, and the fact that the series for F' is the derivative of that for F follow from (12) and (13) when the general theorems on infinite series are taken into account.

Theorem 9. *Suppose that f is periodic with period 2π and is piecewise smooth on any finite interval. Suppose that*

$$\pi a_0 = \int_{-\pi}^{\pi} f(x)\,dx = 0,$$

and define

$$F(x) = \int_{-\pi}^{x} f(t)\,dt.$$

Then

(i) *the Fourier series for F is obtained by integrating that for f term by term, except for the constant term A_0 which is given by*

$$A_0 = -\frac{1}{\pi}\int_{-\pi}^{\pi} xf(x)\,dx;$$

(ii) *F and its Fourier coefficients satisfy the conditions of Theorem 8.*

Proof. If F is to be periodic, the condition $a_0 = 0$ is required. To find A_0, we have (Fig. 10–17)

$$A_0 = \frac{1}{\pi}\int_{-\pi}^{\pi}\int_{-\pi}^{x} f(t)\,dt\,dx = \frac{1}{\pi}\iint f(t)\,dA_{xt}$$

$$= \frac{1}{\pi}\int_{-\pi}^{\pi} f(t)\left[\int_{t}^{\pi} dx\right]dt = \frac{1}{\pi}\int_{-\pi}^{\pi}(\pi - t)f(t)\,dt$$

$$= -\frac{1}{\pi}\int_{-\pi}^{\pi} tf(t)\,dt,$$

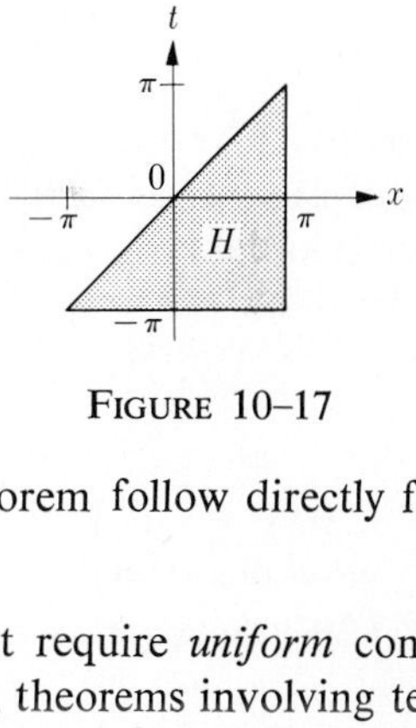

FIGURE 10–17

since $\int_{-\pi}^{\pi} f(t)\,dt = 0$. All other statements in the theorem follow directly from Theorem 8.

Remark. It is interesting that Theorem 9 does not require *uniform* convergence of the derivative series $F'(x) = f(x)$. In general, theorems involving term-by-term integration demand fewer hypotheses than those involving term-by-term differentiation.

———————

EXAMPLE. Use the result of Example 1 in Section 1, which establishes the expansion

$$f(x) = x = 2\sum_{n=1}^{\infty}\frac{(-1)^{n-1}\sin nx}{n}, \qquad -\pi < x < \pi,$$

to obtain the Fourier series for $F(x) = x^2$ on $[-\pi, \pi]$.

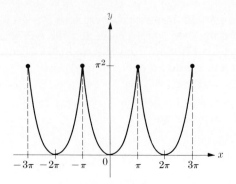

FIGURE 10–18

Solution. The periodic extension F_0 of F satisfies the hypotheses of Theorem 8, and we have

$$F_0'(x) = 4 \sum_{n=1}^{\infty} \frac{(-1)^{n-1} \sin nx}{n}.$$

Therefore

$$F_0(x) = \frac{A_0}{2} + 4 \sum_{n=1}^{\infty} \frac{(-1)^n \cos nx}{n^2},$$

with

$$A_0 = \frac{1}{\pi} \int_{-\pi}^{\pi} F(x)\,dx = \frac{2\pi^2}{3}.$$

Remark. The graph of F_0 is shown in Fig. 10–18. We see that F_0 is continuous but has a corner at each of the points $(\pi \pm 2n\pi,\, \pi^2)$. These corners correspond to the jumps in the periodic extension f_0 of f, the graph of which is shown in Fig. 10–4.

EXERCISES

1. If $h(x) = x/(\sin x)$, show that $h'(x) \geq 0$ for $0 < x \leq \pi$, thereby completing the proof of Lemma 2.

2. Using the results of the example above, find the series for f, where

$$f(x) = \frac{\pi^2 x - x^3}{3} \qquad \text{on } [-\pi, \pi],$$

and show that

$$\sum_{n=1}^{\infty} \frac{1}{n^6} = \frac{\pi^6}{945}.$$

Draw the graph of f_0.

$$\left[\textit{Answer:} \quad f(x) = 4 \sum_{n=1}^{\infty} \frac{(-1)^{n-1} \sin nx}{n^3}. \right]$$

3. Using Theorem 9, find the series for $f(x) = |x|$.

$$\left[\text{Answer:} \quad |x| = \frac{\pi}{2} - \frac{4}{\pi}\left(\frac{\cos x}{1^2} + \frac{\cos 3x}{3^2} + \frac{\cos 5x}{5^2} + \cdots \right). \right]$$

4. Using the method and result of problem 3, find the Fourier series for f if

$$f(x) = \begin{cases} \dfrac{x^2 - \pi x}{2} & \text{for} \quad 0 \leq x \leq \pi, \\[2mm] \dfrac{-x^2 - \pi x}{2} & \text{for} \quad -\pi \leq x \leq 0. \end{cases}$$

Show that

$$\frac{1}{1^6} + \frac{1}{3^6} + \frac{1}{5^6} + \cdots = \frac{\pi^6}{960}.$$

Draw the graph of f_0.

$$\left[\text{Answer:} \quad f(x) = -\frac{4}{\pi}\left(\frac{\sin x}{1^3} + \frac{\sin 3x}{3^3} + \frac{\sin 5x}{5^3} + \cdots \right). \right]$$

5. Using the result of problem 2, find the series for

$$f(x) = \frac{(\pi^2 - x^2)^2}{12}.$$

Draw the graph of f_0.

$$\left[\text{Answer:} \quad f(x) = \frac{2\pi^4}{45} + 4\sum_{n=1}^{\infty} \frac{(-1)^{n-1}}{n^4}\cos nx. \right]$$

6. Find the Fourier series for f and F, given

$$f(x) = |\sin x|, \qquad -\pi \leq x \leq \pi,$$

$$F(x) = \begin{cases} -(1 - \cos x) - \dfrac{2x}{\pi} & \text{for} \quad -\pi \leq x \leq 0, \\[2mm] (1 - \cos x) - \dfrac{2x}{\pi} & \text{for} \quad 0 \leq x \leq \pi. \end{cases}$$

Draw the graph of F_0.

7. Find the Fourier series for the function

$$f(x) = \begin{cases} -\pi - x & \text{for} \quad -\pi \leq x \leq -\dfrac{\pi}{2}, \\[2mm] x & \text{for} \quad -\dfrac{\pi}{2} \leq x \leq \dfrac{\pi}{2}, \\[2mm] \pi - x & \text{for} \quad \dfrac{\pi}{2} \leq x \leq \pi. \end{cases}$$

8. Using the results of problem 4 above and the expansion for $f(x) = x$, as in the example on page 471, find the Fourier series for the function

$$g(x) = \begin{cases} -\dfrac{x^2}{2} & \text{for } -\pi < x \le 0, \\[3mm] \dfrac{x^2}{2} & \text{for } \quad 0 \le x < \pi. \end{cases}$$

Note that $g'(x) = |x|$ on $(-\pi, \pi)$. Is the series obtained by term-by-term differentiation of that for g identical with that for $|x|$ given in problem 3 above? Explain.

9. Complete the proof of Lemma 4.

10. Write out the details which establish the bounds given in (9).

*11. Suppose that f possesses continuous derivatives of all orders for $-\infty < x < \infty$ and is periodic with period 2π. What can be said about the ratios

$$\frac{a_n}{n^k}, \qquad \frac{b_n}{n^k} \qquad \text{as } n \to \infty,$$

where a_n, b_n, are the Fourier coefficients of f, and k is a positive integer?

12. The *Dirichlet conjugate kernel* is defined by

$$\overline{D}_n(u) = \sum_{k=1}^{n} \sin ku.$$

Show that

$$\overline{D}_n(u) = \frac{\cos \tfrac{1}{2}u - \cos (n + \tfrac{1}{2})u}{2 \sin \tfrac{1}{2}u}.$$

7. THE COMPLEX FORM OF FOURIER SERIES

In discussing the complex form of Fourier series we shall employ the terminology and methods of complex vector spaces. The student should review the material in Chapter 8, Sections 4 through 6. Let V_c be the pre-Hilbert space of all complex-valued functions f, where

$$f(x) = f_1(x) + if_2(x) \tag{1}$$

and where f_1 and f_2 are real-valued, piecewise continuous functions on an interval $[a, b]$. We may define on V_c the inner product

$$(f, g) = \int_a^b f(x)\,\overline{g(x)}\,dx.$$

Note that if V were a finite-dimensional space of functions, then V_c would simply be the unitary complexification of V. From the general theory in Chapter 8, Section 4, it follows that both the Schwarz Inequality

$$\|(f, g)\| \le \|f\|\,\|g\|$$

and the "triangle inequality"

$$\|f + g\| \le \|f\| + \|g\|$$

hold in this space. Since the trigonometric set

$$\frac{1}{\sqrt{2\pi}}, \qquad \frac{\cos nx}{\sqrt{\pi}}, \qquad \frac{\sin nx}{\sqrt{\pi}}, \qquad n = 1, 2, \ldots$$

is complete in the space of real, piecewise continuous functions on $[-\pi, \pi]$, it follows that the same set is complete in $V_c[-\pi, \pi]$. An appropriate adjustment yields the complete trigonometric set for the space $V_c[a, b]$.

The definitions of piecewise smooth, normalized, and periodic extension carry over directly to complex-valued functions. For any function f given by (1), we have

$$f'(x) = f_1'(x) + if_2'(x), \qquad \int_a^b f(x)\,dx = \int_a^b f_1(x)\,dx + i\int_a^b f_2(x)\,dx.$$

The definitions of evenness, oddness, and half-range expansions for complex functions are the same as for real-valued functions. The theorems for complex functions on convergence, integration, and differentiation of series are all established in the same way as for real functions.

Although the trigonometric set is a complete orthonormal set for V_c, there are certain problems for which the exponential set, which we now introduce, is more convenient. Using the formulas

$$\cos nx = \frac{e^{inx} + e^{-inx}}{2}, \qquad \sin nx = \frac{-i(e^{inx} - e^{-inx})}{2},$$

we obtain at once

$$a_n \cos nx + b_n \sin nx = c_n e^{inx} + c_{-n} e^{-inx},$$

where

$$c_n = \frac{a_n - ib_n}{2}, \qquad c_{-n} = \frac{a_n + ib_n}{2}, \qquad n = 1, 2, \ldots. \qquad (2)$$

If, in addition, we set

$$c_0 = \frac{a_0}{2}, \qquad (3)$$

we see that if f is represented by the convergent series

$$f(x) = \frac{a_0}{2} + \sum_{n=1}^{\infty} (a_n \cos nx + b_n \sin nx), \qquad (4)$$

then f is also represented by the convergent series

$$f(x) = c_0 + \sum_{n=1}^{\infty} (c_n e^{inx} + c_{-n} e^{-inx}), \qquad (5)$$

where c_n and c_{-n} are given in terms of a_n and b_n by (2) and (3). Moreover, if the series (5) converges to $f(x)$ with the terms grouped as indicated, then the series (4) does also, provided the a_n and b_n are related to c_n and c_{-n} by (2) and (3); that is,

$$a_0 = 2c_0, \qquad a_n = c_n + c_{-n}, \qquad b_n = i(c_n - c_{-n}).$$

Formally, the series (5) can be written in the form

$$f(x) = \sum_{n=-\infty}^{\infty} c_n e^{inx} = \sum_{n=1}^{\infty} c_n e^{inx} + \sum_{n=1}^{\infty} c_{-n} e^{-inx} + c_0$$

but, unless the two series $\sum_{n=1}^{\infty} |c_n|$ and $\sum_{n=1}^{\infty} |c_{-n}|$ are convergent, the rearrangement is not always valid. (See Chapter 9, Section 4.)

We now show that the collection

$$\phi_0 = 1, \; \phi_1 = e^{ix}, \; \ldots, \; \phi_n = e^{inx}, \; \ldots,$$
$$\phi_{-1} = e^{-ix}, \; \ldots, \; \phi_{-n} = e^{-inx}, \; \ldots$$

forms a complex orthogonal set on $[-\pi, \pi]$. We have

$$(\phi_m, \phi_n) = \int_{-\pi}^{\pi} e^{imx} \, \overline{(e^{inx})} \, dx = \int_{-\pi}^{\pi} e^{i(m-n)x} \, dx$$

$$= \left[\frac{e^{i(m-n)x}}{i(m-n)} \right]_{-\pi}^{\pi} = 0 \qquad \text{if } m \neq n,$$

and

$$(\phi_m, \phi_m) = \int_{-\pi}^{\pi} 1 \cdot dx = 2\pi.$$

Since $c_n = \frac{1}{2}(a_n - ib_n)$, we find for $n > 0$

$$c_n = \frac{1}{2\pi} \int_{-\pi}^{\pi} f(x) \, (\cos nx - i \sin nx) \, dx = \frac{1}{2\pi} \int_{-\pi}^{\pi} f(x) e^{-inx} \, dx.$$

Also,

$$c_{-n} = \frac{1}{2\pi} \int_{-\pi}^{\pi} f(x) \, (\cos nx + i \sin nx) \, dx = \frac{1}{2\pi} \int_{-\pi}^{\pi} f(x) e^{inx} \, dx$$

and

$$c_0 = \frac{1}{2\pi} \int_{-\pi}^{\pi} f(x) \, dx = \frac{1}{2\pi} \int_{-\pi}^{\pi} f(x) e^{i \cdot 0 \cdot x} \, dx.$$

Therefore the formula

$$c_k = \frac{1}{2\pi} \int_{-\pi}^{\pi} f(x) e^{-ikx} \, dx \qquad (6)$$

holds for all integers k, positive, negative, or zero. In terms of inner-product notation, we see that

$$c_k = \frac{(f, \phi_k)}{(\phi_k, \phi_k)}, \qquad (\phi_k, \phi_k) = 2\pi,$$

which is the "correct" formula for the Fourier coefficients with respect to any orthogonal set.

Theorem 10. *The set*

$$\left\{ \frac{1}{\sqrt{2\pi}} e^{inx} \right\}, \qquad n = 0, \pm1, \pm2, \ldots,$$

is a complete orthonormal set for V_c on the interval $[-\pi, \pi]$. If f satisfies the hypotheses of Theorem 2 and if the c_k are given by (6), then the series (5) converges to $f(x)$ for each x. Theorems 8 and 9 on the differentiation and integration of series remain valid.

The only point in question is that of completeness, which is equivalent to the question of completeness of the trigonometric set. (See Appendix 1.)

The effect on complex Fourier series of evenness or oddness of a function f is analogous to the effect in the real case. The next theorem describes the various possibilities.

Theorem 11. *Suppose that f satisfies the hypotheses of Theorem 2 and that the c_n are given by (6). Then*

(i)   *f is real if and only if $c_{-n} = \bar{c}_n$ for all n;*
(ii) *f is even if and only if $c_{-n} = c_n$ for all n;*
(iii) *f is even and real if and only if $c_{-n} = c_n = \bar{c}_n$ for all n;*
(iv) *f is odd if and only if $c_{-n} = -c_n$ for all n;*
(v)   *f is odd and real if and only if $c_{-n} = -c_n = \bar{c}_n$ for all n.*

The series (5) is called the **complex Fourier series of f.**

EXAMPLE. If $f = x$ for $-\pi < x < \pi$, find the complex Fourier series for f.

Solution. For any integer $n \neq 0$, we have

$$c_n = \frac{1}{2\pi} \int_{-\pi}^{\pi} xe^{-inx}\, dx = \frac{1}{2\pi}\left[\frac{xe^{-inx}}{-in} \right]_{-\pi}^{\pi} + \frac{1}{2n\pi i} \int_{-\pi}^{\pi} e^{inx}\, dx$$

$$= -\frac{\pi}{2n\pi i}(e^{-in\pi} + e^{in\pi}) + \frac{1}{2n\pi i}\left[\frac{e^{-inx}}{-in} \right]_{-\pi}^{\pi}.$$

Since e^z is periodic with period $2\pi i$, we find

$$e^{-in\pi} = e^{in\pi} = \cos n\pi + i \sin n\pi = (-1)^n.$$

Therefore

$$c_n = \frac{(-1)^n}{-in} = \frac{i}{n}(-1)^n \qquad \text{and} \qquad c_0 = \frac{1}{2\pi}\int_{-\pi}^{\pi} x\, dx = 0.$$

Hence

$$x = \sum_{\substack{n=-\infty \\ n \neq 0}}^{\infty} \frac{i}{n}(-1)^n e^{inx}, \qquad -\pi < x < \pi.$$

Note that the c_n satisfy the conditions of Theorem 11(v) for an odd real function.

EXERCISES

Find the complex Fourier series for each of the following functions.

1. $f(x) = x^2, \quad -\pi < x < \pi$

2. $f(x) = \frac{1}{3}(x^3 - \pi^2 x), \quad -\pi < x < \pi$

3. $f(x) = \begin{cases} -1 & \text{for } -\pi < x < 0 \\ 1 & \text{for } 0 < x < \pi \end{cases}$

4. $f(x) = |x| \quad \text{for } -\pi < x < \pi$

5. $f(x) = e^{ax}, a$ real. [*Hint.* Recall Theorem 23(c) in Section 5 of Chapter 9.]

6. $f(x) = \sin ax, \quad 0 < a < 1$

7. $f(x) = \sinh ax, \quad 0 < a < 1$

8. $f(x) = \cosh ax, \quad 0 < a < 1$

9. $f(x) = 1 + ix, \quad -\pi < x < \pi$

10. $f(x) = 2x + ix^2, \quad -\pi < x < \pi$

11. Show for functions in V_c that

$$\int_{-\pi}^{\pi} |f(x)|^2 \, dx = \pi \left[\frac{|a_0|^2}{2} + \sum_{n=1}^{\infty} (|a_n|^2 + |b_n|^2) \right] = 2\pi \sum_{n=-\infty}^{\infty} |c_n|^2.$$

[*Hint.* Use Parseval's Formula for real functions. See page 459.]

CHAPTER **11**

Implicit Function Theorems. Transformations

---

1. IMPLICIT FUNCTION THEOREMS

An equation such as

$$x^6 + 2y^8 + 7x^2y^2 - 8x + 2y = 0 \tag{1}$$

represents a *relation* between x and y. A pair of numbers which satisfies this equation corresponds to a point in the xy-plane. In general, the totality of points which satisfy an equation of the form

$$F(x, y) = 0 \tag{2}$$

is called the *locus* of the equation. In the study of analytic geometry, our work with loci consisted mostly of the study of arcs or simple smooth curves. However, it is not at all obvious what the locus might be of an equation such as (1) above. Even equations which are quite simple in appearance sometimes have unusual loci. For example, the equation $x^2 + 2y^2 + 9 = 0$ is in the form (2) but has no locus, since the sum of positive quantities can never be zero. The equation $(x - 1)^2 + (y + 2)^2 = 0$ is satisfied only when $x = 1$, $y = -2$, and the locus is a single point. The equation

$$\sin x + \sec y = 0, \tag{3}$$

also in the form of (2), has an interesting locus. Since $|\sin x| \le 1$ and $|\sec y| \ge 1$, the equation can hold only when $\sin x = 1$ and $\sec y = -1$ or when $\sin x = -1$ and $\sec y = 1$. The locus consists of the isolated points

$$(\pi/2 \pm 2m\pi, \pi \pm 2n\pi) \quad \text{and} \quad (3\pi/2 \pm 2m\pi, \pm 2n\pi), \quad m, n = 0, 1, 2, \ldots$$

as shown in Fig. 11–1.

We studied functions defined implicitly in *First Course*, page 142 ff, where we learned how to calculate the derivative. We always assumed that the given ex-

479

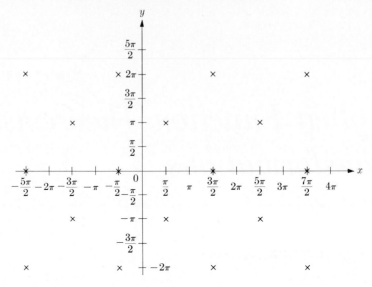

FIGURE 11–1

pressions represented differentiable functions, paying no attention to unusual loci such as (3). If we proceed unthinkingly and apply the chain rule to (3), we obtain

$$\cos x + (\sec y \tan y)\, \frac{dy}{dx} = 0$$

or

$$\frac{dy}{dx} = -\frac{\cos x \cos^2 y}{\sin y}. \tag{4}$$

Equation (4) appears reasonable but is actually an absurd statement, since the locus of (3) is the isolated set described above. A detailed analysis is always required when relations are given implicitly.

Even under the most favorable circumstances, a relation of the form $F(x, y) = 0$ may have a complicated locus, such as the one shown in Fig. 11–2. We shall be interested in studying the behavior of the locus near such particular points as P, Q, R, or S. At each of these points, the locus has a special character, which we now examine. The next definition will be helpful in the discussion which follows.

DEFINITION. By the **local behavior** of a locus or curve, we mean the behavior of the locus in a vicinity or neighborhood (small circle) of a particular point P. The term *local*, which is a technical one in mathematics, implies not only that the property we are describing persists no matter how small the neighborhood of P is, but also that it may be destroyed if a sufficiently large neighborhood is selected. It may also be destroyed if a different point is selected, no matter how close.

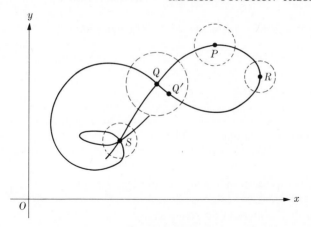

FIGURE 11–2

For example, in Fig. 11–2 we note that the slope of the locus is positive in a small neighborhood of P. This property concerns the local behavior in that a selection of a smaller neighborhood will not change the correctness of the statement, but if the neighborhood selected is sufficiently large the statement is false. The locus in the neighborhood of R also has a special character. The tangent line at R is vertical, and therefore in a neighborhood of R we may think of x as a function of y. However, y is not a function of x since, for each value x_0 nearby, there correspond two values of y (Fig. 11–3). Thus we can describe a property of the local behavior near R by saying that x is a function of y but that y is not a function of x. The statement is false if the neighborhood is sufficiently large. At the point Q the locus intersects itself (Fig. 11–2) and, in a neighborhood of Q, y is not representable as a function of x nor is x representable as a function of y. This statement concerns local behavior because the selection of another point Q', no matter how close to Q, changes the assertion. In a neighborhood of Q', y is representable as a function of x and x is representable as a function of y.

If we are given a specific relation in the form

$$F(x, y) = 0, \qquad (5)$$

we are interested in determining when this equation gives y as a function of x (or x as a function of y). We may also ask when we can solve for one of the variables in terms of the other. In a simple case such as

$$x^2 + 3xy - 2x + 5y - 7 = 0,$$

we merely use elementary algebra to get

$$y = \frac{7 + 2x - x^2}{3x + 5},$$

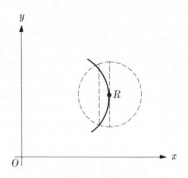

FIGURE 11-3

and y is expressed as a function of x. However, even in cases where we can perform the algebra, certain questions arise. The equation

$$\frac{x^2}{4} + \frac{y^2}{9} - 1 = 0$$

may be solved for y to give

$$y = \pm \tfrac{3}{2}\sqrt{4 - x^2},$$

but still y is not represented as a function of x. If we write $y = +\tfrac{3}{2}\sqrt{4 - x^2}$ and $y = -\tfrac{3}{2}\sqrt{4 - x^2}$, then the entire locus (ellipse) is described by these two functions.

Starting with the relation (5) and a point $P(x_0, y_0)$ on the locus, we suppose that in a neighborhood of P the relation represents y as a function of x. In other words, we assume that the local behavior of the relation allows us to write (Fig. 11–4)

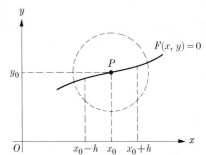

FIGURE 11–4

$$y = f(x) \tag{6}$$

for $x_0 - h < x < x_0 + h$, where h is some positive number. If we substitute (6) into (5), then

$$F(x, f(x)) = 0$$

is an identity for $|x - x_0| < h$. Assuming that all quantities are smooth, we may use the chain rule to get

$$F_x[x, f(x)] + F_y[x, f(x)] \cdot f'(x) = 0.$$

If $F_y \neq 0$, we obtain the formula

$$f'(x) = -\frac{F_x}{F_y}, \tag{7}$$

a result we have used on a number of occasions. See Chapter 4, Section 8.

Implicit function theorems are those which state conditions under which a relation (5) may be put in the form (6), at least in the neighborhood of some point P on the locus. These theorems also decide when the differentiation formula (7) is valid. Before establishing the most important implicit function theorems, we recall several facts which were studied earlier. The first is the Intermediate Value Theorem (*First Course*, page 228).

Intermediate Value Theorem. *Suppose that f is continuous on an interval [a, b] and that f(a) = A, f(b) = B. If C is any number between A and B, there is a number c between a and b such that f(c) = C.*

In other words, a continuous function must take on all intermediate values between any two values it assumes.

A second fact, used many times but never stated as a theorem, concerns the local behavior of a continuous function. Briefly, if a continuous function is positive for some value, it must be positive for all points in some region about this value. Of course, the region may be "very small," but there is one. The next theorem establishes this result for functions in any number of variables.

Theorem 1. *Suppose $f(x_1, x_2, \ldots, x_n)$ is continuous at a point $(x_1^0, x_2^0, \ldots, x_n^0)$, and suppose $f(x_1^0, x_2^0, \ldots, x_n^0) > 0$. Then there is a positive number h such that $f(x_1, x_2, \ldots, x_n)$ is positive for all $(x_1, x_2, \ldots, x_n)$ in the neighborhood*

$$|x_1 - x_1^0| < h, \ |x_2 - x_2^0| < h, \ldots, |x_n - x_n^0| < h. \tag{8}$$

Proof. The domain (8) is an n-dimensional rectangular box of the type used in the definition of limit for functions of several variables. (See Chapter 4, page 121.) Let $\epsilon = f(x_1^0, x_2^0, \ldots, x_n^0)$ and apply the definition of continuity. For any $\epsilon > 0$ there is a $\delta > 0$ such that

$$|f(x_1, x_2, \ldots, x_n) - f(x_1^0, x_2^0, \ldots, x_n^0)| < \epsilon$$

whenever $|x_1 - x_1^0| < \delta, |x_2 - x_2^0| < \delta, \ldots, |x_n - x_n^0| < \delta$. Selecting $\delta = h$, we see that

$$0 = f(x_1^0, x_2^0, \ldots, x_n^0) - \epsilon < f(x_1, x_2, \ldots, x_n)$$
$$< f(x_1^0, x_2^0, \ldots, x_n^0) + \epsilon$$

so long as $(x_1, x_2, \ldots, x_n)$ is in the box (8).

A third fact we shall use in the proof of the first implicit function theorem is the Fundamental Lemma on Differentiation. For functions of two variables, the result is identified with the formula

$$F(x_0 + h, y_0 + k) - F(x_0, y_0)$$
$$= F_x(x_0, y_0)h + F_y(x_0, y_0)k + G_1(h, k)h + G_2(h, k)k, \tag{9}$$

where G_1 and G_2 tend to zero as h, k tend to zero, and $G_1(0, 0) = G_2(0, 0) = 0$. (See Chapter 4, page 127.)

We now prove the first implicit function theorem.

Theorem 2. *Suppose that F, F_x, and F_y are continuous near (x_0, y_0) and suppose that*

$$F(x_0, y_0) = 0, \qquad F_y(x_0, y_0) \neq 0.$$

(a) *Then there are two positive numbers h and k which determine a rectangle R about (x_0, y_0)*

$$R: \ |x - x_0| < h, \qquad |y - y_0| < k,$$

such that for each x with $|x - x_0| < h$ there is a unique number y with $|y - y_0| < k$ which satisfies the equation $F(x, y) = 0$. That is, y is a function of x, and we may write $y = f(x)$. The domain of f contains $|x - x_0| < h$ and the range is in $|y - y_0| < k$.

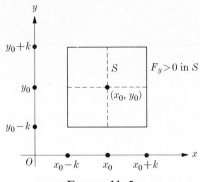

FIGURE 11–5

(b) *The function f determined in (a) and its derivative f' are continuous for* $|x - x_0| < h$. *Furthermore,*

$$F_y[x, f(x)] \neq 0 \qquad and \qquad f'(x) = - \frac{F_x[x, f(x)]}{F_y[x, f(x)]}$$

for $|x - x_0| < h$.

Proof. We may assume that $F_y(x_0, y_0) > 0$; otherwise we simply replace F by $-F$ and repeat the argument. From the definition of the continuity of F_y and from Theorem 1, there must be a number k such that $F_y(x, y) > 0$ in the square region S: $\{|x - x_0| \leq k, |y - y_0| \leq k\}$ (Fig. 11–5). If we fix a value x in $|x - x_0| < k$, then $F(x, y)$, considered as a function of y alone, has a positive slope and therefore is *an increasing function of y in S.* In particular, $F(x_0, y)$ is an increasing function of y. Since $F(x_0, y_0) = 0$ by hypothesis, it follows that

$$F(x_0, y_0 + k) > 0 \qquad and \qquad F(x_0, y_0 - k) < 0.$$

We now apply Theorem 1 to each of these functions and conclude that there is an interval $|x - x_0| < h$ in which

$$F(x, y_0 + k) > 0 \qquad and \qquad F(x, y_0 - k) < 0.$$

We now return to the solution of the equation $F(x, y) = 0$. Fix x in the interval $x_0 - h < x < x_0 + h$ and concentrate on the rectangle R: $|x - x_0| < h$, $|y - y_0| < k$, as shown in Fig. 11–6. Since $F(x, y)$ as a function of y is negative for $y = y_0 - k$ and positive for $y = y_0 + k$ then, according to the Intermediate Value Theorem, there is a value y such that $F(x, y) = 0$. Also, since $F_y > 0$, there cannot be more than one such value. The existence of the function $y = f(x)$ for $|x - x_0| < h$ is thus established.

To see that f is continuous at x_0, we must show that for any $\epsilon > 0$

$$|f(x) - f(x_0)| < \epsilon,$$

provided that x is sufficiently close to x_0. The values of $f(x)$ are restricted by the

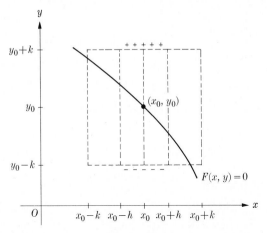

$$\text{Figure } 11\text{–}6$$

choice of the square S—that is, by the size of k. If we select a k' smaller than k and go through the entire process described in the proof, we will obtain the same function f, but it may perhaps be defined on a smaller interval $|x - x_0| < h'$. Selecting $k' = \epsilon$, we see that the choice of $h' = \delta$ in the definition of continuity yields the result. At any other point x_1, we establish the result by constructing the square S and the rectangle R with (x_1, y_1) as center, where $y_1 = f(x_1)$.

To establish part (b), we employ the Fundamental Lemma on Differentiation, Eq. (9). Writing $\Delta f = f(x + \Delta x) - f(x)$ and noting that $G_1(0, 0) = G_2(0, 0) = 0$, we have

$$
\begin{aligned}
0 &= F[x + \Delta x, f(x + \Delta x)] - F[x, f(x)] \\
&= F_x[x, f(x)]\,\Delta x + F_y[x, f(x)]\,\Delta f + G_1(\Delta x, \Delta f)\,\Delta x + G_2(\Delta x, \Delta f)\,\Delta f.
\end{aligned}
$$

Therefore

$$F_x\,\Delta x + F_y\,\Delta f = -G_1\,\Delta x - G_2\,\Delta f$$

and

$$\frac{\Delta f}{\Delta x} = -\frac{F_x + G_1}{F_y + G_2}. \tag{10}$$

Since f is continuous, Δf tends to zero as Δx does. Consequently $G_1(\Delta x, \Delta f)$ and $G_2(\Delta x, \Delta f)$ tend to zero with Δx. The left side of (10) is the difference quotient of f and so tends to $f'(x)$ as the right side tends to $-F_x/F_y$. This is (8) precisely. Furthermore, f' is continuous because of the hypotheses that F_x and F_y are.

Remarks. (i) A geometric interpretation of the theorem is indicated in Fig. 11–7. The theorem states that there exists a rectangle

$$|x - x_0| < h, \qquad |y - y_0| < k$$

such that the part of the locus of $F(x, y) = 0$ which is in this rectangle lies along

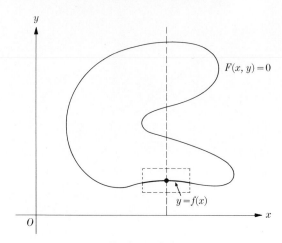

FIGURE 11–7

an arc of the form $y = f(x)$, where f is differentiable. Of course, the line $x = x_0$ may intersect the total locus at several points, as shown in the figure. (ii) The theorem is purely *local* in that nothing is determined about how far the domain of f can be extended. However, the result remains valid no matter how small a rectangle about (x_0, y_0) is selected. (iii) We have here an example of a pure *existence theorem*, in that the proof does not provide us with a method for finding the particular function f. A proof which enables us to determine the actual answer either numerically or analytically is called a *constructive proof*. Constructive proofs, although more desirable than existence proofs, are frequently more complicated. (iv) A corresponding result holds if the variables are interchanged. If $F_x(x_0, y_0) \neq 0$, then the Implicit Function Theorem allows us to write $x = g(y)$, at least locally.

EXAMPLE 1. Apply Theorem 2 to the equation

$$\frac{x^2}{9} + \frac{y^2}{4} - 1 = 0,$$

and find the function f of the theorem whenever possible.

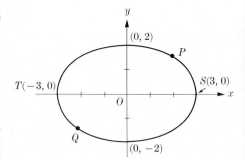

FIGURE 11–8

Solution. The graph of the equation is shown in Fig. 11–8. We see that if P is a point as shown, then $F_y = \frac{1}{2}y > 0$ and y is a function of x. In fact, $y = +\frac{2}{3}\sqrt{9 - x^2}$. At a point Q we have $F_y < 0$ and $y = -\frac{2}{3}\sqrt{9 - x^2}$. However, at points such as S or T, $F_y = 0$ and the theorem fails. On the other hand, we observe that $F_x = \frac{2}{9}x > 0$ at S, and so we can write $x = g(y)$ in a neighborhood of S. We compute, trivially, $g(y) = \frac{3}{2}\sqrt{4 - y^2}$ near S.

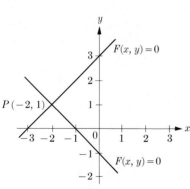

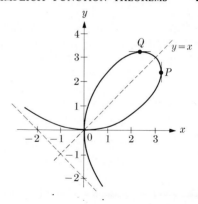

FIGURE 11–9   FIGURE 11–10

EXAMPLE 2. Show that the entire locus of the equation

$$F(x, y) \equiv y^3 + 3x^2y - x^3 + 2x + 3y = 0$$

is a function which is defined for all x.

Solution. We have

$$F_y = 3y^2 + 3x^2 + 3 > 0 \qquad \text{for all } (x, y).$$

Therefore, for each x, $F(x, y)$ is increasing in y for all y. Moreover, $F(x, y) \to -\infty$ as $y \to -\infty$ and $F \to +\infty$ as $y \to +\infty$ for each fixed x. It follows that for each x there is a unique y such that $F(x, y) = 0$. Since the hypotheses of Theorem 2 are satisfied at any point (x_0, y_0) where $F(x_0, y_0) = 0$, the function f determined by the theorem is defined, continuous, and differentiable for all x.

EXAMPLE 3. Discuss the validity of Theorem 2 with regard to the function

$$F(x, y) \equiv x^2 - y^2 + 4x + 2y + 3 = 0.$$

Solution. We have $F_x = 2x + 4$, $F_y = -2y + 2$. At any point on the locus with $y \neq 1$, we may solve for y as a function of x, since $F_y \neq 0$. Also, whenever $x \neq -2$ we may solve for x as a function of y. However, at the point $(-2, 1)$ which is on the locus, we have $F_x = F_y = 0$, and Theorem 2 fails (Fig. 11–9). Formula (8) gives

$$\frac{dy}{dx} = -\frac{F_x}{F_y} = \frac{x + 2}{1 - y},$$

which is indeterminate at $(-2, 1)$. The point $P(-2, 1)$ is a double point of the locus.

EXAMPLE 4. Locate the points on the locus

$$F(x, y) \equiv x^3 + y^3 - 6xy = 0$$

for which Theorem 2 is not applicable.

Solution. This curve is the *folium of Descartes* (Fig. 11–10), which was discussed on page 179. A simple computation yields $F_x = 3x^2 - 6y$, $F_y = 3y^2 - 6x$, and we see that both F_x and F_y vanish at the origin, which is a double point. We cannot obtain y as a function of x at any point where $3y^2 - 6x = 0$. Substituting $x = \frac{1}{2}y^2$ into the equation of the curve, we find $x = 2\sqrt[3]{4}$, $y = 2\sqrt[3]{2}$. This point is designated P in Fig. 11–10. Similarly, setting $F_x = 0$ and solving, we find that we cannot express x as a function of y in a neighborhood of $Q(2\sqrt[3]{2}, 2\sqrt[3]{4})$. Theorem 2 is applicable at all other points on the curve.

Theorem 2 is valid in any number of variables, and the proof introduces no new difficulties. We state the result for reference.

Theorem 2′. *Suppose that $F(x_1, x_2, \ldots, x_n, y)$ and F_y are continuous near the point $(x_1^0, x_2^0, \ldots, x_n^0, y_0)$; suppose also that, for $i = 1, 2, \ldots, n$, each F_{x_i} is continuous near $(x_1^0, x_2^0, \ldots, x_n^0, y_0)$. Let*

$$F(x_1^0, x_2^0, \ldots, x_n^0, y_0) = 0, \qquad F_y(x_1^0, x_2^0, \ldots, x_n^0, y_0) \neq 0.$$

Then there are positive numbers h and k such that
(a) *for each $(x_1, x_2, \ldots, x_n)$ with $|x_i - x_i^0| < h$ for $i = 1, 2, \ldots, n$, there is a unique number y with $|y - y_0| < k$ satisfying $F(x_1, x_2, \ldots, x_n, y) = 0$;*
(b) *if we define f by $f(x_1, x_2, \ldots, x_n) = y$, then f and all first derivatives f_{x_i}, $i = 1, 2, \ldots, n$ are continuous for $|x_i - x_i^0| < h$, and*

$$F_y[x_1, x_2, \ldots, x_n, f(x_1, x_2, \ldots, x_n)] \neq 0,$$

$$f_{x_i} = -\frac{F_{x_i}[x_1, x_2, \ldots, x_n, f(x_1, x_2, \ldots, x_n)]}{F_y[x_1, x_2, \ldots, x_n, f(x_1, x_2, \ldots, x_n)]}$$

if $|x - x_i| < h$, $i = 1, 2, \ldots, n$.

EXAMPLE 5. Using Theorem 2′, can we conclude that the part of the locus of the equation
$$F(x, y, z) = 3x^2 + 2y^2 + z^2 + 2xy + 2xz + 2yz - 9 = 0$$

within some box $|x - 2| < h$, $|y + 1| < h$, $|z + 1| < k$ lies along a surface $z = f(x, y)$ in which f is defined and differentiable for all x, y with $|x - 2| < h$ and $|y + 1| < h$? Solve for z in terms of x and y and discuss.

Solution. Here $x_0 = 2$, $y_0 = -1$, $z_0 = -1$. [In Theorem 2′ we have (x, y, z) instead of (x_1, x_2, y).] Then $F_z = 2z + 2x + 2y$, $F_z(2, -1, -1) = 0$. Consequently, the hypotheses of the theorem *are not satisfied.* Solving for z, we obtain

$$z^2 + 2(x + y)z + (3x^2 + 2y^2 + 2xy - 9) = 0,$$
$$z = -(x + y) \pm \sqrt{9 - 2x^2 - y^2}. \tag{11}$$

The domain of the two functions in (11) is the elliptical region

$$0 \leq 2x^2 + y^2 \leq 9,$$

and we note that the point $(2, -1)$ is on the boundary of this region. Thus there is no box satisfying the conditions. From our knowledge of solid analytic geometry we also observe that the plane tangent to the surface $F(x, y, z) = 0$ at $(2, -1, -1)$ is parallel to the z-axis.

EXERCISES

In each of problems 1 through 10, use Theorem 2 to show that the equation $F(x, y) = 0$ may be represented in the form $y = f(x)$ in a neighborhood of the given point (x_0, y_0). Draw a graph and compute $f'(x_0)$ in each case.

1. $F(x, y) \equiv x + y + x \sin y = 0$; $(x_0, y_0) = (0, 0)$
2. $F(x, y) \equiv y^2 - 2xy + 5x^2 - 16 = 0$; $(x_0, y_0) = (1, 1 - 2\sqrt{3})$
3. $F(x, y) \equiv y^3 + y - x^2 = 0$; $(x_0, y_0) = (0, 0)$
4. $F(x, y) \equiv xe^y - y + 1 = 0$; $(x_0, y_0) = (-1, 0)$
5. $F(x, y) \equiv x^{2/3} + y^{2/3} - 4 = 0$; $(x_0, y_0) = (1, 3\sqrt{3})$
6. $F(x, y) \equiv (x^2 + y^2)^2 - 8(x^2 - y^2) = 0$; $(x_0, y_0) = (\sqrt{3}, 1)$
7. $F(x, y) \equiv xy + \ln(xy) - 1 = 0$; $(x_0, y_0) = (1, 1)$
8. $F(x, y) \equiv x \cos xy$; $(x_0, y_0) = (1, \pi/2)$
9. $F(x, y) \equiv x^5 + y^5 + xy + 4 = 0$; $(x_0, y_0) = (2, -2)$
10. $F(x, y) \equiv 2 \sin x + \cos y - 1 = 0$; $(x_0, y_0) = (\pi/6, 3\pi/2)$

In each of problems 11 through 16, use Theorem 2' to show that the equation $F(x, y, z) = 0$ may be represented in the form $z = f(x, y)$ in a neighborhood of the given point (x_0, y_0, z_0). Find $f_x(x_0, y_0)$ and $f_y(x_0, y_0)$.

11. $F(x, y, z) \equiv x^3 + y^3 + z^3 - 3xyz - 4 = 0$; $(x_0, y_0, z_0) = (1, 1, 2)$
12. $F(x, y, z) \equiv x^4 + y^4 + z^4 - 18 = 0$; $(x_0, y_0, z_0) = (1, 1, 2)$
13. $F(x, y, z) \equiv e^z - z^2 - x^2 - y^2 = 0$; $(x_0, y_0, z_0) = (1, 0, 0)$
14. $F(x, y, z) \equiv z^3 - z - xy \sin z = 0$; $(x_0, y_0, z_0) = (0, 0, 0)$
15. $F(x, y, z) \equiv x + y + z + \cos xyz = 0$; $(x_0, y_0, z_0) = (0, 0, -1)$
16. $F(x, y, z) = x + y + z - e^{xyz} = 0$; $(x_0, y_0, z_0) = (0, \frac{1}{2}, \frac{1}{2})$
17. Do there exist numbers $h > 0$ and $k > 0$ such that all the points satisfying $y^2 - x^3 = 0$ and $|x| < h$, $|y| < k$ (a) lie along an arc $y = f(x)$ where $f(x)$ is defined and smooth for all x with $|x| < h$? (b) lie along an arc $x = g(y)$ where g has the same properties for $|y| < k$?

In problems 18 through 24, in each case plot and discuss the entire locus, indicate the different functions defined implicitly by the equation, and find any points on the locus where $F_y(x_0, y_0) = 0$. At any such points check to see whether $F_x(x_0, y_0) = 0$.

18. $F(x, y) \equiv (x - 3)^4 - (y + 2)^2 = 0$
19. $F(x, y) \equiv xe^y - 2y + 2 = 0$
20. $F(x, y) \equiv 2x^2 + y^2 - x^3 = 0$
21. $F(x, y) \equiv y^3 + x^2y - x^2 = 0$
22. $F(x, y) \equiv y^3 - 3y - x^2 = 0$

23. $F(x, y) \equiv (x^2 + y^2)^2 - 8(x^2 - y^2) = 0$ (lemniscate)

24. $F(x, y) \equiv e^{2xy} - \ln [1/(1 + y^2)] = 0$

In problems 25 and 26, show that $F_z(x_0, y_0, z_0) = 0$ and determine whether or not it is possible to express z as a function of x and y in a neighborhood of (x_0, y_0, z_0). Does the same situation prevail when y and z are interchanged; i.e., can we solve for y in terms of x and z in a neighborhood of (x_0, y_0, z_0)?

25. $F(x, y, z) \equiv 5x^2 + 3y^2 + z^2 - 4xy - 4xz + 2yz - 9 = 0$; $(x_0, y_0, z_0) = (-1, 2, -4)$

26. $F(x, y, z) \equiv 2x^2 + 3y^2 + z^2 + 4xy + 2xz + 4yz - 7 = 0$; $(x_0, y_0, z_0) = (4, -3, 2)$

27. Write out the proof of part (a) of Theorem 2′.

28. Write out the proof of part (b) of Theorem 2′.

29. Find an example of a relation $F(x, y, z) = 0$ such that $F_x(x_0, y_0, z_0) = F_y(x_0, y_0, z_0) = F_z(x_0, y_0, z_0) = 0$, and yet we are able to solve for z in terms of x and y in a neighborhood of (x_0, y_0, z_0).

30. Repeat problem 29, for a relation $F(x_1, x_2, \ldots, x_n, y) = 0$.

2.   IMPLICIT FUNCTION THEOREMS FOR SYSTEMS

Suppose we have the system

$$x^2 + 2xy - 3xu + 4yv = 0, \qquad 4xy + x^3u - 8yv + 2 = 0 \qquad (1)$$

in the four variables $x, y, u,$ and v. It is possible to express u and v as functions of x and y. Solving the first equation for u, we get

$$u = \frac{x^2 + 2xy + 4yv}{3x}, \qquad (2)$$

and we label this equation $u = \phi(x, y, v)$. Next we substitute u from (2) into the second equation in (1) to obtain

$$4xy + x^3 \left(\frac{x^2 + 2xy + 4yv}{3x} \right) - 8yv + 2 = 0.$$

This last equation may be solved for v in terms of x and y. The result is (after a little algebra)

$$v = \frac{x^4 + 2x^3y + 12xy + 6}{24y - 4x^2y}. \qquad (3)$$

We now substitute v from (3) into (2) and find

$$u = \frac{(x^2 + 2xy)(6y - x^2) + (x^4 + 2x^3y + 12xy + 6)y}{3x(6y - x^2)}, \qquad (4)$$

which expresses u as a function of x and y. In other words, starting with the system (1), we obtained a system $u = f(x, y)$ and $v = g(x, y)$ which, in this particular case, consists of the equations (4) and (3).

Suppose that we have a general system of two equations in four unknowns which we write

$$F(x, y, u, v) = 0 \quad \text{and} \quad G(x, y, u, v) = 0. \tag{5}$$

Proceeding in the most elementary manner, as in the example above, we imagine that we can solve the equation $F = 0$ for u in terms of x, y, and v. We write this solution

$$u = \phi(x, y, v).$$

Then, substituting this value of u in the equation $G = 0$, we get a single equation involving x, y, and v only. According to the Implicit Function Theorem (which we suppose it is possible to use), we extract from the relation connecting $x, y,$ and v an explicit function

$$v = g(x, y).$$

Now, continuing to parallel the method of the above example, we substitute for v into ϕ to obtain

$$u = \phi[x, y, g(x, y)],$$

which we write

$$u = f(x, y).$$

In this way, we have expressed u in terms of x and y, and we have expressed v in terms of x and y. We observe that all these results are *local* in character, and thus the equations

$$u = f(x, y) \quad \text{and} \quad v = g(x, y) \tag{6}$$

are valid only in the neighborhood of some point.

Taking (6) into account, we can now write the equations $F = 0$ and $G = 0$ in the form

$$F[x, y, f(x, y), g(x, y)] = 0, \qquad G[x, y, f(x, y), g(x, y)] = 0. \tag{7}$$

Assuming that all functions are smooth and that all operations of differentiation are legitimate, we apply the chain rule in (5) to compute partial derivatives with respect to x. The result is (since u and v are functions of x and y)

$$F_x + F_u u_x + F_v v_x = 0 \quad \text{and} \quad G_x + G_u u_x + G_v v_x = 0. \tag{8}$$

In (8) we treat u_x and v_x as unknowns and the remaining quantities as known. Then a simple application of Cramer's Rule gives

$$u_x = -\frac{\begin{vmatrix} F_x & F_v \\ G_x & G_v \end{vmatrix}}{\begin{vmatrix} F_u & F_v \\ G_u & G_v \end{vmatrix}}, \qquad v_x = -\frac{\begin{vmatrix} F_u & F_x \\ G_u & G_x \end{vmatrix}}{\begin{vmatrix} F_u & F_v \\ G_u & G_v \end{vmatrix}}, \tag{9}$$

provided that the denominator $F_u G_v - F_v G_u$ is not zero. The derivatives u_y and v_y are obtained by similar formulas.

The above development paid no attention either to the feasibility of carrying out the process or to the legitimacy of the various steps. At this time we are more interested in determining when the process leading to equations (6) and (9) is *possible* than we are in deciding when the actual steps can be performed, either analytically or numerically. The next theorem shows the circumstances under which all the various steps are possible—at least theoretically.

Theorem 3A. *Suppose that $F(x, y, u, v)$ and $G(x, y, u, v)$ are continuous and have continuous first derivatives near a point (x_0, y_0, u_0, v_0). Also, suppose that*

$$F(x_0, y_0, u_0, v_0) = 0, \qquad G(x_0, y_0, u_0, v_0) = 0$$

and

$$D_0 = \begin{vmatrix} F_u(x_0, y_0, u_0, v_0) & F_v(x_0, y_0, u_0, v_0) \\ G_u(x_0, y_0, u_0, v_0) & G_v(x_0, y_0, u_0, v_0) \end{vmatrix} \neq 0.$$

Then there are positive numbers h, k_1, and k_2 such that
(a) for each (x, y) with $|x - x_0| < h$, $|y - y_0| < h$, there is a unique solution (u, v) of the equations

$$F(x, y, u, v) = 0, \qquad G(x, y, u, v) = 0$$

with $|u - u_0| < k_1$ and $|v - v_0| < k_2$. We denote these solutions

$$u = f(x, y), \qquad v = g(x, y).$$

(b) The functions f and g are continuous with their first derivatives, and the following formulas hold:

$$f_x = -\frac{1}{D} \begin{vmatrix} F_x & F_v \\ G_x & G_v \end{vmatrix}, \qquad g_x = -\frac{1}{D} \begin{vmatrix} F_u & F_x \\ G_u & G_x \end{vmatrix},$$

$$f_y = -\frac{1}{D} \begin{vmatrix} F_y & F_v \\ G_y & G_v \end{vmatrix}, \qquad g_y = -\frac{1}{D} \begin{vmatrix} F_u & F_y \\ G_u & G_y \end{vmatrix},$$

where $D = F_u G_v - F_v G_u$.

Proof. The proof consists of successive applications of the Implicit Function Theorem of Section 1. In this way the original naïve description of the appropriate steps required for the elimination process can be made legitimate. Since $D_0 \neq 0$, it follows that G_u and G_v cannot both be zero at (x_0, y_0, u_0, v_0). Suppose that $G_v \neq 0$; the proof is similar if $G_u \neq 0$. Then, from Theorem 2', we conclude that there are numbers m and r such that the entire portion of the locus of $G(x, y, u, v) = 0$, for which

$$|x - x_0| < m, \qquad |y - y_0| < m,$$
$$|u - u_0| < m, \qquad \text{and} \qquad |v - v_0| < r$$

may be represented in the form

$$v = H(x, y, u),$$

where H is continuous and differentiable; moreover

$$H_u(x, y, u) = -\frac{G_u[x, y, u, H(x, y, u)]}{G_v[x, y, u, H(x, y, u)]}. \tag{10}$$

(In applying Theorem 2', we replaced h by m, k by r, (x_1, x_2, x_3) by (x, y, u), and the variable y by the variable v. We also used the label H instead of f for the explicitly obtained function.)

Now, let us define

$$K(x, y, u) = F[x, y, u, H(x, y, u)], \tag{11}$$

which merely amounts to a relabeling; the function K is defined for

$$|x - x_0| < m, \qquad |y - y_0| < m, \qquad |u - u_0| < m.$$

Using the chain rule to differentiate (11) with respect to u, we find that

$$K_u = F_u + F_v H_u = F_u + F_v \left(-\frac{G_u}{G_v} \right),$$

in which (10) has been used. Using the values at the point $P_0(x_0, y_0, u_0, v_0)$, we have

$$K_u(x_0, y_0, u_0) = \frac{F_u G_v - F_v G_u}{G_v(x_0, y_0, u_0, v_0)} = \frac{D_0}{G_v(P_0)} \neq 0. \tag{12}$$

Because of (12), we can apply the Implicit Function Theorem to the equation $K(x, y, u) = 0$. We conclude that there are positive numbers m' and r' with $m' \leq m$, $r' \leq m$ such that the locus of $K(x, y, u) = 0$, for which

$$|x - x_0| < m', \qquad |y - y_0| < m' \qquad \text{and} \qquad |u - u_0| < r'$$

is representable in the form

$$u = f(x, y), \qquad \text{with} \qquad |x - x_0| < m', \qquad |y - y_0| < m'.$$

Furthermore, f and its first derivatives are continuous. If we now define

$$v = g(x, y) \equiv H[x, y, f(x, y)] \qquad \text{for} \qquad |x - x_0| < m', \qquad |y - y_0| < m',$$

we see that g and its first derivatives are continuous.

The validity of the formulas for $f_x, \ldots, g_y$ follows at once from the chain rule, as described at the beginning of the section.

Theorem 3A may be generalized to a pair of functions in any number of variables. If we are given

$$F(x_1, x_2, \ldots, x_n, u, v) = 0, \qquad G(x_1, x_2, \ldots, x_n, u, v) = 0,$$

and a point on the locus of F and G at which

$$D_0 = \begin{vmatrix} F_u(x_1^0, x_2^0, \ldots, x_n^0, u^0, v^0) & F_v(x_1^0, x_2^0, \ldots, x_n^0, u^0, v^0) \\ G_u(x_1^0, x_2^0, \ldots, x_n^0, u^0, v^0) & G_v(x_1^0, x_2^0, \ldots, x_n^0, u^0, v^0) \end{vmatrix} \neq 0$$

then, under hypotheses analogous to those given in the above theorem, we may solve for u and v, obtaining

$$u = f(x_1, x_2, \ldots, x_n), \qquad v = g(x_1, x_2, \ldots, x_n) \tag{13}$$

in a neighborhood of $(x_1^0, x_2^0, \ldots, x_n^0, u^0, v^0)$. Furthermore,

$$f_{x_i} = - \frac{\begin{vmatrix} F_{x_i} & F_v \\ G_{x_i} & G_v \end{vmatrix}}{\begin{vmatrix} F_u & F_v \\ G_u & G_v \end{vmatrix}}, \qquad g_{x_i} = \frac{\begin{vmatrix} F_u & F_{x_i} \\ G_u & G_{x_i} \end{vmatrix}}{\begin{vmatrix} F_u & F_v \\ G_u & G_v \end{vmatrix}}. \tag{14}$$

EXAMPLE 1. Show that the locus of the equations

$$F(x, y, u, v) = x^2 - y^2 - u^3 + v^2 + 4 = 0,$$
$$G(x, y, u, v) = 2xy + y^2 - 2u^2 + 3v^4 + 8 = 0,$$

is representable in the form $u = f(x, y)$, $v = g(x, y)$ in a neighborhood of

$$P_0: \{x = 2, \quad y = -1, \quad u = 2, \quad v = 1\}.$$

Find the derivatives u_x, u_y, v_x, v_y at P_0.

Solution. We have

$$F_u = -3u^2, \qquad F_v = 2v, \qquad G_u = -4u, \qquad G_v = 12v^3, \qquad F_x = 2x,$$
$$F_y = -2y, \qquad G_x = 2y, \qquad G_y = 2x + 2y.$$

At the point in question, we find

$$D_0 = F_u G_v - F_v G_u \big|_{P_0} = -128.$$

Since F and G are polynomial expressions (and hence smooth), and since $D_0 \neq 0$, Theorem 3A is applicable. We conclude that u and v are expressible as functions of x and y. A computation yields

$$u_x = \frac{1}{128} \begin{vmatrix} 4 & 2 \\ -2 & 12 \end{vmatrix} = \frac{13}{32}, \qquad v_x = \frac{7}{16}, \qquad u_y = \frac{5}{32}, \qquad v_y = -\frac{1}{16}.$$

EXAMPLE 2. Show that there is a box of the form $|x - 1| < h$, $|y + 1| < k$, $|z - 2| < k(h$ and $k > 0)$ such that the part of the locus of the equations

$$z - 2x - 2y - 2 = 0 \qquad \text{and} \qquad z - x^2 - y^2 = 0$$

in that box lies along the curve determined by a pair of equations of the form $y = f(x)$ and $z = g(x)$. Show that the functions f, g, f', g' are continuous for $|x - 1| < h$. Also, find f and g explicitly. Does a similar box exist about the point $(-1, 1, 2)$?

Solution. Let $F(x, y, z) = z - 2x - 2y - 2$, $G(x, y, z) = z - x^2 - y^2$. Then the Implicit Function Theorem for a general pair of equations is applicable with $x_1 = x$, $u = y$, $v = z$. Since

$$F_y = -2, \quad F_z = 1, \quad G_y = -2y, \quad G_z = 1,$$

at the point $(1, -1, 2)$, we have

$$F_y = -2, \quad F_z = 1, \quad G_y = 2, \quad G_z = 1,$$

and

$$F_u G_v - F_v G_u = F_y G_z - F_z G_y = -4 \neq 0.$$

Therefore such a box exists. To find the explicit functions, we substitute z from $F = 0$ into $G = 0$ and find

$$x^2 + y^2 - 2x - 2y - 2 = 0, \quad z = 2x + 2y + 2$$

or

$$(x - 1)^2 + (y - 1)^2 = 4, \quad z = 2x + 2y + 2.$$

Therefore

$$y = 1 \pm \sqrt{4 - (x - 1)^2}$$

and

$$z = 2x + 2y + 2 \quad \text{for } |x - 1| \leq 2.$$

Since $y = -1$ when $x = 1$, we must take the branch $1 - \sqrt{4 - (x - 1)^2}$. Hence

$$f(x) = 1 - \sqrt{4 - (x - 1)^2},$$
$$g(x) = 2x + 4 - 2\sqrt{4 - (x - 1)^2}$$

for $|x - 1| \leq 2$ or $-1 \leq x \leq 3$. There is no box about the point $(-1, 1, 2)$, since x can never fall below -1 anywhere on the locus.

————————

The implicit function theorems may be extended to cover the situation where we have any number of variables and any number of equations, so long as there are fewer equations than variables. The next theorem, which states the result in full generality, may be established by using a complicated induction argument, of which Theorem 3A is the first step in the inductive process.

Theorem 3. *Suppose that $F_1, F_2, \ldots, F_k$ are functions of the $n + k$ variables $x_1, x_2, \ldots, x_n, u_1, u_2, \ldots, u_k$, and suppose that each F_i and all its first derivatives are continuous in a neighborhood of a point*

$$P_0(x_1^0, x_2^0, \ldots, x_n^0, u_1^0, u_2^0, \ldots, u_k^0).$$

The k equations

$$F_i(x_1^0, x_2^0, \ldots, x_n^0, u_1^0, u_2^0, \ldots, u_k^0) = 0, \qquad i = 1, 2, \ldots, k,$$

are assumed to hold, and the determinant

$$D = \begin{vmatrix} \dfrac{\partial F_1}{\partial u_1} & \dfrac{\partial F_1}{\partial u_2} & \cdots & \dfrac{\partial F_1}{\partial u_k} \\[2mm] \dfrac{\partial F_2}{\partial u_1} & \dfrac{\partial F_2}{\partial u_2} & \cdots & \dfrac{\partial F_2}{\partial u_k} \\ \vdots & & & \\ \dfrac{\partial F_k}{\partial u_1} & \dfrac{\partial F_k}{\partial u_2} & \cdots & \dfrac{\partial F_k}{\partial u_k} \end{vmatrix} \tag{15}$$

evaluated at the point P_0 is assumed to be different from zero. Then there are numbers h and r such that:

(a) *for each $(x_1, x_2, \ldots, x_n)$ with $|x_i - x_i^0| < h$ for $i = 1, 2, \ldots, n$ there is a unique solution $(u_1, u_2, \ldots, u_k)$ of the equations*

$$F_i(x_1, x_2, \ldots, x_n, u_1, u_2, \ldots, u_k) = 0, \qquad i = 1, 2, \ldots, k,$$

for which $|u_j - u_j^0| < r, j = 1, 2, \ldots, k$. The solution is defined by

$$u_j = f_j(x_1, x_2, \ldots, x_n), \qquad j = 1, 2, \ldots, k.$$

(b) *The functions f_j and all their first derivatives are continuous and the determinant D does not vanish for points P in the box described in (a). The derivatives $\partial f_j/\partial x_k$ are obtained by applying Cramer's Rule to each of the n systems*

$$\sum_{j=1}^{k} \frac{\partial F_i}{\partial u_j} \frac{\partial f_j}{\partial x_p} + \frac{\partial F_i}{\partial x_p} = 0, \qquad \begin{array}{l} i = 1, 2, \ldots, k, \\ p = 1, 2, \ldots, n. \end{array} \tag{16}$$

The determinant in (15) is called a **Jacobian determinant** or, simply, a **Jacobian.** Customary notations for the one in (15) are

$$\frac{\partial(F_1, F_2, \ldots, F_k)}{\partial(u_1, u_2, \ldots, u_k)} \quad \text{and} \quad J\left(\frac{F_1, F_2, \ldots, F_k}{u_1, u_2, \ldots, u_k}\right).$$

EXAMPLE 3. Given the functions

$$x_1^2 + 2x_2^2 - 3u_1^2 + 4u_1u_2 - u_2^2 + u_3^3 = 0,$$
$$x_1 + 3x_2 - 4x_1x_2 + 4u_1^2 - 2u_2^2 + u_3^3 = 0,$$
$$x_1^3 - x_2^3 + 4u_1^2 + 2u_2 - 3u_3^2 = 0.$$

Assume that the conditions of Theorem 3 are valid, and use (16) to compute $\partial u_1/\partial x_1$, $\partial u_2/\partial x_1$, $\partial u_3/\partial x_1$.

Solution. We have

$$(-6u_1 + 4u_2)\frac{\partial u_1}{\partial x_1} + (4u_1 - 2u_2)\frac{\partial u_2}{\partial x_1} + 3u_3^2\frac{\partial u_3}{\partial x_1} + 2x_1 = 0,$$

$$8u_1\frac{\partial u_1}{\partial x_1} - 4u_2\frac{\partial u_2}{\partial x_1} + 2u_3\frac{\partial u_3}{\partial x_1} + 1 - 4x_2 = 0,$$

$$8u_1\frac{\partial u_1}{\partial x_1} + 2\frac{\partial u_2}{\partial x_1} - 6u_3\frac{\partial u_3}{\partial x_1} + 3x_1^2 = 0.$$

This linear system of three equations in three unknowns is easily solved by Cramer's Rule. The details are left to the reader. (See Exercise 13.)

EXERCISES

In problems 1 through 6, use Theorem 3 to verify that there is a box $|x - x_0| < h$, $|y - y_0| < k$, $|z - z_0| < k$ such that all the points (x, y, z) in that box which satisfy

$$F(x, y, z) = 0 \quad \text{and} \quad G(x, y, z) = 0$$

lie on the locus of equations of the form $y = f(x)$, $z = g(x)$, where f and g are smooth for $|x - x_0| < h$. In problems 1 and 2, find in symmetric form the equations of the line. In problems 3 through 6, find in symmetric form the equations of the tangent line at (x_0, y_0, z_0). Let P_0 denote (x_0, y_0, z_0).

1. $F(x, y, z) = 2x + y - z - 2$, $G(x, y, z) = x + 2y + z - 1$, $P_0 = (2, -1, 1)$.
2. $F = 3x + 2y - z - 8$, $G = x + y + 2z - 1$, $P_0 = (1, 2, -1)$.
3. $F = x^2 + 2y^2 - z^2 - 2$, $G = 2x - y + z - 1$, $P_0 = (2, 1, -2)$.
4. $F = 2x^2 + y^2 - z^2 + 3$, $G = 3x + 2y + z - 10$, $P_0 = (1, 2, 3)$.
5. $F = x^3 + y^3 + z^3 - 3xyz - 14$, $G = x^2 + y^2 + z^2 - 6$, $P_0 = (2, -1, 1)$.
6. $F = x^2 - xy + 2y^2 - 4xz + 2z^2 - 10$, $G = xyz - 6$, $P_0 = (2, 3, 1)$.

In problems 7 through 10, show that there is a box

$$|x - x_0| < h, \quad |y - y_0| < h, \quad |u - u_0| < k, \quad |v - v_0| < k$$

such that all the points (x, y, u, v) in that box which satisfy the equations

$$F(x, y, u, v) = 0 \quad \text{and} \quad G(x, y, u, v) = 0$$

lie along the locus of the equations $u = f(x, y)$, $v = g(x, y)$, where f and g are smooth for $|x - x_0| < h$ and $|y - y_0| < h$. Find the values of f_x, f_y, g_x, and g_y at (x_0, y_0). Let P_0 denote (x_0, y_0, u_0, v_0).

7. $F = 2x - 3y + u - v$, $G = x + 2y + u + 2v$, $P_0 = (0, 0, 0, 0)$.
8. $F = 2x - y + 2u - v$, $G = 3x + 2y + u + v$, $P_0 = (0, 0, 0, 0)$.
9. $F = x - 2y + u + v - 8$, $G = x^2 - 2y^2 - u^2 + v^2 - 4$,
   $P_0 = (3, -1, 2, 1)$.
10. $F = x^2 - y^2 + uv - v^2 + 3$, $G = x + y^2 + u^2 + uv - 2$,
    $P_0 = (2, 1, -1, 2)$.

11. Show that there is a box (as in problems 1 through 6) about each point, except $(2, -3, 6)$ and $(-10, -15, 30)$ of the locus of

$$z^2 - \tfrac{9}{2}x^2 - 2y^2 = 0 \qquad \text{and} \qquad x + y + z - 5 = 0.$$

Solve explicitly for y and z in terms of x.

12. For what values of (x_0, y_0, u_0, v_0) does there exist a box as in problems 7 through 10 if

$$F = -x + u^2 - v^2, \qquad G = -y + 2uv?$$

Find explicit functions $f(x, y)$ and $g(x, y)$ which are smooth near $x_0 = 3, y_0 = -4$ and which are such that

$$f(x_0, y_0) = -2 \qquad \text{and} \qquad g(x_0, y_0) = 1.$$

13. Complete Example 3 and determine $\partial u_1/\partial x_1, \partial u_2/\partial x_1, \partial u_3/\partial x_1$.

14. For the equations in Example 3, determine

$$\frac{\partial u_1}{\partial x_2}, \qquad \frac{\partial u_2}{\partial x_2}, \qquad \frac{\partial u_3}{\partial x_2}.$$

15. Given that

$$\begin{aligned} u &= f(x, y), & x &= \phi(s, t), \\ v &= g(x, y), & y &= \psi(s, t), \end{aligned}$$

show, by using the chain rule for partial derivatives (see Chapter 4, Section 3) that

$$J\left(\frac{u, v}{x, y}\right) \cdot J\left(\frac{x, y}{s, t}\right) = J\left(\frac{u, v}{s, t}\right).$$

16. Given that

$$u^5 + v^5 + x^5 + 2y = 0, \qquad u^3 + v^3 + y^3 + 2x = 0,$$

find, under the appropriate hypotheses, u_x, u_y, u_{xx}.

17. Given that

$$F(x, y, u, v) = 0, \qquad G(x, y, u, v) = 0,$$

and that the hypotheses of Theorem 3 are satisfied. State conditions under which the relation

$$\frac{\partial u}{\partial x} \frac{\partial y}{\partial u} + \frac{\partial v}{\partial x} \frac{\partial y}{\partial v} = 0$$

is true.

3. TRANSFORMATIONS

We have already introduced the general topic describing a transformation from one set of elements to another. If T is a transformation and P is an element in its domain, we call the element $T(P)$ the **image** of P under T. The totality of images $T(P)$ for P in the domain is called the *range* of T. For a review of the basic facts on transformations, the reader is referred to Chapter 7, Section 3. In that section the primary emphasis was on *linear transformations*.

Many of the transformations we studied in elementary calculus are nonlinear. For example, the transformation from rectangular to polar coordinates in the plane

$$r = \sqrt{x^2 + y^2}, \qquad \theta = \arctan \frac{y}{x},$$

is a nonlinear transformation. Similarly, the transformations from rectangular to cylindrical coordinates and from rectangular to spherical coordinates are nonlinear. Also, it is frequently helpful to make nonlinear changes of variables in both single and multiple integrals. In this section we shall establish some general properties of nonlinear transformations and then show how these results may be applied in the study of multiple integration.

Let D be a domain in the xy-plane and T a transformation from D to a set of elements in another plane, which we denote the uv-plane. We write

$$T: \quad u = f(x, y), \qquad v = g(x, y), \qquad x, y \quad \text{in} \quad D.$$

DEFINITION. We say that the transformation T is **continuously differentiable** in D if and only if f and g are continuous and the first derivatives of f and g are continuous throughout D.

If T is a transformation from a domain in n-dimensional space to n-dimensional space, we write

$$T: \quad u_1 = f_1(x_1, x_2, \ldots, x_n), u_2 = f_2(x_1, x_2, \ldots, x_n), \ldots,$$
$$u_n = f_n(x_1, x_2, \ldots, x_n),$$

with the analogous definition for continuously differentiable transformations.

Just as the inverse relation of a function need not be a function, so the inverse relation of a transformation need not be a transformation. *The inverse of a transformation T is also a transformation if and only if T establishes a one-to-one relation between its domain and its range.* That is, if for each element Q in the range of T, there is only one P in its domain such that $T(P) = Q$, then T has an inverse, and we denote this inverse transformation by T^{-1}. (See Chapter 7, page 322.)

The implicit function theorems are the basis for the following inversion theorem for transformations.

Theorem 4. (Inversion Theorem.) *Suppose*

$$T: \quad u_1 = F_1(x_1, x_2), \qquad u_2 = F_2(x_1, x_2), \qquad x_1, x_2 \quad \text{in} \quad D, \qquad (1)$$

is a continuously differentiable transformation for (x_1, x_2) interior to D. Let $P_0 = (x_1^0, x_2^0)$ be a point in D and let $Q_0 = (u_1^0, u_2^0)$ be the image of P_0 under T. We define J_0 to be the Jacobian

$$J \left(\frac{F_1, F_2}{x_1, x_2} \right)$$

evaluated at P_0, and we suppose that

$$J_0 \neq 0. \qquad (2)$$

Then there exist positive numbers h and k such that whenever $Q = (u_1, u_2)$ is within h of (u_1^0, u_2^0), that is,

$$|u_1 - u_1^0| < h, \qquad |u_2 - u_2^0| < h,$$

there is one and only one point P which satisfies $T(P) = Q$ with $|x_1 - x_1^0| < k$, $|x_2 - x_2^0| < k$. If we define

$$x_1 = f_1(u_1, u_2), \qquad x_2 = f_2(u_1, u_2), \tag{3}$$

then f_1 and f_2 are continuously differentiable for all Q for which

$$|u_1 - u_1^0| < h, \qquad |u_2 - u_2^0| < h.$$

Proof. If we rewrite equations (1) in the form

$$F(u_1, u_2, x_1, x_2) \equiv u_1 - F_1(x_1, x_2) = 0,$$
$$G(u_1, u_2, x_1, x_2) \equiv u_2 - F_2(x_1, x_2) = 0,$$

we see that all the conditions of Theorem 3 are fulfilled. Therefore we may solve for x_1 and x_2 in terms of u_1 and u_2, which is statement (3) precisely. The condition $J_0 \neq 0$ is identical with the condition $D_0 \neq 0$ in Theorem 3.

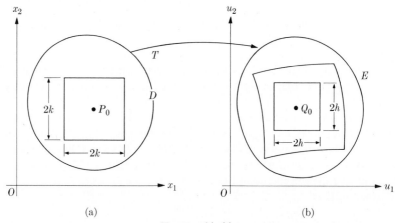

FIGURE 11–11

Remarks. (i) A schematic diagram of Theorem 4 is shown in Fig. 11–11. The transformation T takes the domain D into some region E in the $u_1 u_2$-plane. The inverse mapping is defined in a square of side $2h$ about Q_0.

(ii) Theorem 4 may be extended to n equations of the form

$$u_i = F_i(x_1, x_2, \ldots, x_n), \qquad i = 1, 2, \ldots, n.$$

The inversion theorem states that we may solve (at least locally) for each x_i and obtain $x_i = f_i(u_1, u_2, \ldots, u_n)$, $i = 1, 2, \ldots, n$, with the f_i continuously differentiable.

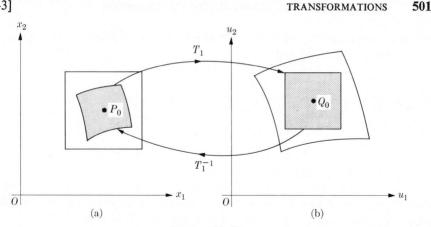

FIGURE 11-12

(iii) We define T_1 to be the transformation T *restricted* to those points P in the square $|x_i - x_i^0| < k$, $i = 1, 2$ which correspond to points Q lying in the image square defined by $|u_i - u_i^0| < h$, $i = 1, 2$. Then T_1 is one-to-one, and its inverse T_1^{-1} is just the transformation $x_i = f_i(u_1, u_2)$, $i = 1, 2$. (See Fig. 11-12.)

(iv) The theorem says nothing about the sizes of h and k, and extreme caution must be used in the applications. It may happen that all other hypotheses of the theorem are satisfied, but if h and k are not taken sufficiently small the inversion from (1) to (3) may not be possible. [See part (d) of the examples below.]

(v) The theorem fails, and we have no information when the Jacobian vanishes. Actually, if $J_0 = 0$ anything can happen, and we exhibit this fact with the following illustrative examples.

(a) The transformation

$$u = x^3, \qquad v = y$$

is one-to-one for all (x, y), since the inversion formulas are

$$x = \sqrt[3]{u}, \qquad y = v.$$

However, the Jacobian

$$J\left(\frac{u, v}{x, y}\right) = 3x^2$$

vanishes along the entire y-axis.

(b) The transformation $u = x^2$, $v = y$ is not one-to-one for all (x, y), since for any $a \neq 0$ we have (a, b) and $(-a, b)$ transforming into (a^2, b). However, the Jacobian

$$J\left(\frac{u, v}{x, y}\right) = 2x$$

vanishes along the entire y-axis, exactly as in Example (a) above, which is one to one.

(c) The transformation

$$u = x^2 - y^2, \qquad v = 2xy \tag{4}$$

has Jacobian

$$J\left(\frac{u, v}{x, y}\right) = 4(x^2 + y^2),$$

which vanishes only at the origin. Setting $w = u + iv$ and $z = x + iy$, we see that

$$u + iv = w = z^2 = (x + iy)^2 = x^2 - y^2 + 2ixy,$$

and so the transformation (4) is the same as the complex transformation

$$w = z^2.$$

Therefore, to each $w \neq 0$ there correspond two numbers z such that $w = z^2$ and the transformation is *not one to one*.

(d) In Example (c) we select for domain D the annular ring between the circles $x^2 + y^2 = 1$ and $x^2 + y^2 = 9$. Then the Jacobian of (4) never vanishes in D. Yet the transformation is not one to one in D. [That is, $(2, 0)$ and $(-2, 0)$ both map into $(4, 0)$.] This example shows that the size of h and k must be suitably chosen before the theorem is valid. The theorem states that the transformation (4) is one to one in any sufficiently small rectangle in the annular ring. Actually, it is one to one within *any* rectangle whose interior is entirely within D.

In discussing transformations it is desirable to find the images of the lines $x = \text{const}$ and $y = \text{const}$, or those of other convenient loci. We illustrate with examples.

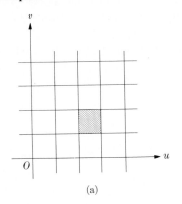

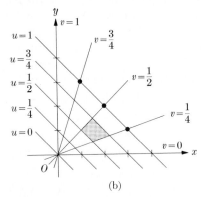

(a)   (b)

FIGURE 11–13

EXAMPLE 1.   Given the transformation

$$x = u - uv, \quad y = uv. \qquad \text{Find} \quad \frac{\partial(x, y)}{\partial(u, v)}$$

and the inverse transformation. In the (x, y)-plane draw the images of the lines $u = 0$, $\frac{1}{4}$, $\frac{1}{2}$, $\frac{3}{4}$, 1 and $v = 0$, $\frac{1}{4}$, $\frac{1}{2}$, $\frac{3}{4}$, 1, and find the image of the square $\frac{1}{2} \leq u \leq \frac{3}{4}$ and $\frac{1}{4} \leq v \leq \frac{1}{2}$ (Fig. 11–13).

Solution. We have

$$\frac{\partial(x, y)}{\partial(u, v)} = \begin{vmatrix} 1 - v & -u \\ v & u \end{vmatrix} = u - uv + uv = u.$$

Solving for u and v, we obtain

$$u = x + y, \qquad v = \frac{y}{x + y}.$$

The line $u = c$ corresponds to $x + y = c$ (if $c \neq 0$) and the line $v = d$ corresponds to $dx = (1 - d)y$. The various lines are drawn and the image of the rectangle is shaded in Fig. 11–13.

EXAMPLE 2. Discuss the transformation

$$u = e^x \cos y, \qquad v = e^x \sin y.$$

Draw the lines $x = \pm k\pi/6$, $y = \pm k\pi/6$ for $k = 0, 1, 2, 3$ and draw their images in the (u, v) plane. Find

$$\frac{\partial(u, v)}{\partial(x, y)}.$$

Solution. The reader will observe that if we set $w = u + iv$ and $z = x + iy$, the transformation becomes $w = e^z$. Introducing polar coordinates (ρ, ϕ) with $\rho > 0$ in the w-plane, we see that we may write

$$\rho = e^x, \qquad \phi = y.$$

Thus the transformation is periodic in y with period 2π. The lines $x = c$ correspond to circles $\rho = e^c$ and lines $y = d$ correspond to rays $\phi = d$. We obtain

$$\frac{\partial(u, v)}{\partial(x, y)} = \begin{vmatrix} e^x \cos y & -e^x \sin y \\ e^x \sin y & e^x \cos y \end{vmatrix} = e^{2x}.$$

The lines and their images are drawn in Fig. 11–14.

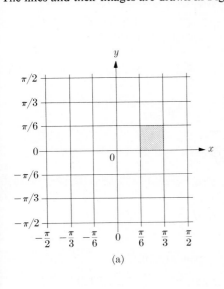

(a)

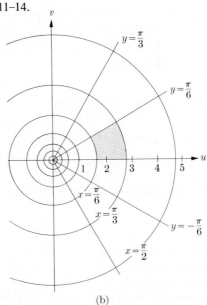

(b)

FIGURE 11–14

EXERCISES

In each of problems 1 through 7, find the Jacobian

$$J\left(\frac{u, v}{x, y}\right)$$

and find the inverse transformation. In the uv-plane, draw the images of the lines $x = \frac{1}{4}$, $\frac{1}{2}, \frac{3}{4}, 1$ and $y = -\frac{1}{2}, -\frac{1}{4}, 0, \frac{1}{4}, \frac{1}{2}$.

1. $u = x, \quad v = y + x^2$ 2. $u = x + xy, \quad v = y, \quad y > -1$

3. $u = 2x - 3y, \quad v = x + 2y$ 4. $u = 2x + 3y, \quad v = x + 2y$

5. $u = x/(1 + x + y), \quad v = y/(1 + x + y), \quad x + y > -1$

6. $u = x^2 - y^2, \quad v = 2xy, x > 0$

7. $u = x \cos(\pi y/2), \quad v = x \sin(\pi y/2), \quad x > 0, \quad -1 < y < 1$

8. Show that the transformation $u = x^2 - y^2$, $v = 2xy$ is one to one if the domain D is any rectangle situated in the half-plane $y > 0$.

9. Given the transformation

$$T: \quad u = \frac{x}{x^2 + y^2}, \qquad v = \frac{y}{x^2 + y^2},$$

show that boundary of any circle in the xy-plane with center at $(0, 0)$ is mapped into the boundary of a circle in the uv-plane. Compare the radii of the two circles.

10. Given the transformation

$$T: \quad u = -x + \sqrt{x^2 + y^2}, \qquad v = -x - \sqrt{x^2 + y^2}.$$

Show that T is not a one-to-one transformation. Decide whether or not T is one to one in the rectangle

$$R_1: \quad 1 \le x \le 3, \qquad -4 \le y \le 4;$$

in the rectangle

$$R_2: \quad -1 \le x \le 1, \qquad 2 \le y \le 6.$$

11. Given the transformation

$$T: \quad u = f(x, y), \qquad v = g(x, y),$$

with Jacobian

$$J\left(\frac{u, v}{x, y}\right) \ne 0$$

at a point P_0. Show that if f and g are continuously differentiable near P_0, then

$$J\left(\frac{u, v}{x, y}\right) \cdot J\left(\frac{x, y}{u, v}\right) = 1.$$

12. Suppose that the transformation

$$T: \quad u = f(x, y), \qquad v = g(x, y)$$

is continuously differentiable and one to one. We have the relation

$$\frac{\partial u}{\partial x}\frac{\partial x}{\partial u} + \frac{\partial u}{\partial y}\frac{\partial y}{\partial u} = 1. \qquad\qquad (*)$$

(a) Denoting the inverse transformations $x = \phi(u, v)$, $y = \psi(u, v)$, differentiate the expression $v = g[\phi(u, v), \psi(u, v)]$ with respect to u to obtain

$$0 = \frac{\partial v}{\partial x}\frac{\partial x}{\partial u} + \frac{\partial v}{\partial y}\frac{\partial y}{\partial u}. \qquad (**)$$

(b) By differentiating (*) and (**) with respect to u (employing the chain rule) and then using Cramer's Rule, find expressions for $\partial^2 x/\partial u^2$ and $\partial^2 y/\partial u^2$.

13. (a) Find the Jacobian of the transformation

$$u = \frac{x}{x^2 + y^2 + z^2}, \qquad v = \frac{y}{x^2 + y^2 + z^2}, \qquad w = \frac{z}{x^2 + y^2 + z^2}.$$

(b) Show that the surface of a sphere in xyz-space with center at the origin transforms into a sphere in uvw-space.

14. (a) Calculate the Jacobian of the transformation

$$u = \cos x \cosh y, \qquad v = \sin x \cosh y, \qquad w = \sinh z.$$

(b) What is the image of the surface $\cosh^2 y - \sinh^2 z = 1$?

4. PRODUCT OF TRANSFORMATIONS. TANGENT LINEAR TRANSFORMATION

The product of two transformations was defined in Chapter 7, Section 3 and, for *linear* transformations, it was shown there that the matrix of the product of two transformations is the product of the matrices. For a nonlinear transformation, the Jacobian occupies the position analogous to that of the determinant of the (square) matrix of a linear transformation. For continuously differentiable transformations, we now establish a result for the product of two nonlinear transformations which corresponds to the result for the product of two linear transformations.

Suppose that T and U are two transformations

$$T: \quad \xi_i = f_i(x_1, x_2, \ldots, x_n), \quad i = 1, 2, \ldots, n, \quad (x_1, x_2, \ldots, x_n) \text{ in } D,$$
$$\tag{1}$$
$$U: \quad u_i = F_i(\xi_1, \xi_2, \ldots, \xi_n), \quad i = 1, 2, \ldots, n, \quad (\xi_1, \xi_2, \ldots, \xi_n) \text{ in } G,$$

and assume that all the image points of D under T are in G. Furthermore, suppose that this image set is a region in G. Then the *product transformation UT* is given by

$$UT: \quad u_i = F_i[f_1(x_1, x_2, \ldots, x_n), f_2(x_1, x_2, \ldots, x_n), \ldots, f_n(x_1, x_2, \ldots, x_n)]$$

for $i = 1, 2, \ldots, n$ and $(x_1, x_2, \ldots, x_n)$ in D.

DEFINITION. The **Jacobian matrix** of a transformation T—that is, the matrix of which the determinant is the Jacobian—is the matrix with its (ij)th element given by $\partial \xi_i/\partial x_j$. We note that the Jacobian matrix of U has $\partial u_i/\partial \xi_j$ for the element in the ith row and jth column.

Using the chain rule, we observe that the (ij)th element of the Jacobian matrix for the product UT is given by the formula

$$\frac{\partial u_i}{\partial x_j} = \sum_{k=1}^{n} \frac{\partial u_i}{\partial \xi_k} \frac{\partial \xi_k}{\partial x_j}. \tag{2}$$

The next theorem is a direct result of formula (2).

Theorem 5.   (a) *The Jacobian matrix of the product UT of two continuously differentiable transformations U and T is the product of the Jacobian matrix of U with the Jacobian matrix of T.*

(b) *The Jacobian of the product of two such transformations is the product of their Jacobians.*

Proof.   Part (a) follows directly from (2) and the definition of the product of two matrices. Part (b) follows from the theorem that the determinant of the product of two square matrices is the product of their determinants (Theorem 9, Chapter 7, page 327).

Corollary 1.   *Suppose that T is a one-to-one transformation and that its Jacobian never vanishes. Then T has an inverse T^{-1}, and the Jacobian matrix of T^{-1} is the inverse of the Jacobian matrix for T. Therefore the Jacobian (determinant) of T^{-1} is the reciprocal of the Jacobian of T.*

Proof.   Let A be the Jacobian matrix of T and B that of T^{-1}. Since $T^{-1}T$ (restricted to D) is the identity, Theorem 5 shows that $BA = I$. Hence $B = A^{-1}$.

Let T be a transformation of the type given in (1) and suppose that

$$P_0(x_1^0, x_2^0, \ldots, x_n^0)$$

is a fixed point in D, the domain of T. The elements of the Jacobian matrix $\partial f_i / \partial x_j$ evaluated at P_0 are denoted by

$$a_{ij} = \frac{\partial f_i}{\partial x_j}(P_0), \qquad i, j = 1, 2, \ldots, n. \tag{3}$$

We now define a special linear transformation (nonhomogeneous) which is associated with the transformation T and the point P_0.

DEFINITION.   The transformation

$$V: \quad v_i = \xi_i^0 + \sum_{j=1}^{n} a_{ij}(x_j - x_j^0), \qquad i = 1, 2, \ldots, n, \tag{4}$$

where the a_{ij} are given by (3) and where $\xi_i^0 = f_i(x_1^0, x_2^0, \ldots, x_n^0), i = 1, 2, \ldots, n$, is called the **tangent linear transformation to** T **at** P_0. Note that the Jacobian matrix of V is the matrix $A = (a_{ij})$, which is constant.

Remark. The tangent linear transformation is obtained by expanding each f_i in a Taylor series about the point P_0 and keeping only the linear terms. In one dimension, V is the equation of the tangent line to the curve $\xi = f(x)$.

Corollary 2. *Suppose that T as given in (1) is a continuously differentiable transformation with Jacobian which does not vanish at a point P_0. Then the tangent linear transformation to T at P_0 given by (4) has an inverse V^{-1} which is a linear transformation. The Jacobian matrix of TV^{-1} evaluated at $Q_0 = V(P_0)$ is the identity matrix.*

Proof. The Jacobian matrix of V is, by definition, that of T evaluated at P_0. The Jacobian matrix of TV^{-1} is the product of the Jacobian matrices. The matrix of $V^{-1}(Q_0)$ is the reciprocal of that of V; since $P_0 = V^{-1}(Q_0)$, the result follows.

EXAMPLE 1. Given the transformation

$$T: \begin{array}{l} u = 2 + 2x + y + 4x^2 + 4xy + y^2, \\ v = -2 - x \quad\;\; + 2x^2 + 3xy + y^2, \end{array}$$

and the point $P_0(-1, 2)$, find the transformations V and TV^{-1} of Corollary 2 above.

Solution. In this case, it is best to expand u and v in a power series about P_0 by setting $x = [(x + 1) - 1]$ and $y = [(y - 2) + 2]$. The result is

$$T: \begin{array}{l} u = 2 + 2(x + 1) + (y - 2) + 4(x + 1)^2 + 4(x + 1)(y - 2) + (y - 2)^2, \\ v = -1 + (x + 1) + (y - 2) + 2(x + 1)^2 + 3(x + 1)(y - 2) + (y - 2)^2. \end{array}$$

At P_0, the coefficients of $x + 1$ and $y - 2$ give the matrix of V. We have

$$V: \begin{array}{l} \xi = 2 + 2(x + 1) + (y - 2), \\ \eta = -1 + (x + 1) + (y - 2). \end{array}$$

Solving for $(x + 1)$ and $(y - 2)$, we find

$$V^{-1}: \begin{array}{l} x + 1 = (\xi - 2) - (\eta + 1), \\ y - 2 = -(\xi - 2) + 2(\eta + 1). \end{array}$$

Therefore a computation yields

$$TV^{-1}: \begin{array}{l} u = 2 + (\xi - 2) + (\xi - 2)^2, \\ v = -1 + (\eta + 1) + (\xi - 2)(\eta + 1). \end{array}$$

EXAMPLE 2. Given the transformations

$$T: \begin{array}{l} x = r \cos \theta, \\ y = r \sin \theta, \end{array} \qquad U: \begin{array}{l} u = 3x - 2y^3, \\ v = x^3 + 5y. \end{array}$$

Verify that the Jacobian matrix of UT is the product of the Jacobian matrices of U and T.

Solution. Denoting the Jacobian matrices of T, U, and UT by $A(T)$, $A(U)$, and $A(UT)$, respectively, we have

$$A(T) = \begin{pmatrix} \cos\theta & -r\sin\theta \\ \sin\theta & r\cos\theta \end{pmatrix}, \qquad A(U) = \begin{pmatrix} 3 & -6y^2 \\ 3x^2 & 5 \end{pmatrix}.$$

For the transformation

$$UT: \quad \begin{aligned} u &= 3r\cos\theta - 2r^3\sin^3\theta, \\ v &= r^3\cos^3\theta + 5r\sin\theta, \end{aligned}$$

we find

$$A(UT) = \begin{pmatrix} 3\cos\theta - 6r^2\sin^3\theta & -3r\sin\theta - 6r^3\sin^2\theta\cos\theta \\ 3r^2\cos^3\theta + 5\sin\theta & -3r^3\cos^2\theta\sin\theta + 5r\cos\theta \end{pmatrix}.$$

The rule for the multiplication of matrices shows that

$$A(U)A(T) = \begin{pmatrix} 3 & -6r^2\sin^2\theta \\ 3r^2\cos^2\theta & 5 \end{pmatrix}\begin{pmatrix} \cos\theta & -r\sin\theta \\ \sin\theta & r\cos\theta \end{pmatrix} = A(UT).$$

EXERCISES

In each of problems 1 through 4, verify that the Jacobian matrix of the product UT of two transformations is the product of the Jacobian matrices of each of the transformations.

1. $T: \begin{aligned} x &= r\cos\theta \\ y &= r\sin\theta \end{aligned}$ $\qquad$ $U: \begin{aligned} u &= x + 2y \\ v &= -x + 3y \end{aligned}$

2. $T: \begin{aligned} r &= \sqrt{x^2 + y^2} \\ \theta &= \arctan(x/y) \end{aligned}$ $\qquad$ $U: \begin{aligned} u &= 2r + \cos\theta \\ v &= r - \sin\theta \end{aligned}$

3. $T: \begin{aligned} r &= x + 2y - z \\ \theta &= 2x - y + 3z \\ \phi &= x^2 + y^2 + z^2 \end{aligned}$ $\qquad$ $U: \begin{aligned} u &= r - 2\theta + 3\phi \\ v &= 2r - \theta^2 + \phi \\ w &= r^2 - \theta^2 + \phi^2 \end{aligned}$

4. $T: \begin{aligned} x &= r\cos\theta \\ y &= r\sin\theta \end{aligned}$ $\qquad$ $U: \begin{aligned} u &= \sqrt{x^2 + y^2} \\ v &= \arctan(y/x) \end{aligned}$

In each of problems 5 through 8, find the tangent linear transformation V to the given transformation T at the point P_0. Also find TV^{-1}.

5. $T: \begin{aligned} u &= x - y - 4xy \\ v &= x + y + 2x^2 - 2y^2 \end{aligned}$ $\qquad\qquad$ $P_0 = (0, 0)$

6. $T: \begin{aligned} u &= 2x - y + (x + y)^2 \\ v &= x + y - (2x - y)^2 \end{aligned}$ $\qquad\qquad$ $P_0 = (0, 0)$

7. $T: \begin{aligned} u &= 1 + x - y + (x - y)(2x + y) \\ v &= -1 + 2x + y + (x - y)^2 + (2x + y)^2 \end{aligned}$ $\qquad$ $P_0 = (0, 0)$

8. $T: \begin{aligned} u &= 1 + x - y + z - x^2 + 2y^2 - z^2 \\ v &= -1 + x - z + xy + yz \\ w &= 2 + y - z + x^2 - z^2 \end{aligned}$ $\qquad$ $P_0 = (0, 0, 0)$

9. Given

$$TV^{-1}: \quad \begin{aligned} x &= u + \frac{u^2 - v^2}{40}, \\ y &= v + \frac{uv}{20}. \end{aligned}$$

Note that the image of the point $(1, 0)$ is $(\frac{41}{40}, 0)$. By expanding x and y in a Taylor series about $(1, 0)$, show that the image of the square $|u - 1| \leq \frac{1}{2}$, $|v| \leq \frac{1}{2}$ is contained in the square $|x - \frac{41}{40}| \leq \frac{43}{40} \cdot \frac{1}{2}$, $|y| \leq \frac{43}{40} \cdot \frac{1}{2}$.

10. Verify that the transformation

$$T: \quad \begin{aligned} s &= x^4 - 2x^2 y^2 + y^4 \\ t &= -x^4 + 2x^2 y^2 \end{aligned}$$

is the product SWU where

$$S: \begin{aligned} s &= u^2 \\ t &= v^2 - 2u^2 \end{aligned} \qquad W: \begin{aligned} u &= \xi \\ v &= \sqrt{\xi^2 + \eta^2} \end{aligned} \qquad U: \begin{aligned} \xi &= x^2 - y^2 \\ \eta &= y^2 \end{aligned}.$$

5. RESULTS FROM THE THEORY OF FUNCTIONS

In order to establish the formula for the change of variables in a multiple integral, we shall need several definitions and theorems which are usually presented in more advanced courses in analysis. While the proofs of the theorems are not essentially more complicated or sophisticated than many of the proofs which we establish in detail, the digression necessary to perform all the required steps would be lengthy. Moreover, the time allotted for the customary calculus course does not make allowance for this additional material.

We state the definitions and theorems for sets and regions in the plane. With minor adjustments the statements are valid for any finite-dimensional Euclidean space. For example, a circle in the plane is translated into a sphere in three-space and into a line segment in one-space.

DEFINITIONS. A set of points F in the plane is **bounded** if it can be enclosed in a sufficiently large circle. A point P is a **limit point** of a set F if every circle containing P contains infinitely many points of F. A set F is **closed** if all the limit points of F are in F. A point P is an **interior point** of a set F if P is the center of a circle which itself is entirely in F. A set G is **open** if every point of G is an interior point.

Remarks. Examples of sets with the above properties are easily found. All points (x, y) satisfying $x^2 + y^2 < R^2$ form an open set $(R > 0)$. All points satisfying $x^2 + y^2 \leq R^2$ form a closed set. The collection of all points (x, y) where x and y are rational numbers forms a set which is neither open nor closed. All the points in the plane form a set which is both open and closed. No set containing only a finite number of points can have a limit point.

The next theorem is usually proved in courses in the Theory of Functions.

Theorem 6. (Heine-Borel.) *Let F be a closed and bounded set in the plane. Suppose that to each point of F we associate a circle $C(P, r_P)$. Call this collection of circles $\mathfrak{F}$. Then there is a finite number of circles of $\mathfrak{F}$ which cover the set F; that is, each point of F is in at least one of this finite number of circles.*

Remarks. (i) The Heine-Borel theorem, which is fundamental in many branches of analysis, is valid for sets in any (finite) number of dimensions. (ii) The collection $\mathfrak{F}$ need not consist of circles. Squares, triangles or any type of neighborhood of each point P would do equally well. (iii) The hypotheses that F is closed and bounded are essential. If either one is dropped the result is false. It is instructive for the student to find examples exhibiting this fact.

The next theorem, which is more useful for our purposes, is an equivalent form of the Heine-Borel Theorem.

Theorem 6'. (Heine-Borel Theorem, Second Form.) *Let $\mathfrak{F}$ be any family of open sets covering the closed and bounded set F; that is, each point of F is in at least one of the sets of $\mathfrak{F}$. Then there is a number $\rho > 0$ such that any circle $C(P, \rho)$ for any P in F is contained in some one of the sets in $\mathfrak{F}$.*

DEFINITION. A function f defined over a set S in the plane is said to be **uniformly continuous on** S if and only if f is continuous on S and, for each $\epsilon > 0$, there is a $\delta > 0$ such that

$$|f(x_1, y_1) - f(x_2, y_2)| < \epsilon \quad \text{whenever} \quad \sqrt{(x_1 - x_2)^2 + (y_1 - y_2)^2} < \delta$$

and $(x_1, y_1), (x_2, y_2)$ are in S.

Remark. The crucial assertion in uniform continuity is the fact that δ *depends only on* ϵ and not on the two selected points $(x_1, y_1), (x_2, y_2)$ in S (so long as the distance between them is less than δ).

Theorem 7. (Uniform Continuity.) *If f is continuous on a bounded and closed set S, then f is uniformly continuous on S.*

The hypotheses that S is closed and bounded are essential. For example, the function $f = 1/[1 - (x^2 + y^2)]$ is continuous on the open set $S: x^2 + y^2 < 1$, but it is not uniformly continuous on S. The theorem on uniform continuity is true for functions in any (finite) number of variables.

6. CHANGE OF VARIABLES IN A MULTIPLE INTEGRAL

One of the principal techniques used in the evaluation of single integrals is the method of substitution. As was shown in *First Course*, Chapter 15, by making a substitution or change of variable of the form $x = g(u)$ we are able to transform an integral

$$\int f(x)\, dx \tag{1}$$

into

$$\int f[g(u)]g'(u)\, du. \tag{2}$$

It sometimes happens that (2) is simpler to evaluate than (1). If we start with a definite integral of the form

$$\int_a^b f(x)\, dx$$

then, after the change of variable $x = g(u)$, we obtain

$$\int_c^d f[g(u)]g'(u)\, du, \tag{3}$$

where $g(c) = a$ and $g(d) = b$. This method is always valid, provided that g and g' are (single valued and) continuous and f is defined and continuous for all the values of $g(u)$ for u on $[c, d]$. (See Morrey, *University Calculus*, Theorem 8–19.) In this section we shall establish a general theorem concerning change of variables in multiple integrals, a special case of which was proved in Chapter 7, Section 5.

We shall employ Taylor's theorem for functions of several variables, established in Chapter 4, Section 10 (Theorem 8, page 165). Since we need this theorem only in the special case where the Taylor series is carried to one term, we state the result in the form of a lemma.

Lemma 1. (Taylor's Theorem with one term.) *Suppose that f is continuously differentiable in a square $|x - a| < h$, $|y - b| < h$. If (x_1, y_1) and (x_2, y_2) are any two points in this square, then*

$$f(x_2, y_2) = f(x_1, y_1) + f_x(\alpha, \beta)(x_2 - x_1) + f_y(\alpha, \beta)(y_2 - y_1),$$

where (α, β) is on the line segment joining the two points.

Remark. Lemma 1 may also be proved by setting

$$\phi(t) = f[x_1 + t(x_2 - x_1),\, y_1 + t(y_2 - y_1)], \qquad 0 \le t \le 1, \tag{4}$$

and using the Theorem of the Mean. (See Exercise 5.)

If T is a *linear* transformation from the xy-plane to the uv-plane, there is a simple relation between the area of a region S in the xy-plane and the area of the image $T(S)$ in the uv-plane. In fact, Theorem 12 of Chapter 7 (page 335) shows that

$$\text{area } (T(S)) = |J_0| \cdot \text{area } (S), \tag{5}$$

where J_0 is the determinant of the transformation matrix. We observe that for linear transformations this determinant is nothing but the Jacobian of T. It is clear that for nonlinear transformations, a formula such as (5) will not hold in general. However, we can show that if the region S is "very small," then a relation similar to (5) holds "approximately." The next lemma establishes this fact for a "small" region. In the statement of the lemma, we use the notation $A(S)$ for the area of a region S and $C(P_0, \delta)$ for the *circle of radius δ with center at P_0*.

Lemma 2.   Suppose that T, given by

$$T: \begin{matrix} u = f(x, y), \\ v = g(x, y), \end{matrix}$$

is continuously differentiable and one-to-one throughout its domain D.   Suppose that $P_0(x_0, y_0)$ is a point interior to D at which the Jacobian J_0 of T is not zero. Then for each $\epsilon > 0$, there is a $\delta > 0$ such that

$$(1 - \epsilon)^2 |J_0| A(E) \le A^-(G) \le A^+(G) \le (1 + \epsilon)^2 |J_0| A(E) \tag{6}$$

whenever E is a region having area* contained in $C(P_0, \delta)$ and G is the image of E under T.

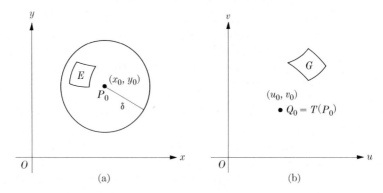

FIGURE 11–15

Proof. Figure 11–15 shows the region E and its image G under T. Let V be the tangent linear transformation

$$V: \begin{matrix} \xi = u_0 + a(x - x_0) + b(y - y_0), \quad a = f_x(P_0), \quad b = f_y(P_0), \\ \eta = v_0 + c(x - x_0) + d(y - y_0), \quad c = g_x(P_0), \quad b = g_y(P_0), \end{matrix}$$

with $u_0 = f(x_0, y_0)$, $v_0 = g(x_0, y_0)$.   Then, from Corollary 2 of Theorem 5, we conclude that V has an inverse (since $J_0 \ne 0$ by hypothesis); we conclude also that the transformation $TV^{-1} \equiv \overline{T}$, which is a mapping from the ξ, η-plane to the u, v-plane (Fig. 11–16), is one-to-one and continuously differentiable on its domain.   The corollary also showed that the Jacobian of $\overline{T}$ is the identity at (ξ_0, η_0).   Note that V is such that $\overline{T}(\xi_0, \eta_0) = (u_0, v_0)$.   We write

$$\overline{T}: \begin{matrix} u = \overline{f}(\xi, \eta), \\ v = \overline{g}(\xi, \eta), \end{matrix} \qquad \overline{P}_0 = (\xi_0, \eta_0) = (u_0, v_0).$$

---

* That is, for which the inner area $A^-(E) = $ the outer area $A^+(E)$. (See *First Course*, Chapter 8, Section 1.)

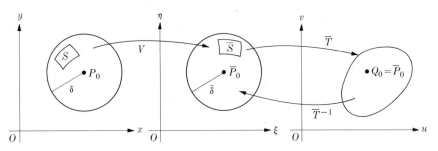

Figure 11–16

Since the transformation V is linear, the image of $C(P_0, \delta)$ in the ξ, η-plane will be contained in a circle $C(\bar{P}_0, \bar{\delta})$ of radius $\bar{\delta}$ about $\bar{P}_0$. (See Fig. 11–16.) We may employ (5) in the *linear* transformation V and, if $\bar{S}$ is the image of S, then*

$$A(\bar{S}) = |J_0| A(S).$$

Since $\bar{T}$ is one-to-one, the inverse of $\bar{T}$ exists, and we write

$$\bar{T}^{-1}: \quad \begin{array}{l} \xi = \bar{F}(u, v) \\ \eta = \bar{G}(u, v) \end{array}.$$

We now apply the results of the Inversion Theorem for smooth transformations with nonvanishing Jacobians. Let $\epsilon > 0$ be given. Then there is a circle $C(\bar{P}_0, \bar{\delta})$ so small that

$$|\bar{f}_\xi(\xi, \eta) - 1| < \frac{\epsilon}{2}, \qquad |\bar{f}_\eta(\xi, \eta)| < \frac{\epsilon}{2},$$

$$|\bar{g}_\xi(\xi, \eta)| < \frac{\epsilon}{2}, \qquad |\bar{g}_\eta(\xi, \eta) - 1| < \frac{\epsilon}{2} \tag{7}$$

for (ξ, η) in $C(\bar{P}_0, \bar{\delta})$, and such that the image of $C(\bar{P}_0, \bar{\delta})$ under $\bar{T}$ is contained in a circle $C(\bar{P}_0, \bar{\bar{\delta}})$, with $\bar{\bar{\delta}}$ so small that

$$|\bar{F}_u(u, v) - 1| < \frac{\epsilon}{2}, \qquad |\bar{F}_v(u, v)| < \frac{\epsilon}{2},$$

$$|\bar{G}_u(u, v)| < \frac{\epsilon}{2}, \qquad |\bar{G}_v(u, v) - 1| < \frac{\epsilon}{2} \tag{8}$$

for all (u, v) in $C(\bar{P}_0, \bar{\bar{\delta}}_0)$. Equations (7) and (8) state essentially that for sufficiently small circles the transformation $\bar{T}$ is practically the identity. Note that if $\bar{T}$ were the identity, then we could take $\epsilon = 0$ in (7) and (8).

Suppose that $\bar{S}$ is a square of side $2h$ located in $C(\bar{P}_0, \bar{\delta})$ (Fig. 11–17). Denote its center by $\bar{P}_1(\xi_1, \eta_1)$ so that $\bar{S}$ is described by

$$|\xi - \xi_1| < h, \ |\eta - \eta_1| < h.$$

---

* Actually $A^-(\bar{S}) = |J_0| A^-(S)$ and $A^+(\bar{S}) = |J_0| A^+(S)$ for any bounded S.

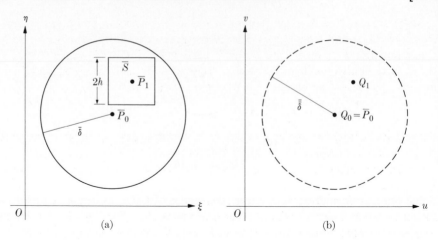

FIGURE 11–17

Let $Q_1 = (u_1, v_1)$ be $\overline{T}(\overline{P}_1)$. Then, applying Taylor's Theorem (Lemma 1) to (ξ, η) and using (7), we have

$$|\bar{f}(\xi, \eta) - \bar{f}(\xi_1, \eta_1)| = |\bar{f}_\xi(\alpha, \beta)(\xi - \xi_1) + \bar{f}_\eta(\alpha, \beta)(\eta - \eta_1)|$$
$$< \left(1 + \frac{\epsilon}{2}\right) h + \frac{\epsilon}{2} h = h(1 + \epsilon), \qquad (9)$$

where (α, β) is on the segment from (ξ_1, η_1) to (ξ, η) (Fig. 11–18).

A similar result holds for $\bar{g}(\xi, \eta)$:

$$|\bar{g}(\xi, \eta) - \bar{g}(\xi_1, \eta_1)| < h(1 + \epsilon). \qquad (10)$$

Inequalities (9) and (10) imply that the image of any region contained in $\overline{S}$ lies in the square

$$\overline{\overline{S}}: \quad |u - u_1| < h(1 + \epsilon),$$
$$|v - v_1| < h(1 + \epsilon)$$

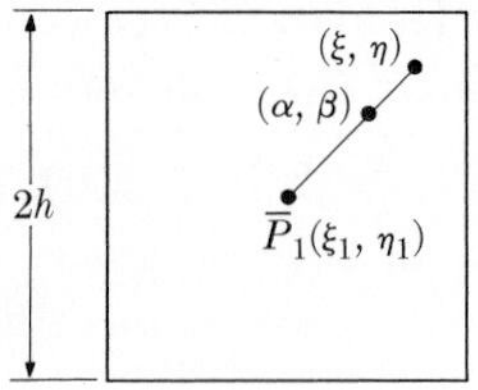

situated in the uv-plane. By using Taylor's Theorem and (8) for $\overline{T}^{-1}$, we get the result that the image of any region in the uv-plane within the square

$$S^*: \quad |u - u_1| < h(1 - \epsilon),$$
$$|v - v_1| < h(1 - \epsilon)$$

maps into a region inside the square

$$\overline{S}: \quad |\xi - \xi_1| < h, \qquad |\eta - \eta_1| < h.$$

FIGURE 11–18

Thus the square $\overline{S}$ itself must map under $\overline{T}$ into a region G which contains S^* and is contained in $\overline{\overline{S}}$. Therefore

$$A(\overline{S})(1 - \epsilon)^2 \le A^-(G) \le A^+(G) \le A(\overline{S})(1 + \epsilon)^2. \qquad (11)$$

More generally, if R is any region in $C(\bar{P}_0, \bar{\delta})$, its area may be approximated from the inside and from the outside by the area of a sum of squares formed by decomposing the $\xi\eta$-plane into the usual network. Since (11) holds for each of the squares, it also holds for the sums. Proceeding to the limit by taking finer and finer subdivisions,† we obtain, for any region R with area $A(R)$,

$$A(R)(1 - \epsilon)^2 \le A^-(G) \le A^+(G) \le A(R)(1 + \epsilon)^2 \tag{12}$$

where, now, G is the image of R under $\bar{T}$. If E is now any region which maps into R under V, we employ the relation $A^\pm(R) = |J_0|A^\pm(E)$ together with (12) to yield the inequality of the lemma. Note that the transformation V followed by $\bar{T}$ is T precisely.

We now state the first fundamental theorem of this section, which gives a formula for the areas of regions connected by continuously differentiable transformations. This result is an extension of Theorem 12, Chapter 7, page 335, which gives the areas of regions connected by linear transformations. The second principal theorem is the general formula for the change of variables in a double integral. This theorem generalizes Theorem 14 of Chapter 7, page 337, which gives the formula for a linear change of variables in a double integral.

Theorem 8. *Suppose that T, given by*

$$T: \quad \begin{aligned} u &= f(x, y), \\ v &= g(x, y), \end{aligned}$$

is a one-to-one, continuously differentiable transformation on its domain D (an open set). Suppose that the Jacobian of T is never zero. If R is any closed, bounded region in D which has area, then its image $T(R)$ has area which is given by

$$A[T(R)] = \iint_R \left| J\left(\frac{u,\ v}{x,\ y}\right) \right| dA_{xy}.$$

Proof. Since u and v are continuously differentiable, it follows that

$$\left| J\left(\frac{u,\ v}{x,\ y}\right) \right|$$

is continuous on R. Since R is closed and bounded, $|J|$ is uniformly continuous. The value of the Jacobian at a point $P(x, y)$ will be denoted $J(x, y)$ or $J(P)$. Let $\epsilon > 0$ be given. From the definition of integral, we can make a subdivision of the plane into rectangles which is so fine that

$$\left| \sum_{i=1}^{n} |J(x'_i, y'_i)|A(F_i) - \iint_R |J(x, y)|\ dA_{xy} \right| < \frac{\epsilon}{3}, \tag{13}$$

---

† Using also the facts (Morrey, *University Calculus*, Sections 8–1 and 8–2) that $A^-(G) \ge \sum A^-(G_i)$ and $A^+(G) \le \sum A^+(G_i)$ whenever G is the union of the G_i and no two of the G_i overlap.

where $F_1, F_2, \ldots, F_n$ are the parts in R of those rectangles which contain points of R and (x_i', y_i') is an arbitrary point in F_i.

We now employ Lemma 2. For each point P in R, we can find a circle $C(P, r_P)$ where r_P is so small that

$$|J(P)|(1 - \eta)A(G) \leq A^-[T(G)] \leq A^+[T(G)] \leq |J(P)|(1 + \eta)A(G) \quad (14)$$

for any region G in C and any desired positive number $0 < \eta < 1$.

Let

$$M = \max |J(P)| \quad \text{for} \quad P \quad \text{on} \quad R.$$

Then we can select $\eta = \epsilon/3MA(R)$.

Define $\mathfrak{F}$ to be the totality of circles $C(P, r_P)$ obtained from Lemma 2 when the above value of η is used. According to the Heine-Borel theorem, there is a number $\rho > 0$ such that any circle $C(P, \rho)$ with P in R is inside some circle of $\mathfrak{F}$. We further refine the subdivision of the plane into rectangles, if necessary, so that each F_i is in some circle $C(P_i', \rho)$ which itself is contained in some $C(P_i, r_{P_i})$ in $\mathfrak{F}$ (Fig. 11–19). Then, because of the uniform continuity of $|J|$ on R, we have, for any $\epsilon' > 0$,

$$\left| \sum_{i=1}^{n} [|J(P_i)|A(F_i) - |J(P_i')|A(F_i)] \right| < \epsilon' \sum_{i=1}^{n} A(F_i) = \epsilon'A(R).$$

From (14), we write for each i,

$$|J(P_i)|A(F_i) - \eta MA(F_i) \leq A^-[T(F_i)] \leq A^+[T(F_i)] \leq |J(P_i)|A(F_i) + \eta MA(F_i).$$

Adding these inequalities for all i, we get

$$\sum_{i=1}^{n} |J(P_i)|A(F_i) - \eta MA(R) \leq A^-[T(R)] \leq A^+[T(R)]$$

$$\leq \sum_{i=1}^{n} |J(P_i)|A(F_i) + \eta MA(R).$$

Now we take $\epsilon' = \epsilon/3A(R)$ and obtain

$$\left| \iint\limits_{R} |J(x, y)|\, dA_{xy} - A^{\pm}[T(R)] \right|$$

$$\leq \left| \iint\limits_{R} |J(x, y)|\, dA_{xy} - \sum_{i=1}^{n} J(P_i')A(F_i) \right|$$

$$+ \left| \sum_{i=1}^{n} J(P_i')A(F_i) - \sum_{i=1}^{n} J(P_i)A(F_i) \right|$$

FIGURE 11–19

$$+ \left| \sum_{i=1}^{n} J(P_i)A(F_i) - A^{\pm}[T(R)] \right| \leq \frac{\epsilon}{3} + \frac{\epsilon}{3} + \frac{\epsilon}{3} = \epsilon.$$

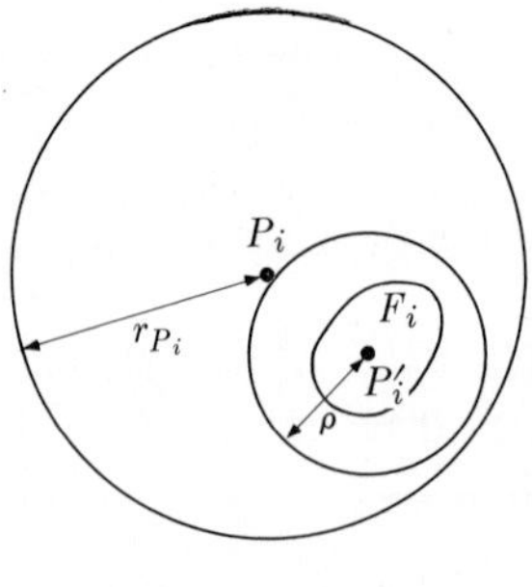

Since ϵ is arbitrary, the theorem follows.

Theorem 9. (Change of Variables Theorem.) *Suppose that T satisfies the hypotheses of Theorem 8 and that $K(u, v)$ is continuous on $T(R)$. If $H(x, y) = K[f(x, y), g(x, y)]$, then H is continuous on R and*

$$\iint\limits_{T(R)} K(u, v) \, dA_{u,v} = \iint\limits_{R} H(x, y) \left| J\left(\frac{u, v}{x, y}\right) \right| dA_{xy}.$$

The proof of Theorem 9 follows from the method of approximation in a way that is similar to the proof of Theorem 14, Chapter 7, page 337.

EXAMPLE 1. Evaluate $\iint_R x \, dA_{xy}$, where R is the region bounded by the curves

$$x = -y^2, \qquad x = 2y - y^2 \quad \text{and} \quad x = 2 - y^2 - 2y.$$

Perform the integration by introducing the new variables u, v:

$$x = u - \frac{(u + v)^2}{4}, \qquad y = \frac{u + v}{2}.$$

Draw R and the corresponding region G in the u, v-plane.

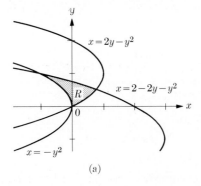

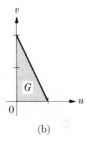

(a)

(b)

FIGURE 11–20

Solution. The equations of the boundary curves for G in the u, v-plane are obtained by substitution. We find

$$x = -y^2 \to u = 0,$$
$$x = 2y - y^2 \to u = u + v \to v = 0,$$
$$x + y^2 = 2 - 2y \to u = 2 - u - v \to 2u + v = 2.$$

The regions R and G are shown in Fig. 11–20. We have

$$J\left(\frac{x, y}{u, v}\right) = \begin{vmatrix} 1 - \dfrac{u + v}{2} & -\dfrac{u + v}{2} \\[2mm] \dfrac{1}{2} & \dfrac{1}{2} \end{vmatrix} = \frac{1}{2}.$$

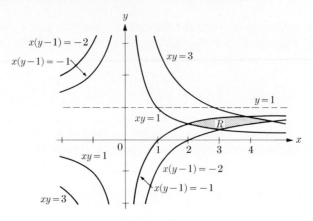

FIGURE 11–21

Therefore

$$\iint_R x \, dA_{xy} = \iint_G \left[u - \frac{(u+v)^2}{4} \right] \cdot \frac{1}{2} \, dA_{uv} = \frac{1}{2} \int_0^1 \int_0^{2-2u} \left[u - \frac{(u+v)^2}{4} \right] dv \, du$$

$$= \frac{1}{2} \int_0^1 \left[uv - \frac{(u+v)^3}{12} \right]_0^{2-2u} du$$

$$= \frac{1}{2} \int_0^1 \left[2u - 2u^2 - \frac{(2-u)^3}{12} + \frac{u^3}{12} \right] du = \frac{1}{48}.$$

EXAMPLE 2. Evaluate $\iint_R x \, dA_{xy}$, where R is the region bounded by the curves $x(1-y) = 1$, $x(1-y) = 2$, $xy = 1$, and $xy = 3$.

Solution. Let $u = x(1-y)$, $v = xy$. Then

$$x = u + v, \qquad y = \frac{v}{u+v}, \qquad J\left(\frac{u,v}{x,y}\right) = x = u + v.$$

Since the Jacobian of the inverse of a one-to-one transformation is the reciprocal of the Jacobian of the transformation, we have

$$J\left(\frac{x,y}{u,v}\right) = \frac{1}{u+v}.$$

Therefore

$$\iint_R x \, dA_{xy} = \iint_{T(R)} (u+v) \left| J\left(\frac{x,y}{u,v}\right) \right| dA_{uv} = \iint_{T(R)} dA_{uv} = \int_1^2 \int_1^3 dv \, du = 2.$$

The region R is shown in Fig. 11–21.

———

The theorems of this section have appropriate generalizations to three or more dimensions. Furthermore, the specialization to one dimension shows that the ordinary method of substitution $x = g(u)$ is valid whenever g is continuously differentiable and $g' \neq 0$.

EXERCISES

1. Show that
$$J\left(\frac{x, y}{r, \theta}\right) = r \qquad \text{if } x = r \cos \theta, \, y = r \sin \theta.$$

2. Show that
$$J\left(\frac{x, y, z}{\rho, \theta, \phi}\right) = \rho^2 \sin \phi \qquad \text{if } x = \rho \cos \theta \sin \phi, \, y = \rho \sin \theta \sin \phi, \, z = \rho \cos \phi.$$

3. Find
$$J\left(\frac{x, y, z}{u, v, w}\right) \qquad \text{given that } x = u(1 - v), \, y = uv, \, z = uvw.$$

4. Find
$$J\left(\frac{x_1, x_2, x_3, x_4}{u_1, u_2, u_3, u_4}\right)$$
if $x_1 = u_1 \cos u_2$, $x_2 = u_1 \sin u_2 \cos u_3$, $x_3 = u_1 \sin u_2 \sin u_3 \cos u_4$, and if $x_4 = u_1 \sin u_2 \sin u_3 \sin u_4$.

5. Show that Lemma 1 can be proved by applying the Theorem of the Mean to the function
$$\phi(t) = f[x_1 + t(x_2 - x_1), \, y_1 + t(y_2 - y_1)], \qquad 0 \le t \le 1.$$

In each of problems 6 through 13, evaluate $\iint_R f(x, y) \, dA_{xy}$, where R is bounded by the curves whose equations are given. Perform the integration by introducing variables u and v as indicated. Draw a graph of R and the corresponding region G in the u, v-plane. Find the inverse of each transformation.

6. $f(x, y) = x^2$; R bounded by $y = 3x$, $x = 3y$ and $x + y = 4$; transformation: $x = 3u + v$, $y = u + 3v$.

7. $f(x, y) = x - y^2$; R bounded by $y = 2$, $x = y^2 - y$, $x = 2y + y^2$; transformation: $x = 2u - v + (u + v)^2$, $y = u + v$.

8. $f(x, y) = y$; R bounded by $x + y - y^2 = 0$, $2x + y - 2y^2 = 1$, $x - y^2 = 0$; transformation: $x = u - v + (u - 2v)^2$, $y = -u + 2v$.

9. $f(x, y) = x^2$; R bounded by $y = -x - x^2$, $y = 2x - x^2$, $y = \frac{1}{2}x - x^2 + 3$; transformation: $x = u - v$, $y = 2u + v - (u - v)^2$.

10. $f(x, y) = (x^2 + y^2)^{-3}$; R bounded by $x^2 + y^2 = 2x$, $x^2 + y^2 = 4x$, $x^2 + y^2 = 2y$, $x^2 + y^2 = 6y$; transformation: $x = u/(u^2 + v^2)$, $y = v/(u^2 + v^2)$.

11. $f(x, y) = 4xy$; R bounded by $y = x$, $y = -x$, $(x + y)^2 + x - y - 1 = 0$; transformation: $x = (u + v)/2$, $y = (-u + v)/2$ (assume that $x + y > 0$).

12. $f(x, y) = y$; R bounded by $x = e^{y/2} - \frac{1}{2}y$, $x = 5 + e^{y/2} - \frac{1}{2}y$, $x = 2y + e^{y/2}$, $3x = y + 3e^{y/2} - 5$; transformation: $u = x - e^{y/2} + (y/2)$, $v = y$.

13. $f(x, y) = x^2 + y^2$; R is the region in the first quadrant bounded by $x^2 - y^2 = 1$, $x^2 - y^2 = 2$, $2xy = 2$, $2xy = 4$; transformation: $u = x^2 - y^2$, $v = 2xy$.

14. Show that the integrals
$$\int_0^a \int_0^{a-y} f(x, y) \, dx \, dy \qquad \text{and} \qquad \int_0^1 \int_0^a f(u - uv, \, uv)u \, du \, dv$$
are equal if $u = x + y$, $v = y/(x + y)$.

15. Show that the integrals

$$\int_0^a \int_0^x f(x, y) \, dx \, dy \quad \text{and}$$

$$\int_0^1 \int_0^{a(1+u)} f\left[\frac{v}{1 + u}, \frac{uv}{1 + u}\right] \frac{v}{(1 + u)^2} \, dv \, du$$

are equal if $x = v/(1 + u)$, $y = uv/(1 + u)$.

16. Evaluate the integral

$$\iiint_R z \, dV_{xyz}$$

where R is the region $x^2 + y^2 \leq z^2$, $x^2 + y^2 + z^2 \leq 1$, $z \geq 0$, by changing to spherical coordinates.

17. By introducing polar coordinates, evaluate

$$\iint_R \frac{dA_{xy}}{(1 + x^2 + y^2)^2}$$

where R is the right-hand loop of the lemmiscate:

$$(x^2 + y^2)^2 - (x^2 - y^2) = 0.$$

18. State and prove Lemma 1 (Taylor's Theorem with one term) for functions of three variables.

19. State and prove Lemma 2 for transformations in three-space.

20. State and prove Theorem 8 for transformations in three-space.

21. Write out a proof of Theorem 9.

CHAPTER 12

Functions Defined by Integrals

---

1. DIFFERENTIATION UNDER THE INTEGRAL SIGN

We recall the elementary integration formula

$$\int_0^1 t^n \, dt = \left[\frac{1}{n+1} t^{n+1} \right]_0^1 = \frac{1}{n+1},$$

valid for any $n > -1$. Since n need not be an integer, we employ the more familiar variable x and write

$$\phi(x) = \int_0^1 t^x \, dt = \frac{1}{x+1}, \qquad x > -1. \tag{1}$$

Suppose we wish to compute the derivative $\phi'(x)$. We can proceed in two ways. Equating the first and last expressions in (1), we have

$$\phi(x) = \frac{1}{x+1}, \qquad \phi'(x) = -\frac{1}{(x+1)^2}.$$

On the other hand, we may try the following procedure:

$$\frac{d}{dx} \phi(x) = \frac{d}{dx} \int_0^1 t^x \, dt = \int_0^1 \frac{d}{dx} (t^x) \, dt = \int_0^1 t^x \ln t \, dt. \tag{2}$$

Is it true that

$$\int_0^1 t^x \ln t \, dt = -\frac{1}{(x+1)^2}, \tag{3}$$

at least for $x > -1$? In this section we shall determine conditions under which a process such as (2) is valid.

Suppose a function ϕ is given by the relation

$$\phi(x) = \int_c^d f(x, t) \, dt, \qquad a \le x \le b,$$

where c and d are constants. If the integration can be performed explicitly, then $\phi'(x)$ can be found by a computation. However, even when the evaluation of the integral is impossible, it sometimes happens that $\phi'(x)$ can be found. The basic formula is given in the next theorem, known as **Leibniz' Rule**.

Theorem 1. *Suppose that ϕ is defined by*

$$\phi(x) = \int_c^d f(x, t)\, dt, \qquad a \leq x \leq b, \tag{4}$$

where c and d are constants. If f and $f_{,1}$ (i.e., $\partial f/\partial x$) are continuous in the rectangle $R: \{a \leq x \leq b, c \leq t \leq d\}$, then

$$\phi'(x) = \int_c^d f_{,1}(x, t)\, dt, \qquad a < x < b. \tag{5}$$

That is, the derivative may be found by differentiating under the integral sign.

Proof. We prove the theorem by showing that the difference quotient $[\phi(x + k) - \phi(x)]/k$ tends to the right side of (5) as k tends to zero. If x is in (a, b) then, from (4), we have

$$\frac{\phi(x + k) - \phi(x)}{k} = \frac{1}{k}\int_c^d f(x + k, t)\, dt - \frac{1}{k}\int_c^d f(x, t)\, dt$$

$$= \frac{1}{k}\int_c^d [f(x + k, t) - f(x, t)]\, dt.$$

Since differentiation and integration are inverse processes, we can write

$$f(x + k, t) - f(x, t) = \int_x^{x+k} f_{,1}(\xi, t)\, d\xi,$$

and so

$$\frac{\phi(x + k) - \phi(x)}{k} = \frac{1}{k}\int_c^d \int_x^{x+k} f_{,1}(\xi, t)\, d\xi\, dt.$$

For convenience we use the subscript notation for the derivative with respect to the first argument. We note that $f_{,1} = \partial f/\partial x$ is uniformly continuous on R, since a function which is continuous on a bounded, closed set is uniformly continuous there. Therefore, if $\epsilon > 0$ is given, there is a $\delta > 0$ such that

$$|f_{,1}(\xi, t) - f_{,1}(x, t)| < \frac{\epsilon}{d - c}$$

for all t in $[c, d]$ and all ξ with $|\xi - x| < \delta$. We now wish to show that

$$\frac{\phi(x + k) - \phi(x)}{k} - \int_c^d f_{,1}(x, t)\, dt \to 0 \qquad \text{as } k \to 0.$$

We write

$$\int_c^d f_{,1}(x, t)\, dt = \frac{1}{k}\int_c^d \int_x^{x+k} f_{,1}(x, t)\, d\xi\, dt,$$

which is true because the integrand on the right does not contain ξ. Substituting this last expression in the one above, we find, for $0 < |k| < \delta$,

$$\left| \frac{\phi(x + k) - \phi(x)}{k} - \int_c^d f_{,1}(x, t)\, dt \right|$$

$$= \left| \int_c^d \left\{ \frac{1}{k} \int_x^{x+k} [f_{,1}(\xi, t) - f_{,1}(x, t)]\, d\xi \right\} dt \right|$$

$$\leq \int_c^d \left| \frac{1}{k} \int_x^{x+k} \frac{\epsilon}{d - c}\, d\xi \right| dt = \frac{\epsilon}{(d - c)} \cdot (d - c) = \epsilon.$$

Since ϵ is arbitrary, the theorem follows.

Theorem 1 shows that the relation (3) is justified for $x > 0$, since the integrand $f(x, t)$ is then continuous in an appropriate rectangle. Later we shall examine more closely the validity of (3) when $-1 < x \leq 0$, in which case the integral is improper. (See *First Course*, Chapter 16, Section 5.)

EXAMPLE 1. Find the value of $\phi'(x)$ if

$$\phi(x) = \int_0^{\pi/2} f(x, t)\, dt; \qquad f(x, t) = \begin{cases} \dfrac{\sin xt}{t}, & t \neq 0, \\ x, & \text{if } t = 0. \end{cases}$$

Solution. Since

$$\lim_{t \to 0} \frac{\sin xt}{t} = x \lim_{t \to 0} \frac{\sin xt}{xt} = x,$$

the integrand is continuous for $0 \leq t \leq \pi/2$ and for all x. Also, we have

$$f_{,1}(x, t) = \begin{cases} \cos xt & \text{if } t \neq 0, \\ 1 = \cos xt & \text{if } t = 0, \end{cases}$$

so $f_{,1}(x, t)$ is continuous everywhere. Therefore

$$\phi'(x) = \int_0^{\pi/2} \cos xt\, dt = -\left[\frac{1}{x} \sin xt \right]_0^{\pi/2} = -\frac{\sin (\pi/2)x}{x}, \qquad x \neq 0.$$

Note that the integral expression for ϕ cannot be evaluated explicitly.

EXAMPLE 2. Evaluate

$$\int_0^1 \frac{du}{(u^2 + 1)^2}$$

by letting

$$\phi(x) = \int_0^1 \frac{du}{u^2 + x} = \frac{1}{\sqrt{x}} \arctan (1/\sqrt{x})$$

and computing $-\phi'(1)$.

Solution.

$$\phi'(x) = -\int_0^1 \frac{du}{(u^2 + x)^2} = \frac{1}{\sqrt{x}} \frac{-\frac{1}{2}x^{-3/2}}{1 + (1/x)} - \frac{1}{2x\sqrt{x}} \arctan \frac{1}{\sqrt{x}}$$

and

$$-\phi'(1) = \int_0^1 \frac{du}{(u^2 + 1)^2} = \frac{1}{2}\left(\frac{1}{2} + \arctan 1\right) = \frac{1}{2}\left(\frac{1}{2} + \frac{\pi}{4}\right).$$

Leibniz' Rule may be extended to the case where the limits of integration also depend on x. We consider a function defined by

$$\phi(x) = \int_{u_0(x)}^{u_1(x)} f(x, t)\, dt, \tag{6}$$

where $u_0(x)$ and $u_1(x)$ are continuously differentiable functions for $a \le x \le b$. Furthermore, the ranges of u_0 and u_1 are assumed to lie between c and d (Fig. 12–1).

To obtain a formula for the derivative $\phi'(x)$, where ϕ is given by an integral such as (6), it is simpler to consider a new integral which is more general than (6). We define

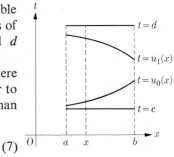

$$F(x, y, z) = \int_y^z f(x, t)\, dt \tag{7}$$

FIGURE 12–1

and obtain the following corollary of Leibniz' Rule.

Theorem 2. *Suppose that f satisfies the conditions of Theorem 1 and that F is defined by (7) with $c \le y, z \le d$. Then*

$$\frac{\partial F}{\partial x} = \int_y^z f_{,1}(x, t)\, dt, \tag{8a}$$

$$\frac{\partial F}{\partial y} = -f(x, y), \tag{8b}$$

$$\frac{\partial F}{\partial z} = f(x, z). \tag{8c}$$

Proof. Formula (8a) is Theorem 1. Formulas (8b) and (8c) are precisely the Fundamental Theorem of Calculus, since taking the partial derivative of F with respect to one variable, say y, implies that x and z are kept *fixed*.

Theorem 3. (General Rule for Differentiation under the Integral Sign.) *Suppose that f and $\partial f/\partial x$ are continuous in the rectangle R: $\{a \le x \le b, c \le t \le d\}$, and suppose that $u_0(x)$, $u_1(x)$ are continuously differentiable for $a \le x \le b$ with* *the range of u_0 and u_1 in $[c, d]$. If ϕ is given by*

$$\phi(x) = \int_{u_0(x)}^{u_1(x)} f(x, t)\,dt,$$

then

$$\phi'(x) = f[x, u_1(x)]\, u_1'(x) - f[x, u_0(x)] \cdot u_0'(x) + \int_{u_0(x)}^{u_1(x)} f_{,1}(x, t)\, dt. \qquad (9)$$

Proof. We observe that

$$F(x, u_0(x), u_1(x)) = \phi(x)$$

in Theorem 2. Applying the Chain Rule, we get

$$\phi'(x) = F_{,1} + F_{,2}u_0'(x) + F_{,3}u_1'(x).$$

Inserting the values of $F_{,1}, F_{,2}$, and $F_{,3}$ from (8), we obtain the desired result (9).

EXAMPLE 3. Find $\phi'(x)$, given that

$$\phi(x) = \int_0^{x^2} \arctan \frac{t}{x^2}\, dt.$$

Solution. We have

$$\frac{\partial}{\partial x}\left(\arctan \frac{t}{x^2}\right) = \frac{-2t/x^3}{1 + (t^2/x^4)} = -\frac{2tx}{t^2 + x^4}.$$

We use formula (9) and find

$$\phi'(x) = (\arctan 1) \cdot (2x) - \int_0^{x^2} \frac{2tx\, dt}{t^2 + x^4}.$$

Setting $t = x^2u$ in the integral on the right, we obtain

$$\phi'(x) = \frac{\pi x}{2} - \int_0^1 \frac{2x^3 u \cdot x^2\, du}{x^4 u^2 + x^4} = \frac{\pi x}{2} - x\int_0^1 \frac{2u\, du}{u^2 + 1} = x\left(\frac{\pi}{2} - \ln 2\right).$$

EXERCISES

In each of problems 1 through 5, express $\phi'(x)$ as a definite integral, using Leibniz' Rule.

1. $\phi(x) = \displaystyle\int_0^1 \frac{\sin xt\, dt}{1 + t}$ 2. $\phi(x) = \displaystyle\int_0^2 \frac{e^{-xt}\, dt}{1 + t^2}$ 3. $\phi(x) = \displaystyle\int_1^2 \frac{e^{-t}\, dt}{1 + xt}$

4. $\phi(x) = \displaystyle\int_0^1 \frac{t^2\, dt}{(1 + xt)^2}$

5. $\phi(x) = \displaystyle\int_0^1 \frac{t^x - 1}{\ln t}\, dt,$ $(f(x, 0) = 0, f(x, 1) = x, x > 0)$

In each of problems 6 through 14, obtain expressions of the form (9) for $\phi'(x)$.

6. $\phi(x) = \displaystyle\int_1^x t^3 \, dt$

7. $\phi(x) = \displaystyle\int_1^{x^2} \cos(t^2) \, dt$

8. $\phi(x) = \displaystyle\int_x^1 e^{t^2} \, dt$

9. $\phi(x) = \displaystyle\int_{x^2}^x \sin(xt) \, dt$

10. $\phi(x) = \displaystyle\int_x^{\tan x} \frac{dt}{1 + xt}$

11. $\phi(x) = \displaystyle\int_{x^2}^{e^x} \tan(xt) \, dt$

12. $\phi(x) = \displaystyle\int_{\sin x}^x \ln(1 + xt) \, dt, \quad 0 < x \le \pi$

13. $\phi(x) = \displaystyle\int_{\cos x}^{1+x^2} \frac{e^{-t} \, dt}{1 + xt}, \quad x > 0$

14. $\phi(x) = \displaystyle\int_{x^2}^{\sin x} e^{xt} \, dt$

15. Given that $\phi(x) = \int_0^{\pi/2} \cos xt \, dt$, obtain $\phi'(x)$ in two ways: (1) by integrating and then differentiating, and (2) by using Leibniz' Rule and then performing the integration.

16. Evaluate

$$\int_0^1 \frac{du}{(u^2 + 1)^3}$$

by using the methods and results of Example 2.

In each of problems 17 through 25, find $\phi'(x)$ by first applying Theorem 1 or 3 and then integrating.

17. $\phi(x) = \displaystyle\int_{\pi/2}^\pi \frac{\cos xt}{t} \, dt$

18. $\phi(x) = \displaystyle\int_1^{x^2} \frac{e^{xt}}{t} \, dt$

19. $\phi(x) = \displaystyle\int_{x^2}^x \frac{\sin xt}{t} \, dt, \quad x > 0$

20. $\phi(x) = \displaystyle\int_x^{x^2} \frac{e^{-xt}}{t} \, dt, \quad x > 0$

21. $\phi(x) = \displaystyle\int_{x^m}^{x^n} \frac{dt}{x + t}, \quad x > 0$

22. $\phi(x) = \displaystyle\int_0^\pi \ln(1 + x \cos t) \, dt, \quad |x| < 1$

23. $\phi(x) = \displaystyle\int_0^1 \frac{x \, dt}{\sqrt{1 - x^2 t^2}}, \quad |x| < 1$

24. $\phi(x) = \displaystyle\int_0^{x^2} f(x, t) \, dt, \quad f(x, t) = \begin{cases} t^{-1} \sin^2 xt & \text{if } t \ne 0 \\ 0 & \text{if } t = 0 \end{cases}$

25. $\phi(x) = \int_0^\pi \ln(1 - 2x \cos t + x^2)\, dt, \quad |x| < 1$

26. Show that if m and n are positive integers, then

$$\int_0^1 t^n (\ln t)^m\, dt = (-1)^m \frac{m!}{(n+1)^{m+1}}.$$

[*Hint*. Differentiate $\int_0^1 x^n\, dx = 1/(n+1)$ and use induction on m. Here we understand that $t^n (\ln t)^m$ is defined to be 0 for $t = 0$, or else that the integral is improper.]

27. Suppose that $\phi(x, y, z) = \int_a^b f(z + x \cos t + y \sin t)\, dt$. Show that $\phi_{zz} = \phi_{xx} + \phi_{yy}$.

2. TESTS FOR CONVERGENCE OF IMPROPER INTEGRALS. THE GAMMA FUNCTION

Suppose that a function f is continuous in the half-open interval $a \le x < b$, and suppose that f tends to infinity as x tends to b. A typical example of such a function is shown in Fig. 12–2. In *First Course*, Chapter 16, Section 5, we defined

$$\int_a^b f(x)\, dx = \lim_{\epsilon \to 0} \int_a^{b-\epsilon} f(x)\, dx$$

whenever the limit exists. If the limit does exist, we say the integral **converges**; otherwise it **diverges**. If the integrand becomes infinite at the left endpoint of an interval, convergence and divergence are defined similarly. In *First Course*, Chapter 16, Section 5, we also discussed integrals in which the interval of integration is infinite. If $f(x)$ is continuous for $a \le x < \infty$, we define

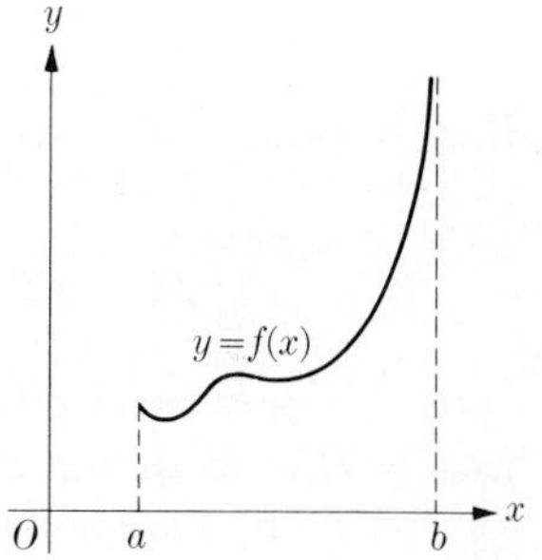

$$\int_a^\infty f(x)\, dx = \lim_{X \to \infty} \int_a^X f(x)\, dx$$

FIGURE 12–2

whenever the limit exists. The terms "convergence" and "divergence" are used in the same way as for finite intervals.

If $f(x)$ becomes infinite at several points in the interval of integration, the interval may be decomposed so that each limit can be calculated separately. Also, the integral of a function f over the infinite interval $-\infty < x < \infty$ is computed by selecting a convenient value C and calculating

$$\lim_{X \to -\infty} \int_X^C f(x)\, dx \quad \text{and} \quad \lim_{Y \to \infty} \int_C^Y f(x)\, dx.$$

The integral $\int_{-\infty}^\infty f(x)\, dx$ is said to converge if *both* limits above exist. For example,

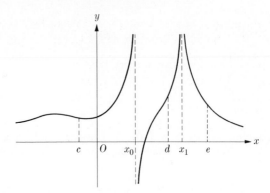

FIGURE 12–3

suppose that we wish to calculate $\int_{-\infty}^{\infty} f(x)\, dx$ for a function f which becomes infinite at the points x_0 and x_1, as shown in Fig. 12–3. We select the convenient values c, d, and e (Fig. 12–3) and evaluate

$$\lim_{X \to -\infty} \int_X^c f(x)\, dx, \qquad \lim_{\epsilon_1 \to 0} \int_c^{x_0-\epsilon_1} f(x)\, dx, \qquad \lim_{\epsilon_2 \to 0} \int_{x_0+\epsilon_2}^d f(x)\, dx,$$

$$\lim_{\epsilon_3 \to 0} \int_d^{x_1-\epsilon_3} f(x)\, dx, \qquad \lim_{\epsilon_4 \to 0} \int_{x_1+\epsilon_4}^e f(x)\, dx, \qquad \lim_{Y \to +\infty} \int_e^Y f(x)\, dx.$$

If all of these limits exist, their sum yields the value of

$$\int_{-\infty}^{\infty} f(x)\, dx.$$

In our previous study of convergence and divergence of integrals, we were usually able to obtain an indefinite integral and then evaluate the limit, either directly or by l'Hôpital's Rule. We now wish to determine methods for testing convergence which are useful even when the integrands are so complicated that we cannot perform the integrations. The following theorem, helpful in establishing convergence tests, is almost a corollary of the Axiom of Continuity. (See *First Course*, page 122.)

Theorem 4.   (a) *Suppose that F is nondecreasing in the half-open interval $a \le x < b$ and that $F(x) \le M$ there. Then* (i) $F(x)$ *tends to a limit L as x tends to b;* (ii) $F(x) \le L$ *in* $[a, b)$; (iii) $L \le M$.
(b) *Similarly, if F is nonincreasing in $[a, b)$ and if $F(x) \ge m$ for $a \le x < b$, then* (i) $F(x) \to l$ *as* $x \to b$; (ii) $F(x) \ge l$; (iii) $l \ge m$.

Proof. We prove (a); the proof of (b) is the same. For each positive integer n, we define $x_n = b - (1/n)(b - a)$. Then $a \le x_n < b$ and $x_n < x_{n+1}$; also, $x_n \to b$ as $n \to \infty$. The numbers $F(x_n)$ form a nondecreasing bounded sequence and, by the Axiom of Continuity, tend to a limit L which is less than or equal to

M; and $F(x_n) \leq L$ for all n. Let $\epsilon > 0$ be given. Then there is an N such that $F(x_n) > L - \epsilon$ for all $n > N$. If x is in the interval (x_{N+1}, b), then $x_{N+1} < x < x_n$ for some n, so that

$$L - \epsilon < F(x_{N+1}) \leq F(x) \leq F(x_n) \leq L,$$

and the theorem is established.

The next theorem, known as the comparison test, is one of the basic tools used in establishing convergence. Note the analogy with the comparison test for series (Chapter 3, Section 3).

Theorem 5. (Comparison Test.) *Suppose that f and g are continuous in the half-open interval $[a, b)$, that $0 \leq |f(x)| \leq g(x)$, and that $\int_a^b g(x)\, dx$ converges. Then $\int_a^b f(x)\, dx$ converges, and*

$$\left| \int_a^b f(x)\, dx \right| \leq \int_a^b g(x)\, dx.$$

The same result holds if b is replaced by $+\infty$, or if $[a, b)$ is replaced by $(a, b]$, or if the interval considered is $(-\infty, b]$.

Proof. We established the result for $[a, b)$; the other cases are proved similarly. We first assume that $f(x) \geq 0$ and define

$$F(X) = \int_a^X f(x)\, dx, \qquad G(X) = \int_a^X g(x)\, dx.$$

Then F and G are nondecreasing on $[a, b)$ and, by hypothesis, $G(X)$ tends to a limit M as $X \to b$. Since, by hypothesis, $F(X) \leq G(X) \leq M$ on $[a, b)$, we find from Theorem 4 that $F(X) \to L \leq M$.

If $f(x)$ is not always nonnegative, we define

$$f_1(x) = \frac{|f(x)| + f(x)}{2}, \qquad f_2(x) = \frac{|f(x)| - f(x)}{2}.$$

Then f_1 and f_2 are continuous on $[a, b)$ and nonnegative there. Furthermore,

$$f_1(x) + f_2(x) = |f(x)| \leq g(x), \qquad f_1(x) - f_2(x) = f(x).$$

Therefore the improper integrals of f_1, f_2, and $|f|$ all exist. From the theorems on limits we conclude that

$$\left| \int_a^b f(x)\, dx \right| = \left| \int_a^b f_1(x)\, dx - \int_a^b f_2(x)\, dx \right| \leq \int_a^b [f_1(x) + f_2(x)]\, dx$$

$$\leq \int_a^b g(x)\, dx.$$

We have the following test for divergence.

Theorem 6. *Suppose that f and g are continuous on* $[a, b)$ *where* $0 \leq g(x) \leq f(x)$, *and suppose that* $\int_a^b g(x)\, dx$ *diverges. Then* $\int_a^b f(x)\, dx$ *diverges. The same result holds if* $[a, b)$ *is replaced by* $[a, +\infty)$, $(a, b]$, *or* $(-\infty, b]$.

Proof. If $\int_a^b f(x)\, dx$ were convergent then, according to Theorem 5, $\int_a^b g(x)\, dx$ would be also.

The comparison tests are useful if we have available a class or several classes of integrals which we know converge or diverge. Then these integrals may be used for comparison purposes. Two such classes are given in the next theorem. The proof is obtained by straightforward integration and evaluation of the resulting limit.

Theorem 7. (a) *The integrals*

$$\int_a^b (b - x)^{-p}\, dx, \qquad \int_a^b (x - a)^{-p}\, dx$$

converge if $p < 1$ *and diverge if* $p \geq 1$.
(b) *The integrals*

$$\int_a^\infty x^{-p}\, dx \qquad and \qquad \int_{-\infty}^{-b} |x|^{-p}\, dx$$

with $a > 0, b > 0$ *converge if* $p > 1$ *and diverge if* $p \leq 1$.

EXAMPLE 1. Test for convergence or divergence

$$\int_0^1 \frac{x^\alpha\, dx}{\sqrt{1 - x^3}},$$

where α is a positive constant.

Solution. Taking

$$f(x) = \frac{x^\alpha}{\sqrt{1 - x^3}}, \qquad g(x) = \frac{1}{\sqrt{1 - x}},$$

we show that $|f(x)| \leq g(x)$ for $0 \leq x \leq 1$. To see this, we observe that

$$f(x) = \frac{x^\alpha}{\sqrt{1 - x^3}} = \frac{x^\alpha}{\sqrt{(1 - x)(1 + x + x^2)}} = \frac{x^\alpha}{\sqrt{1 + x + x^2}}\, g(x).$$

Since $x^\alpha \leq \sqrt{1 + x + x^2}$ for $0 \leq x \leq 1$ so long as $\alpha \geq 0$, we conclude that

$$|f(x)| \leq g(x).$$

However,

$$\int_0^1 g(x)\, dx$$

converges by Theorem 7(a) with $p = \frac{1}{2}$. The Comparison Test shows that the given integral converges for all $\alpha \geq 0$.

EXAMPLE 2. Test for convergence or divergence:

$$\int_1^\infty \frac{\sqrt{x}}{1 + x^{3/2}}\, dx.$$

Solution. For $x \geq 1$, we see that

$$f(x) \equiv \frac{\sqrt{x}}{1 + x^{3/2}} \geq \frac{\sqrt{x}}{x^{3/2} + x^{3/2}} = \frac{1}{2} \cdot \frac{1}{x} \equiv g(x).$$

However, according to Theorem 7(b), $\int_1^\infty g(x)\, dx$ diverges. Therefore the integral diverges.

EXAMPLE 3. Test for convergence or divergence:

$$\int_1^\infty \frac{\sin x}{x^{3/2}}\, dx.$$

Solution. Since

$$\left| \frac{\sin x}{x^{3/2}} \right| \leq \frac{1}{x^{3/2}},$$

we employ Theorem 7(b), with $p = \frac{3}{2}$, and the Comparison Test to conclude that the integral converges.

EXAMPLE 4. Test for convergence or divergence:

$$\int_0^\infty \frac{e^{-x}}{\sqrt{x}}\, dx.$$

Solution. Because the integrand is infinite at $x = 0$, we select a convenient value, say $x = 1$, and break the integral into two parts:

$$\int_0^\infty \frac{e^{-x}}{\sqrt{x}}\, dx = \int_0^1 \frac{e^{-x}}{\sqrt{x}}\, dx + \int_1^\infty \frac{e^{-x}}{\sqrt{x}}\, dx. \tag{1}$$

In the first integral on the right, we have

$$\frac{e^{-x}}{\sqrt{x}} \leq \frac{1}{\sqrt{x}},$$

and $\int_0^1 (1/\sqrt{x})\, dx$ converges by Theorem 7(a) with $p = \frac{1}{2}$. In the second integral on the right, we have

$$\frac{e^{-x}}{\sqrt{x}} \leq e^{-x}.$$

The integral

$$\int_1^X e^{-x}\, dx = [-e^{-x}]_1^X = e^{-1} - e^{-X} \to \frac{1}{e} \quad \text{as } X \to +\infty,$$

so that the second integral converges. Since both integrals on the right in (1) converge, the original integral is convergent.

We shall show that the integral

$$\int_0^\infty t^{x-1}e^{-t}\,dt \tag{2}$$

is convergent for $x > 0$. To do so, we notice that for $0 < x < 1$ the integrand becomes infinite at $t = 0$, and so we treat the integral as in Example 4. We write

$$\int_0^\infty t^{x-1}e^{-t}\,dt = \int_0^1 t^{x-1}e^{-t}\,dt + \int_1^\infty t^{x-1}e^{-t}\,dt. \tag{3}$$

In the first integral on the right, we obtain the inequality

$$t^{x-1}e^{-t} \le t^{x-1}.$$

The integral $\int_0^1 t^{x-1}\,dx$ converges for $x > 0$ (Theorem 7(a), $p < 1$). As for the second integral on the right in (3), we obtain an inequality by computing the maximum value of the function $f(t) = t^{x+1}e^{-t}$. The derivative $f'(t)$ is zero when $t = x + 1$, and the maximum value of $f(t)$ is

$$f(x + 1) = (x + 1)^{x+1}e^{-(x+1)}.$$

Therefore

$$\int_1^\infty t^{x-1}e^{-t}\,dt = \int_1^\infty (t^{x+1}e^{-t})\frac{1}{t^2}\,dt \le (x + 1)^{x+1}e^{-(x+1)}\int_1^\infty \frac{1}{t^2}\,dt.$$

The last integral on the right converges and, consequently, so does (2). This last device can be used to show the convergence of the integral (2) when $x \ge 1$.

DEFINITION. The **Gamma function**, denoted by $\Gamma(x)$, is defined by

$$\Gamma(x) = \int_0^\infty t^{x-1}e^{-t}\,dt, \qquad x > 0.$$

This function, which has many important applications in both mathematics and physics, has a number of interesting properties. The observation that

$$\Gamma(x + 1) = \int_0^\infty t^x e^{-t}\,dt$$

and an integration by parts yield one of the most important properties. We have

$$\int_0^T t^x e^{-t}\,dt = [t^x(-e^{-t})]_0^T + x\int_0^T t^{x-1}e^{-t}\,dt, \qquad x \ge 1. \tag{4}$$

Letting $T \to +\infty$ and using l'Hôpital's Rule in the first term on the right, we obtain

$$\Gamma(x + 1) = x\Gamma(x). \tag{5}$$

(See Exercise 27.) Since

$$\Gamma(1) = \int_0^\infty e^{-t}\, dt = \lim_{X\to\infty} \left[e^{-t}\right]_0^X = 1,$$

we find, by successive application of (5), that

$$\Gamma(2) = 1 \cdot \Gamma(1) = 1, \qquad \Gamma(3) = 2 \cdot \Gamma(2) = 2, \qquad \Gamma(4) = 3 \cdot \Gamma(3) = 1 \cdot 2 \cdot 3,$$

and, by induction, that

$$\Gamma(n + 1) = n!$$

In other words, the Gamma function, which is defined for all real numbers $x > 0$, is a generalization of the "factorial function" defined for positive integers.

The relation (5), which is a **recursion formula**, shows that if the Gamma function is known for any particular value of x, it may be found for all numbers of the form $x + n$, where n is a positive integer. For example, it can be shown that $\Gamma(\frac{3}{2}) = \frac{1}{2}\sqrt{\pi}$. From this fact it is easy to deduce that

$$\Gamma\left(\frac{2k + 1}{2}\right) = \frac{1 \cdot 3 \cdot 5 \cdots (2k - 1)}{2^k}\sqrt{\pi}$$

for any positive integer k.

EXERCISES

In each of problems 1 through 22, test the integral for convergence or divergence.

1. $\displaystyle\int_1^\infty \frac{dx}{(x + 2)\sqrt{x}}$

2. $\displaystyle\int_2^\infty \frac{dx}{(x - 1)\sqrt{x}}$

3. $\displaystyle\int_0^1 \frac{dx}{\sqrt{1 - x^4}}$

4. $\displaystyle\int_0^1 \frac{dx}{\sqrt{(1 - x)^4}}$

5. $\displaystyle\int_0^\infty \frac{dx}{\sqrt{x^3 + 1}}$

6. $\displaystyle\int_2^\infty \frac{dx}{\sqrt{x^3 - 1}}$

7. $\displaystyle\int_0^\infty \frac{x\, dx}{\sqrt{x^4 + 1}}$

8. $\displaystyle\int_1^2 \frac{dx}{x\sqrt{x^2 - 1}}$

9. $\displaystyle\int_3^4 \frac{\sqrt{16 - x^2}}{x^2 - x - 6}\, dx$

10. $\displaystyle\int_{-1}^1 \frac{dx}{\sqrt{1 - x^2}}$

11. $\displaystyle\int_{-1}^1 \frac{dx}{x^2}$

12. $\displaystyle\int_0^{\pi/2} \frac{\sqrt{x}\, dx}{\sin x}$

13. $\displaystyle\int_1^3 \frac{\sqrt{x}}{\ln x}\, dx$

14. $\displaystyle\int_0^\infty \frac{(\arctan x)^2\, dx}{x^2 + 1}$

15. $\displaystyle\int_0^\infty e^{-x^2}\,dx$ 16. $\displaystyle\int_0^\infty x^2 e^{-x^2}\,dx$

17. $\displaystyle\int_0^\infty \frac{dx}{(x+1)\sqrt{x}}$ 18. $\displaystyle\int_0^1 \frac{dx}{\sqrt{x-x^2}}$

19. $\displaystyle\int_0^\infty \frac{\cos x\,dx}{1+x^2}$ 20. $\displaystyle\int_0^\infty e^{-x}\sin x\,dx$

21. $\displaystyle\int_0^\pi \frac{\cos x\,dx}{\sqrt{x}}$ 22. $\displaystyle\int_0^\pi \frac{\sin x}{x\sqrt{x}}\,dx$

23. Show that $\int_2^\infty x^{-1}(\ln x)^{-p}\,dx$ converges if $p>1$ and diverges if $p\le 1$.

24. Show that $\int_0^1 |\ln x|^p\,dx$ converges to $\Gamma(p+1)$ for each $p>0$. [*Hint.* Consider $\int_\epsilon^1 (\ln(1/x))^p\,dx$, set $x=e^{-t}$, and let $\epsilon\to 0$.]

25. Show that $\int_1^\infty e^{-x^2}\,dx = \frac{1}{2}\Gamma(\frac{1}{2}) = \Gamma(\frac{3}{2})$.

26. Show that

$$\int_0^\infty x^p e^{-x^2}\,dx = \tfrac{1}{2}\Gamma\left(\frac{p+1}{2}\right), \qquad p>-1.$$

27. Show how to extend the procedure in Eq. (4) to establish (5) for $x>0$.

3. IMPROPER MULTIPLE INTEGRALS

For functions of one variable we have considered two types of improper integrals: (i) those in which the integrand becomes infinite at some point in the interval of integration, and (ii) those in which the interval of integration becomes infinite.

Double and triple integrals have been defined only for bounded functions and for bounded regions of integration. We now take up the problem of defining a double integral when the integrand f becomes infinite at a single point in a bounded region of integration. For example, suppose we wish to integrate the function

$$f(x,y) = \frac{1}{[(x-1)^2 + y^2]^{1/2}}$$

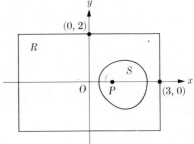

FIGURE 12–4

over the rectangular region R: $|x|\le 3$, $|y|\le 2$. The function f becomes infinite at the point $P(1,0)$, which is in the region of integration. We construct a small region S containing the point P, and we observe that the function f is continuous in the region $R-S$ (Fig. 12–4). Therefore the integral

$$\iint_{R-S} f(x,y)\,dA_{xy} \tag{1}$$

may be defined in the usual way if S is a small circle, square, triangle, or even a region of rather irregular shape, so long as P is interior to S (that is, so long as S contains a closed circle with center at P). For a closed set S we define the **diameter of S** as the maximum distance between any two points of S; we denote this diameter by $d(S)$. Let S_n be a sequence of closed regions, each containing P in its interior, and such that $d(S_n) \to 0$ as $n \to \infty$. Then, intuitively, we define the integral of f over R as the limit L of the integrals taken over $R - S_n$ whenever this limit exists. There are many different ways in which such a sequence S_n may be chosen, and of course the value L must be the same for all possible choices of the $\{S_n\}$. More precisely, we say that if for every $\epsilon > 0$ there is a $\delta > 0$ such that

$$\left| \iint_{R-S} f(x, y) \, dA_{xy} - L \right| < \epsilon$$

whenever S is a closed region containing P in its interior with diameter less than δ, then the **improper integral**

$$\iint_{R} f(x, y) \, dA_{xy}$$

exists and has the value L.

If a function becomes infinite at a finite number of points $P_1, P_2, \ldots, P_k$ within (or on the boundary of) the region of integration but is otherwise continuous, we subdivide R into a number of nonoverlapping pieces so that each portion contains one of the points P_i. (Figure 12–5 shows a typical situation.) Then, if the improper integral over each portion exists, we add the resulting values to obtain the improper integral over R. It can be shown that the value of the integral does not depend on how the region R is subdivided.

In any particular case, it is usually extremely difficult to verify the existence of an improper integral by use of the definition alone. We now establish theorems which are helpful not only in verifying the existence of improper integrals but also in their actual evaluation. Let R be a closed region and P a point in R. We define an **increasing sequence of regions closing down on P** as a sequence of closed regions $H_1, H_2, \ldots, H_n, \ldots$ with these properties: (i) every H_n is in $R - P$; (ii) for each n, $H_n \subset H_{n+1}$; and (iii) if R' is any closed set contained in $R - P$, then there is an integer n such that $R' \subset H_n$. Since H_n is closed and does not contain P, it also does not contain a small circle (of radius r_n, $r_n \to 0$) about P. In Fig. 12–6, for example, we could select H_1 as the portion of R between Γ and Γ_1, H_2 as the portion of R between Γ and Γ_2, H_3 as the portion of R between Γ and Γ_3, and so forth.

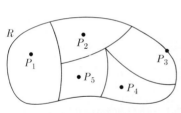

FIGURE 12–5

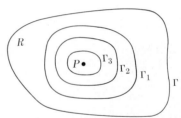

FIGURE 12–6

Theorem 8. (a) *Suppose that* $f(x, y) \geq 0$ *in a closed, bounded region* R, *and suppose that* f *is continuous on* $R - P$, *where* P *is a point of* R. *If there is a number* M *such that*

$$\iint\limits_{R'} f(x, y) \, dA_{xy} \leq M$$

for every closed region $R' \subset R - P$, *then the improper integral of* f *over* R *exists. Moreover, suppose that* $\{H_n\}$ *is an increasing sequence of regions closing down on* P, *as defined above. Then we have*

$$\lim_{n \to \infty} \iint\limits_{H_n} f(x, y) \, dA_{xy} = \iint\limits_{R} f(x, y) \, dA_{xy}.$$

Proof. Let $\{H_n\}$ be any sequence of regions of the type specified. Since $f(x, y) \geq 0$, we see that for each n

$$\iint\limits_{H_{n+1}} f(x, y) \, dA = \iint\limits_{H_n} f(x, y) \, dA + \iint\limits_{H_{n+1} - H_n} f(x, y) \, dA \geq \iint\limits_{H_n} f(x, y) \, dA.$$

Therefore $\iint_{H_n} f(x, y) \, dA_{xy}$ is a nondecreasing sequence bounded by M. From Axiom C (*First Course*, page 122), it follows that there is a number L such that

$$\lim_{n \to \infty} \iint\limits_{H_n} f(x, y) \, dA_{xy} = L. \tag{2}$$

Now, let $\epsilon > 0$ be given. There is an integer N such that

$$L - \epsilon < \iint\limits_{H_{N+1}} f(x, y) \, dA_{xy} \leq \iint\limits_{H_n} f(x, y) \, dA_{xy} \leq L \qquad \text{for } n \geq N + 1.$$

Let R' be any closed region such that

$$H_{N+1} \subset R' \subset R - P.$$

Since R' does not contain P, there is an H_n such that $R' \subset H_n$ for some n (and so for all larger n). Consequently,

$$L - \epsilon < \iint\limits_{H_{N+1}} f(x, y) \, dA \leq \iint\limits_{R'} f(x, y) \, dA \leq \iint\limits_{H_n} f(x, y) \, dA \leq L.$$

Since ϵ is arbitrary, it follows that

$$L = \iint\limits_{R} f(x, y) \, dA.$$

Corollary. *If* f, R, P *are as in Theorem* 8, *if* $\{H_n\}$ *is an increasing sequence closing down on* P, *and if*

$$\iint\limits_{H_n} f(x, y) \, dA_{xy} \leq M$$

for all n, *then the improper integral of* f *over* R *exists and* (2) *holds.*

If an improper integral exists, we say that the integral is **convergent**; if it does not exist, we say that the integral is **divergent**.

EXAMPLE 1. Discuss the convergence or divergence of

$$\iint_R \frac{1}{\sqrt{(x^2 + y^2)^p}} \, dA_{xy}, \qquad p > 0,$$

where R is the disc: $x^2 + y^2 \leq 1$.

Solution. The integrand becomes infinite at the origin. We select for H_n the ring $(1/n) \leq r \leq 1$, where $r = \sqrt{x^2 + y^2}$. Then, changing to polar coordinates, we have

$$\iint_{H_n} \frac{1}{r^p} \, dA = \int_0^{2\pi} \int_{1/n}^1 r^{-p} r \, dr \, d\theta$$

$$= \begin{cases} \dfrac{2\pi}{2 - p} \left(1 - \dfrac{1}{n^{2-p}} \right) & \text{if } p < 2, \\[2ex] 2\pi \ln n & \text{if } p = 2, \\[2ex] \dfrac{2\pi}{p - 2} (n^{p-2} - 1) & \text{if } p > 2. \end{cases}$$

By the Corollary to Theorem 8, the integral converges if $p < 2$. Since $r^{-p} > 0$, it follows that the definition of the existence of improper integral is not satisfied if $p \geq 2$.

So far we have considered integrals in which the integrand becomes infinite at one or several points. However, for functions of two or more variables the integrand may misbehave in many ways. For example, the function

$$f(x, y) = \frac{1}{(x^2 - y)^{1/3}}$$

becomes infinite all along the parabola $C: y = x^2$. It is possible to define an improper integral for functions which are infinite along an arc C. It is necessary to define an increasing sequence of regions closing down on C and to make an appropriate generalization of Theorem 8. (See Exercise 18.) This topic and various extensions to triple integrals will not be discussed in detail.

Suppose that R is an unbounded region in the plane and that f is a function on R. In defining the improper integral of f over R we must be aware that the situation is more complicated than it is for functions of one variable. For integrals along the x-axis, the interval of integration could extend from some point a to $-\infty$, from a point a to $+\infty$, or from $-\infty$ to $+\infty$. As exhibited in Fig. 12–7, unbounded regions in the plane may be of many types. The reader can easily think of many other kinds of unbounded regions.

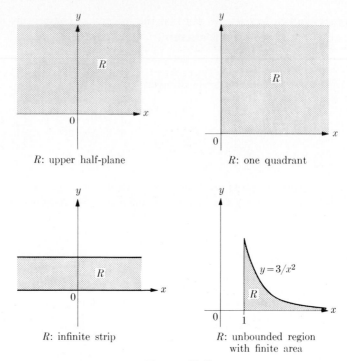

FIGURE 12–7

DEFINITION. Suppose that R is an unbounded region and that f is continuous on R. We say that the **improper integral of f over R exists** if and only if there is a number K such that: for every $\epsilon > 0$ there is a closed bounded region $R_1 \subset R$ with the property that

$$\left| \iint_{R_1} f(x, y) \, dA_{xy} - K \right| < \epsilon$$

for every closed bounded region R' with $R_1 \subset R' \subset R$.

For unbounded regions R, we define an **increasing sequence of regions $\{H_n\}$ filling R** as a sequence of closed bounded regions in R with the properties: (i) for each n, $H_n \subset H_{n+1}$, and (ii) if R' is any bounded region in R, then there is an integer n such that $R' \subset H_n$. With this definition of the sequence $\{H_n\}$, Theorem 8 and its Corollary have obvious analogs for an improper integral taken over an unbounded region R.

EXAMPLE 2. Evaluate $\int_0^\infty e^{-x^2} \, dx$.

Solution. We employ a trick which has become classical in the study of improper multiple integrals but which is seldom useful. First we observe that

$$\int_0^\infty e^{-x^2} \, dx = \lim_{n \to \infty} \int_0^n e^{-x^2} \, dx$$

is a convergent integral. Next we note that

$$\left(\int_0^n e^{-x^2}\, dx\right)^2 = \int_0^n \int_0^n e^{-(x^2+y^2)}\, dx\, dy = \iint_{R_n} e^{-(x^2+y^2)}\, dA,$$

where R_n is the square $0 \le x \le n, 0 \le y \le n$. Let R be the entire first quadrant and let R' be a closed bounded region in R. For n sufficiently large, $R' \subset R_n$ and, denoting

$$M = \left(\int_0^\infty e^{-x^2}\, dx\right)^2,$$

we see that

$$\iint_{R'} e^{-(x^2+y^2)}\, dA \le M.$$

The quantity M is also equal to

$$\lim_{n\to\infty} \iint_{G_n} e^{-(x^2+y^2)}\, dA,$$

where G_n is the quarter circle (polar coordinates): $0 \le \theta \le \pi/2, 0 \le r \le n$. But

$$\iint_{G_n} e^{-(x^2+y^2)}\, dA = \int_0^{\pi/2} \int_0^n e^{-r^2} r\, dr\, d\theta = \frac{\pi}{4}\left[-e^{-r^2}\right]_0^n \to \frac{\pi}{4} \quad \text{as } n \to \infty.$$

Hence $M = \pi/4$ and, taking the square root, we conclude that

$$\int_0^\infty e^{-x^2}\, dx = \frac{\sqrt{\pi}}{2}.$$

Comparison theorems for determining the convergence and divergence of multiple integrals follow the same pattern as do those for integrals of functions of one variable. (See Section 2.)

Theorem 9. (a) *Suppose that f, R, P are as in Theorem 8 and that $|f(x, y)| \le g(x, y)$ on $R - P$. If g is continuous on $R - P$ and the integral of g over R exists, then the integral of f over R does also, and*

$$\left|\iint_R f(x, y)\, dA\right| \le \iint_R g(x, y)\, dA. \tag{3}$$

(b) *Suppose that f and g are continuous on an unbounded region R. If $|f(x, y)| \le g(x, y)$ on R and if the integral of g over R exists, then the integral of f does also, and the inequality (3) holds.*

The proof of Theorem 9 is similar to that of Theorem 5 of the preceding section and is left to the reader. (See Exercise 21.)

Theorem 10.   (a) *Suppose that* f, g, R, P *are as in Theorem* 9 *and that* $0 \le g(x, y) \le f(x, y)$. *If* $\iint_R g(x, y)\, dA$ *diverges, then* $\iint_R f(x, y)\, dA$ *does also.* (b) *If* R *is unbounded and* $\iint_R g(x, y)\, dA$ *diverges, then* $\iint_R f(x, y)\, dA$ *diverges also.*

Theorem 10 follows directly from Theorem 9, since the assumption that $\iint_R f(x, y)\, dA$ converges implies that $\iint_R g\, dA$ does.

EXAMPLE 3. Test for convergence or divergence:

$$\iint_R \frac{\sin xy}{x^2(1 + y^2)}\, dA,$$

where R is the half-infinite strip: $\{1 \le x < \infty, 0 \le y \le 1\}$.

Solution.   We define the domain R_n: $1 \le x \le n$, $0 \le y \le 1$. Setting $f(x, y) = \sin xy / x^2(1 + y^2)$, $g(x, y) = 1/x^2(1 + y^2)$, we have

$$\iint_{R_n} |f(x, y)|\, dA \le \iint_{R_n} g(x, y)\, dA = \int_1^n \int_0^1 \frac{1}{x^2}\, \frac{1}{1 + y^2}\, dy\, dx.$$

However,

$$\iint_{R_n} g(x, y)\, dA = [\arctan y]_0^1 \int_1^n \frac{1}{x^2}\, dx = \frac{\pi}{4}\left[-\frac{1}{x}\right]_1^n \to \frac{\pi}{4} \quad \text{as } n \to \infty.$$

Hence $\iint_R g\, dA$ is convergent, and so $\iint_R f\, dA$ is also.

The entire discussion of this section has an appropriate generalization to functions of three variables and to triple integrals. Improper integrals may be defined for functions $f(x, y, z)$ which become infinite at points, on curves, or on surfaces. Also, integrals over unbounded domains may be defined. (See Exercises 19, 20, 22.) We work an example.

EXAMPLE 4. Test for convergence or divergence:

$$\iiint_R \frac{1}{\sqrt{(x^2 + y^2 + z^2)^p}}\, dV, \qquad p > 0,$$

where R is the unit ball: $x^2 + y^2 + z^2 \le 1$.

Solution.   We define the regions R_n: $(1/n) \le \rho \le 1$ where $\rho = \sqrt{x^2 + y^2 + z^2}$. Then, introducing spherical coordinates

$$x = \rho \cos \theta \sin \phi, \qquad y = \rho \sin \theta \sin \phi, \qquad z = \rho \cos \phi,$$

and computing the Jacobian, we find

$$
\iint_{R_n} \frac{1}{\rho^p}\, dV = \int_0^{2\pi}\!\!\int_0^{\pi}\!\!\int_{1/n}^{1} \frac{1}{\rho^p}\, \rho^2 \sin\phi\, d\rho\, d\phi\, d\theta
$$

$$
= \begin{cases}
\dfrac{4\pi}{3-p}\left(1 - \dfrac{1}{n^{3-p}}\right), & 0 \le p < 3, \\[2ex]
4\pi \ln n, & p = 3, \\[2ex]
\dfrac{4\pi}{p-3}(n^{p-3} - 1), & p > 3.
\end{cases}
$$

We conclude that the integral is convergent for $p < 3$ and divergent for $p \ge 3$.

EXERCISES

In each of problems 1 through 10, the region R is the unit disc: $x^2 + y^2 \le 1$. Test the given integral for convergence or divergence. Specify the choice of the sequence $\{H_n\}$ and describe the set S where the integrand is singular.

1. $\displaystyle \iint_R \frac{x^2\, dA}{(x^2 + y^2)^{3/2}}$

2. $\displaystyle \iint_R \frac{xy\, dA}{(x^2 + y^2)^{3/2}}$

3. $\displaystyle \iint_R \ln\frac{1}{r}\, dA$

4. $\displaystyle \iint_R r^{-2}\left(\ln\frac{2}{r}\right)^{-2} dA$

5. $\displaystyle \iint_R \left(\ln\frac{1}{r}\right)^{-1} dA$

6. $\displaystyle \iint_R \frac{x^2 y^2}{(x^2 + y^2)^3}\, dA$

7. $\displaystyle \iint_R \frac{dA}{\sqrt{1 + x}}$

8. $\displaystyle \iint_R \frac{x\, dA}{(x^2 + y^2)\sqrt{1 - x}}$

9. $\displaystyle \iint_R \frac{dA}{1 - x}$

10. $\displaystyle \iint_R \frac{dA}{\sqrt[3]{(1 - 2x)(1 - 3x)(1 - 4x)}}$

In each of problems 11 through 16, test for convergence or divergence. Specify your choice of $\{H_n\}$.

11. $\displaystyle \iint_R (x^2 + y^2)^{-p/2}\, dA, \quad R: x^2 + y^2 \ge 1; \quad p > 0$

12. $\displaystyle \iint_R \frac{y\, dA}{(x^2 + y^2)(1 - x)^{2/3}}, \quad R: 0 \le x \le 1,\ 0 \le y \le 1$

13. $\displaystyle \iint_R \frac{dA}{(x^2 + y^2)\sqrt{(x - 1)(y - 1)}}, \quad R: x \ge 1,\ y \ge 1$

14. $\displaystyle\iiint_R \rho^{-p}\, dV, \quad \rho = (x^2 + y^2 + z^2)^{1/2}, \quad R: \rho \geq 1; \quad p > 0$

15. $\displaystyle\iiint_R \frac{x\, dV}{\rho^3}, \quad R: 0 \leq \rho \leq 1$

16. $\displaystyle\iiint_R \frac{dV}{\rho^2(1 - x)^{3/2}}, \quad R: 0 \leq \rho \leq 1$

17. Use the result of Example 2 to find $\Gamma(\tfrac{1}{2})$.

*18. (a) Define an improper double integral for a function f which becomes infinite along an arc C contained in a bounded region R. (b) State and prove the appropriate analog of Theorem 8 for improper integrals as defined in (a).

*19. (a) Define an improper triple integral for a function f which is continuous in a bounded region R except at one point where it becomes infinite. (b) State and prove the appropriate analog of Theorem 8 for improper integrals as defined in (a).

20. Define an improper triple integral for a function continuous in an unbounded domain R.

21. Write out a proof of Theorem 9.

22. State and prove the analog of Theorem 9 for triple integrals.

23. State the appropriate generalization of Theorem 9(a) for functions of two variables which become infinite along an arc.

4. FUNCTIONS DEFINED BY IMPROPER INTEGRALS

We recall that the Gamma function is defined by the formula

$$\Gamma(x) = \int_0^\infty t^{x-1} e^{-t}\, dt, \qquad x > 0.$$

Is it possible to differentiate under the integral sign? That is, is the formula (Leibniz' Rule)

$$\Gamma'(x) = \int_0^\infty \frac{\partial}{\partial x}(t^{x-1} e^{-t})\, dt = \int_0^\infty t^{x-1} (\ln t) e^{-t}\, dt$$

valid? Leibniz' Rule, given in Section 1, page 522, was established for proper integrals. Not only does the integral for the Gamma function have an infinite interval of integration, but also its integrand has a singularity at $t = 0$ if $0 < x < 1$. In order to extend Leibniz' Rule to functions given by improper integrals, we must first extend the notion of uniform convergence introduced in Chapter 9, page 397.

Suppose that $F(x, t)$ is continuous for $a \leq x \leq b$ and for $c \leq t < d$. We wish to consider the limit of $F(x, t)$ as t tends to d (from below). The limiting value will depend on x, and we write

$$\lim_{t \to d} F(x, t) = f(x).$$

DEFINITION. We say that $F(x, t) \to f(x)$ **uniformly on** $[a, b]$ **as** $t \to d$ if and only if for each $\epsilon > 0$ there is a $\delta > 0$ such that

$$|F(x, t) - f(x)| < \epsilon \tag{1}$$

for all t in the interval $d - \delta < t < d$ and for all x on $[a, b]$.

Of course the size of δ will depend on ϵ. However, the crucial distinction between uniform limits (called **uniform convergence**) and ordinary limits lies in the fact that for uniform convergence, δ **is independent of** x. In other words, in uniform convergence, inequality (1) holds for all values of x and t in the shaded strip shown in Fig. 12–8.

Uniform convergence for a "continuous variable" may also be defined if the interval $c \leq t < d$ is replaced by $c \leq t < \infty$. We say that $F(x, t) \to f(x)$ **uniformly for** x **on** $[a, b]$ as $t \to +\infty$ if and only if for each $\epsilon > 0$ there is a value T such that

$$|F(x, t) - f(x)| < \epsilon$$

for all $t > T$. Once again, the value of T will depend on ϵ, but the uniformity condition requires that T **not** depend on x.

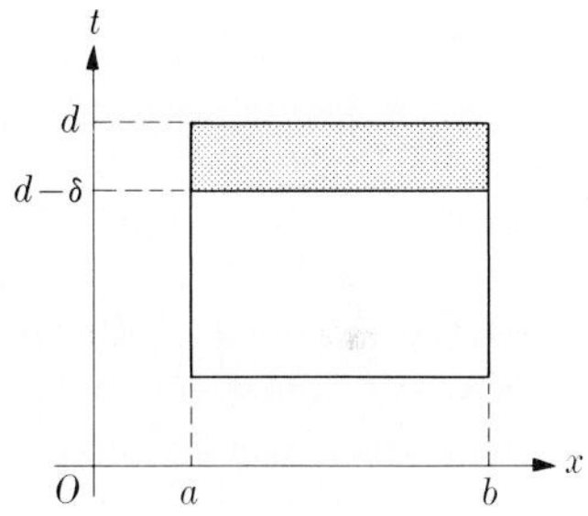

FIGURE 12–8

The next theorems, which are analogs of those for series given in Chapter 9, Section 1, are useful in establishing Leibniz' Rule for improper integrals.

Theorem 11. *Suppose that* $F(x, t)$ *is continuous in* x *on* $[a, b]$ *for each* t, $c \leq t < d$, *and suppose that* $F(x, t) \to \phi(x)$ *uniformly on* $[a, b]$ *as* $t \to d$. *Then* $\phi(x)$ *is continuous on* $[a, b]$. *The same result holds if* d *is replaced by* $+\infty$.

Proof. The proof follows the pattern of that given for Theorem 2 (Chapter 9, page 400). Let x_1, x_2 be two values of x in $[a, b]$. We write $\phi(x_1) - \phi(x_2)$ in the complicated form

$$\phi(x_1) - \phi(x_2) = \phi(x_1) - F(x_1, t) + F(x_1, t) - F(x_2, t) + F(x_2, t) - \phi(x_2).$$

Therefore

$$|\phi(x_1) - \phi(x_2)| \leq |\phi(x_1) - F(x_1, t)| + |F(x_1, t) - F(x_2, t)|$$
$$+ |F(x_2, t) - \phi(x_2)|.$$

Let $\epsilon > 0$ be given. Then, from the uniform convergence of F, the first and third terms on the right may be made $< \epsilon/3$ for all $t \in (d - \delta, d)$. The middle term on the right may be made $< \epsilon/3$ if x_1 and x_2 are close enough, since F is continuous in x. Hence $|\phi(x_1) - \phi(x_2)| < \epsilon$ if $|x_1 - x_2|$ is sufficiently small—that is, if ϕ is continuous.

Theorem 12. *Suppose that the hypotheses of Theorem* 11 *hold and, also, that* $F_x(x, t)$ *is continuous on* $[a, b]$ *for each* t. *If* $F_x(x, t) \to \psi(x)$ *uniformly for* x *on* $[a, b]$ *as* $t \to d$ *or as* $t \to +\infty$, *then* $\psi(x) = \phi'(x)$ *on* (a, b).

The proof follows the pattern of that given for Theorem 4, Chapter 9, page 402, and is left to the student. (See Exercise 20.)

We now see that Leibniz' Rule for improper integrals, given below, is an immediate consequence of the two theorems above.

Theorem 13. *Suppose that* $f(x, \tau)$ *is continuous for* $a \le x \le b$ *and* $c \le \tau < d$; *we define*

$$F(x, t) = \int_c^t f(x, \tau) \, d\tau.$$

If the improper integral

$$\phi(x) = \int_c^d f(x, \tau) \, d\tau$$

exists for $a \le x \le b$, *and if* $F(x, t) \to \phi(x)$ *uniformly for* x *on* $[a, b]$ *as* $t \to d$, *then* ϕ *is continuous on* $[a, b]$. *The same result holds if* d *is replaced by* $+\infty$. (This result is a corollary of Theorem 11.)

Theorem 14. (Leibniz' Rule for improper integrals.) *Suppose that the hypotheses of Theorem* 13 *hold, and suppose also that* f_x *is continuous. If* $F_x(x, t)$ *converges uniformly to* $\psi(x)$ *on* $[a, b]$ *as* $t \to d$ (or $+\infty$), *then*

$$\psi(x) = \phi'(x) = \int_c^d f_x(x, \tau) \, d\tau$$

or, if d *is replaced by* $+\infty$,

$$\psi(x) = \phi'(x) = \int_c^{+\infty} f_x(x, \tau) \, d\tau.$$

Proof. According to Leibniz' Rule, for each $t < d$ (or $< +\infty$), we have

$$F_x(x, t) = \int_c^t f_x(x, \tau) \, d\tau.$$

Hence the result follows from Theorem 12.

EXAMPLE 1. Show that if $\phi(x) = \int_0^\infty e^{-xt} \, dt$, then ϕ and ϕ' are continuous for $x > 0$ and

$$\phi'(x) = \int_0^\infty -te^{-xt} \, dt. \tag{2}$$

Solution. We define

$$F(x, t) = \int_0^t e^{-x\tau} \, d\tau = \frac{1 - e^{-xt}}{x}.$$

Also,

$$F_x(x, t) = \frac{-1 + e^{-xt} + xte^{-xt}}{x^2}.$$

As $t \to \infty$, $F(x, t) \to 1/x$ and $F_x(x, t) \to -1/x^2$. To see that the convergence is uniform, we note that for $h > 0$

$$\left| F(x, t) - \frac{1}{x} \right| = \frac{e^{-xt}}{x} < \frac{e^{-ht}}{h} \qquad \text{for all } x \geq h,$$

$$\left| F_x(x, t) - \left(-\frac{1}{x^2} \right) \right| = \frac{e^{-xt}(1 + xt)}{x^2} \leq \frac{e^{-ht}(1 + ht)}{h^2} \qquad \text{for } x \geq h.$$

Thus the convergence is uniform on any interval $x \geq h$ for positive h. Applying Leibniz' Rule for improper integrals, we conclude that (2) is valid.

In Example 1 it was possible to integrate the expression for $F(x, t)$ and so verify the uniform convergence directly. Since in most instances this direct approach is not possible, it is important to have some indirect tests for uniform convergence. The next theorem, which is a comparison test, is useful in that the direct evaluation of $F(x, t)$ is not required. The test may also be applied to $F_x(x, t)$ and in this way the applicability of Leibniz' Rule for improper integrals may be verified.

Theorem 15. (a) *Suppose that $f(x, t)$ is continuous for $a \leq x \leq b$ and $c \leq t < d$, and suppose that*

$$|f(x, t)| \leq g(t) \qquad \text{for } a \leq x \leq b, \quad c \leq t < d.$$

If $\int_c^d g(t)\, dt$ converges, then the improper integral

$$\phi(x) = \int_c^d f(x, t)\, dt$$

is defined for each x on $[a, b]$ and ϕ is continuous on $[a, b]$.
(b) *Suppose, also, that f_x is continuous for $a \leq x \leq b$, $c \leq t < d$ and that*

$$|f_x(x, t)| \leq g_1(t) \qquad \text{for } a \leq x \leq b, \quad c \leq t < d.$$

If $\int_c^d g_1(t)\, dt$ converges, then

$$\phi'(x) = \int_c^d f_x(x, t)\, dt.$$

That is, Leibniz' Rule holds. The same results hold if d is replaced by $+\infty$.

Proof. (a) That $\phi(x)$ is defined for each x follows from Theorem 5. We define

$$F(x, t) = \int_c^t f(x, \tau)\, d\tau.$$

Then

$$|\phi(x) - F(x, t)| = \left| \int_t^d f(x, \tau)\, d\tau \right| \leq \int_t^d g(\tau)\, d\tau.$$

But since $\int_c^d g(\tau)\, d\tau$ converges, we know that for each $\epsilon > 0$ there is a $\delta > 0$ such that for $t \in (d - \delta, d)$, we have $\left| \int_t^d g(\tau)\, d\tau \right| < \epsilon$. Thus

$$|\phi(x) - F(x, t)| \to 0 \qquad \text{as} \qquad t \to d,$$

uniformly for x on $[a, b]$.

To prove (b), we note that

$$\left| \int_t^d f_x(x, \tau)\, d\tau \right| \le \int_t^d g_1(\tau)\, d\tau \to 0 \qquad \text{as} \qquad t \to d,$$

as above. The convergence is uniform and the result established.

DEFINITION. If

$$F(x, t) = \int_c^t f(x, \tau)\, d\tau$$

and

$$\phi(x) = \int_c^d f(x, \tau)\, d\tau \tag{3}$$

and if $F(x, t) \to \phi(x)$ uniformly for x on $[a, b]$, we say that the **improper integral** (3) **converges uniformly for x on $[a, b]$**.

EXAMPLE 2. Show that the improper integral

$$\phi(x) = \int_1^\infty \frac{\sin t}{x^2 + t^2}\, dt$$

converges uniformly for all x.

Solution. We have

$$\left| \frac{\sin t}{x^2 + t^2} \right| \le \frac{1}{t^2}$$

for all x and all $t \ge 1$. Since $\int_1^\infty (1/t^2)\, dt$ converges, Theorem 15 applies.

EXAMPLE 3. Given the integral

$$\phi(x) = \int_0^\infty \frac{e^{-xt} - e^{-t}}{t}\, dt.$$

Show that the integral for $\phi(x)$ and the integral for $\phi'(x)$ (obtained by differentiation under the integral sign) both converge uniformly for x on $[a, b]$ if $a > 0$. Evaluate $\phi(x)$ by this means.

Solution. Setting $f(x, t) = (e^{-xt} - e^{-t})/t$, we have

$$f_x(x, t) = -e^{-xt}, \qquad |f_x(x, t)| \le \begin{cases} e^{-at}, & a \le 1, \\ e^{-t}, & 1 < a. \end{cases}$$

The last inequality may be written more compactly as

$$|f_x(x, t)| \leq e^{-ht} \qquad \text{where } h = \min (a, 1).$$

Now, using the Theorem of the Mean on f (as a function of x), we find that

$$\left| \frac{e^{-xt} - e^{-t}}{t} \right| = |e^{-\xi t}| \, |1 - x| \qquad \text{for } 0 \leq t \leq 1.$$

Also,

$$\left| \frac{e^{-xt} - e^{-t}}{t} \right| \leq e^{-ht} \qquad \text{for } t \geq 1.$$

Thus the integrals for ϕ and ϕ' converge uniformly. We may apply Leibniz' Rule to get

$$\phi'(x) = \int_0^\infty -e^{-xt} \, dt = \lim_{t \to \infty} \left[\frac{e^{-xt}}{x} \right]_0^t = -\frac{1}{x}.$$

The equation $\phi'(x) = 1/x$ can now be integrated to give $\phi(x) = C - \ln x$. Since

$$\phi(1) = \int_0^\infty \frac{e^{-t} - e^{-t}}{t} \, dt = 0,$$

we obtain

$$\phi(x) = -\ln x = \int_0^\infty \frac{e^{-xt} - e^{-t}}{t} \, dt.$$

EXERCISES

In each of problems 1 through 8, show that the integrals for $\phi(x)$ and $\phi'(x)$ converge uniformly on the given intervals, and find $\phi'(x)$ by Leibniz' Rule.

1. $\phi(x) = \displaystyle\int_0^\infty \frac{e^{-xt} \, dt}{1 + t}, \quad a \leq x, \quad a > 0$

2. $\phi(x) = \displaystyle\int_0^1 \frac{e^{xt}}{\sqrt{t}} \, dt, \quad |x| \leq A, \quad A > 0$

3. $\phi(x) = \displaystyle\int_0^\infty \frac{\cos xt}{1 + t^3} \, dt, \quad |x| \leq A, \quad A > 0$

4. $\phi(x) = \displaystyle\int_0^\infty \frac{e^{-t} \, dt}{1 + xt}, \quad x \geq 0$

5. $\phi(x) = \displaystyle\int_0^1 \frac{\sin xt}{t} (\ln t) \, dt, \quad |x| \leq A, \quad A > 0$

6. $\phi(x) = \displaystyle\int_0^1 \frac{(\ln t)^2}{1 + xt} \, dt, \quad x \geq A, \quad A > -1$

7. $\phi(x) = \displaystyle\int_0^1 \frac{dt}{(1 + xt)\sqrt{1 - t}}, \quad x \geq A, \quad A > -1$

8. $\phi(x) = \displaystyle\int_0^\infty \frac{\sin xt}{t(1 + t^2)}\, dt, \quad |x| \le A, \quad A > 0$

9. Using the complex exponential function (and Theorem 15), show that

$$\int_0^\infty e^{-ax} \cos bx\, dx = \frac{a}{a^2 + b^2},$$

$$\int_0^\infty e^{-ax} \sin bx\, dx = \frac{b}{a^2 + b^2}, \quad a > 0.$$

10. From the fact that

$$\int_0^1 t^x\, dt = \frac{1}{x + 1}, \quad x > -1,$$

deduce that

$$\int_0^1 t^x(-\ln t)^m\, dt = \frac{m!}{(x + 1)^{m+1}}, \quad x > -1.$$

11. From the fact that

$$\int_0^\infty \frac{t^{x-1}}{1 + t}\, dt = \frac{\pi}{\sin \pi x}, \quad 0 < x < 1,$$

deduce that

$$\int_0^\infty \frac{t^{x-1} \ln t}{1 + t}\, dt = \frac{\pi^2 \cos \pi x}{\cos^2 \pi x - 1}.$$

12. Verify that

$$\Gamma'(x) = \int_0^\infty t^{x-1}\, (\ln t)e^{-t}\, dt$$

and, more generally, that

$$\frac{d^n}{dx^n}\, (\Gamma(x)) = \int_0^\infty t^{x-1}\, (\ln t)^n e^{-t}\, dt.$$

13. If

$$\phi(x) = \int_0^\infty \frac{e^{-xt}}{1 + t}\, dt,$$

show that $\phi(x) - \phi'(x) = 1/x$. Justify your steps.

14. If

$$\phi(x) = \int_0^\infty \frac{e^{-xt}\, dt}{1 + t^2},$$

show that $\phi(x) + \phi''(x) = 1/x$. Justify your steps.

15. Given that

$$\phi(x) = \int_0^\infty e^{-t} \left(\frac{1 - \cos xt}{t}\right)\, dt.$$

Find $\phi'(x)$ by Leibniz' Rule. Evaluate the integral for $\phi'(x)$ and then find $\phi(x)$ by integration.

16. Given that

$$\phi(x) = \int_0^1 \frac{t^x - 1}{\ln t}\, dt, \qquad x \ge a, \quad a > -1.$$

Find $\phi'(x)$ by Leibniz' Rule, evaluate, and find $\phi(x)$ by integration.

17. Given that

$$\phi(x) = \int_0^\infty \frac{e^{-t}(1 - \cos xt)}{t^2}\, dt.$$

Find $\phi'(x)$ and $\phi''(x)$ by Leibniz' Rule. Find ϕ' and ϕ by integration.

18. Given that

$$\phi(x) = \int_0^\infty e^{-xt}\, dt = \frac{1}{x},$$

show that

$$\phi^{(n)}(x) = (-1)^n \int_0^\infty t^n e^{-xt}\, dt = (-1)^n \frac{n!}{x^{n+1}}, \qquad x > 0.$$

19. Starting from the formula

$$\int_0^\infty \frac{dt}{t^2 + x} = \frac{\pi}{2} x^{-1/2},$$

deduce the formula

$$\int_0^\infty \frac{dt}{(t^2 + x)^{n+1}} = \frac{1 \cdot 3 \cdot 5 \cdots (2n-1)}{2 \cdot 4 \cdot 6 \cdots 2n} x^{-n-1/2} = \frac{(2n)!}{2^{2n}(n!)^2} x^{-n-1/2}.$$

20. Prove Theorem 12.

*21. Given that

$$P_n(x) = \frac{1}{n!\sqrt{\pi}} \int_{-\infty}^\infty e^{-(1-x^2)t^2} \left(-\frac{d}{dx}\right)^n (e^{-x^2 t^2})\, dt,$$

show that

$$x\frac{d}{dx}(P_n(x)) - \frac{d}{dx}(P_{n-1}(x)) = nP_n(x).$$

$[P_n(x)$ is the Legendre polynomial.]

CHAPTER 13

Vector Field Theory

1. VECTOR FUNCTIONS

A **vector function** $\mathbf{v}(P)$ assigns a specific vector to each element P in a given domain $\mathfrak{D}$. The range of such a function is the collection of vectors which correspond to the points in the domain. In *First Course* (pages 433–442), we discussed vector functions with domain a portion (or all) of the real axis and with range a collection of vectors in a plane. For example, if the domain of a vector function is the interval $a \le t \le b$, then we may represent such a function in the form

$$\mathbf{v}(t) = f(t)\mathbf{i} + g(t)\mathbf{j},$$

where $\mathbf{i}$ and $\mathbf{j}$ are the customary unit vectors. The real-valued functions f and g are defined on the interval $[a, b]$.

If the range of a vector function of one variable is a collection of vectors in three-space, then we can write

$$\mathbf{v}(t) = f(t)\mathbf{i} + g(t)\mathbf{j} + h(t)\mathbf{k},$$

where $\mathbf{i}, \mathbf{j},$ and $\mathbf{k}$ are mutually perpendicular unit vectors in three-space. The functions $f, g,$ and h are ordinary real-valued functions of the variable t. In Chapter 2 (Section 4, page 56), we discussed various properties of vector functions with range in three-space.

We now continue the study of vector functions by considering those with *domain* a portion of the plane or of three-space. A vector function with domain in the xy-plane and with range consisting of vectors in a plane has the representation

$$\mathbf{v}(x, y) = f(x, y)\mathbf{i} + g(x, y)\mathbf{j}. \tag{1}$$

If the range consists of vectors in three-space, we write

$$\mathbf{v}(x, y) = f(x, y)\mathbf{i} + g(x, y)\mathbf{j} + h(x, y)\mathbf{k}.$$

If the domain $\mathfrak{D}$ is a region in three-space, so that we are dealing with functions of three variables, a vector function $\mathbf{u}(P)$ has the representation

$$\mathbf{u}(x, y, z) = f(x, y, z)\mathbf{i} + g(x, y, z)\mathbf{j}$$

when the range is two-dimensional. If the range of such a vector function is a collection of vectors in three-space, we write

$$\mathbf{u}(x, y, z) = f(x, y, z)\mathbf{i} + g(x, y, z)\mathbf{j} + h(x, y, z)\mathbf{k}. \tag{2}$$

The vector functions described above are actually special cases of the general transformations described in Chapter 11, Sections 3 and 4. In fact, we may easily define a vector function with domain $\mathfrak{D}$ in some m-dimensional Euclidean space and with range in some n-dimensional vector space. Of course, the numbers m and n may be different. Although many of the results of this chapter are valid in quite general spaces, we shall state and prove theorems in two and three dimensions only. In this way we can take full advantage of our geometrical insight.

The usual properties of continuity, differentiation, and integration for vector functions of several variables are immediate generalizations of those for vector functions of one variable. A function $\mathbf{v}(x, y)$ as given by (1) is *continuous* if and only if the functions $f(x, y)$ and $g(x, y)$ are. A function $\mathbf{u}(x, y, z)$ as in (2) has partial derivatives if and only if the functions f, g, and h do. For example, we have the formula

$$\frac{\partial \mathbf{u}}{\partial x} = \frac{\partial f(x, y, z)}{\partial x}\mathbf{i} + \frac{\partial g(x, y, z)}{\partial x}\mathbf{j} + \frac{\partial h(x, x, z)}{\partial x}\mathbf{k}.$$

We also use the notation

$$\mathbf{u}_{,1}(x, y, z) = f_{,1}(x, y, z)\mathbf{i} + g_{,1}(x, y, z)\mathbf{j} + h_{,1}(x, y, z)\mathbf{k}.$$

The formulas for partial derivatives with respect to y and z are obvious analogs.

EXAMPLE 1. Find $\partial\mathbf{u}/\partial x$ and $\partial^2\mathbf{u}/\partial x\,\partial y$ if

$$\mathbf{u}(x, y) = (x^2 + 2xy)\mathbf{i} + (1 + x^2 - y^2)\mathbf{j} + (x^3 - xy^2)\mathbf{k}.$$

Solution. We have

$$\frac{\partial \mathbf{u}}{\partial x} = 2(x + y)\mathbf{i} + 2x\mathbf{j} + (3x^2 - y^2)\mathbf{k},$$

$$\frac{\partial^2 \mathbf{u}}{\partial x\,\partial y} = 2\mathbf{i} + 0\cdot\mathbf{j} - 2y\mathbf{k} = 2\mathbf{i} - 2y\mathbf{k}.$$

The formulas for the partial derivatives of the scalar and cross products of vector functions are similar to those for vector functions of one variable. (See Chapter 2, Section 4.) For example, if $\mathbf{u}(x, y, z)$ and $\mathbf{v}(x, y, z)$ are differentiable functions, then the student may verify that

$$\frac{\partial}{\partial x}(\mathbf{u}\cdot\mathbf{v}) = \mathbf{u}\cdot\frac{\partial\mathbf{v}}{\partial x} + \frac{\partial\mathbf{u}}{\partial x}\cdot\mathbf{v}, \qquad \frac{\partial}{\partial x}(\mathbf{u}\times\mathbf{v}) = \mathbf{u}\times\frac{\partial\mathbf{v}}{\partial x} + \frac{\partial\mathbf{u}}{\partial x}\times\mathbf{v}.$$

Analogous formulas hold for derivatives with respect to y and z.

EXAMPLE 2. Given the vector functions

$$\mathbf{u} = (e^x \cos y)\mathbf{i} + (e^y \sin z)\mathbf{j} + (e^z \sin x)\mathbf{k}$$

and

$$\mathbf{v} = (\cos z)\mathbf{i} + (e^{2x} \sin y)\mathbf{j} + e^z\mathbf{k}.$$

Find $\partial(\mathbf{u} \cdot \mathbf{v})/\partial y$. Decide whether the vectors $\partial\mathbf{u}/\partial z$ and $\partial\mathbf{v}/\partial x$ are perpendicular at the point P_0: $x = 1$, $y = 0$, $z = \pi/4$.

Solution. We have

$$\frac{\partial\mathbf{u}}{\partial y} = (-e^x \sin y)\mathbf{i} + (e^y \sin z)\mathbf{j}, \qquad \frac{\partial\mathbf{v}}{\partial y} = e^{2x} \cos y\mathbf{j},$$

$$\frac{\partial}{\partial y}(\mathbf{u} \cdot \mathbf{v}) = \mathbf{u} \cdot \frac{\partial\mathbf{v}}{\partial y} + \frac{\partial\mathbf{u}}{\partial y} \cdot \mathbf{v}$$

$$= e^{2x+y} \cos y \sin z - e^x \sin y \cos z + e^{2x+y} \sin y \sin z.$$

Computing the derivatives

$$\frac{\partial\mathbf{u}}{\partial z} = (e^y \cos z)\mathbf{j} + (e^z \sin x)\mathbf{k}, \qquad \frac{\partial\mathbf{v}}{\partial x} = (2e^{2x} \sin y)\mathbf{j},$$

we find

$$\frac{\partial\mathbf{u}}{\partial z} \cdot \frac{\partial\mathbf{v}}{\partial x} = 2e^{2x+y} \sin y \cos z.$$

At $P_0(1, 0, \pi/4)$, this scalar product is zero, and so the vectors are orthogonal there.

The integration of a vector function is given in terms of the integration of each of its components. If R is a region and $\mathbf{v}(x, y)$ is a vector function with domain containing R, then the integral

$$\iint\limits_{R} \mathbf{v}(x, y)\, dA$$

may be defined by the formula

$$\iint\limits_{R} \mathbf{v}(x, y)\, dA = \left(\iint\limits_{R} f(x, y)\, dA\right)\mathbf{i} + \left(\iint\limits_{R} g(x, y)\, dA\right)\mathbf{j} + \left(\iint\limits_{R} h(x, y)\, dA\right)\mathbf{k}.$$

We see that the definite integral of a vector function is a vector. Similarly, triple integrals and line integrals of a vector function may be defined in terms of the corresponding integrals of each of its components.

The parametric equations

$$x = f(t), \qquad y = g(t), \qquad a \le t \le b \tag{3}$$

represent a curve in the xy-plane. Similarly, the equations

$$x = f(t), \qquad y = g(t), \qquad z = h(t), \qquad a \le t \le b$$

represent a curve in three-space. These curves may also be represented by vector functions. A vector is an equivalence class of directed line segments. If we consider the particular representative directed line segment which has its base at the origin, then the vector function

$$\mathbf{v}(t) = f(t)\mathbf{i} + g(t)\mathbf{j} \tag{4}$$

turns out to be a characterization of the curve given by (3). The heads of the directed line segments representing $\mathbf{v}(t)$ which have their base at the origin trace out the curve (3). This geometric interpretation was discussed in some detail in the study of vector functions of one variable.

The parametric equations

$$x = f(s, t), \qquad y = g(s, t), \qquad z = h(s, t) \tag{5}$$

represent a surface in three-space. The vector function

$$\mathbf{v}(s, t) = f(s, t)\mathbf{i} + g(s, t)\mathbf{j} + h(s, t)\mathbf{k}$$

characterizes the same surface when we consider the collection of those directed line segments with base at the origin which represent $\mathbf{v}$. The heads of these directed line segments form the surface (5).

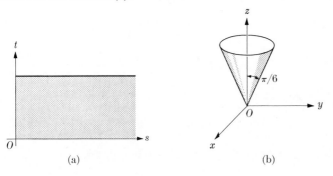

(a)   (b)

FIGURE 13–1

EXAMPLE 3. Sketch the surface represented by the vector function

$$\mathbf{v}(s, t) = (s \cos t)\mathbf{i} + (s \sin t)\mathbf{j} + (\sqrt{3}\, s)\mathbf{k}, \qquad s \geq 0, \quad 0 \leq t \leq 2\pi.$$

Solution. We write

$$x = s \cos t, \qquad y = s \sin t, \qquad z = \sqrt{3}\, s.$$

Squaring and adding to eliminate the parameters s and t, we find

$$3(x^2 + y^2) = z^2,$$

which we recognize as the equation of a cone. The domain of the vector function and the surface represented by it are shown in Fig. 13–1. Since $s \geq 0$, the given locus is only the upper half of the cone.

EXERCISES

In each of problems 1 through 9, find the partial derivatives as indicated.

1. $\mathbf{u}(x, y) = (x^2 + 2xy)\mathbf{i} + (x^3 - y^3)\mathbf{j}$; $\dfrac{\partial \mathbf{u}}{\partial x}, \dfrac{\partial \mathbf{u}}{\partial y}$

2. $\mathbf{u}(x, y) = (x \cos y)\mathbf{i} + (y \sin x)\mathbf{j}$; $\dfrac{\partial \mathbf{u}}{\partial x}, \dfrac{\partial \mathbf{u}}{\partial y}$

3. $\mathbf{u}(x, y, z) = \dfrac{x}{x^2 + y^2}\mathbf{i} + \dfrac{y}{x^2 + z^2}\mathbf{j} + \dfrac{z}{x^2 + y^2}\mathbf{k}$; $\dfrac{\partial \mathbf{u}}{\partial x}, \dfrac{\partial \mathbf{u}}{\partial z}$

4. $\mathbf{u}(x, y, z) = (e^{xy} \ln z)\mathbf{i} + (e^{yz} \ln x)\mathbf{j}$; $\dfrac{\partial \mathbf{u}}{\partial y}, \dfrac{\partial \mathbf{u}}{\partial z}$

5. $\mathbf{v}(s, t) = (s \cos t)\mathbf{i} + (t \sin s)\mathbf{j} + (t^2 - s^2)\mathbf{k}$; $\dfrac{\partial \mathbf{v}}{\partial s}, \dfrac{\partial \mathbf{v}}{\partial t}$

6. $\mathbf{v}(s, t) = (e^s \tan t)\mathbf{i} + (e^{-s} \cos t)\mathbf{j}$; $\dfrac{\partial^2 \mathbf{v}}{\partial s^2}, \dfrac{\partial^2 \mathbf{v}}{\partial t^2}$

7. $\mathbf{v}(r, s, t) = (s^2 - t^2)\mathbf{i} + (t^2 - r^2)\mathbf{j} + (r^2 - s^2)\mathbf{k}$; $\dfrac{\partial^2 \mathbf{v}}{\partial r\, \partial s}, \dfrac{\partial^2 \mathbf{v}}{\partial s\, \partial t}$

8. $\mathbf{u}(x, y) = (x + y)\mathbf{i} + x^2\mathbf{j} + (y - x)\mathbf{k}$; $\dfrac{\partial}{\partial x}\left(|\mathbf{u}(x, y)|^2\right)$

9. $\mathbf{u}(x, y, z) = (\sin xy)\mathbf{i} + (\cos yz)\mathbf{j} + (\sin xz)\mathbf{k}$; $\dfrac{\partial}{\partial y}\left(|\mathbf{u}(x, y, z)|^2\right)$

10. Find $\partial(\mathbf{u} \cdot \mathbf{v})/\partial x$ if
$$\mathbf{u} = (xy)\mathbf{i} + (x - y)\mathbf{j} + (x + y)\mathbf{k},$$
$$\mathbf{v} = (2x - y)\mathbf{i} - x\mathbf{j} + (x + 2y)\mathbf{k}.$$

11. Find $\partial(\mathbf{u} \cdot \mathbf{v})/\partial y$ if
$$\mathbf{u} = \ln(x + y)\mathbf{i} + (x - y)\mathbf{j} + \ln(x - y)\mathbf{k},$$
$$\mathbf{v} = e^{x-y}\mathbf{i} + (x + y)\mathbf{j} + \ln(x + y)\mathbf{k}.$$

12. Find $\partial(\mathbf{u} \times \mathbf{v})/\partial x$ if
$$\mathbf{u} = (1 + x)\mathbf{i} + (y + 2)\mathbf{j} + (x + y)\mathbf{k},$$
$$\mathbf{v} = 2\mathbf{i} + (3 - x + y)\mathbf{j} + (x - 2y)\mathbf{k}.$$

13. Find $\partial(\mathbf{u} \times \mathbf{v})/\partial t$ if
$$\mathbf{u} = (e^t \cos s)\mathbf{i} + (e^{-t} \sin s)\mathbf{j} + e^t\mathbf{k},$$
$$\mathbf{v} = (e^{-t} \sin 2s)\mathbf{i} + (e^t \cos s)\mathbf{j} + 2\mathbf{k}.$$

14. Given the vectors
$$\mathbf{u}(x, y) = (x + y)\mathbf{i} + (2x - y)\mathbf{j},$$
$$\mathbf{v}(x, y) = (x^2 - y)\mathbf{i} + (x + 2y^2)\mathbf{j}.$$

(a) At what values of x and y are $\partial \mathbf{u}/\partial x$ and $\partial \mathbf{v}/\partial x$ orthogonal?
(b) At what values of x and y are $\partial \mathbf{u}/\partial y$ and $\partial \mathbf{v}/\partial y$ orthogonal?
(c) At what values of x and y do both (a) and (b) hold?

15. Given the vectors

$$\mathbf{u}(x, y) = (x - 2y)\mathbf{i} + (x - y)\mathbf{j} + (x + 2y)\mathbf{k},$$
$$\mathbf{v}(x, y) = (2x - y)\mathbf{i} + (x + y)\mathbf{j} + (x + 3y)\mathbf{k}.$$

Find the values of x and y such that

$$\mathbf{u} \cdot (\mathbf{u}_{,1} \times \mathbf{v}_{,2}) = 0.$$

16. Derive a formula for

$$\frac{\partial}{\partial x} [\mathbf{u} \cdot (\mathbf{v} \times \mathbf{w})]$$

in terms of $\partial \mathbf{u}/\partial x$, $\partial \mathbf{v}/\partial x$, and $\partial \mathbf{w}/\partial x$.

17. Describe the surface represented by the vector function

$$\mathbf{v}(s, t) = (3 \cos s \cos t)\mathbf{i} + (3 \cos s \sin t)\mathbf{j} + (3 \sin s)\mathbf{k}$$

for $0 \leq s \leq (\pi/2), 0 \leq t \leq 2\pi$.

18. Describe the surface represented by the vector function

$$\mathbf{v}(s, t) = (s + t)\mathbf{i} + (s + t)\mathbf{j} + (st)\mathbf{k}$$

for $-\infty < s < \infty, -\infty < t < \infty$.

19. Describe the surface represented by the vector function

$$\mathbf{v}(s, t) = (2s \cos t)\mathbf{i} + (3s \sin t)\mathbf{j} + s^2\mathbf{k}$$

for $0 \leq s < \infty, 0 \leq t \leq 2\pi$.

20. Show that if

$$\mathbf{v}(x, y, z) = \frac{\mathbf{w}(x, y, z)}{F(x, y, z)},$$

then

$$\mathbf{v}_{,1} = \frac{F\mathbf{w}_{,1} - \mathbf{w}F_{,1}}{F^2}.$$

2. VECTOR AND SCALAR FIELDS. DIRECTIONAL DERIVATIVE AND GRADIENT

Vector functions occur frequently in applications. The vector velocity of the wind in the atmosphere is an example of a vector function. Other examples of vector functions are the vector velocity of the particles of fluid in a stream and the vector force of gravity exerted by the earth on an object in space.

We may represent a vector function graphically. At each point P of the domain D of a function $\mathbf{v}(P)$, we construct the representative directed line segment of $\mathbf{v}(P)$ having its base at P. If D is a plane region and if the range of $\mathbf{v}(P)$ is a collection of vectors in the plane, the graphical representation appears as in Fig. 13–2. The vectors in Fig. 13–2 form a field of vectors. More precisely, **vector field** is a synonym for vector function. In analogy, an ordinary function f which as-

signs a real number to each point of a region
in the plane or in space is called a **scalar field**
or a **scalar function**.

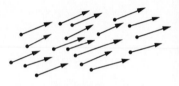

In Section 1 we defined vector fields in terms
of the unit vectors **i**, **j**, and **k**. In other words,
we introduced a rectangular coordinate system
and used this system as a basis for various

FIGURE 13–2

definitions. Many of the most important properties of vector fields are geometric
in character and so are independent of any coordinate system. Therefore, whenever it is possible we shall define properties of vector functions without reference to a particular coordinate system. For example, in Section 1 we defined
the continuity of a vector function $\mathbf{v}(x, y)$ given by

$$\mathbf{v}(x, y) = f(x, y)\mathbf{i} + g(x, y)\mathbf{j}$$

in terms of the continuity of f and g. It is also possible to proceed without such a
reference to coordinate vectors **i** and **j**. Suppose that P_0 is a point in the domain
D of a vector field $\mathbf{v}(P)$. Then we say that $\mathbf{v}(P)$ *is continuous at point P_0* if and
only if for every $\epsilon > 0$ there is a $\delta > 0$ such that

$$|\mathbf{v}(P) - \mathbf{v}(P_0)| < \epsilon$$

whenever $0 < |PP_0| < \delta$. Observe that this definition is valid if the domain of
$\mathbf{v}(P)$ is a two- or three-dimensional region. Furthermore, since $|\mathbf{v}(P) - \mathbf{v}(P_0)|$
is the length of the vector $\mathbf{v}(P) - \mathbf{v}(P_0)$, the definition is applicable for vector
fields which have either two- or three-dimensional ranges.

It is possible to define the definite integral of a scalar or vector field without
reference to a coordinate system. If R is the region of integration, we subdivide
R into a number of smaller regions $R_1, R_2, \ldots, R_n$. Let P_i be any point in R_i.
Then we form the sums

$$\sum_{i=1}^{n} f(P_i)A(R_i) \quad \text{and} \quad \sum_{i=1}^{n} \mathbf{v}(P_i)A(R_i),$$

where $A(R_i)$ is the area (or volume if R is in three-space) of the region R_i. If the
above expressions approach limits under the usual hypotheses imposed for the
definition of integrals, then we say the integral exists. The first sum above will
approach a numerical limit, while the limit of the second sum will be a vector.
Note that no reference to a particular coordinate system is required in these definitions.

Suppose that f is a scalar field, that P is a point in the domain of f, and that $\mathbf{a}$
is a unit vector. We define the **directional derivative of f in the direction of a at the
point P** by

$$\lim_{h \to 0} \frac{f(Q) - f(P)}{h},$$

where Q is a point at distance h from P in the direction of **a**. The relation between P and Q is shown in Fig. 13–3. We denote this directional derivative by

$$D_a f(P).$$

If we hold $\mathbf{a}$ and f fixed and consider P as variable, then $D_a f(P)$ defines a scalar field which is analogous to a partial derivative. In fact, if we introduce a rectangular (x, y, z)-coordinate system and select $\mathbf{a} = \mathbf{i}$, the directional derivative is $\partial/\partial x$. Similarly, $\mathbf{a} = \mathbf{j}$ and $\mathbf{a} = \mathbf{k}$ correspond to partial derivatives with respect to y and z, respectively.

FIGURE 13–3

DEFINITION. A scalar field f is **continuously differentiable** on an open set $\mathfrak{D}$ if and only if f and $D_a f$ are continuous on $\mathfrak{D}$ for each fixed $\mathbf{a}$.

The definition of directional derivative of a vector field is analogous to that of a scalar field.

DEFINITION. The **directional derivative, $D_a \mathbf{w}(P)$, of a vector field $\mathbf{w}$ in the direction of $\mathbf{a}$ at the point P** is given by

$$D_a \mathbf{w}(P) = \lim_{h \to 0} \frac{\mathbf{w}(Q) - \mathbf{w}(P)}{h},$$

where the relation between Q and P is that shown in Fig. 13–3. We say that $\mathbf{w}$ is **continuously differentiable on $\mathfrak{D}$** if and only if $\mathbf{w}$ and $D_a \mathbf{w}$ are continuous on $\mathfrak{D}$ for every fixed $\mathbf{a}$.

Remarks. For fixed $\mathbf{a}$ and $\mathbf{w}$, the directional derivative $D_a \mathbf{w}$ defines a vector field on $\mathfrak{D}$. If a rectangular (x, y, z)-coordinate system is introduced, then selecting $\mathbf{a}$ equal to $\mathbf{i}$, $\mathbf{j}$, and $\mathbf{k}$, in turn, gives the partial derivatives with respect to x, y, and z, respectively. The computation of these derivatives was discussed in Section 1.

While it is convenient to define properties of scalar and vector fields without reference to any coordinate axes, it is usually desirable to use some appropriate coordinate system when specific computations are to be made. Suppose that f is a scalar field in space and that a rectangular (x, y, z)-coordinate system is introduced. Then with f there is associated the unique function f^* defined by the equation

$$f(P) = f^*(x_P, y_P, z_P)$$

where (x_P, y_P, z_P) are the coordinates of P in three-space. Two different coordinate systems will associate two different functions to a given scalar field. If (x, y, z) and (x', y', z') are two rectangular coordinate systems, we know from Chapter 7, Section 8, that the coordinates of a given point P in the two systems are related by equations of the form

$$x_P = a + c_{11}x'_P + c_{12}y'_P + c_{13}z'_P,$$
$$y_P = b + c_{21}x'_P + c_{22}y'_P + c_{23}z'_P,$$
$$z_P = c + c_{31}x'_P + c_{32}y'_P + c_{33}z'_P,$$

in which the matrix $C = (c_{ij})$ is orthogonal. Since we shall restrict ourselves to right-handed systems only, it can be shown that det $C = 1$. (See Section 4, page 596 of this chapter.) If f^* and $f^{*\prime}$ correspond to f under the (x, y, z) and (x', y', z') systems, respectively, then

$$f^{*\prime}(x'_P, y'_P, z'_P) = f^*(a + c_{11}x'_P + c_{12}y'_P + c_{13}z'_P,$$
$$b + c_{21}x'_P + c_{22}y'_P + c_{23}z'_P,$$
$$c + c_{31}x'_P + c_{32}y'_P + c_{33}z'_P).$$

We shall now derive an expression for the directional derivative $D_a f$ in terms of an (x, y, z) coordinate system. Using the mutually orthogonal unit vectors $\mathbf{i}, \mathbf{j}, \mathbf{k}$, the unit vector $\mathbf{a}$ has the representation

$$\mathbf{a} = \lambda\mathbf{i} + \mu\mathbf{j} + \nu\mathbf{k}, \qquad \lambda^2 + \mu^2 + \nu^2 = 1.$$

If the coordinates of P are (x_P, y_P, z_P) and those of Q are $(x_P + \lambda h, y_P + \mu h, z_P + \nu h)$, then the directional derivative is given by

$$D_a f(P) = \lim_{h \to 0} \frac{f^*(x_P + \lambda h, y_P + \mu h, z_P + \nu h) - f^*(x_P, y_P, z_P)}{h}, \quad (1)$$

where f^* is the coordinate function in three-space corresponding to the scalar field f. Recalling the definition of directional derivative as defined on page 138 of Chapter 4, we find

$$D_a f(P) = \lambda f^*_{,1}(x_P, y_P, z_P) + \mu f^*_{,2}(x_P, y_P, z_P) + \nu f^*_{,3}(x_P, y_P, z_P). \quad (2)$$

From this formula, we see that a function f is continuously differentiable if and only if its corresponding function with respect to some given (and therefore every) coordinate system is continuously differentiable.

A similar discussion holds for vector fields. If $\mathbf{u}$ is a vector field in space, then in each coordinate system there corresponds a triple of functions. More precisely, if (x, y, z) is a rectangular coordinate system and $\mathbf{i}, \mathbf{j}, \mathbf{k}$ are the customary unit vectors, then the components of $\mathbf{u}$ with respect to this system are defined by

$$\mathbf{u}(P) = f^*(x_P, y_P, z_P)\mathbf{i} + g^*(x_P, y_P, z_P)\mathbf{j} + h^*(x_P, y_P, z_P)\mathbf{k}.$$

The directional derivative $D_a\mathbf{u}(P)$ in this coordinate system is

$$D_a\mathbf{u}(P) = (D_a f^*)\mathbf{i} + (D_a g^*)\mathbf{j} + (D_a h^*)\mathbf{k}.$$

The directional derivatives of the coordinate functions on the right may be calculated according to formula (2).

While it is important to distinguish between a scalar field f and its corresponding function $f*$ given in a particular coordinate system, we shall, on occasion, drop the * notation. Since almost all computational work is done in some coordinate system, there is little danger of confusion in such a procedure.

EXAMPLE 1. Given the vector field

$$\mathbf{u}(P) = (x^2 - y + z)\mathbf{i} + (2y - 3z)\mathbf{j} + (x + z)\mathbf{k}$$

and the unit vector

$$\mathbf{a} = \frac{1}{\sqrt{6}}(2\mathbf{i} - \mathbf{j} + \mathbf{k}).$$

Find $D_{\mathbf{a}}\mathbf{u}(P)$.

Solution. We have

$$D_{\mathbf{a}}(x^2 - y + z) = \frac{2}{\sqrt{6}}(2x) - \frac{1}{\sqrt{6}}(-1) + \frac{1}{\sqrt{6}}(1) = \frac{1}{\sqrt{6}}(4x + 2),$$

$$D_{\mathbf{a}}(2y - 3z) = \frac{2}{\sqrt{6}}(0) - \frac{1}{\sqrt{6}}(2) + \frac{1}{\sqrt{6}}(-3) = -\frac{5}{\sqrt{6}},$$

$$D_{\mathbf{a}}(x + z) = \frac{2}{\sqrt{6}}(1) - \frac{1}{\sqrt{6}}(0) + \frac{1}{\sqrt{6}}(1) = \frac{3}{\sqrt{6}}.$$

Therefore

$$D_{\mathbf{a}}\mathbf{u}(P) = \frac{1}{\sqrt{6}}(4x + 2)\mathbf{i} - \frac{5}{\sqrt{6}}\mathbf{j} + \frac{3}{\sqrt{6}}\mathbf{k}.$$

Let $f(P)$ be a scalar field, let (x, y, z) be a rectangular coordinate system, and let $\mathbf{a}$ be a unit vector. We define the vector field

$$\mathbf{v}(P) = f_{,1}^*(x_P, y_P, z_P)\mathbf{i} + f_{,2}^*(x_P, y_P, z_P)\mathbf{j} + f_{,3}^*(x_P, y_P, z_P)\mathbf{k}, \tag{3}$$

where $f*$, $\mathbf{i}$, $\mathbf{j}$, and $\mathbf{k}$ have the significance described above. Then we observe that, according to the definition of scalar product, formula (2) may be written

$$D_{\mathbf{a}}f(P) = \mathbf{v}(P) \cdot \mathbf{a}. \tag{4}$$

We now show that the vector $\mathbf{v}(P)$, as given in (4), is unique.

Theorem 1. *Suppose that*

$$\mathbf{v} \cdot \mathbf{a} = \mathbf{w} \cdot \mathbf{a}$$

for every unit vector $\mathbf{a}$ *(in the plane or in space). Then* $\mathbf{v} = \mathbf{w}$.

Proof. The equation $\mathbf{v} \cdot \mathbf{a} = \mathbf{w} \cdot \mathbf{a}$ is equivalent to

$$(\mathbf{v} - \mathbf{w}) \cdot \mathbf{a} = 0. \tag{5}$$

If $\mathbf{v} - \mathbf{w} \neq \mathbf{0}$, we select $\mathbf{a}$ to be a unit vector in the direction of $\mathbf{v} - \mathbf{w}$. Then (5) would not hold.

DEFINITION. The unique vector $\mathbf{v}(P)$ determined by Eq. (4) is called the **gradient of f at P**. We denote this vector by ∇f, which we read "del f." Then equation (4) becomes

$$D_{\mathbf{a}} f(P) = \nabla f \cdot \mathbf{a} \tag{6}$$

for any unit vector $\mathbf{a}$. We observe that since $D_{\mathbf{a}} f(P)$ and $\mathbf{a}$ are defined without reference to a coordinate system, the vector ∇f (which is uniquely determined) has a significance independent of axes.

To calculate ∇f when a rectangular (x, y, z)-system is introduced, we simply use the formula

$$\nabla f = f_{,1}^* \mathbf{i} + f_{,2}^* \mathbf{j} + f_{,3}^* \mathbf{k}.$$

The geometric significance of ∇f is easily seen with reference to the surface $f^*(x, y, z) = \text{const.}$ If P is a point on the surface, the vector $\nabla f(P)$, whenever it is not zero, is perpendicular to the surface. This fact was established in Chapter 4 on page 144. To find a unit vector normal to a surface $f(P) = \text{const}$, we select

$$\mathbf{n} = \frac{\nabla f}{|\nabla f|}$$

evaluated at the desired point.

EXAMPLE 2.   Given the function $f^*(x, y, z) = 3x^2 - y^2 + 2z^2$, find ∇f and a unit normal to the surface $f^*(x, y, z) = 17$ at the point $(1, -2, 3)$.

Solution. We have

$$\nabla f = f_{,1}^* \mathbf{i} + f_{,2}^* \mathbf{j} + f_{,3}^* \mathbf{k} = 6x\mathbf{i} - 2y\mathbf{j} + 4z\mathbf{k},$$
$$\nabla f(1, -2, 3) = 6\mathbf{i} + 4\mathbf{j} + 12\mathbf{k} = 2(3\mathbf{i} + 2\mathbf{j} + 6\mathbf{k}),$$
$$\mathbf{n} = \pm\tfrac{1}{7}(3\mathbf{i} + 2\mathbf{j} + 6\mathbf{k}).$$

The proof of the next theorem is left to the student. (See Exercises 12 and 13.)

Theorem 2. *Suppose that f, g, and u are continuously differentiable scalar fields. Then*

$$\left. \begin{array}{cc} \nabla(f + g) = \nabla f + \nabla g, & \nabla(fg) = f\nabla g + g\nabla f \\[2mm] \nabla\left(\dfrac{f}{g}\right) = \dfrac{1}{g^2}(g\nabla f - f\nabla g) & \text{if } g \neq 0 \end{array} \right\} , \tag{7}$$

$$\nabla f(u) = f'(u)\nabla u. \tag{8}$$

EXERCISES

In each of problems 1 through 6, find ∇f and $D_a f$ at $P_0(x_0, y_0, z_0)$ as given.

1. $f^*(x, y, z) = x^2 - y^2 + 2yz + 2z^2$, $P_0(2, -1, 1)$, $\mathbf{a} = \dfrac{1}{\sqrt{14}}(2\mathbf{i} - \mathbf{j} + 3\mathbf{k})$

2. $f^* = e^x \cos y + xz^2$, $P_0(0, \pi/2, 1)$, $\mathbf{a} = \frac{1}{7}(3\mathbf{i} + 2\mathbf{j} - 6\mathbf{k})$

3. $f^* = \dfrac{x}{(x^2 + y^2 + z^2)^{3/2}}$, $P_0(2, 2, 1)$, $\mathbf{a} = \frac{1}{3}(\mathbf{i} - 2\mathbf{j} + 2\mathbf{k})$

4. $f^* = \ln(x^2 + y^2)$, $P_0(3, 4, 2)$, $\mathbf{a} = \frac{1}{7}(6\mathbf{i} - 3\mathbf{j} - 2\mathbf{k})$

5. $f^* = \tan x + \ln(y + z)$, $P_0(\pi/4, 1, 1)$, $\mathbf{a} = \frac{1}{3}(2\mathbf{i} + 2\mathbf{j} - \mathbf{k})$

6. $f^* = e^{xy} \cos z + e^{yz} \sin x$, $P_0(\pi/6, 0, \pi/3)$, $\mathbf{a} = \dfrac{1}{\sqrt{17}}(3\mathbf{i} + 2\mathbf{j} - 2\mathbf{k})$

In problems 7 through 9, find $D_a \mathbf{u}$ at $P_0(x_0, y_0, z_0)$ as given.

7. $\mathbf{u} = (x^2 + y^2)\mathbf{i} + 2xy\mathbf{j} + 3xz\mathbf{k}$, $P_0(1, 0, 2)$, $\mathbf{a} = \dfrac{1}{\sqrt{21}}(\mathbf{i} + 4\mathbf{j} + 2\mathbf{k})$

8. $\mathbf{u} = \dfrac{x}{x^2 + y^2}\mathbf{i} + \dfrac{y}{y^2 + z^2}\mathbf{j} + \dfrac{z}{x^2 + z^2}\mathbf{k}$, $P_0(1, 1, 2)$, $\mathbf{a} = \dfrac{1}{\sqrt{6}}(2\mathbf{i} + \mathbf{j} - \mathbf{k})$

9. $\mathbf{u} = \cos(xy)\mathbf{i} + \sin(z^2)\mathbf{j} + 2\mathbf{k}$, $P_0(0, \pi/4, 0)$, $\mathbf{a} = \dfrac{1}{\sqrt{29}}(2\mathbf{i} + 5\mathbf{k})$

10. Find the unit vector $\mathbf{a}$ such that $D_a f(P_0)$ is a maximum, given that
$$f^* = 2x^2 - 3y^2 + z^2, \qquad P_0 = (2, 1, 3).$$

11. Find the unit vector $\mathbf{a}$ such that $D_a f(P_0)$ is a minimum, given that
$$f^* = xyz, \qquad P_0 = (1, -3, -2).$$

12. Prove the laws stated in Eqs. (7). [*Hint:* Introduce a coordinate system.]

13. Prove the law (8).

14. Suppose that O is the origin of coordinates, P is a point, and $\mathbf{r}$ is the vector having the directed line segment $\overrightarrow{OP}$ as a representative. We write $\mathbf{r} = \mathbf{v}(\overrightarrow{OP})$. Show that
$$\nabla f(|\mathbf{r}|) = \frac{1}{|\mathbf{r}|} f'(|\mathbf{r}|)\mathbf{r}.$$

In each of problems 15 through 20, find the unit normal to the surface of the given point P_0.

15. $x^2 - 2xy + 2y^2 - z^2 = 9$, $P_0 = (2, -1, 1)$

16. $z = x^2 - y^2$, $P_0 = (3, -2, 5)$

17. $z^2 = x^2 + y^2$, $P_0 = (-3, 4, 5)$

18. $z = e^x \sin y$, $P_0 = (1, \pi/2, e)$

19. $z = \arctan x + \ln(1 + y)$, $P_0(1, 0, \pi/4)$

20. $x^3 + y^3 + z^3 - 3xyz = 14$, $P_0(2, 1, -1)$

3. THE DIVERGENCE OF A VECTOR FIELD

Suppose we have a rectangular coordinate system in space. It is convenient to label the coordinate axes x_1, x_2, x_3 and to use the vectors $\mathbf{e}_1, \mathbf{e}_2, \mathbf{e}_3$ as orthogonal unit vectors. For any scalar function f, we may regard the gradient, ∇, as an **operator**, i.e., as a transformation which takes scalar fields into vector fields. In a given coordinate system we write ∇ in the **symbolic form**

$$\nabla = \frac{\partial}{\partial x_1}\mathbf{e}_1 + \frac{\partial}{\partial x_2}\mathbf{e}_2 + \frac{\partial}{\partial x_3}\mathbf{e}_3. \tag{1}$$

Then we make the convention that ∇f has the meaning

$$\nabla f = \frac{\partial f^*}{\partial x_1}\mathbf{e}_1 + \frac{\partial f^*}{\partial x_2}\mathbf{e}_2 + \frac{\partial f^*}{\partial x_3}\mathbf{e}_3,$$

which is the gradient of f, as it should be.

Suppose now that $\mathbf{w}$ is a vector field in space which has the coordinate representation

$$\mathbf{w}(P) = w_1(x_1, x_2, x_3)\mathbf{e}_1 + w_2(x_1, x_2, x_3)\mathbf{e}_2 + w_3(x_1, x_2, x_3)\mathbf{e}_3. \tag{2}$$

We form the inner product of the symbolic vector ∇ and the vector $\mathbf{w}$, denoted $\nabla \cdot \mathbf{w}$, to get

$$\nabla \cdot \mathbf{w} = \frac{\partial w_1}{\partial x_1} + \frac{\partial w_2}{\partial x_2} + \frac{\partial w_3}{\partial x_3}. \tag{3}$$

The fact that the operator ∇ has a significance independent of the axes suggests that the expression (3) might be independent of the coordinate system. We shall show this to be true.

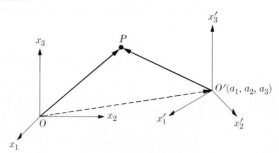

FIGURE 13–4

Let (x_1', x_2', x_3') be another rectangular coordinate system with origin at O' and with unit basis vectors $\mathbf{e}_1', \mathbf{e}_2', \mathbf{e}_3'$. Suppose that O' has coordinates (a_1, a_2, a_3) in the (x_1, x_2, x_3) system. (See Fig. 13–4.) For convenience, we repeat some of the results in Chapter 7, Section 8, showing the relationship between two coordinate systems. If P is any point in space, the vectors $\mathbf{v}(\overrightarrow{OP})$ and $\mathbf{v}(\overrightarrow{O'P})$ have the representations

$$\mathbf{v}(\overrightarrow{OP}) = \sum_{i=1}^{3} x_i\mathbf{e}_i, \qquad \mathbf{v}(\overrightarrow{O'P}) = \sum_{i=1}^{3} x_i'\mathbf{e}_i'.$$

Referring to Fig. 13–4, we see that the relation between the vectors is given by

$$\mathbf{v}(\overrightarrow{OP}) = \mathbf{v}(\overrightarrow{OO'}) + \mathbf{v}(\overrightarrow{O'P}) = \sum_{i=1}^{3} a_i \mathbf{e}_i + \sum_{i=1}^{3} x_i' \mathbf{e}_i'. \tag{4}$$

The connection between the coordinate systems is determined by the constants

$$c_{ij} = \mathbf{e}_i \cdot \mathbf{e}_j',$$

which form the elements of the transformation matrix C. We recall that this matrix is *orthogonal*, i.e., that the elements satisfy the equations

$$\sum_{j=1}^{3} c_{ij} c_{kj} = \delta_{ik} = \begin{cases} 1 & \text{if } i = k \\ 0 & \text{if } i \neq k \end{cases}. \tag{5}$$

Taking the inner product of $\mathbf{e}_k$ with $\mathbf{v}$ in (4), we find

$$\mathbf{e}_k \cdot \mathbf{v}(\overrightarrow{OP}) = \sum_{i=1}^{3} x_i \mathbf{e}_i \cdot \mathbf{e}_k = x_k = a_k + \sum_{j=1}^{3} c_{kj} x_j', \qquad k = 1, 2, 3. \tag{6}$$

Furthermore, if $\mathbf{w}$ is any vector field, it has the two representations

$$\mathbf{w}(P) = \sum_{j=1}^{3} w_j \mathbf{e}_j = \sum_{j=1}^{3} w_j' \mathbf{e}_j'. \tag{7}$$

As we learned in Chapter 7, the relation between w_j' and w_i is obtained by taking the inner product of $\mathbf{w}$ with each $\mathbf{e}_j'$. The result is

$$w_j' = \mathbf{e}_j' \cdot \mathbf{w}(P) = \sum_{k=1}^{3} c_{kj} w_k. \tag{8}$$

We now establish the invariance of the expression $\nabla \cdot \mathbf{w}$.

Theorem 3. *Suppose that $\mathbf{w}$ is a continuously differentiable vector field, that (x_1, x_2, x_3) and (x_1', x_2', x_3') are two rectangular coordinate systems, and that (w_1, w_2, w_3), (w_1', w_2', w_3') are the components of $\mathbf{w}$ in the two systems as given in (7). Then*

$$\frac{\partial w_1}{\partial x_1} + \frac{\partial w_2}{\partial x_2} + \frac{\partial w_3}{\partial x_3} = \frac{\partial w_1'}{\partial x_1'} + \frac{\partial w_2'}{\partial x_2'} + \frac{\partial w_3'}{\partial x_3'}. \tag{9}$$

Proof. Formulas (6) and (8) are the basis for the computations in the proof. From (6) we find

$$\frac{\partial x_k}{\partial x_j'} = c_{kj}, \qquad k, j = 1, 2, 3, \tag{10}$$

and from (8) we have

$$\frac{\partial w_j'}{\partial x_l} = \sum_{k=1}^{3} c_{kj} \frac{\partial w_k}{\partial x_l}, \qquad j, l = 1, 2, 3. \tag{11}$$

Now we use the Chain Rule to write

$$\frac{\partial w'_j}{\partial x'_j} = \sum_{l=1}^{3} \frac{\partial w'_j}{\partial x_l} \frac{\partial x_l}{\partial x'_j}, \qquad j = 1, 2, 3.$$

A substitution from (10) and (11) into the above formula yields

$$\frac{\partial w'_j}{\partial x'_j} = \sum_{l=1}^{3} c_{lj} \frac{\partial w'_j}{\partial x_l} = \sum_{l=1}^{3} \sum_{k=1}^{3} c_{lj} c_{kj} \frac{\partial w_k}{\partial x_l}. \tag{12}$$

Now we add Eqs. (12) for $j = 1, 2, 3$ to obtain

$$\sum_{j=1}^{3} \frac{\partial w'_j}{\partial x'_j} = \sum_{l=1}^{3} \sum_{k=1}^{3} \frac{\partial w_k}{\partial x_l} \sum_{j=1}^{3} c_{lj} c_{kj}$$

$$= \sum_{l=1}^{3} \sum_{k=1}^{3} \frac{\partial w_k}{\partial x_l} \delta_{lk} = \sum_{k=1}^{3} \frac{\partial w_k}{\partial x_k}.$$

The last two equalities above result from the orthogonality of the matrix C, as given in (5).

DEFINITION. If $\mathbf{w}$ is a continuously differentiable vector field, we define its **divergence** as the value of the expression

$$\frac{\partial w_1}{\partial x_1} + \frac{\partial w_2}{\partial x_2} + \frac{\partial w_3}{\partial x_3}.$$

We denote the divergence by

$$\text{div } \mathbf{w} \qquad \text{or} \qquad \nabla \cdot \mathbf{w}.$$

The latter symbol suggests the invariant character of the operator div, a fact established in Theorem 3 above.

The proof of the following simple properties of the divergence operator is left to the student. (See Exercise 15.)

Theorem 4. *If $\mathbf{u}$ and $\mathbf{v}$ are vector fields and f is a scalar field, all continuously differentiable, then*

(a) div $(\mathbf{u} + \mathbf{v}) = $ div $\mathbf{u} + $ div $\mathbf{v}$,

(b) div $(f\mathbf{u}) = f$ div $\mathbf{u} + \nabla f \cdot \mathbf{u}$.

Remarks. (i) It is easy to see that the results of this section may be extended to vectors and scalars in n-dimensional Euclidean space. (See Exercise 21.) (ii) The divergence operator, a differential operator from vector fields to scalar fields, has a geometrical significance. This fact is brought out in the divergence theorem, which is taken up in Section 8 of Chapter 14. The divergence operator is useful for many problems in mechanics, in fluid flow, and in electromagnetism. An application to mechanics is given in Example 2 below, and the connection of the divergence operator with fluid-flow problems is discussed in Section 8 of Chapter 14. The geometric significance of the divergence operator has led to a number of generalizations which have applications in higher analysis and in differential geometry. (iii) The x_1, x_2, x_3 notation is useful in the statements and the proofs of the theorems in that the extension of the results to any number of variables is apparent. However, since the examples and exercises are restricted to three dimensions, we employ the more familiar x, y, z-notation.

EXAMPLE 1. Given the vector

$$\mathbf{w} = \frac{x + z}{x^2 + y^2 + z^2}\mathbf{i} + \frac{y - x}{x^2 + y^2 + z^2}\mathbf{j} + \frac{z - y}{x^2 + y^2 + z^2}\mathbf{k},$$

find div $\mathbf{w}$. Evaluate div $\mathbf{w}$ at $P_0(1, 0, -2)$.

Solution. We have

$$\frac{\partial}{\partial x}\left(\frac{x + z}{x^2 + y^2 + z^2}\right) = \frac{(x^2 + y^2 + z^2) - (x + z)(2x)}{(x^2 + y^2 + z^2)^2}$$

$$= \frac{y^2 + z^2 - x^2 - 2xz}{(x^2 + y^2 + z^2)^2},$$

$$\frac{\partial}{\partial y}\left(\frac{y - x}{x^2 + y^2 + z^2}\right) = \frac{x^2 + z^2 - y^2 + 2xy}{(x^2 + y^2 + z^2)^2},$$

$$\frac{\partial}{\partial z}\left(\frac{z - y}{x^2 + y^2 + z^2}\right) = \frac{x^2 + y^2 - z^2 + 2yz}{(x^2 + y^2 + z^2)^2}.$$

Therefore

$$\text{div } \mathbf{w} = \frac{x^2 + y^2 + z^2 + 2(xy - xz + yz)}{(x^2 + y^2 + z^2)^2}.$$

At P_0 we obtain div $\mathbf{w} = 2/5$.

EXAMPLE 2. Suppose R is the radius of the earth, O is its center, and g is the acceleration due to gravity at the surface of the earth. If P is a point in space near the surface, we denote by $\mathbf{r}$ the vector having the directed line segment $\overrightarrow{OP}$ as a representative. The length $|\mathbf{r}|$ of the vector $\mathbf{r}$ we denote simply by r. From classical physics it is known that the vector field $\mathbf{v}(P)$ due to gravity (called the *gravitational field of the earth*) is given (approximately) by the equation

$$\mathbf{v}(P) = \frac{-gR^2}{r^3}\mathbf{r}.$$

Show that for $r > R$, we have div $\mathbf{v}(P) = 0$.

Solution. We have

$$\text{div } \mathbf{v} = \text{div}\left(\frac{-gR^2}{r^3}\mathbf{r}\right).$$

Using formula (b) of Theorem 4, we find

$$\operatorname{div} \mathbf{v} = \frac{-gR^2}{r^3} \operatorname{div} \mathbf{r} + \nabla \left(\frac{-gR^2}{r^3} \right) \cdot \mathbf{r}$$

$$= -gR^2 \left[\frac{1}{r^3} \operatorname{div} \mathbf{r} + \frac{d}{dr} \left(\frac{1}{r^3} \right) \nabla r \cdot \mathbf{r} \right]$$

$$= -gR^2 (r^{-3} \operatorname{div} \mathbf{r} - 3r^{-4} \nabla r \cdot \mathbf{r}).$$

Introducing a rectangular (x, y, z)-system with origin at O, we may write

$$\mathbf{r} = x\mathbf{i} + y\mathbf{j} + z\mathbf{k},$$

from which we obtain $\operatorname{div} \mathbf{r} = 3$. Also, the length r of $\mathbf{r}$ is given by

$$r = \sqrt{x^2 + y^2 + z^2},$$

and so

$$\frac{\partial r}{\partial x} = \frac{x}{r}, \qquad \frac{\partial r}{\partial y} = \frac{y}{r}, \qquad \frac{\partial r}{\partial z} = \frac{z}{r}.$$

Hence

$$\nabla r = \frac{1}{r} (x\mathbf{i} + y\mathbf{j} + z\mathbf{k}) = \frac{\mathbf{r}}{r}$$

and

$$\nabla r \cdot \mathbf{r} = \frac{\mathbf{r} \cdot \mathbf{r}}{r} = r.$$

We conclude that

$$\operatorname{div} \mathbf{v} = -gR^2 (3r^{-3} - 3r^{-4} \cdot r) = 0.$$

The force field in Example 2 is illustrative of a large class of vector fields which occur in mechanics, electrostatics, fluid dynamics, and so forth. A vector field $\mathbf{v}(P)$ is called *conservative* if there exists a scalar function $u(P)$ of which $\mathbf{v}$ is the gradient. That is, if there is a function u such that

$$\mathbf{v}(P) = \nabla u,$$

we say that u is the **potential function** which corresponds to the conservative force field $\mathbf{v}$. In fact, the potential function u which yields the force field in Example 2 above is

$$u(P) = \frac{gR^2}{r}.$$

A simple computation shows that

$$\nabla u = gR^2 \nabla \left(\frac{1}{r} \right) = -\frac{gR^2}{r^2} \nabla r = -\frac{gR^2}{r^3} \mathbf{r} = \mathbf{v}.$$

Since the result of Example 2 shows that $\nabla \cdot \mathbf{v} = 0$, we conclude that

$$\nabla \cdot \nabla u = \operatorname{div} \nabla u = 0. \tag{13}$$

Equation (13) is known as **Laplace's equation**, and it is often written

$$\nabla^2 u = 0 \quad \text{or} \quad \Delta u = 0.$$

The latter notation is used mainly by mathematicians and the former mainly by scientists in related fields. It is clear that if $\mathbf{v} = \nabla u$ with div $\mathbf{v} = 0$, then u satisfies Laplace's equation, and conversely.

In any rectangular coordinate system we may write

$$\nabla u = \frac{\partial u^*}{\partial x_1}\mathbf{e}_1 + \frac{\partial u^*}{\partial x_2}\mathbf{e}_2 + \frac{\partial u^*}{\partial x_3}\mathbf{e}_3$$

and, therefore, the **Laplacian** has the form

$$\nabla^2 u = \nabla \cdot \nabla u = \frac{\partial^2 u^*}{\partial x_1^2} + \frac{\partial^2 u^*}{\partial x_2^2} + \frac{\partial^2 u^*}{\partial x_3^2}. \tag{14}$$

The Laplacian is a differential operator from scalar fields to scalar fields. The next corollary is an immediate consequence of Theorem 3.

Corollary. *If u and ∇u are continuously differentiable and if (x_1, x_2, x_3) and (x_1', x_2', x_3') are two rectangular coordinate systems, then*

$$\sum_{i=1}^{3} \frac{\partial^2 u^*}{\partial x_i^2} = \sum_{i=1}^{3} \frac{\partial^2 u^{*\prime}}{\partial x_i'^2}.$$

That is, the form of the Laplace operator (14) is the same in every rectangular coordinate system.

EXAMPLE 3. Let $\mathbf{r}(P)$ be the vector from the origin O to a point P in the xy-plane. Define $r = |\mathbf{r}|$. Show that the plane scalar field

$$u(P) = \log r$$

satisfies the Laplace equation.

Solution. We have

$$\nabla u = \nabla(\log r) = \frac{1}{r}\nabla(r) = \frac{1}{r^2}\mathbf{r}.$$

Therefore

$$\nabla u = \text{div } \nabla u = \text{div}\left(\frac{1}{r^2}\mathbf{r}\right) = \frac{1}{r^2}\text{div } \mathbf{r} - 2r^{-3}\nabla r \cdot \mathbf{r}.$$

Since $\mathbf{r}$ is a vector in the plane, we have

$$\mathbf{r} = x\mathbf{i} + y\mathbf{j}, \quad r = \sqrt{x^2 + y^2},$$

and so

$$\text{div } \mathbf{r} = 2 \quad \text{and} \quad \nabla r = \frac{x}{r}\mathbf{i} + \frac{y}{r}\mathbf{j} = \frac{\mathbf{r}}{r}.$$

Therefore

$$\nabla u = \frac{1}{r^2}(2) - 2r^{-3}\frac{\mathbf{r} \cdot \mathbf{r}}{r} = \frac{2}{r^2} - 2r^{-3}(r) = 0.$$

EXERCISES

In each of problems 1 through 10, find div $\mathbf{v}$ in the given coordinate system.

1. $\mathbf{v} = (a_{11}x + a_{12}y + a_{13}z)\mathbf{i} + (a_{21}x + a_{22}y + a_{23}z)\mathbf{j} + (a_{31}x + a_{32}y + a_{33}z)\mathbf{k}$

2. $\mathbf{v} = (x^2 - y^2)\mathbf{i} + (x^2 - z^2)\mathbf{j} + (y^2 - z^2)\mathbf{k}$

3. $\mathbf{v} = (x^2 + 1)\mathbf{i} + (y^2 - 1)\mathbf{j} + z^2\mathbf{k}$

4. $\mathbf{v} = 4xz\mathbf{i} - 2yz\mathbf{j} + (2x^2 - y^2 - z^2)\mathbf{k}$

5. $\mathbf{v} = e^{xz}(\cos yz\mathbf{i} + \sin yz\mathbf{j} - \mathbf{k})$

6. $\mathbf{v} = y \ln (1 + x)\mathbf{i} + z \ln (1 + y)\mathbf{j} + x \ln (1 + z)\mathbf{k}$

7. $\mathbf{v} = \nabla u, \quad u^* = x^3 - 3xy^2$

8. $\mathbf{v} = \nabla u, \quad u^* = a_{11}x^2 + a_{22}y^2 + a_{33}z^2 + 2a_{12}xy + 2a_{13}xz + 2a_{23}yz$

9. $\mathbf{v} = \nabla u, \quad u^* = e^x \cos y + e^y \cos z + e^z \cos x$

10. $\mathbf{v} = 2x\mathbf{i} + 3y\mathbf{j} + (\Delta u)\mathbf{k}, \quad$ where $u = x^3 + y^3 + z^3$

In each of problems 11 through 14, find div $\mathbf{v}$, assuming that $\mathbf{r} = x\mathbf{i} + y\mathbf{j} + z\mathbf{k}$ and that $r = |\mathbf{r}|$.

11. $\mathbf{v} = \mathbf{r}/r^n, \quad n > 0$

12. $\mathbf{v} = f(r)\mathbf{r}$

13. $\mathbf{v} = \nabla u, \quad u = \phi(r)$

14. $\mathbf{v} = \nabla u, \quad u^* = \phi(x + y + z)$

15. Prove Theorem 4.

16. Find div $(f\nabla g - g\nabla f)$

17. Show that div $(\mathbf{a} \times \mathbf{r}) = 0$ for any constant vector $\mathbf{a}$.

18. Show that there is a vector $\mathbf{w}$ such that

$$\text{div } (\mathbf{v} \times \mathbf{a}) = \mathbf{w} \cdot \mathbf{a}$$

for any constant vector $\mathbf{a}$.

19. Suppose that a rigid body is rotating about the z-axis with an angular velocity ω. Show that the velocity of a particle of the body which is at position $P(x, y, z)$ at time t is given by

$$\mathbf{v} = \omega\mathbf{k} \times \mathbf{r}.$$

20. (a) Compute the quantity Δu in cylindrical coordinates given by

$$x = r \cos \theta, \qquad y = r \sin \theta, \qquad z = z.$$

   [*Hint:* Start with formula (14) and use the Chain Rule.]

   (b) Compute Δu in spherical coordinates:

$$x = \rho \cos \theta \sin \phi, \qquad y = \rho \sin \theta \sin \phi, \qquad z = \rho \cos \phi.$$

21. Define the operator ∇ for vectors $\mathbf{v}$ in an n-dimensional space E^n. Write out the statement and proof of Theorem 3 in E^n. Show that the formulas of Theorem 4 are valid for n-dimensional scalar and vector fields.

4. THE CURL OF A VECTOR FIELD

The cross or vector product of two vectors, as defined in Chapter 2, page 51, changes in sign when we shift from a right-handed system of coordinates to a left-handed one. Three-dimensional space is said to have an **orientation** when one of these two systems of coordinates is introduced. We shall use the term **positive orientation** for a right-handed coordinate system and **negative orientation** for a left-handed one.*

If $\mathbf{e}_1$, $\mathbf{e}_2$, $\mathbf{e}_3$ and $\mathbf{e}_1'$, $\mathbf{e}_2'$, $\mathbf{e}_3'$ are two orthonormal bases, we know that there is a relation between them of the form

$$\mathbf{e}_i = \sum_{j=1}^{3} c_{ij}\mathbf{e}_j',$$

where $C = (c_{ij})$ is an orthogonal matrix with $\det C = \pm 1$. If both bases are positively oriented or if both are negatively oriented, then $\det C = +1$. If one system is positively oriented and the other negatively oriented, then $\det C = -1$. We shall always consider positively oriented systems in three-space and, when discussing transformations from one system to another, we shall assume that the new system has the same orientation as the old.

Suppose that $\mathbf{v}$ is a continuously differentiable vector field in space. In the last section we saw that the differential operator div $\mathbf{v}$, which we also write $\nabla \cdot \mathbf{v}$, is independent of the axes chosen and so defines a scalar field. It might be expected —and indeed we shall show it to be the case—that the formal expression $\nabla \times \mathbf{v}$ defines a vector field.

Let $\mathbf{e}_1$, $\mathbf{e}_2$, $\mathbf{e}_3$ be an orthonormal set of unit vectors (positively oriented, of course), and let $\mathbf{u}$ and $\mathbf{v}$ be two vector fields. Then, in terms of the basis vectors, we may write

$$\mathbf{u} = u_1\mathbf{e}_1 + u_2\mathbf{e}_2 + u_3\mathbf{e}_3, \qquad \mathbf{v} = v_1\mathbf{e}_1 + v_2\mathbf{e}_2 + v_3\mathbf{e}_3.$$

We recall (Chapter 2, page 52) that the cross product $\mathbf{u} \times \mathbf{v}$ is given by

$$\mathbf{u} \times \mathbf{v} = (u_2v_3 - u_3v_2)\mathbf{e}_1 + (u_3v_1 - u_1v_3)\mathbf{e}_2 + (u_1v_2 - u_2v_1)\mathbf{e}_3. \tag{1}$$

This formula may be written more compactly in the symbolic form

$$\mathbf{u} \times \mathbf{v} = \begin{vmatrix} \mathbf{e}_1 & \mathbf{e}_2 & \mathbf{e}_3 \\ u_1 & u_2 & u_3 \\ v_1 & v_2 & v_3 \end{vmatrix}, \tag{2}$$

where it is understood that the determinant is expanded in minors according to

---

* It can be shown that it is impossible for a right-handed triple of basis vectors to be deformed continuously into a left-handed one without the vectors becoming linearly dependent at some time during the process. Furthermore, it can be shown that any triple of linearly independent vectors can be deformed continuously (always remaining linearly independent in the process) into either a right-handed set or a left-handed set and, in the light of the preceding sentence, not both.

the first row. Note that (1) and (2) are identical. With the vector operator ∇ given by

$$\nabla = \frac{\partial}{\partial x_1}\,\mathbf{e}_1 + \frac{\partial}{\partial x_2}\,\mathbf{e}_2 + \frac{\partial}{\partial x_3}\,\mathbf{e}_3,$$

we **define** the vector $\nabla \times \mathbf{v}$ by the formula

$$\nabla \times \mathbf{v} = \begin{vmatrix} \mathbf{e}_1 & \mathbf{e}_2 & \mathbf{e}_3 \\ \dfrac{\partial}{\partial x_1} & \dfrac{\partial}{\partial x_2} & \dfrac{\partial}{\partial x_3} \\ v_1 & v_2 & v_3 \end{vmatrix} \tag{3}$$

$$= \left(\frac{\partial v_3}{\partial x_2} - \frac{\partial v_2}{\partial x_3}\right)\mathbf{e}_1 + \left(\frac{\partial v_1}{\partial x_3} - \frac{\partial v_3}{\partial x_1}\right)\mathbf{e}_2 + \left(\frac{\partial v_2}{\partial x_1} - \frac{\partial v_1}{\partial x_2}\right)\mathbf{e}_3.$$

Since $\nabla \times \mathbf{v}$ is defined in terms of a particular coordinate system, it is essential to show that $\nabla \times \mathbf{v}$ is a true vector field (if orientation is preserved).

For any three vectors $\mathbf{b}, \mathbf{c}, \mathbf{d}$, it is not hard to establish the identity

$$(\mathbf{b} \times \mathbf{c}) \cdot \mathbf{d} = \mathbf{b} \cdot (\mathbf{c} \times \mathbf{d}). \tag{4}$$

Using the usual definitions, $b = b_1\mathbf{e}_1 + b_2\mathbf{e}_2 + b_3\mathbf{e}_3$, etc., the student may verify that

$$(\mathbf{b} \times \mathbf{c}) \cdot \mathbf{d} = \begin{vmatrix} b_1 & b_2 & b_3 \\ c_1 & c_2 & c_3 \\ d_1 & d_2 & d_3 \end{vmatrix}.$$

Identity (4) is now an immediate consequence of the properties of determinants. The next theorem establishes the invariance of (3) under a change of basis.

Theorem 5. *Suppose that $\mathbf{v}$ is a continuously differentiable vector field. Then there exists a unique continuous vector field $\mathbf{w}$ such that*

$$\nabla \cdot (\mathbf{v} \times \mathbf{a}) = \mathbf{w} \cdot \mathbf{a}$$

for every constant vector $\mathbf{a}$. In fact, if $\mathbf{e}_1, \mathbf{e}_2, \mathbf{e}_3$ is an orthonormal basis, then $\mathbf{w} = \nabla \times \mathbf{v}$ where the vector $\nabla \times \mathbf{v}$ is given by (3).

Proof. Let $\mathbf{e}_1, \mathbf{e}_2, \mathbf{e}_3$ be an orthonormal basis. We define $\mathbf{a} = a_1\mathbf{e}_1 + a_2\mathbf{e}_2 + a_3\mathbf{e}_3$. Then

$$\mathbf{v} \times \mathbf{a} = (v_2a_3 - v_3a_2)\mathbf{e}_1 + (v_3a_1 - v_1a_3)\mathbf{e}_2 + (v_1a_2 - v_2a_1)\mathbf{e}_3.$$

The formula for the divergence yields

$$\operatorname{div}(\mathbf{v} \times \mathbf{a}) = a_3\frac{\partial v_2}{\partial x_1} - a_2\frac{\partial v_3}{\partial x_1} + a_1\frac{\partial v_3}{\partial x_2} - a_3\frac{\partial v_1}{\partial x_2} + a_2\frac{\partial v_1}{\partial x_3} - a_1\frac{\partial v_2}{\partial x_3}$$

or

$$\nabla \cdot (\mathbf{v} \times \mathbf{a}) = \left(\frac{\partial v_3}{\partial x_2} - \frac{\partial v_2}{\partial x_3}\right)a_1 + \left(\frac{\partial v_1}{\partial x_3} - \frac{\partial v_3}{\partial x_1}\right)a_2 + \left(\frac{\partial v_2}{\partial x_1} - \frac{\partial v_1}{\partial x_2}\right)a_3.$$

Taking into account the identity (4) and the definition of $\nabla \times \mathbf{v}$, we conclude that the last equation states the result of the theorem. The uniqueness of $\mathbf{w}$ follows from Theorem 1.

DEFINITION. For any vector field $\mathbf{v}$ we define **curl v** by the formula

$$\text{curl } \mathbf{v} = \nabla \times \mathbf{v},$$

where $\nabla \times \mathbf{v}$ is given by (3).

EXAMPLE 1. Given the vector field

$$\mathbf{v} = (x_1^2 - x_2^2 + 2x_1x_3)\mathbf{e}_1 + (x_1x_3 - x_1x_2 + x_2x_3)\mathbf{e}_2 + (x_3^2 + x_1^2)\mathbf{e}_3,$$

find curl $\mathbf{v}$. Show that the vectors given by curl $\mathbf{v}$ evaluated at $P_0(1, 2, -3)$ and $P_1(2, 3, 12)$ are orthogonal.

Solution. We have

$$\text{curl } \mathbf{v} = \begin{vmatrix} \mathbf{e}_1 & \mathbf{e}_2 & \mathbf{e}_3 \\ \dfrac{\partial}{\partial x_1} & \dfrac{\partial}{\partial x_2} & \dfrac{\partial}{\partial x_3} \\ x_1^2 - x_2^2 + 2x_1x_3 & x_1x_3 - x_1x_2 + x_2x_3 & x_3^2 + x_1^2 \end{vmatrix}$$

$$= -(x_1 + x_2)\mathbf{e}_1 + (x_3 + x_2)\mathbf{e}_3.$$

At $P_0(1, 2, -3)$, we find curl $\mathbf{v} = -3\mathbf{e}_1 - \mathbf{e}_3 \equiv \mathbf{v}_0$. At $P_1(2, 3, 12)$ we find curl $\mathbf{v} = -5\mathbf{e}_1 + 15\mathbf{e}_3 \equiv \mathbf{v}_1$. Since $\mathbf{v}_0 \cdot \mathbf{v}_1 = 0$, the vectors are orthogonal.

The proofs of the following identities are left to the student. (See Exercises 11 through 15.)

Theorem 6. *Suppose that* $\mathbf{u}$, $\mathbf{v}$, *and* f *are continuously differentiable fields. Then*

(a) curl $(\mathbf{u} + \mathbf{v}) = $ curl $\mathbf{u} + $ curl $\mathbf{v}$,

(b) curl $(f\mathbf{u}) = f$ curl $\mathbf{u} + \nabla f \times \mathbf{u}$,

(c) div $(\mathbf{u} \times \mathbf{v}) = \mathbf{v} \cdot $ curl $\mathbf{u} - \mathbf{u} \cdot $ curl $\mathbf{v}$.

(d) *If* ∇f *is a continuously differentiable vector field, then*

$$\text{curl } \nabla f = \mathbf{0}.$$

(e) *If* $\mathbf{v}$ *is a twice continuously differentiable vector field, then*

$$\text{div curl } \mathbf{v} = 0.$$

We now change to the familiar (x, y, z), $\mathbf{i}$, $\mathbf{j}$, $\mathbf{k}$ notation from that of (x_1, x_2, x_3), $\mathbf{e}_1$, $\mathbf{e}_2$, $\mathbf{e}_3$.

EXAMPLE 2. Given the scalar field $f* = x^2 + y^2 + z^2$ and the vector field

$$\mathbf{u} = (x^2 + y^2)\mathbf{i} + (y^2 + z^2)\mathbf{j} + (z^2 + x^2)\mathbf{k},$$

compute curl $(f\mathbf{u})$.

Solution. We use formula (b) of Theorem 6 and the definition of curl as given in (3). Then

$$\text{curl } \mathbf{u} = -2(z\mathbf{i} + x\mathbf{j} + y\mathbf{k}),$$
$$\nabla f = \quad 2(x\mathbf{i} + y\mathbf{j} + z\mathbf{k}).$$

Therefore

$$\text{curl } (f\mathbf{u}) = f\,\text{curl } \mathbf{u} + \nabla f \times \mathbf{u}.$$

A computation yields

$$\text{curl } (f\mathbf{u}) = -2(x^2 + y^2 + z^2)(z\mathbf{i} + x\mathbf{j} + y\mathbf{k}) + \begin{vmatrix} \mathbf{i} & \mathbf{j} & \mathbf{k} \\ 2x & 2y & 2z \\ x^2 + y^2 & y^2 + z^2 & z^2 + x^2 \end{vmatrix}$$

$$= -2[z(x^2 + 2y^2 + 2z^2) - y(x^2 + z^2)]\mathbf{i}$$
$$- 2[x(2x^2 + y^2 + 2z^2) - z(x^2 + y^2)]\mathbf{j}$$
$$- 2[y(2x^2 + 2y^2 + z^2) - x(y^2 + z^2)]\mathbf{k}.$$

The curl of a vector is intimately connected with the notion of exact differential. We repeat the definition of exact differential as given in Chapter 4, page 184. Suppose that $P(x, y, z)$, $Q(x, y, z)$ and $R(x, y, z)$ are continuously differentiable functions on some region $\mathfrak{D}$. We say that the expression

$$P\,dx + Q\,dy + R\,dz \tag{5}$$

is an **exact differential** if and only if there is a continuously differentiable function $f(x, y, z)$ such that

$$df = P\,dx + Q\,dy + R\,dz. \tag{6}$$

Whenever such a function f exists, the definition of the total differential of a function yields the relations

$$f_x = P, \qquad f_y = Q, \qquad f_z = R. \tag{7}$$

If the region $\mathfrak{D}$ is a rectangular box, then a necessary and sufficient condition for the expression (5) to be an exact differential is that the three equations

$$\frac{\partial Q}{\partial x} = \frac{\partial P}{\partial y}, \qquad \frac{\partial P}{\partial z} = \frac{\partial R}{\partial x}, \qquad \frac{\partial R}{\partial y} = \frac{\partial Q}{\partial z} \tag{8}$$

hold. (See Chapter 4, page 188.)

We may formulate the notion of exact differential in vector terms. We introduce the usual rectangular coordinate system and define the vectors

$$\mathbf{v} = P\mathbf{i} + Q\mathbf{j} + R\mathbf{k}, \qquad \mathbf{r} = x\mathbf{i} + y\mathbf{j} + z\mathbf{k}, \qquad d\mathbf{r} = (dx)\mathbf{i} + (dy)\mathbf{j} + (dz)\mathbf{k}.$$

Suppose that f and ∇f are continuously differentiable on some domain $\mathfrak{D}$ in space.

Then Eqs. (6) and (7) are equivalent to

$$df = \mathbf{v} \cdot d\mathbf{r} \quad \text{and} \quad \nabla f = \mathbf{v}.$$

The necessary and sufficient conditions (8) become the simple vector condition

$$\text{curl } \mathbf{v} = \mathbf{0}. \tag{9}$$

Whenever a vector $\mathbf{v}$ has the form ∇f, we see that (9) is a consequence of Theorem 6(d). Conversely, if (9) holds, then the results on exact differentials in Chapter 4, Section 13, show that there is a function f such that $\nabla f = \mathbf{v}$. We state this conclusion in the following theorem.

Theorem 7. *Suppose that* $\mathbf{v}$ *is a continuously differentiable vector field with* curl $\mathbf{v} = \mathbf{0}$ *in some rectangular parallelepiped* $\mathcal{D}$ *in space. Then there exists a continuously differentiable scalar field* f *in* $\mathcal{D}$ *such that* $\nabla f = \mathbf{v}$. *Any two such fields differ by a constant.*

In Section 6 we shall state the above theorem for more general domains $\mathcal{D}$ (Theorem 13).

If we are given a specific vector field $\mathbf{v}$ with curl $\mathbf{v} = \mathbf{0}$, then the method of determining f such that $\nabla f = \mathbf{v}$ is precisely the one described in Chapter 4, Section 13. We review the process by working an example.

EXAMPLE 3. Given the vector field

$$\mathbf{v} = 2xyz\mathbf{i} + (x^2z + y)\mathbf{j} + (x^2y + 3z^2)\mathbf{k},$$

verify that curl $\mathbf{v} = \mathbf{0}$ and find the function f such that $\nabla f = \mathbf{v}$.

Solution. We have

$$\text{curl } \mathbf{v} = \begin{vmatrix} \mathbf{i} & \mathbf{j} & \mathbf{k} \\ \dfrac{\partial}{\partial x} & \dfrac{\partial}{\partial y} & \dfrac{\partial}{\partial z} \\ 2xyz & (x^2z + y) & (x^2y + 3z^2) \end{vmatrix}$$

$$= (x^2 - x^2)\mathbf{i} + (2xy - 2xy)\mathbf{j} + (2xz - 2xz)\mathbf{k} = \mathbf{0}.$$

We wish to find f such that

$$f_x^* = 2xyz, \quad f_y^* = x^2z + y, \quad f_z^* = x^2y + 3z^2.$$

Integrating the first equation, we get

$$f^* = x^2yz + C(y, z);$$

differentiating f with respect to y and z, we obtain

$$f_y^* = x^2z + C_y(y, z) = x^2z + y, \quad f_z^* = x^2y + C_z(y, z) = x^2y + 3z^2.$$

The equations on the right yield

$$C_y = y, \qquad C_z = 3z^2.$$

Hence $C = \frac{1}{2}y^2 + K(z)$. Differentiation of C with respect to z shows that

$$C_z = K'(z) = 3z^2,$$

so that $K(z) = z^3 + C_1$, where C_1 is a constant. We conclude that

$$C = \frac{1}{2}y^2 + z^3 + C_1$$

and, finally, that

$$f^* = x^2yz + C = x^2yz + \frac{1}{2}y^2 + z^3 + C_1.$$

EXERCISES

In each of problems 1 through 8, find curl $\mathbf{v}$; in case curl $\mathbf{v} = \mathbf{0}$, find the function f such that $\nabla f = \mathbf{v}$.

1. $\mathbf{v} = (2x - y + 3z)\mathbf{i} + (-x + 3y + 2z)\mathbf{j} + (2x + 3y - z)\mathbf{k}$
2. $\mathbf{v} = (2xy + z^2)\mathbf{i} + (2yz + x^2)\mathbf{j} + (2xz + y^2)\mathbf{k}$
3. $\mathbf{v} = e^x(\sin y \cos z\mathbf{i} + \cos y \cos z\mathbf{j} - \sin y \sin z\mathbf{k})$
4. $\mathbf{v} = (x + 2y - z)\mathbf{i} + (x - y + z)\mathbf{j} + (-x + y + 2z)\mathbf{k}$
5. $\mathbf{v} = (x^2 + y^2)^{-1/2}[-xz\mathbf{i} - yz\mathbf{j} + (x^2 + y^2)\mathbf{k}]$
6. $\mathbf{v} = (x^2 + y^2 + z^2)^{-1}(x\mathbf{i} + y\mathbf{j} + z\mathbf{k}), \quad (x, y, z) \neq (0, 0, 0)$
7. $\mathbf{v} = x^2z\mathbf{i} + 0 \cdot \mathbf{j} + xz^2\mathbf{k}$
8. $\mathbf{v} = \dfrac{2x + y + z}{(x + y)(x + z)}\mathbf{i} + \dfrac{x + 2y + z}{(x + y)(y + z)}\mathbf{j} + \dfrac{x + y + 2z}{(y + z)(x + z)}\mathbf{k}$
9. Suppose that $G_1'(x) = g_1(x)$, $G_2'(y) = g_2(y)$, $G_3'(z) = g_3(z)$, and that

$$\mathbf{v} = [x^2(y^3 + z^3) + g_1(x)]\mathbf{i} + [y^2(x^3 + z^3) + g_2(y)]\mathbf{j} + [z^2(x^3 + y^3) + g_3(z)]\mathbf{k}.$$

   Show that curl $\mathbf{v} = \mathbf{0}$ and find f such that $\nabla f = \mathbf{v}$.
10. Verify the identity $(\mathbf{b} \times \mathbf{c}) \cdot \mathbf{d} = \mathbf{b} \cdot (\mathbf{c} \times \mathbf{d})$ by introducing coordinate vectors and calculating each side separately.
11. Prove Theorem 6(a).
12. Prove Theorem 6(b).
13. Prove Theorem 6(c).
14. Prove Theorem 6(d).
15. Prove Theorem 6(e).

In problems 16 through 18, suppose that $\mathbf{r} = x\mathbf{i} + y\mathbf{j} + z\mathbf{k}$, and let $r = |\mathbf{r}|$.
16. Show that

$$\text{curl}\left(\frac{\mathbf{r}}{r}\right) = \mathbf{0}.$$

17. Find curl $[\phi(r)\mathbf{r}]$, where ϕ is a differentiable function.

18. Assuming p and $\mathbf{a}$ constant, find curl $(r^p \mathbf{a} \times \mathbf{r})$.

19. Verify Theorem 6(b) for $f^* = (x^2 + y^2 + z^2)^p$, $\mathbf{u} = z\mathbf{i} + x\mathbf{j} + y\mathbf{k}$.

20. Verify Theorem 6(c) for $\mathbf{u} = y\mathbf{i} + z\mathbf{j} + x\mathbf{k}$, $\mathbf{v} = z\mathbf{i} + x\mathbf{j} + y\mathbf{k}$.

21. Verify Theorem 6(d) with $f^* = (x^2 + y^2 + z^2)^{-1/2}x$.

22. Verify Theorem 6(e) with $\mathbf{v} = (y^2 - z^2)\mathbf{i} + (z^2 - x^2)\mathbf{j} + (x^2 - y^2)\mathbf{k}$.

23. If $\mathbf{v}$ is any vector of the form

$$\mathbf{v} = v_1\mathbf{i} + v_2\mathbf{j} + v_3\mathbf{k},$$

we define the Laplacian of $\mathbf{v}$, denoted $\Delta\mathbf{v}$ or $\nabla^2\mathbf{v}$, by the formula

$$\Delta\mathbf{v} = (\Delta v_1)\mathbf{i} + (\Delta v_2)\mathbf{j} + (\Delta v_3)\mathbf{k}.$$

For any vector field u (sufficiently differentiable), establish the identity

$$\text{curl curl } \mathbf{u} = \text{grad div } \mathbf{u} - \Delta\mathbf{u}. \tag{10}$$

24. Verify (10) with $\mathbf{u} = (y^2 + zx)\mathbf{i} + (z^2 + xy)\mathbf{j} + (x^2 + yz)\mathbf{k}$.

5. LINE INTEGRALS; VECTOR FORMULATION

An arc in three-space is the locus of the equations

$$x = f(t), \quad y = g(t), \quad z = h(t), \quad a \leq t \leq b,$$

provided that f, g, and h are continuous and that no point on the locus corresponds to two different values of t. In terms of vectors and transformations, we define an **arc** as the range of a one-to-one continuous transformation of the form (Fig. 13–5)

$$\mathbf{v}(\overrightarrow{OP}) = \mathbf{r}(t), \quad a \leq t \leq b,$$

with the auxiliary condition that

$$\mathbf{r}(t') \neq \mathbf{r}(t'') \quad \text{if } t' \neq t''.$$

The definition of an arc in a space of any number of dimensions is analogous to its definition in three-space. (Arcs in the plane were defined in *First Course*, page 388.)

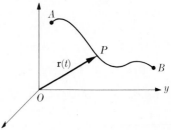

FIGURE 13–5

Suppose that in a rectangular coordinate system an arc C is given by the transformation

$$C: \quad \mathbf{r} = \mathbf{r}(t) = f(t)\mathbf{i} + g(t)\mathbf{j} + h(t)\mathbf{k}, \quad a \leq t \leq b.$$

As t increases from a to b, the point P moves along the arc from the point A, corresponding to $t = a$, to the point B, corresponding to $t = b$. Since the trans-

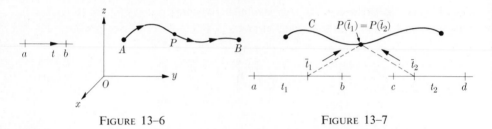

FIGURE 13–6   FIGURE 13–7

formation is one to one, the point P moves along "without doubling back." We say that "P describes the arc in a certain sense." (See Fig. 13–6.)

Any arc C will have many different parametric representations. Suppose that two representations of C are

$$C: \quad x = f(t_1), \qquad y = g(t_1), \qquad z = h(t_1), \qquad a \le t_1 \le b;$$
$$C: \quad x = F(t_2), \qquad y = G(t_2), \qquad z = H(t_2), \qquad c \le t_2 \le d.$$

In vector notation, we write

$$C: \quad \mathbf{r} = \mathbf{r}(t_1), \quad a \le t_1 \le b; \qquad \mathbf{r}(t_1') \ne \mathbf{r}(t_1'') \quad \text{if } t_1' \ne t_1''; \tag{1}$$
$$C: \quad \mathbf{r} = \mathbf{R}(t_2), \quad c \le t_2 \le d; \qquad \mathbf{R}(t_2') \ne \mathbf{R}(t_2'') \quad \text{if } t_2' \ne t_2''. \tag{2}$$

Intuitively we see that as t_1 increases from a to b, the arc C is described in a particular sense. As t_2 increases from c to d, the arc is described either in the same sense or in the opposite sense. (See Fig. 13–7.) There are no other possibilities. We make this conclusion precise in the next theorem, which is stated without proof.

Theorem 8. *Suppose that the arc C is the range of the continuous transformations* (1) *and* (2). *Then there are continuous functions $S_1(t_1)$, $S_2(t_2)$ defined on $[a, b]$ and $[c, d]$, respectively, and with ranges $[c, d]$ and $[a, b]$, respectively, such that*

$$\mathbf{R}[S_1(t_1)] = \mathbf{r}(t_1) \text{ for } t_1 \text{ on } [a, b] \text{ and } \mathbf{r}[S_2(t_2)] = \mathbf{R}(t_2) \text{ for } t_2 \text{ on } [c, d].$$

Either S_1 and S_2 are both increasing or they are both decreasing; moreover, each is the inverse of the other.

As Fig. 13–7 shows, the function $S_1(t_1)$ is the mapping obtained by going from a point $\bar{t}_1$ to the point $P(\bar{t}_1)$ on C and then finding the unique point $\bar{t}_2$ in $[c, d]$ which corresponds to the same point P. Thus we get $\bar{t}_2 = S_1(\bar{t}_1)$. The function S_2 is obtained by reversing the process.

Theorem 8 implies that all parametric representations of an arc C fall into two classes: one class, in which S_1 and S_2 are both increasing, and the other, in which S_1 and S_2 are both decreasing. Two representations in the same class define the same ordering of the points of C in the sense that a point P' on C precedes P'' if and only if $t' < t''$.

DEFINITIONS. A **directed arc** $\vec{C}$ is an arc C together with one of the two order-ings described above. If $\vec{C}$ is a directed arc, we denote the corresponding un-directed arc by C or, if we wish to emphasize the undirected property, by $|\vec{C}|$. A transformation (1) is said to be a **parametric representation** of $\vec{C}$ if and only if it is a representation of $|\vec{C}|$ and establishes the given order on $\vec{C}$. The arc **op-positely directed** to $\vec{C}$ is denoted by $-\vec{C}$.

The definitions and basic properties of line integrals in the plane and in space were taken up in Chapter 4, page 189ff. We shall now see how vector notation may be used to simplify the statements and proofs of some of the elementary theorems on line integrals in space.

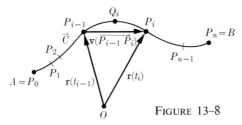

FIGURE 13–8

Let $\vec{C}$ be a directed arc from a point A to a point B. We make a **subdivision of** $\vec{C}$ by arranging the points $A = P_0,\ P_1,\ P_2,\ \ldots,\ P_{n-1},\ P_n = B$ in order along the directed arc. As usual, we define the norm $\|\Delta\|$ of the subdivision as the length of the longest line segment connecting two successive points P_{i-1}, P_i in the subdivision (Fig. 13–8). Suppose that we are given a vector field w de-fined on $|\vec{C}|$. For each i, we select a point Q_i on the subarc from P_{i-1} to P_i (Fig. 13–8) and form the sum of the scalar products

$$\sum_{i=1}^{n} \mathbf{w}(Q_i) \cdot \mathbf{v}(\overrightarrow{P_{i-1}P_i}) = \sum_{i=1}^{n} \mathbf{w}(Q_i) \cdot [\mathbf{r}(t_i) - \mathbf{r}(t_{i-1})]. \tag{3}$$

We may abbreviate this expression by letting $\Delta_i \mathbf{r} = \mathbf{r}(t_i) - \mathbf{r}(t_{i-1})$ and by writing

$$\sum_{i=1}^{n} \mathbf{w}(Q_i) \cdot \Delta_i \mathbf{r}.$$

DEFINITION. Suppose there is a number L with the property that for each $\epsilon > 0$ there is a $\delta > 0$ such that

$$\left| \sum_{i=1}^{n} \mathbf{w}(Q_i) \cdot \Delta_i \mathbf{r} - L \right| < \epsilon$$

for all subdivisions with norm less than δ and for all choices of the Q_i on the arc $\overparen{P_{i-1}P_i}$. Then we say that **the differential $w \cdot dr$ is integrable along** $\vec{C}$. We write

$$L = \int_{\vec{C}} \mathbf{w} \cdot d\mathbf{r}.$$

The next theorem establishes a few of the elementary properties of line integrals in space.

Theorem 9. (a) *There is at most one number L satisfying the conditions of the definition above.*
(b) *When such a number L exists, it is independent of the choice for the origin O.*
(c) *If $\mathbf{w} \cdot d\mathbf{r}$ is integrable along $\vec{C}$, it is integrable along $-\vec{C}$, and*

$$\int_{-\vec{C}} \mathbf{w} \cdot d\mathbf{r} = -\int_{\vec{C}} \mathbf{w} \cdot d\mathbf{r}.$$

Proof. (a) Suppose there are two numbers L_1 and L_2 satisfying the required conditions. We select $\epsilon = (L_2 - L_1)/2$ (supposing that $L_2 > L_1$). Then, if $\|\Delta\|$ is sufficiently small, we have

$$\sum_{i=1}^{n} \mathbf{w}(Q_i) \cdot \Delta_i \mathbf{r} < L_1 + \epsilon = L_1 + \frac{L_2 - L_1}{2} = \frac{L_1 + L_2}{2}. \tag{4}$$

On the other hand, since L_2 is also a limit, we see that

$$\frac{L_1 + L_2}{2} = L_2 - \epsilon < \sum_{i=1}^{n} \mathbf{w}(Q_i) \cdot \Delta_i \mathbf{r}. \tag{5}$$

Since (4) and (5) are contradictory, $L_1 = L_2$.

(b) This statement follows from the fact that each sum, as given on the left side of (3), is independent of O.

(c) If $P_0, P_1, \ldots, P_n$ is a subdivision for $\vec{C}$, then $P_n, P_{n-1}, \ldots, P_0$ is a subdivision for $-\vec{C}$. Therefore a sum of the type (3) for $-\vec{C}$ is

$$\mathbf{w}(Q_n) \cdot \mathbf{v}(\overrightarrow{P_n P_{n-1}}) + \mathbf{w}(Q_{n-1}) \cdot \mathbf{v}(\overrightarrow{P_{n-1} P_{n-2}}) + \cdots$$
$$+ \mathbf{w}(Q_1) \cdot \mathbf{v}(\overrightarrow{P_1 P_0}) = -\sum_{i=1}^{n} \mathbf{w}(Q_i) \cdot \Delta_i \mathbf{r}. \tag{6}$$

The result of part (c) follows by letting $\|\Delta\| \to 0$ in (6).

Suppose that $\vec{C}$ is a directed arc and that $\mathbf{w}$ is a vector field defined on $|\vec{C}|$. We introduce a rectangular coordinate system (x, y, z) and basis vectors $\mathbf{i}, \mathbf{j}, \mathbf{k}$. We now write

$$C: \quad \mathbf{r}(t) = x(t)\mathbf{i} + y(t)\mathbf{j} + z(t)\mathbf{k}, \tag{7}$$
$$\mathbf{w}: \quad \mathbf{w} = P(x, y, z)\mathbf{i} + Q(x, y, z)\mathbf{j} + R(x, y, z)\mathbf{k}. \tag{8}$$

If $\mathbf{r}(t)$ is a continuously differentiable function and $\mathbf{w}$ is a continuous vector field, it can be shown that

$$\int_{\vec{C}} \mathbf{w} \cdot d\mathbf{r} = \int_{\vec{C}} P \, dx + Q \, dy + R \, dz,$$

where the line integral on the right is defined in terms of coordinates as in Chapter 4, Section 14. The evaluation of line integrals is reduced to that of ordinary integrals, as is shown in the next theorem.

Theorem 10. *Suppose that* $\vec{C}$ *has the parametric representation* $\mathbf{r} = \mathbf{r}(t)$, *with* $\mathbf{r}$ *continuously differentiable for* $a \leq t \leq b$. *Suppose that* $\mathbf{w}$ *is continuous on* $|\vec{C}|$. *Then* $\mathbf{w} \cdot d\mathbf{r}$ *is integrable along* $\vec{C}$ *and*

$$\int_{\vec{C}} \mathbf{w} \cdot d\mathbf{r} = \int_a^b \mathbf{w}(t) \cdot \mathbf{r}'(t)\, dt.$$

Furthermore, if $\mathbf{w}$ *is given by* (8), *then*

$$\int_{\vec{C}} \mathbf{w} \cdot d\mathbf{r} = \int_a^b \{P[x(t), y(t), z(t)]x'(t) + Q[x(t), y(t), z(t)]y'(t)$$

$$+ R[x(t), y(t), z(t)]z'(t)\}\, dt.$$

Remark. If $\mathbf{r}$ is continuous and piecewise smooth, we may evaluate the line integral along each smooth subarc and add the results.

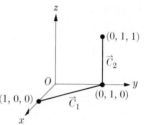

FIGURE 13–9

EXAMPLE. Compute $\int_{\vec{C}} \mathbf{w} \cdot d\mathbf{r}$, where

$$\mathbf{w} = xy\mathbf{i} + xz\mathbf{j} - y\mathbf{k},$$
$$\mathbf{r} = x\mathbf{i} + y\mathbf{j} + z\mathbf{k},$$

and $\vec{C}$ is the directed line segment $\vec{C}_1$ from $(1, 0, 0)$ to $(0, 1, 0)$, followed by $\vec{C}_2$, which is the segment from $(0, 1, 0)$ to $(0, 1, 1)$. (See Fig. 13–9.)

Solution. Along $\vec{C}_1$ we have $z = 0$. Taking $t = y$, we may write the equation of the line segment

$$\mathbf{r}(y) = (1 - y)\mathbf{i} + y\mathbf{j}, \qquad 0 \leq y \leq 1.$$

Then

$$\mathbf{w} = (1 - y)y\mathbf{i} - y\mathbf{k}.$$

Therefore

$$\int_{\vec{C}_1} \mathbf{w} \cdot d\mathbf{r} = \int_0^1 \mathbf{w}(y) \cdot \mathbf{r}'(y)\, dy = \int_0^1 (-y + y^2)\, dy = -\tfrac{1}{6}.$$

Along $\vec{C}_2$ we have $x = 0$ and we take $t = z$. Then we find

$$\mathbf{r}(z) = \mathbf{j} + z\mathbf{k}, \qquad \mathbf{w}(z) = -\mathbf{k}, \qquad \mathbf{r}'(z) = \mathbf{k}.$$

Hence

$$\int_{\vec{C}_2} \mathbf{w} \cdot d\mathbf{r} = \int_0^1 -\mathbf{k} \cdot \mathbf{k}\, dz = -1.$$

The result is

$$\int_{\vec{C}_1} \mathbf{w} \cdot d\mathbf{r} + \int_{\vec{C}_2} \mathbf{w} \cdot d\mathbf{r} = -\tfrac{1}{6} - 1 = -\tfrac{7}{6}.$$

EXERCISES

In each of problems 1 through 10, evaluate $\int_{\overrightarrow{C}} \mathbf{w} \cdot d\mathbf{r}$. Sketch the arc $\overrightarrow{C}$ in each case.

1. $\mathbf{w} = xy\mathbf{i} - y\mathbf{j} + \mathbf{k}$; $\overrightarrow{C}$ is the segment going from $(0, 0, 0)$ to $(1, 1, 1)$.

2. $\mathbf{w} = xy\mathbf{i} - y\mathbf{j} + \mathbf{k}$; $\overrightarrow{C}$ is the arc given by $x = t$, $y = t^2$, $z = t^3$, $0 \le t \le 1$.

3. $\mathbf{w} = x\mathbf{i} - y\mathbf{j} + z\mathbf{k}$; $\overrightarrow{C}$ is the helical path $x = \cos \theta$, $y = \sin \theta$, $z = (1/\pi)\theta$, $0 \le \theta \le 2\pi$.

4. $\mathbf{w} = x\mathbf{i} - y\mathbf{j} + z\mathbf{k}$; $\overrightarrow{C}$ is the segment from $(1, 0, 0)$ to $(1, 0, 2)$.

5. $\mathbf{w} = 2x\mathbf{i} - 3y\mathbf{j} + z^2\mathbf{k}$; $\overrightarrow{C}$ is the path $x = \cos \theta$, $y = \sin \theta$, $z = \theta$, $0 \le \theta \le (\pi/2)$.

6. $\mathbf{w} = 2x\mathbf{i} - 3y\mathbf{j} + z^2\mathbf{k}$; $\overrightarrow{C}$ is the segment from $(1, 0, 0)$ to $(0, 1, \pi/2)$.

7. $\mathbf{w} = y^2\mathbf{i} + x^2\mathbf{j} + 0 \cdot \mathbf{k}$; $\overrightarrow{C}$ is the arc of the parabola $x = t$, $y = t^2$, $z = 0$, $1 \le t \le 2$.

8. $\mathbf{w} = z^2\mathbf{i} + 0 \cdot \mathbf{j} + x^2\mathbf{k}$; $\overrightarrow{C}$ is the segment $\overrightarrow{C}_1 (1, 0, 1)$ to $(2, 0, 1)$ followed by the segment $\overrightarrow{C}_2$ from $(2, 0, 1)$ to $(2, 0, 4)$.

9. $\mathbf{w} = (x^2 - y^2)\mathbf{i} + 2xy\mathbf{j} + 0 \cdot \mathbf{k}$; $\overrightarrow{C}$ is the segment from $(2, 0, 0)$ to $(0, 2, 0)$.

10. $\mathbf{w} = 2yz\mathbf{j} + (z^2 - y^2)\mathbf{k}$; $\overrightarrow{C}$ is the circular arc given by: $x = 0$, $y^2 + z^2 = 4$ going from $(0, 2, 0)$ to $(0, 0, 2)$.

11. Show that

$$\int_{\overrightarrow{C}} \mathbf{w} \cdot d\mathbf{r} = \int_{\overrightarrow{C}} \sqrt{P^2 + Q^2 + R^2} \cos \theta \, ds,$$

where $\mathbf{w} = P\mathbf{i} + Q\mathbf{j} + R\mathbf{k}$, ds is the element of arc along $\overrightarrow{C}$, and θ is the angle made by the vector $\mathbf{r}$ and the vector $\mathbf{w}$.

6. PATH–INDEPENDENT LINE INTEGRALS

A continuous transformation

$$\mathbf{v}(\overrightarrow{OP}) = \mathbf{r}(t); \qquad a \le t \le b$$

is called a **path** in space. Since a path is not necessarily one to one, we see that arcs are special cases of paths. As the examples in Fig. 13–10 show, a path may have loops and multiple intersections. Parametric representations are needed to distinguish the first two paths in Fig. 13–10, both of which start at A and end at B; they are identical in appearance. However, it is intuitively clear that if $\mathbf{r}(t_1)$ is a representation of a path and $t_1 = S(t_2)$ is an increasing function, then the representation $\mathbf{R}(t_2)$ determined by the relation

$$\mathbf{R}(t_2) = \mathbf{r}[S(t_2)]$$

describes the same path as $\mathbf{r}(t_1)$, and in the same order. Also, we shall denote **directed paths** by symbols such as $\overrightarrow{C}$, undirected paths by $|\overrightarrow{C}|$, and oppositely directed paths by $-\overrightarrow{C}$.

The theorems about line integrals which we established for piecewise smooth arcs are equally valid for piecewise smooth paths. As we saw in the preceding

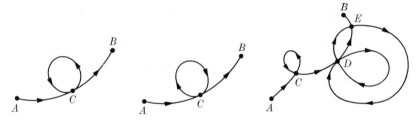

FIGURE 13–10

section, a line integral $\int_{\vec{C}} \mathbf{w} \cdot d\mathbf{r}$ depends not only on the vector field $\mathbf{w}$ and the endpoints A and B of the path but also on the path $\vec{C}$ itself. Whenever the value of a line integral depends only on the vector field $\mathbf{w}$ and on the endpoints A and B of the path but not on $\vec{C}$ itself, we say **the integral is independent of the path**. Such integrals were first discussed on page 198 of Chapter 4. The results given there are now stated in such a way that the extension to higher-dimensional spaces becomes evident.

We say that D is a **connected region in space** if it has the property that any two points in D can be joined by a smooth arc.

Theorem 11. *Suppose that u is a continuously differentiable scalar field on a connected region D and that A and B are in D. Then*

$$\int_{\vec{C}} \nabla u \cdot d\mathbf{r} = u(B) - u(A)$$

for any piecewise smooth path from A to B which is contained in D.

Proof. Since each piecewise smooth path is the finite sum of smooth paths, it is sufficient to prove the theorem for a smooth path. Let (x, y, z) be a rectangular coordinate system with $\mathbf{i}, \mathbf{j}, \mathbf{k}$ the usual basis. The path $\vec{C}$ is given by

$$\mathbf{r} = \mathbf{r}(t) = x(t)\mathbf{i} + y(t)\mathbf{j} + z(t)\mathbf{k}, \qquad a \leq t \leq b,$$

and the vector field ∇u is

$$\nabla u = \frac{\partial u}{\partial x}\mathbf{i} + \frac{\partial u}{\partial y}\mathbf{j} + \frac{\partial u}{\partial z}\mathbf{k}.$$

Then

$$\int_{\vec{C}} \nabla u \cdot d\mathbf{r} = \int_{\vec{C}} \left\{ \frac{\partial u}{\partial x}\frac{dx}{dt} + \frac{\partial u}{\partial y}\frac{dy}{dt} + \frac{\partial u}{\partial z}\frac{dz}{dt} \right\} dt.$$

Along $\vec{C}$ we have $u = u(t)$, and so (noting that $\mathbf{r}(a) = A$, $\mathbf{r}(b) = B$) the above integral is

$$\int_a^b \frac{d}{dt} u[x(t), y(t), z(t)] \, dt = u(B) - u(A),$$

which is the desired result. Note that we have not used the u^* notation.

Theorem 11 shows that, under appropriate hypotheses, every vector field which is the gradient of a scalar field is path-independent. The extension to scalar fields in n variables is immediate. The next theorem shows that, conversely, if a vector field leads to path-independent line integrals in a domain, it must be the gradient of some scalar field.

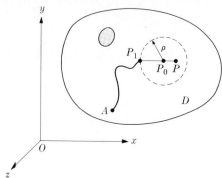

FIGURE 13–11

Theorem 12. *Suppose that* **v** *is a continuous vector field on a domain D, and suppose that for every ordered pair* (A, B) *of points in D, the integral*

$$\int_{\vec{C}} \mathbf{v} \cdot d\mathbf{r}$$

has the same value for every smooth path from A to B with $\vec{C}$ *in D. That is, suppose the integral is path-independent. Then there is a continuously differentiable scalar field u with domain D such that*

$$\mathbf{v} = \nabla u.$$

Proof. Let A be a fixed point in D, and define

$$u(P) = \int_{\vec{C}} \mathbf{v} \cdot d\mathbf{r},$$

where $\vec{C}$ is any smooth path from A to a point P. We shall show that u is the desired scalar field. We introduce the customary rectangular coordinate system and let P_0 be any point in D with coordinates (x_0, y_0, z_0). We construct a sphere with center at P_0 and radius ρ so small that the sphere is entirely in D (Fig. 13–11). Let $\vec{C}_0$ be a directed path in D from A to P_0 which ends with a directed straight line segment $\overrightarrow{P_1 P_0}$, as shown in Fig. 13–11. The coordinates of P_1 are (x_1, y_1, z_1). We suppose for convenience that this segment is parallel to the x-axis. Let P be a point with coordinates $(x_0 + h, y_0, z_0)$, where $|h| < \rho$. Then we have

$$u(x_0 + h, y_0, z_0) = \int_{\vec{C}_1} \mathbf{v} \cdot d\mathbf{r} + \int_{\vec{C}_2} \mathbf{v} \cdot d\mathbf{r}, \tag{1}$$

where $\vec{C}_1$ is the directed path from A to P_1 and where $\vec{C}_2$ is the directed segment

$\overrightarrow{P_1P}$. Now we write

$$\mathbf{v}(x, y, z) = v_1(x, y, z)\mathbf{i} + v_2(x, y, z)\mathbf{j} + v_3(x, y, z)\mathbf{k},$$
$$\mathbf{r} = x\mathbf{i} + y\mathbf{j} + z\mathbf{k},$$

and we observe that, along $\overrightarrow{P_1P}$, $d\mathbf{r} = (dx)\mathbf{i}$. Then, defining

$$\phi(x) = \int_{x_1}^{x} v_1(\xi, y_0, z_0)\, d\xi,$$

we see from (1) that

$$u(x_0 + h, y_0, z_0) - u(x_0, y_0, z_0) = \phi(x_0 + h) - \phi(x_0).$$

We conclude from Leibniz' Rule that

$$u_x(x_0, y_0, z_0) = \lim_{h \to 0} \frac{\phi(x_0 + h) - \phi(x_0)}{h} = \phi'(x_0) = v_1(x_0, y_0, z_0).$$

Since the same arguments work in the y- and z-directions, we obtain $\nabla u(P_0) = \mathbf{v}(P_0)$. But P_0 is an arbitrary point of D, and so the result is established.

The above theorem is intimately connected with Theorem 7 of Section 4, because any vector field $\mathbf{v}$ in a rectangular box with the property that curl $\mathbf{v} = \mathbf{0}$ is the gradient of a scalar function. Thus $\int_{\vec{C}} \mathbf{v} \cdot d\mathbf{r}$ will be independent of the path in such a box. We would like to establish Theorem 7 for more general domains, but in doing so we must exercise extreme care, as the following illustration shows.

We consider the vector field

$$\mathbf{v} = -\frac{y}{x^2 + y^2}\mathbf{i} + \frac{x}{x^2 + y^2}\mathbf{j} + 0 \cdot \mathbf{k}.$$

The reader may easily verify that curl $\mathbf{v} = \mathbf{0}$ for all (x, y, z) so long as $(x, y) \neq (0, 0)$. However, when we select for $\vec{C}$ the circular path

$$\vec{C}: \quad x = \cos t, \quad y = \sin t, \quad z = 0, \quad -\pi \leq t \leq \pi,$$

and for D any region containing $\vec{C}$ and excluding the z-axis, a calculation shows that

$$\int_{\vec{C}} \mathbf{v} \cdot d\mathbf{r} = \int_{\vec{C}} \frac{x\, dy - y\, dx}{x^2 + y^2} = \int_{-\pi}^{\pi} dt = 2\pi.$$

If $\mathbf{v} \cdot d\mathbf{r}$ were an exact differential in D, then $\mathbf{v} = \nabla f$ and the integral would be zero, according to Theorem 7. So we see that *some restriction* on the domain D is essential before Theorem 7 can be extended.

Suppose that A and B are points of a domain D which are connected by two paths $\vec{C}_0$ and $\vec{C}_1$, both lying in D. We shall define formally the concept which states: "$\vec{C}_0$ can be deformed smoothly into $\vec{C}_1$ without going outside D."

DEFINITION. A domain D in the plane or in space is said to be **simply connected** if there exists a vector function $\mathbf{f}$ of the variables t and τ, described as follows: Suppose that $\vec{C}_0$ and $\vec{C}_1$ are smooth paths in D joining the same points A and B. Then $\mathbf{f}$ has these four properties:

(i) $\mathbf{f}(t, \tau)$ is continuous for $a \leq t \leq b$, $0 \leq \tau \leq 1$.

(ii) If $\mathbf{v}(\overrightarrow{OP})$ is the vector (directed line segment) from the origin to a point P, then for $\tau = 0$, $\mathbf{f}(t, 0)$ describes $\vec{C}_0$. That is,

$$\vec{C}_0: \quad \mathbf{v}(\overrightarrow{OP}) = \mathbf{f}(t, 0), \ a \leq t \leq b.$$

(iii) Similarly, for $\tau = 1$,

$$\vec{C}_1: \quad \mathbf{v}(\overrightarrow{OP}) = \mathbf{f}(t, 1), \ a \leq t \leq b.$$

(iv) For each fixed τ on $[0, 1]$, $\mathbf{v}(\overrightarrow{OP}) = \mathbf{f}(t, \tau)$, $a \leq t \leq b$, is a directed path from A to B lying entirely in D.

The definition of simple connectivity in the plane is illustrated in Fig. 13–12, where we denote by $\vec{C}_\tau$ the path $\mathbf{v}(\overrightarrow{OP}) = \mathbf{f}(t, \tau)$ for fixed τ going from A to B. We note that the paths vary smoothly from $\vec{C}_0$ to $\vec{C}_1$ as τ goes from 0 to 1.

It is intuitively clear that the domain D, consisting of the entire plane with a single point (denoted O) removed, is not simply connected. As Fig. 13–13 shows, it is impossible to deform $\vec{C}_0$ into $\vec{C}_1$ without some $\vec{C}_\tau$ passing through O. Similarly, the domain between two concentric circles (Fig. 13–13) is not simply connected.

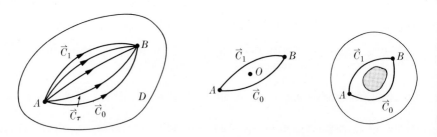

FIGURE 13–12   FIGURE 13–13

In three-space the domain inside a sphere or ellipsoid is simply connected. Also, it can be shown (although we shall not do so) that, on one hand, the region between two concentric spheres *is* simply connected while, on the other, the region inside a torus (that is, the space inside a doughnut) is *not* simply connected. The region between two coaxial cylinders is another example of a region which is not simply connected.

The next theorem, which is an extension of Theorem 7 of Section 4, is proved in the appendix (with a slightly stronger definition of simple connectivity).

Theorem 13. *Suppose that* **v** *is continuously differentiable in a simply connected region D in space and that* curl* **v** $= 0$ *in D. Then there is a continuously differentiable scalar field u on D such that* **v** $= \nabla u$.

A domain D in the plane or in space is said to be **convex** if and only if the line segment $\overline{P_1 P_2}$ lies in D whenever P_1 and P_2 do. Figure 13–14 shows examples of convex and nonconvex domains. The next theorem establishes relationships between convex and simply connected domains.

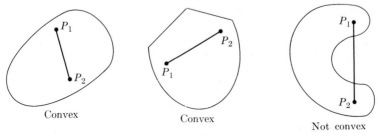

Convex   Convex   Not convex

FIGURE 13–14

Theorem 14. (a) *Any convex domain is simply connected.*
(b) *Suppose D and D' are domains with D' simply connected. If there is a one-to-one twice continuously differentiable transformation with D as its domain and D' as its range, then D is simply connected.*

Proof. (a) Suppose that $\vec{C}_0$ and $\vec{C}_1$ are paths in D. If $\vec{C}_0$: $\mathbf{f}(t, 0)$, $\vec{C}_1$: $\mathbf{f}(t, 1)$, we define
$$\mathbf{f}(t, \tau) = (1 - \tau)\mathbf{f}(t, 0) + \tau \mathbf{f}(t, 1), \qquad 0 \le \tau \le 1.$$

As τ varies from 0 to 1, $\mathbf{f}(t, \tau)$ describes (for each t) the straight line segment joining the points $\mathbf{f}(t, 0)$ and $\mathbf{f}(t, 1)$. By convexity, this segment is in D.
(b) If $\mathbf{f}(t, \tau)$ is the desired function in D' then, under the transformation, it would correspond to a function with similar properties in D.

Remark. The statement and proof of Theorem 14 are valid in a Euclidean space of any dimension.

---

* In the case of n dimensions where

$$\mathbf{v} = \sum_{i=1}^{n} v_i \mathbf{e}_i, \qquad \mathbf{e}_i \text{ orthonormal},$$

this condition is replaced by

$$\frac{\partial v_i}{\partial x_j} = \frac{\partial v_j}{\partial x_i}, \qquad i, j = 1, \ldots, n.$$

This condition is independent of the particular coordinate system. If $n = 2$, the condition reduces to

$$\frac{\partial Q}{\partial x} = \frac{\partial P}{\partial y} \qquad \text{if } v_1 = P \text{ and } v_2 = Q.$$

EXAMPLE. Let D be the set in the plane consisting of all points except the origin and the points on the negative x-axis. Show that D is simply connected.

Solution. The equations $x = r \cos \theta$, $y = r \sin \theta$ set up a one-to-one twice differentiable map of D onto the domain

$$D': \quad r > 0, \qquad -\pi < \theta < \pi,$$

in the (r, θ)-plane. The inverse map is also twice differentiable. The domain D' is a half-infinite strip, and its convexity is easily verified. (See Fig. 13–15; see also Exercise 14.) According to Theorem 14(a), D' is simply connected and then, by part (b), D is also.

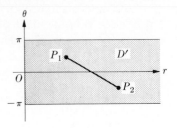

FIGURE 13–15

EXERCISES

In each of problems 1 through 5, verify Theorem 11 by calculating the line integral $\int_{\vec{C}} \nabla u \cdot d\mathbf{r}$ for each of the given paths.

1. $u = x^2 - xy - y^2$;
   $\vec{C}_1$: $x = \cos \theta$, $y = -\sin^2 \theta$, $z = 0$, $0 \le \theta \le \pi/2$;
   $\vec{C}_2$: straight segment from $(1, 0, 0)$ to $(0, 1, 0)$.

2. $u = x^2 - 2y^2 + xz + z^2$;
   $\vec{C}_1$: straight segment from $(1, 0, 1)$ to $(-1, 1, 2)$;
   $\vec{C}_2$: straight segment from $(1, 0, 1)$ to $(1, 0, 2)$ followed by straight segment from $(1, 0, 2)$ to $(-1, 1, 2)$.

3. $u = (x^2 + y^2 + z^2)^{-1/2}$;
   $\vec{C}_1$: $x = \sin \theta \cos \theta$, $y = \sin^2 \theta$, $z = \cos \theta$, $0 \le \theta \le \pi/2$;
   $\vec{C}_2$: straight segment from $(0, 0, 1)$ to $(0, 1, 1)$.

4. $u = z(x^2 + y^2)^{-1/2}$;
   $\vec{C}_1$: $x = 3t$, $y = -4t$, $z = 5t^2$, $1 \le t \le 2$;
   $\vec{C}_2$: $x = 3t$, $y = -4t$, $z = t^4 + 4$, $1 \le t \le 2$.

5. $u = \sin xy + \cos yz + \sin xz$;
   $\vec{C}_1$: straight segment from $(0, \pi/4, 1)$ to $(\pi/2, 1, 0)$;
   $\vec{C}_2$: straight segment from $(0, \pi/4, 1)$ to $(0, 1, \pi/4)$ followed by straight segment from $(0, 1, \pi/4)$ to $(\pi/2, 1, 0)$.

In each of problems 6 through 10, decide whether Theorem 13 is valid in the given domain. If so, find the appropriate scalar field.

6. $\mathbf{v}$: $(2x + 8y - 2z)\mathbf{i} + (2y + 4z + 8x)\mathbf{j} + (2z - 2x + 4y)\mathbf{k}$; D is the interior of the ball $x^2 + y^2 + z^2 \le 1$.

7. $\mathbf{v} = 3(x^2 - yz)\mathbf{i} + 3(y^2 - xz)\mathbf{j} - 3xy\mathbf{k}$; D is the domain between the spheres $x^2 + y^2 + z^2 = 1$, $x^2 + y^2 + z^2 = 9$.

8. $\mathbf{v} = (x^4 - 8y^2 + 2)\mathbf{i} + 2xyz\mathbf{j} + (y^2 - x^2)\mathbf{k}$; D is the interior of the ellipsoid $x^2 + 2y^2 + 3z^2 = 27$.

9. $\mathbf{v} = \dfrac{x}{(x^2 + y^2 + z^2)^{2/3}}\mathbf{i} + \dfrac{y}{(x^2 + y^2 + z^2)^{2/3}}\mathbf{j} + \dfrac{z}{(x^2 + y^2 + z^2)^{2/3}}\mathbf{k};$

   D is the domain between the cylinders

$$x^2 + z^2 = 1 \quad \text{and} \quad x^2 + z^2 = 9.$$

10. $\mathbf{v} = \dfrac{x}{x + y + z}\mathbf{i} + \dfrac{y}{x + y + z}\mathbf{j} + \dfrac{z}{x + y + z}\mathbf{k};$

   D is the parallelepiped: $1 \le x \le 2, \ 1 \le y \le 3, \ 2 \le z \le 4.$

11. Prove that the intersection of any finite number of convex sets is a convex set.

12. Prove, using coordinates, that any half-plane (that is, the part of a plane on one side of a line) is convex.

13. Prove, using coordinates, that any half-space (that is, the portion of three-space on one side of a plane) is convex. Extend the result to n-dimensional Euclidean space.

14. Using the results of problems 11 and 12, show that the domain D' in Fig. 13–15 is convex.

15. Prove that any set of the form $a < x < b, \ c < y < f(x)$ in the plane, where f and f' are smooth in an interval containing $[a, b]$ in its interior, is simply connected. [*Hint.* Set up a transformation from this region onto a rectangle $a < \xi < b$, $0 < \eta < 1$; it is assumed that $f(x) > c$ for $a - h \le x \le b + h$ for some $h > 0$. (See Fig. 13–16.)]

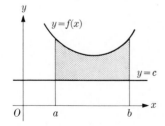

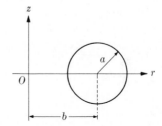

FIGURE 13–16   FIGURE 13–17

16. Use Theorem 13 and the vector field

$$\mathbf{v} = (x^2 + y^2)^{-1}(-y\mathbf{i} + x\mathbf{j} + 0 \cdot \mathbf{k}), \quad \vec{C}: x = \cos\theta, \ y = \sin\theta, \ -\pi \le \theta \le \pi,$$

to show that the plane with the origin removed is not simply connected.

17. A torus is obtained by revolving a circle about an axis in its plane (provided the axis does not intersect the circle). (See Fig. 13–17.) The interior of the torus consists of all points whose cylindrical coordinates (r, θ, z) satisfy

$$(r - b)^2 + z^2 < a^2, \quad 0 < a < b.$$

Using the example of problem 16, show that the interior of a torus in three-space is not simply connected.

18. Show that the set of points (x, y) satisfying

$$\frac{x^2}{a^2} + \frac{y^2}{b^2} < 1$$

is convex. [*Hint.* Consider the function

$$\phi(t) = \frac{[x_1 + t(x_2 - x_1)]^2}{a^2} + \frac{[y_1 + t(y_2 - y_1)]^2}{b^2}, \qquad 0 \le t \le 1.$$

Draw a figure.]

19. Show that the totality of points in the plane which are not on the spiral $r = \theta$, $\theta \ge 0$, r and θ polar coordinates, is simply connected. Draw a figure.

20. Use the results of problems 11 and 12 to show that the interior of every regular polygon in the plane is convex.

CHAPTER 14

The Theorems of Green and Stokes

1. GREEN'S THEOREM

The Fundamental Theorem of Calculus states that differentiation and integration are inverse processes. An appropriate extension of this theorem to double integrals of functions of two variables is known as Green's Theorem. Suppose that $P(x, y)$ and $Q(x, y)$ are smooth (i.e., continuously differentiable) functions defined in some region R of the plane. **A simple closed curve** is a curve that can be obtained as the union of two arcs which have only their endpoints in common. Thus a circle is the union of two half circles. Of course, any two points on a simple closed curve divide it into two arcs in this way. It is intuitively clear that a simple closed curve in the plane divides the plane into two regions, constituting the "interior" and the "exterior" of the curve. This fact is surprisingly hard to prove, and we shall simply assume it here. Actually we do not use this fact in the proofs of any theorems. **A smooth simple closed curve** is one which has a parametric representation $x = x(t)$, $y = y(t)$, $a \leq t \leq b$, in which x, y, x', and y' are continuous and $[x'(t)]^2 + [y'(t)]^2 > 0$ and $x(b) = x(a)$, $x'(b) = x'(a)$, $y(b) = y(a)$, $y'(b) = y'(a)$. If Γ is a smooth simple closed curve which, together with its interior G, is in R, then the basic formula associated with Green's Theorem is

$$\iint_G \left(\frac{\partial Q}{\partial x} - \frac{\partial P}{\partial y} \right) dA = \oint_\Gamma (P \, dx + Q \, dy). \tag{1}$$

The symbol on the right represents the line integral taken in the counterclockwise sense, so that Γ is traversed with the interior of G always on the left.

We shall establish Green's Theorem for regions which have a special shape. Then we shall show how the result for these special regions may be extended to yield the theorem for rather general domains.

Lemma 1. Suppose that G is a region bounded by the straight lines $x = a$, $x = b$, $y = c$, and by an arc (situated above the line $y = c$) with equation $y = f(x)$, $a \leq x \leq b$. Assume that f is either nondecreasing or nonincreasing. If $P(x, y)$

and $Q(x, y)$ are continuously differentiable in a region containing G and its boundary, then

$$\iint\limits_{G} \left(\frac{\partial Q}{\partial x} - \frac{\partial P}{\partial y} \right) dA = \oint\limits_{\partial G} (P \, dx + Q \, dy), \qquad (2)$$

where the symbol ∂G denotes the boundary of G and the line integral is traversed in a counterclockwise sense.

Proof. We shall establish the result by proving separately each of the formulas

$$-\iint\limits_{G} \frac{\partial P}{\partial y} \, dA = \oint\limits_{\partial G} P \, dx, \qquad (2a)$$

$$\iint\limits_{G} \frac{\partial Q}{\partial x} \, dA = \oint\limits_{\partial G} Q \, dy. \qquad (2b)$$

For convenience, suppose that f is nondecreasing. (The proof is the same if f is nonincreasing.) Figure 14–1 shows a typical region G with the arcs directed as shown by the arrows. To prove (2a), we change the double integral to an iterated integral and then employ the Fundamental Theorem of Calculus. We get

$$-\iint\limits_{G} \frac{\partial P}{\partial y} \, dA = - \int_{a}^{b} \int_{c}^{f(x)} \frac{\partial P}{\partial y} \, dy \, dx = - \int_{a}^{b} \{ P[x, f(x)] - P(x, c) \} \, dx. \qquad (3)$$

The right side of (3) may be written as the line integrals

$$\int_{\vec{C}_3} P(x, y) \, dx + \int_{\vec{C}_1} P(x, y) \, dx.$$

Since x is constant along $\vec{C}_2$ and $\vec{C}_4$, we may write

$$\int_{\vec{C}_2} P \, dx = \int_{\vec{C}_4} P \, dx = 0,$$

and so

$$-\iint\limits_{G} \frac{\partial P}{\partial y} \, dA = \int_{\vec{C}_1 + \vec{C}_2 + \vec{C}_3 + \vec{C}_4} P \, dx = \oint\limits_{\partial G} P \, dx.$$

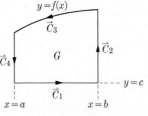

FIGURE 14–1

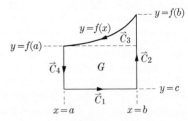

FIGURE 14–2

To establish (2b), we suppose first that f is strictly increasing. Then, referring to Fig. 14–2, we have

$$\iint\limits_{G} \frac{\partial Q}{\partial x} \, dA = \int_{c}^{f(a)} \int_{a}^{b} \frac{\partial Q}{\partial x} \, dx \, dy + \int_{f(a)}^{f(b)} \int_{f^{-1}(y)}^{b} \frac{\partial Q}{\partial x} \, dx \, dy$$

$$= \int_{c}^{f(a)} [Q(b, y) - Q(a, y)] \, dy + \int_{f(a)}^{f(b)} \{ Q(b, y) - Q[f^{-1}(y), y] \} \, dy$$

$$= \int_{c}^{f(b)} Q(b, y) \, dy + \int_{f(a)}^{c} Q(a, y) \, dy + \int_{f(b)}^{f(a)} Q[f^{-1}(y), y] \, dy$$

$$= \int_{\vec{C}_2} Q(x, y) \, dy + \int_{\vec{C}_4} Q(x, y) \, dy + \int_{\vec{C}_3} Q(x, y) \, dy. \tag{4}$$

We use the fact that

$$\int_{\vec{C}_1} Q \, dy = 0$$

to conclude that (2b) holds. If f is nondecreasing rather than strictly increasing, (2b) is still valid, since any portion of $\vec{C}_3$ where f is horizontal makes no contribution to the last integral on the right in (4).

FIGURE 14–3

By means of a simple change of coordinates or other minor adjustment, we see easily that the above lemma holds for all regions of the type shown in Fig. 14–3. The next important step is the observation that the result of the lemma [i.e., Eq. (2)] may be established for any region which can be divided up into a finite number of regions, each of the type considered in the lemma. How this may be done is suggested in Fig. 14–4, which shows a region with a smooth simple closed curve as boundary divided by straight-line segments into a number of regions of special type. It is clear that the double integral over the whole region is the sum of the double integral over the parts. In adding the line integrals we observe that the integrals over the interior segments must cancel, since each segment is directed in opposite senses when considered as the boundary of two adjacent special regions. Unfortunately, it is not true that every region for which formula (2) holds can be divided into special regions in the manner described above.

We shall not attempt to examine the most general type of domain for which formula (2) is valid but shall note some examples of regions to which we may apply the lemma. The shaded domain shown in Fig. 14–5 is bounded by four smooth simple closed curves. It may be subdivided into regions to which the

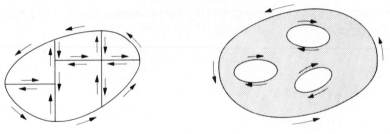

FIGURE 14–4   FIGURE 14–5

lemma applies. It is important to notice in such a case that the line integral as given in (2) must be traversed so that the region G always remains on the left—counterclockwise for the outer boundary curve and clockwise for the three inner boundary curves.

A simple closed curve is **piecewise smooth** if it is made up of a finite number of smooth arcs (i.e., having parametric representations as above) and if these arcs are joined at points called corners. A **corner** is the juncture of two smooth arcs which have limiting tangent lines *making a positive angle* (Fig. 14–6). For a piecewise smooth boundary with corners, it is possible to use the lemma, thus establishing formula (2) for any region which has a boundary consisting of a piecewise smooth simple closed curve. Since polygons have piecewise smooth boundaries, they are included in the collection of regions for which (2) is valid. We now state Green's Theorem in a form which is sufficiently general for most applications.

FIGURE 14–6

Theorem 1. (Green's Theorem in the plane.) *Suppose that G is a region with a boundary consisting of a finite number of piecewise smooth simple closed curves, no two of which intersect. Suppose that $P(x, y)$, $Q(x, y)$ are continuously differentiable functions defined in a region which contains G and ∂G. Then*

$$\iint\limits_{G} \left(\frac{\partial Q}{\partial x} - \frac{\partial P}{\partial y}\right) dA = \oint\limits_{\partial G} (P\, dx + Q\, dy), \tag{5}$$

where the integral on the right is defined to be the sum of the integrals over the boundary curves, each of which is directed so that G is on the left.

The proof of Green's Theorem for all possible regions which may be subdivided into regions of special type has been described above in a discursive manner. In Sections 2 and 3 a proof is given which is sufficiently general to be used in many problems in advanced analysis.

Green's Theorem may be stated in vector form. We write

$$\mathbf{v} = P\mathbf{i} + Q\mathbf{j} + 0 \cdot \mathbf{k}$$

and denote by $\mathbf{r}(t)$ the vector from the origin O of a rectangular coordinate system in the plane to the boundary ∂G of G. We interpret

$$\operatorname{curl} \mathbf{v} = \left(\frac{\partial Q}{\partial x} - \frac{\partial P}{\partial y}\right) \mathbf{k}$$

as the scalar function $[(\partial Q/\partial x) - (\partial P/\partial y)]$ in the $\mathbf{i}$, $\mathbf{j}$-plane. We call this expression the **scalar curl of v**, although we use the same symbol. Then, under the hypotheses of Theorem 1, we have the formula

$$\iint_G \operatorname{curl} \mathbf{v} \, dA = \oint_{\partial G} \mathbf{v} \cdot d\mathbf{r}.$$

It is a simple matter to verify that this formula is identical with (5). However, the vector formulation has the advantage of exhibiting the invariance of the result under a change of coordinates. Furthermore, the nature of the extension to three-space is apparent from the vector formulation.

We now illustrate Green's Theorem in the plane with several examples.

EXAMPLE 1. Verify Green's Theorem when $P(x, y) = 2y$, $Q(x, y) = 3x$ and G is the unit circle: $x^2 + y^2 \leq 1$ (Fig. 14–7).

Solution. We have

$$\frac{\partial Q}{\partial x} - \frac{\partial P}{\partial y} = 3 - 2 = 1.$$

Therefore

$$\iint_G \left(\frac{\partial Q}{\partial x} - \frac{\partial P}{\partial y}\right) dA = \iint_G dA = \pi \cdot 1^2 = \pi.$$

FIGURE 14–7

The equation of ∂G is: $x = \cos \theta$, $y = \sin \theta$, $-\pi \leq \theta \leq \pi$. Hence

$$\oint_{\partial G} (P \, dx + Q \, dy) = \int_{-\pi}^{\pi} \{2(\sin \theta) \, d(\cos \theta) + 3(\cos \theta) \, d(\sin \theta)\}$$

$$= \int_{-\pi}^{\pi} (-2 \sin^2 \theta + 3 \cos^2 \theta) \, d\theta = -2\pi + 3\pi = \pi.$$

EXAMPLE 2. If G is the unit circle as in Example 1, use Green's Theorem to evaluate

$$\int_{\partial G} [(x^2 - y^3) \, dx + (y^2 + x^3) \, dy].$$

Solution. Here, $P = x^2 - y^3$, $Q = y^2 + x^3$. Therefore

$$\int_{\partial G} [(x^2 - y^3) \, dx + (y^2 + x^3) \, dy] = \iint_G 3(x^2 + y^2) \, dA$$

$$= 3 \int_0^{2\pi} \int_0^1 r^2 \cdot r \, dr \, d\theta = \frac{3\pi}{2}.$$

EXAMPLE 3. Let G be the region outside the unit circle which is bounded on the left by the parabola $y^2 = 2(x + 2)$ and on the right by the line $x = 2$. (See Fig. 14–8.) Use Green's Theorem to evaluate

$$\int_{\vec{C}_1} \left(\frac{-y}{x^2 + y^2} \, dx + \frac{x}{x^2 + y^2} \, dy \right),$$

where $\vec{C}_1$ is the oriented outer boundary of G as shown in Fig. 14–8.

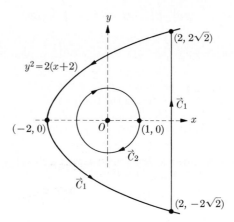

FIGURE 14–8

Solution. We write $P = -y/(x^2 + y^2)$, $Q = x/(x^2 + y^2)$ and observe that P and Q have singularities at the origin. Denoting the boundary of the unit circle oriented *clockwise* by $\vec{C}_2$ and noting that $(\partial Q/\partial x) - (\partial P/\partial y) = 0$, we use Green's Theorem to write

$$\int_{\vec{C}_1 + \vec{C}_2} \left(-\frac{y}{x^2 + y^2} \, dx + \frac{x}{x^2 + y^2} \, dy \right) = 0.$$

Therefore

$$\int_{\vec{C}_1} (P \, dx + Q \, dy) = -\int_{\vec{C}_2} (P \, dx + Q \, dy) = \int_{-\vec{C}_2} (P \, dx + Q \, dy),$$

where $-\vec{C}_2$ is the unit circle oriented *counterclockwise*. Since $x = \cos \theta$, $y = \sin \theta$, $-\pi \le \theta \le \pi$ on the unit circle $-\vec{C}_2$, we obtain

$$\int_{\vec{C}_1} (P \, dx + Q \, dy) = \int_{-\pi}^{\pi} (\sin^2 \theta + \cos^2 \theta) \, d\theta = 2\pi.$$

EXAMPLE 4. Let $\mathbf{v} = -\frac{1}{2} y \mathbf{i} + \frac{1}{2} x \mathbf{j}$ be defined in a region G with area A. Show that

$$A = \int_{\partial G} \mathbf{v} \cdot d\mathbf{r}.$$

Solution. We apply Green's Theorem and obtain

$$\int_{\partial G} \mathbf{v} \cdot d\mathbf{r} = \iint_G \operatorname{curl} \mathbf{v} \, dA = \iint_G (\tfrac{1}{2} + \tfrac{1}{2}) \, dA = A,$$

in which we have used the scalar interpretation of curl **v**.

EXERCISES

In each of problems 1 through 8, verify Green's Theorem.

1. $P(x, y) = -y, \quad Q(x, y) = x; \qquad G: 0 \le x \le 1, \quad 0 \le y \le 1$

2. $P = 0, \ Q = x; \ G$ is the region outside the unit circle, bounded below by the parabola $y = x^2 - 2$ and bounded above by the line $y = 2$.

3. $P = xy, \quad Q = -2xy; \quad G: 1 \le x \le 2, \quad 0 \le y \le 3$

4. $P = e^x \sin y, \quad Q = e^x \cos y; \quad G: 0 \le x \le 1, \quad 0 \le y \le (\pi/2)$

5. $P = \tfrac{2}{3}xy^3 - x^2y, \ Q = x^2y^2; \ G$ is the triangle with vertices at $(0, 0)$, $(1, 0)$, and $(1, 1)$.

6. $P = 0, \ Q = x; \ G$ is the region inside the circle $x^2 + y^2 = 4$ and outside the circles $(x - 1)^2 + y^2 = \tfrac{1}{4}, (x + 1)^2 + y^2 = \tfrac{1}{4}$.

7. $\mathbf{v} = (x^2 + y^2)^{-1}(-y\mathbf{i} + x\mathbf{j}); \ G$ is the region between the circles $x^2 + y^2 = 1$ and $x^2 + y^2 = 4$.

8. $P = 4x - 2y, \quad Q = 2x + 6y; \quad G$ is the ellipse $x = 2 \cos \theta, \quad y = \sin \theta,$ $-\pi \le \theta \le \pi$.

In each of problems 9 through 14, compute the area $A(G)$ of G by using the formula

$$A(G) = \int_{\partial G} x \, dy,$$

which is valid because of Green's Theorem.

9. G is the triangle with vertices at $(1, 1)$, $(4, 1)$, and $(4, 9)$.

10. G is the triangle with vertices at $(2, 1)$, $(3, 4)$, and $(1, 5)$.

11. G is the region bounded by the y-axis, the lines $y = 1$ and $y = 3$, and the curve $x = y^2$.

12. G is the region bounded by the line $y = x + 2$ and the parabola $y = x^2$.

13. G is the region in the first quadrant bounded by the lines $4y = x$ and $y = 4x$ and the hyperbola $xy = 4$.

14. G is the region interior to the ellipse

$$\frac{x^2}{16} + \frac{y^2}{9} = 1.$$

In each of problems 15 through 21, compute $\int_{\partial G} \mathbf{v} \cdot d\mathbf{r}$, using Green's Theorem.

15. $\mathbf{v} = (\tfrac{4}{5}xy^5 + 2y - e^x)\mathbf{i} + (2xy^4 - 4 \sin y)\mathbf{j}; \quad G: 1 \le x \le 2, \quad 1 \le y \le 3$

16. $\mathbf{v} = (2xe^y - x^2y - \frac{1}{3}y^3)\mathbf{i} + (x^2e^y + \sin y)\mathbf{j}$; $G: x^2 + y^2 \leq 1$

17. $\mathbf{v} = 2xy^2\mathbf{i} + 3x^2y\mathbf{j}$; G is the interior of the ellipse

$$\frac{x^2}{a^2} + \frac{y^2}{b^2} = 1.$$

18. $\mathbf{v} = (x^2 + y^2)^y$; G is the interior of the circle $(x - 1)^2 + y^2 = 1$.

19. $\mathbf{v} = (\cosh x - 2) \sin y\mathbf{i} + \sinh x \cos y\mathbf{j}$; $G: 0 \leq x \leq 1, \, 0 \leq y \leq (\pi/2)$

20. $\mathbf{v} = 2 \text{ Arctan } (y/x)\mathbf{i} + \ln (x^2 + y^2)\mathbf{j}$; $G: 1 \leq x \leq 2, \,\, -1 \leq y \leq 1$

21. $\mathbf{v} = -3x^2y\mathbf{i} + 3xy^2\mathbf{j}$; $G: -a \leq x \leq a, \,\, 0 \leq y \leq \sqrt{a^2 - x^2}$

22. Evaluate

$$\int_{\vec{C}} \frac{x \, dy - y \, dx}{x^2 + y^2},$$

where $\vec{C}$ consists of the arc of the parabola $y = x^2 - 1$, $-1 \leq x \leq 2$, followed by the straight segment from $(2, 3)$ to $(-1, 0)$. Do this by applying Green's Theorem to the region G interior to $|\vec{C}|$ and exterior to a small circle of radius ρ centered at the origin.

23. Evaluate

$$\int_{\vec{C}} \frac{x \, dx + y \, dy}{x^2 + y^2}$$

where $\vec{C}$ is the path described in Exercise 22.

24. Suppose that $f(x, y)$ satisfies the Laplace equation ($f_{xx} + f_{yy} = 0$) in a region G. Show that

$$\int_{\partial G^*} (f_y \, dx - f_x \, dy) = 0,$$

where G^* is any region interior to G.

25. If (x^*, y^*) is the location of the center of gravity of a plane region G of uniform density, show that

$$x^* = \frac{\int_{\partial G} x^2 \, dy}{2\int_{\partial G} x \, dy}, \qquad y^* = \frac{\int_{\partial G} y^2 \, dx}{2\int_{\partial G} y \, dx}.$$

26. If $f_{xx} + f_{yy} = 0$ in a region R and v is any smooth function, use the identity $(vf_x)_x = vf_{xx} + v_xf_x$ and a similar one for the derivative with respect to y, to prove that

$$-\int_{\partial G} v(f_y \, dx - f_x \, dy) = \iint_G (v_xf_x + v_yf_y) \, dA,$$

where G is any region interior to R.

2. PARTITION OF UNITY

We define a function ϕ by the formula

$$\phi(s) = \begin{cases} 1, & 0 \le s \le 1, \\ (2s - 1)(s - 2)^2, & 1 \le s \le 2, \\ 0, & 2 \le s. \end{cases} \tag{1}$$

We extend the definition to negative values of s by making ϕ even: $\phi(-s) = \phi(s)$. It is a simple matter to verify that $\phi(1) = 1$, $\phi(2) = 0$ and, therefore, that ϕ is continuous everywhere. Furthermore, since $\phi'(1) = 0$, $\phi'(2) = 0$, we see that ϕ has a continuous first derivative everywhere. Its graph is shown in Fig. 14–9. The function $\phi(s/b)$ with $b > 0$ has the same general behavior as $\phi(s)$ except that the scale is changed. We note that $\phi(s/b)$ is 1 for $|s| \le b$, is between 0 and 1 for $b \le |s| \le 2b$, and vanishes for $|s| \ge 2b$.

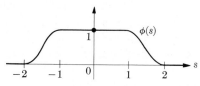

FIGURE 14–9

Functions of two variables with analogous properties are defined by taking products. The function

$$\Phi(x, y) = \phi(x)\phi(y)$$

is 1 in the square S_1: $-1 \le x \le 1$, $-1 \le y \le 1$; is 0 *outside* the square S_2: $-2 \le x \le 2$, $-2 \le y \le 2$; and is between 0 and 1 in the region between S_1 and S_2 (Fig. 14–10). The function Φ and its first partial derivatives are continuous everywhere. A change of scale shows that the function $\Phi(x/b, y/c)$ has the same properties as $\Phi(x, y)$ in a rectangle of width $2b$ and height $2c$.

The function Φ is the basic quantity in performing a decomposition of a region G in the plane. This decomposition will be used to prove Green's Theorem. Let

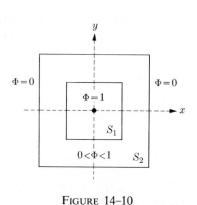

FIGURE 14–10

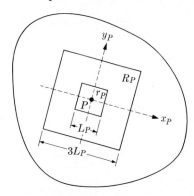

FIGURE 14–11

a bounded region G have for its boundary a finite number of smooth arcs which may form corners at points where they meet. With each point P of G and its boundary ∂G, we associate both a rectangular coordinate system which has P as its origin and a function Φ of the type described above. We consider three cases:

(1) If P is interior to G, we select any two perpendicular lines intersecting at P as coordinate axes (properly oriented, of course). We label these x_P and y_P. Choose any square with center at P, with sides parallel to the axes x_P and y_P, and situated entirely inside G. Denote this square by R_P and the length of one side by $3L_P$. The parallel square with side L_P and center at P we designate r_P (Fig. 14–11). The function

$$\Phi_P(x, y) = \Phi\left(\frac{x_P}{L_P}, \frac{y_P}{L_P}\right) = \phi\left(\frac{x_P}{L_P}\right) \cdot \phi\left(\frac{y_P}{L_P}\right)$$

has the property that it is smooth, is 1 in r_P, and is 0 outside of a square halfway between r_P and R_P. We may set up such a coordinate system and pairs of squares for each point P interior to G. Of course, if P is very near the boundary, the quantity $3L_P$ will be very small; nevertheless, the selections can be made and functions Φ_P formed.

(2) If P is on the boundary of G but interior to a smooth arc, we choose the positive y_P-axis in the direction of the *exterior* normal, as shown in Fig. 14–12. The x_P-axis is then situated along the tangent, properly oriented. We choose a rectangle r_P of width L_P and height H_P parallel to the axes and centered at P. The rectangle of width $3L_P$ and height $3H_P$ is denoted R_P. The numbers L_P and H_P are taken so small that we may express the portion of the boundary in R_P by an equation

$$y_P = f(x_P),$$

where f is smooth. Furthermore, we make the rectangles so small, if necessary, that the conditions

$$|f(x_P)| < H_P \quad \text{for } |x_P| < L_P,$$
$$|f(x_P)| < 3H_P \quad \text{for } |x_P| < 3L_P,$$

are satisfied. In other words, the boundary arc must enter and leave the "sides" of the rectangles* r_P and R_P. With each such boundary point P, we associate the function

$$\Phi_P(x, y) = \phi\left(\frac{x_P}{L_P}\right)\phi\left(\frac{y_P}{H_P}\right),$$

which is 1 in the rectangle r_P and 0 outside a rectangle halfway between r_P and R_P.

---

* That such rectangles can always be found follows from the fact that a smooth arc in the plane has a representation $x_P = x_P(t)$, $y_P = y_P(t)$, $a \leq t \leq b$, in which x_P and y_P are smooth with $[x_P'(t)]^2 + [y_P'(t)]^2 > 0$. If $t = 0$ at P, we have $y_P'(0) = 0$, so $x_P'(0) \neq 0$ and the function x_P has a differentiable inverse T_P; thus we have $x_P(t) = x_P$ and $t = T_P(x_P)$. Hence $y_P = x_P[T_P(x_P)] = f_P(x_P)$ for $|x_P|$ small.

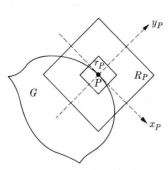

FIGURE 14–12

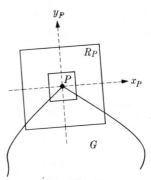

FIGURE 14–13

(3) If P is a corner point of the boundary, we choose the positive y_P-axis along the bisector of the angle between the tangents at P and pointing outside of G; the x_P-axis is selected accordingly (Fig. 14–13). The rectangles r_P and R_P are chosen as in case (2) above, and the function

$$\Phi_P(x, y) = \phi\left(\frac{x_P}{L_P}\right)\phi\left(\frac{y_P}{H_P}\right)$$

is defined as before.

The above description shows that with each point of the region G and its boundary ∂G we may associate a point P and a rectangle (or square) r_P. We recall the Heine-Borel Theorem (Chapter 11, page 510) and, using that theorem, we conclude that a *finite number* of the interiors of the rectangles $\{r_P\}$ cover G and ∂G. We label these covering rectangles

$$r_1, r_2, \ldots, r_k,$$

and we denote their centers $P_1, P_2, \ldots, P_k$. The associated functions are

$$\Phi_{P_1}, \Phi_{P_2}, \ldots, \Phi_{P_k}.$$

Let Q be any point in G. Then Q is in some r_i. According to the way we defined the functions Φ_P, we see that

$$\Phi_{P_i}(Q) = 1,$$

since Φ_{P_i} is identically equal to 1 in all of r_i. Thus, for every point Q of G, we must have

$$\Phi_{P_1}(Q) + \Phi_{P_2}(Q) + \cdots + \Phi_{P_k}(Q) \geq 1.$$

We now define the function

$$\psi_i(Q) = \frac{\Phi_{P_i}(Q)}{\Phi_{P_1}(Q) + \Phi_{P_2}(Q) + \cdots + \Phi_{P_k}(Q)}.$$

Then each $\psi_i(Q)$ is smooth on all the r_i and, since $\Phi_{P_i}(Q)$ is zero outside a rectangle halfway between r_i and R_i, $\psi_i(Q)$ is also.

DEFINITION. The sequence ψ_1, ψ_2, ..., ψ_k is called a **partition of unity.** The term partition of unity comes from the formula

$$\sum_{i=1}^{k} \psi_i(Q) = 1,$$

valid for every point Q not only in G but also in any of the rectangles r_1, r_2, ..., r_k. We have defined a finite sequence of functions which add up to 1 identically, and yet each member of the sequence vanishes, except for a small rectangle about a given point. This form of decomposition, a recently discovered tool, has become extremely useful not only in analysis but also in geometry and topology. It has the virtue of reducing certain types of global problems to local ones. (See page 480.)

EXERCISES

1. Given the function

$$\phi(s) = \begin{cases} 1, & 0 \le s \le 1, \\ (6s^2 - 9s + 4)(2 - s)^3, & 1 \le s \le 2, \\ 0, & 2 \le s < \infty, \end{cases}$$

and $\phi(-s) = \phi(s)$. Show that, for all s, ϕ is twice continuously differentiable and that $0 \le \phi \le 1$.

2. Using the function in Problem 1, construct a twice continuously differentiable function of two variables which is one in a given rectangle, zero outside a larger similarly placed rectangle, and between zero and one in the region between the rectangles.

3. Defining the function $F(x, y) = \phi(x^2 + y^2)$ where ϕ is the function given by (1), show that $F(x, y)$ is one in the unit circle, vanishes outside the circle $x^2 + y^2 \ge 2$, and is between zero and one otherwise.

4. Using the result of Problem 3, find a smooth function which is one in a circle of radius a, vanishes outside a concentric circle of radius $2a$, and is between zero and one otherwise.

5. Same as Problem 4, except that the function is to be twice continuously differentiable. (Use the function in Exercise 1.)

6. Show how to construct a function $G(x, y, z)$ which is smooth, is one inside a rectangular box, is zero outside of a larger box, and is between zero and one in the region between boxes.

7. By considering an expression of the form

$$P(s)(2 - s)^4,$$

where P is a fourth-degree polynomial, show how to construct a function which is one for $|s| \le 1$, zero for $|s| \ge 2$, is between zero and one for $1 \le |s| \le 2$, and is three times continuously differentiable.

8. Show how to construct a smooth function which is one in the unit sphere, vanishes outside the sphere $x^2 + y^2 + z^2 = 2$, and is between zero and one between the two spheres.

9. Show, by sketching the appropriate rectangles, how a partition of unity would be made for the triangle with vertices at $(0, 0)$, $(1, 0)$, $(0, 1)$. Do not construct the functions analytically.

10. Show, by sketching the appropriate rectangles, how a partition of unity would be made for the ellipse $4x^2 + 9y^2 = 36$.

11. Define the function

$$f(x) = \begin{cases} 0, & \text{if } 0 \le x \le 1, \\ e^{1/(x-2)} \cdot e^{1/(x-1)}, & \text{if } 1 < x < 2, \\ 0, & \text{if } 2 \le x. \end{cases}$$

Show that f has (continuous) derivatives of all orders.

12. Let f be the function of Exercise 11. Define the function

$$F(x) = \frac{\displaystyle\int_x^2 f(t)\, dt}{\displaystyle\int_1^2 f(t)\, dt}.$$

Show that F is one for $0 \le x \le 1$, zero for $x \ge 2$, and between zero and one for $1 \le x \le 2$. Furthermore, show that F has derivatives of all orders.

13. Use the function in Problem 12 to find a function with derivatives of all orders which is one in a rectangle, vanishes outside a larger similarly placed rectangle, and is between zero and one in the region between the rectangles.

14. By considering $F(x^2 + y^2)$ where F is the function in Problem 12, show that F is one in a circle, zero outside a larger concentric circle, and between zero and one in the ring between the circles.

15. Describe a partition of unity with circles instead of rectangles.

3. PROOF OF GREEN'S THEOREM

To prove Green's Theorem with the appropriate generality, we must establish the lemma of Section 1 without the hypothesis that f is monotone.

Lemma 1′. Suppose that the hypotheses of Lemma 1 in Section 1 hold without the assumption that f is nonincreasing or nondecreasing. Then formulas (2a) *and* (2b) *of Section 1 are valid.*

Proof. First we observe that the proof of formula (2a), given on page 590, did not use the hypothesis that f is monotone. To establish (2b), we define

$$U(x, y) = \int_c^y Q(x, \eta)\, d\eta.$$

Differentiating and using Leibniz' Rule, we obtain

$$U_x(x, y) = \int_c^y Q_x(x, \eta)\, d\eta, \qquad U_y(x, y) = Q, \qquad U_{yx} = Q_x = U_{xy}.$$

Therefore, using Theorem 11 of Chapter 13, we may write

$$\int_{\partial G} (U_x \, dx + U_y \, dy) = 0 = \int_{\partial G} U_x \, dx + \int_{\partial G} Q \, dy.$$

We conclude [using formula (2a) in the process] that

$$\iint_G Q_x \, dx \, dy = \iint_G U_{xy} \, dy \, dx = -\int_{\partial G} U_x \, dx = \int_{\partial G} Q \, dy.$$

We now prove Green's Theorem. Let $\mathbf{v} = P(x, y)\mathbf{i} + Q(x, y)\mathbf{j}$ be a smooth vector field given in a region containing G. We define

$$\mathbf{v}_i = \psi_i \mathbf{v},$$

where $\psi_1, \psi_2, \ldots, \psi_k$ is a partition of unity, as described in the preceding section. Then it is clear that for all points Q in G,

$$\mathbf{v}(Q) = \sum_{i=1}^{k} \mathbf{v}_i(Q).$$

Green's Theorem will then follow for $\mathbf{v}$ if we prove it for each vector field $\mathbf{v}_i$. Recalling that r_i and R_i are the rectangles associated with ψ_i, we define

$$G_i = G \cap R_i.$$

If P_i, the center of R_i, is interior to G, then so is R_i; in this case $G_i = R_i$. It is apparent that G_i is a region of the special type discussed in Lemmas 1 and 1'. We have

$$\iint_{G_i} \text{curl } \mathbf{v}_i \, dA = \int_{\partial G} \mathbf{v}_i \cdot d\mathbf{r} = 0, \qquad (1)$$

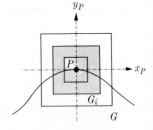

FIGURE 14-14

because $\mathbf{v}_i = \psi_i \mathbf{v}$ is $\mathbf{0}$ near and on the boundary ∂G_i. Therefore (1) holds for all those R_i which have centers in G. Now suppose that P_i is on the boundary of G. We see that $G_i = G \cap R_i$ is again a region of special type, and so Lemma 1' applies. Since $\mathbf{v}_i = \mathbf{0}$ outside G_i and on the part of ∂G_i (see Fig. 14-14) which is interior to G, we have

$$\iint_G \text{curl } \mathbf{v}_i \, dA = \iint_{G_i} \text{curl } \mathbf{v}_i \, dA = \int_{\partial G_i} \mathbf{v}_i \cdot d\mathbf{r} = \int_{\partial G} \mathbf{v}_i \cdot d\mathbf{r}.$$

The result holds in this case also, and so the theorem is established.

4. SURFACE ELEMENTS. PARAMETRIC REPRESENTATION

So far we have interpreted the term "surface in three-space" as the locus in a given rectangular coordinate system of an equation of the form $z = f(x, y)$ or $F(x, y, z) = 0$. The area of a surface of revolution was discussed in Chapter 16,

Section 7 of *First Course*; the definition and computation of areas of more general surfaces was taken up in Section 8 of Chapter 5 in this volume. Now we are interested in studying surfaces which have a structure much more complicated than those we have considered previously.

In order to simplify the study of surfaces, we decompose a given surface into a large number of small pieces and examine each piece separately. It may happen that a particular surface has an unusual and bizarre appearance and yet each piece has a fairly simple structure. In fact, for most surfaces we shall investigate, the "local" behavior will be much like a piece of a sphere, a hyperboloid, a cylinder, or some other smooth surface.

Some surfaces have boundaries and others do not. For example, a hemisphere has a boundary consisting of its equatorial rim. An entire sphere, an ellipsoid, and the surface of a cube are examples of surfaces without boundary.

The general definition given below of a surface without boundary makes use of what we call a **local coordinate system**. That is, at each point P of a surface, we construct a rectangular coordinate system, and we study a piece of the surface containing P by means of this local coordinate system. We make the convention once and for all that a positive orientation of space is selected, and so we allow right-handed coordinate systems only. The cross product of vectors, when it occurs, will be directed in agreement with a right-handed system.

DEFINITION. A surface S is a **smooth surface without boundary** if and only if, for each point P of S, the following conditions hold:

(i) With P there is associated a rectangular local coordinate system (x^P, y^P, z^P) having P as origin.

(ii) There exist two positive numbers r_P and k_P and a function $f^P(x^P, y^P)$ as described in the next two conditions.

(iii) The function f^P is defined and continuously differentiable (i.e., smooth) in a circle $(x^P)^2 + (y^P)^2 < R_P^2$, where R_P is some number larger than r_P. The function f is normalized so that $f^P(0, 0) = f_{,1}^P(0, 0) = f_{,2}^P(0, 0) = 0$. Since f is smooth, this normalization can always be made by adjusting the local coordinate system.

(iv) We denote by Z_P the cylindrical region

$$Z_P: \quad (x^P)^2 + (y^P)^2 < r_P^2,$$
$$-k_P < z^P < k^P.$$

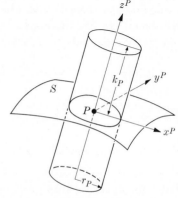

Then r_P and k_P can be selected so small that every point (x^P, y^P, z^P) which satisfies the equation $z^P = f^P(x^P, y^P)$ for $(x^P)^2 + (y^P)^2 < r_P^2$ lies on the surface S and is in Z_P. Conversely, every point of S in the cylinder Z_P is on the locus of the equation $z^P = f^P(x^P, y^P)$. (See Fig. 14–15.)

FIGURE 14–15

From the normalization conditions, it follows that the z^P-axis is normal to the surface S at P and that the tangent plane at P is the $x^P y^P$-coordinate plane. Roughly speaking, a surface is smooth without boundary if we can find a small circle about each point P where the surface has a representation in the form $z = f(x, y)$, with f continuously differentiable in the entire circle. The piece of the surface cut out by the cylinder Z_P must "cover" the circle. We call such a circle a **coordinate patch**.

In the definition of smooth surface *without* boundary, it is essential that we should be able to find a coordinate patch and a function f^P for *every* point P of S. In a hemisphere, for example, all points except those on the bounding equatorial circle satisfy the required conditions. For P on this circle, however, we cannot find a piece of the surface which covers an entire coordinate patch. The next definition describes a smooth surface with a smooth boundary, of which a hemisphere is an example.

DEFINITIONS.   A surface S is a **smooth surface with smooth boundary** if and only if with each point P of S we can associate a region G_P which satisfies either of the following conditions:

(i) G_P is a whole circle, as in the definition of surface without boundary, or
(ii) G_P is that part of a circle $(x^P)^2 + (y^P)^2 < r_P^2$ for which $y^P \leq g^P(x^P)$
(that is, $y^P = g^P$ is a curve in the $x^P y^P$-plane); furthermore,

$$g^P(0) = \frac{d}{dx^P} g^P(0) = 0.$$

(See Fig. 14–16.)   The locus of the equation $z^P = f(x^P, y^P)$ for (x^P, y^P) in G_P is the intersection of S with the cylinder Z_P. If alternative (i) holds, we say that P is an **interior point** of S; if (ii) holds (Fig. 14–16), P is called a **boundary point** of S. The intersection of S with the cylinder Z_P for a boundary point is shown in Fig. 14–17. The region G_P is called a **coordinate half-patch**.

We allow the **boundary of S to be piecewise smooth**, in which case the coordinate half-patch G_P has the appearance of Fig. 14–16 when the boundary point P is interior to a smooth boundary arc, and it has the appearance of Fig. 14–18 when P is at a corner of the boundary.

It is important to be able to decide when a surface is a smooth surface without boundary. Suppose that $F(P)$ is a smooth scalar field defined for P in some open set in three-space. We consider the set S (assumed closed and not empty) of all those points P which satisfy

$$S: \quad F(P) = 0.$$

If the vector field ∇F never vanishes on S, then S is a smooth surface without boundary. To see this, let P be any point on S, choose an (x, y, z) coordinate system with origin at P and with z-axis in the direction of ∇F, which we know is

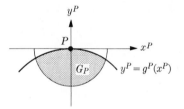

FIGURE 14–16

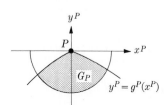

FIGURE 14–18

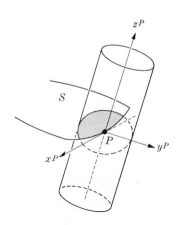

FIGURE 14–17

not **0**. Then $F(P) = F(x, y, z)$ is smooth near the origin and

$$\nabla F(P) = \frac{\partial F^*(0, 0, 0)}{\partial z} \mathbf{k} \neq \mathbf{0}.$$

We may therefore apply the Implicit Function Theorem (see Theorem 2′, page 488) and solve for z in terms of x and y, getting $z = h(x, y)$ in a neighborhood of the origin. The function h satisfies the conditions required of f^P, and the definition of smooth surface without boundary is satisfied.

EXAMPLE. Suppose that a smooth scalar field is defined in all of three-space by the formula

$$F(P) = x^2 + 4y^2 + 9z^2 - 44.$$

Show that the set $S: F(P) = 0$ is a smooth surface.

Solution. We compute the gradient:

$$\nabla F = 2x\mathbf{i} + 8y\mathbf{j} + 18z\mathbf{k}.$$

Then ∇F is zero only at $(0, 0, 0)$. Since S is not void [the point $(\sqrt{44}, 0, 0)$ is on it] and since $(0, 0, 0)$ is not a point of S, the surface is smooth without boundary.

An equation of the form $z - f(x, y) = 0$ is a special case of $F(P) = 0$. Consider the locus of points which satisfy such an equation when (x, y) is in a region G of the xy-plane with a piecewise smooth boundary; this locus forms a surface with boundary. If f is smooth in a region containing G and its boundary ∂G, then the locus is a smooth surface with piecewise smooth boundary. (See Fig. 14–19.)

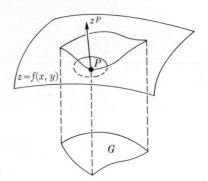

FIGURE 14–19

Under the appropriate assumptions, a system of equations

$$x = x(u, v), \qquad y = y(u, v), \qquad z = z(u, v) \tag{1}$$

represents a portion of a surface. It is convenient to use vector notation, and we write the above system in the form

$$\mathbf{v}(\overrightarrow{OQ}) = \mathbf{r}(u, v), \qquad O \text{ given,} \tag{2}$$

where (u, v) is in G, a region with smooth boundary in the u, v-plane. The quantities u and v are called **parameters**, and the system (1) or (2) which associates a point $Q(x_0, y_0, z_0)$ to each point $T(u_0, v_0)$ is called a **parametric representation** of a portion of a surface. Figure 14–20 shows the region G in the uv-plane, which is the plane of the parameters, and a portion of the surface S in xyz-space. In the vector formulation, we identify the vector $\mathbf{r}$ with the directed line segment from a fixed point O to a point Q on the surface. If $\mathbf{r}$ is one to one and smooth with

$$\mathbf{r}_u \times \mathbf{r}_v \neq \mathbf{0} \qquad \text{for } (u, v) \text{ in } G, \tag{3}$$

the range of (2) is called a **smooth surface element**. The subscripts in (3) denote partial differentiation. We observe that if $\mathbf{r}(u, v) = x(u, v)\mathbf{i} + y(u, v)\mathbf{j} + z(u, v)\mathbf{k}$, then

$$\mathbf{r}_u \times \mathbf{r}_v = \left(\frac{\partial y}{\partial u}\frac{\partial z}{\partial v} - \frac{\partial z}{\partial u}\frac{\partial y}{\partial v}\right)\mathbf{i} + \left(\frac{\partial z}{\partial u}\frac{\partial x}{\partial v} - \frac{\partial x}{\partial u}\frac{\partial z}{\partial v}\right)\mathbf{j} + \left(\frac{\partial x}{\partial u}\frac{\partial y}{\partial v} - \frac{\partial y}{\partial u}\frac{\partial x}{\partial v}\right)\mathbf{k},$$

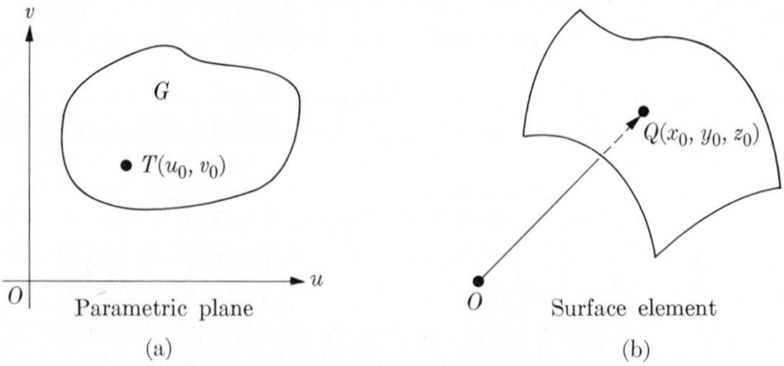

G

$\bullet \, T(u_0, v_0)$

O Parametric plane

(a)

$Q(x_0, y_0, z_0)$

O Surface element

(b)

FIGURE 14–20

which becomes, in Jacobian notation,

$$\mathbf{r}_u \times \mathbf{r}_v = \frac{\partial(y, z)}{\partial(u, v)} \mathbf{i} + \frac{\partial(z, x)}{\partial(u, v)} \mathbf{j} + \frac{\partial(x, y)}{\partial(u, v)} \mathbf{k}.$$

In the preceding definition (page 606) of a smooth surface element, it is clear that each of the loci $z^P = f^P(x^P, y^P)$ for (x^P, y^P) in G_P is a parametric representation of a surface element with

$$x^P = u, \qquad y^P = v, \qquad f^P(x^P, y^P) = f^P(u, v). \tag{4}$$

That is, the parameters u and v are x^P and y^P. In vector symbols, we write

$$\mathbf{v}(\overrightarrow{OQ}) = (x_Q^P - x_0^P)\mathbf{i}^P + (y_Q^P - y_0^P)\mathbf{j}^P + [f^P(x_Q^P, y_Q^P) - z_0^P]\mathbf{k}^P, \tag{5}$$

where $\mathbf{i}^P, \mathbf{j}^P, \mathbf{k}^P$ are unit vectors in the local coordinate system and (x_Q^P, y_Q^P, z_Q^P) and (x_0^P, y_0^P, z_0^P) are the coordinates of Q and O, respectively, in this system. The parametric representation (2) is more general than the special form given in (5).

The next theorem establishes the relation between two parametric representations of the same smooth surface element.

Theorem 2. *Suppose that the transformation (2) defines a smooth surface element for (u, v) in an open region D which contains G and ∂G. Let S and S^* denote the surfaces which are the images under (2) of G and D, respectively. Let (x, y, z) be any rectangular coordinate system. Then*

(a) *if (u_0, v_0) is any point of G, there is a positive number ρ such that the part of S^* corresponding to the square*

$$|u - u_0| < \rho, \qquad |v - v_0| < \rho$$

is of one of the forms

$$z = f(x, y), \qquad x = g(y, z), \qquad or \quad y = h(x, z),$$

where f, g, or h (as the case may be) is smooth near the point (x_0, y_0, z_0) corresponding to (u_0, v_0).

(b) *Suppose that another parametric representation of S and S^* is given by*

$$\mathbf{v}(\overrightarrow{OP}) = \mathbf{r}_1(s, t),$$

with S and S^* the images of G_1 and D_1, respectively, in the (s, t)-plane. Suppose that G_1, D_1, and $\mathbf{r}_1$ have all the properties which G, D, and $\mathbf{r}$ have. Then there is a one-to-one continuously differentiable transformation

$$T: \quad u = U(s, t), \qquad v = V(s, t), \qquad (s, t) \text{ on } D_1,$$

from D_1 to D such that $T(G_1) = G$ and

$$\mathbf{r}[U(s, t), V(s, t)] = \mathbf{r}_1(s, t) \qquad for \ (s, t) \text{ on } D_1.$$

Proof. (a) Since $\mathbf{r}_u \times \mathbf{r}_v \neq \mathbf{0}$ at a point (u_0, v_0), we conclude from the formula

$$\mathbf{r}_u \times \mathbf{r}_v = \frac{\partial(y, z)}{\partial(u, v)}\mathbf{i} + \frac{\partial(z, x)}{\partial(u, v)}\mathbf{j} + \frac{\partial(x, y)}{\partial(u, v)}\mathbf{k} \tag{6}$$

that at least one of the three Jacobians is not zero at (u_0, v_0). Suppose, for instance, that it is

$$\frac{\partial(x, y)}{\partial(u, v)},$$

which is not zero. We set

$$\mathbf{r}(u, v) = X(u, v)\mathbf{i} + Y(u, v)\mathbf{j} + Z(u, v)\mathbf{k},$$

and $x_0 = X(u_0, v_0)$, $y_0 = Y(u_0, v_0)$. Then, from the Implicit Function Theorem, it follows that there are positive numbers h and k such that all numbers x, y, u, v for which

$$|x - x_0| < h, \qquad |y - y_0| < h, \qquad |u - u_0| < k, \qquad |v - v_0| < k,$$

and $x = X(u, v)$, $y = Y(u, v)$ lie along the locus of $u = \phi(x, y)$, $v = \psi(x, y)$, where ϕ and ψ are smooth in the square $|x - x_0| < h$, $|y - y_0| < h$. In this case, the part of S^* near (x_0, y_0, z_0) is the locus of

$$z = Z[\phi(x, y), \psi(x, y)], \qquad |x - x_0| < h, \qquad |y - y_0| < h.$$

The conclusion (a) follows when we select $\rho > 0$ small enough so that the image of the square $|u - u_0| < \rho$, $|v - v_0| < \rho$ lies in the (x, y)-square above.

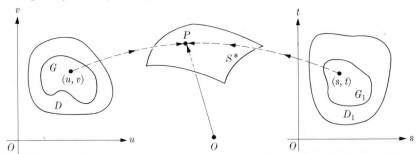

FIGURE 14–21

(b) Since $\mathbf{r}(u, v)$ and $\mathbf{r}_1(s, t)$ are one to one, it follows that to each (s, t) in D_1 there corresponds a unique P on S^* which comes from a unique (u, v) in D (Fig. 14–21). If we define $U(s, t) = u$ and $V(s, t) = v$ by this correspondence, then T is one to one. To see that it is smooth, let (s_0, t_0) be any point in D_1 and let $(u_0, v_0) = T(s_0, t_0)$. Then at least one of the three Jacobians in (6) does not vanish. If, for example, the last one does not vanish, we can solve for u and v in terms of x and y as in part (a). We now set

$$\mathbf{r}_1(s, t) = X_1(s, t)\mathbf{i} + Y_1(s, t)\mathbf{j} + Z_1(s, t)\mathbf{k}.$$

Since there is a one-to-one correspondence between the points P near P_0 and the points (x, y) "below" them, we see that

$$U(s, t) = \phi[X_1(s, t), Y_1(s, t)], \qquad V(s, t) = \psi[X_1(s, t), Y_1(s, t)].$$

Hence U and V are smooth near (s_0, t_0), an arbitrary point of G_1.

Remark. Of course, T^{-1}, the inverse transformation, has the same smoothness properties.

It is desirable to enlarge the class of surfaces we shall consider to include **piece-wise smooth** surfaces. Roughly speaking, a piecewise smooth surface consists of portions of smooth surfaces joined together. The surface of a cube or other poly-hedron is a simple example of a piecewise smooth surface. If two smooth surfaces

$$F(Q) = 0 \qquad \text{and} \qquad G(Q) = 0$$

intersect at a point P_0 where the surfaces have distinct tangent planes, then

$$\nabla F(P_0) \times \nabla G(P_0) \neq \mathbf{0}.$$

Introducing an (x, y, z)-coordinate system, we notice that this condition reduces to

$$\frac{\partial(F, G)}{\partial(y, z)} \mathbf{i} + \frac{\partial(F, G)}{\partial(z, x)} \mathbf{j} + \frac{\partial(F, G)}{\partial(x, y)} \mathbf{k} \neq \mathbf{0} \qquad \text{at } (x_0, y_0, z_0).$$

Therefore at least one of the above three Jacobians is not zero, and so the Implicit Function Theorem (Theorem 3A, page 492) may be employed. We conclude that the part of the intersection of the two surfaces near P_0 is a smooth arc. Such an arc on a piecewise smooth surface is called an **edge**. A point of intersection of two or more edges is called a **vertex**. A more precise definition of piecewise smooth surface is given next.

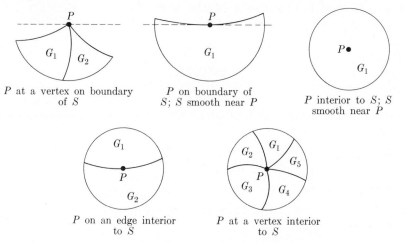

P at a vertex on boundary of S

P on boundary of S; S smooth near P

P interior to S; S smooth near P

P on an edge interior to S

P at a vertex interior to S

FIGURE 14–22

DEFINITION. A surface S is **piecewise smooth** if and only if for each point P of S we can associate a coordinate system (x^P, y^P, z^P), two positive numbers r_P and k_P, a region G_P in the (x^P, y^P)-plane, and a function $f^P(x^P, y^P)$. The region G_P is any one of the three types already described. All points of S in the cylinder Z: $(x^P)^2 + (y^P)^2 < r_P^2$, $-k_P < z^P < k_P$ lie on the locus of $z^P = f^P(x^P, y^P)$ with (x^P, y^P) in G_P. The function f^P is continuous, normalized so that $f^P(0, 0) = 0$, and piecewise smooth in the circle

$$(x^P)^2 + (y^P)^2 < R_P^2$$

for some $R_P > r_P$. We say that f^P is piecewise smooth on G_P if and only if G_P can be divided up into regions $G_1, G_2, \ldots, G_k$, as shown in Fig. 14–22, the indicated arcs in each G_P being smooth. The function f^P is representable in the form

$$f^P(x^P, y^P) = f^{P_i}(x^P, y^P),$$

where f^{P_i} is smooth on an open set containing G_i.

We show that the surface of a cube is a piecewise smooth surface without boundary. If P is on one of the faces, we may take the z^P-axis as the outer normal to that face (Fig. 14–23); we select $k_P = l$ (the edge of the cube) and r_P smaller than the radius of the largest circle on that face with center at P. If P is a vertex, the z^P-axis may be taken along the ray CP, C being the center of the cube, r_P and k_P being taken small enough so that the part of the cylinder of radius r_P and axis the line CP for which $-k_P < z^P < k_P$ intersects only the edges and faces adjacent to P (Fig. 14–23). If P is on an edge, the construction is analogous.

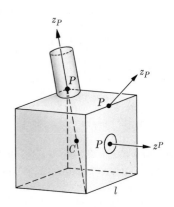

FIGURE 14–23

EXERCISES

In each of problems 1 through 4, a scalar field F is given in terms of a given coordinate system. Decide whether or not the locus of $F(P) = 0$ is a smooth surface without boundary.

1. $F(P) = 4x^2 + 9y^2 - 2z^2 - 8$
2. $F(P) = x^2 - 3y^2 - (z - 2)^2 - 4$
3. $F(P) = x^2 + 4y^2 - z^2$
4. $F(P) = x^2 + 2y^2 - z^2 - 6x + 2z + 8$
5. Given the surfaces

$$F \equiv x^2 + y^2 + z^2 - 25 = 0, \qquad G \equiv x^2 + 4y^2 + 4z^2 - 52 = 0,$$

show that the tangent planes at any intersection point are distinct. Describe a piecewise smooth surface composed of portions of $F = 0$ and $G = 0$.

6. Let $f(x, y) = 0$ be the equation of a smooth simple closed arc in the x, y-plane. Show that the cylinder $f(x, y, z) \equiv f(x, y) = 0$ is a smooth surface without boundary.

7. Let S be the surface of a regular tetrahedron. Sketch, as in Fig. 14–22, the various types of regions G_P corresponding to a point on a face, a point on an edge, and a point at a vertex.

8. Let S be the surface of a pyramid with a square base. Sketch, as in Fig. 14–22, the various types of regions G_P corresponding to a point on a face, a point on an edge, and a point at a vertex.

9. Let S be the *lateral surface* of a pyramid with a regular pentagon as a base. Show that this surface is a piecewise smooth surface with piecewise smooth boundary. Sketch, as in Fig. 14–22, the various types of regions which may occur, both interior to the surface and at boundary points.

5. AREA OF A SURFACE. SURFACE INTEGRALS

Suppose that for (x, y) in a region G of the xy-plane, the function $z = f(x, y)$ represents a portion S of a surface in three-space. The area of a surface was defined in Chapter 5, Section 8; we saw there that if f has continuous first derivatives, the area $A(S)$ can be calculated by the formula

$$A(S) = \iint_G \sqrt{1 + [f_{,1}(x, y)]^2 + [f_{,2}(x, y)]^2} \, dA.$$

In general, a surface in three-space does not have a representation of the form $z = f(x, y)$, and we now take up the problem of defining and computing the area of a surface which is given in the parametric form*

$$\mathbf{v}(\overrightarrow{OQ}) = \mathbf{r}(u, v), \qquad (u, v) \text{ in } G. \tag{1}$$

Suppose that σ is a smooth surface element represented by (1) with r and G satisfying all the conditions required in the definition of such an element. If v is held constant, equal to v_0, say, then the locus of (1) is a smooth curve on the surface element σ. Therefore the vector $\mathbf{r}_u(u, v_0)$ is a vector tangent to the curve on the surface. Similarly, the vector $\mathbf{r}_v(u_0, v)$ is tangent to the smooth curve on the surface element σ obtained when u is set equal to the constant u_0 (Fig. 14–24). The vectors $\mathbf{r}_u(u_0, v_0)$ and $\mathbf{r}_v(u_0, v_0)$ lie in the plane tangent to the surface point P_0 corresponding to $\mathbf{r}(u_0, v_0)$. Since the cross prod-

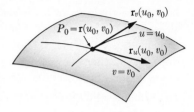

FIGURE 14–24

---

* A brief study of the terms used in Section 4 is sufficient for an understanding of this section.

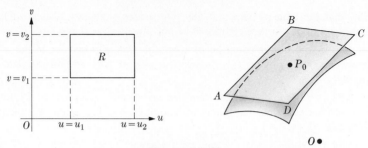

FIGURE 14–25

uct of two vectors is orthogonal to each of them, it follows that *the vector* $\mathbf{r}_u(u_0, v_0) \times \mathbf{r}_v(u_0, v_0)$ *is a vector normal to the surface at* P_0.

For convenience, we write $\mathbf{a} = \mathbf{r}_u(u_0, v_0)$, $\mathbf{b} = \mathbf{r}_v(u_0, v_0)$, and we consider the **tangent linear transformation** defined by

$$\mathbf{v}(\overrightarrow{OP}) = \mathbf{r}(u_0, v_0) + (u - u_0)\mathbf{a} + (v - v_0)\mathbf{b}. \tag{2}$$

We notice that the image of the rectangle $R: u_1 \le u \le u_2, v_1 \le v \le v_2$ under (2) is a parallelogram $ABCD$ in the plane tangent to the surface at $P_0 = \mathbf{r}(u_0, v_0)$ (Fig. 14–25). The points A, B, C, and D are determined by

$$\mathbf{v}(\overrightarrow{OA}) = \mathbf{r}(u_0, v_0) + (u_1 - u_0)\mathbf{a} + (v_1 - v_0)\mathbf{b},$$
$$\mathbf{v}(\overrightarrow{OB}) = \mathbf{r}(u_0, v_0) + (u_2 - u_0)\mathbf{a} + (v_1 - v_0)\mathbf{b},$$
$$\mathbf{v}(\overrightarrow{OC}) = \mathbf{r}(u_0, v_0) + (u_1 - u_0)\mathbf{a} + (v_2 - v_0)\mathbf{b},$$
$$\mathbf{v}(\overrightarrow{OD}) = \mathbf{r}(u_0, v_0) + (u_2 - u_0)\mathbf{a} + (v_2 - v_0)\mathbf{b}.$$

Using vector subtraction, we find that

$$\mathbf{v}(\overrightarrow{AB}) = (u_2 - u_1)\mathbf{a}, \qquad \mathbf{v}(\overrightarrow{AC}) = (v_2 - v_1)\mathbf{b}.$$

Denoting the area of R by $A(R)$, we use the definition of cross product to obtain

$$\text{Area } ABCD = |\mathbf{v}(\overrightarrow{AB}) \times \mathbf{v}(\overrightarrow{AC})| = |(u_2 - u_1)(v_2 - v_1)\mathbf{a} \times \mathbf{b}|$$
$$= A(R)|\mathbf{r}_u(u_0, v_0) \times \mathbf{r}_v(u_0, v_0)|.$$

In defining the area of a surface element, we would expect that for a very small piece near P_0, the area of $ABCD$ would be a good approximation to the (as yet undefined) area of the portion of the surface which is the image of R. We use this fact to define the area of a surface element F. We subdivide the region G in the uv-plane into a number of subregions $G_1, G_2, \ldots, G_k$ and define the norm $\|\Delta\|$ as the maximum diameter of any of the subregions. We define

$$\text{Area of } F = \lim_{\|\Delta\| \to 0} \sum_{i=1}^{n} A(G_i)|\mathbf{r}_u(u_i, v_i) \times \mathbf{r}_v(u_i, v_i)|,$$

where (u_i, v_i) is any point in G_i and where the limit has the usual interpretation as given in the definition of integral. Therefore the definition of area of F is

$$A(F) = \iint\limits_{G} |\mathbf{r}_u(u, v) \times \mathbf{r}_v(u, v)| \, dA_{uv}. \tag{3}$$

We immediately raise the following question concerning the use of parameters. Suppose that the same smooth surface element F has another parametric representation

$$\mathbf{r}' = \mathbf{r}'(s, t) \qquad \text{for } (s, t) \text{ in } G'.$$

Is it true that the formula

$$A(F) = \iint\limits_{G'} |\mathbf{r}'_s(s, t) \times \mathbf{r}'_t(s, t)| \, dA_{st} \tag{4}$$

gives the same value as formula (3)? Because of the rule for change of variable in a multiple integral (Theorem 9, Chapter 11, page 517) and the rule for the product of Jacobians

$$\frac{\partial(y, x)}{\partial(s, t)} = \frac{\partial(y, x)}{\partial(u, v)} \frac{\partial(u, v)}{\partial(s, t)},$$

it follows that (3) and (4) yield the same value. In fact, we have

$$|\mathbf{r}'_s(s, t) \times \mathbf{r}'_t(s, t)| = |\mathbf{r}_u(u, v) \times \mathbf{r}_v(u, v)| \cdot \left|\frac{\partial(u, v)}{\partial(s, t)}\right|.$$

If the representation is the simple one $z = f(x, y)$ discussed at the beginning of the section, we may set $x = u$, $y = v$, $z = f(u, v)$, and find

$$\frac{\partial(y, z)}{\partial(u, v)} = -f_u, \qquad \frac{\partial(z, x)}{\partial(u, v)} = -f_v, \qquad \frac{\partial(x, y)}{\partial(u, v)} = 1.$$

Then (3) becomes

$$A(F) = \iint\limits_{G} \sqrt{1 + (f_u)^2 + (f_v)^2} \, dA_{uv}, \tag{5}$$

which is the formula we established in Section 8, Chapter 5.

Unfortunately, not every surface can be covered by a single smooth surface element, and so (5) cannot be used exclusively for the computation of surface area. In fact, it can be shown that even a simple surface such as a sphere cannot be part of a single smooth surface element. (See, however, Example 2 below.) Consequently, to find the area of a portion of a smooth surface F, we must show that F is the union of a finite number of nonoverlapping pieces of the surface F_1, $F_2, \ldots, F_k$, each F_i being in some surface element σ_i. The area $A(F_i)$ in each element may be found by means of formula (3). The area $A(F)$ is then the sum of the areas $A(F_i)$. It is not difficult to see that two different decompositions of F, each into a finite number of pieces, yield the same result for $A(F)$.

If F is a portion of a surface contained in a smooth surface element σ given by (1), then F is the image of some region G in the uv-plane. Suppose that $f(P)$ is a continuous scalar field defined for P on F. We can define the integral of f taken over the portion of the surface element F. We define

$$\iint_F f(Q)\, dS = \iint_G f[Q(u, v)] \cdot |\mathbf{r}_u \times \mathbf{r}_v|\, dA_{uv}. \tag{6}$$

In other words, we define the integral over a surface in terms of a double integral taken over a plane region G.

In general, if F is not contained in a single surface element, we make a subdivision $F_1, F_2, \ldots, F_k$ of F, each F_i being in a single smooth surface element σ_i. If σ_i is given by $\mathbf{v}(\overline{OQ}) = \mathbf{r}^{(i)}(u, v)$, and if f is continuous on F, we define

$$\iint_F f(Q)\, dS = \sum_{i=1}^{k} \iint_{F_i} f(Q)\, dS. \tag{7}$$

It can be shown that two different subdivisions yield the same value for the integral in (7). Therefore the definition is consistent. Usually a number of parametric representations are needed to evaluate the integrals on the right in (7).

If a surface element σ has the representation $z = \phi(x, y)$ in a given coordinate system, then (6) becomes

$$\iint_F f(Q)\, dS = \iint_G f[x, y, \phi(x, y)]\sqrt{1 + \phi_x^2 + \phi_y^2}\, dA_{xy}, \tag{8}$$

with corresponding formulas in the cases $x = \phi(y, z)$ or $y = \phi(x, z)$. The evaluation of the integral on the right in (8) follows the usual rules for evaluation of double integrals which we studied earlier. Some examples illustrate the technique.

EXAMPLE 1. Evaluate $\iint_F z^2\, dS$, where F is the part of the lateral surface of the cylinder $x^2 + y^2 = 4$ between the planes $z = 0$ and $z = x + 3$.

Solution. (See Fig. 14–26.) If we transform to cylindrical coordinates r, θ, z, then F lies on the surface $r = 2$. We may choose θ and z as parametric coordinates on F and write

F: $x = 2\cos\theta$, $\quad y = 2\sin\theta$, $\quad z = z$, $\quad (\theta, z)$ in G,

G: $-\pi \le \theta \le \pi$, $\quad 0 \le z \le 3 + 2\cos\theta$.

See Fig. 14–27. The element of surface area dS is given by

$$dS = |\mathbf{r}_\theta \times \mathbf{r}_z|\, dA_{\theta z},$$

and since

$$\mathbf{r}_\theta \times \mathbf{r}_z = \frac{\partial(y, z)}{\partial(\theta, z)}\mathbf{i} + \frac{\partial(z, x)}{\partial(\theta, z)}\mathbf{j} + \frac{\partial(x, y)}{\partial(\theta, z)}\mathbf{k}$$

$$= (2\cos\theta)\mathbf{i} + (2\sin\theta)\mathbf{j} + 0 \cdot \mathbf{k},$$

we have

$$dS = 2\, dA_{\theta z}.$$

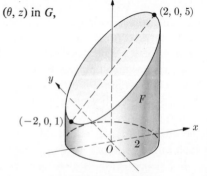

FIGURE 14–26

Then

$$\iint_F z^2 \, dS = 2 \iint_G z^2 \, dA_{\theta z} = 2 \int_{-\pi}^{\pi} \int_0^{3+2\cos\theta} z^2 \, dz \, d\theta$$

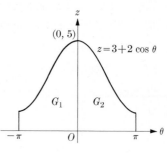

$$= \frac{2}{3} \int_{-\pi}^{\pi} (3 + 2\cos\theta)^3 \, d\theta$$

$$= \frac{2}{3} \int_{-\pi}^{\pi} (27 + 54\cos\theta + 36\cos^2\theta$$

$$+ 8\cos^3\theta) \, d\theta$$

$$= \tfrac{2}{3}(54\pi + 36\pi) = 60\pi. \tag{9}$$

FIGURE 14–27

Remark. Strictly speaking, the whole of F is not a smooth surface element, since the transformation from G to F shows that the points $(-\pi, z)$ and (π, z) of G are carried into the same points on F. For a smooth surface element, the condition that the transformation be one to one is therefore violated. However, if we subdivide G into G_1 and G_2 as shown in Fig. 14–27, the images are smooth surface elements. The evaluation as given in (9) is unaffected.

Surface integrals can be used to make approximate computations of various physical quantities. The center of mass and the moment of inertia of a thin curvilinear plate are sometimes computable in the form of surface integrals. The potential due to a distribution of an electric charge over a surface may be expressed in the form of an integral. (See Exercises 18 through 21 at the end of this section.) With each surface F we may associate a mass (assuming it to be made of a thin material), and this mass may be given by a density function δ. The density will be assumed continuous but not necessarily constant. The total mass $M(F)$ of a surface F is given by

$$M(F) = \iint_F \delta(P) \, dS.$$

The formula for the moment of inertia of F about the z-axis becomes

$$I_z = \iint_F \delta(Q)(x_Q^2 + y_Q^2) \, dS,$$

with corresponding formulas for I_x and I_y. The formula for the center of mass is analogous to those which we studied earlier. (See Chapter 5, Section 7, and also Example 3 below.)

EXAMPLE 2. Find the moment of inertia about the x-axis of the part of the surface of the unit sphere $x^2 + y^2 + z^2 = 1$ which is above the cone $z^2 = x^2 + y^2$. Assume $\delta = $ const.

Solution. (See Fig. 14–28.) In spherical coordinates (ρ, ϕ, θ), the portion F of the surface is given by $\rho = 1$. Then (ϕ, θ) are parametric coordinates and

$$F: \quad x = \sin\phi\cos\theta, \qquad y = \sin\phi\sin\theta, \qquad z = \cos\phi, \qquad \text{with } (\phi, \theta) \text{ in } G:$$

$$G: \quad 0 \le \phi \le \pi/4, \qquad 0 \le \theta \le 2\pi.$$

We compute $\mathbf{r}_\phi \times \mathbf{r}_\theta$, getting

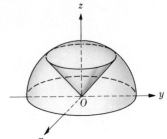

$$\mathbf{r}_\phi \times \mathbf{r}_\theta = \frac{\partial(y, z)}{\partial(\phi, \theta)}\mathbf{i} + \frac{\partial(z, x)}{\partial(\phi, \theta)}\mathbf{j} + \frac{\partial(x, y)}{\partial(\phi, \theta)}\mathbf{k}$$

$$= (\sin^2\phi\cos\theta)\mathbf{i} + (\sin^2\phi\sin\theta)\mathbf{j}$$

$$+ (\sin\phi\cos\phi)\mathbf{k}.$$

We obtain

$$dS = |\mathbf{r}_\phi \times \mathbf{r}_\theta| \, dA_{\phi\theta} = \sin\phi \, dA_{\phi\theta}.$$

Figure 14–28

Then

$$I_z = \iint_G (y^2 + z^2)\,\delta\sin\phi\,dA_{\phi\theta} = \delta\int_0^{\pi/4}\int_0^{2\pi}(\sin^2\phi\sin^2\theta + \cos^2\phi)\sin\phi\,d\theta\,d\phi$$

$$= \pi\delta\int_0^{\pi/4}(\sin^2\phi + 2\cos^2\phi)\sin\phi\,d\phi$$

$$= \pi\delta\int_0^{\pi/4}(1 + \cos^2\phi)\sin\phi\,d\phi = \pi\delta\left[-\cos\phi - \tfrac{1}{3}\cos^3\phi\right]_0^{\pi/4}$$

$$= \frac{\pi\delta}{12}(16 - 7\sqrt{2}).$$

Remark. The parametric representation of F by spherical coordinates does not fulfill the conditions required for such representations, as given in Theorem 2, since $|\mathbf{r}_\phi \times \mathbf{r}_\theta| = \sin\phi$ vanishes for $\phi = 0$. However, the same surface, with a small hole cut out around the z-axis, is of the required type. A limiting process in which the size of the hole tends to zero yields the above result for the moment of inertia I_z.

EXAMPLE 3. Let R be the region in three-space bounded by the cylinder $x^2 + y^2 = 1$ and the planes $z = 0$, $z = x + 2$. Evaluate $\iint_F x\,dS$, where F is the surface made up of the entire boundary of R (Fig. 14–29). If F is made of thin material of uniform density δ, find its mass. Also, find $\bar{x}$, the x-coordinate of the center of mass.

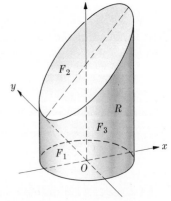

Solution. As shown in Fig. 14–29, the surface is composed of three parts: F_1, the circular base in the xy-plane; F_2, the lateral surface of the cylinder; and F_3, the part of the plane $z = x + 2$ inside the cylinder. We have

$$\iint_{F_1} x\,dS = \iint_{G_1} x\,dA_{xy} = 0,$$

Figure 14–29

where

$$G_1: \quad x^2 + y^2 \le 1, \quad z = 0$$

is a circle. On F_3, we see that $z = f(x, y) = x + 2$, so that $dS = \sqrt{2}\, dA_{xy}$ and

$$\iint_{F_3} x\, dS = \sqrt{2} \iint_{G_1} x\, dA_{xy} = 0.$$

On F_2, we choose coordinates (θ, z) as in Example 1 and obtain $x = \cos\theta$, $y = \sin\theta$, $z = z$, $dS = dA_{\theta z}$. We write

$$\iint_{F_2} x\, dS = \iint_{G_2} \cos\theta\, dA_{\theta z} = \int_{-\pi}^{\pi} \int_0^{2+\cos\theta} \cos\theta\, dz\, d\theta,$$

where

$$G_2: \quad -\pi \le \theta \le \pi, \quad 0 \le z \le 2 + \cos\theta.$$

Therefore

$$\iint_{F_2} x\, dS = \int_{-\pi}^{\pi} (2\cos\theta + \cos^2\theta)\, d\theta = \pi.$$

We conclude that

$$\iint_F x\, dS = \iint_{F_1+F_2+F_3} x\, dS = \pi.$$

We observe that the surface F is a piecewise smooth surface without boundary. There are no vertices on F, but there are two smooth edges, namely the intersections of $x^2 + y^2 = 1$ with the planes $z = 0$ and $z = x + 2$. To compute the mass $M(F)$, we have

$$M(F) = \iint_F \delta\, dS = \delta A(F) = \delta[A(F_1) + A(F_2) + A(F_3)].$$

Clearly, $A(F_1) = \pi$, $A(F_3) = \pi\sqrt{2}$. To find $A(F_2)$, we write

$$A(F_2) = \iint_{G_2} dA_{\theta z} = \int_{-\pi}^{\pi} \int_0^{2+\cos\theta} dz\, d\theta = \int_{-\pi}^{\pi} (2 + \cos\theta)\, d\theta = 4\pi.$$

Therefore $M(F) = \delta[\pi + 4\pi + \pi\sqrt{2}] = \pi(5 + \sqrt{2})\,\delta$. We use the formula

$$\bar{x} = \frac{\displaystyle\iint_F \delta x\, dA}{M(F)} \quad \text{to get} \quad \bar{x} = \frac{\delta\pi}{\delta\pi(5 + \sqrt{2})} = \frac{1}{5 + \sqrt{2}}.$$

EXERCISES

In each of problems 1 through 9, evaluate

$$\iint_F f(x, y, z)\, dS.$$

1. $f(x, y, z) = x$, F is the part of the plane $x + y + z = 1$ in the first octant.
2. $f(x, y, z) = x^2$, F is the part of the plane $z = x$ inside the cylinder $x^2 + y^2 = 1$.

3. $f(x, y, z) = x^2$, F is the part of the cone $z^2 = x^2 + y^2$ between the planes $z = 1$ and $z = 2$.

4. $f(x, y, z) = x^2$, F is the part of the cylinder $z = x^2/2$ cut out by the planes $y = 0$, $x = 2$, and $y = x$.

5. $f(x, y, z) = xz$, F is the part of the cylinder $x^2 + y^2 = 1$ between the planes $z = 0$ and $z = x + 2$.

6. $f(x, y, z) = x$, F is the part of the cylinder $x^2 + y^2 = 2x$ between the lower and upper nappes of the cone $z^2 = x^2 + y^2$.

7. $f(x, y, z) = 1$, F is the part of the vertical cylinder erected on the spiral $r = \theta$, $0 \le \theta \le \pi/2$ (polar coordinates in the xy-plane), bounded below by the xy-plane and above by the cone $z^2 = x^2 + y^2$.

8. $f(x, y, z) = x^2 + y^2 - 2z^2$, F is the surface of the sphere $x^2 + y^2 + z^2 = a^2$.

9. $f(x, y, z) = x^2$, F is the total boundary of the region R in three-space bounded by the cone $z^2 = x^2 + y^2$ and the planes $z = 1$, $z = 2$. (See problem 3.)

In each of problems 10 through 14, find the moment of inertia of F about the indicated axis, assuming that $\delta = $ const.

10. The surface F of problem 3; x-axis.

11. The surface F of problem 6; x-axis.

12. The surface F of problem 7; z-axis.

13. The total surface of the tetrahedron L bounded by the coordinate planes and the plane $x + y + z = 1$, having vertices $B(1, 0, 0)$, $E(0, 1, 0)$, $R(0, 0, 1)$, and $S(0, 0, 0)$; y-axis.

14. The torus $(r - b)^2 + z^2 = a^2$, $0 < a < b$; z-axis. Note that

$$r^2 = x^2 + y^2$$

and that we may introduce parameters

$$x = (b + a \cos \phi) \cos \theta,$$
$$y = (b + a \cos \phi) \sin \theta,$$
$$z = a \sin \phi,$$

with the torus swept out by $0 \le \phi \le 2\pi$, $0 \le \theta \le 2\pi$. (See Fig. 14-30).

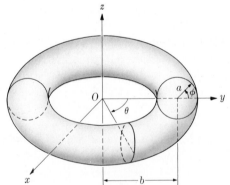

FIGURE 14-30

In each of problems 15 through 17, find the center of mass, assuming $\delta = $ const.

15. F is the surface of Example 2.   16. F is the surface of problem 13.

17. F is the part of the sphere $x^2 + y^2 + z^2 = 4a^2$ inside the cylinder $x^2 + y^2 = 2$

The electrostatic potential $V(Q)$ at a point Q due to a distribution of charge (with charge density δ) on a surface F is given by

$$V(Q) = \iint\limits_{F} \frac{\delta(P)\, dS}{|PQ|},$$

where $|PQ|$ is the distance from Q in space not on F to a point P on the surface F. In problems 18 through 21, find $V(Q)$ at the point given, assuming δ constant.

18. $Q = (0, 0, 0)$, F is the part of the cylinder $x^2 + y^2 = 1$ between the planes $z = 0$ and $z = 1$.

19. $Q = (0, 0, c)$; F is the surface of the sphere $x^2 + y^2 + z^2 = a^2$. Do two cases: (i), $c > a > 0$ and (ii), $a > c > 0$.

20. $Q = (0, 0, c)$; F is the upper half of the sphere $x^2 + y^2 + z^2 = a^2$; $0 < c < a$.

21. $Q = (0, 0, 0)$; F is the surface of problem 3.

6. ORIENTABLE SURFACES

Suppose S is a smooth surface element represented by the equation

$$\mathbf{v}(\overrightarrow{OQ}) = \mathbf{r}(u, v) \qquad \text{with } (u, v) \text{ in } G, \tag{1}$$

where G is a region in the (u, v)-plane with piecewise smooth boundary ∂G. According to the definition of smooth surface element, we know that $\mathbf{r}_u \times \mathbf{r}_v \neq \mathbf{0}$ for (u, v) in a domain D containing G and its boundary. Therefore we can define the **unit normal to the surface** S by the formula (Fig. 14–31)

$$\mathbf{n} = \frac{\mathbf{r}_u \times \mathbf{r}_v}{|\mathbf{r}_u \times \mathbf{r}_v|}, \qquad (u, v) \text{ in } G. \tag{2}$$

Whenever S is a smooth surface element, the vector $\mathbf{n}$ is a continuous function of u and v. We say that a smooth surface S is an **orientable surface** if coordinate patches can be chosen for points of S so that the unit normal $\mathbf{n}(P)$ as defined above is a continuous function of P over the entire surface S. Thus every smooth surface element is orientable. If $\mathbf{n}$ is a unit normal to a surface, then $-\mathbf{n}$ is a unit normal pointing in the opposite direction. Therefore each orientable surface has two possible **orientations**. If an orientable surface is oriented by taking $\mathbf{n}$ as in (2), we say that the orientation is *positive*; the oppositely directed normal gives a surface a *negative* orientation.

It is easy to see that surfaces such as spheres, ellipsoids, right circular (infinite) cylinders, and so forth, are all orientable. However, there are smooth surfaces for which there is no way to choose a continuously varying normal $\mathbf{n}(P)$ over the

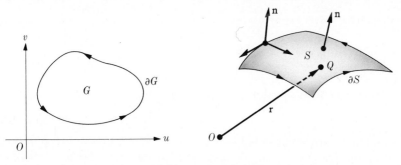

FIGURE 14–31

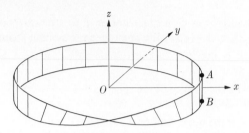

FIGURE 14–32

whole surface. One such surface is the **Möbius** strip drawn in Fig. 14–32. The student can make a model of such a surface from a long, narrow, rectangular strip of paper by giving one end a half-twist and then gluing the ends together. A Möbius strip can be represented parametrically on a rectangle $R: 0 \leq \theta \leq 2\pi$, $-h \leq s \leq h$ (Fig. 14–33) by the equations

$$x = \left(a + s \sin \frac{\theta}{2}\right) \cos \theta, \qquad y = \left(a + s \sin \frac{\theta}{2}\right) \sin \theta, \qquad z = s \cos \frac{\theta}{2}. \quad (3)$$

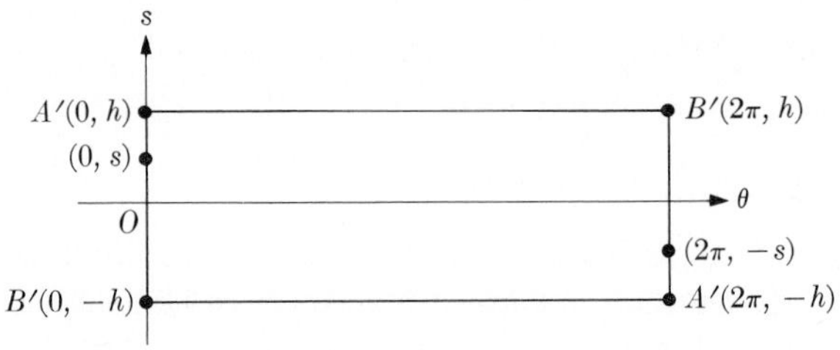

FIGURE 14–33

To obtain the unit normal to the surface, we compute $\mathbf{r}_\theta \times \mathbf{r}_s$ from the formula

$$\mathbf{r}_\theta \times \mathbf{r}_s = \frac{\partial(y, z)}{\partial(\theta, s)} \mathbf{i} + \frac{\partial(z, x)}{\partial(\theta, s)} \mathbf{j} + \frac{\partial(x, y)}{\partial(\theta, s)} \mathbf{k}.$$

We find

$$\frac{\partial(y, z)}{\partial(\theta, s)} = \left(a + s \sin \frac{\theta}{2}\right) \cos \frac{\theta}{2} \cos \theta + \frac{s}{2} \sin \theta,$$

$$\frac{\partial(z, x)}{\partial(\theta, s)} = \left(a + s \sin \frac{\theta}{2}\right) \cos \frac{\theta}{2} \sin \theta - \frac{s}{2} \cos \theta,$$

$$\frac{\partial(x, y)}{\partial(\theta, s)} = - \left(a + s \sin \frac{\theta}{2}\right) \sin \frac{\theta}{2},$$

and

$$|\mathbf{r}_\theta \times \mathbf{r}_s|^2 = \left(a + s \sin \frac{\theta}{2}\right)^2 + \frac{1}{4} s^2.$$

Therefore $\mathbf{r}_\theta \times \mathbf{r}_s$ is never $\mathbf{0}$, so that if we define $\mathbf{n}(\theta, s)$ by (2), we see that $\mathbf{n}(\theta, s)$ is continuous. However, $\mathbf{n}(0, 0) = \mathbf{i}$, $\mathbf{n}(2\pi, 0) = -\mathbf{i}$, and the points $(0, 0)$, $(2\pi, 0)$ in the parametric plane correspond to the same point on the surface.

The transformation (3) is one to one, except that the points of R given by $(0, s)$ and $(2\pi, -s)$, $-h \le s \le h$, are always carried into the same point on the surface (Fig. 14–33). The Möbius strip, often called a "one-sided surface," is not a smooth surface element (defined in Section 4). If a pencil line is drawn down the center of the strip (corresponding to $s = 0$) then, after one complete trip around, the line is on the "opposite side." Two circuits are needed to "close up" the curve made by the pencil line. The Möbius strip is not an orientable surface.

A smooth surface element S which, according to (1) is the image of a plane region G, has a boundary (denoted ∂S) which is the image of the boundary ∂G of G. The closed curve ∂G, which is piecewise smooth, can be made into a directed curve by traversing it in a given direction. We say that ∂G is **positively directed** when we travel along it so that the interior of G is always on the left. We write $\partial \vec{G}$ for a positively directed closed curve. In Fig. 14–31 the curve ∂G is shown positively directed. The closed curve ∂S, the image of ∂G, becomes a directed curve in space, with the direction the one induced by $\partial \vec{G}$. We say the curve $\partial \vec{S}$ is **positively directed** when its direction corresponds to the positively directed curve $\partial \vec{G}$. Geometrically the curve $\partial \vec{S}$ is directed so that when one walks along $\partial \vec{S}$ in an upright position with his head in the direction of the positive normal $\mathbf{n}$ to the surface, then the surface is on his left. In terms of right- and left-handed coordinate systems, if $\mathbf{t}$ is tangent to $\partial \vec{S}$ pointing in the positive direction, if $\mathbf{n}$ is perpendicular to $\mathbf{t}$ and in the direction of the positive normal, and if $\mathbf{b}$ is perpendicular to the vectors $\mathbf{t}$ and $\mathbf{n}$ and pointing toward the surface, then the triple $\mathbf{t}$, $\mathbf{b}$, and $\mathbf{n}$ correspond to a right-handed system (Fig. 14–34).

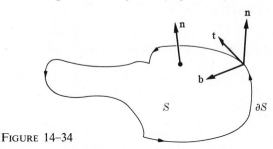

FIGURE 14–34

We wish to extend the notion of orientable surface to *piecewise smooth surfaces*. Such surfaces have edges, and so a continuous unit normal vector field cannot be defined over the entirety of such a surface. However, a piecewise smooth surface F can be subdivided into a finite number of smooth surface elements $F_1, F_2, \ldots, F_k$. Each such surface element may be positively oriented and each boundary ∂F_i may be positively directed. Suppose that γ_{ij} is a smooth arc which is the common boundary of two surface elements F_i and F_j. If the positive direction of γ_{ij} as part of $\partial \vec{F}_i$ is the opposite of the positive direction of γ_{ij} as part of $\partial \vec{F}_j$ for all arcs γ_{ij}, we say that the surface F is an **orientable, piecewise smooth surface**. This condition can be shown to be equivalent to the possibility of defining the unit normal $\mathbf{n}(Q)$ on each smooth surface element and associating with each point P of the surface a cartesian coordinate system (x^P, y^P, z^P), a function f^P, a domain G_P, and a cylinder r_P as in the definition of a piecewise smooth surface. The normal $\mathbf{n}(Q)$ is the positive one on each smooth part of the element $z^P = f^P(x^P, y^P)$ for (x^P, y^P) on G_P. (See also Appendix 3.)

Figure 14–35 shows a piecewise smooth, orientable surface and its decomposition into smooth surface elements. The boundary $\partial\vec{F}$ is a positively directed, closed, piecewise smooth curve. Those boundary arcs of $F_1, F_2, \ldots, F_k$ which are traversed only once comprise $\partial\vec{F}$. From the discussion of the last paragraph, it follows that if a piecewise smooth surface is orientable according to one de-

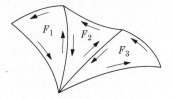

FIGURE 14–35

composition into smooth surface elements, then it is orientable according to any other such decomposition. It can be shown that there is no way to subdivide a Möbius strip into smooth elements, with two adjacent elements always having oppositely directed common boundary arcs. In other words, a Möbius strip is not orientable even if it is treated as a piecewise smooth surface. On the other hand, the surface of a cube, which is piecewise smooth, is an orientable surface. We select the "outward" pointing normal on each face and traverse the boundary of any face in a counterclockwise direction as we view it from outside the cube. It is easily verified that all edges are traversed twice, once in each direction.

7. STOKES' THEOREM

Let $\vec{S}$ be a smooth, positively oriented surface element and (x, y, z) a fixed rectangular coordinate system. We represent $\vec{S}$ parametrically:

$$\vec{S}: \quad \mathbf{r}(u, v) = x(u, v)\mathbf{i} + y(u, v)\mathbf{j} + z(u, v)\mathbf{k}.$$

Then the unit normal $\mathbf{n}(u, v)$ is given by

$$\mathbf{n}(u, v) = \frac{1}{|\mathbf{r}_u \times \mathbf{r}_v|}\left[\frac{\partial(y, z)}{\partial(u, v)}\mathbf{i} + \frac{\partial(z, x)}{\partial(u, v)}\mathbf{j} + \frac{\partial(x, y)}{\partial(u, v)}\mathbf{k}\right].$$

If $\mathbf{v}(x, y, z)$ is a continuous vector field defined on S with coordinate functions

$$\mathbf{v}(x, y, z) = v_1(x, y, z)\mathbf{i} + v_2(x, y, z)\mathbf{j} + v_3(x, y, z)\mathbf{k},$$

we can compute the scalar product $\mathbf{v} \cdot \mathbf{n}$:

$$\mathbf{v} \cdot \mathbf{n} = \frac{1}{|\mathbf{r}_u \times \mathbf{r}_v|}\left[v_1\frac{\partial(y, z)}{\partial(u, v)} + v_2\frac{\partial(z, x)}{\partial(u, v)} + v_3\frac{\partial(x, y)}{\partial(u, v)}\right].$$

Since $\mathbf{v} \cdot \mathbf{n}$ is a continuous function on S, we may define the integral

$$\iint\limits_{S} \mathbf{v} \cdot \mathbf{n}\, dS.$$

The surface element dS can be computed in terms of the parameters (u, v) according to the formula

$$dS = |\mathbf{r}_u \times \mathbf{r}_v|\, dA_{uv},$$

and so

$$\iint\limits_{S} \mathbf{v} \cdot \mathbf{n}\, dS = \iint\limits_{G} \left[v_1 \frac{\partial(y, z)}{\partial(u, v)} + v_2 \frac{\partial(z, x)}{\partial(u, v)} + v_3 \frac{\partial(x, y)}{\partial(u, v)} \right] dA_{uv}, \tag{1}$$

where G is the domain in the (u, v)-plane which has $\vec{S}$ for its image.

If $\mathbf{r}_1(s, t)$ is another smooth representation of S, it follows from Theorem 2 that there is a one-to-one smooth transformation

$$u = U(s, t), \qquad v = V(s, t)$$

such that

$$\mathbf{r}[U(s, t), V(s, t)] = \mathbf{r}_1(s, t).$$

According to the rule for multiplying Jacobians,

$$\frac{\partial(y, z)}{\partial(s, t)} = \frac{\partial(y, z)}{\partial(u, v)} \frac{\partial(u, v)}{\partial(s, t)}$$

[and the analogous equations for (z, x) and (x, y)], we see that the representation $\mathbf{r}_1(s, t)$ gives the same orientation as the representation $\mathbf{r}(u, v)$ if and only if

$$\frac{\partial(u, v)}{\partial(s, t)} > 0.$$

In such a case, if we replace (u, v) by (s, t) in (1) and integrate over G_1 [the region in the (s, t) plane whose image is S], we obtain the formula for $\iint_S \mathbf{v} \cdot \mathbf{n}\, dS$ in terms of the parameters s and t.

When S is a piecewise smooth orientable surface, we can represent it as the union of a number of smooth surface elements $S_1, S_2, \ldots, S_k$. Then we define

$$\iint\limits_{S} \mathbf{v} \cdot \mathbf{n}\, dS = \sum_{i=1}^{k} \iint\limits_{S_i} \mathbf{v} \cdot \mathbf{n}_i\, dS_i.$$

Each integral on the right is evaluated as in (1).

EXAMPLE 1. Let R be the region bounded by the cylinder $x^2 + y^2 = 1$ and the planes $z = 0$ and $z = x + 2$ (Fig. 14–36). Let S be the entire boundary of R. Find the value of $\iint_S \mathbf{v} \cdot \mathbf{n}\, dS$ where $\mathbf{n}$ is the outward directed unit normal on S and

$$\mathbf{v} = 2x\mathbf{i} - 3y\mathbf{j} + z\mathbf{k}.$$

Solution. The surface S is piecewise smooth and we label the smooth portions S_1, S_2, and S_3, as shown in Fig. 14–36. On S_1 we have

$$\mathbf{n} = -\mathbf{k}, \qquad \mathbf{v} \cdot \mathbf{n}\, dS = -z\, dA_{xy} = 0,$$

as S_1 is in the plane $z = 0$. Therefore we have $\iint_{S_1} \mathbf{v} \cdot \mathbf{n}\, dS = 0$. On S_3, we have $z = x + 2$ and

$$\mathbf{n} = \frac{1}{\sqrt{2}}(-\mathbf{i} + \mathbf{k}),$$

$$\mathbf{v} \cdot \mathbf{n} = \frac{1}{\sqrt{2}}(-2x + z), \qquad dS = \sqrt{2}\, dA_{xy},$$

$$\mathbf{v} \cdot \mathbf{n}\, dS = (-2x + z)\, dA_{xy}.$$

We obtain $\iint_{S_3} \mathbf{v} \cdot \mathbf{n}\, dS = \iint_{S_1} (-x + 2)\, dA_{xy}$. We see that $\iint_{S_1} x\, dA_{xy} = 0$, since the integrand is an odd function. Also, $2\iint_{S_1} dA_{xy} = 2\pi$.

On S_2, we select cylindrical coordinates

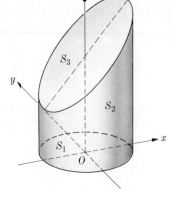

FIGURE 14–36

$$S_2: \quad x = \cos\theta, \quad y = \sin\theta, \quad z = z, \quad (\theta, z) \text{ in } F_2;$$
$$F_2: \quad -\pi \le \theta \le \pi, \quad 0 \le z \le 2 + \cos\theta.$$

We find

$$\mathbf{n} = (\cos\theta)\mathbf{i} + (\sin\theta)\mathbf{j}, \qquad \mathbf{v} \cdot \mathbf{n} = 2\cos^2\theta - 3\sin^2\theta.$$

Therefore

$$\iint_{S_2} \mathbf{v} \cdot \mathbf{n}\, dS = \iint_{F_2} (2\cos^2\theta - 3\sin^2\theta)\, dA_{\theta z} = \int_{-\pi}^{\pi} \int_0^{2+\cos\theta} (2 - 5\sin^2\theta)\, dz\, d\theta$$

$$= \int_{-\pi}^{\pi} (4 - 10\sin^2\theta + 2\cos\theta - 5\sin^2\theta\cos\theta)\, d\theta$$

$$= 8\pi - 10\pi + 0 + 0 = -2\pi.$$

Hence

$$\iint_S \mathbf{v} \cdot \mathbf{n}\, dS = \iint_{S_1} \mathbf{v} \cdot \mathbf{n}\, dS + \iint_{S_2} \mathbf{v} \cdot \mathbf{n}\, dS + \iint_{S_3} \mathbf{v} \cdot \mathbf{n}\, dS = 0.$$

Suppose that $\vec{S}$ is an oriented, piecewise smooth surface with the boundary $\partial\vec{S}$ positively directed and made up of a finite number of smooth arcs $\vec{C}_1, \vec{C}_2, \ldots, \vec{C}_k$. If $\mathbf{v}$ is a continuous vector field defined on and near $\partial\vec{S}$, we define

$$\int_{\partial\vec{S}} \mathbf{v} \cdot d\mathbf{r} = \sum_{i=1}^{k} \int_{\vec{C}_i} \mathbf{v} \cdot d\mathbf{r}.$$

With the aid of this definition we can state Stokes' Theorem.

Theorem 3. (Stokes' Theorem.) *Suppose that S is a bounded, closed, orientable, piecewise smooth surface and that $\mathbf{v}$ is a smooth vector field defined on an open set containing $|\vec{S}|$. Then*

$$\iint_{\vec{S}} (\operatorname{curl}\mathbf{v}) \cdot \mathbf{n}\, dS = \int_{\partial\vec{S}} \mathbf{v} \cdot d\mathbf{r}. \tag{2}$$

Corollary. *Suppose that* $\vec{S}$ *is a bounded, closed, orientable, piecewise smooth surface without boundary and that* **v** *is a smooth vector field defined on an open set containing* $|\vec{S}|$. *Then*

$$\iint\limits_{\vec{S}} (\text{curl } \mathbf{v}) \cdot \mathbf{n} \, dS = 0. \tag{3}$$

A general form of Stokes' Theorem is proved in Appendix 3, using Theorem 4 below and the partition of unity. However, Theorem 4, proved in this section, establishes Stokes' Theorem for those special piecewise smooth surface elements associated with a given point P.

Lemma. *Suppose that* $f(u, v)$ *is smooth on an open set* D *containing a piecewise smooth, closed, bounded region* G. *Then there exists a sequence* $f_1, f_2, \dots, f_n, \dots$ *such that each* f_n, f_{nu}, f_{nv} *is smooth on* G *and* $f_n \to f, f_{nu} \to f_u, f_{nv} \to f_v$ *uniformly on* G.

This Lemma is proved in Appendix 3.

Theorem 4. (Stokes' Theorem for surface elements.) *Suppose that* (x, y, z) *is a right-handed coordinate system in space, that* G *is a region of one of the types in Fig. 14–22, and that* $f(x, y)$ *is continuous and piecewise smooth on* G. *Assume that* $f(x, y) = f_i(x, y)$ *on* G_i *where* f_i *is smooth on an open set* D_i *containing* G_i, $i = 1, 2, \dots, k$. *If* $\vec{S}$ *is the locus of* $z = f(x, y)$ *oriented by choosing* **n** *as the positively directed normal, then* (2) *holds.*

Proof. We shall prove the result for each G_i. Then the theorem will follow for G by addition, since the arcs ending at O are described in opposite senses when considered as parts of the boundary of two adjacent G_i (Fig. 14–37). The representation of f is equivalent to the selection of (x, y) as the parameters (u, v). We have

$$x = u, \qquad y = v, \qquad z = f(u, v) \qquad \text{for } (u, v) \text{ in } G_i, \tag{4}$$

and

$$\frac{\partial(y, z)}{\partial(u, v)} = -f_u, \qquad \frac{\partial(z, x)}{\partial(u, v)} = -f_v, \qquad \frac{\partial(x, y)}{\partial(u, v)} = 1. \tag{5}$$

We first assume that f, f_x, f_y are smooth on G_i. We choose parametric representations $x = X(t), y = Y(t)$ of the arcs $\vec{C}$ of ∂G_i and set $z(t) = f[X(t), Y(t)]$. The vector field **v** is represented in coordinate functions by

$$\mathbf{v}(x, y, z) = P(x, y, z)\mathbf{i} + Q(x, y, z)\mathbf{j} + R(x, y, z)\mathbf{k}.$$

Therefore

$$\int_{\partial \vec{S}_i} \mathbf{v} \cdot d\mathbf{r} = \int_{\partial \vec{S}_i} (P \, dx + Q \, dy + R \, dz)$$

$$= \int_{\partial \vec{G}_i} [(P + Rf_x) \, dx + (Q + Rf_y) \, dy], \tag{6}$$

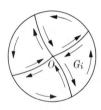

FIGURE 14–37

in which we have set $dz = f_x \, dx + f_y \, dy$. Applying Green's Theorem to the integral on the right in (6), we obtain

$$
\iint_{\partial \vec{S}_i} \mathbf{v} \cdot d\mathbf{r} = \iint_{G_i} [Q_x + Q_z f_x + Rf_{xy} + (R_x + R_z f_x)f_y - P_y - P_z f_y - Rf_{xy}
$$

$$
- (R_y + R_z f_y)f_x] \, dA_{xy}
$$

$$
= \iint_{G_i} [(Q_x + Q_z f_x) - (P_y + P_z f_y) + (R_x f_y - R_y f_x)] \, dA_{xy}. \tag{7}
$$

Using (4) and (5), we see that

$$
(\text{curl } \mathbf{v}) \cdot \mathbf{n} \, dS = [(R_y - Q_z)(-f_x) + (P_z - R_x)(-f_y) + (Q_x - P_y) \cdot 1] \, dA_{xy}. \tag{8}
$$

The result follows by comparing (7) and (8). When $f = f_i$ on G_i and f_i is smooth on D_i, we may approximate to f on G_i by f_n as described in the lemma above. The result holds for each n, and holds in the limit because of the uniform convergence and because the second derivatives f_{xy} which appear in the proof do not appear on either side of the final result (2).

EXAMPLE 2. Verify Stokes' Theorem, given that

$$
\mathbf{v} = y\mathbf{i} + z\mathbf{j} + x\mathbf{k}
$$

and that $\vec{S}$ is the part of the surface of the cylinder $x^2 + y^2 = 1$ between the planes $z = 0$ and $z = x + 2$, oriented with $\mathbf{n}$ pointing outward (Fig. 14–38).

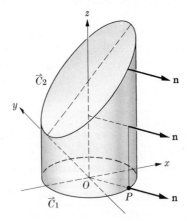

FIGURE 14–38

Solution. An examination of Fig. 14–38 shows that the curves $\vec{C}_1$ and $\vec{C}_2$ must be directed as exhibited. We choose cylindrical coordinates and write

$$
\vec{S}: \quad x = \cos\theta, \quad y = \sin\theta, \quad z = z, \quad \text{for } (\theta, z) \text{ on } F;
$$

$$
F: \quad -\pi \le \theta \le \pi, \quad 0 \le z \le 2 + \cos\theta.
$$

We think of S as made up of two smooth surface elements corresponding to F_1 and F_2, as shown in Fig. 14–39. We must make this subdivision because the representation of

$\vec{S}$ by F is not one to one. The formulas

$$\frac{\partial(y, z)}{\partial(\theta, z)} = \cos \theta,$$

$$\frac{\partial(z, x)}{\partial(\theta, z)} = \sin \theta,$$

$$\frac{\partial(x, y)}{\partial(\theta, z)} = 0,$$

$$\mathbf{n} = (\cos \theta)\mathbf{i} + (\sin \theta)\mathbf{j}$$

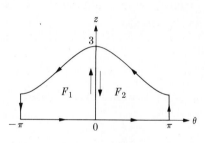

FIGURE 14–39

show that the representation agrees with the given orientation. Therefore the orientation of the boundary of $\vec{S}$ is determined by the orientations of the boundaries of F_1 and F_2. We note that when we form the boundary integral, those parts taken over vertical segments cancel. Now

$$\text{curl } \mathbf{v} = -\mathbf{i} - \mathbf{j} - \mathbf{k}, \quad \mathbf{n} = (\cos \theta)\mathbf{i} + (\sin \theta)\mathbf{j}, \quad \text{and} \quad dS = dA_{\theta z}.$$

We obtain

$$\iint\limits_{\vec{S}} (\text{curl } \mathbf{v} \cdot \mathbf{n})\, dS = \int_{-\pi}^{\pi} \int_{0}^{2+\cos \theta} (-\cos \theta - \sin \theta)\, dz\, d\theta$$

$$= -\int_{-\pi}^{\pi} [(2 \cos \theta + \cos^2 \theta) + (2 + \cos \theta)\sin \theta]\, d\theta = -\pi. \quad (9)$$

For the boundary integral, we have

$$\int_{\partial \vec{S}} \mathbf{v} \cdot d\mathbf{r} = \int_{\vec{C}_1} \mathbf{v} \cdot d\mathbf{r} - \int_{\vec{C}_2} \mathbf{v} \cdot d\mathbf{r},$$

in which the integrals on the right are taken in a counterclockwise direction. We find:

On $\vec{C}_1$: $\mathbf{v} = (\sin \theta)\mathbf{i} + (\cos \theta)\mathbf{k}, \quad d\mathbf{r} = (-\sin \theta\mathbf{i} + \cos \theta\mathbf{j})\, d\theta.$

On $\vec{C}_2$: $\mathbf{v} = (\sin \theta)\mathbf{i} + (2 + \cos \theta)\mathbf{j} + (\cos \theta)\mathbf{k},$
$$d\mathbf{r} = (-\sin \theta\mathbf{i} + \cos \theta\mathbf{j} - \sin \theta\mathbf{k})\, d\theta.$$

Taking the scalar products, we get

$$\int_{\vec{C}_1} \mathbf{v} \cdot d\mathbf{r} = \int_{-\pi}^{\pi} (-\sin^2 \theta)\, d\theta = -\pi, \quad (10)$$

$$\int_{\vec{C}_2} \mathbf{v} \cdot d\mathbf{r} = \int_{-\pi}^{\pi} (-\sin^2 \theta + 2 \cos \theta + \cos^2 \theta - \sin \theta \cos \theta)\, d\theta = 0. \quad (11)$$

A comparison of (9) with (10) and (11) verifies Stokes' Theorem.

EXERCISES

In each of problems 1 through 7, compute $\iint_{\vec{S}} \mathbf{v} \cdot \mathbf{n}\, dS.$

1. $\mathbf{v} = (x + 1)\mathbf{i} - (2y + 1)\mathbf{j} + z\mathbf{k}$; $\vec{S}$ is the triangle with vertices $(1, 0, 0)$, $(0, 1, 0)$, and $(0, 0, 1)$, with $\mathbf{n}$ pointing away from the origin.

2. $\mathbf{v} = x\mathbf{i} + y\mathbf{j} + z\mathbf{k}$; $\vec{S}$ is the part of the paraboloid $2z = x^2 + y^2$ inside the cylinder $x^2 + y^2 = 2x$ with $\mathbf{n} \cdot \mathbf{k} > 0$ (**n** pointing upward).

3. $\mathbf{v} = x^2\mathbf{i} + y^2\mathbf{j} + z^2\mathbf{k}$; $\vec{S}$ is the part of the cone

$$z^2 = x^2 + y^2 \qquad \text{for which} \qquad 1 \le z \le 2,$$

with $\mathbf{n} \cdot \mathbf{k} > 0; \mathbf{n} \cdot \mathbf{i} > 0$.

4. $\mathbf{v} = xy\mathbf{i} + xz\mathbf{j} + yz\mathbf{k}$; $\vec{S}$ is the part of the cylinder $y^2 = 2 - x$ cut out by the cylinders $y^2 = z$ and $y = z^3$.

5. $\mathbf{v} = y^2\mathbf{i} + z\mathbf{j} - x\mathbf{k}$; $\vec{S}$ is the part of the cylinder $y^2 = 1 - x$ between the planes $z = 0$ and $z = x$; $x \ge 0$, with $\mathbf{n} \cdot \mathbf{i} > 0$.

6. $\mathbf{v} = 2x\mathbf{i} - y\mathbf{j} + 3z\mathbf{k}$; $\vec{S}$ is the part of the cylinder $z^2 = x$ to the left of the cylinder $y^2 = 1 - x$ and $\mathbf{n} \cdot \mathbf{i} > 0$, **i** pointing to the right.

*7. $\mathbf{v} = x\mathbf{i} + y\mathbf{j} - 2z\mathbf{k}$; $\vec{S}$ is the part of the cylinder $x^2 + y^2 = 2x$ between the two nappes of the cone $z^2 = x^2 + y^2$, **n** pointing outward.

In each of problems 8 through 13, verify Stokes' Theorem.

8. $\mathbf{v} = z\mathbf{i} + x\mathbf{j} + y\mathbf{k}$; $\vec{S}$ is the part of the paraboloid $z = 1 - x^2 - y^2$ for which $z \ge 0$ and $\mathbf{n} \cdot \mathbf{k} > 0$.

9. $\mathbf{v} = y^2\mathbf{i} + xy\mathbf{j} - 2xz\mathbf{k}$; $\vec{S}$ is the hemisphere $x^2 + y^2 + z^2 = a^2$, $z \ge 0$ with $\mathbf{n} \cdot \mathbf{k} > 0$.

10. $\mathbf{v} = -yz\mathbf{i}$; $\vec{S}$ is the part of the sphere $x^2 + y^2 + z^2 = 4$ outside the cylinder $x^2 + y^2 = 1$, **n** pointing outward.

11. $\mathbf{v} = -z\mathbf{j} + y\mathbf{k}$; $\vec{S}$ is the part of the vertical cylinder $r = \theta$ (cylindrical coordinates), $0 \le \theta \le \pi/2$, bounded below by the (x, y)-plane and above by the cone $z^2 = x^2 + y^2$, $\mathbf{n} \cdot \mathbf{i} > 0$ for $\theta > 0$.

12. $\mathbf{v} = y\mathbf{i} + z\mathbf{j} + x\mathbf{k}$; $\vec{S}$ is the part of the surface $z^2 = 4 - x$ to the right of the cylinder $y^2 = x$, $\mathbf{n} \cdot \mathbf{i} > 0$, **i** pointing to the right.

13. $\mathbf{v} = z\mathbf{i} - x\mathbf{k}$; $\vec{S}$ is the part of the cylinder $r = 2 + \cos \theta$ above the (x, y)-plane and below the cone $z^2 = x^2 + y^2$, **n** pointing outward.

In each of problems 14 through 16, compute $\int_{\partial \vec{S}} \mathbf{v} \cdot d\mathbf{r}$, using Stokes' Theorem.

14. $\mathbf{v} = r^{-3}\mathbf{r}, \mathbf{r} = x\mathbf{i} + y\mathbf{j} + z\mathbf{k}, r = |\mathbf{r}|$; $\vec{S}$ is the surface of Example 2.

15. $\mathbf{v} = (e^x \sin y)\mathbf{i} + (e^x \cos y - z)\mathbf{j} + y\mathbf{k}$; $\vec{S}$ is the surface of Exercise 3.

16. $\mathbf{v} = (x^2 + z)\mathbf{i} + (y^2 + x)\mathbf{j} + (z^2 + y)\mathbf{k}$; $\vec{S}$ is the part of the sphere $x^2 + y^2 + z^2 = 1$ above the cone $z^2 = x^2 + y^2$; $\mathbf{n} \cdot \mathbf{k} > 0$.

17. Show that if $\vec{S}$ is given by $z = f(x, y)$ for $x^2 + y^2 \le 1$, where f is smooth and if $\mathbf{v} = (1 - x^2 - y^2)\mathbf{w}(x, y, z)$, where **w** is any smooth vector field defined on an open set containing $|\vec{S}|$, then $\iint_{\vec{S}} (\text{curl } \mathbf{v} \cdot \mathbf{n}) \, dS = 0$.

18. Suppose that $\mathbf{v} = r^{-3}(y\mathbf{i} + z\mathbf{j} + x\mathbf{k})$, where $r = |\mathbf{r}| = |x\mathbf{i} + y\mathbf{j} + z\mathbf{k}|$ and $\vec{S}$ is the unit sphere with **n** directed outward. Show by direct calculation that

$$\iint_S (\text{curl } \mathbf{v}) \cdot \mathbf{n} \, dS = 0.$$

8. THE DIVERGENCE THEOREM

Stokes' Theorem, which relates an integral over a surface in space to a line integral over the boundary of the surface, is a generalization of Green's Theorem. Another type of generalization, known as the Divergence Theorem, establishes a connection between an integral over a three-dimensional domain and an integral over the surface which forms the boundary of the domain. It can be shown that the Divergence Theorem and the theorems of Green and Stokes are all special cases of a general formula which connects an integral over a set of points in some n-dimensional space with another integral over the boundary of that set of points.

Theorem 5. (The Divergence Theorem.) *Suppose that a bounded domain G in three-space is bounded by a piecewise smooth, orientable surface without boundary and suppose that* **v** *is a smooth vector field defined on an open set containing G and* ∂G. *Then*

$$\iiint_G \operatorname{div} \mathbf{v} \, dV = \iint_{\partial G} \mathbf{v} \cdot \mathbf{n} \, dS, \tag{1}$$

where the boundary ∂G *is oriented by taking* **n** *as the exterior normal.*

The proof of the Divergence Theorem (with an added condition that the boundary is not too irregular) is given in Appendix 4. However, certain special cases which give the principal content of the result will be established in this section.

Theorem 6. *Suppose that* **v** *and G are such that there exists an* (x, y, z) *coordinate system in which* $\mathbf{v}(x, y, z) = R(x, y, z)\mathbf{k}$ *and G is of the form*

$$G: \quad x^2 + y^2 < r_0^2, \quad c < z < f(x, y),$$

where f is piecewise smooth (as indicated in Fig. 14–40). If R and R_z *are continuous on an open set containing G and* ∂G, *then* (1) *holds.*

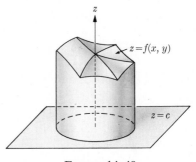

FIGURE 14–40

Proof. We note that $\operatorname{div} \mathbf{v} = R_z$ and, therefore, that

$$\iiint_G \operatorname{div} \mathbf{v} \, dV = \iiint_G R_z \, dV_{xyz} = \iint_{C(0, r_0)} \int_c^{f(x, y)} R_z \, dz \, dA_{xy},$$

where $C(0, r_0)$ is the circle with center at the origin and radius r_0. Upon performing the integration with respect to z, we find

$$\iiint_G \operatorname{div} \mathbf{v} \, dV = \iint_{C(0, r_0)} \{ R[x, y, f(x, y)] - R(x, y, c) \} \, dA_{xy}. \tag{2}$$

Let $\partial G = S_1 \cup S_2 \cup S_3$, where S_1 is the circle in the plane $z = c$, S_2 is the lateral surface of the cylinder, and S_3 is the top surface $z = f(x, y)$. We have

$$\text{On } S_2: \quad \mathbf{n} \cdot \mathbf{k} = 0, \qquad \iint_{S_2} \mathbf{v} \cdot \mathbf{n} \, dS = 0. \tag{3}$$

$$\text{On } S_1: \quad \mathbf{n} = -\mathbf{k}, \qquad \iint_{S_1} \mathbf{v} \cdot \mathbf{n} \, dS = - \iint_{C(0,r)} R(x, y, c) \, dA_{xy}. \tag{4}$$

$$\text{On } S_3: \quad \mathbf{n} = (1 + f_x^2 + f_y^2)^{-1/2}(-f_x \mathbf{i} - f_y \mathbf{j} + \mathbf{k}),$$
$$dS = (1 + f_x^2 + f_y^2)^{1/2} \, dA_{xy};$$
$$\iint_{S_3} \mathbf{v} \cdot \mathbf{n} \, dS = \iint_{C(0,r)} R[x, y, f(x, y)] \, dA_{xy}. \tag{5}$$

A comparison of (2) with (3), (4), and (5) yields the result.

Theorem 7. *Suppose that the hypotheses of Theorem 6 hold, except that* $\mathbf{v}$ *has the form*

$$\mathbf{v} = P(x, y, z)\mathbf{i} + Q(x, y, z)\mathbf{j}$$

with P, Q smooth on an open set containing G and ∂G. Then (1) holds.

Proof. We define the functions $U(x, y, z)$, $V(x, y, z)$ by the formulas

$$U(x, y, z) = \int_c^z Q(x, y, t) \, dt, \qquad V(x, y, z) = - \int_c^z P(x, y, t) \, dt.$$

Also, we set

$$\mathbf{w} = U\mathbf{i} + V\mathbf{j} \quad \text{and} \quad -R = V_x - U_y, \quad \mathbf{u} = R\mathbf{k}.$$

Then $\mathbf{w}$ is smooth and R, R_z are continuous, so that

$$\text{curl } \mathbf{w} = -V_z \mathbf{i} + U_z \mathbf{j} + (V_x - U_y)\mathbf{k} = \mathbf{v} - \mathbf{u},$$
$$\text{div } \mathbf{v} = P_x + Q_y = R_z = \text{div } \mathbf{u}. \tag{6}$$

We apply the corollary to Stokes' Theorem and obtain

$$\iint_{\partial G} (\text{curl } \mathbf{w}) \cdot \mathbf{n} \, dS = 0 = \iint_{\partial G} (\mathbf{v} - \mathbf{u}) \cdot \mathbf{n} \, dS.$$

Therefore

$$\iint_{\partial G} \mathbf{v} \cdot \mathbf{n} \, dS = \iint_{\partial G} \mathbf{u} \cdot \mathbf{n} \, dS = \iiint_G \text{div } \mathbf{u} \, dV; \tag{7}$$

the last equality holds because Theorem 6 may be applied to $\mathbf{u} = R\mathbf{k}$. Taking (6) and (7) into account, we get

$$\iint_{\partial G} \mathbf{v} \cdot \mathbf{n} \, dS = \iiint_G \text{div } \mathbf{v} \, dV.$$

It is clear that Theorems 6 and 7 hold for cylindrical domains G formed by the projection of a piece of the surface $z = f(x, y)$ on a plane $z = c$ parallel to the xy-plane, so long as the projection is a piecewise smooth region R, as shown in Fig. 14–41. The proof is the same. Also, the Divergence Theorem holds for cylindrical domains which are parallel to the x- or y-axes. By addition, the result is valid for smooth vector fields $\mathbf{v}$ defined over regions G which are the sum of cylindrical domains of the kind just described. These regions may be quite general. However, the proof given in Appendix 4, which uses the partition of unity, avoids the difficulty of describing such regions in detail.

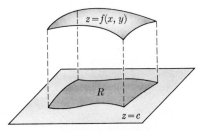
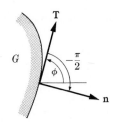

FIGURE 14–41   FIGURE 14–42

The Divergence Theorem has a straightforward extension to n-dimensional space. Integrals over an n-dimensional volume and integrals over an $(n - 1)$-dimensional "hypersurface" can be defined in analogy with integrals in three dimensions and with integrals on two-dimensional surfaces. Also, the scalar product of two vectors in n-dimensional space has been defined. The appropriate Stokes formula is

$$\overbrace{\int \cdots \int}^{n \text{ integrals}}_{G} \operatorname{div} \mathbf{v} \, dV = \overbrace{\int \cdots \int}^{n - 1 \text{ integrals}}_{\partial G} \mathbf{v} \cdot \mathbf{n} \, dS \tag{8}$$

where, if $\mathbf{v}$ is given in terms of basis vectors $\mathbf{e}_1, \ldots, \mathbf{e}_n$ by the relation

$$\mathbf{v} = \sum_{i=1}^{n} P_i \mathbf{e}_i,$$

then

$$\operatorname{div} \mathbf{v} = \sum_{i=1}^{n} \frac{\partial P_i}{\partial x_i}. \tag{9}$$

It can be shown that (9) is independent of the particular basis chosen.

In two dimensions the Divergence Theorem is a direct consequence of Green's Theorem. To see this we choose the usual coordinate system in the (xy)-plane and suppose that $\mathbf{v} = P\mathbf{i} + Q\mathbf{j}$. We define $\mathbf{u} = Q\mathbf{i} - P\mathbf{j}$. As Fig. 14–42 shows, the exterior normal $\mathbf{n}$ to a region G makes an angle of $-\pi/2$ with the tangent vector $\mathbf{T}$, directed so that the interior of G is always on the left as we proceed around the boundary. If ϕ is the angle the vector $\mathbf{T}$ makes with the positive x-

direction, we may write

$$\mathbf{T} = (\cos \phi)\mathbf{i} + (\sin \phi)\mathbf{j}, \qquad \mathbf{n} = (\sin \phi)\mathbf{i} - (\cos \phi)\mathbf{j},$$
$$\mathbf{v} \cdot \mathbf{T} = \mathbf{u} \cdot \mathbf{n}, \qquad \text{div } \mathbf{u} = Q_x - P_y.$$

The relation (1) applied to $\mathbf{u}$ is, when translated in terms of $\mathbf{v}$ (or P, Q), a restatement of Green's Theorem.

EXAMPLE 1. Verify the Divergence Theorem, given that G is the domain between the concentric spheres S_1 and S_2 of radius 1 and 2, respectively, and center at O; the vector $\mathbf{v}$ is $\mathbf{v} = \mathbf{r}/r^3$, with $\mathbf{r} = \mathbf{v}(\overrightarrow{OP})$, P a point in the domain, and $r = |\mathbf{r}|$.

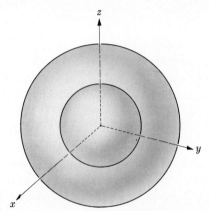

FIGURE 14–43

Solution. (See Fig. 14–43.) Let $\vec{S}_1$ and $\vec{S}_2$ be the spheres, both oriented with $\mathbf{n}$ pointing outward from O. Then

$$\iint_{\partial G} \mathbf{v} \cdot \mathbf{n} \, dS = \iint_{\vec{S}_2} \mathbf{v} \cdot \mathbf{n} \, dS - \iint_{\vec{S}_1} \mathbf{v} \cdot \mathbf{n} \, dS,$$

since the normal on $\vec{S}_1$, as part of ∂G, points toward O when the Divergence Theorem is used. According to Example 2 in Section 3 of Chapter 13, we easily find that

$$\text{div } \mathbf{v} = 0 \qquad \text{and hence} \qquad \iiint_G \text{div } \mathbf{v} \, dV = 0.$$

Now, on both $\vec{S}_1$ and $\vec{S}_2$, we have $\mathbf{n} = r^{-1}\mathbf{r}$, and so

$$\iint_{S_2} \mathbf{v} \cdot \mathbf{n} \, dS - \iint_{S_1} \mathbf{v} \cdot \mathbf{n} \, dS = \tfrac{1}{4} A(S_2) - 1 \cdot A(S_1) = 0.$$

EXAMPLE 2. Given that G is the domain inside the cylinder $x^2 + y^2 = 1$ and between the planes $z = 0$ and $z = x + 2$, and given that

$$\mathbf{v} = (x^2 + ye^z)\mathbf{i} + (y^2 + ze^x)\mathbf{j} + (z^2 + xe^y)\mathbf{k},$$

use the Divergence Theorem to evaluate $\iint_{\partial G} \mathbf{v} \cdot \mathbf{n} \, dS$.

Solution. We have div $\mathbf{v} = 2x + 2y + 2z$. Therefore, letting F denote the unit circle, we obtain (Fig. 14–38)

$$\iint_{\partial G} \mathbf{v} \cdot \mathbf{n}\, dS = \iiint_G 2(x + y + z)\, dV = 2\iint_F \left[(x + y)z + \tfrac{1}{2}z^2\right]_0^{x+2} dA_{xy}$$

$$= \iint_F \left[2y(x + 2) + 2x^2 + 4x + (x^2 + 4x + 4)\right] dA_{xy}$$

$$= \iint_F (3x^2 + 4)\, dA_{xy} = \int_0^1 \int_0^{2\pi} \left[(3r^3 \cos^2 \theta) + 4r\right] dr\, d\theta$$

$$= \pi \int_0^1 (3r^3 + 8r)\, dr = \tfrac{19}{4}\pi.$$

The Divergence Theorem has an important physical interpretation in connection with problems in fluid flow. We suppose that a fluid (liquid or gas) is flowing through a region in space. In general, the density ρ and the velocity vector $\mathbf{u}$ will depend not only on the point P in space but also on the time t. We select a point P in space, a value of t, and a small plane surface σ through P (Fig. 14–44). For a moment we suppose that ρ and $\mathbf{u}$ are constant. Then, after a short time Δt, all the particles on σ at time t would be in the shaded region σ' shown in Fig. 14–44. The particles sweep out a small cylindrical region as they travel from σ to σ'. The total mass of fluid flowing across σ in time Δt is just enough to fill up the oblique cylinder between σ and σ'. We denote this cylindrical region by G, its boundary by ∂G, and the outward normal on the boundary by $\mathbf{n}$. If $\mathbf{u} \cdot \mathbf{n} > 0$ on σ, then the fluid flow is *out* of G across σ, while if $\mathbf{u} \cdot \mathbf{n} < 0$ on σ, the flow is *into* G across σ. The amount of fluid leaving (or entering, if the value is negative) G across σ is then given by

$$(\rho\mathbf{u})\, \Delta t \cdot \mathbf{n}A(\sigma).$$

The net *rate* of flow of mass out of G across σ (per unit time) is just

$$\rho\mathbf{u} \cdot \mathbf{n}A(\sigma).$$

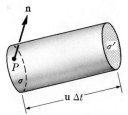

FIGURE 14–44

If ρ and $\mathbf{u}$ are continuous on ∂G, a piecewise smooth boundary, and if we divide up ∂G into small, smooth surface elements such as σ and add the results, we obtain (in the limit)

$$\iint_{\partial G} (\rho\mathbf{u}) \cdot \mathbf{n}\, dS$$

for the rate of flow of the total amount of mass out of G. We denote by $M(G, t)$ the total mass in the region G at time t. Using the definition of density, we may write

$$M(G, t) = \iiint_G \rho(P, t)\, dV_P$$

and, using Leibniz' Rule (extended to triple integrals), we find

$$\frac{d}{dt} M(G, t) = \iiint\limits_{G} \rho_t(P, t) \, dV_P.$$

Since $(d/dt)M$ is the amount of mass flowing into G, we obtain

$$\iiint\limits_{G} \rho_t(P, t) \, dV_P = - \iint\limits_{\partial G} (\rho \mathbf{u}) \cdot \mathbf{n} \, dS.$$

If ρ and $\mathbf{u}$ are smooth, we may apply the Divergence Theorem to the boundary integral and get the equation

$$\iiint\limits_{G} [\rho_t(P, t) + \operatorname{div}(\rho \mathbf{u})] \, dV = 0.$$

Since G is arbitrary, we can divide the above equation by $V(G)$, the volume of G, and let G shrink to a point P. The result is the **equation of continuity**

$$\rho_t + \operatorname{div}(\rho \mathbf{u}) = 0.$$

If the fluid is an incompressible liquid, so that $\rho = \text{const}$, the equation of continuity becomes

$$\operatorname{div} \mathbf{u} = 0.$$

EXERCISES

In each of problems 1 through 10, verify the Divergence Theorem by computing separately each side of Eq. (1).

1. $\mathbf{v} = xy\mathbf{i} + yz\mathbf{j} + zx\mathbf{k}$; G is bounded by the coordinate planes and the plane $x + y + z = 1$.

2. $\mathbf{v} = x^2\mathbf{i} - y^2\mathbf{j} + z^2\mathbf{k}$; G is bounded by: $x^2 + y^2 = 4$, $z = 0$, $z = 2$.

3. $\mathbf{v} = 2x\mathbf{i} + 3y\mathbf{j} - 4z\mathbf{k}$; G is the sphere $x^2 + y^2 + z^2 \leq 4$.

4. $\mathbf{v} = x^2\mathbf{i} + y^2\mathbf{j} + z^2\mathbf{k}$; G is bounded by: $y^2 = 2 - x$, $z = 0$, $z = x$.

5. $\mathbf{v} = x\mathbf{i} + y\mathbf{j} + z\mathbf{k}$; G is the domain outside $x^2 + y^2 = 1$ and inside $x^2 + y^2 + z^2 = 4$.

6. $\mathbf{v} = x\mathbf{i} - 2y\mathbf{j} + 3z\mathbf{k}$; G is bounded by $y^2 = x$ and $z^2 = 4 - x$.

7. $\mathbf{v} = r^{-3}(z\mathbf{i} + x\mathbf{j} + y\mathbf{k})$; G is the domain outside $x^2 + y^2 + z^2 = 1$ and inside $x^2 + y^2 + z^2 = 4$.

8. $\mathbf{v} = 3x\mathbf{i} - 2y\mathbf{j} + z\mathbf{k}$; G is bounded by $x^2 + z^2 = 4$, $y = 0$, $x + y + z = 3$.

9. $\mathbf{v} = 2x\mathbf{i} + y\mathbf{j} + z\mathbf{k}$; G is bounded by $z = x^2 + y^2$ and $z = 2x$.

10. $\mathbf{v} = x\mathbf{i} + y\mathbf{j} + z\mathbf{k}$; G is bounded by $x^2 + y^2 = 4$ and $x^2 + y^2 - z^2 = 1$.

In each of problems 11 through 13, evaluate $\iint_{\partial G} \mathbf{v} \cdot \mathbf{n} \, dS$, using the Divergence Theorem.

11. $\mathbf{v} = ye^z\mathbf{i} + (y - ze^x)\mathbf{j} + (xe^y - z)\mathbf{k}$; G is the interior of the torus $(r - b)^2 + z^2 \le a^2$, $0 < a < b$; r, z cylindrical coordinates.

12. $\mathbf{v} = x^3\mathbf{i} + y^3\mathbf{j} + z^3\mathbf{k}$; G is the sphere $x^2 + y^2 + z^2 \le 1$.

13. $\mathbf{v} = x^3\mathbf{i} + y^3\mathbf{j} + z\mathbf{k}$; G is bounded by: $x^2 + y^2 = 1$, $z = 0$, and $z = x + 2$.

14. Suppose that G is a region in three-space with a boundary ∂G for which the Divergence Theorem is applicable. Prove the following formula for integration by parts if u and v are smooth on an open set D containing G and ∂G:

$$\iiint_G u \operatorname{div} \mathbf{v} \, dV = \iint_{\partial G} u\mathbf{v} \cdot \mathbf{n} \, dS - \iiint_G \nabla u \cdot \mathbf{v} \, dV.$$

15. Suppose that D and G are as in Problem 14 and that u, ∇u, and v are smooth in D. Let $\partial/\partial n$ denote the directional derivative on ∂G in the direction of $\mathbf{n}$. Show that

$$\iiint_G v\nabla^2 u \, dV = \iint_{\partial G} v \frac{\partial u}{\partial n} \, dS - \iiint_G \nabla v \cdot \nabla u \, dV.$$

16. Suppose that u satisfies Laplace's equation in a region G of the type described in Problem 14. Show that

$$\iint_{\partial G} \frac{\partial u}{\partial n} \, dS = 0.$$

[*Hint.* Use the formula in Problem 15.]

17. Suppose that u satisfies Laplace's equation in a region G of the type described in Problem 14. Show that if $u = 0$ on ∂G, then $u \equiv 0$ in G. [*Hint:* Set $v = u$ in the formula of Problem 15.]

18. Evaluate $\iint_S \mathbf{v} \cdot \mathbf{n} \, dS$, where S is the surface of the torus

$$S: \quad (r - 3)^2 + z^2 = 1;$$

$\mathbf{n}$ is the outward normal on S; $\mathbf{R} = (x - 3)\mathbf{i} + y\mathbf{j} + z\mathbf{k}$; $R = |\mathbf{R}|$, $\mathbf{v} = R^{-3}\mathbf{R}$. [*Hint.* Use the Divergence Theorem, with G as the part of the interior of the torus which is outside a small sphere of radius ρ and center at $(3, 0, 0)$.]

Some Special Types of Differential Equations

1. DEFINITIONS AND EXAMPLES

The equation

$$\frac{dy}{dx} = 2x \tag{1}$$

is an example of a differential equation. The function $y = x^2 + c$ (where c is any constant) is a solution of this equation. In general, any equation which contains derivatives is called a **differential equation**. Other examples of differential equations are

$$2x\frac{d^2y}{dx^2} + \left(\frac{dy}{dx}\right)^2 = \frac{1}{y} + 2x,$$

$$\frac{d^4y}{dx^4} + x\left(\frac{dy}{dx}\right)^2 - y^3 = \tan x, \tag{2}$$

$$x^2\frac{\partial z}{\partial x} + 2\frac{\partial z}{\partial y} - 3xy + 4z \sec y = 0.$$

The first two equations contain ordinary derivatives; from the way they are written, we see that x is the independent variable and that y is a function of x. The third equation in (2) involves partial derivatives of the variable z in terms of the independent variables x and y.

DEFINITIONS. An **ordinary differential equation** is an equation which involves one unknown function, say y, and one or more derivatives of y taken with respect to an independent variable x. The equation (1) and the first two equations in (2) are ordinary differential equations. A **partial differential equation** is one which contains partial derivatives.

We shall restrict our attention entirely to the problem of finding solutions of ordinary differential equations. (The systematic study of partial differential equations is usually taken up in more advanced courses.) When we use the term "differential equation" we shall always mean ordinary differential equation.

636

The order of a differential equation is the order of the highest derivative which appears. The first equation in (2) is of the second order, and the second equation in (2) is of the fourth order.

A differential equation in the form

$$c_0(x)\frac{d^n y}{dx^n} + c_1(x)\frac{d^{n-1}y}{dx^{n-1}} + \cdots + c_{n-1}(x)\frac{dy}{dx} + c_n(x)y = f(x) \qquad (3)$$

is said to be a linear differential equation. The functions $c_0(x)$, $c_1(x)$, $\ldots$, $c_n(x)$, $f(x)$ may be arbitrary functions of x, and their character does not affect the linearity of the equation. Equation (3) is of the nth order. The equations

$$x^2 \frac{d^2 y}{dx^2} + 2\frac{dy}{dx} - x^3 y = \tan x, \qquad \sqrt{x}\,\frac{dy}{dx} + 8y = 2,$$

$$\frac{d^3 y}{dx^3} + 2x^4 \frac{dy}{dx} + 8y = 6x$$

are examples of linear differential equations. On the other hand, the equations

$$\frac{d^2 y}{dx^2} + y\frac{dy}{dx} = 1, \qquad \left(\frac{dy}{dx}\right)^2 + 2y = 8x,$$

$$\left(\frac{dy}{dx}\right)\left(\frac{d^2 y}{dx^2}\right) + 8y = x^2, \qquad \frac{dy}{dx} + y^2 = 2$$

are all examples of nonlinear differential equations.

To solve a differential equation means to find all the sufficiently differentiable functions y such that, for each x (in some set), the numerical values of y and its derivatives satisfy the given equation. In the following sections, we shall often be content to find relations which define the desired solutions y implicitly. Some particularly simple differential equations may be solved by inspection. For example, the equation

$$\frac{dy}{dx} = f(x) \qquad (4)$$

has the solution

$$y = \int f(x)\,dx + c.$$

While it is desirable to evaluate the indefinite integral of f and so obtain an explicit answer, we call $\int f(x)\,dx + c$ a solution even when the actual integration cannot be performed. When the integration is not carried out we say that the solution has been obtained except for a *quadrature*. The equation

$$\frac{d^2 y}{dx^2} = f(x) \qquad (5)$$

may be integrated to give

$$\frac{dy}{dx} = F(x) + C,$$

where $F(x) = \int f(x)\,dx$. Integrating once more, we find

$$y = G(x) + Cx + D,$$

in which G is an antiderivative of F, and C and D are constants. Another example of an equation which we can solve easily is the linear first-order equation

$$\frac{dy}{dx} = ay + b, \qquad a \neq 0 \tag{6}$$

with a and b constant. We write

$$\frac{dy}{y + (b/a)} = a\,dx, \qquad y + \frac{b}{a} > 0,$$

and obtain, upon integration,

$$\ln\left(y + \frac{b}{a}\right) = ax + C.$$

Therefore

$$y + \frac{b}{a} = e^{ax+C} = De^{ax}, \qquad \text{where* } D = e^C.$$

We note that the solutions of Eqs. (4) and (6), which are both of the first order, contain one constant of integration, while the solution of (5), a second-order equation, contains two arbitrary constants of integration.

The motion of one or several particles is described by equations which involve both the velocity and the acceleration of the individual particles. Since velocity and acceleration are defined in terms of derivatives, these equations of motion are differential equations. Newton's law states that the force acting on a particle is proportional to the acceleration; therefore it is easy to see how differential equations arise in mechanics and many branches of science and engineering. The subject of differential equations has been studied extensively by mathematicians, physicists, and engineers for several centuries. The material presented in the next sections is intended as a brief introduction to the subject.

2. EQUATIONS WITH VARIABLES SEPARABLE

In order to give the student practice in solving differential equations, we shall consider a few types of equations which can be solved by special devices. From the point of view of the theory of differential equations, those equations which are amenable to these tricks are not particularly interesting. Some of these methods, however, are useful in more advanced topics in differential equations. Moreover, in studying these special methods, the reader will understand more clearly the reasons for the various classifications of differential equations.

---

* The interested reader can show (recalling that $d \ln |u| = u^{-1}\,du$) that $y = (-b/a) + De^{ax}$ is a solution of (6) even if D is negative or 0, and that all the solutions are obtained in this way.

If a first-order differential equation is of the form

$$\frac{dy}{dx} = f(x)g(y), \qquad g(y) \neq 0,$$

we say that the **variables in the equation are separable.** That is, we may write

$$\frac{dy}{g(y)} = f(x)\, dx,$$

and the solution, except for the quadratures involved, is

$$\int \frac{dy}{g(y)} = \int f(x)\, dx + C.$$

EXAMPLE 1. Solve

$$\frac{dy}{dx} = \frac{x \cos y}{e^x \sin y}, \qquad y \neq n\pi, \qquad (n + \tfrac{1}{2})\, \pi, n = 0, \pm 1, \pm 2, \cdots.$$

Solution. We can separate variables by writing

$$\frac{\sin y\, dy}{\cos y} = xe^{-x}\, dx,$$

from which we conclude that

$$-\ln \cos y = -(x + 1)e^{-x} + C.$$

Occasionally differential equations of the first order are given in terms of differentials. Thus an expression such as

$$P(x, y)\, dx + Q(x, y)\, dy = 0$$

is equivalent to the differential equation

$$\frac{dy}{dx} = -\frac{P(x, y)}{Q(x, y)}.$$

No distinction will be made between equations with derivatives and those with differentials.

EXAMPLE 2. Solve: $(1 + x^2)(1 + y^2)\, dx - xy\, dy = 0$.

Solution. We have

$$\frac{(1 + x^2)\, dx}{x} = \frac{y\, dy}{1 + y^2},$$

and the variables are separated. The solution is

$$\ln x + \tfrac{1}{2}x^2 = \tfrac{1}{2} \ln (1 + y^2) + C. \tag{2}$$

Remarks. Equation (2) yields the solution in implicit form. We shall frequently perform the necessary algebra to obtain an explicit result. Writing

$$\ln (1 + y^2) = 2 \ln x + x^2 - 2C,$$

we find

$$1 + y^2 = e^{2 \ln x + x^2 - 2C}$$

$$= e^{2 \ln x} \cdot e^{x^2} \cdot e^{-2C}$$

$$= Dx^2 e^{x^2},$$

where we have written $D = e^{-2C}$. Hence the explicit solution to Example 2 is

$$y = \pm \sqrt{Dx^2 e^{x^2} - 1}.$$

Since at present we restrict ourselves to real solutions, the domain of the solution above consists of those x which satisfy the inequality $x^2 e^{x^2} \geq 1/D$.

Equations which fall into the "variables separable" class are of an extremely special character. It is interesting to compare equations of type (1) with an equation of the form

$$\frac{dy}{dx} = f(x) + g(y). \tag{3}$$

The variables are not separable in this case and, in fact, there is no known way to reduce every such equation to a simple problem in quadratures.

EXERCISES

In the following exercises, solve the given differential equations.

1. $x^2 \, dy - \cos^2 y \, dx = 0$

2. $(1 - x)y^2 \, dx + x \, dy = 0$

3. $y^2 \, dx + 2y^2 \, dy = dy$

4. $\dfrac{dy}{dx} + 2y = 3$

5. $e^y \, dx + x^2(2 + e^y) \, dy = 0$

6. $\dfrac{dy}{dx} = \dfrac{y}{x}$

7. $\dfrac{dy}{dx} = xy$

8. $\sqrt{a^2 - x^2} \, dy - y\sqrt{a^2 - y^2} \, dx = 0$

9. $e^{2x+y} \, dx - 2e^{x-y} \, dy = 0$

10. $\dfrac{dy}{dx} = xe^{x+y}$

11. $(xy^2 + x) \, dx + (x^2 y + y) \, dy = 0$

12. $(xy^2 + 2y^2) \, dx + (xy^2 - x) \, dy = 0$

13. $2y \, dy + 4x^3\sqrt{4 - y^4} \, dx = 0$

14. $(1 + x^2)\dfrac{dy}{dx} + 2xy = x$

15. $\dfrac{\ln y}{\ln x} dy - \dfrac{x^4}{y^2} dx = 0$

3. HOMOGENEOUS FUNCTIONS

A function $f(x, y)$ is said to be **homogeneous of degree** n if and only if

$$f(tx, ty) = t^n f(x, y) \qquad \text{for all } t > 0 \text{ and all } (x, y) \neq (0, 0).$$

For example, the function

$$f(x, y) = x^3 + 2x^2 y - 3xy^2 + y^3$$

is homogeneous of degree 3, since

$$f(tx, ty) = (tx)^3 + 2(tx)^2(ty) - 3(tx)(ty)^2 + (ty)^3$$
$$= t^3(x^3 + 2x^2 y - 3xy^2 + y^3) = t^3 f(x, y).$$

The function

$$f(x, y) = \sqrt{x^2 + y^2 - 8xy}$$

is homogeneous of degree 1, and the function

$$f(x, y) = \frac{x^2 + 2y^2}{8xy + 4y^2}$$

is homogeneous of degree zero. The function $f(x, y) = 1 + x^2 + y^2$ is not homogeneous.

A function f of the k variables $x_1, x_2, \ldots, x_k$ is homogeneous of degree n if and only if

$$f(tx_1, tx_2, \ldots, tx_k) = t^n f(x_1, x_2, \ldots, x_k).$$

Note that n is not restricted to positive values. The function

$$f = (x_1^2 + x_2^2 + x_3^2)^{-1/3}$$

is homogeneous of degree $-\frac{2}{3}$.

RULE. *A differential equation of the form*

$$P(x, y)\, dx + Q(x, y)\, dy = 0, \tag{1}$$

*in which P and Q are homogeneous of the **same degree** is reducible to the variables separable case.*

To establish the rule, we assume that P and Q are homogeneous of degree n and set

$$y(x) = xv(x),$$

where v is a new unknown function. Since

$$dy = x\, dv + v\, dx,$$

Eq. (1) becomes

$$P(x, vx)\, dx + Q(x, vx)(x\, dv + v\, dx) = 0$$

or

$$x^n P(1, v)\, dx + x^n Q(1, v)(x\, dv + v\, dx) = 0 \qquad \text{if } x > 0.$$

Therefore
$$[P(1, v) + vQ(1, v)]\, dx + xQ(1, v)\, dv = 0 \qquad \text{if } x > 0$$
and
$$[P(-1, -v) + vQ(-1, -v)]\, dx + xQ(-1, -v)\, dv = 0 \qquad \text{if } x < 0.$$

In either case, we have
$$\frac{dx}{x} = -\frac{Q(\pm1, \pm v)}{P(\pm1, \pm v) + vQ(\pm1, \pm v)}\, dv,$$

and the variables are separated.

EXAMPLE. Solve: $(x^2 - xy + y^2)\, dx + x^2\, dy = 0$.

Solution. $P = x^2 - xy + y^2$ and $Q = x^2$ are homogeneous of degree 2. Setting $y = xv$, we obtain
$$x^2(1 - v + v^2)\, dx + x^2(x\, dv + v\, dx) = 0$$
or
$$\frac{dv}{1 + v^2} = -\frac{dx}{x}.$$

Integrating, we find
$$\arctan v = -\ln(x) + C \qquad \text{and} \qquad v = \tan(C - \ln|x|).$$

Finally, setting $v = y/x$, we get
$$y = x \tan(C - \ln|x|).$$

Differential equations in which both P and Q are homogeneous of the same degree rarely occur in actual practice.

EXERCISES

In each of problems 1 through 10, decide whether or not the given function is homogeneous. If so, find the degree.

1. $f(x, y) = x^4 + 8x^3y - 2x^2y^2 + y^4$ 2. $f(x, y) = 8x^2$

3. $f(x, y) = y^3 + 1$ 4. $f(x, y, z) = x + y + z$

5. $f(x, y) = \dfrac{(x^2 + y^2 - xy)^{1/2}}{x^3 + y^3 + x^2y}$ 6. $f(x, y) = x^2 + 2xy + x(x + y)$

7. $f(x, y, z) = xyz + 4$ 8. $f(x, y) = \dfrac{x^2(x + y)}{x^4 + x^3y + x^2(x^2 + y^2)}$

9. $f(x, y) = \ln\dfrac{y}{x} + \arctan\dfrac{x}{y}$ 10. $f(x, y) = x \ln y - y \ln x$

In each of problems 11 through 22, solve the differential equation.

11. $(x + 2y)\, dx + (-2x + y)\, dy = 0$ 12. $(2x - y)\, dx + (x - 2y)\, dy = 0$

13. $(2x + y)\, dx - y\, dy = 0$

14. $(2x^2 - y^2)\, dx - xy\, dy = 0$

15. $\dfrac{dy}{dx} = \dfrac{3x^2 + 6xy - y^2}{5x^2 + 2xy + y^2}$

16. $\dfrac{dy}{dx} = \dfrac{x^2 + y^2}{2xy}$

17. $x\dfrac{dy}{dx} = y - \sqrt{x^2 - y^2}$

18. $\dfrac{dy}{dx} = \dfrac{y}{x} + \sin\dfrac{y}{x}$

19. $(x\sqrt{x^2 + y^2} - y^2)\, dx + xy\, dy = 0$

20. $[x + (x - y)e^{y/x}]\, dx + xe^{y/x}\, dy = 0$

21. $\left(\dfrac{1}{x} - \dfrac{y}{x^2}\, e^{y/x}\right) dx + \left(\dfrac{1}{x}\, e^{y/x} - \dfrac{1}{y}\right) dy = 0$

22. $\left(\dfrac{1}{x - y} + \dfrac{y}{x^2 + y^2}\right) dx + \left(\dfrac{1}{y - x} - \dfrac{x}{x^2 + y^2}\right) dy = 0$

23. If $f(x)$ and $g(y)$ are each homogeneous of degree zero, show that the equation

$$\frac{dy}{dx} = f(x) + g(y)$$

may be solved explicitly.

24. Given the equation

$$(a_1x + b_1y + c_1)\, dx + (a_2x + b_2y + c_2)\, dy = 0,$$

with $a_1, b_1, c_1, a_2, b_2, c_2$ constant, show how h and k may be selected so that the change in variables $x' = x + h$ and $y' = y + k$ reduces the equation to one of homogeneous type. Express h and k in terms of the $a_i, b_i,$ and c_i. Are any conditions necessary?

4. EXACT DIFFERENTIALS

Suppose that the differential equation

$$P(x, y)\, dx + Q(x, y)\, dy = 0 \tag{1}$$

satisfies the condition

$$\frac{\partial P}{\partial y} = \frac{\partial Q}{\partial x}.$$

Then the left side of (1) is an exact differential and there is a function $f(x, y)$ such that

$$df = P\, dx + Q\, dy.$$

Methods for finding the function f were discussed in Chapter 4, Section 13. (See also Chapter 13, Section 4.) Once the function f is obtained, we see that any differentiable function y which satisfies the equation

$$f(x, y) = c,$$

where c is any constant, is a solution of (1); this follows immediately from the Chain Rule. In general, the relation (2) will only give y as a function of x in implicit form.

EXAMPLE. Solve

$$(3x^2 + 4xy - y^2 - 2)\,dx + (2x^2 - 2xy + 3y^2 + 3)\,dy = 0.$$

Solution. We have

$$\frac{\partial}{\partial y}(3x^2 + 4xy - y^2 - 2) = 4x - 2y = \frac{\partial}{\partial x}(2x^2 - 2xy + 3y^2 + 3).$$

Then

$$f(x, y) = \int [3x^2 + 4xy - y^2 - 2]\,dx + C(y)$$

$$= x^3 + 2x^2y - xy^2 - 2x + C(y)$$

and

$$f_y(x, y) = 2x^2 - 2xy + C'(y) = 2x^2 - 2xy + 3y^2 + 3.$$

Therefore

$$C'(y) = 3y^2 + 3, \qquad C(y) = y^3 + 3y + C_1.$$

We conclude that if y is any differentiable function satisfying the equation

$$f(x, y) \equiv x^3 + 2x^2y - xy^2 - 2x + y^3 + 3y = C \tag{3}$$

then y satisfies the differential equation. The solution of y in terms of x is expressed implicitly by (3).

EXERCISES

In the following exercises, solve the given differential equations.

1. $(x + 2y - 2)\,dx + (2x - y + 3)\,dy = 0$
2. $(x + y - 4)\,dx + (x + y + 2)\,dy = 0$
3. $(y - 3x^2 + 2)\,dx + (x - y^2 + 2y)\,dy = 0$
4. $(3x^2 - 4xy + y^2 - 3y)\,dx + (2xy - 2x^2 + 6y^2 - 3x)\,dy = 0$
5. $(2x^3 + 6xy^2 - 2y^3 + 4x + 3y)\,dx + (6x^2y - 6xy^2 - 4y^3 + 3x - 2y)\,dy = 0$
6. $(2x \sin y + e^x \cos y)\,dx + (x^2 \cos y - e^x \sin y)\,dy = 0$

7. $\arcsin y\,dx + \dfrac{x + 2\sqrt{1 - y^2}\cos y}{\sqrt{1 - y^2}}\,dy = 0$

8. $(2x + ye^{xy})\,dx + (\cos y + xe^{xy})\,dy = 0$

9. $\left(\dfrac{1}{x - y} + \dfrac{x}{x^2 + y^2}\right)dx + \left(\dfrac{1}{y - x} + \dfrac{y}{x^2 + y^2}\right)dy = 0$

10. $e^x(x^2 + y^2 + 2x)\,dx + 2ye^x\,dy = 0$

11. $(x \ln y + y \ln x + y)\,dx + \left(\dfrac{x^2}{2y} + x \ln x\right)dy = 0$

12. $\arctan(y^2)\,dx + \dfrac{2xy}{1 + y^4}\,dy = 0$

5. INTEGRATING FACTORS

It sometimes happens that an expression

$$P(x, y)\, dx + Q(x, y)\, dy$$

is not an exact differential, but that a function $I(x, y)$ can be found so that

$$I(x, y)(P\, dx + Q\, dy) \tag{1}$$

is an exact differential. If $I \neq 0$, then the solutions of $P\, dx + Q\, dy = 0$ and $I(P\, dx + Q\, dy) = 0$ are the same.

DEFINITION. If a nonvanishing function $I(x, y)$ makes the expression (1) an exact differential, it is called an **integrating factor**.

There are no general rules for finding integrating factors, and usually none can be found. However, there are a few simple cases where known expressions can be employed as integrating factors. We illustrate with two examples, and remark that the exercises at the end of this section were concocted so that integrating factors for them can be found.

EXAMPLE 1. Solve $x^2\, dy + (xy + x^2 + 1)\, dx = 0$.

Solution. We observe that the equation is not an exact differential. However, we note that

$$x^2\, dy + xy\, dx = x(x\, dy + y\, dx) = x\, d(xy).$$

Therefore, if we divide the original equation by x, we obtain

$$x\, dy + y\, dx + \left(x + \frac{1}{x}\right) dx = 0.$$

We may write this equation

$$d(xy) + \left(x + \frac{1}{x}\right) dx = 0,$$

and so the solution is

$$xy + \tfrac{1}{2}x^2 + \ln|x| = C.$$

The integrating factor we employed is $I(x, y) = 1/x$.

In Example 1, the key to the solution was the observation that the combination $x\, dy + y\, dx$ is $d(xy)$. There are a few other simple differentials which, when we recognize their form in an equation, may lead to an integrating factor. Some of these differentials are:

$$\frac{x\, dy - y\, dx}{x^2} = d\left(\frac{y}{x}\right), \qquad \frac{x\, dy - y\, dx}{x^2 + y^2} = d\left(\arctan\frac{y}{x}\right)$$

$$x\, dx + y\, dy = \tfrac{1}{2}\, d(x^2 + y^2).$$

EXAMPLE 2. Solve

$$x \, dy - y \, dx = 2x^2 \ln y \, dy.$$

Solution. We observe that

$$\frac{x \, dy - y \, dx}{x^2} = d\left(\frac{y}{x}\right),$$

and so we try $I(x, y) = 1/x^2$ as an integrating factor. The result is

$$\frac{x \, dy - y \, dx}{x^2} - 2 \ln y \, dy = 0, \tag{2}$$

which is exact. To obtain the solution, we may employ the methods of Section 4 or proceed in the following way. Equation (2) may be written

$$d\left(\frac{y}{x}\right) = 2 \ln y \, dy$$

and, upon integration of both sides, we find

$$\frac{y}{x} - 2(y \ln y - y) = C.$$

EXERCISES

In the following exercises, solve the given differential equations.

1. $x \, dy - y \, dx = (x^2 + y^2) \, dx$
2. $x \, dy - y \, dx = (x^2 y + x^2 y^3) \, dy$
3. $x \, dy - y \, dx = (x^2 + y^2) \, dy$
4. $x \, dy - (y + x^3 e^{2x}) \, dx = 0$
5. $ye^{-x/y} \, dx - (xe^{-x/y} + y^3) \, dy = 0$
6. $x \cos^2 y \, dx + 2 \csc x \, dy = 0$
7. $dx + (x \tan y - 2 \sec y) \, dy = 0$
8. $x \, dx + y \, dy + (x^2 + y^2)(y \, dx - x \, dy) = 0$
9. $(2 - xy)y \, dx + (2 + xy)x \, dy = 0$
10. $(\sqrt{x^2 - y^2} - y) \, dx + x \, dy = 0$. [*Hint.* Compute $d \arcsin (y/|x|)$ for $x > 0$ and for $x < 0$.]

6. THE GENERAL LINEAR FIRST-ORDER EQUATION

The most general linear equation of the first order has the form

$$c_0(x)\frac{dy}{dx} + c_1(x)y = f(x). \tag{1}$$

Dividing this equation by $c_0(x)$, we obtain

$$\frac{dy}{dx} + P(x)y = Q(x), \tag{2}$$

which is equivalent to (1) so long as $c_0(x) \neq 0$. It turns out that every such equation has an integrating factor I which is a function of x alone. To see this, we multiply (2) by I, getting

$$I\frac{dy}{dx} + IPy = Q(x)I. \tag{3}$$

If we write

$$I\frac{dy}{dx} + \frac{dI}{dx}y \equiv \frac{d}{dx}(Iy) = Q(x)I, \tag{4}$$

then (3) and (4) will be identical if and only if

$$\frac{dI}{dx} = IP(x). \tag{5}$$

Equation (5) is of the variables separable type, and we find

$$\frac{dI}{I} = P(x)\,dx, \quad \ln I = \int P(x)\,dx.$$

Therefore the function

$$I = e^{\int P\,dx} \tag{6}$$

puts (3) into the form (4). However, (4) is easily integrated. We see at once that

$$\int \frac{d}{dx}(Iy) = \int QI\,dx,$$

and so

$$e^{\int P\,dx}y = \int Qe^{\int P\,dx}\,dx + C.$$

The explicit solution is

$$y = e^{-\int P\,dx}\int^x Q(\xi)e^{\int^\xi P\,dx}\,d\xi + Ce^{-\int P\,dx}.$$

It is much easier to remember the form of the integrating factor (6) and work through the development of the solution than it is to try to remember the formula for the explicit answer. An example shows the method.

EXAMPLE. Solve

$$\frac{dy}{dx} + 2xy = 2x^3.$$

Solution. We have $P = 2x$ and $\int P\,dx = x^2$, and we take $I = e^{x^2}$ as an integrating factor. We write

$$e^{x^2}\frac{dy}{dx} + 2xe^{x^2}y = 2x^3e^{x^2},$$

and so

$$\frac{d}{dx}(e^{x^2}y) = 2x^3e^{x^2}.$$

Therefore

$$e^{x^2}y = \int 2x^3 e^{x^2}\, dx + C = (x^2 - 1)e^{x^2} + C$$

and

$$y = (x^2 - 1) + Ce^{-x^2}.$$

EXERCISES

Solve the following differential equations.

1. $\dfrac{dy}{dx} - \dfrac{y}{x} = x^n, \quad n \neq 0$

2. $\dfrac{dy}{dx} - \dfrac{y}{x} = 1$

3. $\dfrac{dy}{dx} - y = (3x^2 + 4x - 3)e^x$

4. $\dfrac{dy}{dx} - 2y = e^x$

5. $x\dfrac{dy}{dx} + y = e^{2x} - \sin x$

6. $x\dfrac{dy}{dx} + 2y = e^x$

7. $\dfrac{dy}{dx} - \dfrac{xy}{x^2 - 1} = x$

8. $\dfrac{dy}{dx} + y \sin x = xe^{\cos x}$

9. $\dfrac{dy}{dx} + \dfrac{3y}{2x - 3} = \dfrac{x}{\sqrt{2x - 3}}$

10. $\dfrac{dy}{dx} - ry = \cos sx, \quad r, s$ constant

11. $(2y - x^2 + 1)\, dx + (x^2 - 1)\, dy = 0$

12. $(3xy + 2e^{x^3})x\, dx - dy = 0$

13. $[y \cos 2x + 2(\sin 2x)^{3/2}]\, dx + \sin 2x\, dy = 0$

14. $(x^2 + 1)y' - 2xy = 6x^2 + 2x$

15. $(a^2 - x^2)\dfrac{dy}{dx} + 2ay = (a^2 - x^2)^2$

7. EQUATIONS OF BERNOULLI AND CLAIRAUT

The **equation of Bernoulli**

$$\frac{dy}{dx} + P(x)y = Q(x)y^n, \qquad n \neq 1 \tag{1}$$

can always be reduced to a linear equation and so solved by the method described in Section 6. To see this, we divide both sides of (1) by y^n, obtaining

$$y^{-n}\frac{dy}{dx} + P(x)y^{1-n} = Q(x). \tag{2}$$

If we set $z = y^{1-n}$ and $dz/dx = (1 - n)y^{-n}(dy/dx)$, then (2) becomes

$$\frac{1}{1 - n}\frac{dz}{dx} + P(x)z = Q(x),$$

which is a linear equation. Once z is determined, the substitution $y = z^{1/(1-n)}$ yields the solution.

EXAMPLE 1. Solve

$$(1 + x^2)\frac{dy}{dx} + xy = x^3 y^3.$$

Solution. Upon division by $1 + x^2$, this equation is a Bernoulli equation with $n = 3$. Setting $z = y^{1-n} = y^{-2}$ and dividing by y^3, we find

$$y^{-3}\frac{dy}{dx} + \frac{x}{1 + x^2}y^{-2} = \frac{x^3}{1 + x^2}$$

or

$$-2y^{-3}\frac{dy}{dx} - \frac{2x}{1 + x^2}y^{-2} = \frac{-2x^3}{1 + x^2}.$$

Therefore

$$\frac{dz}{dx} - \frac{2x}{1 + x^2}z = -\frac{2x^3}{1 + x^2}. \tag{3}$$

To obtain the integrating factor, we first calculate

$$\int -\frac{2x}{1 + x^2}\,dx = -\ln(1 + x^2).$$

Then the integrating factor is $e^{-\ln(1+x^2)} = (1 + x^2)^{-1}$. Multiplying (3) by this integrating factor, we can write

$$\frac{d}{dx}[(1 + x^2)^{-1}z] = -\frac{2x^3}{(x^2 + 1)^2} = -\frac{2x(x^2 + 1)}{(x^2 + 1)^2} + \frac{2x}{(x^2 + 1)^2}$$

and

$$(1 + x^2)^{-1}z = -\ln(x^2 + 1) - \frac{1}{x^2 + 1} + C,$$

$$y^{-2} = z = -(x^2 + 1)\ln(x^2 + 1) + C(x^2 + 1) - 1,$$

$$y = 1/[-(x^2 + 1)\ln(x^2 + 1) + C(x^2 + 1) - 1]^{1/2}.$$

Any equation in the special form

$$x\frac{dy}{dx} - y + f\left(\frac{dy}{dx}\right) = 0 \tag{4}$$

is called a **Clairaut equation**. If f is a differentiable function of its argument, this equation can always be solved by an interesting and useful trick. We write

$$p = \frac{dy}{dx},$$

and Clairaut's equation becomes

$$y = px + f(p). \tag{5}$$

Differentiation of (5) with respect to x yields a second-order equation. However, because of the special form of Clairaut's equation, this differential equation can be integrated twice to give a solution to the original problem. To see this, we differentiate (5), obtaining

$$\frac{dy}{dx} = p + x\frac{dp}{dx} + f'(p)\frac{dp}{dx}$$

or, since dy/dx is p, the equation

$$[x + f'(p)]\frac{dp}{dx} = 0.$$

The only possible solutions are those functions which satisfy

$$\frac{dp}{dx} = 0 \quad \text{or} \quad x + f'(p) = 0.$$

In the first alternative

$$p = C,$$

and substitution of this fact into Clairaut's equation (5) gives the solution

$$y = Cx + f(C). \tag{6}$$

In the second alternative $x = -f'(p)$, and substitution of this relation into (5) yields

$$y = f(p) - pf'(p). \tag{7}$$

If p is treated as a parameter, then the two equations

$$x = -f'(p), \qquad y = f(p) - pf'(p) \tag{8}$$

determine a curve which is a solution of Clairaut's equation.

A striking feature of Clairaut's equation is the relation between Eq. (6) and the original equation. We see that *replacing p by C in Clairaut's equation always determines a solution.*

The solution (8) does not involve a constant of integration and, except to note that such solutions are called singular, we shall not discuss the particular features of these special curves.

———————

EXAMPLE 2. Solve $$y = x\frac{dy}{dx} + \cos\left(\frac{dy}{dx}\right).$$

Solution. This is a Clairaut equation with $f(p) = \cos p$. One solution is

$$y = Cx + \cos C.$$

The singular solution is

$$x = \sin p, \qquad y = \cos p + p \sin p.$$

Eliminating the parameter in the singular solution, we find that

$$y = \begin{cases} \sqrt{1 - x^2} + x(\arcsin x + 2n\pi), & \text{or} \\ -\sqrt{1 - x^2} + x(\pi - \arcsin x + 2n\pi) \end{cases}, \quad |x| \leq 1, \ n = 0, \pm 1, \pm 2, \ldots.$$

EXERCISES

In each of problems 1 through 14, solve the differential equation. Find the singular solution whenever appropriate and eliminate the parameter in this solution, if possible.

1. $\dfrac{dy}{dx} + \dfrac{2y}{x} = x^2 y^4$

2. $\dfrac{dy}{dx} - y = \dfrac{\sin 2x}{y^2}$

3. $\dfrac{ds}{dt} = \dfrac{t^2 + s^2}{2st}$

4. $x\dfrac{dy}{dx} + y = y^2 \ln x$

5. $(y + xy^2)\, dx - dy = 0$

6. $(1 + x^2)\, dy - [(1 + x)y + x^2 y^3]\, dx = 0$

7. $(x^2 - y^3)\, dx + 3xy^2\, dy = 0$

8. $x(1 + x^2)\, dy - [(1 + x^2)y + xy^2]\, dx = 0$

9. $(y - xy^3)\, dx + x\, dy = 0$

10. $x\dfrac{dy}{dx} - y = 3\left(\dfrac{dy}{dx}\right)^2$

11. $2x\dfrac{dy}{dx} - 2y - \left(\dfrac{dy}{dx}\right)^2 = 0$

12. $y = x\dfrac{dy}{dx} + \sin 2\left(\dfrac{dy}{dx}\right)$

13. $y = x\dfrac{dy}{dx} + \ln\left(\dfrac{dy}{dx}\right)$

14. $y = x\dfrac{dy}{dx} + \sqrt{1 + (dy/dx)^2}$

15. By following the method of Clairaut's equation, solve

$$y = 2px - xp^2, \qquad \text{where } p = dy/dx.$$

8. SECOND-ORDER EQUATIONS HAVING A SPECIAL FORM

A completely general second-order differential equation has the form

$$F\left(x,\ y,\ \frac{dy}{dx},\ \frac{d^2y}{dx^2}\right) = 0, \tag{1}$$

where F is an arbitrary function of its arguments. We are not in a position to develop the theory of such general differential equations. If we assume that the Implicit Function Theorem holds with respect to the last argument of the function F in (1), we can solve for d^2y/dx^2 and write the general second-order equation in the form

$$\frac{d^2y}{dx^2} = f\left(x,\ y,\ \frac{dy}{dx}\right). \tag{2}$$

Equation (2) looks better but is still too general to handle by elementary means. However, we are able to treat four types of special cases of Eq. (2).

Type I: $\dfrac{d^2y}{dx^2} = f(x).$

Type III: $\dfrac{d^2y}{dx^2} = f(y).$

Type II: $\dfrac{d^2y}{dx^2} = f\left(x, \dfrac{dy}{dx}\right).$

Type IV: $\dfrac{d^2y}{dx^2} = f\left(y, \dfrac{dy}{dx}\right).$

Before taking up methods for solving each of these types, we should realize how special they are. For example, the simple *linear* second-order equation

$$\frac{d^2y}{dx^2} + C_1(x)\frac{dy}{dx} + C_2(x)y = f(x)$$

is not one of the four types listed above.

Type I. Equations of the form $d^2y/dx^2 = f(x)$.

These have already been discussed. We obtain

$$\frac{dy}{dx} = F(x) + C_1, \qquad y = G(x) + C_1x + C_2,$$

where F is a particular antiderivative of f and G is one of F.

Type II. Equations of the form $d^2y/dx^2 = f(x, dy/dx)$.

Here we set $dy/dx = p$ and, since $d^2y/dx^2 = dp/dx$, the equation becomes

$$\frac{dp}{dx} = f(x, p).$$

If this first-order equation can be solved, so that

$$p = G(x, C_1),$$

then a second integration of the equation

$$\frac{dy}{dx} = G(x, C_1)$$

can be performed to yield

$$y = F(x, C_1) + C_2,$$

in which F is an antiderivative of G.

EXAMPLE 1. Solve

$$x\frac{d^2y}{dx^2} + 2\frac{dy}{dx} = x^3, \qquad x \neq 0.$$

Solution. Since

$$\frac{d^2y}{dx^2} = x^2 - \frac{2}{x}\frac{dy}{dx},$$

the equation is of Type II. Setting $p = dy/dx$, we obtain

$$\frac{dp}{dx} + \frac{2}{x}p = x^2,$$

which is a linear first-order equation. The function $I = x^2$ is an integrating factor, and so

$$\frac{d}{dx}(x^2 p) = x^4 \quad \text{or} \quad x^2 p = \tfrac{1}{5}x^5 + C_1.$$

Therefore

$$\frac{dy}{dx} = \frac{1}{5}x^3 + C_1 x^{-2} \quad \text{and} \quad y = \frac{x^4}{20} - C_1 x^{-1} + C_2.$$

Type III. Equations of the form $d^2y/dx^2 = f(y)$.

Here we set $p = dy/dx$, and the equation becomes

$$\frac{dp}{dx} = f(y).$$

Now, by the Chain Rule, we may write

$$\frac{dp}{dx} = \frac{dp}{dy}\frac{dy}{dx} = \frac{dp}{dy}p,$$

and so

$$p\frac{dp}{dy} = f(y),$$

which is an equation with variables separable. Integrating once, we obtain

$$\tfrac{1}{2}p^2 = F(y) + C_1,$$

where F is an antiderivative of f. Then we have

$$\frac{dy}{dx} = \pm\sqrt{2F(y) + 2C_1},$$

which again is a variables separable equation. Therefore

$$x = \pm\int \frac{dy}{\sqrt{2C_1 + 2F(y)}}, \quad C_1 + F(y) > 0.$$

Many problems in differential equations concern solutions which satisfy additional auxiliary conditions. When these conditions are just sufficient to determine specific values for the constants of integration which arise, we say that we have a **particular solution** to the differential equation. We illustrate with an example.

EXAMPLE 2. If a particle of mass m pounds is moving with an acceleration of a ft/sec^2 then, according to Newton's Law, the force F is given by the formula $F = ma$, where F is measured in *poundals*. Suppose that a particle of mass m moving along the positive x-axis is repelled with a force $F = 4mx^{-3}$ poundals by a particle situated at the origin.

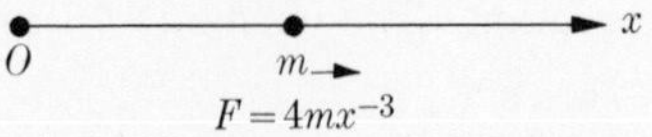

FIGURE 15–1 $F = 4mx^{-3}$

(See Fig. 15–1.) Find the motion of the particle if, at time $t = 0$, $x = 2$ and the velocity $v = -\sqrt{3}$. Distance is measured in feet and time in seconds.

Solution. Since $a = d^2x/dt^2$, we have

$$4mx^{-3} = m\frac{d^2x}{dt^2}, \qquad x(0) = 2, \qquad v(0) = -\sqrt{3}.$$

The equation

$$\frac{d^2x}{dt^2} = 4x^{-3}$$

is of Type III. Writing $v = dx/dt$, we see that

$$\frac{d^2x}{dt^2} = \frac{dv}{dt} = \frac{dv}{dx}\frac{dx}{dt} = v\frac{dv}{dx} = 4x^{-3}.$$

Integrating, we find (using $\frac{1}{2}C_1^2$ instead of C_1 for a constant of integration)

$$v^2 = C_1^2 - 4x^{-2} = x^{-2}(C_1^2x^2 - 4).$$

Now we employ the auxiliary conditions to find C_1. Substituting $x = 2$, $v = -\sqrt{3}$ when $t = 0$, we obtain

$$3 = \tfrac{1}{4}(4C_1^2 - 4) \qquad \text{or} \qquad C_1^2 = 4.$$

Hence

$$v = \pm \frac{2}{x}\sqrt{x^2 - 1}, \qquad x \geq 1$$

but, since $v = -\sqrt{3}$ when $x = 2$, the negative sign must be chosen. Therefore

$$\frac{dx}{dt} = -\frac{2}{x}\sqrt{x^2 - 1}$$

and

$$\frac{x\,dx}{\sqrt{x^2 - 1}} = -2\,dt \qquad \text{or} \qquad \sqrt{x^2 - 1} = -2t + C_2.$$

Since $x = 2$ when $t = 0$, we get

$$x^2 = 1 + (\sqrt{3} - 2t)^2 \qquad \text{or} \qquad x = \sqrt{1 + (\sqrt{3} - 2t)^2}.$$

The particle starts at the location $x = 2$ and moves to the left until, at time $t = \sqrt{3}/2$ sec, it reaches the location $x = 1$; at this point v is zero. Then it begins moving to the right and continues to do so with increasing speed.

Type IV. Equations of the form $d^2y/dx^2 = f(y, dy/dx)$.

Here we proceed as in equations of Type III. We write

$$p = \frac{dy}{dx}, \qquad \frac{d^2y}{dx^2} = p\frac{dp}{dy} = f(y, p).$$

The first-order equation

$$\frac{dp}{dy} = \frac{1}{p} f(y, p)$$

may or may not be solvable by one of the methods we studied for first-order equations. If it is, we have

$$p = F(y, C_1),$$

and then the equation

$$\frac{dy}{dx} = F(y, C_1)$$

is of the variables separable type. The solution is

$$x = \int \frac{dy}{F(y, C_1)} + C_2,$$

provided that $F \neq 0$.

EXAMPLE 3. Find the particular solution of the equation

$$y\frac{d^2y}{dx^2} + 4y^2 - \frac{1}{2}\left(\frac{dy}{dx}\right)^2 = 0$$

for which $y = 1$ and $dy/dx = -\sqrt{8}$ when $x = 0$.

Solution. This equation is of Type IV. Setting

$$p = \frac{dy}{dx}, \qquad \frac{d^2y}{dx^2} = p\frac{dp}{dy},$$

we obtain

$$yp\frac{dp}{dy} + 4y^2 - \frac{1}{2}p^2 = 0 \qquad \text{and} \qquad \frac{dp}{dy} - \frac{1}{2y}p = -4yp^{-1},$$

which we recognize as a Bernoulli equation with $n = -1$. Thus we find (after setting $q = p^2$)

$$2p\frac{dp}{dy} - \frac{1}{y}p^2 = -8y \qquad \text{or} \qquad \frac{dq}{dy} - \frac{1}{y}q = -8y.$$

The integrating factor for the linear equation in q is y^{-1}. Hence

$$\frac{d}{dy}(y^{-1}q) = -8 \qquad \text{or} \qquad y^{-1}q = -8y + C_1.$$

Substituting back, we get

$$p^2 = C_1 y - 8y^2.$$

For $x = 0$, we have $y = 1$ and $p = -\sqrt{8}$. Therefore $C_1 = 16$ and, for convenience, we write $C_1 = 16a^2$ where $a = \pm 1$. Then

$$\frac{dy}{dx} = \sqrt{8}\, a\sqrt{2y - y^2}.$$

The fact that $dy/dx = -\sqrt{8}$ when $y = 1$ means that a must be selected as -1. Separation of variables yields

$$\frac{dy}{\sqrt{1 - (y - 1)^2}} = -\sqrt{8}\, dx \quad \text{or} \quad \arcsin{(y - 1)} = -\sqrt{8}\, x + C_2.$$

Finally, since $y = 1$ when $x = 0$, we conclude that $C_2 = 0$, and the particular solution is

$$y = 1 - \sin{(x\sqrt{8})}.$$

EXERCISES

In each of problems 1 through 14, solve the given differential equation.

1. $\dfrac{d^2 y}{dx^2} = e^{2x} + \cos 2x$

2. $\dfrac{d^2 y}{dx^2} = \dfrac{1}{x}$

3. $\dfrac{d^2 y}{dx^2} = \dfrac{x}{\sqrt{1 + x^2}}$

4. $\dfrac{d^2 y}{dx^2} + 2 \dfrac{dy}{dx} = 0$

5. $\dfrac{d^2 y}{dx^2} - 2 \dfrac{dy}{dx} = e^{2x}$

6. $x \dfrac{d^2 y}{dx^2} + 2 \dfrac{dy}{dx} = x$

7. $\dfrac{d^2 y}{dx^2} - 4 = \left(\dfrac{dy}{dx}\right)^2$

8. $\dfrac{d^2 y}{dx^2} - 4 \dfrac{dy}{dx} = \left(\dfrac{dy}{dx}\right)^2$

9. $\dfrac{d^2 y}{dx^2} + k^2 y = 0$

10. $\dfrac{d^2 y}{dx^2} + y \dfrac{dy}{dx} = 0$

11. $y \dfrac{d^2 y}{dx^2} = \left(\dfrac{dy}{dx}\right)^2 + 4$

12. $(x + 2) \dfrac{d^2 y}{dx^2} + \dfrac{dy}{dx} = x^2$

13. $\dfrac{d^2 y}{dx^2} = \dfrac{1}{y^2}$

14. $\dfrac{d^2 y}{dx^2} = e^y$

In each of problems 15 through 20, find the solution of the given differential equation for which y and dy/dx have their given values for the specified value of x.

15. $\dfrac{d^2 y}{dx^2} = \dfrac{3}{2} y^2$, $y = 4$ and $\dfrac{dy}{dx} = 8$ when $x = 2$

16. $\dfrac{d^2 y}{dx^2} = 2y^3 + 8y$, $y = 2$ and $\dfrac{dy}{dx} = -8$ for $x = \dfrac{\pi}{4}$

17. $\dfrac{d^2 y}{dx^2} + \dfrac{dy}{dx} = \sin x$, $y = -3$ and $\dfrac{dy}{dx} = \dfrac{1}{2}$ when $x = \dfrac{\pi}{2}$

18. $\dfrac{d^2 y}{dx^2} + y = 0$, $y = -1$ and $\dfrac{dy}{dx} = -\sqrt{3}$ when $x = \dfrac{\pi}{6}$

19. $\dfrac{d^2 y}{dx^2} = y$, $y = 4$ and $\dfrac{dy}{dx} = -5$ when $x = 0$

20. $\dfrac{d^2 s}{dt^2} = 100 - \left(\dfrac{ds}{dt}\right)^2$, $s = 0$ and $\dfrac{ds}{dt} = 26$ when $t = 0$

9. APPLICATIONS

Differential equations may be applied in such disciplines as mechanics, physics, and engineering as well as in various branches of mathematics. In this section we restrict our attention to some applications to geometric problems and to those simple problems in mechanics which employ elementary forms of Newton's laws. In the applications to be considered we must first determine the differential equation which describes the situation and then find a solution which takes into account the various conditions prescribed in the particular application. Since each problem seems unlike all others, at least in appearance, it is difficult to give general rules of procedure which would cover large classes of applications. All we can do is study a number of illustrative examples and then solve some typical exercises.

EXAMPLE 1. Find the equation of the curve such that its slope at any point $P(x, y)$ is equal to the difference of the squares of the distances of $P(x, y)$ from the points $(1, 0)$ and $(4, 0)$.

Solution. See Fig. 15–2, in which the distances from P to $(1, 0)$ and $(4, 0)$ are labeled d_1 and d_2, respectively. The conditions of the problem state that

$$\frac{ay}{dx} = d_1^2 - d_2^2,$$

which is a differential equation. Therefore

$$\frac{dy}{dx} = (x - 1)^2 + y^2 - (x - 4)^2 - y^2$$

$$= 6x - 15.$$

Integrating, we find

$$y = 3x^2 - 15x + C.$$

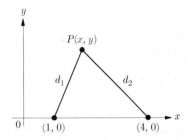

FIGURE 15–2

The solution consists of a family of parabolas.

EXAMPLE 2. Find the equation in polar coordinates of the curve such that the tangent of the angle between the radius vector and the tangent to the curve is equal to the square of the radius vector.

Solution. Let $P(r, \theta)$ be a typical point on the curve and denote by ψ the angle between the radius vector and the tangent. (See Fig. 15–3.) The condition of the problem states that

$$\tan \psi = r^2.$$

In the study of curves in polar coordinates we learned (*First Course*, page 414) that

$$\cot \psi = \pm \frac{1}{r} \frac{dr}{d\theta},$$

and so the condition of the problem yields the differential equation

$$r\frac{d\theta}{dr} = \pm r^2 \quad \text{or} \quad r\frac{dr}{d\theta} = \pm 1.$$

The solution is the family of curves

$$r^2 = \pm 2\theta + C.$$

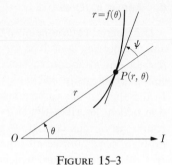

Suppose we are given a relation

FIGURE 15–3

$$\phi(x, y, c) = 0, \tag{1}$$

in which c is an arbitrary constant. In general, Eq. (1) will represent a family of
curves, one for each value of the constant c. Suppose we wish to find a first-order
differential equation which is satisfied by every curve of the family. If (1) can be
solved for c, giving

$$f(x, y) = c,$$

then such a differential equation is obtained by differentiating with respect to x
or y. We write

$$f_x \, dx + f_y \, dy = 0,$$

which is the desired equation. If it is not convenient to solve for c first, we may
differentiate (1) with respect to x, getting

$$\phi_x(x, y, c) + \phi_y(x, y, c)\frac{dy}{dx} = 0. \tag{2}$$

Elimination of c between (1) and (2) yields the differential equation. In the applications it is usually essential that the constant c appearing in the family of
curves be absent from the differential equation.

EXAMPLE 3. The equation

$$\frac{x^2}{c^2 + 4} + \frac{y^2}{c^2} = 1 \tag{3}$$

represents a family of confocal ellipses. Find a first-order differential equation satisfied
by all such ellipses.

Solution. Multiplying by $c^2(c^2 + 4)$, we obtain

$$c^2x^2 + (c^2 + 4)y^2 = c^2(c^2 + 4).$$

Differentiating with respect to x, we find

$$c^2x + (c^2 + 4)y\frac{dy}{dx} = 0 \quad \text{or} \quad c^2 = -\frac{4y(dy/dx)}{x + y(dy/dx)}.$$

Substituting in (3), we get (after some algebraic work)

$$\frac{x^2[x + y(dy/dx)]}{4x} - \frac{y^2[x + y(dy/dx)]}{4y(dy/dx)} = 1.$$

A simplification yields

$$xy\left(\frac{dy}{dx}\right)^2 + (x^2 - y^2 - 4)\frac{dy}{dx} - xy = 0.$$

We could have obtained the same result by first solving (3) for c^2 and then differentiating with respect to x.

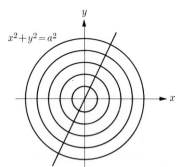

$x^2 + y^2 = a^2$

FIGURE 15–4

Suppose that $\phi(x, y, c) = 0$ represents a family of curves. An **orthogonal trajectory** to this family is a curve which intersects every member of the family at right angles. For example, the family of curves $x^2 + y^2 = a^2$ has as an orthogonal trajectory any line through the origin (Fig. 15–4). We can use the methods of differential equations to find the orthogonal trajectories of a given family. The general procedure depends on the known fact that two curves which intersect at right angles have slopes at the point of intersection which are the negative reciprocals of each other. Therefore, to find the orthogonal trajectories of a family $\phi(x, y, c) = 0$, we first find the differential equation which the family satisfies (eliminating c in the process), and we then replace dy/dx by $-dx/dy$ in this equation. The solution of the new differential equation gives the orthogonal trajectories. An example illustrates the process.

EXAMPLE 4. Find the orthogonal trajectories of the family of parabolas $y^2 = cx$. Sketch these trajectories.

Solution. We have

$$2y\frac{dy}{dx} = c \qquad \text{or} \qquad y = 2x\frac{dy}{dx}$$

as the differential equation of the parabolas. For the orthogonal trajectories, we write

$$y = -2x\frac{dx}{dy} \qquad \text{or} \qquad \frac{dy}{dx} = -\frac{2x}{y} \qquad \text{or} \qquad 2x\,dx + y\,dy = 0.$$

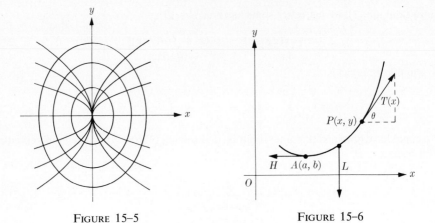

FIGURE 15–5   FIGURE 15–6

The solution is

$$2x^2 + y^2 = 2c^2 \qquad \text{or} \qquad \frac{x^2}{c^2} + \frac{y^2}{2c^2} = 1.$$

The orthogonal trajectories are confocal ellipses. These ellipses and the family of parabolas are sketched in Fig. 15–5.

We give an example from mechanics.

EXAMPLE 5. Each foot of a uniform cable weighs w pounds. Assuming that the cable is perfectly flexible and that it is suspended from its ends, find the equation of the curve which the cable forms.

Solution. To find the differential equation we use the fact that the forces on any object in equilibrium must balance. The forces on the cable are those due to gravity and to the tension in the cable which prevents it from falling. Let $A(a, b)$ be the lowest point of the cable, and let $P(x, y)$ be any point on the cable with $x > a$, for example (Fig. 15–6). The horizontal component of the tension in the cable will be a constant which we denote by H. The vertical pull of gravity, directed downward, is denoted by $\mathbf{L}(x)$. If θ is the angle of the tangent at P, we decompose the tension vector $\mathbf{T}$ (which is tangent to the cable) into its vertical and horizontal components. Writing $L(x)$, $T(x)$, and H for the magnitudes of the corresponding vectors, we have

$$T(x) \sin \theta = L(x), \qquad T(x) \cos \theta = H.$$

Since $\tan \theta = dy/dx$, we obtain

$$\frac{dy}{dx} = \frac{L(x)}{H}.$$

The downward pull of gravity L is due to the weight of the cable alone. Letting $s(x)$ denote the length of the cable, we have

$$L(x) = ws(x), \qquad \frac{dy}{dx} = \frac{w}{H} s(x) = ks(x) \quad \text{if} \quad k = \frac{w}{H}.$$

Differentiating the equation $dy/dx = ks$ and taking into account the relation $ds = \sqrt{1 + (dy/dx)^2}\, dx$, we find

$$\frac{d^2y}{dx^2} = k\sqrt{1 + (dy/dx)^2}. \tag{4}$$

This equation is of Type II among those considered in Section 8. To find y, let $p = dy/dx$; then (4) becomes

$$\frac{dp}{\sqrt{1 + p^2}} = k\, dx. \tag{5}$$

If we set $p = \sinh u$, so that $\sqrt{1 + p^2} = \cosh u$, $dp = \cosh u\, du$, we can integrate (5) and obtain

$$\operatorname{argsinh} p = kx + C_1 \qquad \text{or} \qquad p = \sinh(kx + C_1).$$

Since $p = dy/dx = 0$ when $x = a$, we get $C_1 = -ka$ and therefore

$$\frac{dy}{dx} = \sinh k(x - a), \qquad y = \frac{1}{k}\cosh k(x - a) + C_2.$$

Setting $y = b$ when $x = a$, we have

$$y = b + \frac{1}{k}[\cosh k(x - a) - 1]$$

for the equation of the flexible cable. It is a remarkable fact that every flexible cable with no external force except that due to gravity assumes the shape of a curve given by the hyperbolic cosine function.

EXERCISES

1. Find the equation of the curve such that the tangent to the curve at $P(x, y)$, the line OP joining the origin to P, and the x-axis bound an isosceles triangle having base along the x-axis.

2. Find the equation in polar coordinates of a curve if the angle from the radius vector to the tangent is half the angle from the x-axis to the radius vector.

3. Find the equation of a locus which is such that the tangent to the curve at any point $P = (x, y)$ on it and the line joining P to the origin make complementary angles with the positive x-axis.

4. Show that if the normal to a curve always passes through a fixed point the curve is a circle (or part of one).

5. Find the function $y(x)$ such that for any $X(>a)$ the area bounded by the line $x = a$, $x = X$, and $y = 0$ and the curve $y = y(x)$ is $Ks(X)$ where $s(X)$ is the length of the arc $y = y(x)$ for $a \le x \le X$.

6. Find the equation of a curve in which the perpendicular from the origin to the tangent is equal to the abscissa of the point of contact.

In problems 7 through 11, find the differential equation of the first order for each of the given families of curves.

7. $y = Cx + e^x$ 8. $y = x + Ce^x$

9. $x^2y + x = Cy^2$ 10. $y = \sin(x + C)$

11. The family of all parabolas having focus at the origin and the x-axis as axis.

In each of problems 12 through 15, find the orthogonal trajectories of the given family of curves, and sketch.

12. $x^2 + y^2 = C$ 13. $2x^2 + 3y^2 = C$

14. $x^{2/3} + y^{2/3} = C^{2/3}$ 15. $x^2 + y^2 = Cx$

16. Suppose in Example 5 that the cable is that of a suspension bridge and that the cable supports a load weighing W lb/ft. (a) Find the equation of the curve along which the cable hangs, if the weight of the cable may be neglected. (b) Write the differential equation of the curve of the cable, taking into account its weight w per foot of length.

17. A heavy object is dropped from rest in a resisting medium in which the resistance is directly proportional to the velocity. Determine its motion.

18. A heavy object is thrown downward into the medium of Problem 17 with a velocity v_0. Determine its motion.

19. A body falls from rest through the air. If air resistance is proportional to the square of the velocity, determine its motion.

20. A body is projected vertically upward with a velocity of v_0. If air resistance is as in Problem 19, determine its motion.

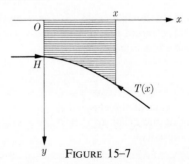

y FIGURE 15–7

21. Find the shape of the arch of a stone-arch bridge if the resultant stress at any point of the arch due to the weight of the masonry above is directed along the tangent to the arch at that point. (See Fig. 15–7.) Assume that the masonry is uniform in density and that the surface of the road is horizontal.

22. A particle of mass m moving along a straight line is acted on by a force of magnitude $km|x|$ directed toward the origin and by a resistance force of magnitude $lm|dx/dt|$ and directed oppositely to the motion. Write the differential equation of the motion.

CHAPTER 16

Linear Differential Equations

1. OPERATORS

We consider the linear differential equation

$$a_0(x)\frac{d^n y}{dx^n} + a_1(x)\frac{d^{n-1}y}{dx^{n-1}} + \cdots + a_{n-1}(x)\frac{dy}{dx} + a_n(x)y = f(x), \qquad (1)$$

in which the functions $a_0, a_1, \ldots, a_n$ are defined and continuous on some interval I of the x-axis. If y and its first n derivatives are continuous on I, then the left side of (1) is a continuous function of x.

DEFINITIONS. We define M to be the set of all functions with at least n continuous derivatives on an interval I. We define N to be the set of all functions continuous on I.

We may interpret the left side of (1) as a transformation which takes an element of M into an element of N. This transformation is from one set (or space) of functions into another. Such a transformation is called an **operator**. An operator involving derivatives of the functions in its domain is called a **differential operator**. The left side of (1) is an example of a differential operator.

Recalling the terminology and results in Chapter 7, we note that M and N are linear vector spaces. An operator from a set in one vector space to a set in another is called a **linear operator** if it is a linear transformation as defined in Chapter 7, page 339. That is, L is a linear operator if

$$L(c_1 f_1 + c_2 f_2 + \cdots + c_k f_k) = c_1 L(f_1) + c_2 L(f_2) + \cdots + c_k L(f_k)$$

for any functions $f_1, f_2, \ldots, f_k$ in the domain of L and any numbers $c_1, c_2, \ldots, c_k$. The only distinction between linear operators and linear transformations lies in the domain of definition. An operator may be defined for any set of elements in a vector space, while a linear transformation, as discussed in Chapter 7, is always defined for an entire linear space (or subspace).

We note that the left side of Eq. (1) *is a linear operator.* In fact, the basis for calling a differential equation linear is just the fact that it is linear when considered as a transformation. This statement holds for linear partial differential equations as well.

If $f(x)$ is continuous on $[a, b]$ and c is a fixed point in this interval, then the relation

$$F(x) = \int_c^x f(t)\, dt$$

may be treated as an operator relation. We see that for each function f in the space of continuous functions on $[a, b]$ there is associated a function F. We write

$$L(f) = \int_c^x f(t)\, dt,$$

which indicates that L is the transformation taking f into F. Such an operator is called an **integral operator**, and the student may verify at once that it is linear.

We shall now develop some elementary algebra for operators. Suppose L_1 and L_2 are two operators with domains of functions $\mathfrak{D}_1$ and $\mathfrak{D}_2$, respectively. If c_1 and c_2 are any numbers, we define the operator L by the relation

$$L = c_1 L_1 + c_2 L_2. \tag{2}$$

We must proceed with caution in defining L, since a function y in $\mathfrak{D}_1$ may not be in $\mathfrak{D}_2$. In such a case $L_1 y$ has a meaning but $L_2 y$ does not. Thus the operator L in (2) has for its domain $\mathfrak{D}$ only those functions y which are in *both* $\mathfrak{D}_1$ and $\mathfrak{D}_2$. We call $\mathfrak{D}$ the intersection of the domains $\mathfrak{D}_1$ and $\mathfrak{D}_2$, and we use the notation $\mathfrak{D} = \mathfrak{D}_1 \cap \mathfrak{D}_2$. Figure 16–1 shows $\mathfrak{D}$ schematically, with each point considered as a function. We define L more precisely than in (2) by writing

$$L(y) = c_1 L_1(y) + c_2 L_2(y) \quad \text{for } y \text{ in } \mathfrak{D} = \mathfrak{D}_1 \cap \mathfrak{D}_2.$$

The operator L is called a **linear combination** of the operators L_1 and L_2. The product $L_1 L_2$ of two operators is defined by the relation

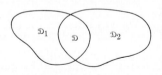

$$L_1 L_2(y) = L_1[L_2(y)]. \tag{3}$$

FIGURE 16–1

Again we must proceed with caution. The operator is defined only if y is in the domain $\mathfrak{D}_2$ and if the function $L_2(y)$ is in the domain of L_1. Of course, there may not be any such y, in which case we say the domain of $L_1 L_2$ is the empty set. Similarly, the domain of $c_1 L_1 + c_2 L_2$ may be the empty set if $\mathfrak{D}_1$ and $\mathfrak{D}_2$ have no functions in common.

If y is a differentiable function of x, then dy/dx is a function of x. In other words, the simple process of differentiation is an operator process.

DEFINITION. The operator whose domain $\mathfrak{D}$ consists of all functions y with one continuous derivative which carries y into dy/dx is denoted by D. We define the operator D^2 by the product rule

$$D^2 = D \cdot D,$$

as in (3). Of course D^2 is defined only for the subclass of $\mathfrak{D}$ consisting of functions with two continuous derivatives. The operators $D^3, D^4, \ldots$ are defined by

$$D^3 = D \cdot D^2, \; D^4 = D \cdot D^3, \ldots.$$

Remarks. We see that

$$D(y) = \frac{dy}{dx}, \; D^2(y) = \frac{d^2 y}{dx^2}, \ldots, \; D^n(y) = \frac{d^n y}{dx^n}, \ldots,$$

observing that as n increases, the corresponding domain of D^n becomes a smaller and smaller portion of the domain $\mathcal{D}$ in the above definition.

According to the rule for forming linear combinations, we may define the operator

$$P(D) = c_0 D^n + c_1 D^{n-1} + \cdots + c_{n-1} D + c_n D^0,$$

where $D^0 = 1$ and where $c_0, c_1, \ldots, c_n$ are any numbers. The differential operator $P(D)$, when written out, is simply

$$P(D)(y) = c_0 \frac{d^n y}{dx^n} + c_1 \frac{d^{n-1} y}{dx^{n-1}} + \cdots + c_{n-1} \frac{dy}{dx} + c_n y.$$

DEFINITION. The ordinary polynomial

$$P(r) = c_0 r^n + c_1 r^{n-1} + \cdots + c_{n-1} r + c_n,$$

corresponding to the operator $P(D)$, is called the **auxiliary polynomial**. The auxiliary polynomial plays an important part in obtaining solutions of linear differential equations with constant coefficients.

EXAMPLE 1. Write the equation $(D^2 + D^1 + 4D^0)(y) = x$ in traditional notation.

Solution. We have

$$D^2 y = \frac{d^2 y}{dx^2}, \qquad D^1 y = \frac{dy}{dx}, \qquad D^0 y = y.$$

The equation is

$$\frac{d^2 y}{dx^2} + \frac{dy}{dx} + 4y = x \qquad \text{or} \qquad y'' + y' + 4y = x.$$

EXAMPLE 2. Given $L_1 = 2D^1 + 3D^0, L_2 = 3D^1 - D^0$, find $(L_1 L_2)(y)$ and $(L_2 L_1)(y)$ in traditional notation.

Solution. Letting $u = L_2(y)$, we have $L_1[L_2(y)] = L_1(u)$. Since

$$u = 3D^1(y) - D^0(y) = 3y' - y,$$

we find

$$(L_1 L_2)(y) = L_1 u = 2D(u) + 3u = 2\frac{du}{dx} + 3u$$

$$= 2(3y'' - y') + 3(3y' - y) = 6y'' + 7y' - 3y.$$

Similarly, if we let $v = L_1(y) = 2y' + 3y$, then

$$(L_2L_1)(y) = 3\frac{dv}{dx} - v = 3(2y'' + 3y') - (2y' + 3y)$$
$$= 6y'' + 7y' - 3y.$$

Note that $(L_1L_2)(y) = (L_2L_1)(y)$.

The algebra of operators may be extended so that expressions of the type

$$L = a_1(x)L_1 + a_2(x)L_2$$

may be formed. It is clear that if L_1 and L_2 are linear differential operators with domains which have a nonempty intersection, then L will be a linear differential operator defined for a nonempty domain $\mathfrak{D}$. Similarly, the product of operators with variable coefficients may be defined as in (3).

EXAMPLE 3. Given $L_1 = 2xD^1 + D^0$ and $L_2 = x^2D^1 - 2xD^0$, find $(L_1L_2)(y)$ and $(L_2L_1)(y)$ in traditional notation.

Solution. Let $u = L_2y$. Then $u = x^2y' - 2xy$ and

$$(L_1L_2)(y) = L_1u = 2xu' + u = 2x(x^2y' - 2xy)' + x^2y' - 2xy$$
$$= 2x^3y'' + x^2y' - 6xy.$$

Similarly, letting $v = L_1(y) = 2xy' + y$, we get

$$(L_2L_1)(y) = L_2v = x^2(2xy' + y)' - 2x(2xy' + y)$$
$$= 2x^3y'' - x^2y' - 2xy.$$

Note that $(L_1L_2)(y) \neq (L_2L_1)(y)$.

Examples 2 and 3 exhibit a basic difference between operators with constant coefficients and those with variable coefficients. Differential equations with constant coefficients are much easier to solve than are those with variable coefficients, a fact which is intimately connected with the next theorem.

Theorem 1. *Suppose that L_1 and L_2 are polynomial differential operators:*

$$L_1 = a_0D^m + a_1D^{m-1} + \cdots + a_{m-1}D + a_mD^0,$$
$$L_2 = b_0D^n + b_1D^{n-1} + \cdots + b_{n-1}D + b_nD^0,$$

where the a_i and b_i are constants. Then if c_1 and c_2 are any constants, we have

$$c_1L_1 + c_2L_2 = c_2L_2 + c_1L_1 \qquad \text{and} \qquad L_1L_2 = L_2L_1.$$

Proof. The first conclusion is immediate from the definition of linear combination of operators. To prove the second result, we set

$$u = L_2(y) = b_0 D^n(y) + b_1 D^{n-1}(y) + \cdots + b_n(y)$$

and compute

$$(L_1 L_2)(y) = L_1(u).$$

Applying each of the terms in L_1 to u, we may write

$$a_0 D^m u = a_0 b_0 D^{m+n}(y) + a_0 b_1 D^{m+n-1}(y) + \cdots + a_0 b_n D^m y,$$
$$a_1 D^{m-1} u = a_1 b_0 D^{m+n-1}(y) + a_1 b_1 D^{m+n-2}(y) + \cdots + a_1 b_n D^{m-1} y,$$
$$\vdots \qquad\qquad \vdots$$
$$a_m u = a_m b_0 D^n(y) + a_m b_1 D^{n-1}(y) + \cdots + a_m b_n y.$$

The quantity $L_1(u)$ is obtained by adding these equations. The expression for $(L_1 L_2)(y)$ is simply the product of the auxiliary polynomials with r replaced by D. However, $(L_2 L_1)(y)$ is also the product of the same two polynomials. Therefore $L_1 L_2 = L_2 L_1$.

———————

EXAMPLE 4. Show that

$$[(D - 2)(D - 3)](y), \qquad [(D - 3)(D - 2)](y) \qquad \text{and} \qquad (D^2 - 5D + 6)(y)$$

are all identical.

Solution. We may use Theorem 1 or proceed directly. Letting $u = (D - 3)y$, we find

$$[(D - 2)(D - 3)](y) = (D - 2)(u) = (D - 2)\left(\frac{dy}{dx} - 3y\right)$$
$$= \frac{d^2 y}{dx^2} - 3\frac{dy}{dx} - 2\frac{dy}{dx} + 6y = (D^2 - 5D + 6)(y).$$

Similarly, $[(D - 3)(D - 2)](y)$ is easily shown to give the same value.

———————

Remark. We shall change notation and replace D^0 by 1. For example, $2D^0 y$ will be written $2y$.

EXERCISES

In each of problems 1 through 4, write the equation in traditional notation.

1. $(2D^2 - D^1 + D^0)(y) = e^x$
2. $(D^3 - 2D^0)(y) = x^2$
3. $(D^3 + 2D^2 - D^1)(y) = 0$
4. $(2D^3 - D^2 - D^1 + D^0)(y) = 0$

In each of problems 5 through 8, write the equation in operator notation (as above).

5. $y'' - 3y' + 2y = 0$
6. $y''' + y'' - 2y = 0$
7. $2y''' + y' - 3y = 0$
8. $y''' - 2y'' + 3y' - 2y = x^2$

In each of problems 9 through 16, find $(L_1 L_2)(y)$ and $(L_2 L_1)(y)$ and write them in traditional notation. Check the validity of the operator equation $L_1 L_2 = L_2 L_1$.

9. $L_1 = D + 1$, $L_2 = D - 1$ 10. $L_1 = L_2 = D - 2$

11. $L_1 = D + 2$, $L_2 = D - 1$ 12. $L_1 = D^2 + D - 2$, $L_2 = D + 1$

13. $L_1(y) = (x + 1)y' - y$, $L_2(y) = y' + xy$

14. $L_1(y) = 2xy' + 3y$, $L_2(y) = xy' - 2y$

15. $L_1(y) = y'' - 3xy' + 2y$, $L_2 = y'' + 3xy' + 2x^2 y$

16. $L_1(y) = y''' - 4y$, $L_2(y) = xy''' + 4y'$

17. (a) Show that if $L_1, L_2, \ldots, L_n$ are differential operators with constant coefficients, then

$$L_{i_1} L_{i_2} \ldots L_{i_n} = L_{j_1} L_{j_2} \ldots L_{j_n}$$

where $i_1, i_2, \ldots, i_n$ is one permutation of the numbers $1, 2, \ldots, n$ and $j_1, j_2, \ldots,$ j_n is another permutation of the same numbers.

(b) What can be said if *one* of the operators has variable coefficients and all the remaining have constant coefficients?

2. LINEAR DIFFERENTIAL EQUATIONS OF THE SECOND ORDER WITH CONSTANT COEFFICIENTS

In this section we show how to solve the second-order equation

$$(D^2 + a_1 D + a_2)y = 0, \tag{1}$$

in which a_1 and a_2 are real constants. The auxiliary polynomial equation is

$$r^2 + a_1 r + a_2 = 0,$$

and we denote the roots of this quadratic equation by r_1 and r_2. We consider three cases which depend on the nature of the roots.

Case I. The roots r_1 and r_2 are real and unequal. We may write Eq. (1) in the form

$$(D - r_1)(D - r_2)y = 0. \tag{2}$$

If y is a solution of (2), we set $u = (D - r_2)y$, and then

$$(D - r_1)u = 0 \quad \text{or} \quad \frac{du}{dx} = r_1 u.$$

We separate variables and obtain

$$u = ce^{r_1 x}.$$

Then y satisfies the equation

$$(D - r_2)y = ce^{r_1 x}$$

or

$$\frac{dy}{dx} - r_2 y = ce^{r_1 x}.$$

This linear equation has $e^{-r_2 x}$ as an integrating factor. Therefore

$$\frac{d}{dx}(e^{-r_2 x}y) = ce^{(r_1-r_2)x}, \tag{3}$$

from which we conclude that

$$ye^{-r_2 x} = \frac{c}{r_1 - r_2} e^{(r_1-r_2)x} + C_2, \qquad r_1 \neq r_2.$$

The solution of (1) is

$$y = C_1 e^{r_1 x} + C_2 e^{r_2 x}, \qquad r_1 \neq r_2 \tag{4}$$

in which we have set $c/(r_1 - r_2)$ equal to C_1.

Case II. The roots are real and equal. The argument is the same as in Case I until we get to Eq. (3). Then, since $r_1 = r_2$, we replace (3) by

$$\frac{d}{dx}(e^{-r_2 x}y) = c$$

and we integrate, to obtain

$$y = (C_1 x + C_2)e^{rx} \qquad \text{with } r = r_1 = r_2. \tag{5}$$

Case III. The roots r_1 and r_2 are complex conjugates. In Chapter 9, Section 5, we defined e^z for complex values of z, and we showed that if x is real, then the formula

$$\frac{d}{dx}e^{ax} = ae^{ax}$$

holds for complex constants a as well as for real ones. Hence the analysis for Case I applies equally well when r_1 and r_2 are complex. Writing

$$r_1 = \alpha + i\beta, \qquad r_2 = \alpha - i\beta, \qquad \beta \neq 0$$

where α and β are real, we obtain from (4) the solution

$$y = C_1 e^{(\alpha+i\beta)x} + C_2 e^{(\alpha-i\beta)x}, \tag{6}$$

where C_1 and C_2 are constants. In Eq. (1) the coefficients a_1 and a_2 are real, and we are interested in real solutions $y(x)$. On the other hand, (6) gives the appearance of a complex-valued function. If we write (6) in the form

$$y = C_1 e^{\alpha x}(\cos \beta x + i\sin \beta x) + C_2 e^{\alpha x}(\cos \beta x - i\sin \beta x)$$
$$= D_1 e^{\alpha x}\cos \beta x + D_2 e^{\alpha x}\sin \beta x$$

or

$$y = e^{\alpha x}(D_1 \cos \beta x + D_2 \sin \beta x), \tag{7}$$

the solution will be real whenever D_1 and D_2 are. We observe that

$$D_1 = C_1 + C_2, \qquad D_2 = i(C_1 - C_2),$$

and therefore C_1 and C_2 will be complex, in general.

The preceding analysis leads to the following theorem.

Theorem 2. *The function $y(x)$ is a solution of Eq. 2 if and only if*
(i), *y is given by (4) when $r_1 \neq r_2$ and r_1, r_2 are real;*
(ii), *y is given by (5) when $r_1 = r_2$; and*
(iii), *y is given by (7) when r_1 and r_2 are complex.*

EXAMPLE 1. Solve the equation

$$\frac{d^2y}{dx^2} - 3\frac{dy}{dx} + 2y = 0.$$

Solution. The equation can be written in the form

$$(D^2 - 3D + 2)y = 0.$$

The auxiliary polynomial equation is

$$r^2 - 3r + 2 = (r - 1)(r - 2) = 0.$$

We have Case I, and the solution is

$$y = C_1e^x + C_2e^{2x}$$

where C_1 and C_2 are arbitrary constants.

EXAMPLE 2. Solve

$$y'' - 2y' + 2y = 0.$$

Solution. The roots of the auxiliary polynomial are $1 \pm i$. Using (7) with $\alpha = 1$, $\beta = 1$, we obtain

$$y = e^x(D_1 \cos x + D_2 \sin x).$$

EXAMPLE 3. Find the solution of the equation

$$y'' + 6y' + 9y = 0$$

which satisfies the conditions that $y = 2$ when $x = 0$ and $y = 0$ when $x = 1$.

Solution. We have

$$r^2 + 6r + 9 = 0 \quad \text{and} \quad r_1 = r_2 = -3.$$

Therefore

$$y = (C_1x + C_2)e^{-3x}.$$

Substituting $y = 2$, $x = 0$, we get $C_2 = 2$. Then, for $y = 0$, $x = 1$, we have

$$0 = (C_1 + 2)e^{-3} \quad \text{or} \quad C_1 = -2.$$

The result is

$$y = (-2x + 2)e^{-3x}.$$

EXERCISES

In each of problems 1 through 10, solve the given differential equation.

1. $y'' - 4y = 0$ 2. $y'' + y' - 2y = 0$

3. $(D^2 - 4D + 3)(y) = 0$ 4. $(D^2 - 3D + 1)(y) = 0$

5. $y'' + 2y' + 2y = 0$ 6. $(D^2 + 4D + 4)(y) = 0$

7. $(D^2 - 2D + 3)(y) = 0$ 8. $y'' - 8y' + 16y = 0$

9. $y'' + 6y' + 10y = 0$ 10. $y'' + y' + y = 0$

In each of problems 11 through 14, find the solution which satisfies the given initial conditions.

11. $y'' - y' - 2y = 0$, $y(0) = 1$, $y'(0) = -2$

12. $(D^2 + D - 6)(y) = 0$, $y(0) = -1$, $y'(0) = 3$

13. $(D^2 + 4)(y) = 0$, $y(0) = 3$, $y'(0) = -2$

14. $y'' - 6y' + 9y = 0$, $y(0) = -1$, $y'(0) = 2$.

3. NONHOMOGENEOUS EQUATIONS

If L is any linear operator of the form

$$L \equiv a_0(x)D^n + a_1(x)D^{n-1} + \cdots + a_{n-1}(x)D + a_n(x),$$

then the equation $L(y) = 0$, in which the right side is zero, is said to be a **homogeneous** differential equation. The equation $L(y) = f(x)$ with $f(x)$ not identically zero is called a **nonhomogeneous** differential equation.

In Section 2 we considered homogeneous second-order equations with constant coefficients. In this section we shall show how to solve nonhomogeneous second-order equations of the form

$$(D^2 + a_1D + a_2)y = f(x) \qquad (1)$$

in which a_1 and a_2 are constants and $f(x)$ is a given function.

The first step in obtaining a solution of (1) is to find a solution of the same equation with $f = 0$. That is, we find a solution of the corresponding homogeneous equation.

DEFINITION. The solution of the corresponding homogeneous equation is called the **complementary function**.

The solution of nonhomogeneous equations is obtained by means of the following general principle.

PRINCIPLE 1. *Any solution of a nonhomogeneous linear equation $L(y) = f$ is the sum of a particular solution of the nonhomogeneous equation and a solution of the corresponding homogeneous equation (complementary function).*

This principle is easily established by letting y_1 be a particular solution and y any solution of $L(y) = f$; then, since L is linear,

$$L(y - y_1) = L(y) - L(y_1) = f - f = 0,$$

and so the function $v = y - y_1$ is a complementary function. Therefore

$$y = y_1 + v,$$

and the principle is established.

We exhibit the technique for finding particular solutions when the function f has various forms. We proceed on a case-by-case basis.

Case I. The function f has the form ae^{bx}, where b is not a root of the auxiliary polynomial. In this case we observe that

$$L(ce^{bx}) = c(b^2 + a_1b + a_2)e^{bx}$$

and so, if we select

$$c = \frac{a}{b^2 + a_1b + a_2}, \tag{2}$$

then $y_1 = ce^{bx}$ is a particular solution of

$$L(y_1) = ae^{bx}.$$

Notice that if b were a root of the auxiliary polynomial, the selection of c as in (2) would be impossible.

———————

EXAMPLE 1. Solve the equation

$$y'' + y' - 6y = 2e^{-x}.$$

Solution. The auxiliary polynomial $r^2 + r - 6 = 0$ has roots $2, -3$ and, since $b = -1$, b is not a root. We select

$$c = \frac{2}{1 + 1 \cdot (-1) + (-6)} = -3,$$

and a particular solution is $y_1 = -3e^{-x}$. The complementary function is

$$v = C_1e^{2x} + C_2e^{-3x},$$

and so the general solution is

$$y = C_1e^{2x} + C_2e^{-3x} - 3e^{-x}.$$

———————

Case II. The function f is a polynomial. To obtain a particular solution y_1 of Eq. (1), we observe that if y_1 is a polynomial, then $L(y_1)$ will be a polynomial of the same degree (*unless* $a_2 = 0$, when the equation can be solved anyway by the methods of Chapter 15, Sections 8 and 6). *Thus, if f is a polynomial and $a_2 \neq 0$, a particular solution can be found by setting y_1 equal to a general polynomial of the* *same degree as f with unknown coefficients.* The coefficients are then obtained by computing $L(y_1)$, setting the result equal to f, and equating coefficients of like powers of x. An example illustrates the procedure.

EXAMPLE 2. Solve the equation $(D^2 + D - 2)y = x^2$.

Solution. Writing $(D + 2)(D - 1)y = x^2$, we see that the complementary function is $v = C_1 e^{-2x} + C_2 e^x$. To get a particular solution, we set y_1 equal to *a general polynomial of the second degree.* That is,

$$y_1 = ax^2 + bx + c.$$

Then

$$y_1' = 2ax + b, \qquad y_1'' = 2a,$$

$$L(y_1) = 2a + 2ax + b - 2ax^2 - 2bx - 2c$$

$$= (-2a)x^2 + (2a - 2b)x + (2a + b - 2c).$$

We set this equal to $x^2(=f)$ and equate coefficients. The result is

$$-2a = 1, \qquad 2a - 2b = 0, \qquad 2a + b - 2c = 0,$$

or

$$a = -\tfrac{1}{2}, \qquad b = -\tfrac{1}{2}, \qquad c = -\tfrac{3}{4}.$$

The desired solution is

$$y = v + y_1 = C_1 e^{-2x} + C_2 e^x - \tfrac{1}{2}(x^2 + x + \tfrac{3}{2}).$$

The methods described in the two cases above come under the general heading of the **method of undetermined coefficients**. We now describe another method, which sometimes works even when f does not satisfy the conditions of either Case I or Case II.

Suppose that v is a solution of the homogeneous equation $L(v) = 0$ and we wish to find a solution which satisfies $L(y) = f$. We set

$$y = vw$$

where w is to be determined. Once we find w, the problem is solved. To compute $L(y)$ we write

$$y' = vw' + v'w, \qquad y'' = vw'' + 2v'w' + v''w$$

and

$$L(y) = vw'' + (2v' + a_1 v)w' + L(v)w.$$

Since we assumed that v is a solution of the homogeneous equation, we have $L(v) = 0$, and therefore

$$L(y) = vw'' + (2v' + a_1 v)w'.$$

If y is to be a solution, we must have

$$vw'' + (2v' + a_1 v)w' = f(x).$$

This equation can be solved by the methods of Sections 6 and 8 in Chapter 15, or we can multiply the above equation by v and get

$$\frac{d}{dx}(v^2 w') + a_1(v^2 w') = vf(x).$$

Now $e^{a_1 x}$ is an integrating factor, and we obtain

$$\frac{d}{dx}(e^{a_1 x}v^2 w') = e^{a_1 x}vf$$

and

$$e^{a_1 x}v^2 w' = C + \int e^{a_1 x}vf(x)\,dx.$$

Since v is a known function, we can integrate once more to find w, and the problem is solved. An explicit result can be obtained only when all the integrations can be performed.

Remark. The student should not try to memorize the last formula above. It is better to proceed by setting $y = vw$ and making the necessary computations in each problem.

EXAMPLE 3. Solve $(D^2 + D - 2)y = xe^x$.

Solution. Writing $(D + 2)(D - 1)y = xe^x$, we see that $v = e^x$ is a solution of the homogeneous equation. We set

$$y = e^x w$$

and compute

$$y' = e^x(w' + w), \qquad y'' = e^x(w'' + 2w' + w)$$
$$L(y) = e^x(w'' + 3w') = xe^x.$$

We now solve the equation

$$w'' + 3w' = x$$

or, since e^{3x} is an integrating factor,

$$\frac{d}{dx}(e^{3x}w') = xe^{3x}.$$

The result is

$$e^{3x}w' = C + \int xe^{3x}\,dx = C + \tfrac{1}{3}xe^{3x} - \tfrac{1}{9}e^{3x}$$

and

$$w' = Ce^{-3x} + \tfrac{1}{3}x - \tfrac{1}{9}.$$

Another integration yields

$$w = C_1 e^{-3x} + C_2 + \tfrac{1}{6}x^2 - \tfrac{1}{9}x$$

and

$$y = e^x w = C_1 e^{-2x} + C_2 e^x + (\tfrac{1}{6}x^2 + \tfrac{1}{9}x)e^x.$$

PRINCIPLE 2. *If $f(x) = c_1 f_1(x) + c_2 f_2(x)$ and if $y_1(x)$, $y_2(x)$ are solutions of $L(y_1) = f_1(x)$, $L(y_2) = f_2(x)$, respectively, then $y = c_1 y_1(x) + c_2 y_2(x)$ is a solution of $L(y) = f$.*

To establish the principle, we merely have to observe that for linear operators

$$L(y) = L(c_1y_1 + c_2y_2) = c_1L(y_1) + c_2L(y_2) = c_1f_1(x) + c_2f_2(x) = f(x).$$

The principle also holds if f is the finite sum of any number of functions. If f is the sum of a polynomial and several functions of the type in Case I, the principle yields a quick method for finding a solution.

The developments of this section work equally well when $f(x)$ is a complex-valued function of the real variable x. Of course, in such a case the solution will be complex even when a_1 and a_2 are real. That is, if

$$f(x) = f_1(x) + if_2(x)$$

and if y satisfies the equation

$$L(y) = y'' + a_1y' + a_2y = f, \qquad a_1, a_2 \text{ real,}$$

we may write $y(x) = y_1(x) + iy_2(x)$ and, therefore,

$$L(y) = L(y_1 + iy_2) = L(y_1) + iL(y_2) = f_1(x) + if_2(x).$$

Two complex numbers are equal if and only if their real and imaginary parts are. Hence,

$$L(y_1) = f_1(x) \qquad \text{and} \qquad L(y_2) = f_2(x).$$

The fact that complex solutions can be separated into real and imaginary parts is employed in the next example.

———————

EXAMPLE 4. Solve $y'' + 4y = e^x \cos x$.

Solution. The complementary function is $v = C_1 \cos 2x + C_2 \sin 2x$. We could proceed by the method of undetermined coefficients, but instead we employ the following trick. We notice that $e^x \cos x$ is the real part of $e^{(1+i)x}$. The equation

$$y'' + 4y = e^{(1+i)x}$$

can be solved by the method of Case I. The solution will be complex, but the real part of the solution will give the answer to the original problem. We set

$$y_1 = ce^{(1+i)x}$$

and compute

$$y_1'' + 4y_1 = (4 + 2i)ce^{(1+i)x} = e^{(1+i)x}$$

Therefore $c = 1/(4 + 2i) = (2 - i)/10$, and we write

$$y_1 = \frac{2 - i}{10} e^{(1+i)x} = \frac{2 - i}{10} e^x(\cos x + i \sin x)$$

$$= \frac{e^x}{10} [(2 \cos x + \sin x) + i(2 \sin x - \cos x)].$$

The general solution of the given equation is

$$y = C_1 \cos 2x + C_2 \sin 2x + \frac{e^x}{10}(2 \cos x + \sin x).$$

EXERCISES

In each of problems 1 through 14, solve the given differential equation.

1. $y'' + 2y' = e^x$

2. $y'' + y' - 2y = e^{2x} - 2$

3. $y'' + 2y' + 2y = 2x + 3 + e^{-x}$

4. $y'' + 2y' + y = x + e^x$

5. $y'' - 4y' + 4y = x^2 - e^{-2x}$

6. $(D^2 - D)(y) = e^x$

7. $(D - 1)^2(y) = e^x$

8. $(D^2 - 1)(y) = e^{-x}$

9. $y'' - y = \cos x$

10. $(D^2 + 4)(y) = \cos x$

11. $(D^2 + 2D + 2)(y) = \sin 2x$

12. $(D^2 + 2D + 5)(y) = e^x \sin x$

13. $y'' + y = \sin x$

14. $y'' - 2y' + 2y = e^x \cos x$

In problems 15 through 20, find in each case that solution which satisfies the initial conditions $y(0) = y'(0) = 0$.

15. $y'' - y = x$

16. $(D^2 - D - 2)(y) = e^{3x}$

17. $(D^2 + D - 6)(y) = -4e^x$

18. $y'' + y = x$

19. $(D^2 + 4)(y) = e^x$

20. $y'' + 2y' = x$

4. LINEAR EQUATIONS OF THE SECOND ORDER WITH VARIABLE COEFFICIENTS

It is not usually possible to solve the second-order equation

$$a_0(x)\frac{d^2y}{dx^2} + a_1(x)\frac{dy}{dx} + a_2(x)y = f(x)$$

explicitly, as was possible for the general linear equation of the first order. We shall restrict ourselves to some special methods which work in particular cases.

If $v(x)$ is a nonzero solution of the homogeneous second-order equation, then the substitution

$$y = v(x)w(x),$$

as illustrated in the preceding section, leads to a solution. Of course, the method depends on our knowing in advance one solution of the homogeneous equation.

EXAMPLE 1. Given the equation.

$$L(y) \equiv xy'' - (2x + 1)y' + (x + 1)y = x^2 e^x,$$

verify that $v = e^x$ is a solution of the homogeneous equation and find a solution of $L(y) = x^2 e^x$.

Solution. Since $v'' = v' = v = e^x$, we see that

$$L(v) = e^x(x - 2x - 1 + x + 1) = 0.$$

Setting $y = e^x w$, we obtain

$$y' = e^x(w' + w), \qquad y'' = e^x(w'' + 2w' + w),$$
$$L(y) = e^x(xw'' - w') = x^2 e^x$$

and

$$xw'' - w' = x^2 \qquad \text{or} \qquad \frac{d}{dx}(x^{-1}w') = 1.$$

Therefore

$$x^{-1}w' = x + C_1 \qquad \text{and} \qquad w' = x^2 + C_1 x.$$

Another integration yields

$$w = \tfrac{1}{3}x^3 + \tfrac{1}{2}C_1 x^2 + C_2$$

and

$$y = e^x w = \frac{x^3}{3}e^x + \frac{x^2}{2}C_1 e^x + C_2 e^x.$$

Whenever two solutions of the homogeneous equation (denoted by y_1 and y_2) can be found, it is possible to solve the nonhomogeneous equation by a method known as **variation of parameters.** This method depends on the following theorem.

Theorem 3. *Suppose that y_1 and y_2 are solutions, in some interval containing x_0, of the equation*

$$y'' + a_1(x)y' + a_2(x)y = 0;$$

define

$$W(x) = y_1 y_2' - y_2 y_1'.$$

Then

$$W(x) = W(x_0)e^{-A_1(x)} \qquad \text{where} \qquad A_1(x) = \int_{x_0}^{x} a_1(t)\, dt.$$

Proof. Since y_1 and y_2 are solutions of the homogeneous equation, we have

$$y_1'' = -a_1 y_1' - a_2 y_1, \qquad y_2'' = -a_1 y_2' - a_2 y_2.$$

Therefore

$$W' = y_1 y_2'' + y_1' y_2' - y_2 y_1'' - y_1' y_2' = y_1 y_2'' - y_2 y_1''$$
$$= -y_1(a_1 y_2' + a_2 y_2) + y_2(a_1 y_1' + a_2 y_1) = -a_1 W.$$

Integrating the differential equation $W' = -a_1 W$ from x_0 to x, we obtain the conclusion of the theorem.

DEFINITIONS. The function W defined in Theorem 3 is called the **Wronskian** of the two solutions. We recall the definitions that two functions are **linearly dependent** if and only if one is a constant multiple of the other; otherwise the functions are **linearly independent**.

It is clear from the definition of W that if y_1 and y_2 are linearly dependent on an interval, so that $y_1 = cy_2$, then $W(x) = 0$ on that interval. Moreover, either $W \equiv 0$ or $W(x)$ is never zero in any interval on which $a_1(x)$ is continuous.

Method of variation of parameters. We assume that y_1 and y_2 are solutions of the homogeneous equation and set

$$y = y_1 v_1 + y_2 v_2, \tag{1}$$

where v_1 and v_2 are to be determined in such a way that y is the desired solution. If we compute $L(y) = f$, taking (1) for y, then in general we will get *one second-order equation* in the *two* functions v_1 and v_2. We now show that if we impose the condition

$$y_1 v_1' + y_2 v_2' = 0, \tag{2}$$

then the equation $L(y) = f$ will not involve second derivatives of v_1 and v_2. To see this, we compute

$$y' = y_1' v_1 + y_1 v_1' + y_2' v_2 + y_2 v_2' = y_1' v_1 + y_2' v_2$$

if (2) is taken into account. Then

$$y'' = y_1' v_1' + y_2' v_2' + y_1'' v_1 + y_2'' v_2.$$

A simple computation shows that

$$L(y) \equiv y'' + a_1 y' + a_2 y = y_1' v_1' + y_2' v_2' + v_1 L(y_1) + v_2 L(y_2).$$

Since $L(y_1) = 0$ and $L(y_2) = 0$ by hypothesis, we conclude that $L(y) = f$ if and only if

$$y_1' v_1' + y_2' v_2' = f.$$

This equation combined with (2) leads to the system

$$y_1 v_1' + y_2 v_2' = 0, \qquad y_1' v_1' + y_2' v_2' = f \tag{3}$$

for the two unknown functions v_1', v_2'. The determinant of these two equations is the Wronskian W. So if y_1 and y_2 are linearly independent, then W never vanishes and the system (3) can be solved for v_1', v_2' in terms of *known* functions of x. An integration yields v_1 and v_2 and, with it, the solution y as given by (1). An explicit solution can be obtained only when the integrations can be carried out.

EXAMPLE 2. Solve $y'' + y = \sec^3 x$ by the method of variation of parameters.

Solution. The equation $y'' + y = 0$ has the general solution $y = C_1 \cos x + C_2 \sin x$. We select $y_1 = \sin x$, $y_2 = \cos x$ and observe that y_1 and y_2 are linearly independent. The solution is

$$y = (\sin x)v_1 + (\cos x)v_2.$$

Then, according to (3), the equations for v_1' and v_2' are

$$(\sin x)v_1' + (\cos x)v_2' = 0, \qquad (\cos x)v_1' - (\sin x)v_2' = \sec^3 x.$$

Solving for v_1', v_2', we find

$$v_1' = \sec^2 x, \qquad v_2' = (\cos x)^{-3}(-\sin x).$$

Integrating these equations, we obtain

$$v_1 = \tan x + C_1, \qquad v_2 = -\tfrac{1}{2}\sec^2 x + C_2.$$

Therefore

$$y = C_1 \sin x + C_2 \cos x + \frac{\sin^2 x}{\cos x} - \frac{1}{2 \cos x}$$

$$= C_1 \sin x + (C_2 - 1) \cos x + \tfrac{1}{2}\sec x.$$

EXERCISES

In each of problems 1 through 6, show that x or e^x is a solution of the homogeneous equation and then solve the equation by the method of Example 1.

1. $(1 + x^2)y'' + 2xy' - 2y = 0.$
2. $(1 - x)y'' + xy' - y = 0.$ *Answer:* $y = C_1 x + C_2 e^x.$
3. $xy'' - (x + 3)y' + 3y = 0$
4. $(x - 3)y'' - (4x - 9)y' + (3x - 6)y = 0$
5. $(x^2 + 1)y'' - 2xy' + 2y = 0$
6. $xy'' - (2x - 1)y' + (x - 1)y = 0$

In each of problems 7 through 11, solve the equation by the method of variation of parameters.

7. $y'' - y = xe^x$
8. $y'' + y' - 6y = xe^{2x}$
9. $y'' + 2y' + y = xe^{-x}$
10. $y'' + 2y' + 5y = xe^{-x} \cos 2x$
11. $(1 - x)y'' + xy' - y = (1 - x)^2.$ [*Hint.* Use the results of Exercise 2.]

5. LINEAR EQUATIONS OF HIGHER ORDER WITH CONSTANT COEFFICIENTS. SPECIAL CASES

In this section we discuss homogeneous equations of the form

$$(D - r)^n y = 0 \tag{1}$$

and nonhomogeneous equations of the form

$$(D - r)^n y = e^{sx}(c_0 + c_1 x + \cdots + c_p x^p) \tag{2}$$

where r, s, c_0, c_1, $\ldots$, c_p may be complex numbers. The first result, which yields solutions of (1), depends on two simple and useful lemmas.

Lemma 1. *The following identity holds for any function u with n derivatives:*

$$(D - r)^n(e^{rx}u) = e^{rx}D^n(u). \tag{3}$$

Proof. We proceed by induction. For $n = 1$, we have

$$(D - r)(e^{rx}u) = \frac{d}{dx}(e^{rx}u) - r(e^{rx}u)$$

$$= e^{rx}D(u) + ure^{rx} - re^{rx}u = e^{rx}D(u).$$

Now *assume* that $(D - r)^{n-1}(e^{rx}u) = e^{rx}D^{n-1}(u)$. We wish to show that (3) holds. We have

$$(D - r)^n(e^{rx}u) = (D - r)(D - r)^{n-1}(e^{rx}u) = (D - r)e^{rx}D^{n-1}(u),$$

in which the induction hypothesis has been used. Therefore

$$(D - r)^n(e^{rx}u) = D(e^{rx}D^{n-1}u) - re^{rx}D^{n-1}u$$

$$= e^{rx}D^n u + re^{rx}D^{n-1}u - re^{rx}D^{n-1}u$$

$$= e^{rx}D^n u,$$

and the induction is complete.

Lemma 2. *If $D^n(u) = 0$, then u is a polynomial of degree $n - 1$, and conversely.* The proof may be established by induction and is left to the reader.

Theorem 4. *The function y is a solution of the equation*

$$(D - r)^n y = 0$$

if and only if there are constants $c_0, c_1, \ldots, c_{n-1}$ such that

$$y = e^{rx}(c_0 + c_1 x + \cdots + c_{n-1}x^{n-1}).$$

Proof. Let $y = e^{rx}u$. Then

$$(D - r)^n y = (D - r)^n(e^{rx}u) = e^{rx}D^n(u) = 0.$$

Therefore $D^n(u) = 0$ and, by Lemma 2, u is a polynomial of degree $\leq n - 1$. To exhibit the utility of Theorem 4, we observe that a solution of

$$[D - (3 + 2i)]^4(y) = 0$$

may be calculated instantaneously. The result is

$$y = e^{(3+2i)x}(c_0 + c_1 x + c_2 x^2 + c_3 x^3).$$

Before determining the general character of solutions of (2), we establish a lemma concerning the products of polynomials and exponentials.

Lemma 3. Suppose that $r_1, r_2, \ldots, r_k$ are distinct complex numbers and that $Q_1(x), Q_2(x), \ldots, Q_k(x)$ are polynomials. If

$$\sum_{j=1}^{k} Q_j(x)e^{r_j x} = 0 \tag{4}$$

for all real x, then the $Q_j(x)$ vanish identically.

Proof. Since the series e^z converges for all complex z, the left side of (4) is represented by its Maclaurin series for all real x. Therefore the series converges for all complex values and the sum must be zero for every complex number. If $k = 1$, then clearly $Q_1(x) \equiv 0$, since there is but one term and $e^{r_1 x}$ is never zero. Hence, let $k \geq 2$ and, by rearranging the order if necessary, assume that

$$0 \leq |r_1| \leq |r_2| \leq \cdots \leq |r_k|.$$

We shall suppose that $Q_k(x) \not\equiv 0$ and reach a contradiction. We write

$$r_j = |r_j|e^{i\theta_j}, \quad j = 1, 2, \ldots, k$$

and note that $|r_k| > 0$, since all the r_j are distinct. We make a change of variable in (4) by setting

$$t = e^{i\theta_k}x.$$

Then

$$\sum_{j=1}^{k} Q_j(x)e^{r_j x} = \sum_{j=1}^{k} Q_j(e^{-i\theta_k}t)e^{r_j(e^{-i\theta_k}t)}.$$

To simplify the notation, we write $R_j(t) \equiv Q_j(e^{-i\theta_k}t)$ and $s_j = e^{-i\theta_k}r_j$. Then we have $R_k(t) \not\equiv 0$ by hypothesis and so (4) becomes

$$\sum_{j=1}^{k} R_j(t)e^{s_j t} \equiv 0. \tag{5}$$

We observe that $|s_j| = |r_j|$ for each j and, since s_k is real, that $s_k = |r_k|$. We arranged the order so that s_k is larger than $|s_j|, j = 1, 2, \ldots, k - 1$. We set

$$s_j = a_j + ib_j, \quad j = 1, 2, \ldots, k,$$

and we shall reach a contradiction by using the strict inequality $a_j < a_k, j = 1, 2, \ldots, k - 1$. (See Fig. 16-2.) Suppose, to be specific, that

$$R_k(t) = c_0 + c_1 t + \cdots + c_p t^p, \quad c_p \neq 0, \quad p \geq 0.$$

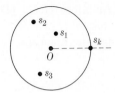

FIGURE 16-2

In (5) we divide by $c_p e^{s_k t} t^p$ and obtain

$$1 \equiv \frac{1}{|c_p|}\left[\left|c_{p-1}t^{-1} + c_{p-2}t^{-2} + \cdots + c_0 t^{-p} + \sum_{j=1}^{k-1} t^{-p}R_j(t)e^{ib_j t}e^{-(s_k-a_j)t}\right|\right]$$

and

$$1 \leq \frac{1}{|c_p|} \left[|c_{p-1}t^{-1}| + \cdots + |c_0 t^{-p}| + \sum_{j=1}^{k-1} |t|^{-p} |R_j(t)| e^{-(s_k - a_j)t} \right]. \tag{6}$$

L'Hôpital's Rule shows that when we multiply an exponential in t with negative exponent by a polynomial in t of arbitrary degree, the resulting expression tends to zero as $t \to +\infty$. Since $s_k - a_j > 0$, we conclude that each term on the right in (6) tends to zero. This yields a contradiction and $Q_k(x) \equiv 0$.

Remark. It may happen that one of the r_j in (4) is zero.

Theorem 5. *The equation*

$$(D - r)^n y = e^{rx}(b_0 + b_1 x + \cdots + b_p x^p), \qquad b_p \neq 0 \tag{7}$$

has a unique particular solution of the form

$$y = x^n e^{rx} R_p(x), \tag{8}$$

where R_p is a polynomial of degree p. In fact,

$$R_p(x) = \frac{b_0}{1 \cdot 2 \cdots n} + \frac{b_1 x}{2 \cdot 3 \cdots (n)(n+1)} + \cdots$$
$$+ \frac{b_p x^p}{(p+1)(p+2)\cdots(p+n)}.$$

Proof. Set $y = e^{rx}u$. We use Lemma 1 to obtain

$$(D - r)^n(y) = (D - r)^n(e^{rx}u) = e^{rx}D^n(u).$$

If (7) is to hold, we must have

$$D^n(u) = (b_0 + b_1 x + \cdots + b_p x^p).$$

It is readily verified by integrating n times that $u = x^n R_p(x)$ is a solution of this equation. Hence (8) is the desired solution. To show that it is unique, we suppose that $y^* = x^n e^{rx} R_p^*(x)$ is also a solution. Then the function $v = y - y^*$ satisfies the equation

$$(D - r)^n(v) = (D - r)^n(y - y^*) = (D - r)^n(y) - (D - r)^n(y^*) = 0.$$

According to Theorem 5,

$$v = y - y^* = e^{rx}\{x^n[R_p(x) - R_p^*(x)]\} = e^{rx}(c_0 + c_1 x + \cdots + c_{n-1} x^{n-1}).$$

The polynomial in braces is $\equiv 0$ or of degree at least n, while the polynomial on the right is of degree $\leq n - 1$. Therefore all coefficients must be zero and $R_p \equiv R_p^*$.

EXAMPLE 1. Solve $(D - 1)^3(y) = e^x(1 + 2x + 3x^2)$.

Solution I. The complementary function is

$$v = e^x(c_0 + c_1x + c_2x^2).$$

By Theorem 5, a particular solution is given by

$$y = e^x\left(\frac{x^3}{1 \cdot 2 \cdot 3} + \frac{2x^4}{2 \cdot 3 \cdot 4} + \frac{3x^5}{3 \cdot 4 \cdot 5}\right).$$

The answer is

$$y = e^x(c_0 + c_1x + c_2x^2 + \tfrac{1}{6}x^3 + \tfrac{1}{12}x^4 + \tfrac{1}{20}x^5).$$

Solution II. Setting $y = e^x u$, we obtain

$$(D - 1)^3(y) = e^x D^3 u = e^x(1 + 2x + 3x^2).$$

The student can easily solve

$$D^3(u) = 1 + 2x + 3x^2$$

to obtain the solution.

Theorem 5 gives the form of a particular solution when the exponential on the right side has a coefficient which is a root of the auxiliary polynomial. The next theorem yields the form of a particular solution when the coefficient on the right is not a root of this polynomial.

Theorem 6. *The equation*

$$(D - r)^n(y) = e^{sx}Q_p(x), \qquad r \neq s \tag{9}$$

in which r and s are complex numbers and Q_p is a polynomial of degree p, has a unique particular solution y of the form

$$y = e^{sx}R_p(x),$$

where R_p is a polynomial of degree p.

Proof. Setting $y = e^{rx}u$, we see that y is a solution of (9) if and only if

$$(D - r)^n y = (D - r)^n(e^{rx}u) = e^{rx}D^n(u) = e^{sx}Q_p(x).$$

If we set $a = s - r$, then we must solve the equation

$$D^n(u) = e^{ax}Q_p(x).$$

We integrate with respect to x, getting

$$D^{n-1}(u) = \int Q_p(x)e^{ax}\,dx = Q_p(x)\frac{e^{ax}}{a} - \frac{1}{a}\int Q'_p(x)e^{ax}\,dx,$$

in which the last equality is obtained by integrating by parts. The quantity Q'_p is a polynomial of degree $p - 1$. We integrate by parts p more times and find

$$D^{n-1}(u) = \frac{e^{ax}}{a}[Q_p(x) - a^{-1}Q'_p(x) + \cdots + (-1)^p a^{-p} Q_p^{(p)}(x)] + K_0, \quad (10)$$

where K_0 is a constant of integration. The function inside the brackets is a polynomial of degree p. We repeat the process by integrating (10) with respect to x. We get

$$D^{n-2}(u) = \frac{e^{ax}}{a^2}[A_p(x)] + K_1 + K_0 x$$

where $A_p(x)$ is a polynomial of degree p and K_0, K_1 are constants. Continuing, we finally obtain

$$u = e^{ax}R_p(x) + c_0 + c_1 x + \cdots + c_{n-1}x^{n-1}$$

where R_p is a polynomial of degree p and the constants of integration are relabeled $c_0, c_1, \ldots, c_{n-1}$. We have found the particular solution

$$y = (c_0 + c_1 x + \cdots + c_{n-1}x^{n-1})e^{rx} + e^{sx}R_p(x).$$

A solution of the desired form is obtained by setting $c_0 = c_1 = \cdots = c_{n-1} = 0$. To establish the uniqueness of the solution of this form, suppose $y^* = e^{sx}R_p^*(x)$ is also a solution. Then

$$(D - r)^n(y - y^*) = 0,$$

so that

$$y - y^* = e^{sx}[R_p(x) - R_p^*(x)] = e^{rx}(c_0 + c_1 x + \cdots + c_{n-1}x^{n-1}).$$

Since $s \neq r$, we conclude from Lemma 3 that $R_p = R_p^*$ and $c_0 = c_1 = \cdots = c_{n-1} = 0$.

Remark. To find a particular solution of Eq. (9), it is a good idea to set $y = e^{rx}u$ with

$$u = e^{ax}(b_0 + b_1 x + \cdots + b_p x^p), \qquad a = s - r,$$

and then use the method of undetermined coefficients. The quantities b_i are determined by equating coefficients after calculating $D^n(u)$ and setting the result equal to $e^{ax}Q_p(x)$. We illustrate with an example.

EXAMPLE 2. Solve $(D - 1)^3 y = e^{3x}(2 - 3x + x^2)$.

Solution. The complementary function is $v = e^x(c_0 + c_1 x + c_2 x^2)$. To obtain a particular solution, we set $y_1 = e^x u$. Then

$$e^x D^3(u) = e^{3x}(2 - 3x + x^2) \qquad \text{or} \qquad D^3(u) = e^{2x}(2 - 3x + x^2).$$

Let $u_1 = e^{2x}(b_0 + b_1x + b_2x^2)$. Then

$$D(u_1) = e^{2x}(b_1 + 2b_2x) + (b_0 + b_1x + b_2x^2) \cdot 2e^{2x}$$
$$= e^{2x}[(2b_0 + b_1) + 2(b_1 + b_2)x + 2b_2x^2].$$
$$D^2(u_1) = e^{2x}[(4b_0 + 4b_1 + 2b_2) + 4(b_1 + 2b_2)x + 4b_2x^2].$$
$$D^3(u_1) = e^{2x}[(8b_0 + 12b_1 + 12b_2) + 8(b_1 + 3b_2)x + 8b_2x^2]$$
$$= e^{2x}(2 - 3x + x^2).$$

Equating coefficients, we find

$$8b_0 + 12b_1 + 12b_2 = 2, \qquad 8(b_1 + 3b_2) = -3, \qquad 8b_2 = 1$$

or

$$b_2 = \tfrac{1}{8}, \qquad b_1 = -\tfrac{3}{4}, \qquad b_0 = \tfrac{19}{16}.$$

The general solution is

$$y = v + y_1 = e^x(c_0 + c_1x + c_2x^2) + e^{3x}(\tfrac{19}{16} - \tfrac{3}{4}x + \tfrac{1}{8}x^2).$$

EXERCISES

Solve the following differential equations.

1. $D^3(y) = 0$
2. $(D - 1)^2(y) = 0$
3. $(D + 1)^3(y) = 0$
4. $(D - 1)^2(y) = e^x$
5. $(D + 1)^2(y) = (2x + 1)e^{-x}$
6. $(D - 2)^2(y) = x^2e^{2x}$
7. $(D - 1)^3(y) = (x^2 - 3)e^x$
8. $(D + 2)^3(y) = (2x^2 - 3x + 1)e^{-2x}$
9. $\dfrac{d^2y}{dx^2} - 2\dfrac{dy}{dx} + y = e^{2x}$
10. $\dfrac{d^2y}{dx^2} + 4\dfrac{dy}{dx} + 4y = xe^{-x}$
11. $(D + 1)^2(y) = (x + 1)e^x$
12. $(D - 2)^2(y) = 3x^2 - 2x + 1$
13. $(D - 1)^2(y) = (x^2 + 2x - 1)e^{2x}$
14. $(D - 1)^3(y) = xe^{2x}$
15. $(D + 1)^3(y) = xe^x$
16. $(D + 1)^3(y) = x^2 - x + 1$
17. $D^3(y) = x^2e^x$
18. $D^3(y) = (2x^2 + 3)e^{-x}$

6. LINEAR DIFFERENTIAL EQUATIONS WITH CONSTANT COEFFICIENTS. GENERAL CASE

The most general linear differential equation with constant coefficients has the form

$$L(y) \equiv (D - r_1)^{n_1}(D - r_2)^{n_2} \ldots (D - r_k)^{n_k}(y) = f(x), \tag{1}$$

where $r_1, r_2, \ldots, r_k$ are distinct complex numbers; $n_1, n_2, \ldots, n_k$ are positive integers; and $f(x)$ is a given function of x.

Theorem 7. *Suppose that $f(x) \equiv 0$ in (1) so that the equation is homogeneous. Then y is a solution of (1) if and only if it has the form*

$$y = P_1(x)e^{r_1x} + P_2(x)e^{r_2x} + \cdots + P_k(x)e^{r_kx} \tag{2}$$

where each P_i is a polynomial of degree $n_i - 1$ or less.

Proof. We observe that the operator L is unchanged if the order of the factors is changed (Theorem 1, page 666). Consider any one of the terms of (2), say $P_2(x)e^{r_2 x}$. We rearrange the order of the left side of (1) so that $(D - r_2)^{n_2}$ is applied first. Then

$$(D - r_2)^{n_2}[e^{r_2 x}P_2(x)] = e^{r_2 x}D^{n_2}[P_2(x)] = 0,$$

since P_2 is a polynomial of degree less than n_2. Thus each term of (2) satisfies (1) with $f \equiv 0$, and so (2) itself does. Now, to show that all solutions of (1) with $f \equiv 0$ are of the form (2), we define

$$u_1 = (D - r_1)^{n_1}(y),$$
$$u_2 = (D - r_2)^{n_2}(u_1) = (D - r_2)^{n_2}(D - r_1)^{n_1}(y),$$
$$\vdots$$
$$u_{k-1} = (D - r_{k-1})^{n_{k-1}}(u_{k-2}),$$
$$u_k = (D - r_k)^{n_k}(u_{k-1}) = L(y).$$

Assuming that y is a solution of (1) with $f \equiv 0$, we obtain

$$(D - r_k)^{n_k}(u_{k-1}) = 0$$

so that, according to Theorem 4,

$$u_{k-1} = e^{r_k x}P_k^{(1)}(x)$$

where $P_k^{(1)}$ is a polynomial of degree less than n_k. Now we write

$$u_{k-1} = (D - r_{k-1})^{n_{k-1}}(u_{k-2}) = e^{r_k x}P_k^{(1)}(x).$$

This equation is in the form (9) of Section 5. The form of u_{k-2}, according to Theorem 6, is

$$u_{k-2} = e^{r_{k-1} x}P_{k-1}^{(2)}(x) + e^{r_k x}P_k^{(2)}(x),$$

where $P_{k-1}^{(2)}(x)$ is a polynomial of degree less than n_{k-1} and $P_k^{(2)}(x)$ is a polynomial of the same degree as $P_k^{(1)}(x)$. Continuing in this way, we find

$$u_{k-2} = (D - r_{k-2})^{n_{k-2}}(u_{k-3})$$

and

$$u_{k-3} = e^{r_{k-2} x}P_{k-2}^{(3)}(x) + e^{r_{k-1} x}P_{k-1}^{(3)}(x) + e^{r_k x}P_k^{(3)}(x).$$

After k steps we get an expression for y of the form (2).

EXAMPLE 1. Solve $(D^2 - 1)^3(y) = 0$.

Solution. We write

$$(D - 1)^3(D + 1)^3(y) = 0$$

and use Theorem 7. The result is

$$y = e^x(c_1 + c_2 x + c_3 x^2) + e^{-x}(c_4 + c_5 x + c_6 x^2).$$

EXAMPLE 2. Solve $(D - 1 - i)^2(D + 2)^3(D - 3i)(y) = 0$.

Solution. We use Theorem 7 to write

$$y = e^{(1+i)x}(c_0 + c_1x) + e^{-2x}(c_2 + c_3x + c_4x^2) + c_5e^{3ix}.$$

The next theorem yields the form of particular solutions of (1) when

$$f(x) = e^{sx}Q_p(x)$$

where Q_p is a polynomial. There are two cases, according as s is or is not a root of the auxiliary polynomial of L.

Theorem 8. (a) *If* $f(x) = e^{sx}Q_p(x)$ *where* $s \neq r_i$ *for any* r_i, $i = 1, 2, \ldots k$, *and if* Q_p *is a polynomial of degree* p, *then* (1) *has a unique particular solution* y_1 *of the form*

$$y_1 = e^{sx}R_p(x), \tag{3}$$

where R_p *is a polynomial of degree* p.
(b) *If* $f(x) = e^{r_ix}Q_p(x)$ *where* Q_p *is a polynomial of degree* p, *then* (1) *has a unique particular solution* y_1 *of the form*

$$y_1 = x^n{}_ie^{r_ix}R_p(x), \tag{4}$$

where R_p *is a polynomial of degree* p.

Proof. (a) We define

$$u_1 = (D - r_1)^{n_1}(y_1)$$

where y_1 is a particular solution of (1). In the same way we define $u_2, u_3, \ldots, u_k$ by the relations

$$u_2 = (D - r_2)^{n_2}(u_1),$$
$$u_3 = (D - r_3)^{n_3}(u_2),$$
$$\vdots$$
$$u_k = (D - r_k)^{n_k}(u_{k-1}) = L(y) = f = e^{sx}Q_p(x).$$

The equation

$$(D - r_k)^{n_k}(u_{k-1}) = e^{sx}Q_p(x), \qquad s \neq r_k,$$

has been solved in Theorem 6. A particular solution is $u_{k-1} = e^{sx}R_p^{(1)}(x)$, where $R_p^{(1)}$ is a polynomial of degree p. Continuing in this way, we solve

$$u_{k-1} = (D - r_{k-1})^{n_{k-1}}(u_{k-2}) = e^{sx}R_p^{(1)}(x)$$

and get a solution of the same form for u_{k-2}. Finally, we find an expression of the form (3) as the solution of (1). If there were two solutions of the form (3), the difference would satisfy (1) with $f(x) \equiv 0$. Thus the difference would be of the form (2). Now, employing Lemma 3 of the last section, we conclude that the solutions must coincide.

(b) After rearranging the factors in (1), if necessary, we may assume that $i = k$ in the expression (4). We let

$$u = (D - r_k)^{n_k}(y)$$

and write

$$L(y) \equiv (D - r_1)^{n_1}(D - r_2)^{n_2} \ldots (D - r_{k-1})^{n_{k-1}}(u) = e^{r_k x}Q_p(x). \qquad (5)$$

The solution of this equation was obtained in part (a), since $r_1, r_2, \ldots, r_k$ are all distinct. Therefore a particular solution of (5) has the form

$$u = e^{r_k x}S_p(x),$$

where S_p is a polynomial of degree p. Now we must solve

$$u = (D - r_k)^{n_k}(y) = e^{r_k x}S_p(x).$$

However, this was done in Theorem 5, and the solution has the specified form (4). The uniqueness follows from the same argument as that given in part (a).

───────

EXAMPLE 3. Find the form of the most general solution of the equation (but do not solve)

$$L(y) \equiv (D - 1)^3(D - 2)^4(D - 3)^3(y) = e^x(2 - 3x + 4x^2)$$
$$+ e^{2x}(3 - 2x + x^2 - x^3) + e^{3x}x^3$$
$$+ e^{4x}(1 - 3x^2) + (x - x^3)e^{5x}.$$

Solution. The complementary function is

$$v = e^x(c_0 + c_1x + c_2x^2) + e^{2x}(c_3 + c_4x + c_5x^2 + c_6x^3)$$
$$+ e^{3x}(c_7 + c_8x + c_9x^2).$$

A particular solution of

$$L(y_1) = e^x(2 - 3x + 4x^2)$$

has the form [by Theorem 8, part (b)]

$$y_1 = x^3e^x(a + bx + cx^2).$$

Particular solutions of

$$L(y_2) = e^{2x}(3 - 2x + x^2 - x^3) \quad \text{and} \quad L(y_3) = e^{3x}x^3$$

have the forms [by Theorem 8, part (b)]

$$y_2 = x^4e^{2x}(d + ex + fx^2 + gx^3), \qquad y_3 = x^3e^{3x}(h + jx + kx^2 + lx^3).$$

According to part (a) of Theorem 8, the equations

$$L(y_4) = e^{4x}(1 - 3x^2) \quad \text{and} \quad L(y_5) = e^{5x}(x - x^3)$$

have the forms

$$y_4 = e^{4x}(m + nx + px^2), \qquad y_5 = e^{5x}(q + rx + sx^2 + tx^3),$$

respectively. The constants a, b, c, $\ldots$, s, t are not arbitrary but would be obtained by the method of undetermined coefficients, as described previously and as demonstrated in Example 4 below. The form of the general solution is, according to Principle 2 in Section 3 on page 674,

$$y = v + y_1 + y_2 + y_3 + y_4 + y_5.$$

As an aid in the solution of equations by the method of undetermined coefficients, we prove the following useful formula, which is an extension of the identity

$$(D - r)^n(e^{rx}u) = e^{rx}D^n(u) \tag{6}$$

established on page 680.

Lemma. If P is any polynomial, then

$$P(D)(e^{rx}u) = e^{rx}P(D + r)u. \tag{7}$$

Proof. Any polynomial $P(t)$ may be written as a polynomial of the same degree in powers of $(t - a)$ for any constant a (Taylor expansion about $t = a$). Thus we may write

$$P(D) = a_0 + a_1(D - r) + \cdots + a_n(D - r)^n = \sum_{j=0}^{n} a_j(D - r)^j.$$

Using (6) above, we see that

$$P(D)(e^{rx}u) = \sum_{j=0}^{n} a_j(D - r)^j(e^{rx}u) = e^{rx}\sum_{j=0}^{n} a_jD^j(u).$$

But

$$P(D + r) = a_0 + a_1D + \cdots + a_nD^n = \sum_{j=0}^{n} a_jD^j,$$

and so (7) is established.

EXAMPLE 4. Solve $(D - 1)^3(D - 2)^3(y) = x^2e^{3x}$.

Solution. The complementary function is

$$v = e^x(C_0 + C_1x + C_2x^2) + e^{2x}(C_3 + C_4x + C_5x^2). \tag{8}$$

To obtain a particular solution y_1, we write $y_1 = e^{3x}u_1$ where $u_1 = a + bx + cx^2$. Using the lemma, we see that

$$(D - 1)^3(D - 2)^3(y_1) = e^{3x}(D + 2)^3(D + 1)^3(u_1) = e^{3x}x^2.$$

Thus we must solve

$$(D + 2)^3(D + 1)^3(u_1) = x^2.$$

Expanding and multiplying, we obtain

$$(D + 2)^3 = D^3 + 6D^2 + 12D + 8, \qquad (D + 1)^3 = D^3 + 3D^2 + 3D + 1,$$

and

$$(D + 2)^3(D + 1)^3 = D^6 + 9D^5 + 33D^4 + 63D^3 + 66D^2 + 36D + 8.$$

By differentiating, we find

$$8u_1 = 8(a + bx + cx^2), \qquad 36D(u_1) = 36(b + 2cx),$$
$$66D^2(u_1) = 132c, \qquad D^3(u_1) = D^4(u_1) = D^5(u_1) = D^6(u_1) = 0.$$

Therefore

$$(8a + 36b + 132c) + (8b + 72c)x + 8cx^2 = x^2.$$

Equating coefficients, we get

$$8c = 1, \qquad 8b + 72c = 0, \qquad 8a + 36b + 132c = 0$$

or

$$c = \tfrac{1}{8}, \qquad b = -\tfrac{9}{8}, \qquad a = 3.$$

The particular solution is

$$y_1 = e^{3x}(3 - \tfrac{9}{8}x + \tfrac{1}{8}x^2),$$

and the general solution is

$$y = v + y_1,$$

i.e., the sum of the complementary function (8) and the particular solution y_1.

EXERCISES

Solve the following differential equations.

1. $\dfrac{d^2y}{dx^2} - 4\dfrac{dy}{dx} + 3y = 0$ 2. $\dfrac{d^2y}{dx^2} + \dfrac{dy}{dx} - 6y = 0$

3. $\dfrac{d^3y}{dx^3} - \dfrac{d^2y}{dx^2} - 2\dfrac{dy}{dx} = 0$ 4. $\dfrac{d^2y}{dx^2} - 3\dfrac{dy}{dx} + y = 0$

5. $\dfrac{d^2y}{dx^2} - 4y = 2e^{2x}$ 6. $\dfrac{d^2y}{dx^2} + \dfrac{dy}{dx} - 2y = xe^x$

7. $(D^3 - 2D^2 - D + 2)(y) = 0$ 8. $(D^3 - 6D + 5)(y) = 0$

9. $(D^3 - 3D + 2)(y) = 2e^x$ 10. $\dfrac{d^4y}{dx^4} - 5\dfrac{d^2y}{dx^2} + 4y = 0$

11. $\dfrac{d^3y}{dx^3} - 2\dfrac{d^2y}{dx^2} + \dfrac{dy}{dx} = 1$ 12. $\dfrac{d^2y}{dx^2} - \dfrac{dy}{dx} - 6y = xe^x$

13. $\dfrac{d^4y}{dx^4} - \dfrac{d^3y}{dx^3} = xe^x$ 14. $\dfrac{d^2y}{dx^2} - 7\dfrac{dy}{dx} + 12y = xe^{2x} + 2xe^{3x}$

15. $\dfrac{d^2y}{dx^2} + 2\dfrac{dy}{dx} - 3y = xe^x + x^2$ 16. $(D^3 - D^2 - D + 1)(y) = xe^x$

17. $(D^4 - 8D^2 + 16)(y) = xe^{2x}$ 18. $D(D - 1)^3(y) = (x^2 + 2x)e^x$

19. $(D + 1)(D - 2)^3(y) = x^2e^{2x} + xe^x$ 20. $D^2(D + 1)^3(y) = x^2e^{-x}$

7. COMPLEX ROOTS IN EQUATIONS WITH REAL COEFFICIENTS

We consider the linear equation

$$L(y) \equiv (a_0 D^n + a_1 D^{n-1} + \cdots + a_{n-1} D + a_n) y = f(x), \tag{1}$$

in which $a_0, a_1, \ldots, a_{n-1}, a_n$ are *real* constants. The auxiliary polynomial equation

$$P(r) = a_0 r^n + a_1 r^{n-1} + \cdots + a_{n-1} r + a_n = 0 \tag{2}$$

has n roots, some of which may be complex. If $a + bi$ is a root of (2), it is easy to see that $a - bi$ is also. For we write

$$P(a + bi) = Q(a, b) + iR(a, b), \tag{3}$$

where $Q(a, b)$ and $R(a, b)$ are real polynomials obtained by substituting $a + bi$ for r in (2) and then separating all the real and imaginary terms which occur. Thus, if $P(a + bi) = 0$, then *both*

$$Q(a, b) = 0 \quad \text{and} \quad R(a, b) = 0.$$

The reader may verify that

$$P(a - bi) = Q(a, b) - iR(a, b),$$

so that $a - bi$ is a root whenever $a + bi$ is. If $a + bi$ and $a - bi$ are roots, then $P(r)$ has the factor $[r - (a + bi)][r - (a - bi)] = (r - a)^2 + b^2$, which is a real quadratic factor. Upon dividing P by this factor, we see that the quotient will have *real* coefficients. Therefore any other complex roots will occur in conjugate pairs. *We conclude that if $a + bi$ is a root of multiplicity n, then $a - bi$ is also.*

The terms in the complementary function corresponding to a pair of complex roots with multiplicity n have the form

$$e^{(a+bi)x}(K_0 + K_1 x + \cdots + K_{n-1} x^{n-1})$$
$$+ e^{(a-bi)x}(K_0' + K_1' x + \cdots + K_{n-1}' x^{n-1})$$

where $K_0, \ldots, K_{n-1}, K_0', \ldots, K_{n-1}'$ are arbitrary *complex* constants. These terms in the complementary function may be written in the alternative form

$$e^{ax}[(C_0 + C_1 x + \cdots + C_{n-1} x^{n-1}) \cos bx$$
$$+ (C_0' + C_1' x + \cdots + C_{n-1}' x^{n-1}) \sin bx], \tag{4}$$

where $C_0, \ldots, C_{n-1}, C_0', \ldots, C_{n-1}'$ are arbitrary constants. If the constants happen to be real, the contribution to the complementary function will be real.

If $f \equiv 0$ in (1) and if the coefficients are real, it is possible to obtain real solutions when the contributions due to complex roots are written in the form (4).

In the case of a nonhomogeneous equation, suppose that

$$f(x) = e^{ax}[P_1(x) \cos bx + P_2(x) \sin bx], \tag{5}$$

where P_1 and P_2 are polynomials of degree not exceeding p and one (or both) of them is of degree p. We use the fact that $\cos bx = \frac{1}{2}(e^{ibx} + e^{-ibx})$, $\sin bx = (1/2i)(e^{ibx} - e^{-ibx})$ to write (5) in the exponential form

$$f(x) = P_1^*(x)e^{(a+bi)x} + P_2^*(x)e^{(a-bi)x},$$

in which $P_1^* = \frac{1}{2}(P_1 + iP_2)$, $P_2^* = \frac{1}{2}(P_1 - iP_2)$. Since f is now in the appropriate form, we may use the results of Section 6 to obtain solutions of $Ly = f$. We state the results in the form of a theorem.

Theorem 9. *If f is in the form (5) and the coefficients in (1) are real, then the equation $L(y) = f$ has a unique particular solution of the form*

$$e^{ax}[Q_1(x)\cos bx + Q_2(x)\sin bx]$$

whenever $a \pm bi$ are not roots of the auxiliary polynomial. The equation $L(y) = f$ has a unique particular solution of the form

$$x^n e^{ax}[Q_1(x)\cos bx + Q_2(x)\sin bx]$$

whenever $a \pm bi$ are roots of the auxiliary polynomial of multiplicity n. The polynomials Q_1 and Q_2 are of degree not exceeding p.

Remarks. The degree of Q_1 and Q_2 will depend on the degrees of P_1^* and P_2^* and, since these are combinations of P_1 and P_2, it is not always the case that P_1^* and P_2^* will have the same degrees as P_1 and P_2.

EXAMPLE 1. Solve $(D - 1)^2(D^2 - 2D + 2)^3(y) = 0$.

Solution. We note that $D^2 - 2D + 2 = [D - (1 + i)][D - (1 - i)]$. Hence

$$y = e^x(C_0 + C_1 x) + e^{x+ix}(K_0 + K_1 x + K_2 x^2) + e^{x-ix}(K_3 + K_4 x + K_5 x^2),$$

which we may write

$$y = e^x(C_0 + C_1 x) + e^x[(C_2 + C_3 x + C_4 x^2)\cos x + (C_5 + C_6 x + C_7 x^2)\sin x].$$

We have selected

$$C_2 = K_0 + K_3, \qquad C_3 = K_1 + K_4, \qquad C_4 = K_2 + K_5,$$
$$C_5 = i(K_0 - K_3), \qquad C_6 = i(K_1 - K_4), \qquad C_7 = i(K_2 - K_5).$$

EXAMPLE 2. Find the form of the solution of the equation (but do not solve)

$$(D - 2)^3(D^2 - 2D + 5)^2(y) = xe^x + x^2 e^{2x} + x^3 e^x \sin x.$$

Solution. The auxiliary polynomial has the root 2 with multiplicity 3 and the roots $1 \pm 2i$ with multiplicity 2. The complementary function is

$$v = e^{2x}(C_0 + C_1 x + C_2 x^2) + e^x[(C_3 + C_4 x)\cos 2x + (C_5 + C_6 x)\sin 2x].$$

The particular solution is of the form

$$y_1 = e^x(a + bx) + x^3 e^{2x}(c + dx + ex^2)$$
$$+ e^x[(f + gx + hx^2 + jx^3) \cos x + (k + lx + mx^2 + px^3) \sin x],$$

where the constants $a, b, \ldots, p$ all have definite values which can be found by the method of undetermined coefficients.

If the given function f is of the form

$$e^{ax}P(x) \cos bx \qquad \text{or} \qquad e^{ax}Q(x) \sin x,$$

it may be expeditious to use the idea of Example 4, Section 3. That is, we replace $\cos bx$ by e^{ibx}, use the techniques of the preceding section, and then take the real part of the solution (since $\cos bx$ is the real part of e^{ibx}). If $\sin bx$ occurs, we take the imaginary part. The next example illustrates the method.

EXAMPLE 3. Solve $(D - 2)(D^2 - 2D + 5)^2(y) = xe^x \cos 2x$.

Solution. The complementary function is

$$v = C_0 e^{2x} + e^x[(C_1 + C_2 x) \cos 2x + (C_3 + C_4 x) \sin 2x]. \tag{6}$$

To obtain a particular solution, we observe that the right side of the differential equation is the real part of $xe^{(1+2i)x}$. So we find a particular solution of

$$P(D)(Y) \equiv (D - 2)(D^2 - 2D + 5)^2(Y) = xe^{(1+2i)x}$$

and take its real part. We set

$$Y = e^{(1+2i)x}u.$$

Then, using the lemma of Section 6, we obtain

$$P(D)(Y) \equiv (D - 2)(D - 1 + 2i)^2(D - 1 - 2i)^2(Y) = e^{(1+2i)x}P(D + 1 + 2i)(u)$$
$$= e^{(1+2i)x}(D - 1 + 2i)(D + 4i)^2 D^2(u) = xe^{(1+2i)x}.$$

Therefore

$$(D - 1 + 2i)(D + 4i)^2 D^2(u) = x$$

or

$$[D^3 + (-1 + 10i)D^2 - 8(4 + i)D + 16(1 - 2i)]D^2(u) = x.$$

Since $1 + 2i$ is a root of multiplicity 2, we select

$$u = x^2(a + bx) = ax^2 + bx^3$$

and compute

$$D^2(u) = 2a + 6bx,$$
$$[D^3 + (-1 + 10i)D^2 + \cdots]D^2(u) = 16(1 - 2i)(2a + 6bx) - 8(4 + i)6b = x.$$

Equating coefficients, we find

$$96b(1 - 2i) = 1, \qquad 32(1 - 2i)a = 48(4 + i)b$$

and

$$b = \frac{1 + 2i}{480}, \qquad a = -\frac{16 - 13i}{1600}.$$

We conclude that

$$Y = e^{(1+2i)x}u = e^{(1+2i)x}\left(\frac{-16 + 13i}{1600}x^2 + \frac{1 + 2i}{480}x^3\right).$$

Taking the real part of Y, we get the particular solution

$$y_1 = x^2 e^x\left[\left(-\frac{1}{100} + \frac{x}{480}\right)\cos 2x - \left(\frac{13}{1600} + \frac{x}{240}\right)\sin 2x\right].$$

The final result is $y = y_1 + v$, where v is the complementary function (6).

EXERCISES

In each of problems 1 through 4, write the form of the solution of the given equation but do not find the coefficients.

1. $(D^2 - 1)^2(y) = x^2 e^x + e^{2x}(2x + 3)\sin 3x$
2. $(D^2 + 1)^3(y) = x^3 - 2x^2 e^x + x^2 \cos x - 2x \sin x$
3. $(D^2 - 2D + 2)^3(y) = x^2 e^{-x} + xe^x + e^x(x^2 \cos x - x^3 \sin x)$
4. $(D - 1)^2(D^2 + 2D + 5)^3(y) = x \cos x + x^2 e^x + e^{-x}(x^2 \cos 2x + x^3 \sin 2x)$

In problems 5 through 17, in each case solve the given equation.

5. $\dfrac{d^2y}{dx^2} + y = 2x$

6. $\dfrac{d^2y}{dx^2} + 4y = \sin 2x$

7. $\dfrac{d^4y}{dx^4} + 2\dfrac{d^2y}{dx^2} + y = 0$

8. $\dfrac{d^4y}{dx^4} + 8\dfrac{d^2y}{dx^2} + 16 = e^x$

9. $(D^4 + 5D^2 + 4)(y) = x$

10. $(D^2 - 2D + 5)(y) = x^2$

11. $(D^3 + D^2 + D + 1)(y) = \cos 2x$

12. $(D^2 - D - 2)(y) = 2\cos x$

13. $(D^3 + 8)(y) = 4x^3 - 2x^2$

14. $(D^3 - D^2 + D - 1)(y) = xe^x$

*15. $(D^4 + 8D^2 - 9)(y) = \cos 3x + e^{2x}$

*16. $(D^2 + 4)^2(y) = 2\cos 2x - 6x \sin 2x + 6 \sin x$

*17. $(D^2 - 2D + 2)^2(y) = e^x(2x \cos x - 6 \sin x) + xe^x$

8. THE EULER-CAUCHY EQUATION

The equation

$$x^n \frac{d^n y}{dx^n} + a_1 x^{n-1} \frac{d^{n-1}y}{dx^{n-1}} + \cdots + a_{n-1}x\frac{dy}{dx} + a_n y = f(x), \qquad (1)$$

where $a_1, a_2, \ldots, a_n$ are constants, is called the **Euler-Cauchy equation.** By a simple change of variable, this equation may be transformed into an equation with constant coefficients of the type we have considered in the preceding sections. For $x > 0$, we let

$$x = e^t, \qquad y(x) = y(e^t) = Y(t) = Y(\ln x). \qquad (2)$$

Differentiating, we find that

$$\frac{dy}{dx} = Y'(\ln x) \cdot x^{-1} \qquad \text{or} \qquad x\frac{dy}{dx} = Y'.$$

Using this last result with y replaced by $x(dy/dx)$ and Y replaced by Y', we obtain

$$x\frac{d}{dx}\left(x\frac{dy}{dx}\right) = Y'' \qquad \text{or} \qquad x^2\frac{d^2y}{dx^2} = D(D - 1)(Y)$$

where we have let D stand for the operator d/dt. By induction it is easily established that

$$x^n\frac{d^ny}{dx^n} = D(D - 1)\cdots(D - n + 1)(Y), \qquad n = 1, 2, \ldots.$$

Thus the change of variable (2) transforms the Euler-Cauchy equation into one with constant coefficients.

———————

EXAMPLE 1. Solve $x^2y'' + xy' + y = x$.

Solution. Letting $x = e^t$, denoting d/dt by D, and letting $y(x) = Y(t)$, we see that this equation becomes

$$[D(D - 1) + D + 1](Y) = e^t \qquad \text{or} \qquad (D^2 + 1)(Y) = e^t.$$

The complementary function is

$$V = C_1 \cos t + C_2 \sin t,$$

and a particular solution is

$$Y_1 = \tfrac{1}{2}e^t.$$

Therefore

$$Y = C_1 \cos t + C_2 \sin t + \tfrac{1}{2}e^t$$

or

$$y = C_1 \cos (\ln x) + C_2 \sin (\ln x) + \tfrac{1}{2}x, \qquad x > 0.$$

———————

The transformation (2) shows that if r is real, then

$$e^{rt} = x^r, \qquad x > 0.$$

So if all the roots of the auxiliary polynomial of the transformed equation are real and distinct, the solutions of the homogeneous equation (1) will be linear combinations of $x^{r_1}, x^{r_2}, \ldots, x^{r_n}$, where $r_1, r_2, \ldots, r_n$ are the n real and distinct roots. This statement is still true if some of the roots are complex, so long as all of them are distinct. However, if r is complex, we must interpret x^r by the relation

$$x^r = e^{r \ln x}, \qquad x > 0.$$

The reader may easily verify that if x^r is substituted for y in (1), then x^r is a factor of the left side. Thus when the auxiliary polynomial has distinct roots, the substitution $y = x^r$ may be used instead of (2) to obtain a solution. But if the auxiliary polynomial has multiple roots, it is better to make the substitution (2), as in Example 1.

EXAMPLE 2. Solve $x^2 y'' + xy' - s^2 y = 0$, where s is a nonzero constant.

Solution. Setting $y = x^r$, we obtain

$$y' = rx^{r-1}, \qquad y'' = r(r-1)x^{r-2}.$$

Therefore

$$x^2 y'' + xy' - s^2 y = x^r[r(r-1) + r - s^2] = x^r(r^2 - s^2) = 0,$$

and so

$$r = \pm s.$$

The solution is

$$y = C_1 x^s + C_2 x^{-s}.$$

Note that the method fails if $x = 0$.

EXERCISES

In each of problems 1 through 6, solve by assuming a solution of the form x^r. If complex values of r occur, write the result in a form which is real if the constants are.

1. $x^2 y'' - 2xy' + 2y = 0$
2. $x^2 y'' + 2xy' - s(s+1)y = 0$
3. $x^3 y''' + 2x^2 y'' + xy' - y = 0$
4. $x^3 y''' + 3x^2 y'' = 0$
5. $x^2 y'' + xy' + 4y = 0$
6. $x^2 y'' - xy' + 5y = 0$

In each of problems 7 through 13, solve the given equation.

7. $x^2 y'' + 4xy' + 2y = 2 \ln x$
8. $x^2 y'' - xy' + y = x^2 \ln x$
9. $x^2 y'' - xy' + y = x \ln x$
10. $x^3 y''' + 2x^2 y'' - xy' + y = x^2 \ln x$
11. $x^3 y''' + xy' - y = x \ln x$
12. $x^3 y''' + 4x^2 y'' + xy' - y = 2x^{-1} \ln x$
13. $x^2 y'' + xy' + 4y = \cos (2 \ln x)$

9. LINEAR SYSTEMS WITH CONSTANT COEFFICIENTS

In this section and in subsequent ones we shall use t as the independent variable and x, y, z, x_1, x_2, etc., as dependent variables. A second-order equation of the form

$$a_0(t)\frac{d^2 x}{dt^2} + a_1(t)\frac{dx}{dt} + a_2 x = f(t) \tag{1}$$

may be transformed into an equivalent system of two equations of the first order for two unknown functions. We set

$$y = \frac{dx}{dt} \tag{2a}$$

and observe that (1) becomes

$$a_0(t)\frac{dy}{dt} + a_1(t)y + a_2x = f(t). \tag{2b}$$

Equations (2a) and (2b) form a system of two first-order equations for the two unknown functions x and y. A solution of this system is equivalent to a solution of (1). The process of reducing a higher-order equation to a system of the first order works quite generally. For example, the nth-order equation

$$\frac{d^n x}{dt^n} = f\left(t, \ x, \ \frac{dx}{dt}, \ \dots, \ \frac{d^{n-1}x}{dt^{n-1}}\right)$$

is equivalent to the following system of n equations of the first order for the n unknown functions $x, u_1, u_2, \dots, u_{n-1}$:

$$\frac{dx}{dt} = u_1,$$

$$\frac{du_1}{dt} = u_2,$$

$$\vdots \tag{3}$$

$$\frac{du_{n-2}}{dt} = u_{n-1},$$

$$\frac{du_{n-1}}{dt} = f(t, \ x, \ u_1, \ u_2, \ \dots, \ u_{n-1}).$$

Systems of simultaneous differential equations occur frequently in applications. Even when such systems are not of the first order, they frequently may be transformed into an equivalent first-order system. For example, the motion of a particle subject to various forces in three-space is governed, according to Newton's Law, by a system of second-order equations of the form

$$\frac{d^2 x}{dt^2} = P(x, y, z),$$

$$\frac{d^2 y}{dt^2} = Q(x, y, z), \tag{4}$$

$$\frac{d^2 z}{dt^2} = R(x, y, z),$$

where P, Q, and R are the components of the force. If we introduce the variables

$$u = \frac{dx}{dt}, \qquad v = \frac{dy}{dt}, \qquad w = \frac{dz}{dt}, \tag{5}$$

then the system (4) becomes

$$\frac{du}{dt} = P(x, y, z), \qquad \frac{dv}{dt} = Q(x, y, z), \qquad \frac{dw}{dt} = R(x, y, z). \qquad (6)$$

The first-order system (5), (6) of six equations for the unknowns x, y, z, u, v, w is equivalent to the second-order system (4).

A general first-order system for n unknown functions x_1, x_2, $\ldots$, x_n may be written

$$\frac{dx_1}{dt} = f_1(t, x_1, x_2, \ldots, x_n),$$

$$\frac{dx_2}{dt} = f_2(t, x_1, x_2, \ldots, x_n),$$

$$\vdots \qquad\qquad\qquad\qquad (7)$$

$$\frac{dx_n}{dt} = f_n(t, x_1, x_2, \ldots, x_n),$$

where $f_1, f_2, \ldots, f_n$ are *given* functions of their arguments. We may use vector notation and simplify the form of (7). Regarding x_1, x_2, $\ldots$, x_n as a vector $\mathbf{x}$ in V_n and $f_1, f_2, \ldots, f_n$ as a vector $\mathbf{f}$, we write (7) in vector notation

$$\frac{d\mathbf{x}}{dt} = \mathbf{f}(t, \mathbf{x}).$$

It is natural to ask when we can expect a system as general as (7) to have a solution. We state without proof the following existence theorem, which shows that the restrictions on $\mathbf{f}$ are quite mild.

Theorem 10. *Suppose that $f_1, f_2, \ldots, f_n$ and the derivatives $(\partial f_i/\partial x_j)$, $i, j = 1$, $2, \ldots, n$ are continuous in the box*

$$|t - t^0| < k, \qquad |x_i - x_i^0| < k, \qquad i = 1, 2, \ldots, n,$$

where t^0, x_1^0, x_2^0, $\ldots$, x_n^0 are given numbers. Then, for some positive number h there exists a solution vector $\mathbf{x}(t) = (x_1(t), x_2(t), \ldots, x_n(t))$ of the system (7) which is defined for t in the interval

$$|t - t^0| < h$$

and which satisfies the conditions $x_i(t^0) = x_i^0$, $i = 1, 2, \ldots, n$ (or, in vector form, $\mathbf{x}(t^0) = \mathbf{x}^0$). The solution is unique in the sense that if $\mathbf{x}^*(t)$ is another solution such that $\mathbf{x}^*(t^0) = \mathbf{x}^0$, then $\mathbf{x}^*(t) = \mathbf{x}(t)$ in their common interval of definition.

The most *general linear first-order system* of n differential equations for n unknown functions has the form

$$\frac{dx_i}{dt} = \sum_{j=1}^{n} a_{ij}(t)x_j + b_i(t), \qquad i = 1, 2, \ldots, n. \qquad (8)$$

If all the $b_i(t)$ are zero, the system is called *homogeneous*; otherwise it is called *nonhomogeneous*. The system (8) is in the same form as (7) with

$$f_i = \sum_{j=1}^{n} a_{ij}(t)x_j + b_i(t).$$

Since

$$\frac{\partial f_i}{\partial x_j} = a_{ij}(t), \qquad i, j = 1, 2, \ldots, n,$$

the Existence Theorem 10 applies to (8) whenever the $a_{ij}(t)$ and $b_i(t)$ are continuous.

If all the a_{ij} are constant, we say the system (8) is a **linear system with constant coefficients.** *For the remainder of this section we shall assume that the system is of this type.* We begin by solving a system of two equations for two unknown functions. We write

$$\frac{dx}{dt} = ax + by + e(t),$$

$$\frac{dy}{dt} = cx + dy + f(t), \tag{9}$$

where a, b, c, and d are constants. If $b = c = 0$, the system breaks up into two separate first-order linear equations which are easily solved by the method of Chapter 15, Section 6. If one of them, say b, is not zero, we may solve the system (9) by a method known as **elimination by differentiation.** To exhibit the method, we differentiate the first equation in (9), getting

$$\frac{d^2x}{dt^2} = a\frac{dx}{dt} + b\frac{dy}{dt} + e'(t).$$

We substitute dy/dt from the second equation of (9) into this second-order equation to obtain

$$\frac{d^2x}{dt^2} = a\frac{dx}{dt} + b[cx + dy + f(t)] + e'(t). \tag{10}$$

Now y may be eliminated by solving the first equation of (9) for y and inserting its value into (10). The result is

$$\frac{d^2x}{dt^2} = a\frac{dx}{dt} + bcx + d\left[\frac{dx}{dt} - ax - e(t)\right] + bf(t) + e'(t)$$

or

$$\frac{d^2x}{dt^2} - (a + d)\frac{dx}{dt} + (ad - bc)x = e'(t) - de(t) + bf(t).$$

This second-order equation with constant coefficients may be solved by the methods described in Section 3. Once x is determined, y may be found by substitution in the first equation of (9).

EXAMPLE 1. Solve the system

$$\frac{dx}{dt} = 2x + 3y + 2e^{2t}, \qquad \frac{dy}{dt} = x + 4y + 3e^{2t}.$$

Solution. We use the method of elimination by differentiation. The derivative of the second equation is

$$\frac{d^2y}{dt^2} = \frac{dx}{dt} + 4\frac{dy}{dt} + 6e^{2t}.$$

Substitution of dx/dt from the first equation yields

$$\frac{d^2y}{dt^2} = (2x + 3y + 2e^{2t}) + 4\frac{dy}{dt} + 6e^{2t}.$$

Since $x = (dy/dt) - 4y - 3e^{2t}$, we see that

$$\frac{d^2y}{dt^2} = 6\frac{dy}{dt} - 5y + 2e^{2t},$$

and x has been eliminated. By the methods we learned for solving a second-order equation with constant coefficients, we find

$$y = C_1e^t + C_2e^{5t} - \tfrac{2}{3}e^{2t}.$$

Therefore

$$x = \frac{dy}{dt} - 4y - 3e^{2t} = -3C_1e^t + C_2e^{5t} - \tfrac{5}{3}e^{2t}.$$

Remarks. (i) The solution in the above example may be written in the vector form

$$x\mathbf{i} + y\mathbf{j} = C_1e^t(\mathbf{i} - 3\mathbf{j}) + C_2e^{5t}(\mathbf{i} + \mathbf{j}) - e^{2t}(\tfrac{2}{3}\mathbf{i} + \tfrac{5}{3}\mathbf{j}).$$

(ii) A solution $x = x(t)$, $y = y(t)$ of a system of two equations may be interpreted geometrically as the parametric equations of a curve in the plane. More generally, a solution $\mathbf{x} = \mathbf{x}(t)$ of three or four or more equations consists of the parametric equations of a curve in the space with a corresponding number of dimensions.

EXAMPLE 2. Solve the system

$$\frac{dx}{dt} = 2x - 2y - 4, \qquad \frac{dy}{dt} = 4x - 2y - 2,$$

and find the particular solution which satisfies $x(0) = -1$, $y(0) = -4$. Interpret the result geometrically.

Solution. Using the method of elimination by differentiation, we find

$$\frac{d^2x}{dt^2} + 4x = -4.$$

The solution is $x(t) = C_1 \cos 2t + C_2 \sin 2t - 1$. Substitution of $x(0) = -1$ yields $C_1 = 0$. The solution for y becomes

$$y = C_2 \sin 2t - C_2 \cos 2t - 3,$$

and $y(0) = -4$ gives $C_2 = 1$. We obtain

$$x(t) = \sin 2t - 1, \qquad y(t) = \sin 2t - \cos 2t - 3.$$

The parameter may be eliminated, and the relation between x and y is

$$2x^2 - 2xy + y^2 - 2x + 4y + 4 = 0.$$

The solution curve traces out an ellipse in the xy-plane.

EXERCISES

In each of the systems 1 through 7, solve by the method of differentiation and elimination.

1. $\dfrac{dx}{dt} = y, \qquad \dfrac{dy}{dt} = x$

2. $\dfrac{dx}{dt} = y, \qquad \dfrac{dy}{dt} = -x$

3. $\dfrac{dx}{dt} = -y, \qquad \dfrac{dy}{dt} = -3x + 2y$

4. $\dfrac{dx}{dt} = x + 2y, \qquad \dfrac{dy}{dt} = -x + 3y$

5. $\dfrac{dx}{dt} = -x + y + 2e^{-2t}, \qquad \dfrac{dy}{dt} = 5x + 3y - e^{-2t}$

6. $\dfrac{dx}{dt} = y + te^{t}, \qquad \dfrac{dy}{dt} = -x + 2y + (t + 1)e^{t}$

7. $\dfrac{dx}{dt} = x + y, \qquad \dfrac{dy}{dt} = y - z, \qquad \dfrac{dz}{dt} = -2y$

In each of problems 8 through 10, solve as above and find that solution for which $x(0) = y(0) = 0$. Interpret the result geometrically.

8. $\dfrac{dx}{dt} = -x - y - e^{t}, \qquad \dfrac{dy}{dt} = 3x + 3y + 2e^{t}$

9. $\dfrac{dx}{dt} = -x + 2y + (2t + 3), \qquad \dfrac{dy}{dt} = -2x + 3y + (t + 1)$

10. $\dfrac{dx}{dt} = -x + y + 2e^{t}, \qquad \dfrac{dy}{dt} = -5x + 3y - e^{t}$

10. MATRIX METHODS FOR FIRST-ORDER SYSTEMS WITH CONSTANT COEFFICIENTS

The method of elimination by differentiation described in Section 9 will be extended in the next section to handle general linear systems. We now present another standard method for the solution of first-order systems, one which makes use of the theory of eigenvalues of matrices. (See Chapter 8, Section 1.)

We first consider the homogeneous system

$$\frac{dx_1}{dt} = a_{11}x_1 + a_{12}x_2 + \cdots + a_{1n}x_n,$$

$$\frac{dx_2}{dt} = a_{21}x_1 + a_{22}x_2 + \cdots + a_{2n}x_n, \tag{1}$$

$$\vdots$$

$$\frac{dx_n}{dt} = a_{n1}x_1 + a_{n2}x_2 + \cdots + a_{nn}x_n,$$

where the coefficients a_{ij} are all constant. On the basis of our experience we look for a solution in the form

$$x_1 = e^{\lambda t}x_1^0, \ x_2 = e^{\lambda t}x_2^0, \ \ldots, \ x_n = e^{\lambda t}x_n^0 \tag{2}$$

or, in vector notation,

$$\mathbf{x} = e^{\lambda t}\mathbf{x}^0,$$

where λ and $\mathbf{x}^0$ are constants which are at our disposal. By direct substitution in (1), we see that the vector $\mathbf{x}$ is a solution of (1) if and only if

$$\lambda e^{\lambda t}x_i^0 = \sum_{j=1}^{n} a_{ij}e^{\lambda t}x_j^0, \qquad i = 1, 2, \ldots, n. \tag{3}$$

Dividing by $e^{\lambda t}$, we observe that (3) holds if and only if

$$(a_{11} - \lambda)x_1^0 + a_{12}x_2^0 + \cdots + a_{1n}x_n^0 = 0,$$
$$a_{21}x_1^0 + (a_{22} - \lambda)x_2^0 + \cdots + a_{2n}x_n^0 = 0, \tag{4}$$
$$\vdots$$
$$a_{n1}x_1^0 + a_{n2}x_2^0 + \cdots + (a_{nn} - \lambda)x_n^0 = 0.$$

The system of n linear, algebraic equations (4) for the n unknowns $x_1^0, x_2^0, \ldots, x_n^0$ is homogeneous. Therefore it has a nonzero solution if and only if the determinant of the coefficients vanishes (Chapter 6, page 300). That is, the determinant

$$\begin{vmatrix} a_{11} - \lambda & a_{12} & \cdots & a_{1n} \\ a_{21} & a_{22} - \lambda & \cdots & a_{2n} \\ \vdots & \vdots & & \vdots \\ a_{n1} & a_{n2} & \cdots & a_{nn} - \lambda \end{vmatrix} \tag{5}$$

must be zero. We denote the matrix (a_{ij}) by A and, as usual, the $n \times n$ identity matrix by I. The vanishing of (5) then becomes

$$\det \|A - \lambda I\| = 0. \tag{6}$$

We now recall the study of eigenvalues of matrices given in Chapter 8, Section 1. We call

$$P(\lambda) = (-1)^n \det \|A - \lambda I\|$$

the **characteristic polynomial** associated with the first-order system (1).   The n roots of the polynomial equation $P(\lambda) = 0$ are called the **eigenvalues of the system**.   Each root determines a set of numbers $x_1^0, x_2^0, \ldots, x_n^0$ and thereby a solution (2) of the system (1).   Two solutions corresponding to two different eigenvalues λ_1, λ_2 of (6) are linearly independent.   (In this connection, see Chapter 8, page 392.)   Thus, if all n roots of (6) are distinct, there will be n linearly independent solutions.   Then it follows from the uniqueness portion of Existence Theorem 10 that every solution of the homogeneous system will be a linear combination of these n linearly independent solutions.

———————

EXAMPLE 1. Use the method of this section to solve the system

$$\frac{dx_1}{dt} = 15x_1 - 32x_2 + 25x_3,$$

$$\frac{dx_2}{dt} = 8x_1 - 17x_2 + 14x_3, \tag{7}$$

$$\frac{dx_3}{dt} = 2x_1 - 4x_2 + 4x_3.$$

Solution. The characteristic polynomial equation is

$$\begin{vmatrix} 15 - \lambda & -32 & 25 \\ 8 & -17 - \lambda & 14 \\ 2 & -4 & 4 - \lambda \end{vmatrix} = 0$$

or, when we multiply the second row by 2 and subtract from the first, and then multiply the third row by 4 and subtract from the second, we have

$$\begin{vmatrix} -1 - \lambda & 2 + 2\lambda & -3 \\ 0 & -1 - \lambda & -2 + 4\lambda \\ 2 & -4 & 4 - \lambda \end{vmatrix} = 0.$$

An expansion in terms of the first column yields

$$(\lambda + 1)^2(4 - \lambda) - 4(\lambda + 1)(4\lambda - 2) + 4(\lambda + 1)(4\lambda - 2) - 6(\lambda + 1) = 0$$

and

$$\lambda^3 - 2\lambda^2 - \lambda + 2 = 0 \quad \text{or} \quad (\lambda + 1)(\lambda - 1)(\lambda - 2) = 0.$$

For $\lambda = -1$, the system (4) becomes

$$16x_1^0 - 32x_2^0 + 25x_3^0 = 0,$$
$$8x_1^0 - 16x_2^0 + 14x_3^0 = 0,$$
$$2x_1^0 - 4x_2^0 + 5x_3^0 = 0.$$

A solution is

$$x_1^0 = 2c_1, \qquad x_2^0 = c_1, \qquad x_3^0 = 0.$$

The corresponding solution of the system (7) is

$$x_1 = 2c_1e^{-t}, \qquad x_2 = c_1e^{-t}, \qquad x_3 = 0.$$

For $\lambda = 1$ and $\lambda = 2$ we obtain in the same way

$$\begin{pmatrix} x_1^0 \\ x_2^0 \\ x_3^0 \end{pmatrix} = c_2 \begin{pmatrix} 1 \\ 2 \\ 2 \end{pmatrix}, \qquad \begin{pmatrix} x_1^0 \\ x_2^0 \\ x_3^0 \end{pmatrix} = c_3 \begin{pmatrix} 3 \\ 2 \\ 1 \end{pmatrix}.$$

Every solution of the system (7) must be of the form

$$\begin{pmatrix} x_1 \\ x_2 \\ x_3 \end{pmatrix} = c_1e^{-t} \begin{pmatrix} 2 \\ 1 \\ 0 \end{pmatrix} + c_2e^{t} \begin{pmatrix} 1 \\ 2 \\ 2 \end{pmatrix} + c_3e^{2t} \begin{pmatrix} 3 \\ 2 \\ 1 \end{pmatrix}.$$

We now consider methods for finding a particular solution of a nonhomogeneous system

$$\frac{dx_i}{dt} = \sum_{j=1}^{n} a_{ij}x_j + b_i(t). \tag{8}$$

Suppose we have found n linearly independent solutions of the corresponding homogeneous system. We denote these complementary functions (in vector form) by

$$\mathbf{v}^1, \ \mathbf{v}^2, \ \ldots, \ \mathbf{v}^n. \tag{9}$$

Then we can use the method of variation of parameters in a way similar to that for a single equation (Section 4) to obtain a particular solution of (8). Suppose the vector solution $\mathbf{v}^i$ has components

$$\mathbf{v}^i = (v_1^i, v_2^i, \ldots, v_n^i).$$

We set

$$x_i = \sum_{k=1}^{n} z_k v_i^k, \qquad i = 1, 2, \ldots, n, \tag{10}$$

and try to select the z_k which are functions of t so that $\mathbf{x} = (x_1, x_2, \ldots, x_n)$ is a particular solution. We require that $\mathbf{x}$ satisfy

$$\frac{dx_i}{dt} - \sum_{j=1}^{n} a_{ij}x_j = b_i. \tag{11}$$

In addition, differentiation of (10) gives

$$\frac{dx_i}{dt} - \sum_{k=1}^{n} z_k \frac{dv_i^k}{dt} = \sum_{k=1}^{n} v_i^k \frac{dz_k}{dt}. \tag{12}$$

Because $\mathbf{v}^k$ is a solution of the homogeneous system

$$\frac{dv_i^k}{dt} = \sum_{j=1}^{n} a_{ij}v_j^k, \qquad k = 1, 2, \ldots, n,$$

it is easy to verify that the left sides of (11) and (12) are identical. Therefore x_i as given by (10) will be a particular solution if the right sides of (11) and (12) are equal—that is, if

$$\sum_{k=1}^{n} v_i^k \frac{dz_k}{dt} = b_i(t), \qquad i = 1, 2, \ldots, n. \tag{13}$$

If the equations (13) can be solved for (dz_k/dt) and the resulting expressions integrated to determine the z_k, then the formulas (10) yield a particular solution of the nonhomogeneous system. A general solution of the nonhomogeneous system is the sum of the particular solution with a linear combination of the n complementary functions.

EXAMPLE 2. Find a particular solution of the system

$$\frac{dx_1}{dt} = 15x_1 - 32x_2 + 25x_3 + (4t - 55)e^t,$$

$$\frac{dx_2}{dt} = 8x_1 - 17x_2 + 14x_3 + (2t - 31)e^t,$$

$$\frac{dx_3}{dt} = 2x_1 - 4x_2 + 4x_3 - 7e^t.$$

Solution. The corresponding homogeneous system was solved in Example 1. Therefore, the vectors

$$\mathbf{v}^1 = \begin{pmatrix} 2 \\ 1 \\ 0 \end{pmatrix} e^{-t}, \qquad \mathbf{v}^2 = \begin{pmatrix} 1 \\ 2 \\ 2 \end{pmatrix} e^t, \qquad \mathbf{v}^3 = \begin{pmatrix} 3 \\ 2 \\ 1 \end{pmatrix} e^{2t}$$

form a linearly independent set of complementary functions. The system (13) becomes

$$\begin{pmatrix} 2e^{-t} & e^t & 3e^{2t} \\ e^{-t} & 2e^t & 2e^{2t} \\ 0 & 2e^t & e^{2t} \end{pmatrix} \begin{pmatrix} \dfrac{dz_1}{dt} \\ \dfrac{dz_2}{dt} \\ \dfrac{dz_3}{dt} \end{pmatrix} = \begin{pmatrix} (4t - 55)e^t \\ (2t - 31)e^t \\ -7e^t \end{pmatrix}$$

Solving these, we obtain

$$\frac{dz_1}{dt} = (2t - 17)e^{2t}, \qquad \frac{dz_2}{dt} = 0, \qquad \frac{dz_3}{dt} = -7e^{-t}$$

and

$$z_1 = (t - 9)e^{2t} + c_1, \qquad z_2 = c_2, \qquad z_3 = 7e^{-t} + c_3.$$

Setting the constants equal to zero, we get the particular solution

$$x_1 = z_1 v_1^1 + z_2 v_1^2 + z_3 v_1^3 = e^t(2t - 18 + 21) = e^t(2t + 3),$$
$$x_2 = z_1 v_2^1 + z_2 v_2^2 + z_3 v_2^3 = e^t(t - 9 + 14) = e^t(t + 5),$$
$$x_3 = z_1 v_3^1 + z_2 v_3^2 + z_3 v_3^3 = 7e^t.$$

By subtracting $3\mathbf{v}^2$ from this solution—which we may do, since $\mathbf{v}^2$ satisfies the homogeneous system—we find the "simpler" particular solution

$$(x_1^*, x_2^*, x_3^*) = e^t(2t, t - 1, 1).$$

In case the characteristic polynomial equation has multiple roots, or if the $b_i(t)$ are of the form $e^{st}Q_j(t)$, where s is one of the eigenvalues of the system, the method of elimination by differentiation presented in the next section leads to a solution. In fact, the method of the next section leads to a solution in any case and applies to equations which have a quite general form.

EXERCISES

In each of problems 1 through 6, solve by the method of this section.

Answer:

1. $\dfrac{dx_1}{dt} = 2x_1 - x_2 + 3x_3$

$\dfrac{dx_2}{dt} = -x_1 + x_2 - x_3$

$\dfrac{dx_3}{dt} = x_2 - x_3$

$$\begin{pmatrix} x_1^{(1)} & x_1^{(2)} & x_1^{(3)} \\ x_2^{(1)} & x_2^{(2)} & x_2^{(3)} \\ x_3^{(1)} & x_3^{(2)} & x_3^{(3)} \end{pmatrix} = \begin{pmatrix} e^{-t} & e^t & 4e^{2t} \\ 0 & -2e^t & -3e^{2t} \\ -e^{-t} & -e^t & -e^{2t} \end{pmatrix}$$

Answer:

2. $\dfrac{dx_1}{dt} = -x_2 + x_3$

$\dfrac{dx_2}{dt} = -3x_1 - 2x_2 - 3x_3$

$\dfrac{dx_3}{dt} = x_1 + x_2$

$$\begin{pmatrix} x_1^{(1)} & x_1^{(2)} & x_1^{(3)} \\ x_2^{(1)} & x_2^{(2)} & x_2^{(3)} \\ x_3^{(1)} & x_3^{(2)} & x_3^{(3)} \end{pmatrix} = \begin{pmatrix} e^{-2t} & e^{-t} & e^t \\ e^{-2t} & 0 & -e^t \\ -e^{-2t} & -e^{-t} & 0 \end{pmatrix}$$

Answer:

3. $\dfrac{dx_1}{dt} = -2x_1 - x_2 - x_3$

$\dfrac{dx_2}{dt} = -x_1 - x_3$

$\dfrac{dx_3}{dt} = 3x_1 + x_2 + 2x_3$

$$\begin{pmatrix} x_1^{(1)} & x_1^{(2)} & x_1^{(3)} \\ x_2^{(1)} & x_2^{(2)} & x_2^{(3)} \\ x_3^{(1)} & x_3^{(2)} & x_3^{(3)} \end{pmatrix} = \begin{pmatrix} e^{-t} & 1 & 0 \\ 0 & -1 & e^t \\ -e^{-t} & -1 & -e^t \end{pmatrix}$$

Answer:

4. $\dfrac{dx_1}{dt} = -3x_1 + 3x_2 - 2x_3$

$\dfrac{dx_2}{dt} = -4x_1 + 4x_2 - 2x_3$

$\dfrac{dx_3}{dt} = -3x_1 + 3x_2 - 2x_3$

$$\begin{pmatrix} x_1^{(1)} & x_1^{(2)} & x_1^{(3)} \\ x_2^{(1)} & x_2^{(2)} & x_2^{(3)} \\ x_3^{(1)} & x_3^{(2)} & x_3^{(3)} \end{pmatrix} = \begin{pmatrix} e^{-2t} & 1 & e^{t} \\ e^{-2t} & 1 & 2e^{t} \\ e^{-2t} & 0 & e^{t} \end{pmatrix}$$

Answer:

5. $\dfrac{dx_1}{dt} = -5x_1 + 6x_2 + 2x_3$

$\dfrac{dx_2}{dt} = -4x_1 + 5x_2 + 2x_3$

$\dfrac{dx_3}{dt} = -4x_1 + 4x_2 + 3x_3$

$$\begin{pmatrix} x_1^{(1)} & x_1^{(2)} & x_1^{(3)} \\ x_2^{(1)} & x_2^{(2)} & x_2^{(3)} \\ x_3^{(1)} & x_3^{(2)} & x_3^{(3)} \end{pmatrix} = \begin{pmatrix} 2e^{-t} & e^{t} & e^{3t} \\ e^{-t} & e^{t} & e^{3t} \\ e^{-t} & 0 & e^{3t} \end{pmatrix}$$

6. $\dfrac{dx_1}{dt} = 2x_1 - 2x_2 + 2x_3 + x_4$

$\dfrac{dx_2}{dt} = -x_1 + 3x_2 \qquad + 3x_4$

$\dfrac{dx_3}{dt} = \qquad 4x_3 - 2x_4$

$\dfrac{dx_4}{dt} = \qquad 2x_3 - x_4$

In each of problems 7 through 11, find a particular solution of the nonhomogeneous system of which the homogeneous part is the same as that in the given problem. We set $b_1 = f$, $b_2 = g$, $b_3 = h$. Use the answers given.

7. Problem 1 with $f(t) = -5$, $g(t) = -t + 5$, $h(t) = -2t + 2$
8. Problem 2 with $f(t) = 0$, $g(t) = 11e^{2t}$, $h(t) = 5e^{2t}$
9. Problem 3 with $f(t) = 5\cos t - 4\sin t$, $g(t) = 5\cos t - 4\sin t$, $h(t) = -11\cos t + 2\sin t$.
10. Problem 4 with $f(t) = 4t + 10$, $g(t) = 4t + 13$, $h(t) = 4t + 11$
11. Problem 5 with $f(t) = e^{t}(-2t + 5)$, $g(t) = e^{t}(-2t + 3)$, $h(t) = e^{t}(-2t + 3)$

11. GENERAL LINEAR SYSTEMS WITH CONSTANT COEFFICIENTS

Suppose that $R(\lambda)$, a polynomial in λ of degree n, is given by

$$R(\lambda) = r_0\lambda^n + r_1\lambda^{n-1} + \cdots + r_{n-1}\lambda + r_n$$

where $r_0, r_1, \ldots, r_n$ are numbers. Denoting d/dt by D, we recall from Section 1 that $R(D)$ may be interpreted as an operator acting on functions $x = x(t)$. We write

$$R(D)x \equiv r_0 D^n x + r_1 D^{n-1}x + \cdots + r_{n-1}Dx + r_n x.$$

If, for each $i, j = 1, 2, \ldots, n$, the quantity

$$a_{ij}(\lambda)$$

is a polynomial in λ with constant coefficients, then the equations

$$\sum_{j=1}^{n} a_{ij}(D)x_j = b_i(t), \qquad i = 1, 2, \ldots, n \tag{1}$$

form a system of n differential equations with constant coefficients for the n unknown functions $x_1, x_2, \ldots, x_n$. Since each $a_{ij}(\lambda)$ may be *a polynomial of any degree whatsoever*, the system (1) consists of n differential equations, each of which may be of any order.

In this section we present a systematic method for solving any system of the form (1) in which the $a_{ij}(D)$ are *polynomial operators with constant coefficients*. This method depends on the possibility of reducing a matrix whose individual elements are *polynomials $a_{ij}(\lambda)$* to a triangular form. As in the case of the reduction of matrices with numerical entries (Chapter 6, Sections 7 and 8), we show that this reduction can always be performed by means of certain *elementary transformations operating on rows only*. These transformations are of three types:

    (a) interchanging two rows;
    (b) multiplying a row by a nonzero *constant*;
    (c) multiplying one row by a *polynomial* and subtracting the result from another row.

It is important to observe that each of these elementary transformations is reversible.

We begin by exhibiting the method in a particular case.

––––––––––

EXAMPLE 1. Reduce the matrix

$$A = \begin{pmatrix} \lambda - 2 & 3 & 3 \\ 1 & \lambda + 1 & 2 \\ -1 & 1 & \lambda \end{pmatrix}$$

to triangular form, using the elementary transformations above.

Solution. Interchanging the first and second rows and then the second and third rows, we get

$$A \cong \begin{pmatrix} 1 & \lambda + 1 & 2 \\ -1 & 1 & \lambda \\ \lambda - 2 & 3 & 3 \end{pmatrix}.$$

Now, adding the first row to the second row and then multiplying the first row by $\lambda - 2$ and subtracting from the third row, we find

$$A \cong \begin{pmatrix} 1 & \lambda + 1 & 2 \\ 0 & \lambda + 2 & \lambda + 2 \\ 0 & 5 + \lambda - \lambda^2 & 7 - 2\lambda \end{pmatrix} = A^1.$$

We would like to replace $5 + \lambda - \lambda^2$ by a zero. To do this, we observe that

$$\lambda^2 - \lambda - 5 = (\lambda + 2)(\lambda - 3) + 1,$$

so that if we multiply the second row of A^1 by $\lambda - 3$ and add the result to the third row, we obtain

$$A \cong A^1 \cong \begin{pmatrix} 1 & \lambda + 1 & 2 \\ 0 & \lambda + 2 & \lambda + 2 \\ 0 & -1 & \lambda^2 - 3\lambda + 1 \end{pmatrix} \cong \begin{pmatrix} 1 & \lambda + 1 & 2 \\ 0 & -1 & \lambda^2 - 3\lambda + 1 \\ 0 & \lambda + 2 & \lambda + 2 \end{pmatrix}.$$

The last matrix was obtained by interchanging the second and third rows. Multiplying the second row by $\lambda + 2$ and adding to the third row, we conclude finally that

$$A \cong \begin{pmatrix} 1 & \lambda + 1 & 2 \\ 0 & -1 & \lambda^2 - 3\lambda + 1 \\ 0 & 0 & (\lambda + 2)(\lambda^2 - 3\lambda + 2) \end{pmatrix}.$$

This last matrix is in the desired triangular form.

——————

The results of Example 1 will now be used to solve a system of differential equations. We replace λ by D in the matrix A and consider the system

$$
\begin{aligned}
(D - 2)x_1 + 3x_2 + 3x_3 &= b_1(t), \\
x_1 + (D + 1)x_2 + 2x_3 &= b_2(t), \\
-x_1 + x_2 + Dx_3 &= b_3(t).
\end{aligned}
\tag{2}
$$

To solve this system we must observe the effect of elementary row transformations on the equations of the system.

(a) The interchange of two rows corresponds to the interchange of two equations. The set of solutions of the system remains the same.

(b) The multiplication of a row by a nonzero constant corresponds to the multiplication of an equation by that constant. The set of solutions of the system is unaffected.

(c) The multiplication of any row by a polynomial $Q(D)$ and the subtraction of the result from another row corresponds to *the application of the differential operator $Q(D)$ to a particular equation* and the subtraction of the result from another equation of the system. Since each transformation is reversible,* the system obtained is equivalent to the original system.

——————

EXAMPLE 2. Use the results of Example 1 to show how to solve the system (2).

——————

* That is, any sufficiently differentiable solution of either system is a solution of the other.

Solution. The augmented matrix of the system is defined as for linear algebraic equations. (See Chapter 6, Section 8.) In this case, we have

$$\begin{pmatrix} D-2 & 3 & 3 & \vdots & b_1 \\ 1 & D+1 & 2 & \vdots & b_2 \\ -1 & 1 & D & \vdots & b_3 \end{pmatrix}.$$

Proceeding through the various row operations of Example 1, we note that the augmented matrix becomes

$$\begin{pmatrix} 1 & D+1 & 2 & \vdots & b_2 \\ 0 & -1 & (D^2-3D+1) & \vdots & b_1 - b_2 + (D-3)b_3 \\ 0 & 0 & (D+2)(D^2-3D+2) & \vdots & (D+2)b_1 - (D+1)b_2 + (D^2-D-5)b_3 \end{pmatrix}.$$

The corresponding system equivalent to (2) is

$$x_1 + (D+1)x_2 + 2x_3 = b_2(t),$$
$$-x_2 + (D^2 - 3D + 1)x_3 = b_1 - b_2 + (D-3)b_3,$$
$$(D+2)(D^2 - 3D + 2)x_3 = (D+2)b_1 - (D+1)b_2 + (D^2 - D - 5)b_3.$$

The third equation may be solved for $x_3(t)$. Then substitution of the result in the second equation yields x_2. Finally, once x_2 and x_3 are known, the first equation gives the answer for x_1. Of course, the functions b_1, b_2, b_3 must be such that the differentiations in the third equation can be performed.

EXAMPLE 3. Solve the system

$$Dx_1 + x_2 - x_3 = e^{-t}(t+2),$$
$$x_1 + (D+1)x_2 + x_3 = e^{-t}(2t+1),$$
$$x_1 - x_2 + (D+2)x_3 = e^{-t}(t).$$

Solution. We form the augmented matrix and interchange the first and third rows, getting

$$\begin{pmatrix} D & 1 & -1 & \vdots & e^{-t}(t+2) \\ 1 & D+1 & 1 & \vdots & e^{-t}(2t+1) \\ 1 & -1 & D+2 & \vdots & e^{-t}t \end{pmatrix} \cong \begin{pmatrix} 1 & -1 & D+2 & \vdots & e^{-t}t \\ 1 & D+1 & 1 & \vdots & e^{-t}(2t+1) \\ D & 1 & -1 & \vdots & e^{-t}(t+2) \end{pmatrix}.$$

Subtracting the first row from the second and then "multiplying" the first row by D and subtracting from the third, we find

$$\cong \begin{pmatrix} 1 & -1 & D+2 & \vdots & e^{-t}t \\ 0 & D+2 & -D-1 & \vdots & e^{-t}(t+1) \\ 0 & D+1 & -(D+1)^2 & \vdots & e^{-t}(2t+1) \end{pmatrix}.$$

Now we subtract the third row from the second to obtain

$$\cong \begin{pmatrix} 1 & -1 & D+2 & \vdots & e^{-t}t \\ 0 & 1 & D(D+1) & \vdots & -e^{-t}t \\ 0 & D+1 & -(D+1)^2 & \vdots & e^{-t}(2t+1) \end{pmatrix}.$$

Finally, "multiplying" the second row by $D + 1$ and subtracting the result from the third row, we get the desired augmented matrix:

$$\begin{pmatrix} 1 & -1 & D + 2 & \vdots & e^{-t}t \\ 0 & 1 & D(D + 1) & \vdots & -e^{-t}t \\ 0 & 0 & -(D + 1)^3 & \vdots & e^{-t}(2t + 2) \end{pmatrix} = A^1.$$

To get a solution, we first solve the equation

$$-(D + 1)^3 x_3 = e^{-t}(2t + 2).$$

Setting $x_3 = e^{-t}w$, we observe that this equation is equivalent to

$$-D^3 w = 2t + 2,$$

which has the solution

$$w = C_1 + C_2 t + C_3 t^2 - \tfrac{1}{3}t^3 - \tfrac{1}{12}t^4$$

or

$$x_3 = e^{-t}w = e^{-t}(C_1 + C_2 t + C_3 t^2 - \tfrac{1}{3}t^3 - \tfrac{1}{12}t^4).$$

From the second row of the augmented matrix A^1, we have

$$x_2 = -D(D + 1)x_3 - e^{-t}t = e^{-t}[C_2 + 2C_3(t - 1) + t - \tfrac{1}{3}t^3].$$

The first row of A^1 yields

$$x_1 = x_2 - (D + 2)x_3 + e^{-t}t$$
$$= e^{-t}[-C_1 - C_2 t - C_3(t^2 + 2) + 2t + t^2 + \tfrac{1}{3}t^3 + \tfrac{1}{12}t^4].$$

The above examples illustrate the following general theorem.

Theorem 11. *Any system of differential equations of the form*

$$\sum_{j=1}^{n} a_{ij}(D)x_j = b_i(t), \qquad i = 1, 2, \ldots, n \tag{3}$$

where the $a_{ij}(D)$ are any polynomials, is equivalent to one of the same form in which the a_{ij} are identically zero for $i > j$ (i.e., below the diagonal). We assume that the functions $b_i(t)$ are sufficiently differentiable so that all elementary row operations may be performed.

Proof. It is sufficient to show that any $n \times n$ matrix with polynomial elements can be reduced to triangular form by elementary transformations of types (a), (b), and (c). To do this, we proceed by induction on n. For $n = 1$, there is nothing to prove. Suppose the result holds for all square $k \times k$ matrices. We wish to establish the result for $(k + 1) \times (k + 1)$ matrices. Setting $n = k + 1$, we do the following:

(i) By interchanging rows, we arrange that $a_{11}(\lambda)$ is of degree smaller than (or equal to) that of any of the polynomials in the first column.

(ii) We divide $a_{11}(\lambda)$ into each polynomial $a_{j1}(\lambda)$ of the first column ($j = 2$, $3, \ldots, n$), obtaining a quotient $q_{j1}(\lambda)$ and a remainder $r_{j1}(\lambda)$. For each j, $r_{j1}(\lambda)$ is a polynomial of degree lower than $a_{11}(\lambda)$.

(iii) We now multiply the first row by $q_{j1}(\lambda)$ and subtract the result from the jth row. This leaves the polynomial $r_{j1}(\lambda)$ in the first column of the jth row. Do this for all $j = 2, 3, \ldots, n$.

Repeat steps (i), (ii), and (iii). The degree of the entry in the upper left-hand corner must be reduced each time we complete a cycle. In a finite number of steps we arrive at a polynomial (or constant) entry in the upper left-hand corner with all the remaining elements of the first-column zeros. By the induction hypothesis, the $k \times k$ matrix obtained by deleting the first row and first column can already be put in the desired form. Hence the induction process is completed.

Remarks. (i) The consistency of systems of m equations in n unknowns with $m \neq n$ may be investigated by the method of elementary row operations. The results are analogous to those obtained for linear algebraic equations in which the number of equations is different from the number of unknowns. (See Chapter 6, Section 8.)

(ii) The **order of any equation** in the system (3) is the highest exponent to which D is raised in the terms of that equation. The **order of the system** (3) is *the degree of the polynomial in λ formed by the determinant of the matrix $a_{ij}(\lambda)$.* Generally speaking, a solution of the system (3) will have as many constants of integration as the order of the system.

(iii) Interchanging two columns of the matrix of a system corresponds to a relabeling of the unknowns. If we add this operation to the three elementary row operations, it can be shown that the reduction of a system as in Theorem 11 can be carried out without increasing the order of the system.

EXERCISES

Solve the following systems.

1. $\quad (D - 1)x_1 \qquad\quad - 2x_2 \quad - x_3 = 0$
   $\qquad\quad -x_1 + (D - 3)x_2 \quad - 2x_3 = 0$
   $\qquad\qquad\qquad\quad 2x_2 + Dx_3 = 0$

2. $\quad (D + 2)x_1 \qquad\quad - 5x_2 \qquad\quad + x_3 = 2$
   $\qquad\qquad x_1 + (D - 3)x_2 \qquad\quad + x_3 = -1$
   $\qquad\qquad 2x_1 \qquad\quad - 3x_2 + (D - 1)x_3 = 1$

3. $(D - 4)x_1 \qquad\qquad - x_2 \qquad\quad - 2x_3 = e^{-t}$
   $\qquad\quad x_1 + (D - 1)x_2 \qquad\quad + x_3 = -2e^{-t}$
   $\qquad\quad 3x_1 \qquad\qquad + x_2 + (D + 1)x_3 = -e^{-t}$

4. $\quad (D + 3)x_1 \qquad\quad + 2x_2 \qquad\quad + 2x_3 = e^t$
   $\qquad\quad -x_1 + (D - 1)x_2 \qquad\qquad - x_3 = -e^t$
   $\qquad\quad -4x_1 \qquad\qquad - 2x_2 + (D - 3)x_3 = 2e^t$

5. $(D - 1)x_1 \qquad\quad - 2x_2 \;\; - x_3 = 2e^{2t}$
        $-x_1 + (D - 3)x_2 \; - 2x_3 = e^{2t}$
                  $2x_2 + Dx_3 = -e^{2t}$

6. $(D - 1)x_1 \qquad\quad - 4x_2 \qquad\quad - 16x_3 = 2te^{-t}$
     $- 4x_1 + (D - 7)x_2 \qquad\quad - 12x_3 = (t + 1)e^{-t}$
        $x_1 \qquad\quad + 2x_2 + (D + 4)x_3 = -te^{-t}$

7. $(D - 1)x_1 - (D - 1)x_2 \qquad\qquad = e^t$
         $x_1 + (D - 2)x_2 - (D - 1)x_3 = -2e^t$
   $-(D - 2)x_1 + (D - 3)x_2 + (D + 2)x_3 = 2e^t$

8. $(4D^2 + 4D)x_1 + (13D + 4)x_2 + 9x_3 = 2t + 1$
   $(3D^2 + 3D)x_1 + (10D + 3)x_2 + 7x_3 = t$
   $(2D^2 + \;\; D)x_1 + \;\; (6D + 1)x_2 + 5x_3 = -t$

CHAPTER 17

Series Solutions of Differential Equations

1. SOLUTION BY INFINITE SERIES

Linear equations and linear systems with constant coefficients can frequently be solved explicitly. For linear equations with variable coefficients and for nonlinear equations, it is generally true that explicit solutions in terms of elementary functions do not exist. In this chapter we describe a method for obtaining solutions of differential equations in the form of convergent infinite series. The process is particularly useful in those cases in which explicit solutions cannot be found. We first work two simple examples and then state a theorem which confirms the validity of the method.

EXAMPLE 1. Express the general solution of the equation

$$L(y) \equiv y'' + xy' + y = 0$$

as a Maclaurin series.

Solution. We write the solution in the form

$$y = a_0 + a_1 x + a_2 x^2 + \cdots + a_n x^n + \cdots = \sum_{n=0}^{\infty} a_n x^n. \tag{1}$$

The solution is determined when the coefficients a_i, $i = 0, 1, 2, \ldots$ are found and when the convergence of the series is verified. Assuming that the series (1) has a positive radius of convergence, we may differentiate term by term and obtain

$$y' = a_1 + 2a_2 x + 3a_3 x^2 + \cdots + na_n x^{n-1} + \cdots = \sum_{n=1}^{\infty} na_n x^{n-1}$$

and

$$y'' = 2a_2 + 3 \cdot 2a_3 x + 4 \cdot 3a_4 x^2 + \cdots + n(n-1)a_n x^{n-2} + \cdots$$

$$= \sum_{n=2}^{\infty} n(n-1)a_n x^{n-2}.$$

Using the series for y, y', and y'', we find

$$L(y) = \sum_{n=2}^{\infty} n(n-1)a_n x^{n-2} + x \sum_{n=1}^{\infty} na_n x^{n-1} + \sum_{n=0}^{\infty} a_n x^n = 0. \qquad (2)$$

The above series may be rewritten so that corresponding powers of x can be combined. In a moment we shall see the importance of this rearrangement. If we replace n by $k + 2$, the first series in (2) becomes

$$\sum_{k=0}^{\infty} (k+2)(k+1)a_{k+2}x^k.$$

For the second series in (2), we replace n by k, allow the sum to begin with $k = 0$, and move the x inside the sum sign. The result is

$$\sum_{k=0}^{\infty} ka_k x^k.$$

The last series in (2) is left unaltered except that the index n is replaced by k. With these changes in notation, we find

$$L(y) = \sum_{k=0}^{\infty} [(k+2)(k+1)a_{k+2} + (k+1)a_k]x^k = 0. \qquad (3)$$

Now we state *the basic tool in determining solutions in infinite series: A power series can vanish identically if and only if the coefficient of every power of x vanishes* (a consequence of the Corollary to Theorem 9, Chapter 9). Applying this fact to (3), we obtain

$$a_{k+2} = -\frac{1}{k+2} a_k, \qquad k = 0, 1, 2, \ldots .$$

This formula, called a **recurrence relation**, is typical of solutions determined by infinite series. It enables us to express $a_2, a_3, \ldots, a_n, \ldots$ in terms of a_0 and a_1. In particular, selecting $k = 2, 4, 6, \ldots$, we find

$$a_2 = -\frac{a_0}{2}, \qquad a_4 = -\frac{a_2}{4} = (-1)^2 \frac{a_0}{2 \cdot 4},$$

$$a_6 = -\frac{a_4}{6} = (-1)^3 \frac{a_0}{2 \cdot 4 \cdot 6}, \ldots.$$

For $k = 1, 3, 5, \ldots$, we have

$$a_3 = -\frac{a_1}{3}, \qquad a_5 = -\frac{a_3}{5} = (-1)^2 \frac{a_1}{3 \cdot 5}, \qquad a_7 = (-1)^3 \frac{a_1}{3 \cdot 5 \cdot 7}, \ldots.$$

The general expressions for a_{2k} and a_{2k+1} are

$$a_{2k} = (-1)^k \frac{a_0}{2 \cdot 4 \cdot 6 \cdots 2k} = (-1)^k \frac{a_0}{2^k k!},$$

$$a_{2k+1} = (-1)^k \frac{a_1}{1 \cdot 3 \cdot 5 \cdots (2k+1)} = (-1)^k \frac{2^k k! a_0}{(2k+1)!}.$$

We conclude that the solution is given by

$$y = a_0\left[1 - \frac{x^2}{2} + \cdots + \frac{(-1)^k x^{2k}}{2^k \cdot k!} + \cdots\right]$$
$$+ a_1\left[x - \frac{x^3}{1\cdot 3} + \cdots + \frac{(-1)^k 2^k k! x^{2k+1}}{(2k+1)!} + \cdots\right].$$

Note that a_0 and a_1 are constants of integration and that, by the ratio test, each series is convergent for all values of x.

EXAMPLE 2. Solve Legendre's equation

$$L(y) \equiv (1 - x^2)y'' - 2xy' + p(p+1)y = 0$$

by the method of power series; the number p is a constant but not necessarily an integer.

Solution. We set $y = \sum_{n=0}^{\infty} a_n x^n$ and, as in Example 1, we find

$$y' = \sum_{n=0}^{\infty} na_n x^{n-1}, \qquad y'' = \sum_{n=0}^{\infty} (n+2)(n+1)a_{n+2}x^n.$$

Also,

$$-2xy' = \sum_{n=0}^{\infty} (-2na_n x^n),$$

$$-x^2 y'' = \sum_{n=0}^{\infty} - (n+2)(n+1)a_{n+2}x^{n+2} = \sum_{k=0}^{\infty} - k(k-1)a_k x^k.$$

Inserting the various series into the expression for $L(y)$ and changing the summation index to k throughout, we find

$$L(y) = \sum_{k=0}^{\infty} \left\{(k+2)(k+1)a_{k+2} + [p(p+1) - k(k+1)]a_k\right\}x^k = 0.$$

Using the fact that the coefficient of x^k vanishes for every k, we get the recurrence formula

$$a_{k+2} = \frac{(k-p)(k+p+1)}{(k+2)(k+1)}a_k.$$

Evaluating $a_2, a_4, a_6, \ldots$ in terms of a_0, and evaluating $a_3, a_5, a_7, \ldots$ in terms of a_1, we obtain

$$y = a_0\left[1 - \frac{p(p+1)}{2!}x^2 + \frac{p(p-2)(p+1)(p+3)}{4!}x^4 - + \cdots\right]$$
$$+ a_1\left[x - \frac{(p-1)(p+2)}{3!}x^3 + \frac{(p-1)(p-3)(p+2)(p+4)}{5!}x^5 - + \cdots\right].$$

If p is an even integer the series in the first bracket is finite, the expression being a polynomial of degree p. If p is an odd integer the series in the second bracket is a polynomial, all terms beyond the pth being zero. These polynomials (when p is an integer) are precisely the Legendre polynomials studied in Chapter 10, Section 5.

The validity of the results in the two examples above depends on the following theorem, which we state without proof.

Theorem 1. *Suppose that* $f(x, u_0, u_1, \ldots, u_{n-1})$ *is an analytic function of the* $n + 1$ *variables* $x, u_0, u_1, u_2, \ldots, u_{n-1}$. *That is, the* $(n + 1)$-*tuple power series expansion of* f *converges to* f *for*

$$|x - x^0| < k, \qquad |u_i - u_i^0| < k, \quad i = 0, 1, 2, \ldots, n - 1,$$

where k *is some positive number and* $x^0, u_0^0, u_1^0, \ldots, u_{n-1}^0$ *are given numbers. Then the unique solution of the equation*

$$\frac{d^n y}{dx^n} = f(x, y, y', y'', \ldots, y^{(n-1)})$$

specified in the existence theorem of Section 9, Chapter 16 (Theorem 10), is represented by a power series which converges in some interval $|x - x^0| < h$ *with* $h > 0$. *This solution satisfies* $y^{(i)}(x^0) = u_i^0, i = 0, 1, 2, \ldots, n - 1$.

A corresponding theorem holds for first-order systems of the form

$$\frac{dy_i}{dx} = f_i(x, y_1, y_2, \ldots, y_n), \qquad i = 1, 2, \ldots, n$$

when the functions $f_i, i = 1, 2, \ldots, n$ are analytic functions of all their arguments.

The above theorem does not give any information about the size of h in comparison with the size of k. However, such information is available in the case of *linear* differential equations. The next theorem, stated without proof, is extremely useful in establishing the convergence of power series solutions of linear equations with variable coefficients.

Theorem 2. *Suppose that* y *is a solution of the linear equation*

$$\frac{d^n y}{dx^n} + a_1(x)\frac{d^{n-1} y}{dx^{n-1}} + \cdots + a_{n-1}(x)\frac{dy}{dx} + a_n(x)y = b(x)$$

in which $a_1, a_2, \ldots, a_n, b$ *are all analytic functions of* x *in a neighborhood of* x^0. *Let* $h > 0$ *be the smallest radius of convergence of the power series expansions of* $a_1, a_2, \ldots, a_n, b$. *Then the power series expansion of the solution* y *has a radius of convergence at least equal to* h.

A corresponding theorem holds for linear first-order systems in the form

$$\frac{dy_i}{dx} = \sum_{k=1}^{n} a_{ik}(x)y_k + b_i(x), \qquad i = 1, 2, \ldots, n.$$

This theorem is verified in Example 1, since the linear equation

$$y'' = -xy' - y$$

has polynomial coefficients on the right side. Hence the series expansion for the solution converges for all values of x. If we solve Legendre's equation as given in Example 2, for y'', we obtain

$$y'' = \frac{2x}{1 - x^2} y' - \frac{p(p + 1)}{1 - x^2} y.$$

Since the Maclaurin expansions of both $2x/(1 - x^2)$ and $p(p + 1)/(1 - x^2)$ have radius 1, it follows that the solution y as given in Example 2 converges at least for $|x| < 1$.

For both linear and nonlinear equations it is frequently quite difficult or impossible to obtain a general formula for the nth term in the power series expansion of the solution. If we seek only the first few terms of the Taylor series of a *particular* solution, then we just need to know y and its first few derivatives at x^0, the value about which the Taylor expansion is taken. In such cases the method exhibited in the next example is valuable. Of course, the convergence of the series must be considered separately.

EXAMPLE 3. Find the terms out to that in x^4 of the Maclaurin series for the particular solution of

$$y' = x^2 + y^2$$

which satisfies the condition $y = 1$ when $x = 0$.

Solution. Substituting $y = 1$, $x = 0$ in the differential equation, we find $y'(0) = 1$. We differentiate the differential equation, getting

$$y'' = 2x + 2yy'.$$

Substituting $x = 0$, $y = 1$, $y' = 1$ in this equation, we obtain $y''(0) = 2$. Repeating the process once again, we see that

$$y''' = 2 + 2yy'' + 2y'^2$$

and

$$y'''(0) = 2 + 2 \cdot 2 + 2 \cdot 1 = 8.$$

A continuation yields

$$y^{(4)} = 2yy''' + 6y'y''$$

and

$$y^{(4)}(0) = 2(8) + 6(1)(2) = 28.$$

Therefore

$$y = 1 + x + \frac{2x^2}{2!} + 8\frac{x^3}{3!} + 28\frac{x^4}{4!} + \cdots$$

$$= 1 + x + x^2 + \tfrac{4}{3}x^3 + \tfrac{7}{6}x^4 + \cdots. \tag{4}$$

Remarks. In Example 3 it is impossible to find an expression for the coefficient of x^n, for arbitrary n. However, from Theorem 1 we know the series (4) converges for $|x| < h$ with some $h > 0$. The method of repeated differentiation of the differential equation to obtain a succession of terms in the Taylor expansion (i.e., the method of Example 3) is particularly useful for nonlinear equations. If the method of Examples 1 and 2 were employed for Example 3, we would have to compute $(\sum_{n=0}^{\infty} a_n x^n)^2$ and combine like powers of x; this technique is quite cumbersome when only a few terms of the expansion are needed.

EXERCISES

In each of problems 1 through 9, find the general solution by the method of power series. Find a formula for the general term as in Example 1.

1. $y'' - y = 0$ 　　　　　　　　　　　　2. $y'' + y = 0$
3. $y'' + xy' + 2y = 0$ 　　　　　　　　4. $y'' + 2xy' + 2y = 0$
5. $y'' + 2xy + 4y = 0$ 　　　　　　　　6. $y'' - xy' + y = 0$
7. $y'' - (x - 1)y = 0$; set $y = \sum_{n=0}^{\infty} a_n(x - 1)^n$. [*Hint.* To make certain formulas hold for all n, define $a_k = 0$ for $k < 0$.]
8. $y'' - xy' - 2y = 1 + x$ 　　　　　　9. $y'' - x^2 y = 0$
10. Obtain the recurrence relation for the coefficients in the solution of

$$(x^2 - 3x + 2)y'' + (x^2 - 2x - 1)y' + (x - 3)y = 0$$

and find a_2, a_3, a_4, a_5. For what values of x will the series surely converge?

11. For what values of x will the series for the solutions of the equation

$$(2 - x)y'' + (x - 1)y' - y = 0$$

surely converge? Show that the recurrence relation is

$$a_{n+2} = \frac{(n + 1)^2 a_{n+1} - (n - 1)a_n}{2(n + 1)(n + 2)}.$$

Show that if the a_n satisfy this relation for every n, then

$$a_n = \frac{(a_0 + a_1)}{2 \cdot n!} \qquad \text{for all } n \geq 2.$$

What is the general solution?

In each of problems 12 through 20, find the terms of the series solution out to and including the term in $(x - a)^n$ for the given n and a.

12. $y' = xy^2 + 1$; 　　$y = 1$ for $x = 1$; 　　$a = 1$, $n = 4$
13. $y' = \sin(xy) + x$; 　$y = 1$ for $x = 0$; 　$a = 0$, $n = 4$
14. $y' = \sin(x^2 + y)$; 　$y = \pi/2$ for $x = 0$; 　$a = 0$, $n = 4$
15. $y'' + 2xy' + y = 0$; 　$y = 0$, $y' = 1$ for $x = 0$; 　$a = 0$, $n = 5$

16. $y'' = x^2 - y^2; \quad y = 1, \quad y' = 0 \quad \text{for} \quad x = 0; \quad a = 0, \quad n = 5$

17. $(1 + x^2)y'' + x(y' + y) = 0; \quad y = 0, \quad y' = 1 \quad \text{for} \quad x = 0; \quad a = 0, \quad n = 4$

18. $\dfrac{dy_1}{dx} = xy_1 + 2y_2;$

$\dfrac{dy_2}{dx} = -2y_1 + xy_2; \quad y_1 = 0, \quad y_2 = 1 \quad \text{for} \quad x = 0; \quad a = 0, \quad n = 3$

[Find expansions for y_1 and y_2.]

19. $\dfrac{dy_1}{dx} = y_1^2 + y_2^2;$

$\dfrac{dy_2}{dx} = y_2^2 - y_1; \quad y_1 = y_2 = 1 \quad \text{for} \quad x = 0; \quad a = 0, \quad n = 2$

20. $y''' = xy + yy'; \quad y = 0, \quad y' = 1, \quad y'' = 2 \quad \text{for} \quad x = 0; \quad a = 0, \quad n = 5$

2. SOLUTION BY GENERALIZED POWER SERIES

The theorems of Section 1 are useful for obtaining power series expansions so long as the required conditions of analyticity are satisfied. However, it happens that some of the most important differential equations of mathematical physics do not satisfy the hypotheses in Theorems 1 and 2. The coefficient in a simple linear equation such as

$$y'' + \frac{1}{x^2} y = 0$$

has a singularity at $x = 0$. Therefore, while a Taylor series solution exists about any value $a \neq 0$, no ordinary power series solution can be obtained about $x = 0$.

For *linear* equations with coefficients which are singular, it is sometimes possible to obtain expansions in what are called generalized power series. These are series which have the form

$$x^\rho \sum_{n=0}^{\infty} a_n x^n, \qquad a_0 \neq 0$$

in which the number ρ is not necessarily an integer. We first state the theorem appropriate for second-order linear equations and then illustrate the procedure with examples.

Theorem 3. (a) *Suppose that $g_1(x)$ and $g_2(x)$ are given by Maclaurin series, convergent for $|x| < R$ for some $R > 0$. Then the equation*

$$y'' + \frac{g_1(x)}{x} y' + \frac{g_2(x)}{x^2} y = 0 \tag{1}$$

will have at least one solution of the form

$$y = x^\rho \sum_{n=0}^{\infty} a_n x^n, \qquad a_0 \neq 0, \tag{2}$$

where ρ is a constant, not necessarily an integer. In particular, ρ is a root of the quadratic equation

$$\rho(\rho - 1) + g_1(0)\rho + g_2(0) = 0. \tag{3}$$

(b) *If the roots ρ_1 and ρ_2 of (3) are such that $\rho_1 - \rho_2$ is not an integer, then there is a solution of the form (2) for each of the roots ρ_1 and ρ_2. If $\rho_1 - \rho_2$ is an integer then there is at least one solution corresponding to the larger root. Moreover, the series (2) converges for $|x| < R$, at least.*

Remarks. (i) The theorem is valid if x is replaced by $x - a$ throughout.
(ii) The quadratic equation (3) is called the **indicial equation** corresponding to the differential equation (1).

We first work two examples and then discuss the significance of the indicial equation.

EXAMPLE 1. Find solutions in generalized series about $x = 0$ of the equation

$$L(y) \equiv 2x^2 y'' - 3xy' + (3 + x^2)y = 0.$$

Solution. We recognize that the above equation is in the form (1) with $g_1(x) = -\frac{3}{2}$ and $g_2(x) = \frac{1}{2}(3 + x^2)$. The indicial equation is

$$\rho(\rho - 1) - \tfrac{3}{2}\rho + \tfrac{3}{2} = 0,$$

and the roots $\rho_1 = \frac{3}{2}$, $\rho_2 = 1$. Since $\rho_1 - \rho_2$ is not an integer there are two solutions in generalized power series. We select

$$y = x^{3/2} \sum_{n=0}^{\infty} a_n x^n = \sum_{n=0}^{\infty} a_n x^{n+3/2}, \qquad a_0 \neq 0.$$

Taking derivatives, we get

$$y' = \sum_{n=0}^{\infty} (n + \tfrac{3}{2})a_n x^{n+1/2}, \qquad y'' = \sum_{n=0}^{\infty} (n + \tfrac{3}{2})(n + \tfrac{1}{2})a_n x^{n-1/2}.$$

Substituting in the expression for $L(y)$, we find

$$L(y) \equiv \sum_{n=0}^{\infty} 2(n + \tfrac{3}{2})(n + \tfrac{1}{2})a_n x^{n+3/2} - \sum_{n=0}^{\infty} 3(n + \tfrac{3}{2})a_n x^{n+3/2}$$

$$+ \sum_{n=0}^{\infty} 3a_n x^{n+3/2} + \sum_{n=0}^{\infty} a_n x^{n+7/2} = 0.$$

By changing the index of summation, we may replace the series

$$\sum_{n=0}^{\infty} a_n x^{n+7/2} \qquad \text{by} \qquad \sum_{n=2}^{\infty} a_{n-2} x^{n+3/2}$$

Introducing the convention that $a_{-1} = a_{-2} = 0$, we may replace $n = 2$ by $n = 0$ as the lower limit in the last sum above. In this way $L(y)$ becomes

$$L(y) \equiv x^{3/2} \sum_{n=0}^{\infty} \{[2(n + \tfrac{3}{2})(n + \tfrac{1}{2}) - 3(n + \tfrac{3}{2}) + 3]a_n + a_{n-2}\}x^n = 0.$$

For each integer n the quantity in braces must vanish. For $n = 0$, this quantity is identically zero, and we shall see later the relation of the indicial equation to this fact. The resulting recurrence relation for all $n \geq 1$ is (after some simplification)

$$a_n = -\frac{a_{n-2}}{n(2n + 1)}.$$

We conclude that $a_1 = a_3 = \cdots = a_{2n+1} = \cdots = 0$. For even values of $n(=2k)$, we obtain

$$a_{2k} = \frac{(-1)^k a_0}{2 \cdot 4 \cdots (2k) \cdot 5 \cdot 9 \cdots (4k + 1)} = \frac{(-1)^k a_0}{2^k k! \cdot 5 \cdot 9 \cdots (4k + 1)}.$$

The solution, which we denote y_1, is

$$y_1 = a_0 x^{3/2}\left[1 - \frac{x^2}{2 \cdot 5} + \frac{x^4}{2^2 \cdot 2!(5)(9)} - \cdots + \frac{(-1)^k x^{2k}}{2^k k! \cdot 5 \cdot 9 \cdots (4k + 1)} + \cdots\right].$$

Starting with a solution in the form

$$y = x \sum_{n=0}^{\infty} b_n x^n, \qquad b_0 \neq 0,$$

corresponding to the root $\rho_2 = 1$, and going through a series of calculations similar to the ones just performed, we obtain the second solution,

$$y_2 = b_0 x\left[1 - \frac{x^2}{2 \cdot 3} + \frac{x^4}{2^2 \cdot 2! \cdot 3 \cdot 7} - \cdots + \frac{(-1)^k x^{2k}}{2^{2k} \cdot k! \cdot 3 \cdot 7 \cdots (4k - 1)} + \cdots\right].$$

Every solution of $L(y) = 0$ is a linear combination of y_1 and y_2.

EXAMPLE 2. Find a solution of the form

$$y = x^\rho \sum_{n=0}^{\infty} a_n x^n$$

of the equation

$$L(y) \equiv x^2 y'' + (1 + 3x)xy' - (1 + 6x)y = 0.$$

Solution. The equation is in the form (1) with

$$g_1(x) = 1 + 3x \qquad \text{and} \qquad g_2(x) = -(1 + 6x).$$

The indicial equation (3) of Theorem 3 is

$$\rho(\rho - 1) + 1 \cdot \rho - 1 = 0 \qquad \text{or} \qquad \rho^2 - 1 = 0.$$

Since the roots $\rho_1 = 1$, $\rho_2 = -1$ differ by an integer, we try $\rho_1 = +1$, the larger root. Setting

$$y = x \sum_{n=0}^{\infty} a_n x^n = \sum_{n=0}^{\infty} a_n x^{n+1},$$

we find

$$xy' = \sum_{n=0}^{\infty} (n+1)a_n x^{n+1}, \qquad x^2 y' = \sum_{n=0}^{\infty} na_{n-1} x^{n+1}, \qquad a_{-1} = 0,$$

$$x^2 y'' = \sum_{n=0}^{\infty} (n+1)na_n x^{n+1}, \qquad xy = \sum_{n=0}^{\infty} a_{n-1} x^{n+1}.$$

Therefore

$$L(y) = x \sum_{n=0}^{\infty} \left\{ [(n+1)^2 - 1]a_n + 3(n-2)a_{n-1} \right\} x^n = 0. \tag{4}$$

The quantity in braces vanishes for $n = 0$ because of the indicial equation. Equation (4) yields the recurrence relation

$$a_n = -\frac{3(n-2)}{n(n+2)} a_{n-1}, \qquad n \geq 1.$$

From the recurrence formula, we find

$$a_1 = -\frac{3(1-2)}{1(1+2)} a_0 = a_0, \qquad a_2 = 0, \; a_3 = 0, \ldots, \; a_n = 0, \ldots.$$

Selecting $a_0 = 1$, we get the solution

$$y_1(x) = x + x^2.$$

A method for finding the second solution is described later in Example 4. (See page 726.)

It is worthwhile to see how the indicial equation (3), as given in Theorem 3, is related to the derivation of series expansions in the form (2). For the equations in Examples 1 and 2, the indicial equation is the item corresponding to $n = 0$ in the sequence of recurrence relations which arise in solutions by generalized series. We now show that this result is true quite generally for second-order equations in the form (1), i.e., for an equation given by

$$x^2 y'' + xg_1(x)y' + g_2(x)y = 0 \tag{5}$$

with $g_1(x)$ and $g_2(x)$ analytic functions. Inserting the series $x^\rho \sum_{n=0}^{\infty} a_n x^n$ and its derivatives in (5), we obtain

$$x^\rho \left\{ \sum_{n=0}^{\infty} (n+\rho)(n+\rho-1)a_n x^n + g_1(x) \sum_{n=0}^{\infty} (n+\rho)a_n x^n \right.$$

$$\left. + g_2(x) \sum_{n=0}^{\infty} a_n x^n \right\} = 0. \tag{6}$$

Since $g_1(x)$ and $g_2(x)$ are analytic, they have the Maclaurin expansions

$$g_1(x) = b_0 + b_1 x + b_2 x^2 + \cdots, \qquad g_2(x) = c_0 + c_1 x + c_2 x^2 + \cdots,$$

with $b_0 = g_1(0)$, $c_0 = g_2(0)$, ... If these series are substituted in (6) and then the various terms are collected in corresponding powers of x, we know that for each n the coefficient of x^n must vanish. The term corresponding to $n = 0$ is just

$$[\rho(\rho - 1) + g_1(0)\rho + g_2(0)]a_0 = 0,$$

which leads to the indicial equation, since $a_0 \neq 0$.

The next example shows a method for finding a second solution when the roots of the indicial equation differ by an integer.

EXAMPLE 3. Solve the equation

$$L(y) \equiv x^2y'' + 6xy' + (6 - x^2)y = 0.$$

Solution. The indicial equation is

$$\rho(\rho - 1) + 6\rho + 6 = \rho^2 + 5\rho + 6 = (\rho + 2)(\rho + 3) = 0.$$

The roots are $\rho_1 = -2$, $\rho_2 = -3$, and we select a series with the form

$$y = x^{-2} \sum_{n=0}^{\infty} a_n x^n, \qquad a_0 \neq 0.$$

Then we find (with $a_{-1} = a_{-2} = 0$) that

$$xy' = \sum_{n=0}^{\infty} (n - 2)a_n x^{n-2}, \qquad x^2y'' = \sum_{n=0}^{\infty} (n - 2)(n - 3)a_n x^{n-2},$$

$$x^2y = \sum_{n=0}^{\infty} a_{n-2}x^{n-2}.$$

Substitution of these series in the equation $L(y) = 0$ yields

$$L(y) = \sum_{n=0}^{\infty} \left\{ [(n - 2)(n + 3) + 6]a_n - a_{n-2} \right\} x^{n-2} = 0.$$

To check the indicial equation we note that the quantity in braces vanishes for $n = 0$. For $n \geq 1$ we get the recurrence relation

$$a_n = \frac{a_{n-2}}{n(n + 1)}.$$

Hence $a_1 = a_3 = \cdots = a_{2k+1} = \cdots = 0$ and

$$a_2 = \frac{a_0}{2 \cdot 3}, \ldots, a_{2k} = \frac{a_0}{(2k + 1)!}.$$

The solution is

$$y = a_0 x^{-2}\left(1 + \frac{x^2}{3!} + \frac{x^4}{5!} + \cdots\right) = a_0 x^{-3}\left(x + \frac{x^3}{3!} + \frac{x^5}{5!} + \cdots\right).$$

We select $a_0 = 1$ and observe that the last series in parentheses is sinh x. Therefore, denoting the solution by y_1, we conclude that

$$y_1 = x^{-3} \sinh x.$$

With the aid of the solution y_1, we are able to find a second solution y_2. The method proceeds by setting

$$y_2 = y_1 v$$

and then determining v ($\neq$ constant) so that $L(y_2) = 0$. (Recall Section 4, Chapter 16.) We have

$$y_2' = y_1 v' + y_1' v, \qquad y_2'' = y_1 v'' + 2y_1' v' + y_1'' v.$$

Computing $L(y_2)$, we find

$$
\begin{aligned}
L(y_2) &= x^2(y_1 v'' + 2y_1' v' + y_1'' v) + 6x(y_1 v' + y_1' v) + (6 - x^2)y_1 v = 0 \\
&= v[x^2 y_1'' + 6xy_1' + (6 - x^2)y_1] + x^2 y_1 v'' + v'(2x^2 y_1' + 6xy_1) = 0 \\
&= x^2 y_1 v'' + v'(2x^2 y_1' + 6xy_1) = 0.
\end{aligned}
$$

We must solve the equation

$$\frac{v''}{v'} = -\frac{2y_1'}{y_1} - \frac{6}{x}.$$

An integration shows that

$$\ln |v'| = \ln C - 2 \ln |y_1| - 6 \ln |x|$$

or

$$v' = \frac{C}{y_1^2 x^6} = C \operatorname{csch}^2 x.$$

A second integration yields

$$v = -C \coth x + C_1$$

and, consequently,

$$y_2 = y_1 v = -Cx^{-3} \cosh x + C_1 x^{-3} \sinh x.$$

Selecting $C = -1$, $C_1 = 0$, we get $y_2 = x^{-3} \cosh x$. Then any solution of $L(y) = 0$ must be of the form

$$y = Ay_1 + By_2 = x^{-3}(A \sinh x + B \cosh x).$$

Remark. The method carried out in Example 3 for finding a second solution y_2 may be performed for any equation of the form

$$y'' + \frac{g_1(x)}{x} y' + \frac{g_2(x)}{x^2} y = 0.$$

Then if $y_2 = y_1 v$, the function v must satisfy the equation

$$\frac{v''}{v'} = -\left(\frac{2y_1'}{y_1} + \frac{g_1(x)}{x} \right).$$

If this equation can be integrated explicitly twice, the second solution y_2 may be found in explicit form whenever a first solution y_1 is known.

We have not exhausted all possible types of solutions of the indicial equation. For example, the roots may coincide. The next theorem, stated without proof, describes in more detail the types of solutions which occur when the roots coincide or differ by an integer.

Theorem 4. *Suppose that the hypotheses of Theorem 3(a) hold. We have:*

(a) *If the roots of the indicial equation are both equal to ρ, then the equation* (1) *has two solutions y_1 and y_2 with the form*

$$y_1(x) = x^\rho \sum_{n=0}^\infty a_n x^n, \qquad a_0 \neq 0,$$

$$y_2(x) = y_1(x) \ln x + x^{\rho+1} \sum_{n=0}^\infty b_n x^n.$$

(b) *If the roots are ρ_1 and $\rho_2 = \rho_1 - k$, where k is a positive integer, then equation* (1) *has two solutions of the form*

$$y_1(x) = x^{\rho_1} \sum_{n=0}^\infty a_n x^n, \qquad a_0 \neq 0,$$

$$y_2(x) = \alpha y_1(x) \ln x + x^{\rho_2} \sum_{n=0}^\infty b_n x^n.$$

The number α is a constant which may be zero.

———————

EXAMPLE 4. Find a second solution to the equation in Example 2.

Solution. The first solution is $y_1(x) = x + x^2$. Using Theorem 4, we set

$$y_2(x) = \alpha(x + x^2) \ln x + x^{-1} \sum_{n=0}^\infty b_n x^n$$

(since $\rho_1 = 1$, $\rho_2 = -1$). We now compute $L(y_2)$:

$$xy_2' = \alpha(x + 2x^2) \ln x + \alpha(x + x^2) + \sum_{n=0}^\infty (n - 1)b_n x^{n-1},$$

$$x^2 y_2'' = 2\alpha x^2 \ln x + \alpha(x + 3x^2) + \sum_{n=0}^\infty (n - 1)(n - 2)b_n x^{n-1},$$

$$x^2 y_2' = \alpha(x^2 + 2x^3) \ln x + \alpha(x^2 + x^3) + \sum_{n=0}^\infty (n - 2)b_{n-1} x^{n-1}, \qquad b_{-1} = 0,$$

$$xy_2 = \alpha(x^2 + x^3) \ln x + \sum_{n=0}^\infty b_{n-1} x^{n-1},$$

$$L(y_2) = \alpha(2x + 7x^2 + 3x^3) + \sum_{n=0}^\infty \{n(n - 2)b_n + 3(n - 4)b_{n-1}\} x^{n-1} = 0.$$

We obtain the following equations by setting the coefficients of x^n equal to zero:

$$n = 1: \quad -b_1 - 9b_0 = 0 \quad \text{or} \quad b_1 = -9b_0,$$

$$n = 2: \quad -6b_1 + 2\alpha = 0 \quad \text{or} \quad \alpha = +3b_1 = -27b_0,$$

$$n = 3: \quad 3b_3 - 3b_2 + 7\alpha = 0 \quad \text{or} \quad b_3 = b_2 + 63b_0,$$

$$n = 4: \quad (4)(2)b_4 + 3\alpha = 0 \quad \text{or} \quad b_4 = \frac{81}{2 \cdot 4} b_0,$$

$$n > 4: \quad b_n = -\frac{3(n-4)}{n(n-2)} b_{n-1}.$$

From this recurrence formula, we calculate

$$b_5 = -\frac{3 \cdot 1}{5 \cdot 3} b_4, \qquad b_6 = -\frac{3 \cdot 2}{6 \cdot 4} b_5 = \frac{(-3)^2 \cdot 2!}{3 \cdot 5 \cdot 4 \cdot 6} b_4$$

$$b_7 = -\frac{3 \cdot 3}{5 \cdot 7} b_6 = \frac{(-3)^3 \cdot 3!}{3 \cdot 5 \cdot 4 \cdot 6 \cdot 5 \cdot 7} b_4, \ldots$$

$$b_{4+m} = \frac{(-3)^m m! b_4}{3 \cdot 4 \cdots (2+m) \cdot 5 \cdot 6 \cdots (4+m)} = \frac{(-3)^{4+m} m! 3!}{(2+m)!(4+m)!} b_0.$$

The solution y_2 is

$$y_2(x) = (b_2 - 27b_0 \ln x)(x + x^2)$$

$$+ b_0 x^{-1} \left[1 - 9x + 63x^3 + \sum_{m=0}^{\infty} \frac{(-3)^{4+m} m! 3! x^{4+m}}{(2+m)!(4+m)!} \right].$$

Of course, we may set $b_2 = 0$, $b_0 = 1$, and then every solution will be a linear combination of y_1 and y_2.

EXERCISES

In each of problems 1 through 10, find a first solution y_1, using the methods of Examples 2 and 3; obtain an expression for y_1 in closed form by recognizing the series, and then find a second solution by the method of Example 3.

1. $xy'' + 2y' + xy = 0$

2. $x(1 - x)y'' + 2(1 - 2x)y' - 2y = 0$

3. $(1 + x)x^2 y'' + (4 + 6x)xy' + (2 + 6x)y = 0$

4. $x^2 y'' + (6 - x)xy' + (6 - 3x)y = 0$

5. $(2 - x)x^2 y'' + (5 - \frac{9}{2}x)xy' + (1 - 3x)y = 0$

6. $(1 - x)x^2 y'' + (1 - 3x)xy' - y = 0$

7. $x^2 y'' + (2 - 2x)xy' - 2xy = 0$

8. $x^2 y'' - (2 + x)xy' + (2 + x)y = 0$

9. $x(2 - x)y'' - (2 + x)y' + y = 0$

10. $x^2 y'' - (4 - x)xy' + (6 - 2x)y = 0$

In problems 11 through 15, solve in terms of generalized series. For what values of x does the series surely converge?

11. $2x^2y'' - (1 + x)xy' - 2xy = 0$

12. $2x^2y'' + (3 - 2x)xy' - (1 + 4x)y = 0$

13. $2x^2y'' + (-7 + 2x)xy' + (7 - 5x)y = 0$

14. The confluent hypergeometric equation $xy'' + (p - x)y' - qy = 0$; assume that p is not 0 or an integer.

15. Gauss' hypergeometric equation, c not an integer:

$$x(1 - x)y'' + [c - (a + b + 1)x]y' - aby = 0.$$

In problems 16 through 21, solve using the methods of Examples 2 and 4.

16. $x^2y'' + (4 - x)xy' + (2 - x)y = 0$

17. $x^2y'' - (3 - 2x)xy' + (3 - 6x)y = 0$

18. $x^2y'' - (1 + x)xy' + (1 + x)y = 0$

19. $x^2y'' + (1 - x)xy' + xy = 0$

20. $(1 - x)x^2y'' + (3 - 5x)xy' + (1 - 4x)y = 0$

21. $x^2y'' - (1 + x)xy' + 3xy = 0$

3. BESSEL FUNCTIONS

The differential equation

$$x^2y'' + xy' + (x^2 - p^2)y = 0, \tag{1}$$

in which p is a constant, is known as **Bessel's equation**. This equation arises in many problems in mathematical physics and engineering, and the various properties of its solutions have been studied extensively.

Dividing Bessel's equation by x^2, we obtain

$$y'' + \frac{1}{x}y' + \frac{x^2 - p^2}{x^2}y = 0,$$

and we notice that Theorem 3 of the last section is applicable to this equation. To obtain solutions we form the indicial equation $\rho(\rho - 1) + \rho - p^2 = 0$ and observe that the roots are $\rho_1 = p$, $\rho_2 = -p$. (The constant p is not necessarily an integer.) We assume $p \geq 0$ and set

$$y = x^p \sum_{n=0}^{\infty} a_nx^n = \sum_{n=0}^{\infty} a_nx^{n+p}, \qquad a_0 \neq 0.$$

We find

$$y' = \sum_{n=0}^{\infty} (n + p)a_nx^{n+p-1}, \qquad y'' = \sum_{n=0}^{\infty} (n + p)(n + p - 1)a_nx^{n+p-2},$$

$$L(y) = x^p \sum_{n=0}^{\infty} \{[(n + p)(n + p - 1) + (n + p) - p^2]a_n + a_{n-2}\}x^n = 0.$$

We have set $a_{-2} = a_{-1} = 0$ as usual. Setting the coefficient of x^n equal to zero, we obtain the recurrence relation

$$n(n + 2p)a_n = -a_{n-2}, \qquad n \geq 1.$$

Therefore $a_1 = a_3 = \cdots = a_{2k+1} = \cdots = 0$ and

$$a_2 = \frac{(-1)a_0}{2 \cdot 2(1 + p)}, \qquad a_4 = \frac{-a_2}{2^2 \cdot 2(2 + p)} = \frac{(-1)^2 a_0}{2^4 \cdot 2!(p + 1)(p + 2)}, \dots,$$

$$a_{2k} = \frac{(-1)^k a_0}{2^{2k} k!(p + 1)(p + 2) \cdots (p + k)}.$$

A solution of Bessel's equation is

$$y = a_0 x^p \left[1 - \frac{x^2}{2^2 \cdot 1!(p + 1)} + \frac{x^4}{2^4 \cdot 2!(p + 1)(p + 2)} + \cdots \right.$$

$$\left. + \frac{(-1)^k x^{2k}}{2^{2k} \cdot k!(p + 1)(p + 2) \cdots (p + k)} + \cdots \right]$$

$$= a_0 x^p \sum_{k=0}^{\infty} \frac{(-1)^k x^{2k}}{2^{2k} k!(p + 1) \cdots (p + k)}. \tag{2}$$

With the aid of the gamma function, which was introduced in Chapter 12, page 532, the solution (2) may be written in a simplified manner. We recall that one of the principal properties of the gamma function is the recursion formula

$$\Gamma(r + 1) = r\Gamma(r).$$

This relation allows us to write

$$\Gamma(p + 2) = (p + 1)\Gamma(p + 1), \qquad \Gamma(p + 3) = (p + 2)(p + 1)\Gamma(p + 1)$$

and, for any positive integer k,

$$\Gamma(p + k + 1) = (p + k)(p + k - 1) \cdots (p + 2)(p + 1)\Gamma(p + 1).$$

Multiplying numerator and denominator of (2) by $\Gamma(p + 1)$, we can now write the solution of Bessel's equation in the form

$$y = a_0 \Gamma(p + 1) x^p \sum_{k=0}^{\infty} \frac{(-1)^k x^{2k}}{2^{2k} k! \Gamma(p + k + 1)}. \tag{3}$$

DEFINITION. The function $J_p(x)$, defined by the infinite series

$$J_p(x) = \sum_{k=0}^{\infty} \frac{(-1)^k (x/2)^{2k+p}}{k! \Gamma(p + k + 1)},$$

is called the **Bessel function of order** p.

In terms of the Bessel function, the solution (3) of Bessel's equation is

$$y = a_0 2^p \Gamma(p + 1) J_p(x).$$

Since $a_0 2^p \Gamma(p + 1)$ is a constant, we see that $J_p(x)$ is a solution of Bessel's equation (1).

If p is positive and not an integer, then the methods of the previous section may be used to show that $J_{-p}(x)$ is a second linearly independent solution of (1). (See Exercise 1.) Thus *when p is not an integer, every solution of (1) is of the form*

$$y = A J_p(x) + B J_{-p}(x).$$

If p is a positive integer, it is easy to see that

$$J_{-p}(x) = (-1)^p J_p(x).$$

A second linearly independent solution can be obtained, however, by applying Theorem 4 of the last section.

For $p \geq 0$ the infinite series defining the Bessel function converges for all values of x. (See Exercise 3.) This fact is easily verified by use of the ratio test. Furthermore, for p a nonnegative integer, we observe that

$$J_p(x) = J_p(-x), \qquad p \text{ even,}$$
$$J_p(x) = -J_p(-x), \qquad p \text{ odd.}$$

(See Exercise 2.) Since convergent power series may be differentiated term by term, the proof of the next theorem is routine.

Theorem 5. *We have*

$$J_p'(x) = J_{p-1}(x) - \frac{p}{x} J_p(x). \tag{4}$$

Proof. The relation (4) may be written

$$\frac{d}{dx}\left(x^p J_p(x)\right) = x^p J_{p-1}(x).$$

Then

$$x^p J_p(x) = \sum_{k=0}^{\infty} \frac{(-1)^k x^{2p+2k}}{2^{2k+p} k! \Gamma(p + k + 1)}$$

and

$$\frac{d}{dx}\left(x^p J_p(x)\right) = \sum_{k=0}^{\infty} \frac{(-1)^k 2(p + k) x^{2p+2k-1}}{2^{2k+p} k! \Gamma(p + k + 1)}$$

$$= x^p \sum_{k=0}^{\infty} \frac{(-1)^k x^{2k+(p-1)}}{2^{2k+(p-1)} k! \Gamma(k + p)}$$

$$= x^p J_{p-1}(x).$$

In a similar manner we may easily establish the relation

$$J'_p(x) = \frac{p}{x} J_p(x) - J_{p+1}(x). \tag{5}$$

(See Exercise 4.)

The Bessel function for $p = \frac{1}{2}$ is particularly interesting, since it is expressible in terms of elementary functions. To see this we write

$$y'' + \frac{1}{x} y' + \left(1 - \frac{1}{4x^2}\right) y = 0 \tag{6}$$

and make the substitution

$$y = x^{-1/2} v.$$

Then we find

$$y' = -\tfrac{1}{2} x^{-3/2} v + x^{-1/2} v', \qquad y'' = \tfrac{3}{4} x^{-5/2} v - x^{-3/2} v' + x^{-1/2} v''.$$

A substitution in (6) yields the equation

$$v'' + v = 0,$$

which has the general solution

$$v = A \sin x + B \cos x.$$

Since $J_{1/2}(0) = 0$ while $J_{-1/2}(x) \to \infty$ as $x \to 0$, we conclude (using the fact that $\Gamma(\tfrac{3}{2}) = \tfrac{1}{2}\sqrt{\pi}$) that

$$J_{1/2}(x) = \sqrt{\frac{2}{\pi}} \frac{\sin x}{\sqrt{x}}, \qquad J_{-1/2}(x) = \sqrt{\frac{2}{\pi}} \frac{\cos x}{\sqrt{x}}.$$

With the aid of the recursion formulas (4) and (5), all Bessel functions of order $m + \frac{1}{2}$ where m is an integer can be obtained in terms of elementary functions. [See Exercise 5(c).]

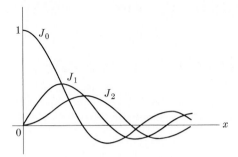

FIGURE 17–1

The graphs of $J_0(x)$, $J_1(x)$, and $J_2(x)$ are shown in Fig. 17–1. The oscillatory character shown in the graphs and verified for Bessel functions of half-integer order can be established quite generally. Let p be any constant, and make the same change of variables in Bessel's equation which we used in the half-integer case:

$$y = x^{-1/2} v.$$

Bessel's equation becomes

$$v'' + \left(1 - \frac{4p^2 - 1}{4x^2}\right) v = 0. \tag{7}$$

The solution of this equation is $v = \sqrt{x}\, J_p(x)$. However, for large values of x, Eq. (7) behaves like solutions of $v'' + v = 0$, that is, like $\sin x$ and $\cos x$. Thus $J_p(x)$ must also have the same oscillatory character and, furthermore, must die out in such a way that the effect of the growth term $\sqrt{x}$ is canceled. The facts indicated by this heuristic argument may be established rigorously.

The most extensive and complete treatment of Bessel functions is given in the book, *Treatise on the Theory of Bessel Functions*, by G. N. Watson. This text devotes over 800 pages entirely to the study of these functions.

EXERCISES

1. If $p > 0$ is not an integer, show that

$$J_{-p}(x) = x^{-p} \sum_{k=0}^{\infty} \frac{(-1)^k x^{2k}}{2^{2k-p} k! \Gamma(-p + k + 1)}$$

is a second solution of Bessel's equation which is independent of $J_p(x)$.

2. If p is a positive integer, show that
   (a) $J_{-p}(x) = (-1)^p J_p(x)$. [*Hint:* Use the fact that $\Gamma(r) \to \infty$ as $r \to$ a negative integer.]
   (b) $J_p(x) = J_p(-x)$, if p is even,
   (c) $J_p(x) = -J_p(-x)$, if p is odd.
   It is assumed that $J_{-p}(x)$ is defined for positive integers p as $\lim_{q \to p} J_{-q}(x)$.

3. Show that for all p, the power series $x^{-p} J_p(x)$ converges for all values of x. [*Hint:* Use 2(a) above for p a negative integer.]

4. Prove the recursion formula (5):

$$J_p'(x) = \frac{p}{x} J_p(x) - J_{p+1}(x).$$

5. (a) Using the recursion formulas (4) and (5), show that

$$J_{p-1}(x) + J_{p+1}(x) = \frac{2p}{x} J_p(x). \tag{8}$$

   (b) Establish the recursion formula

$$J_p'(x) = \tfrac{1}{2}\{J_{p-1}(x) - J_{p+1}(x)\}.$$

   (c) Using (8), find expressions in terms of elementary functions for

$$J_{3/2}(x), \qquad J_{5/2}(x), \qquad J_{-3/2}(x).$$

6. (a) Find a solution of the **modified Bessel equation**

$$y'' + \frac{1}{x} y' - \left(1 + \frac{p^2}{x^2}\right) y = 0.$$

(b) If the solution found in part (a) is denoted by $y_1(x)$, show that the arbitrary constant in the solution may be chosen so that

$$y_1(x) = J_p(ix)/(i^p).$$

7. Express $J_4(x)$ in terms of $J_0(x)$ and $J_1(x)$. [*Hint.* Use recursion formula (8).]

8. Show that

$$\frac{d}{dx}\left(J_p(x)J_{p+1}(x)\right) = xJ_p^2(x) - xJ_{p+1}^2(x) - J_p(x)J_{p+1}(x).$$

9. If the change of variable $v = x^{p/2}J_p(\sqrt{x})$ is made, show that v satisfies the equation

$$v'' + \frac{(1-p)}{x}v' + \frac{1}{4x}v = 0.$$

10. If the change of variable $v = x^p J_p(x)$ is made, show that v satisfies the equation

$$v'' + \frac{2p-1}{x}v' + v = 0.$$

11. If the change of variable $v = J_p(\sqrt{x})$ is made, show that v satisfies the equation

$$v'' + \frac{1}{x}v' + \frac{x-p^2}{4x^2}v = 0.$$

*12. Show that if p is an integer, then

$$J_{p+1/2}(x) = \frac{(-1)^p\sqrt{2}\,x^{p+1/2}}{\sqrt{\pi}}\left(\frac{1}{x}\frac{d}{dx}\right)^p\left(\frac{\sin x}{x}\right).$$

[*Hint.* Use the recurrence formula and induction on p. We interpret the symbol

$$\left(\frac{1}{x}\frac{d}{dx}\right)^p \quad \text{as} \quad \frac{1}{x}\frac{d}{dx}\left[\left(\frac{1}{x}\frac{d}{dx}\right)^{p-1}\right]$$

for all integers $p > 1$.]

13. From the formula ($p > 0$)

$$J_p(x) = \frac{x^p}{2^{p-1}\Gamma(p)}\int_0^{\pi/2} J_0(x\sin\theta)\cos^{2p-1}\theta\sin\theta\,d\theta$$

show that

$$J_p'(x) = \frac{x}{p}J_p(x) - \frac{x^p}{2^{p-1}\Gamma(p)}\int_0^{\pi/2} J_1(x\sin\theta)\cos^{2p-1}\theta\sin^2\theta\,d\theta.$$

*14. Use the results of Exercise 13 to show that

$$J_p(x) = \frac{x^{p-1}}{2^{p-2}\Gamma(p-1)}\int_0^{\pi/2} J_1(x\sin\theta)\cos^{2p-3}\theta\sin^2\theta\,d\theta.$$

Appendix 1

We show that the trigonometric set

$$\frac{1}{\sqrt{2\pi}}, \qquad \frac{1}{\sqrt{\pi}} \cos nx, \qquad \frac{1}{\sqrt{\pi}} \sin nx, \qquad n = 1, 2, \ldots \qquad (1)$$

is complete.

Lemma 1. Suppose that the periodic extension of f is piecewise smooth and normalized on $[-\pi, \pi]$. Then there is a sequence $\{f_n\}$ of continuous 2π-periodic functions which converges in the mean-square sense to f on $[-\pi, \pi]$.

Proof. Let $2h$ be the smallest of the $X_i - X_{i-1}$. For each n, define (see Fig. 1) $f_n(x) = f(x)$ on $[X_{i-1} + h/n, X_i - h/n]$ for each i; on each interval $[X_{i-1}, X_{i-1} + h/n]$ or $[X_i - h/n, X_i]$, define f_n as that linear function coinciding with $f(x)$ at the ends of each such interval. Then each f_n is continuous and periodic. Also, if $|f(x)| \leq M$, this is true of each $f_n(x)$. Thus $|f(x) - f_n(x)| \leq 2M$ on each such interval, the sum of their lengths being $\leq 2\pi/n$. Hence

$$\int_{-\pi}^{\pi} |f(x) - f_n(x)|^2 \, dx \leq \frac{4M^2 \cdot 2\pi}{n}.$$

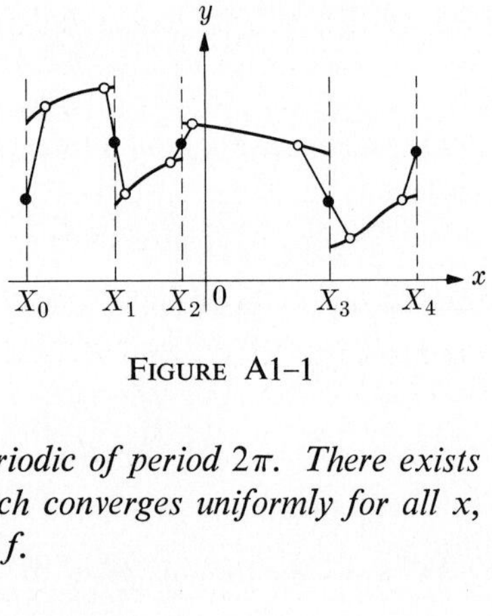

FIGURE A1-1

Lemma 2. Suppose that f is continuous and periodic of period 2π. There exists a sequence $\{f_n\}$ of smooth periodic functions which converges uniformly for all x, and hence in the mean-square sense on $[-\pi, \pi]$ to f.

Proof. Define

$$F(x) = \int_{-\pi}^{x} f(t) \, dt, \qquad L = \int_{-\pi}^{\pi} f(t) \, dt.$$

Since f is periodic, we see that

$$F(x + 2\pi) - F(x) = \int_{x}^{x+2\pi} f(t) \, dt = \int_{-\pi}^{\pi} f(t) \, dt = L.$$

Since F is smooth $[F'(x) = f(x)$ everywhere$]$, we see that

$$f_h(x) = \frac{F(x + h) - F(x)}{h} = \frac{1}{h} \int_{x}^{x+h} f(t) \, dt, \qquad h > 0,$$

734

is periodic and smooth. Since f is continuous on $[-\pi, \pi]$, it is uniformly continuous there and hence everywhere. So, let $\epsilon > 0$ be given. There is a $\delta > 0$ such that $|f(t') - f(t'')| < \epsilon$ whenever $|t' - t''| < \delta$. Hence, if $0 < h < \delta$, we have (for each fixed x)

$$f_n(x) - f(x) = \frac{1}{h} \int_x^{x+h} [f(t) - f(x)]\, dt, \qquad |f_n(x) - f(x)| < \epsilon.$$

If we take $f_n = f_h$, where $h = 1/n$, we see that f_n converges uniformly to f.

We now prove the theorem. We recall that, in any Euclidean vector space, we have (Corollary to Theorem 7 of Chapter 8)

$$\left\| \sum_{j=1}^n \mathbf{v}_j \right\| \le \sum_{j=1}^n \|\mathbf{v}_j\|.$$

We use this inequality for our function space V of Chapter 10, Section 4.

Theorem. *The trigonometric set* (1) *is complete.*

Proof. Suppose $f \in V$. According to Lemma 1, there is, for each n, a continuous 2π-periodic function g_n such that $\|g_n - f\| < 1/3n$. From Lemma 2, we conclude for each n that there is a smooth 2π-periodic function h_n such that $\|h_n - g_n\| < 1/3n$. But, since the partial sums of the series for h_n converge uniformly to h_n for each fixed n, it follows that there is such a partial sum, which we call f_n, such that $\|f_n - h_n\| < 1/3n$. Thus

$$\|f - f_n\| \le \|f - g_n\| + \|g_n - h_n\| + \|h_n - f_n\| < \frac{1}{n}.$$

Accordingly, finite linear combinations of the elements in the set (1) are dense in V, which is the desired result.

Remark. The completeness of the normalized Legendre polynomials follows from Lemma 1 and the famous theorem of Weierstrass to the effect that any function f continuous on a bounded closed interval can be approximated uniformly on that interval by polynomials.

Appendix 2

We establish Theorem 13 in Chapter 13 under an additional hypothesis concerning the definition of simple connectivity. We say a domain is **strongly simply connected** if and only if it is simply connected as defined on page 584 and if the derivatives $\mathbf{f}_t, \mathbf{f}_\tau, \mathbf{f}_{t\tau}$, and $\mathbf{f}_{\tau t}$ of the function $\mathbf{f}$ given in the definition of simple connectivity are all continuous.

Theorem. *Suppose that* **v** *is continuously differentiable in a strongly simply connected region D in space and that* curl **v** $= 0$ *in D. Then there is a continuously differentiable scalar field u on D such that* **v** $= \nabla u$.

Proof. In the light of Theorem 12 on page 582, it is sufficient to prove that $\int_C \mathbf{v} \cdot d\mathbf{r}$ is independent of the path. It is convenient to use the summation notation, and so we denote the coordinates of a rectangular system by (x_1, x_2, x_3). We let $\vec{C}_0$, $\vec{C}_1$, and $\vec{C}_\tau$ be paths from A to B as in the definition of simple connectivity, and we write

$$\mathbf{v}(x_1, x_2, x_3) = v_1(x_1, x_2, x_3)\mathbf{i} + v_2(x_1, x_2, x_3)\mathbf{j} + v_3(x_1, x_2, x_3)\mathbf{k},$$
$$\mathbf{f}(t, \tau) = f_1(t, \tau)\mathbf{i} + f_2(t, \tau)\mathbf{j} + f_3(t, \tau)\mathbf{k}.$$

We define

$$\phi(\tau) = \int_{\vec{C}_\tau} \mathbf{v} \cdot d\mathbf{r} = \int_a^b \sum_{i=1}^3 v_i[f_1(t, \tau), \ f_2(t, \tau), \ f_3(t, \tau)] \frac{\partial f_i}{\partial t} \, dt.$$

Then, using Leibniz' Rule (and the Chain Rule), we obtain

$$\phi'(\tau) = \int_a^b \sum_{i=1}^3 \left\{ v_i(f_1, f_2, f_3) \frac{\partial^2 f_i}{\partial t \, \partial \tau} + \sum_{j=1}^3 \frac{\partial v_i}{\partial x_j} \frac{\partial f_i}{\partial t} \frac{\partial f_j}{\partial \tau} \right\} dt.$$

Integrating by parts the terms in the first sum above, we eliminate the second derivatives of the f_i and find

$$\phi'(\tau) = \left[\sum_{i=1}^3 v_i \frac{\partial f_i}{\partial \tau} \right]_{t=a}^{t=b} + \int_a^b \sum_{i,j=1}^3 \frac{\partial v_i}{\partial x_j} \left(\frac{\partial f_i}{\partial t} \frac{\partial f_j}{\partial \tau} - \frac{\partial f_i}{\partial \tau} \frac{\partial f_j}{\partial t} \right) dt. \qquad (1)$$

Denoting the coordinates of A and B by (x_1^0, x_2^0, x_3^0) and (x_1^1, x_2^1, x_3^1), respectively, we observe that for all τ in $[0, 1]$

$$f_i(a, \tau) = x_i^0 \quad \text{and} \quad f_i(b, \tau) = x_i^1, \quad i = 1, 2, 3.$$

Thus the first term on the right in (1) vanishes. Moreover, if we interchange the indices i and j in the second sum in (1), we get

$$\phi'(\tau) = \int_a^b \sum_{i,j=1}^n \left(\frac{\partial v_i}{\partial x_j} - \frac{\partial v_j}{\partial x_i} \right) \frac{\partial f_i}{\partial t} \frac{\partial f_j}{\partial \tau} \, dt.$$

The condition that curl **v** $= 0$ is equivalent to the condition

$$\frac{\partial v_i}{\partial x_j} - \frac{\partial v_j}{\partial x_i} = 0, \quad i, j = 1, 2, 3$$

and so $\phi'(\tau) = 0$. Hence $\phi(\tau)$ is constant and therefore independent of the path.

Remark. This proof generalizes to *n*-space if we replace the condition curl $\mathbf{v} = \mathbf{0}$ by the condition

$$\frac{\partial v_i}{\partial x_j} - \frac{\partial v_j}{\partial x_i} = 0, \qquad i, j = 1, 2, \ldots, n \tag{2}$$

which can be shown to be independent of the axes. However, the equations (2) cannot be expressed in terms of vector operators for $n > 3$. More complicated objects called *alternating tensors* or *exterior differential forms* are employed. The specific details are usually discussed in courses in differential geometry.

Appendix 3

We prove Stokes' Theorem as stated on page 624 of Chapter 14. To do so we require two lemmas.

Lemma 1.   Suppose that $T: u = U(s, t)$, $v = V(s, t)$ *is a one-to-one smooth transformation from an open set* D_1 *onto an open set* D *which carries a piecewise smooth region* G_1 *contained in* D_1 *onto a region* G *of the same type contained in* D. *Assume that*

$$\frac{\partial(u, v)}{\partial(s, t)} > 0 \qquad \text{for all } (s, t) \quad \text{on} \quad D_1.$$

If $s = S(r)$, $t = T(r)$, $a \le r \le b$ *is a representation of a smooth directed arc* $\vec{C}_1$ *of* ∂G_1, *then*

$$u = U[S(r), T(r)], \qquad v = V[S(r), T(r)], \qquad a \le r \le b$$

is a representation of the corresponding directed arc $\vec{C}$ *of* ∂G.

Proof. Let $P_1(s_0, t_0)$ be a point on $\vec{C}_1$ and $P(u_0, v_0)$ be the corresponding point on $\vec{C}$. That is, $u_0 = U(s_0, t_0)$, $v_0 = V(s_0, t_0)$. Furthermore, suppose that $s_0 = S(r_0)$, $t_0 = T(r_0)$, where $a < r_0 < b$ (Fig. A3–1). We choose new coor-

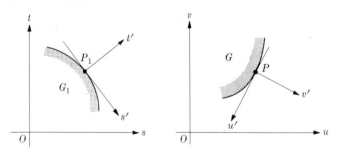

FIGURE A3–1

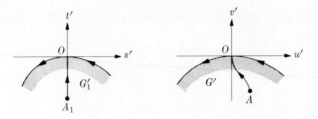

FIGURE A3–2

dinate systems (s', t') in the (s, t)-plane and (u', v') in the (u, v)-plane as shown in Fig. A3–1, which are obtained by a translation and rotation of coordinates. The regions G_1 and G are transformed into regions G_1' and G' under these changes, and we write $G_1' = \sigma(G_1)$, $G' = \tau(G)$. The regions D_1 and D go into D_1' and D', respectively. We observe that $\partial G_1'$ is tangent to the s'-axis at the origin and $\partial G'$ is tangent to the u'-axis at the origin (Fig. A3–2). It follows that at $(0, 0)$ we have

$$\frac{\partial v'}{\partial s'} = 0, \qquad \frac{\partial u'}{\partial s'} \neq 0, \qquad \frac{\partial (u', v')}{\partial (s', t')} = \frac{\partial u'}{\partial s'} \frac{\partial v'}{\partial t'} > 0.$$

Since a segment $\overline{A_1 O}$ of the negative t'-axis is in G_1', it is carried into an arc $\overset{\frown}{AO}$ in G'; this shows that

$$\frac{\partial v'}{\partial t'} > 0 \qquad \text{and so} \qquad \frac{\partial u'}{\partial s'} > 0.$$

The last inequality states that as one moves to the left along $\partial G_1'$ (Fig. A3–2), the image point moves to the left along $\partial G'$.

The second lemma we prove is the one stated on page 625.

Lemma 2. *Suppose that $f(u, v)$ is smooth on an open set D containing a piecewise smooth, closed, bounded region G. Then there exists a sequence $f_1, f_2, \ldots, f_n, \ldots$ such that each f_n, f_{nu}, f_{nv} is smooth on G and $f_n \to f, f_{nu} \to f_u, f_{nv} \to f_v$ uniformly on G.*

Proof. Let P be any point of G. Since G is closed and contained in D, which is open, it follows from the Heine-Borel Theorem (Second Form, page 510) that there is a $\rho > 0$ such that the circle $C(P, \rho)$ with center at P and radius ρ is in D. We let G_0 be the set of all P in G such that $C(P, \rho/2)$ is in D; it can be shown that G_0 is a closed set. For (u_0, v_0) in G, it is then clear that the square

$$|u - u_0| \leq h, \qquad |v - v_0| \leq h$$

is in G_0 so long as $0 < h < \rho/2\sqrt{2}$. We now define the function

$$f_h(u, v) = \frac{1}{4h^2} \int_{u-h}^{u+h} \int_{v-h}^{v+h} f(s, t) \, dA_{st}$$

for all (u, v) in G with $0 < h < \rho/2\sqrt{2}$. Holding v fixed and applying Leibniz' Rule for differentiating under the integral sign, we obtain

$$f_{h,1}(u, v) = \frac{1}{4h^2} \int_{v-h}^{v+h} [f(u + h, t) - f(u - h, t)] \, dt$$

$$= \frac{1}{4h^2} \int_{v-h}^{v+h} \int_{u-h}^{u+h} f_{,1}(s, t) \, dA_{st}.$$

Applying Leibniz' Rule again, first with respect to u and then with respect to v, we get

$$f_{h,1,1}(u, v) = \frac{1}{4h^2} \int_{v-h}^{v+h} [f_{,1}(u + h, t) - f_{,1}(u - h, t)] \, dt,$$

$$f_{h,1,2}(u, v) = \frac{1}{4h^2} \int_{u-h}^{u+h} [f_{,1}(s, v + h) - f_{,1}(s, v - h)] \, ds.$$

A similar argument shows that

$$f_{h,2}(u, v) = \frac{1}{4h^2} \int_{u-h}^{u+h} \int_{v-h}^{v+h} f_{,2}(s, t) \, dA_{st}$$

and that $f_{h,2}$ is smooth for $0 < h < \rho/2\sqrt{2}$.

Since $f_{,1}$ and $f_{,2}$ are continuous on G_0, a closed, bounded set, they are uniformly continuous there. Hence for any $\epsilon > 0$, there is a δ with $0 < \delta < \rho/2\sqrt{2}$ such that

$$|f_{,k}(s, t) - f_{,k}(u, v)| < \epsilon$$

if

$$|s - u| < \delta, \qquad |t - v| < \delta$$

with (u, v) on G and $k = 1, 2$. Therefore if $0 < h < \delta$ and if $k = 1$ or 2, then

$$|f_{h,k}(u, v) - f_{,k}(u, v)| = \frac{1}{4h^2} \left| \int_{u-h}^{u+h} \int_{v-h}^{v+h} [f_{,k}(s, t) - f_{,k}(u, v)] \, dA_{st} \right| < \epsilon.$$

The sequence $f_1, f_2, \ldots, f_n, \ldots$ is obtained by letting $h = h_n \to 0$.

Theorem. (Stokes' Theorem.) *Suppose that S is a bounded, closed, orientable, piecewise smooth surface and that $\mathbf{v}$ is a smooth vector field defined on an open set containing $|\vec{S}|$. Then*

$$\iint_{\vec{S}} (\operatorname{curl} \mathbf{v}) \cdot \mathbf{n} \, dS = \int_{\partial \vec{S}} \mathbf{v} \cdot d\mathbf{r}.$$

Proof. With each point P of S we may associate a rectangular coordinate system (x^P, y^P, z^P) and two cylindrical domains: A large cylinder Γ_P given by

$$\Gamma_P: \quad (x^P)^2 + (y^P)^2 < 9r_P^2, \qquad -3k_P < z^P < 3k_P$$

where $r_P > 0$, $k_P > 0$; and a small cylinder γ_P given by

$$\gamma_P: \quad (x^P)^2 + (y^P)^2 < r_P^2, \qquad -k_P < z^P < k_P.$$

With the point P we also associate: (i) a domain G_P in the (x^P, y^P) plane which is one of the types in Fig. 14–22 with r_P replaced by $3r_P$; (ii) a piecewise smooth function $f^P(x^P, y^P)$ on G_P such that $f^P(0, 0) = 0$; (iii) the function f^P satisfies the inequality

$$|f^P(x^P, y^P)| < \begin{cases} k_P & \text{if } (x^P)^2 + (y^P)^2 < r_P^2, \\ 3k_P & \text{if } (x^P, y^P) \text{ is in } G_P; \end{cases}$$

(iv) all the points of $|\vec{S}|$ in Γ_P are on the locus of

$$z^P = f^P(x^P, y^P) \qquad \text{for } (x^P, y^P) \text{ in } G_P;$$

(v) at each point Q of S where f^P is smooth, $\mathbf{n}(Q)$ is the positively directed normal to $\vec{S}$. (vi) The numbers r_P and k_P are reduced in size, if necessary, so that the vector field $\mathbf{v}$ is smooth in each cylinder Γ_P.

We now define

$$\alpha_P(Q) = \phi\left[\frac{\sqrt{(x_Q^P)^2 + (y_Q^P)^2}}{r_P}\right] \phi\left(\frac{z_Q^P}{k_P}\right),$$

where ϕ is the function defined on page 597 at the beginning of the section on the partition of unity. The triple (x_Q^P, y_Q^P, z_Q^P) are the coordinates of Q in the system centered at the point P. Then each $\alpha_P(Q)$ is smooth everywhere and

$$\alpha_P(Q) = \begin{cases} 1 & \text{on } \gamma_P, \\ 0 & \text{outside, on, and near the boundary of } \Gamma_P. \end{cases}$$

From the Heine-Borel Theorem (page 510), it follows that a finite number of the γ_P cover an open set containing $|\vec{S}|$. Call these cylinders $\gamma_1, \gamma_2, \ldots \gamma_k$, each γ_i being centered at P_i. Let Γ_i be the corresponding large cylinders and define

$$\psi_i(Q) = \frac{\alpha_i(Q)}{\alpha_1(Q) + \alpha_2(Q) + \cdots + \alpha_k(Q)}.$$

Therefore $\sum_{i=1}^k \psi_i(Q) = 1$ for all Q on an open set D containing $|\vec{S}|$. We define

$$\mathbf{v}_i = \psi_i \mathbf{v}$$

and, since $\mathbf{v}_i = \mathbf{0}$ outside Γ_i, we may write

$$\iint_{\vec{S}} (\text{curl } \mathbf{v}_i) \cdot \mathbf{n} \, dS = \iint_{\vec{S}_i} (\text{curl } \mathbf{v}_i) \cdot \mathbf{n} \, dS = \int_{\partial \vec{S}_i} \mathbf{v}_i \cdot d\mathbf{r}_i$$

$$= \int_{\partial \vec{S}} \mathbf{v}_i \cdot d\mathbf{r},$$

in which we use the fact that in Chapter 14, Section 7, the theorem was proved for the special domains S_i. The result follows by addition.

The corollary to Stokes' Theorem on page 625 follows, since if $\vec{S}$ has no boundary, then every G_i is a full circle, and so $\mathbf{v}_i = \mathbf{0}$ near ∂G_i for each i. Hence

$$\int_{\partial S_i} \mathbf{v} \cdot d\mathbf{r} = 0$$

for each i.

Appendix 4

PROOF OF THE DIVERGENCE THEOREM

We prove the Divergence Theorem (Theorem 5, page 629) for regions which have boundaries that are not too irregular. We suppose that G is a region in three-space, the boundary of which consists of a finite number of disjoint piecewise smooth surfaces without boundary, each of which has the following additional property: With each point P of such a surface S is associated a coordinate system (x^P, y^P, z^P), a function f^P, and a cylindrical region γ_P, as in the definition of a piecewise smooth surface, in such a way that the part of G in γ_P consists of those points of γ_P for which $z^P < f^P(x^P, y^P)$. In this case we say that G has a **regular boundary**. If G has a regular boundary it follows that ∂G (i.e., each boundary surface) may be oriented by defining $\mathbf{n}(Q)$ as the outward-pointing normal; if Q is on a smooth part of the locus of $z^P = f^P(x^P, y^P)$ as chosen above, $\mathbf{n}(Q)$ is the positive normal.

Theorem. (The Divergence Theorem.) *Suppose that a bounded domain G in three-space has a regular boundary ∂G and that $\mathbf{v}$ is a smooth vector field defined on an open set containing G and ∂G. Then*

$$\iiint_G \text{div } \mathbf{v} \, dV = \iint_{\partial G} \mathbf{v} \cdot \mathbf{n} \, dS,$$

where the boundary ∂G is oriented by taking $\mathbf{n}$ as the exterior normal.

Proof. With each point P of ∂G, we associate a coordinate system (x^P, y^P, z^P), a function f^P, cylindrical regions Γ_P and γ_P, and a function α_P as in the proof of Stokes' Theorem. (See Appendix 3.) We impose the additional condition, in line with the definition above, that the intersection of G with Γ_P is the part of Γ_P for which $z^P < f^P(x^P, y^P)$. For each point P in G itself, we choose (x^P, y^P, z^P) as any right-handed system with origin at P, and define the cylinders Γ_P, γ_P as before with $k_P = r_P$. The quantity r_P is so small that Γ_P is in G. For these interior points P, we define

$$\alpha_P(x^P, y^P, z^P) = \phi\left[r_P^{-1}\sqrt{(x^P)^2 + (y^P)^2}\right] \cdot \phi[r_P^{-1}z^P].$$

From the Heine-Borel Theorem (page 510), we conclude that $G \cup \partial G$ is covered by a finite number of the γ_P. Denote them by $\gamma_1, \gamma_2, \ldots \gamma_k$, each γ_i being centered at some P_i; let Γ_i and α_i be the corresponding Γ_P and α_P. We then define

$$\psi_i(Q) = \frac{\alpha_i(Q)}{\alpha_1(Q) + \alpha_2(Q) + \cdots + \alpha_k(Q)}$$

and we notice that the ψ_i form a smooth partition of unity on an open set D containing G and its boundary. Setting $\mathbf{v}_i = \psi_i\mathbf{v}$, we have

$$\mathbf{v} = \mathbf{v}_1 + \mathbf{v}_2 + \cdots + \mathbf{v}_k,$$

and it is sufficient to prove the Divergence Theorem for each $\mathbf{v}_i$. If P_i and hence Γ_i is interior to G, then $\mathbf{v}_i = \mathbf{0}$ outside Γ_i as well as on and near its boundary. Applying Theorems 6 and 7 of Chapter 14 to $\mathbf{v}_i$ in Γ_i, we find that

$$\iiint_G \operatorname{div} \mathbf{v}_i \, dV = \iiint_{\Gamma_i} \operatorname{div} \mathbf{v}_i \, dV = \iint_{\partial\Gamma_i} \mathbf{v}_i \cdot \mathbf{n}_i \, dS = 0 = \iint_{\partial G} \mathbf{v}_i \cdot \mathbf{n} \, dS.$$

When P_i is on ∂G, we still have $\mathbf{v}_i = \mathbf{0}$ outside Γ_i, so that Theorems 6 and 7 of Chapter 14 applied to Γ_i yield

$$\iiint_G \operatorname{div} \mathbf{v}_i \, dV = \iiint_{\Gamma_i} \operatorname{div} \mathbf{v}_i \, dV = \iint_{\partial\Gamma_i} \mathbf{v}_i \cdot \mathbf{n} \, dS = \iint_{\partial G} \mathbf{v}_i \cdot \mathbf{n} \, dS.$$

The last equality holds because $\mathbf{v}_i = \mathbf{0}$ on that portion of $\partial\Gamma_i$ which is not part of ∂G.

Introduction to
a Short Table of Integrals

Let us recall some of the methods of integration with which the student should be familiar. These devices together with the integrals listed below enable the student to perform expeditiously any integration which is required in order to work the problems in this text.

(1) Substitution in a table of integrals.

EXAMPLE: Letting $u = \tan x$, $du = \sec^2 x\, dx$, we find that $\int e^{\tan x} \sec^2 x\, dx = \int e^u\, du = e^u + C = e^{\tan x} + C$.

(2) Certain trigonometric and hyperbolic integrals. We illustrate with trigonometric integrals; the corresponding hyperbolic forms are treated similarly.

(a) $\int \sin^m u \cos^n u\, du$.

(i) n an odd positive integer, m arbitrary: Factor out $\cos u\, du$ and express the remaining cosines in terms of sines.

EXAMPLE:

$$\int \sin^4 2x \cos^3 2x\, dx = \tfrac{1}{2}\int \sin^4 2x(1 - \sin^2 2x) \cdot (2 \cos 2x\, dx)$$

$$= \tfrac{1}{2}\int (\sin^4 2x - \sin^6 2x)d(\sin 2x) = \tfrac{1}{10}\sin^5 2x - \tfrac{1}{14}\sin^7 2x + C.$$

(ii) m an odd positive integer, n arbitrary: Factor out $\sin u\, du$ and express the remaining sines in terms of cosines.

(iii) m and n both even integers ≥ 0: Reduce the degree of the expression by the substitutions

$$\sin^2 u = \frac{1 - \cos 2u}{2}, \qquad \cos^2 u = \frac{1 + \cos 2u}{2}.$$

EXAMPLE:

$$\int \sin^4 u\, du = \tfrac{1}{4}\int (1 - \cos 2u)^2\, du$$

$$= \tfrac{1}{4}\int (1 - 2 \cos 2u)\, du + \tfrac{1}{8}\int (1 + \cos 4u)\, du$$

$$= \frac{3u}{8} - \frac{\sin 2u}{4} + \frac{\sin 4u}{32} + C.$$

743

(b) $\int \tan^m u \sec^n u \, du.$

(i)   n an even positive integer, m arbitrary.   Factor out $\sec^2 u \, du$ and express the remaining secants in terms of $\tan u$.

EXAMPLE:

$$\int \frac{\sec^4 u \, du}{\sqrt{\tan u}} = \int (\tan u)^{-1/2}(1 + \tan^2 u) \cdot (\sec^2 u \, du)$$

$$= \int [(\tan u)^{-1/2} + (\tan u)^{3/2}] d\,(\tan u)$$

$$= 2 \,(\tan u)^{1/2} + \tfrac{2}{5}\,(\tan u)^{5/2} + C.$$

(ii)   m an odd positive integer, n arbitrary.   Factor out $\sec u \tan u \, du$ and express the remaining tangents in terms of the secants.

EXAMPLE:

$$\int \frac{\tan^3 u \, du}{\sqrt[3]{\sec u}} = \int (\sec u)^{-4/3} \tan^2 u \cdot (\sec u \tan u \, du)$$

$$= \int [(\sec u)^{2/3} - (\sec u)^{-4/3}] d\,(\sec u)$$

$$= \tfrac{3}{5}\,(\sec u)^{5/3} + 3\,(\sec u)^{-1/3} + C.$$

(c) $\int \cot^m u \csc^n u \, du.$ These are treated like those in (b).

(3)   Trigonometric and hyperbolic substitutions.

(a)   If $\sqrt{a^2 - u^2}$ occurs (or $a^2 - u^2$ occurs in the denominator), set $u = a \sin \theta$, $\theta = \arcsin (u/a)$, $du = a \cos \theta \, d\theta$, $\sqrt{a^2 - u^2} = a \cos \theta$.

EXAMPLE:

$$\int \sqrt{a^2 - u^2} \, du = a^2 \int \cos^2 \theta \, d\theta = \frac{a^2}{2} \int (1 + \cos 2\theta) \, d\theta$$

$$= \frac{a^2}{2} (\theta + \sin \theta \cos \theta) + C$$

$$= \frac{a^2}{2} \arcsin \frac{u}{a} + u\sqrt{a^2 - u^2} + C.$$

(b)   If $\sqrt{a^2 + u^2}$ occurs, set $u = a \tan \theta$, etc.

EXAMPLE:

$$\int u^3 \sqrt{a^2 + u^2} \, du = a^5 \int \tan^3 \theta \sec^3 \theta \, d\theta.$$

The last integral is of type (2)(b)(ii) above.

(c)   If $\sqrt{u^2 - a^2}$ occurs, set $u = a \sec \theta$, etc.

EXAMPLE: $\displaystyle\int \frac{\sqrt{u^2 - a^2}}{u} \, du = \int \frac{a \tan \theta}{a \sec \theta} \cdot a \sec \theta \tan \theta \, d\theta$

$$= \int a \tan^2 \theta \, d\theta = a \int (\sec^2 \theta - 1) \, d\theta$$

$$= a (\tan \theta - \theta) + C$$

$$= \sqrt{u^2 - a^2} - a \operatorname{arcsec} (u/a) + C.$$

A hyperbolic substitution is sometimes more effective:

EXAMPLE: If we let $u = a \sinh v$, then

$$\int \sqrt{a^2 + u^2} \, du = \int a^2 \cosh^2 v \, dv = \frac{a^2}{2} \int (1 + \cosh 2v) \, dv$$

$$= \frac{a^2}{2} (v + \sinh v \cosh v) + C$$

$$= \frac{a^2}{2} \operatorname{argsinh} \left(\frac{u}{a}\right) + u\sqrt{a^2 + u^2} + C.$$

(4) **Integrals involving quadratic functions.** Complete the square in the quadratic function and introduce a simple change of variable to reduce the quadratic to one of the forms $a^2 - u^2$, $a^2 + u^2$, or $u^2 - a^2$.

EXAMPLE:

$$\int \frac{(2x - 3) \, dx}{x^2 + 2x + 2} = \int \frac{(2x - 3) \, dx}{(x + 1)^2 + 1}.$$

Let $u = x + 1$. Then $x = u - 1$, $dx = du$, and

$$\int \frac{(2x - 3) \, dx}{(x + 1)^2 + 1} = \int \frac{(2u - 5) \, du}{u^2 + 1} = \ln (u^2 + 1) - 5 \arctan u + C$$

$$= \ln (x^2 + 2x + 2) - 5 \arctan (x + 1) + C.$$

(5) **Integration by parts:** $\int u \, dv = uv - \int v \, du.$

EXAMPLE: If we let $u = x$ and $v = e^x$, then the formula for integration by parts gives

$$\int xe^x \, dx = \int u \, dv = uv - \int v \, du = xe^x - \int e^x \, dx = xe^x - e^x + C.$$

(6) **Integration of rational functions** (quotients of polynomials). If the degree of the numerator $\geq$ that of the denominator, divide out, thus expressing the given function as a polynomial plus a "proper fraction." Each proper fraction can be expressed as a sum of simpler "proper partial fractions":

$$\frac{P(x)}{Q(x)} = \frac{P_1(x)}{Q_1(x)} + \cdots + \frac{P_n(x)}{Q_n(x)},$$

in which no two Q_i have common factors and each Q_i is of the form $(x - a)^k$ or $(ax^2 + bx + c)^k$; of course $Q = Q_1 \cdot Q_2 \cdots Q_n$. Each of these fractions can be expressed uniquely in terms of still simpler fractions as follows:

$$\frac{P_i(x)}{(x - a)^k} = \frac{A_1}{x - a} + \frac{A_2}{(x - a)^2} + \cdots + \frac{A_k}{(x - a)^k},$$

$$\frac{P_i(x)}{(ax^2 + bx + c)^k} = \frac{A_1 x + B_1}{ax^2 + bx + c} + \frac{A_2 x + B_2}{(ax^2 + bx + c)^2} + \cdots$$
$$+ \frac{A_k x + B_k}{(ax^2 + bx + c)^k}.$$

Each of these simplest fractions can be integrated by methods already described. The constants are obtained by multiplying up the denominators and either equating coefficients of like powers of x in the resulting polynomials or by substituting a sufficient number of values of x to determine the coefficients.

EXAMPLE 1: Integrate

$$\int \frac{x^2 + 2x + 3}{x(x - 1)(x + 1)} \, dx.$$

According to the results above, there exist constants A, B, and C such that

$$\frac{x^2 + 2x + 3}{x(x - 1)(x + 1)} = \frac{A}{x} + \frac{B}{x - 1} + \frac{C}{x + 1}.$$

Multiplying up, we see that we must have

$$A(x - 1)(x + 1) + Bx(x + 1) + Cx(x - 1) \equiv x^2 + 2x + 3.$$

The constants are most easily found by substituting $x = 0, 1$, and -1 in turn in this identity, yielding $A = -3$, $B = 3$, $C = 1$.

EXAMPLE 2: Integrate

$$\int \frac{3x^2 + x - 2}{(x - 1)(x^2 + 1)}.$$

There are constants A, B, and C such that

$$\frac{3x^2 + x - 2}{(x - 1)(x^2 + 1)} = \frac{A}{x - 1} + \frac{Bx + C}{x^2 + 1},$$

or

$$A(x^2 + 1) + Bx(x - 1) + C(x - 1) \equiv 3x^2 + x - 2.$$

Setting $x = 1, 0$, and -1 in turn, we find that $A = 1$, $C = 3$, $B = 2$.

EXAMPLE 3: Show how to break up the fraction

$$\frac{2x^6 - 3x^5 + x^4 - 4x^3 + 2x^2 - x + 1}{(x - 2)^3 (x^2 + 2x + 2)^2} \equiv \frac{P(x)}{Q(x)}$$

into simplest partial fractions. Do not determine the constants.

Solution. $\dfrac{P(x)}{Q(x)} = \dfrac{P_1(x)}{(x-2)^3} + \dfrac{P_2(x)}{(x^2+2x+2)^2}$

$$= \dfrac{A_1}{x-2} + \dfrac{A_2}{(x-2)^2} + \dfrac{A_3}{(x-2)^3} + \dfrac{A_4 x + A_5}{x^2 + 2x + 2}$$

$$+ \dfrac{A_6 x + A_7}{(x^2 + 2x + 2)^2}.$$

(7) Three rationalizing substitutions.

(a) If the integrand contains a single irrational expression of the form $(ax + b)^{p/q}$, the substitution $z = (ax + b)^{1/q}$ will convert the integral into that of a rational function of z.

EXAMPLE: If we let $z = (x + 1)^{1/3}$ so that $x = z^3 - 1$ and $dx = 3z^2\, dz$, then

$$\int \frac{\sqrt[3]{x+1}}{x}\, dx = \int \frac{z}{z^3 - 1} \cdot 3z^2\, dz$$

$$= \int 3\, dz + \int \frac{3\, dz}{(z-1)(z^2 + z + 1)}.$$

(b) If a single irrational expression of the form $\sqrt{a^2 - x^2}$, $\sqrt{a^2 + x^2}$ or $\sqrt{x^2 - a^2}$ occurs with an odd power of x outside, the substitution $z = \sqrt{a^2 - x^2}$ (or etc.) reduces the given integral to that of a rational function.

EXAMPLE. If we let $z = \sqrt{a^2 - x^2}$, then

$$x^2 = a^2 - z^2, \qquad x\, dx = -z\, dz,$$

$$\int \frac{\sqrt{a^2 - x^2}}{x^3}\, dx = -\int \frac{z^2\, dz}{(z^2 - a^2)^2}.$$

(c) In case a given integrand is a rational function of trigonometric functions, the substitution

$$t = \tan\,(\theta/2), \qquad \theta = 2\arctan t, \qquad d\theta = \frac{2\, dt}{1 + t^2},$$

$$\cos\theta = \frac{1 - t^2}{1 + t^2}, \qquad \sin\theta = \frac{2t}{1 + t^2},$$

reduces the integral to one of a rational function of t.

EXAMPLE: Making these substitutions, we obtain

$$\int \frac{d\theta}{5 - 4\cos\theta} = \int \frac{(2\, dt)/(1 + t^2)}{5 - 4[(1 - t^2)/(1 + t^2)]} = \int \frac{2\, dt}{1 + 9t^2}.$$

A Short Table of Integrals

The constant of integration is omitted.

Elementary formulas

1. $\int u^n \, du = \dfrac{u^{n+1}}{n+1}, \qquad n \neq -1$

2. $\int \dfrac{du}{u} = \ln |u|$

3. $\int e^u \, du = e^u$

4. $\int a^u \, du = \dfrac{a^u}{\ln a}, \qquad a > 0, \quad a \neq 1$

5. $\int \sin u \, du = -\cos u$

6. $\int \cos u \, du = \sin u$

7. $\int \sec^2 u \, du = \tan u$

8. $\int \csc^2 u \, du = -\cot u$

9. $\int \sec u \tan u \, du = \sec u$

10. $\int \csc u \cot u \, du = -\csc u$

11. $\int \sinh u \, du = \cosh u$

12. $\int \cosh u \, du = \sinh u$

13. $\int \operatorname{sech}^2 u \, du = \tanh u$

14. $\int \operatorname{csch}^2 u \, du = -\coth u$

15. $\int \operatorname{sech} u \tanh u \, du = -\operatorname{sech} u$

16. $\int \operatorname{csch} u \coth u \, du = -\operatorname{csch} u$

17. $\int \dfrac{du}{\sqrt{a^2 - u^2}} = \arcsin \dfrac{u}{a}, \qquad a > |u|$

18. $\int \dfrac{du}{a^2 + u^2} = \dfrac{1}{a} \arctan \dfrac{u}{a}, \qquad a \neq 0$

19. $\int \dfrac{du}{u\sqrt{u^2 - a^2}} = \dfrac{1}{a} \operatorname{arcsec} \dfrac{u}{a}, \qquad |u| > a$

20. $\int \dfrac{du}{\sqrt{a^2 + u^2}} = \begin{cases} \operatorname{argsinh} \dfrac{u}{a} \\[2mm] \ln(u + \sqrt{a^2 + u^2}) \end{cases}$

21. $\int \dfrac{du}{a^2 - u^2} = \begin{cases} \dfrac{1}{a} \operatorname{argtanh} \dfrac{u}{a}, & |u| < a \\[3mm] \dfrac{1}{a} \operatorname{argcoth} \dfrac{u}{a}, & |u| > a \end{cases}$

22. $\int \dfrac{du}{u\sqrt{a^2 - u^2}} = -\dfrac{1}{a} \operatorname{argsech} \dfrac{u}{a}, \qquad 0 < u < a$

23. $\displaystyle\int \frac{du}{|u|\sqrt{u^2 + a^2}} = -\frac{1}{a}\operatorname{argcsch}\frac{u}{a}, \qquad u \neq 0$

24. $\displaystyle\int \frac{du}{\sqrt{u^2 - a^2}} = \begin{cases} \operatorname{argcosh}\dfrac{u}{a} \\[2mm] \ln|u + \sqrt{u^2 - a^2}| \end{cases} \qquad |u| > a > 0$

Algebraic forms

25. $\displaystyle\int \frac{u\,du}{a + bu} = \frac{u}{b} - \frac{a}{b^2}\ln(a + bu)$
26. $\displaystyle\int \frac{du}{u(a + bu)} = \frac{1}{a}\ln\left|\frac{u}{a + bu}\right|$

27. $\displaystyle\int \frac{u\,du}{(a + bu)^2} = \frac{a}{b^2}\left(\frac{1}{a + bu} + \frac{1}{a}\ln|a + bu|\right)$

28. $\displaystyle\int \frac{du}{u(a + bu)^2} = \frac{1}{a(a + bu)} + \frac{1}{a^2}\ln\left|\frac{u}{a + bu}\right|$

29. $\displaystyle\int \frac{du}{u\sqrt{a + bu}} = \frac{1}{\sqrt{a}}\ln\left|\frac{\sqrt{a + bu} - \sqrt{a}}{\sqrt{a + bu} + \sqrt{a}}\right|$

30. $\displaystyle\int u\sqrt{a + bu}\,du = \frac{2(3bu - 2a)\sqrt{(a + bu)^3}}{15b^2}$

31. $\displaystyle\int \frac{\sqrt{a + bu}}{u}\,du = 2\sqrt{a + bu} + a\int \frac{du}{u\sqrt{a + bu}}$

32. $\displaystyle\int \frac{u\,du}{\sqrt{a + bu}} = \frac{2(bu - 2a)}{3b^2}\sqrt{a + bu}$

33. $\displaystyle\int \sqrt{a^2 - u^2}\,du = \frac{u}{2}\sqrt{a^2 - u^2} + \frac{a^2}{2}\operatorname{Arcsin}\frac{u}{a}, \qquad |u| < a$

34. $\displaystyle\int \sqrt{u^2 \pm a^2}\,du = \frac{u}{2}\sqrt{u^2 \pm a^2} \pm \frac{a^2}{2}\ln\left|u + \sqrt{u^2 \pm a^2}\right|$

35. $\displaystyle\int \frac{\sqrt{a^2 \pm u^2}}{u}\,du = \sqrt{a^2 \pm u^2} - a\ln\left|\frac{a + \sqrt{a^2 \pm u^2}}{u}\right|$

36. $\displaystyle\int \frac{\sqrt{u^2 - a^2}}{u}\,du = \sqrt{u^2 - a^2} - a\operatorname{Arccos}\frac{a}{u}, \qquad 0 < a < |u|$

Trigonometric forms

37. $\displaystyle\int \tan u\,du = -\ln|\cos u|$
38. $\displaystyle\int \sec u\,du = \ln|\sec u + \tan u|$

39. $\displaystyle\int \sin^2 u\,du = \tfrac{1}{2}u - \tfrac{1}{4}\sin 2u$

40. $\displaystyle\int \sin^n u \, du = -\frac{\sin^{n-1} u \cos u}{n} + \frac{n-1}{n}\int \sin^{n-2} u \, du$

41. $\displaystyle\int \cos^n u \, du = \frac{\cos^{n-1} u \sin u}{n} + \frac{n-1}{n}\int \cos^{n-2} u \, du$

42. $\displaystyle\int \frac{du}{\sin^n u} = -\frac{\cos u}{(n-1)\sin^{n-1} u} + \frac{n-2}{n-1}\int \frac{du}{\sin^{n-2} u}, \qquad n \neq 1$

43. $\displaystyle\int \sin mu \sin nu \, du = \frac{\sin(m-n)u}{2(m-n)} - \frac{\sin(m+n)u}{2(m+n)}, \qquad m \neq \pm n$

44. $\displaystyle\int \cos mu \cos nu \, du = \frac{\sin(m-n)u}{2(m-n)} + \frac{\sin(m+n)u}{2(m+n)}, \qquad m \neq \pm n$

45. $\displaystyle\int \sin mu \cos nu \, du = -\frac{\cos(m-n)u}{2(m-n)} - \frac{\cos(m+n)u}{2(m+n)}, \qquad m \neq \pm n$

46. $\displaystyle\int u^n \sin u \, du = -u^n \cos u + n\int u^{n-1} \cos u \, du$

47. $\displaystyle\int u^n \cos u \, du = u^n \sin u - n\int u^{n-1} \sin u \, du$

48. $\displaystyle\int \operatorname{Arcsin} u \, du = u \operatorname{Arcsin} u + \sqrt{1-u^2}$

49. $\displaystyle\int \operatorname{Arccos} u \, du = u \operatorname{Arccos} u - \sqrt{1-u^2}$

50. $\displaystyle\int u \operatorname{Arcsin} u \, du = \tfrac{1}{4}\left[(2u^2 - 1)\operatorname{Arcsin} u + u\sqrt{1-u^2}\right]$

Logarithmic and exponential forms

51. $\displaystyle\int \ln u \, du = u(\ln|u| - 1)$

52. $\displaystyle\int (\ln u)^2 \, du = u(\ln u)^2 - 2u \ln|u| + 2u$

53. $\displaystyle\int u^n \ln u \, du = \frac{u^{n+1}}{n+1}\ln|u| - \frac{u^{n+1}}{(n+1)^2}, \qquad n \neq -1$

54. $\displaystyle\int u^n e^u \, du = u^n e^u - n\int u^{n-1} e^u \, du$

55. $\displaystyle\int e^{au} \sin bu \, du = \frac{e^{au}(a \sin bu - b \cos bu)}{a^2 + b^2}$

56. $\displaystyle\int e^{au} \cos bu \, du = \frac{e^{au}(a \cos bu + b \sin bu)}{a^2 + b^2}$

Answers to Odd-Numbered Exercises

CHAPTER 1

SECTION 1

1. $\overline{AB} = \sqrt{30}, \overline{AC} = \sqrt{21}, \overline{BC} = \sqrt{19}$
3. $\overline{AB} = \sqrt{17}, \overline{AC} = \sqrt{18}, \overline{BC} = \sqrt{35}$; right triangle
5. $\overline{AB} = \sqrt{21}, \overline{AC} = \sqrt{51}, \overline{BC} = \sqrt{126}$
7. $(3, \frac{5}{2}, 2)$ 9. $(-1, 2, -\frac{1}{2})$
11. $\frac{1}{2}\sqrt{61}, \frac{1}{2}\sqrt{61}, \frac{1}{2}\sqrt{10}$ 13. $\frac{1}{2}\sqrt{74}, \frac{1}{2}\sqrt{74}, \frac{1}{2}\sqrt{26}$
15. $P_2(4, 5, -6), Q(1, 3, 0)$ 17. Not on line
19. Not on line
21. Straight line parallel to y-axis   27. $6x + 4y - 8z - 5 = 0$; plane

SECTION 2

1. $1, -5, 2; 1/\sqrt{30}, -5/\sqrt{30}, 2/\sqrt{30}$ 3. $7, 1, 2; 7/\sqrt{54}, 1/\sqrt{54}, 2/\sqrt{54}$
5. $P_2(5, 7, 6)$ 7. $P_2(0, 4, 2)$
9. Yes   11. No   13. Yes   15. No
17. Yes   19. $\sqrt{2}/3$ 21. $41/3\sqrt{190}$ 23. $4\sqrt{10}$

SECTION 3

1. $\dfrac{x-2}{3} = \dfrac{y-3}{6} = \dfrac{z-4}{2}$; $(5, 9, 6)$; $(-1, -3, 2)$

3. $\dfrac{x-1}{-2} = \dfrac{y-2}{3} = \dfrac{z+1}{-3}$; $(-1, 5, -4)$; $(5, -4, 5)$

5. $\dfrac{x-1}{2} = \dfrac{y}{1} = \dfrac{z+1}{-3}$ 7. $\dfrac{x-4}{2} = \dfrac{y}{-1} = \dfrac{z}{-3}$

9. $\dfrac{x-3}{2} = \dfrac{y+1}{0} = \dfrac{z+2}{0}$ 11. Perpendicular

13. Not perpendicular

15. $\dfrac{x-4}{1} = \dfrac{y}{-2} = \dfrac{z-2}{0}; \dfrac{x-3}{0} = \dfrac{y-1}{-1} = \dfrac{z-4}{2}; \dfrac{x-2}{1} = \dfrac{y-5}{-3} = \dfrac{z}{2}$

17. $(-\frac{3}{7}, -\frac{13}{7}, 0)$; $(-\frac{5}{3}, 0, \frac{13}{3})$; $(0, -\frac{5}{2}, -\frac{3}{2})$

19. $x = 3 - t, y = 1 + 3t, z = 5 + t$

21. $A'B'$: $\dfrac{x-2}{5} = \dfrac{y}{0} = \dfrac{z-6}{2}$; $A'C'$: $\dfrac{x-2}{3} = \dfrac{y}{0} = \dfrac{z-6}{1}$;

$B'C'$: $\dfrac{x+3}{8} = \dfrac{y}{0} = \dfrac{z-4}{3}$

751

SECTION 4

1. $4x + 2y - z = 4$
3. $4x - z = 7$
5. $2x - z = 1$
7. $9x + y - 5z = 16$
9. $2x + 3y - 4z + 11 = 0$
11. $3x - 2z = 10$

13. $\dfrac{x + 2}{2} = \dfrac{y - 3}{3} = \dfrac{z - 1}{1}$
15. $\dfrac{x + 1}{1} = \dfrac{y}{0} = \dfrac{z + 2}{2}$

17. $3x + 2y - z = 0$
19. $x - 2y - 3z + 5 = 0$

21. $\dfrac{x - 2}{3} = \dfrac{y + 1}{-2} = \dfrac{z - 3}{4}$
23. $\dfrac{x - 1}{2} = \dfrac{y + 2}{-1} = \dfrac{z}{4}$

25. $2x - 2y - z = 4$
27. $2x - 3y - 5z = 7$
29. $2x - 2y - z = 6$
33. $aA + bB + cC = 0$

SECTION 5

1. $\frac{8}{21}$
3. $\sqrt{14/17}$
5. $x = 11 + 5t, y = -19 - 8t, z = -t$
7. $x = -1 - 2t, y = t, z = 3$
9. $(3, 2, -1)$
11. $(-\frac{3}{5}, 0, -\frac{6}{5})$
13. 1
15. $8/\sqrt{29}$
17. $10x - 17y + z + 25 = 0$
19. $14x + 8y - 13z + 15 = 0$
21. $y + z = 1$
23. $(\frac{91}{57}, \frac{31}{57}, \frac{25}{57})$
25. No intersection
27. $x = 3 + 4t, y = -1 + 5t, z = 2 - t$
29. $x = 29t, y = 2 + t, z = 4 - 22t$

SECTION 6

1. $x^2 + y^2 + z^2 - 4x - 2z - 11 = 0$
3. $x^2 + y^2 + z^2 - 6x + 4y - 12z = 0$
5. Sphere; $C(-1, 0, 2), r = 2$
7. No locus
9. Sphere; $C(3, -2, -1), r = 2$
11. $x^2 + y^2 + z^2 + 5x + 7y - 12z + 41 = 0$
13. Plane
15. Plane
17. Parabolic cylinder
19. Elliptic cylinder
21. Circular cylinder
23. Circular cylinder
25. Circle
27. None

SECTION 7

1. Ellipsoid
3. Ellipsoid (oblate)
5. Elliptic hyperboloid of two sheets
7. Elliptic hyperboloid of one sheet
9. Elliptic paraboloid; axis: x-axis
11. Circular paraboloid; axis: y-axis
13. Circular cone about y-axis
15. Elliptic hyperboloid of one sheet
17. Hyperbolic paraboloid

SECTION 8

1. Sphere
3. Elliptic hyperboloid of one sheet
5. Elliptic paraboloid
7. Hyperbolic paraboloid
9. Circular cone
11. Hyperbolic paraboloid
13. Ellipsoid
15. Elliptic hyperboloid of one sheet

SECTION 9

1. (a) $\left(3\sqrt{2}, \frac{\pi}{4}, 7\right)$; (b) $(4\sqrt{5}, \tan^{-1} 2, 2)$; (c) $(\sqrt{13}, \tan^{-1}(-\frac{3}{2}), 1)$

3. (a) $\left(2\sqrt{3}, \frac{\pi}{4}, \cos^{-1}\left(\frac{1}{\sqrt{3}}\right)\right)$; (b) $\left(2\sqrt{3}, -\frac{\pi}{4}, \cos^{-1}\left(\frac{-1}{\sqrt{3}}\right)\right)$;

(c) $\left(2\sqrt{2}, \frac{2\pi}{3}, \frac{\pi}{4}\right)$

5. (a) $\left(4, \frac{\pi}{3}, 0\right)$; (b) $\left(1, \frac{2\pi}{3}, -\sqrt{3}\right)$; (c) $\left(\frac{7}{2}, \frac{\pi}{2}, 7\sqrt{3}/2\right)$

7. $r^2 + z^2 = 9$ 9. $r^2 = 4z$ 11. $r^2 = z^2$

13. $r^2 \cos 2\theta = 4$ 15. $r = 4 \sin \theta$ 17. $\rho = 4 \cos \phi$

19. $\phi = \pi/4$ 21. $\rho = \pm 2/(1 \mp \cos \phi)$

SECTION 10

1. Vertices of region: $(0, 0, 0)$, $(0, 0, 2)$, $(0, 4, 0)$, $(4, 0, 0)$
3. Vertices of region: $(0, 0, 0)$, $(0, 0, -2)$,$(0, 2, 0)$, $(4, 0, 0)$
5. Infinite pentagonal prism bounded by $z \geq 0$ $(0, 0, 0)$, $(2, 0, 0)$, $(0, 2, 0)$, $(2, \frac{3}{2}, 0)$, $(\frac{5}{3}, 2, 0)$
7. Half infinite strip in the plane $y = 4$ satisfying the inequality $z \geq -2$ and situated above the segment joining $P(5, 4, -2)$ and $Q(3, 4, -2)$.

CHAPTER 2

SECTION 1

1. $-i - 2j + 4k$ 3. $-10j - 7k$

5. $4i - 4k$ 7. $\frac{1}{\sqrt{29}} (3i + 2j - 4k)$

9. $\frac{1}{\sqrt{21}} (2i - 4j - k)$ 11. $B = (3, 3, -4)$

13. $A = (-1, -2, 0)$ 15. $A = (\frac{3}{2}, 0, 3)$, $B = (\frac{5}{2}, -2, 5)$
17. $A = (\frac{7}{4}, -\frac{3}{4}, \frac{7}{2})$, $B = (\frac{3}{4}, \frac{1}{4}, \frac{3}{2})$ 19. $4i - 5j + k$

SECTION 2

1. $-7/2\sqrt{21}$ 3. $-\sqrt{15/28}$ 5. $2/\sqrt{574}$

7. $-\frac{2}{7}$ 9. $-23/\sqrt{50}$ 11. 96

13. 35 15. $\frac{1}{7}(2i - 6j + 3k)$ 17. $\frac{1}{\sqrt{62}} (3i - 2j + 7k)$

19. $k = 2; h = \frac{2}{5}$ 21. $k = 3; h = \frac{42}{145}$ 25. $3g + 5h = 0$
27. $4g - 9h = 0$ 29. $g = -1, h = 1$

SECTION 3

1. $i - 7j + 5k$ 3. $-15i + 5j - 6k$
5. 0 7. $7\sqrt{3}/2; x - y - z = 0$
9. $3\sqrt{35}/2; 11x + 5y + 13z = 30$ 11. $\sqrt{421}/2; 9x + 12y + 14z = 32$

13. $107/\sqrt{1038}$

15. $\dfrac{x+1}{2} = \dfrac{y-3}{-11} = \dfrac{z-2}{-7}$

17. $\dfrac{x-1}{1} = \dfrac{y+2}{-1} = \dfrac{z-3}{1}$

19. $\dfrac{x-3}{-2} = \dfrac{y}{1} = \dfrac{z-1}{3}$

21. $\dfrac{x+2}{2} = \dfrac{y-1}{11} = \dfrac{z+1}{8}$

23. $8x + 14y + 13z + 37 = 0$

25. $2x - z + 1 = 0$

27. $5x - 3y - z = 6$

29. $4x - 3y - z + 9 = 0$

31. $x - y - z + 6 = 0$

33. $\dfrac{x-3}{2} = \dfrac{y+2}{1} = \dfrac{z}{2}$

SECTION 4

1. $\mathbf{f}'(t) = 2\mathbf{i} + 2t\mathbf{j} + 3\mathbf{k}; \mathbf{f}''(t) = 2\mathbf{j}$

3. $\mathbf{f}'(t) = 2e^{2t}\mathbf{i} - 2e^{-2t}\mathbf{j}; \mathbf{f}''(t) = 4e^{2t}\mathbf{i} + 4e^{-2t}\mathbf{j}$

5. $\mathbf{f}'(t) = -2(\sin 2t)\mathbf{i} - \dfrac{1}{t^2}\mathbf{j} + 2(\cos 2t)\mathbf{k}$

7. $f'(t) = 9t^2 + 2t + 2$

$\mathbf{f}''(t) = -4(\cos 2t)\mathbf{i} + \dfrac{2}{t^3}\mathbf{j} - 4(\sin 2t)\mathbf{k}$

9. $f'(t) = 1$

11. 14

13. $\tfrac{3}{8}\pi\sqrt{\pi^2 + 8} + \ln(\pi + \sqrt{\pi^2 + 8}) - \tfrac{3}{2}\ln 2$

15. $\mathbf{r}'(t) = (\sin t + t\cos t)\mathbf{i} + (\cos t - t\sin t)\mathbf{j} + \mathbf{k}; |\mathbf{r}'(t)| = \sqrt{2 + t^2};$

$\mathbf{r}''(t) = (2\cos t - t\sin t)\mathbf{i} - (2\sin t + t\cos t)\mathbf{j}$

SECTION 5

1. $(3t^2\mathbf{i} - \mathbf{j} + 2\mathbf{k})/\sqrt{9t^4 + 5}$

3. $(-e^{-2t}\mathbf{i} + e^{2t}\mathbf{j} + t\mathbf{k})/\sqrt{e^{-4t} + e^{4t} + t^2}$

5. $\dfrac{[(\cos t + \sin t)\mathbf{i} + e^t(2\cos t - \sin t)\mathbf{j} - e^{-2t}\mathbf{k}]}{[(1 + 2\cos t \sin t) + e^{2t}(2\cos t - \sin t)^2 + e^{-4t}]^{1/2}}$

7. $\mathbf{T} = \dfrac{1}{\sqrt{6}}(\mathbf{i} - \mathbf{j} + 2\mathbf{k}); \mathbf{B} = \mathbf{N} = 0; \kappa = 0; \dfrac{x-4}{1} = \dfrac{y}{-1} = \dfrac{z-10}{2};$

osculating plane not defined

9. $\mathbf{T} = \dfrac{1}{\sqrt{3}}(\mathbf{i} + \mathbf{j} + \mathbf{k}); \mathbf{N} = \dfrac{1}{\sqrt{2}}(-\mathbf{i} + \mathbf{j}); \mathbf{B} = (-\mathbf{i} - \mathbf{j} + 2\mathbf{k}); \kappa = \sqrt{2}/3;$

$\dfrac{x-1}{1} = \dfrac{y}{1} = \dfrac{z-1}{1}; x + y - 2z + 1 = 0$

11. $\mathbf{T} = \dfrac{1}{\sqrt{5}}(\mathbf{j} + 2\mathbf{k}); \mathbf{N} = \mathbf{i}; \mathbf{B} = \dfrac{1}{\sqrt{5}}(2\mathbf{j} - \mathbf{k}); \kappa = \tfrac{1}{10}; \dfrac{x-2}{0} = \dfrac{y}{1} = \dfrac{z}{2};$

$2y - z = 0$

13. $\mathbf{v} = \mathbf{i} + 6\mathbf{j} + 18\mathbf{k}; \mathbf{T} = \mathbf{v}/19; \mathbf{a} = 3\mathbf{j} + 18\mathbf{k}; \mathbf{N} = -\tfrac{1}{19}(6\mathbf{i} + 17\mathbf{j} - 6\mathbf{k});$

$a_T = 18; a_N = 3; R = \tfrac{361}{3}$

15. $\mathbf{v} = \mathbf{i} + 2\mathbf{j} - 2\mathbf{k}; \mathbf{T} = \mathbf{v}/3$

$\mathbf{a} = 2\mathbf{i} + 4\mathbf{k}; \mathbf{N} = (2\mathbf{i} + \mathbf{j} + 2\mathbf{k})/3$

$a_T = -2; a_N = 36; R = \tfrac{1}{4}$

CHAPTER 3

SECTION 1

1. $\frac{3}{4}$ 3. $\frac{3}{2}$ 5. $\frac{2}{3}$ 7. 0   9. 3
11. 4   13. ∞ 15. 0   17. $\ln 3 - \ln 2$
19. 0   21. 0   23. -1 25. 0   27. 0
29. 2   31. 1   33. 0   35. 0   37. 1
39. 1   41. 1   43. e

SECTION 2

1. $\frac{53}{99}$ 3. $515{,}000/1111$ 5. $\frac{3718}{999}$ 7. $\frac{9}{4}$

9. $\frac{1}{4} + \frac{4}{11} + \frac{3}{7} + \frac{8}{17} + \frac{1}{2}$ 11. $e - \dfrac{e^2}{8} + \dfrac{e^3}{27} - \dfrac{e^4}{64} + \dfrac{e^5}{125}$

SECTION 3

1. Convergent   3. Convergent   5. Divergent
7. Convergent   9. Divergent   11. Divergent
13. Convergent   15. Convergent   17. Divergent
19. Convergent   21. Convergent   23. Convergent
25. Convergent   27. Convergent   29. Convergent

SECTION 4

1. Divergent   3. Divergent
5. Absolutely convergent   7. Conditionally convergent
9. Absolutely convergent   11. Absolutely convergent
13. Divergent   15. Absolutely convergent
17. Absolutely convergent   19. Conditionally convergent
21. Absolutely convergent   23. Divergent
25. Conditionally convergent   27. Conditionally convergent

SECTION 5

1. Converges for $-1 < x < 1$ 3. Converges for $-\frac{1}{2} < x < \frac{1}{2}$
5. Converges for $-1 < x < 1$ 7. Converges for $-2 \le x - 1 \le 2$
9. Converges for $-1 \le x + 2 < 1$ 11. Converges for $-\infty < x < \infty$
13. Converges for $-\frac{2}{3} < x \le \frac{2}{3}$ 15. Converges for $-1 < x - 1 < 1$
17. Converges for $-3 \le x + 4 \le 3$ 19. Converges for $-8 < x - 2 < 8$
21. Converges for $-\frac{4}{3} < x < \frac{4}{3}$ 23. Converges for $-1 \le x < 1$
25. Converges for $-1 < x \le 1$ 27. Converges for $-\frac{3}{2} \le x \le \frac{3}{2}$
29. (a) converges for $|x| < 1$ 31. Converges for $|x| < 1$

SECTION 6

1. $\displaystyle\sum_{n=0}^{\infty} \frac{x^n}{n!}$ 3. $\displaystyle\sum_{n=1}^{\infty} \frac{(-1)^{n-1}x^n}{n}$ 5. $\displaystyle\sum_{n=0}^{\infty} (n+1)x^n$

7. $1 + \displaystyle\sum_{n=1}^{\infty} \frac{(\frac{1}{2})(-\frac{1}{2})(-\frac{3}{2})\cdots(-n+\frac{3}{2})x^n}{n!}$

9. $\ln 3 + \displaystyle\sum_{n=1}^{\infty} \frac{(-1)^{n-1}(x-3)^n}{3^n \cdot n}$

11. $\dfrac{1}{2} - \dfrac{\sqrt{3}}{2}\left(x - \dfrac{\pi}{3}\right) - \dfrac{1}{2}\dfrac{\left(x - \dfrac{\pi}{3}\right)^2}{2!} + \dfrac{\sqrt{3}}{2}\dfrac{\left(x - \dfrac{\pi}{3}\right)^3}{3!} + \cdots$

13. $2 + \dfrac{1}{4}(x - 4) + 2\displaystyle\sum_{n=2}^{\infty}(-1)^{n-1}\dfrac{1\cdot 3\cdots(2n-3)(x-4)^n}{2\cdot 4\cdots(2n)\cdot 4^n}$

15. $\cos\left(\dfrac{1}{2}\right) - x\sin\left(\dfrac{1}{2}\right) - \dfrac{x^2}{2!}\cos\left(\dfrac{1}{2}\right) + \dfrac{x^3}{3!}\sin\left(\dfrac{1}{2}\right) + \dfrac{x^4}{4!}\cos\left(\dfrac{1}{2}\right) + \cdots$

17. $1 - x^2 + \dfrac{x^4}{2}$ 19. $1 - x^2 + x^4$

21. $1 + x - \dfrac{x^3}{3} - \dfrac{x^4}{6}$ 23. $x + \dfrac{x^3}{2\cdot 3} + \dfrac{3x^5}{2\cdot 4\cdot 5}$

25. $\frac{1}{2}x^2 + \frac{1}{12}x^4 + \frac{1}{45}x^6$ 27. $1 - \frac{1}{2}x^2 + \frac{5}{24}x^4$

29. $2 + 2\sqrt{3}\left(x - \dfrac{\pi}{3}\right) + 7\left(x - \dfrac{\pi}{3}\right)^2 + \dfrac{23\sqrt{3}}{3}\left(x - \dfrac{\pi}{3}\right)^3$

SECTION 7

1. 0.81873   3. 1.22140   5. 0.87758   7. 0.1823
9. 0.36788   11. 1.01943   13. 0.96905   15. 1.97435
17. 0.95635   19. −0.22314   21. 0.017452   23. 0.99619

SECTION 8

1. $\displaystyle\sum_{n=0}^{\infty}\dfrac{(-1)^n x^{2n}}{(2n+1)!}$

3. $\displaystyle\sum_{n=1}^{\infty}\dfrac{(-1)^{n-1}x^{2n-1}}{(2n)!}$

5. $x + \displaystyle\sum_{n=1}^{\infty}\dfrac{1\cdot 3\cdot 5\cdots(2n-1)}{2\cdot 4\cdot 6\cdots(2n)}\cdot\dfrac{x^{2n+1}}{2n+1}$

7. $\displaystyle\sum_{n=0}^{\infty}\dfrac{x^{2n+1}}{2n+1}$

9. $2\displaystyle\sum_{n=0}^{\infty}\dfrac{x^{2n+1}}{2n+1}$

11. $\displaystyle\sum_{n=0}^{\infty}(-1)^n(n+1)x^{2n+1}$

13. $\dfrac{1}{2} - \dfrac{1}{2}\displaystyle\sum_{n=0}^{\infty}\dfrac{(-1)^n(2x)^{2n}}{(2n)!}$

SECTION 9

1. $1 - \dfrac{3}{2}x + \dfrac{1\cdot 3\cdot 5}{2^2\cdot 2!}x^2 - \dfrac{1\cdot 3\cdot 5\cdot 7}{2^3\cdot 3!}x^3 + \dfrac{1\cdot 3\cdot 5\cdot 7\cdot 9}{2^4\cdot 4!}x^4$

3. $1 - \dfrac{2}{3}x^2 + \dfrac{2\cdot 5}{3^2\cdot 2!}x^4 - \dfrac{2\cdot 5\cdot 8}{3^3\cdot 3!}x^6 + \dfrac{2\cdot 5\cdot 8\cdot 11}{3^4\cdot 4!}x^8$

5. $1 + 7x^3 + 21x^6 + 35x^9 + 35x^{12} + 21x^{15} + 7x^{18} + x^{21}$

7. $\dfrac{1}{3^3} - \dfrac{3}{3^4}x^{1/2} + \dfrac{3\cdot 4}{3^5\cdot 2!}x - \dfrac{3\cdot 4\cdot 5}{3^6\cdot 3!}x^{3/2} + \dfrac{3\cdot 4\cdot 5\cdot 6}{3^7\cdot 4!}x^2$

9. 0.90452   11. 0.48540   13. 1.31790
15. 0.32939   17. 0.50826   19. 0.69315

SECTION 10

1. $x - x^2 + \dfrac{5}{6} x^3 - \dfrac{5}{6} x^4 + \dfrac{101}{120} x^5$ 3. $1 - x + \dfrac{x^2}{2} - \dfrac{x^3}{2} + \dfrac{13}{24} x^4 - \dfrac{13}{24} x^5$

5. $x - \frac{1}{24}x^3 + \frac{1}{24}x^4 - \frac{71}{1920}x^5$ 7. $x - x^2 + \frac{23}{24}x^3 - \frac{11}{12}x^4 + \frac{563}{640}x^5$

9. $1 + x + x^2 + \frac{2}{3}x^3 + \frac{1}{2}x^4 + \frac{3}{10}x^5$ 11. $1 + x - \frac{1}{3}x^3 + \frac{1}{6}x^4 + \frac{3}{10}x^5$

13. $x - x^2 + \frac{2}{3}x^3 - \frac{2}{3}x^4 + \frac{13}{15}x^5$ 15. $x - \frac{1}{3}x^3 + \frac{1}{5}x^5$

17. $1 + \frac{1}{3}x + \frac{11}{9}x^2 + \frac{41}{81}x^3 + \frac{8}{243}x^4$ 19. $1 + \frac{3}{2}x^2 + \frac{1}{2}x^3 + \frac{3}{8}x^4$

CHAPTER 4

SECTION 1

1. $f_{,1} = 2x + 2y^2 - 2; f_{,2} = 4xy$

3. $f_{,1} = 3x^2y - 6xy^2 + 2y; f_{,2} = x^3 - 6x^2y + 2x$

5. $f_{,1} = x/\sqrt{x^2 + y^2}; f_{,2} = y/\sqrt{x^2 + y^2}$

7. $f_{,1} = 2x/(x^2 + y^2); f_{,2} = 2y/(x^2 + y^2)$

9. $f_{,1} = -y/(x^2 + y^2); f_{,2} = x/(x^2 + y^2)$

11. $f_{,1} = ye^{x^2+y^2}(2x^2 + 1); f_{,2} = xe^{x^2+y^2}(2y^2 + 1)$

13. $f_{,1}(1, 2) = \infty; f_{,2}(1, 2) = \infty$

15. $f_{,1} = \dfrac{1}{2} e^{\sqrt{2}/2}(4 + \sqrt{2}); f_{,2} = \dfrac{\pi}{2} e^{\sqrt{2}/2}$

17. $f_{,1} = 32 \ln 2; f_{,2} = 32(1 + \ln 2)$

19. $f_{,1} = 2xy - 4xz + 3yz + 2z^2; f_{,2} = x^2 + 3xz - 2yz;$
    $f_{,3} = -2x^2 + 3xy - y^2 + 4xz$

21. $f_{,1} = (yz + y \cot xy - 2z \tan 2xz)f(x, y, z)$
    $f_{,2} = (xz + x \cot xy)f(x, y, z)$
    $f_{,3} = (xy - 2x \tan 2xz)f(x, y, z)$

23. $\dfrac{\partial w}{\partial x} = \dfrac{y^2 - x^2}{x(x^2 + y^2)}; \dfrac{\partial w}{\partial y} = \dfrac{x^2 - y^2}{y(x^2 + y^2)}$

25. $\dfrac{\partial w}{\partial y} = \dfrac{1}{x} \cos\left(\dfrac{y}{x}\right) e^{\sin(y/x)}; \dfrac{\partial w}{\partial x} = -\dfrac{y}{x^2} \cos\left(\dfrac{y}{x}\right) e^{\sin(y/x)}$

SECTION 2

1. $\dfrac{\partial w}{\partial x} = \dfrac{1 - 6x}{12w}; \dfrac{\partial w}{\partial y} = -\dfrac{1 + 4y}{12w}$

3. $\dfrac{\partial w}{\partial x} = \dfrac{y - x - w}{x + w}; \dfrac{\partial w}{\partial y} = \dfrac{x - 3y}{w + x}$

5. $\dfrac{\partial w}{\partial r} = \dfrac{w(r^2 + s^2) + 2r \coth rw}{-r(r^2 + s^2) + \operatorname{csch} rw}; \dfrac{\partial w}{\partial s} = \dfrac{2s \cosh rw}{1 - r(r^2 + s^2) \sinh rw}$

7. $\dfrac{\partial w}{\partial x} = \dfrac{2w \tan 2xw - y \cot xy - yw}{x(y - 2 \tan 2xw)}; \dfrac{\partial w}{\partial y} = \dfrac{w + \cot xy}{2 \tan 2xw - y}$

9. $\dfrac{\partial w}{\partial x} = \dfrac{z(y + 2x + w)}{yz + 3w^2 - xz}; \dfrac{\partial w}{\partial y} = \dfrac{z(w - x - z)}{xz - yz - 3w^2}$

$\dfrac{\partial w}{\partial z} = \dfrac{yw - xw - xy - 2yz - x^2}{xz - yz - 3w^2}$

11. $\dfrac{\partial w}{\partial x} = \dfrac{ye^{x(y-w)} + w^2}{y^2 e^{w(y-x)} - xw - 1}$; $\dfrac{\partial w}{\partial y} = \dfrac{xe^{y(x-w)} - yw - 1}{y^2 - (1 + xw)e^{w(x-y)}}$

SECTION 3

1. $\dfrac{\partial f}{\partial s} = 10s$; $\dfrac{\partial f}{\partial t} = 10t$
3. $\dfrac{\partial f}{\partial s} = 4s(s^2 + t^2)$; $\dfrac{\partial f}{\partial t} = 4t(s^2 + t^2)$

5. $\dfrac{\partial f}{\partial s} = \dfrac{y(2y - x)}{(x^2 + y^2)^{3/2}}$; $\dfrac{\partial f}{\partial t} = \dfrac{-y(2x + y)}{(x^2 + y^2)^{3/2}}$

7. $\dfrac{\partial f}{\partial s} = 10x + 13y - 4z$; $\dfrac{\partial f}{\partial t} = -5x + y + 2z$

9. $\dfrac{df}{dr} = 4(2r^3 + 3r^2 - 6r + 1)$

11. $\dfrac{\partial f}{\partial r} = 2r(3u^2 + 4u - 3)$; $\dfrac{\partial f}{\partial s} = -2s(3u^2 + 4u - 3)$; $\dfrac{\partial f}{\partial t} = 2t(3u^2 + 4u - 3)$

13. $\dfrac{\partial z}{\partial r} = 0$; $\dfrac{\partial z}{\partial \theta} = -4$
15. $\dfrac{dw}{dt} = \dfrac{1}{8}\sqrt{2}\pi(4 + \sqrt{2})$

17. $\dfrac{\partial z}{\partial r} = 0$; $\dfrac{\partial z}{\partial \theta} = \dfrac{1}{2}$
19. $\dfrac{\partial w}{\partial s} = 0$; $\dfrac{\partial w}{\partial t} = -16$

SECTION 4

1. $\dfrac{290\pi}{\sqrt{13}}$
3. $\dfrac{25}{R} = \dfrac{T}{15}$

5. $\dfrac{40\pi\sqrt{3} - 63}{252}$
7. $-\dfrac{2\sqrt{10\pi} + 1}{15}$

11. (a) $\begin{cases} \dfrac{\partial z}{\partial r} = \dfrac{\partial z}{\partial x}\cos\theta + \dfrac{\partial z}{\partial y}\sin\theta; \\[2mm] \dfrac{\partial z}{\partial \theta} = r\left[-\dfrac{\partial z}{\partial x}\sin\theta + \dfrac{\partial z}{\partial y}\cos\theta\right] \end{cases}$
17. $1 + 2\sqrt{3} - \dfrac{14 + \sqrt{3}}{\sqrt{79 - 10\sqrt{3}}}$

SECTION 5

1. $d_\theta f = 6\cos\theta + 8\sin\theta$
3. $d_\theta f = \dfrac{-3\cos\theta + 4\sin\theta}{25}$

5. $d_\theta f = \dfrac{\cos\theta - \sqrt{3}\sin\theta}{2}$

7. $d_\theta f = 2\cos\theta + \sin\theta$; max if $\cos\theta = \dfrac{2}{\sqrt{5}}$, $\sin\theta = \dfrac{1}{\sqrt{5}}$

9. $d_\theta f = \dfrac{1}{2}\cos\theta + \dfrac{1}{2}\sqrt{3}\sin\theta$; max if $\theta = \pi/3$

11. $D_\mathbf{a} f = 7\lambda + 4\mu + 2\nu$ where $\mathbf{a} = \lambda\mathbf{i} + \mu\mathbf{j} + \nu\mathbf{k}$

13. $D_\mathbf{a} f = -\lambda$ where $\mathbf{a} = \lambda\mathbf{i} + \mu\mathbf{j} + \nu\mathbf{k}$

15. $\dfrac{11}{3}$

17. $2/\sqrt{6}$

19. $d_\theta T = 400 \cos \theta - 300 \sin \theta$; $\tan \theta = \frac{4}{3}$; slope of curve is $\frac{4}{3}$

21. $\nabla f = -4\mathbf{i} - \frac{4}{5}\mathbf{j}$

23. $(2e^2 + 2e^{-1} - e^{-2})\mathbf{i} + (-3e^2 + e^{-1} + e^{-2})\mathbf{j} - (4e^2 + e^{-2})\mathbf{k}$

25. $D_\mathbf{a} f = \dfrac{1}{\sqrt{14}}[-3e^2 \cos 1 - e^2 \sin 1 + 2e]$; $D_\mathbf{a} f = e\sqrt{1 + e^2}$ is maximum

27. $D_\mathbf{a} f = \dfrac{3 - \cos 1 - 6 \sin 1 + 2 \cos 2}{\sqrt{14}}$;

  $D_\mathbf{a} f = \sqrt{\cos^2 1 + \cos^2 2 + (1 - 2 \sin 1)^2}$ max

SECTION 6

1. $4x - 4y - z - 6 = 0$; $\dfrac{x - 2}{4} = \dfrac{y + 1}{-4} = \dfrac{z - 6}{-1}$

3. $x - 2y + z - 2 = 0$; $\dfrac{x - 2}{-1} = \dfrac{y + 1}{2} = \dfrac{z + 2}{-1}$

5. $ex - z = 0$; $\dfrac{x - 1}{e} = \dfrac{y - (\pi/2)}{0} = \dfrac{z - e}{-1}$

7. $3x - 4y + 25z = 25 (\ln 5 - 1)$; $\dfrac{x + 3}{3} = \dfrac{y - 4}{-4} = \dfrac{z - \ln 5}{25}$

9. $2x + 2y + 3z = 3$; $\dfrac{x - 2}{2} = \dfrac{y - 1}{2} = \dfrac{z + 1}{3}$

11. $4x - 3z = 25$; $\dfrac{x - 4}{4} = \dfrac{y + 2}{0} = \dfrac{z + 3}{-3}$

13. $3x + 6y + 2z = 36$; $\dfrac{x - 4}{3} = \dfrac{y - 1}{6} = \dfrac{z - 9}{2}$

19. $\dfrac{x - 4}{4} = \dfrac{y + 2}{3} = \dfrac{z - 20}{20}$ 21. $\dfrac{x - 4}{3} = \dfrac{y + 3}{4} = \dfrac{z - 16}{24}$

SECTION 7

1. $df = -0.17$; $\Delta f = -0.1689$

3. $df = \dfrac{\pi}{2} (\pi - 5)$; $\Delta f = -\sqrt{3} + \sin \dfrac{13\pi^2}{2}$

5. $df = -0.45$; $\Delta f = -0.456035$ 7. $df = 0.05$; $\Delta f = 0.0499$

9. $df = -0.08$; $\Delta f = -0.080384$ 11. $300x^{-1}h + 700y^{-1}k + 400z^{-1}l$

13. $V \approx 4309.92$ 15. 1.4%

17. 3.3π 19. $1 + \dfrac{5\pi\sqrt{3}}{18} \% \approx 2.5\%$

21. 0.111427

23. $(12t^5 + 15t^4 + 48t^3 - 15t^2 + 12t - 35)\, dt$

25. $4r(3r^2 - s^2 + 6s^{-2})\, dr + 4s(3s^2 - r^2 - 6r^2 s^{-4})\, ds$

SECTION 8

1. $\dfrac{dy}{dx} = \dfrac{2x + 3y + 2}{-3x + 8y + 6}$

3. $\dfrac{dy}{dx} = -\dfrac{2x + (1 + x^2 + y^2)ye^{xy}}{2y + (1 + x^2 + y^2)xe^{xy}}$

5. $\dfrac{dy}{dx} = -\dfrac{y}{x}$

7. $\dfrac{dy}{dx} = -\dfrac{3x(x^2 + y^2)^{3/2} - y}{3y(x^2 + y^2)^{3/2} + x}$

9. $\dfrac{\partial w}{\partial x} = -\dfrac{3x^2 + 6xw + 2}{3x^2 - y^2 + 4yw - 3}$

11. $\dfrac{\partial w}{\partial x} = -\dfrac{yw \cos(xyw) + 2x}{xy \cos(xyw) + 2w}$

13. $\dfrac{\partial w}{\partial y} = \dfrac{2y + 3x - 3}{2w - 4x + 3z}$

15. $\dfrac{dy}{dx} = -\dfrac{2x + 2}{y}$; $\dfrac{dz}{dx} = -2(x + 2)$

17. $\dfrac{dy}{dx} = \dfrac{4z + 2xy}{4z^2 - 3y^2}$; $\dfrac{dz}{dx} = -\dfrac{3y + 2xz}{4z^2 - 3y^2}$

19. $\dfrac{\partial u}{\partial x} = \dfrac{u}{2(u^2 + v^2)}$; $\dfrac{\partial u}{\partial y} = \dfrac{v}{2(u^2 + v^2)}$

$\dfrac{\partial v}{\partial x} = -\dfrac{v}{2(u^2 + v^2)}$; $\dfrac{\partial v}{\partial y} = \dfrac{u}{2(u^2 + v^2)}$

21. $\dfrac{\partial u}{\partial x} = \dfrac{2xv}{u + v}$; $\dfrac{\partial u}{\partial y} = \dfrac{1}{2(u + v)}$; $\dfrac{\partial v}{\partial x} = \dfrac{2xu}{u + v}$; $\dfrac{\partial v}{\partial y} = -\dfrac{1}{2(u + v)}$

23. $\dfrac{\partial u}{\partial x} = 0$; $\dfrac{\partial u}{\partial y} = \dfrac{y}{u}$; $\dfrac{\partial v}{\partial x} = -\dfrac{x}{v}$; $\dfrac{\partial v}{\partial y} = 0$

25. $\dfrac{\partial u}{\partial x} = \dfrac{g_v}{f_u g_v - f_v g_u}$; $\dfrac{\partial u}{\partial y} = \dfrac{-f_v}{f_u g_v - f_v g_u}$; $\dfrac{\partial v}{\partial x} = \dfrac{-g_u}{f_u g_v - f_v g_u}$; $\dfrac{\partial v}{\partial y} = \dfrac{f_u}{f_u g_v - f_v g_u}$

SECTION 9

1. $f_{,1,2} = f_{,2,1} = -2$

3. $-2x + 4y$

5. $e^{rs}[(rs + 1)\sin r \cos s + r \cos r \cos s - s \sin r \sin s - \cos r \sin s]$

7. $(t^2 - s^2)/(s^2 + t^2)^2$

9. $u_{xy} = u_{yx} = -2xy/(x^2 + y^2 + z^2)^2$; $u_{xz} = u_{zx} = -2xz/(x^2 + y^2 + z^2)^2$

11. $u_{xy} = u_{yx} = -3z$; $u_{xz} = u_{zx} = -3y$

13. $u_{xy} = u_{yx} = e^{xy}[xy(x^2 + z^2) + z^2](x^2 + z^2)^{-3/2}$
    $u_{xz} = u_{zx} = ze^{xy}[3x - y(x^2 + z^2)](x^2 + z^2)^{-5/2}$

17. $\dfrac{\partial^2 z}{\partial r^2} = 2 \cos 2s$

19. $z_{rr} = 12r^2 - 24rs - 12s^2$

21. $F_{yy} + 2F_{yz}\dfrac{\partial z}{\partial y} + F_{zz}\left(\dfrac{\partial z}{\partial y}\right)^2 + F_z\dfrac{\partial^2 z}{\partial y^2}$

25. $\dfrac{\partial^2 u}{\partial r \, \partial s} = F_x\dfrac{\partial^2 x}{\partial r \, \partial s} + F_y\dfrac{\partial^2 y}{\partial r \, \partial s} + F_{xx}\dfrac{\partial x}{\partial r}\dfrac{\partial x}{\partial s}$

$+ F_{xy}\left(\dfrac{\partial x}{\partial r}\dfrac{\partial y}{\partial s} + \dfrac{\partial x}{\partial s}\dfrac{\partial y}{\partial r}\right) + F_{yy}\dfrac{\partial y}{\partial r}\dfrac{\partial y}{\partial s}$

SECTION 10

1. $10 + 13(x - 2) + 4(y - 1) + 6(x - 2)^2 + 2(x - 2)(y - 1) + 2(y - 1)^2$
   $+ (x - 2)^3 + (x - 2)(y - 1)^2$

3. $x + y - \frac{1}{6}(x + y)^3$

5. $1 + (x + y) + \frac{1}{2}(x + y)^2 + \frac{1}{6}(x + y)^3$

7. $1 - \frac{1}{2}x^2 - \frac{1}{2}y^2 + \frac{1}{24}x^4 + \frac{1}{4}x^2y^2 + \frac{1}{24}y^4$

9. $3 + 4(x - 1) + 2(y - 1) + z + (x - 1)^2 + z^2$
$$+ 2(x - 1)(y - 1) + (y - 1)z$$

11. $\phi''(0) = 2\lambda^2 + 8\lambda\mu + 2\mu^2$. No.

13. $\displaystyle\sum_{i=0}^{k} \sum_{j=0}^{i} \frac{i!k!}{i!j!(k - i)!(i - j)!} A^i B^{i-i} C^{k-i}$

SECTION 11

1. Rel min at $(2, -1)$ 3. Rel min at $(1, -2)$

5. Rel min at $(\frac{27}{2}, 5)$; saddle point at $(\frac{3}{2}, 1)$

7. No rel max or min; $(1, \frac{2}{3}), (-1, -\frac{4}{3})$ saddle points.

9. Rel max at $\left(\dfrac{\pi}{3} \pm 2n\pi, \dfrac{\pi}{3} \pm 2m\pi\right)$ Rel min at $\left(-\dfrac{\pi}{3} \pm 2n\pi, -\dfrac{\pi}{3} \pm 2m\pi\right)$

Test fails at $(\pi \pm 2n\pi, \pi \pm 2m\pi)$

11. Rel max at $(0, 0)$

13. Test fails; critical points are along the lines $(x, n\pi)$

15. $(-\frac{3}{2}, -\frac{1}{2}, -\frac{1}{4})$ 17. $x = -\frac{9}{31}, y = \frac{55}{62}, z = \frac{18}{31}, t = -\frac{41}{31}$

19. $\dfrac{17}{\sqrt{10}}$ 21. $\frac{1}{2}\sqrt{10}$ 23. $\dfrac{6}{\sqrt{3}}, \dfrac{4}{\sqrt{3}}, \dfrac{8}{\sqrt{3}}$

25. Length = width = 4 times height

27. Take $S: 0 \le \alpha < \pi/2, 0 < x < P\cos\alpha/(1 + \cos\alpha)$. Max A is $P^2(2 - \sqrt{3})/4$ and occurs when $\alpha = \pi/6, x = P(2 - \sqrt{3})/2$. Max of A on the boundary is $P^2/16$, which is less than $P^2(2 - \sqrt{3})/4$.

SECTION 12

1. $\frac{8}{7}$ 3. $d^2/(a^2 + b^2 + c^2)$

5. Min $= \frac{134}{75}$; occurs at $(\frac{16}{15}, \frac{1}{3}, -\frac{11}{15})$

7. $x = \pm\frac{1}{2}(1 - \sqrt{5})\sqrt{50 - 10\sqrt{5}}, y = \mp\sqrt{50 - 10\sqrt{5}}$

9. 2, 2, 1

11. $h = 2\sqrt{5}$; $H = \dfrac{V}{25\pi} - \frac{2}{3}\sqrt{5}$

13. $\frac{17}{23}$; occurs at $x = \frac{10}{23}, y = \frac{1}{23}, z = -\frac{1}{23}, t = \frac{17}{23}$

15. Closest: $(\pm\frac{1}{2}\sqrt{2}, \pm\frac{1}{2}\sqrt{2})$; farthest: $(\pm\sqrt{2}, \mp\sqrt{2})$

17. Max dimensions: length $\frac{100}{3}$, width = height $= \frac{50}{3}$

19. If C is the cost (in dollars) per square ft of material, then length = width = $\frac{1}{3}\sqrt{D/C}$; height $= \frac{1}{2}\sqrt{D/C}$

21. $x = aA/(a + b + c), y = bA/(a + b + c), z = cA/(a + b + c)$

SECTION 13

1. Exact; $(x^4/4) + x^3y + (y^4/4) + C$ 3. Exact; $2xy - \ln x + \ln y + C$

5. Not exact   7. Exact; $e^{x^2}\sin y + C$

9. Exact; $\arctan(y/x) + C$ 11. Not exact

13. Not exact   15. Not exact

17. Exact; $e^x \sin y \cos z + C$

19. (b) $f(x, y, z, t) = x^3 + z^3 - t^3 + y^2 + 2xz - yt + 3x - 2y + 4t + C$

SECTION 15

1. $\frac{7}{2}$ 3. $\frac{231}{64}$ 5. $\pi + 8$ 7. $4 + 2\ln\left(\frac{5}{4}\right)$

9. $\arcsin\left(\frac{4}{5}\right)$ 11. $508\sqrt{3}/11$ 13. $\frac{64}{105}$

15. $\frac{8}{9^4}\left\{(37)^{3/2}\left[\frac{1}{9}(37)^3 - \frac{3}{7}(37)^2 + \frac{3}{5}(37) + \frac{2}{3}\right]\right.$

$$\left. - (10)^{3/2}\left[\frac{1}{9}(10)^3 - \frac{3}{7}(10)^2 + \frac{3}{5}(10) + \frac{2}{3}\right]\right\}$$

17. $\frac{1}{2}(\sin 7 + \sin 1) + 2(\cos 3 - \cos 7)$

19. 24   21. $\frac{1}{2}\ln 17$ 23. 0

SECTION 16

1. $\frac{101}{3}$ 3. $-e + \cos 1$

5. $-\frac{7}{5}$ 7. $\sin 3 - \cos 2 + e^{-6}\cos 6$

9. -9 11. $e^{-6}\cos 1 + \sin 2 - \cos 3$

13. $\frac{1}{2}\ln\frac{a^2 + b^2}{c^2 + d^2}$ 15. $(d^2 + e^2 + f^2)^{-1/2} - (a^2 + b^2 + c^2)^{-1/2}$

CHAPTER 5

SECTION 1

1. 0.969   3. 3.5355   5. 3.8140   7. 3.2148   9. 3.4036

SECTION 2

1. Smallest $= 0$; largest $= 180$ 3. Smallest $= 48$; largest $= 1680$

5. Smallest $= -27\pi\sqrt{2}$; largest $= 27\pi\sqrt{2}$ 7. Smallest $= 9$; largest $= 9\sqrt{10}$

SECTION 3

1. $\frac{423}{4}$ 3. $-\frac{20975}{14}$

5. $\frac{18}{7}\sqrt{3} - \frac{16}{21}\sqrt{2} - \frac{559}{8}$ 7. 0

9. $\frac{1}{4}(2\cos 4 - \cos 8 - 1)$ 11. $40\frac{88}{105}$

13. $\frac{\pi}{4}(\sqrt{3} - 1)$ 15. $\frac{2\pi - 3\sqrt{3}}{48}$

17. $2e^{1/\sqrt{2}} - e\sqrt{2}$ 19. 0

21. $(2\sqrt{2} - 1)/3$ 23. $32\sqrt{2}/15$

25. $\frac{64}{3}$ 27. $\frac{8}{15}$ 29. 8π 31. $\frac{16}{5}$

SECTION 4

1. $\frac{5}{12}$ 3. $\frac{2}{3} + 4\ln\left(\frac{3}{2}\right)$ 5. $\frac{9}{8}(3\pi + 2)$ 7. $\frac{9}{20}$

9. $\frac{2}{5}$ 11. $\frac{207}{10}$ 13. $2a^3$

SECTION 6

1. $4\pi/3$ 3. 2π 5. $64(3\pi - 4)/9$ 7. $\sqrt{2} - 1$

9. $\frac{64}{3}$ 11. 4π 13. $32\pi(8 - 3\sqrt{3})/3$

15. $\frac{64}{9}$ 17. $9\pi/2$ 19. $4\pi\sqrt{3}$ 21. $\pi/4$

23. $(3\pi - 4)a^3/9$ 25. $[\sqrt{2} + \ln(1 + \sqrt{2})]a^3/3$

SECTION 7

1. $I = ma^2/3, R = a/\sqrt{3}$
3. $I = 9m/35; R = 3/\sqrt{35}$
5. $I = 2ma^2/5, R = a\sqrt{2/5}$
7. $I = m(\pi^2 - 4)/2; R = \sqrt{(\pi^2 - 4)/2}$
9. $47m/14$
11. $5m/14$
13. $3ma^2/10$
15. $\pi ma^2/8$

17. $I = \rho\left(\dfrac{\pi}{8} - \dfrac{1}{5}\right), \ m = \rho\left(\dfrac{\pi}{4} - \dfrac{1}{3}\right)$
19. $\bar{x} = \frac{235}{112}, \bar{y} = \frac{25}{32}$

21. $\bar{x} = \frac{5}{9}, \bar{y} = \frac{4}{7}$
23. $\bar{x} = \bar{y} = 5a/8$
25. $\bar{x} = \frac{5}{3}, \bar{y} = 0$
27. $\bar{x} = -2/(8\sqrt{3} - 1), \bar{y} = 0$

SECTION 8

1. $\dfrac{552 - 1600\sqrt{10}}{1215}$
3. $8a^2$
5. $2\pi\sqrt{2}$

7. $9\sqrt{2}$
9. $\pi\sqrt{2}/2$
11. $8(20 - 3\pi)/9$
13. $2a^2(\pi - 2)/3$
15. 16
17. $a^2(\pi + 6\sqrt{3} - 12)/3$

SECTION 9

1. $4\pi b^2 a/3$
3. $64\pi/15$
5. $2\pi a^3/21$
7. $4\pi/9$
9. $269\pi/24$
11. $V = 256\pi\sqrt{2}/105; \bar{x} = 256\sqrt{2}/105\pi$

SECTION 10

1. $\frac{1}{8}$
3. $\frac{109}{1008}$
5. $(16 - 3\pi)/3$
7. $\frac{3}{2}$
9. $\frac{1}{3}$
11. $abc^2/24$
13. $a^4/840$
15. $243\pi/2$
17. $27a^5(2\pi + 3\sqrt{3})/2$

SECTION 11

1. $2\pi a^3\delta(2 - \sqrt{2})/6$
3. $\pi k(b^4 - a^4)$
5. $12\pi ka^5\sqrt{3}/5$
7. $128ka^5(15\pi - 26)/225$
9. $7\pi ka^4/6$
11. $7\pi\delta a^3/6$
13. $2\delta a^3(3\pi + 20 - 16\sqrt{2})/9$
15. $k\pi a^4/4$

SECTION 12

1. $2\delta a^5/3$
3. $ka^6/90$

5. $32\delta\left(\dfrac{8}{27} - \dfrac{1}{5} + \dfrac{1}{21}\right) = \dfrac{4352\delta}{945}$
7. $\dfrac{5\pi\delta}{16}$

9. $4\pi k(b^6 - a^6)/9$
11. $4ka^6/9$
13. $4\pi abc\delta(a^2 + b^2)/15$
15. $4\pi\delta[(c^2 - a^2)^{3/2}(2c^2 + 3a^2) - (b^2 - a^2)^{3/2}(2b^2 + 3a^2)]/15$
17. $\bar{x} = \bar{z} = 2a/5; \bar{y} = a/5$
19. $\bar{y} = \bar{z} = 0; \bar{x} = \frac{8}{7}$
21. $\bar{x} = \bar{y} = 0; \bar{z} = \frac{1}{3}$
23. $\bar{y} = \bar{z} = 0; \bar{x} = 2a$
25. $\bar{x} = \bar{y} = 0; \bar{z} = 3(2 + \sqrt{2})a/16$
27. $\bar{x} = \bar{y} = 0; \bar{z} = 9a/7$

29. $\bar{x} = \bar{y} = 0; \ \bar{z} = \dfrac{4(1591 - 288\sqrt{3})a}{7(391 - 192\sqrt{3})}$
31. $\bar{y} = \bar{z} = 0; \bar{x} = a\sqrt{2}/2$

CHAPTER 6

SECTION 1

1. $(-1, 1, 0)$
3. $(1, -1, 2)$
5. $(2, 1, -1, -2)$
7. $(1, 1, -1, -1)$
9. $(1, 2, -1, 2)$
11. $(2, -1, 0, 3, -2)$

SECTION 2

1. $A = \begin{pmatrix} -4 & -7 \\ 7 & -6 \end{pmatrix}$
   3. $A = \begin{pmatrix} 5 & 1 & 9 \\ 4 & 1 & 7 \end{pmatrix}$

5. $A = \begin{pmatrix} 0 & -2 & 1 \\ -2 & 2 & 0 \\ 0 & 2 & -1 \end{pmatrix}$
   7. $A = \begin{pmatrix} 3 & 2 \\ 1 & 4 \end{pmatrix}$

11. $A = \begin{pmatrix} -4 & 1 & -2 \\ 1 & 3 & -4 \end{pmatrix}$, $B = \begin{pmatrix} 3 & 0 & 1 \\ 0 & -2 & 3 \end{pmatrix}$

13. $A = \begin{pmatrix} 2 & -1 & 1 \\ 1 & 2 & 0 \end{pmatrix}$, $B = \begin{pmatrix} 1 & 1 & 2 \\ -1 & 1 & -2 \end{pmatrix}$

15. $AB = \begin{pmatrix} 1 & -4 \\ 7 & 7 \end{pmatrix}$, $BA = \begin{pmatrix} 0 & -5 \\ 7 & 8 \end{pmatrix}$

17. $AB = \begin{pmatrix} 0 & 1 \\ 0 & 1 \end{pmatrix}$, $BA = \begin{pmatrix} 0 & 1 \\ 0 & 1 \end{pmatrix}$

19. $AB = \begin{pmatrix} 5 & 3 & -4 \\ -6 & 11 & -1 \end{pmatrix}$

21. $AB = \begin{pmatrix} 0 & 0 & 0 \\ 0 & 0 & 0 \\ 0 & 0 & 0 \end{pmatrix}$, $BA = \begin{pmatrix} 0 & 0 & 0 \\ 0 & 0 & 0 \\ 1 & 1 & 0 \end{pmatrix}$

SECTION 3

17. 65   19. 6   21. 72   23. 45

25. $\displaystyle\sum_{j=0}^{4} \sum_{i=0}^{4-j} (3i + 2j)$

SECTION 4

1. -2 3. 24

SECTION 5

1. 14   3. 21   5. -21
7. -69 9. 297   11. -22

SECTION 6

1. $(1, -2, -1)$ 3. $(-\frac{73}{19}, \frac{21}{19}, -\frac{43}{19})$ 5. $(2, -1, 3)$
7. $(1, \frac{2}{3}, -\frac{1}{3})$ 9. $(2, -1, \frac{1}{2}, 1)$

SECTION 7

1. rank $= 2$ 3. rank $= 3$ 5. rank $= 3$ 7. rank $= 4$
9. rank $= 3$ 11. rank $= 4$ 13. rank $= 4$ 15. rank $= 4$

SECTION 8

  1. $(-1, 1, 0)$ 3. $(2, -4, 7)$

  5. $r = 2$; $r^* = 3$; inconsistent

  7. $r = r^* = 3$; $(4 - 3x_3, 2 - x_3, x_3, 5x_3 - 4)$

  9. $(0, 0, 0)$ 11. $(\frac{5}{3}x_4, -x_4, -\frac{2}{3}x_4, x_4)$

 13. $(2x_3, x_3, x_3, 1)$; infinitely many solutions   15. $r = 3$; $r^* = 4$; inconsistent

SECTION 9

  1. $(1, -2, 3)$ 3. $(3, \frac{1}{2}, -\frac{1}{3})$ 5. $(4, -2, 0, 1)$ 7. $(-1, -3, -2)$

CHAPTER 7

SECTION 1

  1. If $\bar{P}: (x_1, \ldots, x_n)$, then

$$x_j = \frac{px_j'' + qx_j'}{p + q}, \qquad j = 1, \ldots, n$$

  3. (a) If the hyperplanes have equations

$$a_{11}x_1 + a_{12}x_2 + a_{13}x_3 + a_{14}x_4 = p_1,$$
$$a_{21}x_1 + a_{22}x_2 + a_{23}x_3 + a_{24}x_4 = p_2,$$
$$a_{31}x_1 + a_{32}x_2 + a_{33}x_3 + a_{34}x_4 = p_3,$$
$$a_{41}x_1 + a_{42}x_2 + a_{43}x_3 + a_{44}x_4 = p_4,$$

       they will have a single point of intersection if and only if the determinant of the coefficients a_{ij} is not zero.

  5. (a) If the (directed) lines have direction cosines $(\lambda_1, \ldots, \lambda_n)$ and $(\mu_1, \ldots, \mu_n)$, then $\cos \theta = \lambda_1\mu_1 + \cdots + \lambda_n\mu_n$.

      (b) A line

$$\frac{x_1 - x_1'}{a_1} = \cdots = \frac{x_n - x_n'}{a_n}$$

      and a hyperplane

$$A_1x_1 + \cdots + A_nx_n + B = 0$$

      are perpendicular if and only if the sequences $(a_1, \ldots, a_n)$ and $(A_1, \ldots, A_n)$ are proportional.

  7. The intersection with $x_n = c$ has the equations: $\begin{cases} x_n = c, \\ x_1^2 + \cdots + x_{n-1}^2 = c^2. \end{cases}$

  9. No. Because, for example, the sum of the polynomials $x^2 + x + 2$ and $-x^2 - 1$ is not of even degree.

 11. Yes   13. No   17. No

 19. $\begin{pmatrix} 1 & 0 \\ 0 & 0 \end{pmatrix}$, $\begin{pmatrix} 0 & 1 \\ 0 & 0 \end{pmatrix}$, $\begin{pmatrix} 0 & 0 \\ 1 & 0 \end{pmatrix}$, $\begin{pmatrix} 0 & 0 \\ 0 & 1 \end{pmatrix}$

 21. Yes   23. Yes

SECTION 2

  1. $\mathbf{v} = 3\mathbf{v}_1 - 2\mathbf{v}_2$, $r = 2$

  3. All linearly independent, $r = 2$

  5. $\mathbf{v} = 2\mathbf{v}_1 + \mathbf{v}_2 - 2\mathbf{v}_3$, $r = 3$

  7. $\mathbf{v}$ not in space, $\mathbf{v}_2 = 2\mathbf{v}_1 - \mathbf{v}_3$, $r = 2$

9. $\mathbf{v}_3 = 2\mathbf{v}_1 - \mathbf{v}_2$, $r = 2$; $\mathbf{v} = c_1\mathbf{v}_1 + c_2\mathbf{v}_2 + c_3\mathbf{v}_3$;
$c_1 = 1 - 2c_3$, $c_2 = -2 + c_3$
11. $\mathbf{v}_2 = 2\mathbf{v}_1$, $\mathbf{v}_1$, $\mathbf{v}_3$, $\mathbf{v}_4$ independent
13. $\mathbf{v}_1$, $\mathbf{v}_2$, $\mathbf{v}_3$ form basis; $\mathbf{v}_4 = -\frac{2}{15}\mathbf{v}_1 + \frac{3}{15}\mathbf{v}_2 - \frac{11}{15}\mathbf{v}_3$
15. $\mathbf{v}_1$, $\mathbf{v}_2$, $\mathbf{v}_3$ form a basis; $\mathbf{v}_4 = 2\mathbf{v}_1 - \mathbf{v}_2 + 3\mathbf{v}_3$; $\mathbf{v}_5 = -\mathbf{v}_1 + 2\mathbf{v}_2 - 2\mathbf{v}_3$
19. One choice is $(1, 2, 0)$, $(0, 2, 1)$
21. One choice is $(-4, 1, -3, 0, 0)$, $(-5, -1, 0, 3, 0)$ and $(1, 0, 0, 0, 1)$
23. One choice is $(\frac{4}{5}, -\frac{3}{5}, 1, 0, 0)$, $(-\frac{3}{5}, -\frac{4}{5}, 0, 1, 0)$ and $(0, 1, 0, 0, 1)$

SECTION 3

1. $(6, -2)$; $y_1 - 3y_2 = 12$; $4y_1^2 + 9y_2^2 = 36$
3. $(1, 3)$; $(-3, 1)$; $y_1^2 + y_2^2 = 2$
5. $-y_1 + 3y_2 + 3 = 0$; $2y_1^2 - 6y_1y_2 + 5y_2^2 = 4$
7. $y_1 + 2y_2 = 0, \frac{5}{2}, 5$; $-2y_1 + y_2 = 0, \frac{5}{2}, 5$
9. $y_1 + y_3 = 0$; $y_1 - 2y_2 + 5y_3 = 0$; $y_1 + 4y_2 - y_3 = 6$
11. $T: y_1 = x_1 + x_2, y_2 = 2x_2$
13. $T: y_1 = x_1 - x_2 + 2x_3, y_2 = 2x_2 - x_3, y_3 = x_3$

SECTION 4

5. If we take
$$A = \begin{pmatrix} a & b \\ c & d \end{pmatrix}, \qquad B = \begin{pmatrix} x & y \\ z & w \end{pmatrix},$$
then
$$\det(A + B) \leq \det A + \det B \Leftrightarrow dx + aw - cy - bz \leq 0$$

7. $\begin{pmatrix} 2 & -1 \\ -1 & 1 \end{pmatrix}$

9. $\frac{1}{5}\begin{pmatrix} 2 & 1 \\ -1 & 2 \end{pmatrix}$

$-$11. $\begin{pmatrix} -1 & 0 & 1 \\ 1 & 1 & 0 \\ 3 & 1 & -1 \end{pmatrix}$

13. $\begin{pmatrix} 34 & 15 & -5 & -4 \\ 4 & 2 & 0 & -1 \\ -11 & -5 & 2 & 1 \\ -7 & -3 & 1 & 1 \end{pmatrix}$

SECTION 5

1. $\frac{5}{6}$

3. $\frac{248}{15}$

5. $\frac{10}{3}$

7. 2

9. $\frac{3771}{2}$

11. 4π

13. 18π

SECTION 6

1. $\xi_1 = \frac{1}{2}(x_1 - x_2)$, $\xi_2 = \frac{1}{2}(x_1 + x_2)$
3. $\xi_1 = \frac{1}{3}(2x_1 + x_2)$, $\xi_2 = \frac{1}{3}(x_1 + 2x_2)$
5. $\mathbf{w}_1 = \mathbf{v}_1 + \mathbf{v}_2$, $\mathbf{w}_2 = \mathbf{v}_1 + 2\mathbf{v}_2$
7. $\xi_1 = \frac{1}{3}(2x_1 + x_2 - x_3)$, $\xi_2 = \frac{1}{3}(x_1 + 2x_2 + x_3)$, $\xi_3 = \frac{1}{3}(x_1 - x_2 + x_3)$

9. $\begin{pmatrix} 2 & 0 \\ 0 & -2 \end{pmatrix}$

11. $\frac{1}{5}\begin{pmatrix} -1 & -12 \\ -2 & 26 \end{pmatrix}$

13. $\begin{pmatrix} -7 & -6 \\ 15 & 11 \end{pmatrix}$

SECTION 7

1. $\dfrac{1}{2\sqrt{19}}$

3. $-\dfrac{10}{\sqrt{570}}$

[*Note:* The answers for problems 5–14 are not unique.]

5. $v_1 = \dfrac{1}{\sqrt{3}} (e_1 - e_2 + e_3)$, $\quad v_2 = \dfrac{1}{\sqrt{6}} (2e_1 + e_2 - e_3)$

7. $v_1 = \frac{1}{3}(2e_1 - 2e_2 + e_3)$, $\quad v_2 = \frac{1}{3}(e_1 + 2e_2 + 2e_3)$

9. $v_1 = \dfrac{1}{\sqrt{15}} (2e_1 - e_2 + e_3 + 3e_4)$, $\quad v_2 = \dfrac{1}{\sqrt{15}} (e_1 - e_2 + 3e_3 - 2e_4)$

11. $v_1 = \frac{1}{2}(1, -1, -1, 1)$, $\quad v_2 = \frac{1}{2}(-1, -1, 1, 1)$, $\quad v_3 = \frac{1}{2}(1, 1, 1, 1)$

13. $v_1 = \dfrac{1}{\sqrt{3}} (1, 1, 0, 0, 1, 0)$, $\quad v_2 = \frac{1}{2}(1, 0, -1, 0, -1, -1)$

17. $g = \frac{28}{15}$, $\quad h = \frac{7}{15}$

SECTION 8

1. $\begin{pmatrix} \frac{1}{2}\sqrt{3} & -\frac{1}{2} \\ \frac{1}{2} & \frac{1}{2}\sqrt{3} \end{pmatrix}$

3. $\begin{pmatrix} \frac{2}{3} & \frac{1}{3} & -\frac{2}{3} \\ -\frac{2}{3} & \frac{2}{3} & -\frac{1}{3} \\ \frac{1}{3} & \frac{2}{3} & \frac{2}{3} \end{pmatrix}$

5. $\dfrac{1}{2} \begin{pmatrix} 1 & -1 & 1 & -1 \\ 1 & 1 & 1 & 1 \\ -1 & 1 & 1 & -1 \\ -1 & -1 & 1 & 1 \end{pmatrix}$

7. $v_1 = \dfrac{1}{\sqrt{14}} (-2, 1, -3, 0)$, $\quad v_2 = \dfrac{1}{2\sqrt{35}} (3, 9, 1, 7)$

9. $2x'^2 + 3y'^2 - z'^2 = 6$

11. $B = \begin{pmatrix} 18 & 0 & 0 \\ 0 & 27 & 0 \\ 0 & 0 & -9 \end{pmatrix}$

13. $\begin{pmatrix} 15 & 0 & 0 \\ 0 & 8 & 0 \\ 0 & 0 & 1 \end{pmatrix}$

15. $\begin{pmatrix} \dfrac{1}{\sqrt{14}} & \dfrac{2}{\sqrt{14}} & \dfrac{3}{\sqrt{14}} \\[2mm] -\dfrac{1}{\sqrt{3}} & -\dfrac{1}{\sqrt{3}} & \dfrac{1}{\sqrt{3}} \\[2mm] \dfrac{5}{\sqrt{42}} & -\dfrac{4}{\sqrt{42}} & \dfrac{1}{\sqrt{42}} \end{pmatrix}$

17. $\begin{pmatrix} \dfrac{1}{\sqrt{7}} & \dfrac{2}{\sqrt{7}} & \dfrac{-1}{\sqrt{7}} & \dfrac{1}{\sqrt{7}} \\[2mm] -\dfrac{1}{\sqrt{3}} & \dfrac{1}{\sqrt{3}} & \dfrac{1}{\sqrt{3}} & 0 \\[2mm] \dfrac{1}{\sqrt{2}} & 0 & \dfrac{1}{\sqrt{2}} & 0 \\[2mm] \dfrac{1}{\sqrt{42}} & \dfrac{2}{\sqrt{42}} & \dfrac{-1}{\sqrt{42}} & \dfrac{-6}{\sqrt{42}} \end{pmatrix}$

CHAPTER 8

SECTION 1

1. Eigenvalues: 3, 2

Eigenvectors: $(-2, 1)$, $(-1, 1)$

$H = \begin{pmatrix} -2 & -1 \\ 1 & 1 \end{pmatrix}$

3. Eigenvalues: 3, 9

Eigenvectors: $(1, -1)$, $(1, -4)$

$H = \begin{pmatrix} 1 & 1 \\ -1 & -4 \end{pmatrix}$

5. Eigenvalues: $1, 2, 3$

$$H = \begin{pmatrix} 1 & 0 & 0 \\ -1 & 1 & 0 \\ -2 & 2 & 1 \end{pmatrix}$$

7. Eigenvalues: $8, 4, -2$

$$H = \begin{pmatrix} 1 & 1 & 1 \\ 0 & -2 & 0 \\ -1 & -3 & -3 \end{pmatrix}$$

SECTION 2

1. $e_1' = \dfrac{1}{\sqrt{21}}(4e_1 - e_2 + 2e_3)$

3. $e_1' = \dfrac{1}{\sqrt{30}}(3e_1 + 4e_2 + 2e_3 + e_4)$

(orthonormal)

$e_2' = \dfrac{1}{\sqrt{105}}(-6e_1 + 2e_2 + e_3 + 8e_4)$

5. $w_1 = -2e_1 + 3e_2 + e_3 + 3e_4$

7. $Q = (x_1')^2 + 4(x_2')^2 - 2(x_3')^2;\quad X = HX';\quad H = \dfrac{1}{3}\begin{pmatrix} 2 & 2 & 1 \\ 1 & -2 & 2 \\ -2 & 1 & 2 \end{pmatrix}$

9. $Q = 4(x_2')^2 + (x_3')^2;\quad X = HX';$

$$H = \begin{pmatrix} 0 & \dfrac{2}{\sqrt{6}} & \dfrac{1}{\sqrt{3}} \\ \dfrac{1}{\sqrt{2}} & \dfrac{-1}{\sqrt{6}} & \dfrac{1}{\sqrt{3}} \\ \dfrac{1}{\sqrt{2}} & \dfrac{1}{\sqrt{6}} & \dfrac{-1}{\sqrt{3}} \end{pmatrix}$$

11. $Q = 3(x_2')^2 - 2(x_3')^2;\quad X = HX';$

$$H = \begin{pmatrix} \dfrac{2}{\sqrt{6}} & \dfrac{1}{\sqrt{3}} & 0 \\ \dfrac{-1}{\sqrt{6}} & \dfrac{1}{\sqrt{3}} & \dfrac{1}{\sqrt{2}} \\ \dfrac{1}{\sqrt{6}} & \dfrac{-1}{\sqrt{3}} & \dfrac{1}{\sqrt{2}} \end{pmatrix}$$

13. $Q = 6(x_1')^2 + 6(x_2')^2 + 2(x_3')^2;\quad X = HX';$

$$H = \begin{pmatrix} \dfrac{1}{\sqrt{2}} & 0 & -\dfrac{1}{\sqrt{2}} \\ 0 & 1 & 0 \\ \dfrac{1}{\sqrt{2}} & 0 & \dfrac{1}{\sqrt{2}} \end{pmatrix}$$

SECTION 3

1. (a) $(3 + i)$,   (b) $\frac{1}{10} + \frac{7}{10}i$,   (c) $-i$,   (d) $-\frac{2}{5} - \frac{4}{5}i$

3. $1 + 4i$; -3; $-6 + 2i$; $\frac{1}{4} + \frac{3}{4}i$

5. $1 - 3i$; $5 + i$; $-8 - 4i$; $-\frac{1}{2} + i$

7. $-12 - i$ 11. No solution

13. No solution   15. No solution

17. (a) $\dfrac{1}{\sqrt{2}}$, $\dfrac{\pi}{4}$;   (b) 1, $\dfrac{\pi}{2}$;   (c) $\dfrac{1}{2}$, $\dfrac{\pi}{3}$;   (d) 4, π

19. (a) Circle: $x^2 + y^2 = 1$;   (b) disc: $x^2 + y^2 < 4$;
    (c) $x^2 + y^2 > 1$ (outside of circle $x^2 + y^2 = 1$).

21. (a) Disc: $(x - 1)^2 + y^2 < 1$;   (b) circle: $(x + 2)^2 + y^2 = 4$, or $|z + 2| = 2$

23. $z = \pm(1 + \sqrt{3}\,i)$ 25. $z = \dfrac{\sqrt{3} + i}{2}, \dfrac{-\sqrt{3} + i}{2}, -i$

27. $z = -2, 1 \pm \sqrt{3}\,i$ 29. $z = \pm\dfrac{1 + i}{\sqrt{2}}, \pm\dfrac{1 - i}{\sqrt{2}}$

31. $z = \cos\dfrac{2k\pi}{5} + i\sin\dfrac{2k\pi}{5}$, $k = 0, 1, 2, 3, 4$

33. $z = \pm 1, \dfrac{1 \pm \sqrt{3}\,i}{2}, \dfrac{-1 \pm \sqrt{3}\,i}{2}$

SECTION 4

1. $-\dfrac{1}{4}\begin{pmatrix} 2 - i & -2 - i \\ -3 + i & 1 + i \end{pmatrix}$

3. $\dfrac{1}{6}\begin{pmatrix} i & 1 - 3i & 1 \\ -(2 + 3i) & 2i & 3 + 2i \\ 1 & 2i & -i \end{pmatrix}$ 5. $\dfrac{1}{2}\begin{pmatrix} 20 - 4i & -7 - 3i & 1 + 2i \\ -8 + 6i & 4 & -1 - i \\ -4i & -1 + i & 1 \end{pmatrix}$

7. v_1, v_2, and v are independent

9. v_1 and v_2 independent; $v = (2 + i)v_1 + (-1 + i)v_2$

11. $\begin{pmatrix} \xi_1 \\ \xi_2 \end{pmatrix} = \dfrac{1}{4}\begin{pmatrix} -1 + 3i & -2i \\ -1 - 3i & 2 \end{pmatrix}\begin{pmatrix} x_1 \\ x_2 \end{pmatrix}$

13. $v_3 = ie_1 + e_2 + (1 - i)e_3$ 15. $v_3 = -2e_1 + e_2 + 2e_3$

17. $B = \dfrac{1}{10}\begin{pmatrix} -12 + 16i & -14 + 22i \\ 11 - 3i & 2 + 4i \end{pmatrix}$

19. $B = \dfrac{1}{10}\begin{pmatrix} 23 - 19i & -17 - 9i & -4 + 2i \\ -19 + 7i & 11 + 17i & 2 - 6i \\ -2 + 26i & 8 + 16i & 6 + 12i \end{pmatrix}$

†21. $w_1 = e_1 + ie_2 - e_3$
     $w_2 = (-3 + i)e_1 + 2e_2 - (3 + i)e_3$

†23. $w_1 = e_1 - (2 + i)e_2$, $w_2 = e_3$, $w_3 = e_4$

---

† A dagger before a problem number indicates that the answer given is not unique.

SECTION 5

1. $k = \dfrac{3(1 + 3i)}{5}$ 3. $-\dfrac{3(5 - 3i)}{17}$ 5. $k = -7(1 + 2i)/5$

†7. $e_1' = \dfrac{1}{\sqrt{2}}(e_1 + ie_2)$

$e_2' = \frac{1}{2}[e_1 - ie_2 + (1 + i)e_3]$

9. $e_1' = \dfrac{1}{2\sqrt{2}}[e_1 + (Hi)e_2 - (2 - i)e_3]$

$e_2' = \dfrac{1}{2\sqrt{3}}[(-1 + i)e_1 + (2 - 2i)e_2 - (1 + i)e_3]$

11. $k_1 = -\dfrac{1 - i}{8}$, $k_2 = \dfrac{3 + 3i}{8}$

SECTION 6

1. $\lambda = 0, 2$; $H = \begin{pmatrix} -i & i \\ 1 & 1 \end{pmatrix}$

3. $\lambda = 4 + i, 2 - i$; $H = \begin{pmatrix} 1 & -i \\ 1 + i & 1 + i \end{pmatrix}$

5. $\lambda = 2 + 3i$; double root; no H

7. $\lambda = 3, 1$; $H = \dfrac{1}{\sqrt{2}}\begin{pmatrix} i & 1 \\ 1 & i \end{pmatrix}$

9. $\lambda = 4, -2$; $H = \dfrac{1}{\sqrt{6}}\begin{pmatrix} 1 & -2 - i \\ 2 - i & 1 \end{pmatrix}$

11. $\lambda = 6, 4, -4$; $H = \dfrac{1}{2}\begin{pmatrix} -\sqrt{2}\,i & i & i \\ \sqrt{2} & 1 & 1 \\ 0 & -1 + i & 1 - i \end{pmatrix}$

CHAPTER 9

SECTION 1

1. Yes 3. Yes 5. Yes
7. Yes 9. Yes 11. Yes
13. No 17. f_n converges uniformly $\Leftrightarrow \beta > 2\alpha$
21. f_n' does not; F_n does 23. f_n' does not; F_n does not

SECTION 2

1. $(0 <)h < 1$ 3. $0 < h < 1$ 5. $0 < h$
7. $0 < h < 2$ 9. $0 < h < 1$ 11. $0 < h \le 1$
13. $h > 0$ 17. $0 < h < h_0, h_0 \ln h_0 = 1$

SECTION 3

1. $f(x) = \displaystyle\sum_{n=0}^{\infty} (n + 1)x^{2n}$ 3. $f(x) = \displaystyle\sum_{n=0}^{\infty} \dfrac{ex^{2n}}{n!}$

5. $f(x) = \displaystyle\sum_{n=0}^{\infty} \frac{(-1)^n (n+1)(n+2)x^{3n}}{2! \cdot 3^{3n}}$

7. $f(x) = \displaystyle\sum_{n=0}^{\infty} \frac{(-1)^n (3x^2)^{2n+1}}{\pi^{2n+1} \cdot (2n+1)!}$

9. $f(x) = \displaystyle\sum_{n=0}^{\infty} \frac{x^{2n}}{(n+1)!}$

SECTION 4

3. Converges   7. Diverges

9. $1 - 2x + \frac{3}{2}x^2 - 3x^3 + \frac{85}{24}x^4 - \frac{49}{12}x^5 + \cdots$

11. $x + \frac{1}{2}x^2 + \frac{1}{3}x^3 + 0 \cdot x^4 + \frac{3}{40}x^5 + \cdots$

13. $1 + x + (\frac{1}{2}x^2 - \frac{1}{2}y^2) + (\frac{1}{6}x^3 - \frac{1}{2}xy^2) + \cdots$

15. $1 - x + (\frac{1}{2}x^2 + \frac{1}{2}y^2) - (\frac{1}{6}x^3 + \frac{1}{2}xy^2) + \cdots$

17. $1 + \cdots$

†19. Error $< (0.2)^5 \left(\dfrac{1}{116} + \dfrac{25}{24^2} + \dfrac{10}{57} + \dfrac{15}{28} + \dfrac{10}{9} + \dfrac{25}{16} \right) < 0.00111$

†21. Error $< (0.2)^5 \left(1 + 1 + \dfrac{1}{2} + \dfrac{1}{6} + \dfrac{1}{24} + \dfrac{1}{120} \right) < 0.00073$

SECTION 5

13. (a) $e(\cos 1 + i \sin 1)$;   (b) $-\frac{1}{2}(\sqrt{3} \cdot \cosh 1 + i \sinh 1)$;
    (c) $\frac{1}{2}(\cosh 2 + i\sqrt{3})$

15. (a) $\ln 4 + \pi i$;   (b) $\frac{1}{2} \ln 2 + \dfrac{\pi i}{4}$;   (c) $- \dfrac{\pi i}{2}$

17. (a) $\frac{1}{2}(1 - \cos 2x \cosh 2y + i \sin 2x \sinh 2y)$

    (b) $\dfrac{\sin 2x + i \sinh 2y}{\cos 2x + \cosh 2y}$

19. (a) $\ln 3 + \frac{1}{2} \ln 2 + \dfrac{\pi i}{4}$

    (b) $e^{\sin (x^2-y^2) \cosh (2xy)} (\cos \theta + i \sin \theta)$, $\theta = \cos (x^2 - y^2) \sinh (2xy)$

21. $1 + \dfrac{1}{2}z + \displaystyle\sum_{n=1}^{\infty} \frac{(-1)^{n-1} 1 \cdot 3 \cdots (2n-3)}{2 \cdot 4 \cdots (2n)} z^n$, $|z| < 1$

CHAPTER 10

SECTION 1

1. $f(x) = \displaystyle\sum_{k=1}^{\infty} \frac{\sin (2k-1)x}{2k-1}$ 3. $f(x) = \dfrac{\pi^2}{3} - 4 \displaystyle\sum_{n=1}^{\infty} \frac{(-1)^{n-1} \cos nx}{n^2}$

5. $f(x) = \dfrac{2}{\pi} \displaystyle\sum_{k=1}^{\infty} \frac{\sin (2k-1)x}{2k-1} - \dfrac{2}{\pi} \displaystyle\sum_{k=1}^{\infty} \frac{\sin (4k-2)x}{2k-1}$

7. $f(x) = \dfrac{1}{2} - \dfrac{2}{\pi} \displaystyle\sum_{k=1}^{\infty} \frac{(-1)^{k-1} \cos (2k-1)x}{2k-1}$

9. $f(x) = \dfrac{2}{\pi} + \dfrac{4}{\pi} \displaystyle\sum_{k=1}^{\infty} \frac{(-1)^{k-1} \cos 2kx}{4k^2 - 1}$

11. $f(x) = \dfrac{2 \sinh \pi}{\pi} \left[\dfrac{1}{2} + \displaystyle\sum_{n=1}^{\infty} \dfrac{(-1)^n (\cos nx - n \sin nx)}{n^2 + 1} \right]$

13. $f(x) = \frac{1}{2} - \frac{1}{2} \cos 2x$

15. $f(x) = -\dfrac{\pi}{4} + 2 \displaystyle\sum_{k=1}^{\infty} \dfrac{\sin (2k - 1)x}{2k - 1} - \dfrac{2}{\pi} \displaystyle\sum_{k=1}^{\infty} \dfrac{\cos (2k - 1)x}{(2k - 1)^2}$

19. $f(x) = 2 \displaystyle\sum_{n=1}^{\infty} \dfrac{(-1)^{n-1} \sin nx}{n} + \dfrac{\pi^2}{3} - 4 \displaystyle\sum_{n=1}^{\infty} \dfrac{(-1)^{n-1} \cos nx}{n^2}$

SECTION 2

1. $f(x) = \dfrac{1}{2} + \dfrac{2}{\pi} \left(\cos x - \dfrac{\cos 3x}{3} + \dfrac{\cos 5x}{5} - \cdots \right)$

3. $f(x) = \dfrac{2}{\pi} - \dfrac{4}{\pi} \displaystyle\sum_{k=1}^{\infty} \dfrac{\cos 2kx}{4k^2 - 1}$

5. $f(x) = \dfrac{\pi}{2} - \dfrac{4}{\pi} \displaystyle\sum_{k=1}^{\infty} \dfrac{\cos (2k - 1)x}{(2k - 1)^2}$

7. $f(x) = \dfrac{\pi^3}{4} + \displaystyle\sum_{k=1}^{\infty} \dfrac{3\pi}{2k^2} \cos 2kx$

$\qquad + \dfrac{6}{\pi} \displaystyle\sum_{k=1}^{\infty} \left[\dfrac{4}{(2k - 1)^4} - \dfrac{\pi^2}{(2k - 1)^2} \right] \cos (2k - 1)x$

9. $f(x) = \dfrac{2}{\pi} \displaystyle\sum_{k=1}^{\infty} \dfrac{\sin (2k - 1)x}{2k - 1} + \dfrac{2}{\pi} \displaystyle\sum_{k=1}^{\infty} \dfrac{\sin (4k - 2)x}{2k - 1}$

11. $f(x) = - \displaystyle\sum_{k=1}^{\infty} \dfrac{8k}{\pi(4k^2 - 1)} \sin 2kx$ 13. $f(x) = 2 \displaystyle\sum_{n=1}^{\infty} \dfrac{(-1)^{n-1} \sin nx}{n}$

15. $f(x) = 2 \displaystyle\sum_{n=1}^{\infty} (-1)^{n-1} \left(\dfrac{\pi^2}{n} - \dfrac{6}{n^3} \right) \sin nx$

SECTION 3

1. $f(x) = \dfrac{4}{\pi} \displaystyle\sum_{k=1}^{\infty} \dfrac{\sin (2k - 1)x}{2k - 1}$

3. $f(x) = \dfrac{2}{\pi} - \dfrac{4}{\pi} \displaystyle\sum_{n=1}^{\infty} \dfrac{\cos nx}{4n^2 - 1}$

5. $f(x) = \dfrac{1}{2} - \dfrac{2}{\pi} \displaystyle\sum_{n=1}^{\infty} \dfrac{(-1)^n}{n} \sin \dfrac{n\pi x}{2} - \dfrac{4}{\pi^2} \displaystyle\sum_{k=1}^{\infty} \dfrac{1}{(2k - 1)^2} \cos \dfrac{(2k - 1)\pi x}{2}$

7. $f(x) = \dfrac{1}{2} + \dfrac{4}{\pi^2} \displaystyle\sum_{k=1}^{\infty} \dfrac{\cos (2k - 1)x}{(2k - 1)^2}$

9. $f(x) = - \dfrac{4}{\pi} \displaystyle\sum_{k=1}^{\infty} \dfrac{\sin k\pi x}{k} - \dfrac{8}{\pi^3} \displaystyle\sum_{k=1}^{\infty} \left[- \dfrac{\pi^2}{2k - 1} + \dfrac{4}{(2k - 1)^3} \right] \sin \dfrac{(2k - 1)\pi x}{2}$

11. $f(x) = \dfrac{2}{\pi} \displaystyle\sum_{k=1}^{\infty} \dfrac{\sin 2k\pi x}{k}$

13. $f(x) = \dfrac{1}{4} + \dfrac{1}{\pi} \displaystyle\sum_{k=1}^{\infty} (-1)^{k-1} \dfrac{\cos (2k-1)x}{2k-1}$

$\qquad + \dfrac{1}{\pi} \displaystyle\sum_{k=1}^{\infty} \dfrac{5 \sin (2k-1)x + \sin (4k-2)x}{2k-1}$

SECTION 4

7. $g(x) = 1 + \frac{4}{3} \sin 3x$

9. $g(x) = \dfrac{\sinh \pi}{\pi} \left(-\dfrac{4}{5} \sin 2x + \dfrac{2}{17} \cos 4x + \dfrac{5}{13} \sin 5x \right)$

11. $x = 2 \displaystyle\sum_{n=1}^{\infty} (-1)^n \dfrac{\sin nx}{n}, \qquad \displaystyle\int_{-1}^{1} x^2 \, dx = \dfrac{2\pi^3}{3}$

13. $|x| - \dfrac{\pi}{2} = -\dfrac{4}{\pi} \displaystyle\sum_{k=1}^{\infty} \dfrac{\cos (2k-1)x}{(2k-1)^2}, \qquad \displaystyle\int_{-\pi}^{\pi} \left(|x| - \dfrac{\pi}{2} \right)^2 = \dfrac{\pi^3}{6}$

15. $g(x) = \dfrac{3}{2} x$ $\qquad\qquad\qquad$ 17. $g(x) = \dfrac{12x}{\pi^2}$

SECTION 5

1. $g_3 = \dfrac{\sqrt{14}}{4} (5x^3 - 3x)$

9. $g_0 = 1, \qquad g_1 = \sqrt{3}\,(2x - 1), \qquad g_2 = \sqrt{5}\,(6x^2 - 6x + 1)$

11. $g_0 = \dfrac{1}{\sqrt{\pi}}, \qquad g_1 = \sqrt{\dfrac{2}{\pi}} \cos x, \qquad g_2 = \sqrt{\dfrac{2\pi}{\pi^2 - 8}} \left(\sin x - \dfrac{2}{\pi} \right)$

SECTION 6

1. $x^2 = \dfrac{\pi^2}{3} + \displaystyle\sum_{\substack{n=-\infty \\ n \neq 0}}^{\infty} \dfrac{(-1)^n}{n^2} e^{inx}$

3. $f(x) = \dfrac{2}{\pi i} \displaystyle\sum_{k=1}^{\infty} \dfrac{1}{2k-1} \left[e^{(2k-1)ix} - e^{-(2k-1)ix} \right]$

5. $e^{ax} = \displaystyle\sum_{n=-\infty}^{\infty} \dfrac{(-1)^n \sinh \pi a}{\pi (a - in)} e^{inx}$

7. $\sinh ax = \dfrac{\sinh \pi a}{\pi} \displaystyle\sum_{n=-\infty}^{\infty} \dfrac{(-1)^n in}{a^2 + n^2} e^{inx}$ $\qquad$ 9. $f(x) = 1 - \displaystyle\sum_{\substack{n=-\infty \\ n \neq 0}}^{\infty} (-1)^n \dfrac{e^{inx}}{n}$

CHAPTER 11

SECTION 1

In problems 1 through 10, $F(x_0, y_0) = 0$.

1. $F_y(x_0, y_0) = 1 \neq 0,\ f'(x_0) = -1$

3. $F_y(x_0, y_0) = 1,\ f'(x_0) = 0$

5. $F_y(x_0, y_0) = 2/3\sqrt{3}$, $f'(x_0) = -\sqrt{3}$

7. $F_y(x_0, y_0) = 2$, $f'(x_0) = -1$

9. $F_y(x_0, y_0) = 82$, $f'(x_0) = -\frac{39}{41}$

In problems 11 through 16, $F(x_0, y_0, z_0) = 0$.

11. $F_z(x_0, y_0, z_0) = 9$, $f_x(x_0, y_0) = f_y(x_0, y_0) = \frac{1}{3}$

13. $F_z(x_0, y_0, z_0) = 1$, $f_x = 2$, $f_y = 0$

15. $F_z = 1$, $f_x = f_y = -1$ 17. (a) No   (b) No

19. $F_y = 0$ at $(2e^{-2}, 2)$, $F_x \neq 0$ there

21. $F_y = 0$ at $(0, 0)$, $F_x(0, 0) = 0$

23. $F_y = 0$ at $(0, 0)$ and $(\pm 2\sqrt{2}, 0)$

$\qquad F_x(0, 0) = 0$, $F_x(\pm 2\sqrt{2}, 0) \neq 0$

25. Cannot solve for z in a full box about $(-1, 2)$. Can solve for y in a box about $x = -1$, $z = -4$.

SECTION 2

In problems 1 through 6, $F(x_0, y_0, z_0) = G(x_0, y_0, z_0) = 0$.

1. $F_y G_z - F_z G_y = 3 \neq 0$ at P_0

Line: $\dfrac{x-1}{1} = \dfrac{y}{-1} = \dfrac{z}{1}$

3. $F_y G_z - F_z G_y = 8 \neq 0$ at P_0

Tan line: $\dfrac{x-2}{2} = \dfrac{y-1}{1} = \dfrac{z+2}{-3}$

5. $F_y G_z - F_z G_y = 12 \neq 0$ at P_0

Tan line: $\dfrac{x-2}{2} = \dfrac{y+1}{1} = \dfrac{z-1}{-3}$

In problems 7 through 10, $F = G = 0$ at P_0.

7. $D_0 = F_u G_v - F_v G_u = 3 \neq 0$

$f_x = -\frac{5}{3}$, $g_x = \frac{1}{3}$, $f_y = \frac{4}{3}$, $g_y = -\frac{5}{3}$

9. $D_0 = 6$ at P_0

$f_x = \frac{2}{3}$, $f_y = \frac{4}{3}$, $g_x = -\frac{5}{3}$, $g_y = \frac{2}{3}$

11. $F_y G_z - F_z G_y \equiv D = -4y - 2z$

$D = F = G = 0 \Leftrightarrow P = (2, -3, 6)$ or $(-10, -15, 30)$

13. $\dfrac{\partial u_1}{\partial x_1} = \dfrac{1}{D} \begin{vmatrix} -3x_1^2 + 12x_2 - 3 & 2 - 12u_2 \\ -4x_1 - 3x_1^2 u_3 & 8u_1 - 4u_2 + 2u_3 \end{vmatrix}$

$\dfrac{\partial u_2}{\partial x_1} = \dfrac{1}{D} \begin{vmatrix} 32u_1 & -3x_1^2 + 12x_2 - 3 \\ -12u_1 + 8u_2 + 8u_1 u_3 & -4x_1 - 3x_1^2 u_3 \end{vmatrix}$

$\dfrac{\partial u_3}{\partial x_1} = \dfrac{1}{6u_3} \left[8u_1 \dfrac{\partial u_1}{\partial x_1} + 2 \dfrac{\partial u_2}{\partial x_1} + 3x_1^2 \right]$

$D = \begin{vmatrix} 32u_1 & 2 - 12u_2 \\ -12u_1 + 8u_2 + 8u_1 u_3 & 8u_1 - 4u_2 + 2u_3 \end{vmatrix}$

17. The result will hold if, also,

$$J \frac{F, G}{x, y} \neq 0,$$

so that the equations define one-to-one transformations from (u, v) to (x, y).

SECTION 3

1. $J\left(\dfrac{u, v}{x, y}\right) = 1$; T^{-1}: $\begin{aligned} x &= u \\ y &= v - u^2 \end{aligned}$

3. $\dfrac{\partial(u, v)}{\partial(x, y)} = 7$; T^{-1}: $\begin{aligned} x &= (2u + 3v)/7 \\ y &= (-u + 2v)/7 \end{aligned}$

5. $J\left(\dfrac{u, v}{x, y}\right) = (1 + x + y)^{-3}$; T^{-1}: $\begin{aligned} x &= u/(1 - u - v) \\ y &= v/(1 - u - v) \end{aligned}$, $u + v < 1$

7. $\dfrac{\partial(u, v)}{\partial(x, y)} = \dfrac{\pi x}{2}$; T^{-1}: $x = \sqrt{u^2 + v^2}$, $y = \dfrac{2}{\pi} \arctan \dfrac{v}{u}$

9. T^{-1}: $x = \dfrac{u}{u^2 + v^2}$, $y = \dfrac{v}{u^2 + v^2}$, $x^2 + y^2 = v^2 \Leftrightarrow u^2 + v^2 = \dfrac{1}{r^2}$

13. (a) $\dfrac{\partial(u, v, w)}{\partial(x, y, z)} = (x^2 + y^2 + z^2)^{-3}$

(b) T^{-1}: $x = \dfrac{u}{u^2 + v^2 + w^2}$, $y = \dfrac{v}{u^2 + v^2 + w^2}$, $z = \dfrac{w}{u^2 + v^2 + w^2}$

$x^2 + y^2 + z^2 = r^2 \Leftrightarrow u^2 + v^2 + w^2 = 1/r^2$

SECTION 4

5. V: $\xi = x - y$, $\eta = x + y$;

TV^{-1}: $\begin{aligned} u &= \xi + \xi^2 - \eta^2 \\ v &= \eta + 2\xi\eta \end{aligned}$

7. V: $\xi = 1 + x - y$, $\eta = -1 + 2x + y$

TV^{-1}: $\begin{aligned} u &= 1 + (\xi - 1) + (\xi - 1)(\eta + 1) \\ v &= -1 + (\eta + 1) + (\xi - 1)^2 + (\eta + 1)^2 \end{aligned}$

9. TV^{-1}: $\begin{aligned} x &= \tfrac{41}{40} + \tfrac{21}{20}(u - 1) + \tfrac{1}{40}[(u - 1)^2 - v^2] \\ y &= \tfrac{21}{20}v + \tfrac{1}{20}(u - 1)v \end{aligned}$

If $|u - 1| \leq \tfrac{1}{2}$ and $|v| \leq \tfrac{1}{2}$, then

$-\tfrac{21}{20} \cdot \tfrac{1}{2} - \tfrac{1}{160} \leq x - \tfrac{41}{40} \leq \tfrac{21}{20} \cdot \tfrac{1}{2} + \tfrac{1}{160}$

$-\tfrac{21}{20} \cdot \tfrac{1}{2} - \tfrac{1}{80} \leq y \leq \tfrac{21}{20} \cdot \tfrac{1}{2} + \tfrac{1}{80}$

SECTION 6

3. $J = u^2 v$

7. 4; T^{-1}: $\begin{aligned} u &= (x + y - y^2)/3 \\ v &= (-x + 2y - y^2)/3 \end{aligned}$ 9. 4; T^{-1}: $\begin{aligned} u &= (x + y + x^2)/3 \\ v &= (-2x + y + x^2)/3 \end{aligned}$

11. $-\tfrac{1}{105}$; T^{-1}: $u = x - y$, $v = x + y$

13. $\dfrac{1}{2}$; T^{-1}: $x = \sqrt{\dfrac{\sqrt{u^2 + v^2} + v}{2}}$, $y = \dfrac{v}{2x}$, $x > 0$

17. $\dfrac{\pi}{4} - \dfrac{1}{2}$

CHAPTER 12

SECTION 1

1. $\phi'(x) = \displaystyle\int_0^1 \dfrac{t \cos xt}{1 + t} \, dt$ 3. $\phi'(x) = \displaystyle\int_1^2 -\dfrac{t}{(1 + xt)^2} e^{-t} \, dt$

5. $\phi'(x) = \displaystyle\int_0^1 t^x \, dt, \; x > 0$

7. $\phi'(x) = 2x \cos (x^4)$

9. $\phi'(x) = \sin (x^2) - 2x \sin (x^3) + \displaystyle\int_{x^2}^x t \cos (xt) \, dt$

11. $\phi'(x) = e^x \tan (xe^x) - 2x \tan (x^3) + \displaystyle\int_{x^2}^e t \sec^2 (xt) \, dt$

13. $\phi'(x) = \dfrac{2x}{1 + x + x^3} e^{-(1+x^2)} + \dfrac{\sin x}{1 + x \cos x} e^{-\cos x}$
$$+ \int_{\cos x}^{1+x^2} - \frac{t}{(1 + xt)^2} e^{-t} \, dt$$

15. $\phi'(x) = \dfrac{\pi}{2x} \cos \left(\dfrac{\pi x}{2}\right) - \dfrac{1}{x^2} \sin \left(\dfrac{\pi x}{2}\right)$

17. $\phi'(x) = \dfrac{1}{x}\left[\cos (\pi x) - \cos \left(\dfrac{\pi x}{2}\right)\right]$ 19. $\phi'(x) = \dfrac{1}{x}[2 \sin (x^2) - 3 \sin (x^3)]$

21. $\phi'(x) = \dfrac{x^{n-1} + nx^{n-2}}{1 + x^{n-1}} - \dfrac{x^{m-1} + mx^{m-2}}{1 + x^{m-1}}, \; x > 0$

23. $\phi'(x) = \dfrac{1}{\sqrt{1 - x^2}}$ 25. 0

SECTION 2

1. Converges	3. Converges	5. Converges	7. Diverges
9. Diverges	11. Diverges	13. Diverges	15. Converges
17. Converges	19. Converges	21. Converges	

SECTION 3

1. Converges	3. Converges	5. Diverges	7. Converges
9. Converges	11. Converges if $p > 2$, diverges if $p \le 2$		13. Converges
15. Converges	17. $\Gamma(\tfrac{1}{2}) = \sqrt{\pi}$		

SECTION 4

1. $\phi'(x) = \displaystyle\int_0^\infty \dfrac{-te^{-xt}}{1 + t} \, dt$ 3. $\phi'(x) = - \displaystyle\int_0^\infty \dfrac{t \sin (xt) \, dt}{1 + t^3}$

5. $\phi'(x) = \displaystyle\int_0^1 (\ln t) \cdot \cos (xt) \, dt$ 7. $\phi'(x) = - \displaystyle\int_0^1 \dfrac{t \, dt}{(1 + xt)^2 \sqrt{1 - t}}$

15. $\phi(x) = \tfrac{1}{2} \ln (1 + x^2)$

17. $\phi'(x) = \text{Arctan } x$
   $\phi(x) = x \text{ Arctan } x - \tfrac{1}{2} \ln (1 + x^2)$

CHAPTER 13

SECTION 1

1. $\mathbf{u}_x = (2x + 2y)\mathbf{i} + 3x^2\mathbf{j}$
   $\mathbf{u}_y = 2x\mathbf{i} - 3y^2\mathbf{j}$

3. $\mathbf{u}_x = \dfrac{-x^2 + y^2}{(x^2 + y^2)^2}\mathbf{i} - \dfrac{2xy}{(x^2 + z^2)^2}\mathbf{j} - \dfrac{2xz}{(x^2 + y^2)^2}\mathbf{k}$

$\quad \mathbf{u}_z = -\dfrac{2yz}{(x^2 + z^2)^2}\mathbf{j} + \dfrac{1}{x^2 + y^2}\mathbf{k}$

5. $\mathbf{v}_s = (\cos t)\mathbf{i} + (t \cos s)\mathbf{j} - 2s\mathbf{k}$

$\quad \mathbf{v}_t = (-s \sin t)\mathbf{i} + (\sin s)\mathbf{j} + 2t\mathbf{k}$

7. $\mathbf{v}_{rs} = \mathbf{0}, \ \mathbf{v}_{st} = \mathbf{0}$

9. $\dfrac{\partial}{\partial y}(|\mathbf{u}|^2) = x \sin(2xy) - z \sin(2yz)$

11. $\dfrac{\partial}{\partial y}(\mathbf{u} \cdot \mathbf{v}) = \dfrac{e^{x-y}}{x + y} - e^{x-y} \ln(x + y) - 2y - \dfrac{\ln(x + y)}{x - y} + \dfrac{\ln(x - y)}{x + y}$

13. $\dfrac{\partial}{\partial t}(\mathbf{u} \times \mathbf{v}) = (-2e^{-t} \sin s - 2e^{2t} \cos s)\mathbf{i} + (-2e^{t} \cos s)\mathbf{j}$

$\qquad\qquad\qquad\qquad + (2e^{2t} \cos^2 s + 2e^{-2t} \sin s \sin 2s)\mathbf{k}$

15. $\mathbf{u} \cdot (\mathbf{u}_{,1} \times \mathbf{v}_2) = 0 \Leftrightarrow y = 0$

17. S is the hemisphere $x^2 + y^2 + z^2 = 9, \ z \ge 0$

19. S is the cone $z^2 = \dfrac{x^2}{4} + \dfrac{y^2}{9}$

SECTION 2

1. $\nabla f = 4\mathbf{i} + 4\mathbf{j} + 2\mathbf{k}, \ D_\mathbf{a} f = \dfrac{10}{\sqrt{14}}$

3. $\nabla f = \frac{1}{81}(\mathbf{i} - 4\mathbf{j} - 2\mathbf{k}), \ D_\mathbf{a} f = \frac{5}{243}$

5. $\nabla f = 2\mathbf{i} + \frac{1}{2}\mathbf{j} + \frac{1}{2}\mathbf{k}, \ D_\mathbf{a} f = \frac{3}{2}$

7. $D_\mathbf{a}\mathbf{u} = \dfrac{2}{\sqrt{21}}(\mathbf{i} + 4\mathbf{j} + 6\mathbf{k})$

9. $D_\mathbf{a}\mathbf{u} = \mathbf{0}$

11. $\mathbf{a} = \frac{1}{7}(6\mathbf{i} - 2\mathbf{j} - 3\mathbf{k})$

15. $\mathbf{n} = \pm \dfrac{1}{\sqrt{26}}(3\mathbf{i} - 4\mathbf{j} - \mathbf{k})$

17. $\mathbf{n} = \pm \dfrac{1}{5\sqrt{2}}(3\mathbf{i} - 4\mathbf{j} + 5\mathbf{k})$

19. $\mathbf{n} = \pm\frac{1}{3}(-\mathbf{i} - 2\mathbf{j} + 2\mathbf{k})$

SECTION 3

1. $a_{11} + a_{22} + a_{33}$ 3. $2x + 2y + 2z$

5. $e^{xz}(2z \cos yz - x)$ 7. 0   9. 0

11. $r^{-n}(3 - n)$ 13. $\phi''(r) + 2r^{-1}\phi'(r)$

SECTION 4

1. $\text{curl } \mathbf{v} = \mathbf{i} + \mathbf{j}$ 3. $f^*(x, y, z) = e^x \sin y \cos z + C$

5. $\text{curl } \mathbf{v} = \dfrac{2}{\sqrt{x^2 + y^2}}(y\mathbf{i} - x\mathbf{j})$ 7. $\text{curl } \mathbf{v} = (x^2 - z^2)\mathbf{j}$

9. $f^* = \frac{1}{3}(x^3y^3 + x^3z^3 + y^3z^3) + G_1(x) + G_2(y) + G_3(z) + C$

17. $\text{curl }[\phi(r)\mathbf{r}] = \mathbf{0}$

SECTION 5

1. $\frac{5}{6}$ 3. 2   5. $-\frac{5}{2}+\frac{\pi^3}{24}$ 7. $\frac{137}{10}$ 9. $\frac{8}{3}$

SECTION 6

In problems 1 through 5, we give the common answer.

1. -2 3. 0   5. $2-\dfrac{1}{\sqrt{2}}$

7. Valid; $f^* = x^3 - 3xyz + y^3 + C$
9. Domain not simply connected but have $f^* = \frac{3}{2}(x^2 + y^2 + z^2)^{1/3} + C$ in D.

CHAPTER 14

SECTION 1

In problems 1 through 8, we give the value of $\iint_G (Q_x - P_y)\, dA_{xy}$.

1. 2   3. $-\frac{27}{2}$ 5. $\frac{1}{4}$ 7. 0
9. 12   11. $\frac{26}{3}$ 13. $4\ln 4$ 15. $-\frac{988}{5}$
17. 0   19. 2   21. $\dfrac{3\pi a^4}{4}$ 23. 0

SECTION 4

1. Yes   3. No
5. Surfaces intersect along the circles $y^2 + z^2 = 9$, $x = \pm 4$
   At any point (x_0, y_0, z_0) on this intersection
   $$\nabla F \times \nabla G = \pm 48(-z_0\mathbf{j} + y_0\mathbf{k})$$
   $$|\nabla F \times \nabla G| = 48 \cdot \sqrt{y_0^2 + z_0^2} = 144 \neq 0$$

SECTION 5

1. $\dfrac{\sqrt{3}}{6}$ 3. $\dfrac{15\pi\sqrt{2}}{4}$ 5. 2π

7. $\frac{1}{3}[(1 + \pi^2/4)^{3/2} - 1]$ 9. $\dfrac{\pi}{4}(15\sqrt{2} + 17)$

11. $\dfrac{1024\delta}{45}$ 13. $\dfrac{(2 + \sqrt{3})\delta}{6}$

15. $\bar{z} = \dfrac{2 + \sqrt{2}}{4}$, $\bar{x} = \bar{y} = 0$ 17. $\bar{x} = \dfrac{4a}{3(\pi - 2)}$, $\bar{y} = \bar{z} = 0$

19. Case (i) $\dfrac{4\pi a^2 \delta}{c}$, Case (ii) $4\pi\delta a$ 21. $2\pi\delta$

SECTION 7

1. 0   3. $\dfrac{15\pi}{2}$ 5. $\frac{4}{15}$ 7. $\frac{64}{3}$
In problems 8 through 13, we give the value of $\int_{\partial S} \mathbf{v} \cdot d\mathbf{r}$.

9. 0   11. $2\left(\dfrac{\pi^2}{4} - 1\right)$ 13. 0   15. 0

SECTION 8

In problems 1 through 10, we give the value of $\iiint_G \operatorname{div} \mathbf{v}\, dV$.

1. $\frac{1}{8}$ 3. $\frac{4}{3}\pi$ 5. $12\pi\sqrt{3}$ 7. 0   9. 2π 11. 0   13. 5π

CHAPTER 15

SECTION 2

1. $\tan y + x^{-1} = C$

3. $x = -y^{-1} - 2y + C$

5. $-x^{-1} = 2e^{-y} - y + C$

7. $\ln|y| = \frac{1}{2}x^2 + C$

9. $e^x + e^{-2y} = C$

11. $(x^2 + 1)(y^2 + 1) = C$

13. $y^2 = 2\sin(C - x^4)$

15. $\dfrac{y^3}{3}\ln|y| - \dfrac{y^3}{9} = \dfrac{x^5}{5}\ln|x| - \dfrac{x^5}{25} + C$

SECTION 3

1. Degree 4   3. Not homogeneous   5. Degree -2

7. Not homogeneous   9. Degree 0   11. $2\operatorname{Arctan}(y/x) - \frac{1}{2}\ln(x^2 + y^2) = C$

13. $(y - 2x)^2(x + y) = C$ 15. $(x - y)(3x + y) = C(x + y)$

17. $\ln|x| + \arcsin\left(\dfrac{y}{|x|}\right) = C$ 19. $x\ln|x| + \sqrt{x^2 + y^2} = Cx$

21. $y = Cx$

23. $f(x)$ and $g(y)$ are constants, at least if $x > 0$, etc.

SECTION 4

1. $\dfrac{x^2}{2} + 2xy - 2x - \dfrac{y^2}{2} + 3y = C$ 3. $-x^3 - \dfrac{y^3}{3} + xy + y^2 + 2x = C$

5. $\dfrac{x^4}{2} + 3x^2y^2 - 2xy^3 - y^4 + 2x^2 + 3xy - y^2 = C$

7. $x\arcsin y + 2\sin y = C$

9. $(x - y)^2(x^2 + y^2) = C$ 11. $\frac{1}{2}x^2\ln|y| + xy\ln|x| = C$

SECTION 5

1. $\arctan(y/x) = x + C$ 3. $\arctan(y/x) = y + C$ 5. $e^{-x/y} + y^2/2 = C$

7. $x\sec y - 2\tan y = C$ 9. $2(xy)^{-1} + \ln|x| - \ln|y| = C$

SECTION 6

1. $y = x^{n+1}/n + Cx$ 3. $y = e^x(x^3 + 2x^2 - 3x + C)$

5. $y = (2x)^{-1}(e^{2x} + 2\cos x + C)$ 7. $y = x^2 - 1 + C\sqrt{x^2 - 1}$

9. $y = (2x - 3)^{-3/2}(4x^3 - 9x^2 + 6C)/6$

11. $y = (1 - x)(x + C)/(1 + x)$ 13. $y = (\sin 2x)^{-1/2}(\cos 2x + C)$

15. $y = \frac{1}{3}(a - x)(a + x)^2 + \dfrac{C(a - x)}{(a + x)}$

SECTION 7

1. $y = (x^3 + Cx^6)^{-1/3}$ 3. $s^2 = t^2 + Ct$

5. $y = (1 - x + Ce^{-x})^{-1}$ 7. $y = (Cx - x^2)^{1/3}$

9. $y^{-2} = (2x + Cx^2)$ 11. $y = Cx - \dfrac{C^2}{2}$; sing. sol. $y = \dfrac{x^2}{2}$

13. $y = Cx + \ln C$; sing. sol.: $y = -1 + \ln(-1/x)$

15. $y = 2Cx^{1/2} - C^2$ or $y = x$

SECTION 8

1. $y = \frac{1}{4}e^{2x} - \frac{1}{4}\cos 2x + C_1x + C_2$

3. $y = \frac{1}{2}[x\sqrt{1 + x^2} + \ln(x + \sqrt{1 + x^2})] + C_1x + C_2$

5. $y = \frac{1}{4}(2x - 1)e^{2x} + C_1e^{2x} + C_2$ 7. $y = \ln|\sec(2x + C_1)| + C_2$

9. $y = C_1 \sin kx + C_2 \cos kx$ 11. $y = (2/C_1) \cosh(C_1x + C_2)$

13. $C_2 \pm x = \frac{2}{3}(-y)^{3/2}$

or

$C_2 \pm x = C_1\sqrt{(y - C_1^2)^2 - C_1^4} + C_1^3 \operatorname{argcosh}[(y - C_1^2)/C_1^2]$

or

$C_2 \pm x = -C_1\sqrt{C_1^4 - (y + C_1^2)^2} - C_1^3 \arcsin[(y + C_1^2)/C_1^2]$

15. $y = 4(3 - x)^{-2}$ 17. $y = -\frac{1}{2}(\cos x + \sin x + 5)$

19. $y = 4 \cosh x - 5 \sinh x$

SECTION 9

1. $xy = C$ 3. $x^2 - y^2 = C, \quad y/x > 0$

5. $y = k \cosh(C \pm x/k)$ 7. $x\dfrac{dy}{dx} + e^x - xe^x - y = 0$

9. $(2xy + 1)y = x(xy + 2)\dfrac{dy}{dx}$ 11. $y\left(\dfrac{dy}{dx}\right)^2 + 2x\dfrac{dy}{dx} - y = 0$

13. $y^2 = Cx^3$ 15. $x^2 + y^2 = Cy$

17. $s = \dfrac{g}{k}t - \dfrac{g}{k^2}(1 - e^{-kt})$ (s = distance fallen, k = const in resistance law)

19. $s = \dfrac{1}{k}\ln\cosh(akt), \quad a = \sqrt{g/k}$

21. $y = b \cosh ax, \quad a = \sqrt{\rho/H}$ (ρ = density)

CHAPTER 16

SECTION 1

1. $2y'' - y' + y = e^x$ 3. $\dfrac{d^3y}{dx^3} + 2\dfrac{d^2y}{dx^2} - \dfrac{dy}{dx} = 0$

5. $(D^2 - 3D^1 + 2D^0)(y) = 0$ or $(D^2 - 3D + 2)(y) = 0$

7. $(2D^3 + D - 3)(y) = 0$ 9. $(L_1L_2)(y) = (L_2L_1)(y) = y'' - y$

11. $(L_1L_2)(y) = (L_2L_1)(y) = y'' + y' - 2y$

13. $(L_1L_2)(y) = (x + 1)y'' + (x^2 + x - 1)y' + y$

$(L_2L_1)(y) = (x + 1)y'' + (x^2 + x)y' - xy$

15. $(L_1L_2)(y) = y^{(4)} + (8 - 7x^2)y'' + (7x - 6x^3)y' + (4 - 8x^2)y$

$(L_2L_1)(y) = y^{(4)} - (4 + 7x^2)y'' - (3x + 6x^3)y' + 4x^2y$

SECTION 2

1. $y = C_1e^{2x} + C_2e^{-2x}$ 3. $y = C_1e^x + C_2e^{3x}$

5. $y = e^{-x}(C_1 \cos x + C_2 \sin x)$ 7. $y = e^x[C_1 \cos(x\sqrt{2}) + C_2 \sin(x\sqrt{2})]$

9. $y = e^{-3x}(C_1 \cos x + C_2 \sin x)$ 11. $y = \frac{4}{3}e^{-x} - \frac{1}{3}e^{2x}$

13. $y = -\sin 2x + 3 \cos 2x$

SECTION 3

1. $y = (e^x/3) + C_1 + C_2e^{-2x}$

3. $y = x + 1/2 + e^{-x} + e^{-x}(C_1 \cos x + C_2 \sin x)$

5. $y = \frac{1}{4}x^2 + \frac{1}{2}x + \frac{3}{8} - \frac{1}{16}e^{-2x} + (C_1x + C_2)e^{2x}$

7. $y = e^x(\frac{1}{2}x^2 + C_1x + C_2)$ 9. $y = -\frac{1}{2}\cos x + C_1e^x + C_2e^{-x}$

11. $y = -\frac{1}{5}\cos 2x - \frac{1}{10}\sin 2x + e^{-x}(C_1\cos x + C_2\sin x)$
13. $y = -\frac{1}{2}x\cos x + C_1\cos x + C_2\sin x$
15. $y = -x + \cosh x$ 17. $y = e^x - \frac{1}{5}e^{-3x} - \frac{4}{5}e^{2x}$
19. $y = \frac{1}{5}e^x - \frac{1}{5}\cos 2x - \frac{1}{10}\sin 2x$

SECTION 4

1. $y = C_1x + C_2(1 + x\arctan x)$ 3. $y = C_1e^x + C_2(x^3 + 3x^2 + 6x + 6)$
5. $y = C_1x + C_2(x^2 - 1)$ 7. $y = (\frac{1}{4}x^2 - \frac{1}{4}x + C_1)e^x + C_2e^{-x}$
9. $y = (\frac{1}{6}x^3 + C_1x + C_2)e^{-x}$ 11. $y = x^2 + x + 1 + C_1x + C_2e^x$

SECTION 5

1. $y = C_0 + C_1x + C_2x^2$ 3. $y = (C_0 + C_1x + C_2x^2)e^{-x}$
5. $y = e^{-x}(\frac{1}{3}x^3 + \frac{1}{2}x^2 + C_1x + C_0)$
7. $y = e^x(\frac{1}{60}x^5 - \frac{1}{2}x^3 + C_2x^2 + C_1x + C_0)$
9. $y = e^{2x} + e^x(C_1x + C_0)$ 11. $y = \frac{1}{4}xe^x + e^{-x}(C_1x + C_0)$
13. $y = (x - 1)^2e^{2x} + (C_0 + C_1x)e^x$
15. $y = e^x(\frac{1}{8}x - \frac{3}{16}) + e^{-x}(C_0 + C_1x + C_2x^2)$
17. $y = e^x(x^2 - 6x + 12) + C_0 + C_1x + C_2x^2$

SECTION 6

1. $y = C_1e^{3x} + C_2e^x$ 3. $y = C_1 + C_2e^{2x} + C_3e^{-x}$
5. $y = (\frac{1}{2}x + C_1)e^{2x} + C_2e^{-2x}$ 7. $y = C_1e^x + C_2e^{2x} + C_3e^{-x}$
9. $y = (\frac{1}{3}x^2 + C_1x + C_2)e^x + C_3e^{-2x}$
11. $y = x + C_1 + (C_2 + C_3x)e^x$
13. $y = e^x(\frac{1}{2}x^2 - 3x + C_1) + C_2 + C_3x + C_4x^2$
15. $y = e^x(\frac{1}{8}x^2 - \frac{1}{16}x + C_1) + C_2e^{-3x} - (\frac{14}{27} + \frac{4}{9}x + \frac{1}{3}x^2)$
17. $y = e^{2x}(\frac{1}{96}x^3 - \frac{1}{64}x^2 + C_1x + C_2) + (C_3 + C_4x)e^{-2x}$
19. $y = e^{2x}(\frac{1}{180}x^5 - \frac{1}{108}x^4 + \frac{1}{81}x^3 + C_1x^2 + C_2x + C_3)$
      $+ C_4e^{-x} + e^x(-\frac{1}{2}x - \frac{5}{4})$

SECTION 7

1. $y = e^x(ax^4 + bx^3 + cx^2 + C_1x + C_2) + e^{-x}(C_3 + C_4x)$
      $+ e^{2x}[(d + e_1x)\cos 3x + (f + gx)\sin 3x]$
3. $y = e^{-x}(a + bx + cx^2) + e^x(d + e_1x)$
      $+ e^x\cos x(fx^3 + gx^4 + hx^5 + jx^6 + C_1 + C_2x + C_3x^2)$
      $+ e^x\sin x(kx^3 + lx^4 + mx^5 + nx^6 + C_4 + C_5x + C_6x^2)$
5. $y = 2x + C_1\cos x + C_2\sin x$
7. $y = (C_1 + C_2x)\cos x + (C_3 + C_4x)\sin x$
9. $y = \frac{1}{4}x + C_1\cos 2x + C_2\sin 2x + C_3\cos x + C_4\sin x$
11. $y = -\frac{1}{15}\cos 2x - \frac{2}{15}\sin 2x + C_1e^{-x} + C_2\cos x + C_3\sin x$
13. $y = -\frac{3}{8} - \frac{1}{4}x^2 + \frac{1}{2}x^3 + C_1e^{-2x} + e^x[C_2\cos(x\sqrt{3}) + C_3\sin(x\sqrt{3})]$
15. $y = \frac{1}{39}e^{2x} + (-\frac{1}{60}x + C_1)\sin 3x + C_2\cos 3x + C_3e^{-x} + C_4e^x$
17. $y = xe^x + e^x\cos x(-\frac{1}{12}x^3 + C_1x + C_2) + e^x\sin x(x^2 + C_3x + C_4)$

SECTION 8

1. $y = C_1x + C_2x^2$ 3. $y = C_1x + C_2\cos(\ln x) + C_2\sin(\ln x)(x > 0)$
5. $y = C_1\cos(2\ln x) + C_2\sin(2\ln x)$ 7. $y = \ln x - \frac{3}{2} + C_1x^{-1} + C_2x^{-2}$
9. $y = x[\frac{1}{6}(\ln x)^3 + C_1\ln x + C_2]$
11. $y = x[\frac{1}{24}(\ln x)^4 + C_1(\ln x)^2 + C_2\ln x + C_3]$
13. $y = [C_1\cos(2\ln x) + (C_2 + \frac{1}{4}\ln x)\sin(2\ln x)]$

SECTION 9

1. $x = C_1e^{-t} + C_2e^t$, $y = -C_1e^{-t} + C_2e^t$

3. $x = C_1e^{-t} - C_2e^{3t}$, $y = C_1e^{-t} + 3C_2e^{3t}$

5. $x = C_1e^{4t} + C_2e^{-2t} + \frac{11}{6}te^{-2t}$
   $y = 5C_1e^{4t} - C_2e^{-2t} + (-\frac{11}{6}t - \frac{1}{6})e^{-2t}$

7. $x' = C_1e^{-t} + C_2e^t + C_3e^{2t}$, $y = -2C_1e^{-t} + C_3e^{2t}$, $z = -4C_1e^{-t} - C_3e^t$

9. $x = (8 - 3t)e^t - 2t - 8$, $2y = (13 - 6t)e^t - 4t - 13$

SECTION 10

7. $x_1 = t + 1$, $x_2 = 2t - 1$, $x_3 = 1$

9. $x_1 = \cos t - 2\sin t$, $x_2 = 2\cos t + \sin t$, $x_3 = 3\cos t$

11. $x_1 = te^t$, $x_2 = (t - 1)e^t$, $x_3 = (t + 1)e^t$

SECTION 11

1. $x_1 = -C_1e^{2t} + e^t(-C_2 + \frac{1}{2}C_3 - C_3t)$
   $x_2 = -C_1e^{2t} - \frac{1}{2}e^t(C_2 + C_3 + C_3t)$
   $x_3 = C_1e^{2t} + (C_2 + C_3t)e^t$

3. $x_1 = -\frac{2}{3}e^{-t} - C_1e^{2t} - (C_2 + C_3 + C_3t)e^t$
   $x_2 = e^{-t} + (C_2 + 2C_3 + C_3t)e^t$
   $x_3 = \frac{2}{3}e^{-t} + C_1e^{2t} + (C_2 + C_3t)e^t$

5. $x_1 = e^{2t}(3t + C_1) + e^t(2C_2 - 3C_3 + 2C_3t)$
   $x_2 = e^{2t}(3t + C_1) + e^t(C_2 + C_3t)$
   $x_3 = e^{2t}(1 - 3t - C_1) - e^t(2C_2 - 2C_3 + 2C_3t)$

7. $x_1 = C_3e^{-t} + e^t(-\frac{3}{4}t^2 - \frac{9}{4}t + \frac{17}{8} - \frac{3}{2}C_1t + \frac{3}{4}C_1 + \frac{3}{2}C_2)$
   $x_2 = C_3e^{-t} + e^t(-\frac{3}{4}t^2 - \frac{13}{4}t + \frac{17}{8} - \frac{3}{2}C_1t - \frac{1}{4}C_1 + \frac{3}{2}C_2)$
   $x_3 = C_3e^{-t} + e^t(-\frac{1}{4}t^2 - \frac{5}{4}t + \frac{17}{8} - \frac{1}{2}C_1t - \frac{1}{4}C_1 + \frac{1}{2}C_2)$

CHAPTER 17

SECTION 1

1. $y = a_0 \sum_{k=0}^{\infty} \frac{x^{2k}}{(2k)!} + a_1 \sum_{k=0}^{\infty} \frac{x^{2k+1}}{(2k + 1)!}$

3. $y = a_0 \sum_{k=0}^{\infty} \frac{(-1)^k x^{2k}}{1 \cdot 3 \cdots (2k - 1)} + a_1 \sum_{k=0}^{\infty} \frac{(-1)^k x^{2k+1}}{2^k \cdot k!}$

5. $y = a_0 \sum_{k=0}^{\infty} \frac{(-2x^2)^k}{1 \cdot 3 \cdots (2k - 1)} + a_1 \sum_{k=0}^{\infty} \frac{(-1)^k x^{2k+1}}{k!}$

7. $y = a_0\left[1 + \sum_{k=1}^{\infty} \frac{1 \cdot 4 \cdots (3k - 2)}{(3k)!} x^{3k}\right]$
   $+ a_1\left[x + \sum_{k=1}^{\infty} \frac{2 \cdot 5 \cdots (3k - 1)}{(3k + 1)!} x^{3k+1}\right]$

9. $y = a_0\left(1 + \frac{x^4}{3 \cdot 4} + \frac{x^8}{3 \cdot 4 \cdot 7 \cdot 8} + \frac{x^{12}}{3 \cdot 4 \cdot 7 \cdot 8 \cdot 11 \cdot 12} + \cdots\right)$
   $+ a_1\left(x + \frac{x^5}{4 \cdot 5} + \frac{x^9}{4 \cdot 5 \cdot 8 \cdot 9} + \frac{x^{13}}{4 \cdot 5 \cdot 8 \cdot 9 \cdot 12 \cdot 13} + \cdots\right)$

11. $|x| < 2$. The general solution is

$$y = a_0 + a_1 x + \frac{a_0 + a_1}{2}(e^x - 1 - x)$$

13. $y = 1 + x^2 + \frac{5}{24}x^4 + \cdots$

15. $y = x - \frac{1}{2}x^3 + \frac{7}{40}x^5 + \cdots$

17. $y = x - \frac{1}{6}x^3 - \frac{1}{12}x^4 + \cdots$

19. $y_1 = 1 + 2x + 2x^2 + \cdots, \quad y_2 = 1 - x^2 + \cdots$

SECTION 2

1. $y_1 = x^{-1}\sin x, \quad y_2 = x^{-1}\cos x$

3. $y_1 = x^{-1}(1 + x)^{-1}, \quad y_2 = x^{-2}$

5. $y_1 = x^{-1/2}(1 - x/2)^{-1}, \quad y_2 = x^{-1}(1 - x/2)$

7. $y_1 = x^{-1}(e^{2x} - 1), \quad y_2 = x^{-1}$

9. $y_1 = x^2(2 - x)^{-1}, \quad y_2 = (2 - x)^{-1}$

11. $\left.\begin{array}{l} y_1 = x^{3/2}\displaystyle\sum_{n=0}^{\infty} \dfrac{(2n + 5)x^n}{5 \cdot 2^n \cdot n!} \\[4mm] y_2 = 1 + 2x - \displaystyle\sum_{n=2}^{\infty} \dfrac{(n + 1)x^n}{1 \cdot 3 \cdots (2n - 3)} \end{array}\right\}$; all x

13. $\left.\begin{array}{l} y_1 = x^{7/2}\left[1 + \displaystyle\sum_{n=1}^{\infty} \dfrac{(-2x)^n}{7 \cdot 9 \cdots (2n + 5)}\right] \\[4mm] y_2 = xe^{-x} \end{array}\right\}$; all x

15. $y_1 = 1 + \dfrac{a \cdot b}{c} \cdot \dfrac{x}{1!} + \dfrac{a(a + 1)b(b + 1)}{c(c + 1)} \dfrac{x^2}{2!} + \cdots$

$y_2 = x^{1-c}\left[1 + \dfrac{(1 - c + a)(1 - c + b)}{2 - c} \dfrac{x^2}{2!}\right.$

$+ \dfrac{(1 - c + a)(2 - c + a)(1 - c + b)(2 - c + b)}{(2 - c)(3 - c)} \dfrac{x^3}{3!} + \cdots\Big]$; $|x| < 1$

17. $y_1 = x^3; \quad y_2 = -4x^3 \ln x + x\left[1 - 4x - \displaystyle\sum_{n=3}^{\infty} \dfrac{2!(-2x)^n}{(n - 2) \cdot n!}\right]$

19. $y_1 = 1 - x; \quad y_2 = (1 - x)\ln x + x\left[3 - \displaystyle\sum_{n=1}^{\infty} \dfrac{x^n}{n(n + 1)(n + 1)!}\right]$

21. $y_1 = x^2 - \frac{1}{3}x^3; \quad y_2 = -3(x^2 - \frac{1}{3}x^3)\ln x$

$$+ 1 + 3x - \tfrac{7}{3}x^3 + \sum_{n=4}^{\infty} \frac{3!(n - 4)!}{(n - 2)!n!}x^n$$

SECTION 3

5. (c) $J_{3/2}(x) = \sqrt{2/\pi}\,(x^{-3/2}\sin x - x^{-1/2}\cos x)$

$J_{5/2}(x) = \sqrt{2/\pi}\,(3x^{-5/2}\sin x - 3x^{-3/2}\cos x - x^{-1/2}\sin x)$

$J_{-3/2}(x) = \sqrt{2/\pi}\,(-x^{-3/2}\cos x - x^{-1/2}\sin x)$

7. $J_4(x) = (48x^{-3} - 8x^{-1})J_1 - (24x^{-2} - 1)J_0$

Index